PROCESS MODELING, SIMULATION, AND CONTROL FOR CHEMICAL ENGINEERS

McGraw-Hill Chemical Engineering Series

Building the Literature of a Profession

Fifteen prominent chemical engineers first met in New York more than 60 years ago to plan a continuing literature for their rapidly growing profession. From industry came such pioneer practitioners as Leo H. Baekeland, Arthur D. Little, Charles L. Reese, John V. N. Dorr, M. C. Whitaker, and R. S. McBride. From the universities came such eminent educators as William H. Walker, Alfred H. White, D. D. Jackson, J. H. James, Warren K. Lewis, and Harry A. Curtis. H. C. Parmelee, then editor of *Chemical and Metallurgical Engineering*, served as chairman and was joined subsequently by S. D. Kirkpatrick as consulting editor.

After several meetings, this committee submitted its report to the McGraw-Hill Book Company in September 1925. In the report were detailed specifications for a correlated series of more than a dozen texts and reference books which have since become the McGraw-Hill Series in Chemical Engineering and which became the cornerstone of the chemical engineering curriculum.

From this beginning there has evolved a series of texts surpassing by far the scope and longevity envisioned by the founding Editorial Board. The McGraw-Hill Series in Chemical Engineering stands as a unique historical record of the development of chemical engineering education and practice. In the series one finds the milestones of the subject's evolution: industrial chemistry, stoichiometry, unit operations and processes, thermodynamics, kinetics, and transfer operations.

Chemical engineering is a dynamic profession, and its literature continues to evolve. McGraw-Hill and its consulting editors remain committed to a publishing policy that will serve, and indeed lead, the needs of the chemical engineering profession during the years to come.

The Series

Bailey and Ollis: *Biochemical Engineering Fundamentals*
Bennett and Myers: *Momentum, Heat, amd Mass Transfer*
Beveridge and Schechter: *Optimization: Theory and Practice*
Brodkey and Hershey: *Transport Phenomena: A Unified Approach*
Carberry: *Chemical and Catalytic Reaction Engineering*
Constantinides: *Applied Numerical Methods with Personal Computers*
Coughanowr and Koppel: *Process Systems Analysis and Control*
Douglas: *Conceptual Design of Chemical Processes*
Edgar and Himmelblau: *Optimization of Chemical Processes*
Fahien: *Fundamentals of Transport Phenomena*
Finlayson: Nonlinear Analysis in Chemical Engineering
Gates, Katzer, and Schuit: *Chemistry of Catalytic Processes*
Holland: *Fundamentals of Multicomponent Distillation*
Holland and Liapis: *Computer Methods for Solving Dynamic Separation Problems*
Katz, Cornell, Kobayashi, Poettmann, Vary, Elenbaas, and Weinaug: *Handbook of Natural Gas Engineering*
King: *Separation Processes*
Luyben: *Process Modeling, Simulation, and Control for Chemical Engineers*
McCabe, Smith, J. C., and Harriott: *Unit Operations of Chemical Engineering*
Mickley, Sherwood, and Reed: *Applied Mathematics in Chemical Engineering*
Nelson: *Petroleum Refinery Engineering*
Perry and Chilton (Editors): *Chemical Engineers' Handbook*
Peters: *Elementary Chemical Engineering*
Peters and Timmerhaus: *Plant Design and Economics for Chemical Engineers*
Probstein and Hicks: *Synthetic Fuels*
Reid, Prausnitz, and Sherwood: *The Properties of Gases and Liquids*
Resnick: *Process Analysis and Design for Chemical Engineers*
Satterfield: *Heterogeneous Catalysis in Practice*
Sherwood, Pigford, and Wilke: *Mass Transfer*
Smith, B. D.: *Design of Equilibrium Stage Processes*
Smith, J. M.: *Chemical Engineering Kinetics*
Smith, J. M., and Van Ness: *Introduction to Chemical Engineering Thermodynamics*
Treybal: *Mass Transfer Operations*
Valle-Riestra: *Project Evolution in the Chemical Process Industries*
Van Ness and Abbott: *Classical Thermodynamics of Nonelectrolyte Solutions: with Applications to Phase Equilibria*
Van Winkle: *Distillation*
Volk: *Applied Statistics for Engineers*
Walas: *Reaction Kinetics for Chemical Engineers*
Wei, Russell, and Swartzlander: *The Structure of the Chemical Processing Industries*
Whitwell and Toner: *Conservation of Mass and Energy*

Also available from McGraw-Hill

Schaum's Outline Series in Civil Engineering

Each outline includes basic theory, definitions, and hundreds of solved problems and supplementary problems with answers.

Current List Includes:

Advanced Structural Analysis
Basic Equations of Engineering
Descriptive Geometry
Dynamic Structural Analysis
Engineering Mechanics, 4th edition
Fluid Dynamics
Fluid Mechanics & Hydraulics
Introduction to Engineering Calculations
Introductory Surveying
Reinforced Concrete Design, 2d edition
Space Structural Analysis
Statics and Strength of Materials
Strength of Materials, 2d edition
Structural Analysis
Theoretical Mechanics

Available at Your College Bookstore

PROCESS MODELING, SIMULATION, AND CONTROL FOR CHEMICAL ENGINEERS

Second Edition

William L. Luyben

Process Modeling and Control Center
Department of Chemical Engineering
Lehigh University

McGraw-Hill Publishing Company

New York   St. Louis   San Francisco   Auckland   Bogotá   Caracas   Hamburg
Lisbon   London   Madrid   Mexico   Milan   Montreal   New Delhi
Oklahoma City   Paris   San Juan   São Paulo   Singapore   Sydney   Tokyo   Toronto

This book was set in Times Roman.
The editors were Lyn Beamesderfer and John M. Morriss;
the production supervisor was Friederich W. Schulte.
The cover was designed by John Hite.
New drawings were done by Oxford Illustrators Ltd.
Project supervision was done by Harley Editorial Services.
R. R. Donnelley & Sons Company was printer and binder.

PROCESS MODELING, SIMULATION, AND CONTROL FOR CHEMICAL ENGINEERS

234567890 DOC DOC 9543210

ISBN 0-07-039159-9

Library of Congress Cataloging-in-Publication Data

Luyben, William L.
    Process modeling, simulation, and control for chemical engineers
    William L. Luyben.—2nd ed.
        p.   cm.
    Bibliography: p.
    Includes index.
    ISBN 0-07-039159-9
    1. Chemical processes—Mathematical models. 2. Chemical process-Data
processing. 3. Chemical process control. ί. Title.
TP155.7.L88   1989
660.2′81—dc19   88-32134

ABOUT THE AUTHOR

William L. Luyben received his B.S. in Chemical Engineering from the Pennsylvania State University where he was the valedictorian of the Class of 1955. He worked for Exxon for five years at the Bayway Refinery and at the Abadan Refinery (Iran) in plant technical service and design of petroleum processing units. After earning a Ph.D. in 1963 at the University of Delaware, Dr. Luyben worked for the Engineering Department of DuPont in process dynamics and control of chemical plants. In 1967 he joined Lehigh University where he is now Professor of Chemical Engineering and Co-Director of the Process Modeling and Control Center.

Professor Luyben has published over 100 technical papers and has authored or coauthored four books. Professor Luyben has directed the theses of over 30 graduate students. He is an active consultant for industry in the area of process control and has an international reputation in the field of distillation column control. He was the recipient of the Eckman Education Award in 1975 and the Instrumentation Technology Award in 1969 from the Instrument Society of America.

Overall, he has devoted over 35 years to his profession as a teacher, researcher, author, and practicing engineer.

This book is dedicated to
Robert L. Pigford and Page S. Buckley,
two authentic pioneers
in process modeling
and process control

CONTENTS

Part II Computer Simulation

Part III   Time-Domain Dynamics and Control

Part IV Laplace-Domain Dynamics and Control

Part VI Multivariable Processes

Part VII Sampled-Data Control Systems

PREFACE

The first edition of this book appeared over fifteen years ago. It was the first chemical engineering textbook to combine modeling, simulation, and control. It also was the first chemical engineering book to present sampled-data control. This choice of subjects proved to be popular with both students and teachers and of considerable practical utility.

During the ten-year period following publication, I resisted suggestions from the publisher to produce a second edition because I felt there were really very few useful new developments in the field. The control hardware had changed drastically, but the basic concepts and strategies of process control had undergone little change. Most of the new books that have appeared during the last fifteen years are very similar in their scope and content to the first edition. Basic classical control is still the major subject.

However, in the last five years, a number of new and useful techniques have been developed. This is particularly true in the area of multivariable control. Therefore I feel it is time for a second edition.

In the area of process control, new methods of analysis and synthesis of control systems have been developed and need to be added to the process control engineer's bag of practical methods. The driving force for much of this development was the drastic increase in energy costs in the 1970s. This led to major redesigns of many new and old processes, using energy integration and more complex processing schemes. The resulting plants are more interconnected. This increases control loop interactions and expands the dimension of control problems. There are many important processes in which three, four, or even more control loops interact.

As a result, there has been a lot of research activity in multivariable control, both in academia and in industry. Some practical, useful tools have been developed to design control systems for these multivariable processes. The second edition includes a fairly comprehensive discussion of what I feel are the useful techniques for controlling multivariable processes.

Another significant change over the last decade has been the dramatic increase in the computational power readily available to engineers. Most calculations can be performed on personal computers that have computational horsepower equal to that provided only by mainframes a few years ago. This means that engineers can now routinely use more rigorous methods of analysis and synthesis. The second edition includes more computer programs. All are suitable for execution on a personal computer.

In the area of mathematical modeling, there has been only minor progress. We still are able to describe the dynamics of most systems adequately for engineering purposes. The trade-off between model rigor and computational effort has shifted toward more precise models due to the increase in computational power noted above. The second edition includes several more examples of models that are more rigorous.

In the area of simulation, the analog computer has almost completely disappeared. Therefore, analog simulation has been deleted from this edition. Many new digital integration algorithms have been developed, particularly for handling large numbers of "stiff" ordinary differential equations. Computer programming is now routinely taught at the high school level. The second edition includes an expanded treatment of iterative convergence methods and of numerical integration algorithms for ordinary differential equations, including both explicit and implicit methods.

The second edition presents some of the material in a slightly different sequence. Fifteen additional years of teaching experience have convinced me that it is easier for the students to understand the time, Laplace, and frequency techniques if both the dynamics and the control are presented together for each domain. Therefore, openloop dynamics and closedloop control are both discussed in the time domain, then in the Laplace domain, and finally in the frequency domain. The z domain is discussed in Part VII.

There has been a modest increase in the number of examples presented in the book. The number of problems has been greatly increased. Fifteen years of quizzes have yielded almost 100 new problems.

The new material presented in the second edition has come from many sources. I would like to express my thanks for the many useful comments and suggestions provided by colleagues who reviewed this text during the course of its development, especially to Daniel H. Chen, Lamar University; T. S. Jiang, University of Illinois—Chicago; Richard Kermode, University of Kentucky; Steve Melsheimer, Cignson University; James Peterson, Washington State University; and R. Russell Rhinehart, Texas Tech University. Many stimulating and useful discussions of multivariable control with Bjorn Tyreus of DuPont and Christos Georgakis of Lehigh University have contributed significantly. The efforts and suggestions of many students are gratefully acknowledged. The "LACEY" group (Luyben, Alatiqi, Chiang, Elaahi, and Yu) developed and evaluated much of the new material on multivariable control discussed in Part VI. Carol Biuckie helped in the typing of the final manuscript. Lehigh undergraduate and graduate classes have contributed to the book for over twenty years by their questions, youthful enthusiasm, and genuine interest in the subject. If the ultimate gift that a teacher can be given is a group of good students, I have indeed been blessed. Alhamdulillahi!

William L. Luyben

PROCESS MODELING, SIMULATION, AND CONTROL FOR CHEMICAL ENGINEERS

CHAPTER

1

INTRODUCTION

This chapter is an introduction to process dynamics and control for those students who have had little or no contact or experience with real chemical engineering processes. The objective is to illustrate where process control fits into the picture and to indicate its relative importance in the operation, design, and development of a chemical engineering plant.

This introductory chapter is, I am sure, unnecessary for those practicing engineers who may be using this book. They are well aware of the importance of considering the dynamics of a process and of the increasingly complex and sophisticated control systems that are being used. They know that perhaps 80 percent of the time that one is "on the plant" is spent at the control panel, watching recorders and controllers (or CRTs). The control room is the nerve center of the plant.

1.1 EXAMPLES OF THE ROLE OF PROCESS DYNAMICS AND CONTROL

Probably the best way to illustrate what we mean by process dynamics and control is to take a few real examples. The first example describes a simple process where dynamic response, the time-dependent behavior, is important. The second example illustrates the use of a single feedback controller. The third example discusses a simple but reasonably typical chemical engineering plant and its conventional control system involving several controllers.

Example 1.1. Figure 1.1 shows a tank into which an incompressible (constant-density) liquid is pumped at a variable rate F_0 (ft^3/s). This inflow rate can vary with

1

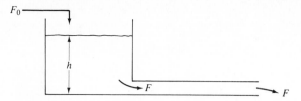

FIGURE 1.1
Gravity-flow tank.

time because of changes in operations upstream. The height of liquid in the vertical cylindrical tank is h (ft). The flow rate out of the tank is F (ft^3/s).

Now F_0, h, and f will all vary with time and are therefore functions of time t. Consequently we use the notation $F_{0(t)}$, $h_{(t)}$, and $F_{(t)}$. Liquid leaves the base of the tank via a long horizontal pipe and discharges into the top of another tank. Both tanks are open to the atmosphere.

Let us look first at the steadystate conditions. By steadystate we mean, in most systems, the conditions when nothing is changing with time. Mathematically this corresponds to having all time derivatives equal to zero, or to allowing time to become very large, i.e., go to infinity. At steadystate the flow rate out of the tank must equal the flow rate into the tank. In this book we will denote steadystate values of variables by an overscore or bar above the variables. Therefore at steadystate in our tank system $\bar{F}_0 = \bar{F}$.

For a given $\bar{F}$, the height of liquid in the tank at steadystate would also be some constant $\bar{h}$. The value of $\bar{h}$ would be that height that provides enough hydraulic pressure head at the inlet of the pipe to overcome the frictional losses of liquid flowing down the pipe. The higher the flow rate $\bar{F}$, the higher $\bar{h}$ will be.

In the steadystate design of the tank, we would naturally size the diameter of the exit line and the height of the tank so that at the maximum flow rate expected the tank would not overflow. And as any good, conservative design engineer knows, we would include in the design a 20 to 30 percent safety factor on the tank height.

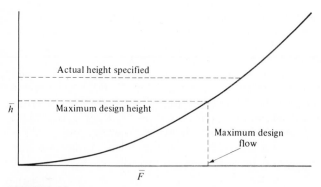

FIGURE 1.2
Steadystate height versus flow.

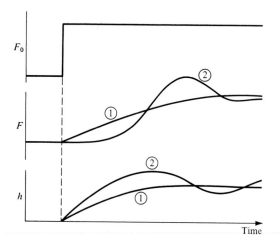

Time **FIGURE 1.3**

Since this is a book on control and instrumentation, we might also mention that a high-level alarm and/or an interlock (a device to shut off the feed if the level gets too high) should be installed to guarantee that the tank would not spill over. The tragic accidents at Three Mile Island, Chernobyl, and Bhopal illustrate the need for well-designed and well-instrumented plants.

   The design of the system would involve an economic balance between the cost of a taller tank and the cost of a bigger pipe, since the bigger the pipe diameter the lower is the liquid height. Figure 1.2 shows the curve of $\bar{h}$ versus $\bar{F}$ for a specific numerical case.

   So far we have considered just the traditional steadystate design aspects of this fluid flow system. Now let us think about what would happen dynamically if we changed F_0. How will $h_{(t)}$ and $F_{(t)}$ vary with time? Obviously F eventually has to end up at the new value of F_0. We can easily determine from the steadystate design curve of Fig. 1.2 where h will go at the new steadystate. But what paths will $h_{(t)}$ and $F_{(t)}$ take to get to their new steadystates?

   Figure 1.3 sketches the problem. The question is which curves (1 or 2) represent the actual paths that F and h will follow. Curves 1 show gradual increases in h and F to their new steadystate values. However, the paths could follow curves 2 where the liquid height rises above its final steadystate value. This is called "overshoot." Clearly, if the peak of the overshoot in h is above the top of the tank, we would be in trouble.

   Our steadystate design calculations tell us nothing about what the dynamic response to the system will be. They tell us where we will start and where we will end up but not how we get there. This kind of information is what a study of the dynamics of the system will reveal. We will return to this system later in the book to derive a mathematical model of it and to determine its dynamic response quantitatively by simulation.

Example 1.2. Consider the heat exchanger sketched in Fig. 1.4. An oil stream passes through the tube side of a tube-in-shell heat exchanger and is heated by condensing steam on the shell side. The steam condensate leaves through a steam trap (a device

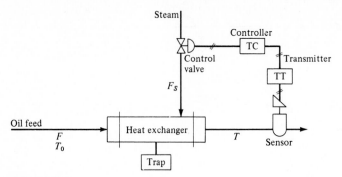

FIGURE 1.4

that only permits liquid to pass through it, thus preventing "blow through" of the steam vapor). We want to control the temperature of the oil leaving the heat exchanger. To do this, a thermocouple is inserted in a thermowell in the exit oil pipe. The thermocouple wires are connected to a "temperature transmitter," an electronic device that converts the millivolt thermocouple output into a 4- to 20-milliampere "control signal." The current signal is sent into a temperature controller, an electronic or digital or pneumatic device that compares the desired temperature (the "setpoint") with the actual temperature, and sends out a signal to a control valve. The temperature controller opens the steam valve more if the temperature is too low or closes it a little if the temperature is too high.

We will consider all the components of this temperature control loop in more detail later in this book. For now we need only appreciate the fact that the automatic control of some variable in a process requires the installation of a sensor, a transmitter, a controller, and a final control element (usually a control valve). Most of this book is aimed at learning how to decide what type of controller should be used and how it should be "tuned," i.e., how should the adjustable tuning parameters in the controller be set so that we do a good job of controlling temperature.

Example 1.3. Our third example illustrates a typical control scheme for an entire simple chemical plant. Figure 1.5 gives a simple schematic sketch of the process configuration and its control system. Two liquid feeds are pumped into a reactor in which they react to form products. The reaction is exothermic, and therefore heat must be removed from the reactor. This is accomplished by adding cooling water to a jacket surrounding the reactor. Reactor effluent is pumped through a preheater into a distillation column that splits it into two product streams.

Traditional steadystate design procedures would be used to specify the various pieces of equipment in the plant:

Fluid mechanics. Pump heads, rates, and power; piping sizes; column tray layout and sizing; heat-exchanger tube and shell side baffling and sizing

Heat transfer. Reactor heat removal; preheater, reboiler, and condenser heat transfer areas; temperature levels of steam and cooling water

Chemical kinetics. Reactor size and operating conditions (temperature, pressure, catalyst, etc.)

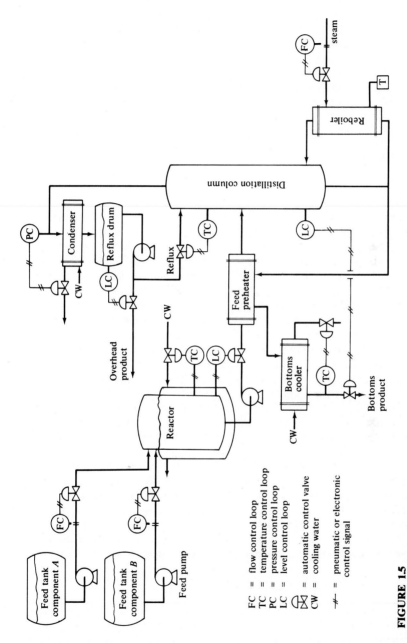

FIGURE 15
Typical chemical plant and control system.

FC = flow control loop
TC = temperature control loop
PC = pressure control loop
LC = level control loop
⊠ = automatic control valve
CW = cooling water
⊬ = pneumatic or electronic
        control signal

5

Thermodynamics and mass transfer. Operating pressure, number of plates and reflux ratio in the distillation column; temperature profile in the column; equilibrium conditions in the reactor

But how do we decide how to control this plant? We will spend most of our time in this book exploring this important design and operating problem. All our studies of mathematical modeling, simulation, and control theory are aimed at understanding the dynamics of processes and control systems so that we can develop and design better, more easily controlled plants that operate more efficiently and more safely.

For now let us say merely that the control system shown in Fig. 1.5 is a typical conventional system. It is about the minimum that would be needed to run this plant automatically without constant operator attention. Notice that even in this simple plant with a minimum of instrumentation the total number of control loops is 10. We will find that most chemical engineering processes are multivariable.

1.2 HISTORICAL BACKGROUND

Most chemical processing plants were run essentially manually prior to the 1940s. Only the most elementary types of controllers were used. Many operators were needed to keep watch on the many variables in the plant. Large tanks were employed to act as buffers or surge capacities between various units in the plant. These tanks, although sometimes quite expensive, served the function of filtering out some of the dynamic disturbances by isolating one part of the process from upsets occurring in another part.

With increasing labor and equipment costs and with the development of more severe, higher-capacity, higher-performance equipment and processes in the 1940s and early 1950s, it became uneconomical and often impossible to run plants without automatic control devices. At this stage feedback controllers were added to the plants with little real consideration of or appreciation for the dynamics of the process itself. Rule-of-thumb guides and experience were the only design techniques.

In the 1960s chemical engineers began to apply dynamic analysis and control theory to chemical engineering processes. Most of the techniques were adapted from the work in the aerospace and electrical engineering fields. In addition to designing better control systems, processes and plants were developed or modified so that they were easier to control. The concept of examining the many parts of a complex plant together as a single unit, with all the interactions included, and devising ways to control the entire plant is called *systems engineering*. The current popular "buzz" words *artificial intelligence* and *expert systems* are being applied to these types of studies.

The rapid rise in energy prices in the 1970s provided additional needs for effective control systems. The design and redesign of many plants to reduce energy consumption resulted in more complex, integrated plants that were much more interacting. So the challenges to the process control engineer have continued to grow over the years. This makes the study of dynamics and control even more vital in the chemical engineering curriculum than it was 30 years ago.

1.3 PERSPECTIVE

Lest I be accused of overstating the relative importance of process control to the main stream of chemical engineering, let me make it perfectly clear that the tools of dynamic analysis are but one part of the practicing engineer's bag of tools and techniques, albeit an increasingly important one. Certainly a solid foundation in the more traditional areas of thermodynamics, kinetics, unit operations, and transport phenomena is essential. In fact, such a foundation is a prerequisite for any study of process dynamics. The mathematical models that we derive are really nothing but extensions of the traditional chemical and physical laws to include the time-dependent terms. Control engineers sometimes have a tendency to get too wrapped up in the dynamics and to forget the steadystate aspects. Keep in mind that if you cannot get the plant to work at steadystate you cannot get it to work dynamically.

An even greater pitfall into which many young process control engineers fall, particularly in recent years, is to get so involved in the fancy computer control hardware that is now available that they lose sight of the process control objectives. All the beautiful CRT displays and the blue smoke and mirrors that computer control salespersons are notorious for using to sell hardware and software can easily seduce the unsuspecting control engineer. Keep in mind your main objective: to come up with an effective control system. How you implement it, in a sophisticated computer or in simple pneumatic instruments, is of much less importance.

You should also appreciate the fact that fighting your way through this book will not in itself make you an expert in process control. You will find that a lot remains to be learned, not so much on a higher theoretical level as you might expect, but more on a practical-experience level. A sharp engineer can learn a tremendous amount about process dynamics and control that can never be put in a book, no matter how practically oriented, by climbing around a plant, talking with operators and instrument mechanics, tinkering in the instrument shop, and keeping his or her eyes open in the control room.

You may question, as you go through this book, the degree to which the dynamic analysis and controller design techniques discussed are really used in industry. At the present time 70 to 80 percent of the control loops in a plant are usually designed, installed, tuned, and operated quite successfully by simple, rule-of-thumb, experience-generated techniques. The other 20 to 30 percent of the loops are those on which the control engineer makes his money. They require more technical knowledge. Plant testing, computer simulation, and detailed controller design or process redesign may be required to achieve the desired performance. These critical loops often make or break the operation of the plant.

I am confident that the techniques discussed in this book will receive wider and wider applications as more young engineers with this training go to work in chemical plants. This book is an attempt by an old dog to pass along some useful engineering tools to the next generation of pups. It represents over thirty years of experience in this lively and ever-challenging area. Like any "expert," I've learned from my successes, but probably more from my failures. I hope this book helps you to have many of the former and not too many of the latter. Remember the old saying: "If you are making mistakes, but they are new ones, you are getting smarter."

1.4 MOTIVATION FOR STUDYING PROCESS CONTROL

Some of the motivational reasons for studying the subjects presented in this book are that they are of considerable practical importance, they are challenging, and they are fun.

1. *Importance.* The control room is the major interface with the plant. Automation is increasingly common in all degrees of sophistication, from single-loop systems to computer-control systems.
2. *Challenging.* You will have to draw on your knowledge of all areas of chemical engineering. You will use most of the mathematical tools available (differential equations, Laplace transforms, complex variables, numerical analysis, etc.) to solve real problems.
3. *Fun.* I have found, and I hope you will too, that process dynamics is fun. You will get the opportunity to use some simple as well as some fairly advanced mathematics to solve real plant problems. There is nothing quite like the thrill of working out a controller design on paper and then seeing it actually work on the plant. You will get a lot of satisfaction out of going into a plant that is having major control problems, diagnosing what is causing the problem and getting the whole plant lined out on specification. Sometimes the problem is in the process, in basic design, or in equipment malfunctioning. But sometimes it is in the control system, in basic strategy, or in hardware malfunctioning. Just your knowledge of what a given control device *should do* can be invaluable.

1.5 GENERAL CONCEPTS

I have tried to present in this book a logical development. We will begin with fundamentals and simple concepts and extend them as far as they can be gainfully extended. First we will learn to derive mathematical models of chemical engineering systems. Then we will study some of the ways to solve the resulting equations, usually ordinary differential equations and nonlinear algebraic equations. Next we will explore their openloop (uncontrolled) dynamic behavior. Finally we will learn to design controllers that will, if we are smart enough, make the plant run automatically the way we want it to run: efficiently and safely.

Before we go into details in the subsequent chapters, it may be worthwhile at this point to define some very broad and general concepts and some of the terminology used in dynamics and control.

1. *Dynamics.* Time-dependent behavior of a process. The behavior with no controllers in the system is called the *openloop* response. The dynamic behavior with feedback controllers included with the process is called the *closedloop* response.

2. *Variables.*

    a. *Manipulated variables.* Typically flow rates of streams entering or leaving a process that we can change in order to control the plant.

    b. *Controlled variables.* Flow rates, compositions, temperatures, levels, and pressures in the process that we will try to control, either trying to hold them as constant as possible or trying to make them follow some desired time trajectory.

    c. *Uncontrolled variables.* Variables in the process that are not controlled.

    d. *Load disturbances.* Flow rates, temperatures, or compositions of streams entering (but sometimes leaving) the process. We are not free to manipulate them. They are set by upstream or downstream parts of the plant. The control system must be able to keep the plant under control despite the effects of these disturbances.

Example 1.4. For the heat exchanger shown in Fig. 1.4, the load disturbances are oil feed flow rate F and oil inlet temperature T_0. The steam flow rate F_s is the manipulated variable. The controlled variable is the oil exit temperature T.

Example 1.5. For a binary distillation column (see Fig. 1.6), load disturbance variables might include feed flow rate and feed composition. Reflux, steam, cooling water, distillate, and bottoms flow rates might be the manipulated variables. Controlled variables might be distillate product composition, bottoms product composition, column pressure, base liquid level, and reflux drum liquid level. The uncontrolled variables would include the compositions and temperatures on all the trays. Note that one physical stream may be considered to contain many variables:

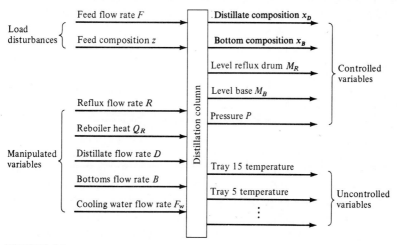

FIGURE 1.6

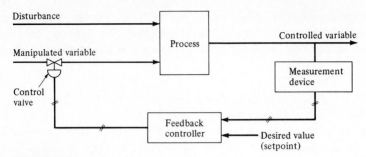

FIGURE 1.7
Feedback control loop.

its flow rate, its composition, its temperature, etc., i.e., all its intensive and extensive properties.

3. *Feedback control.* The traditional way to control a process is to measure the variable that is to be controlled, compare its value with the desired value (the setpoint to the controller) and feed the difference (the error) into a feedback controller that will change a manipulated variable to drive the controlled variable back to the desired value. Information is thus "fed back" from the controlled variable to a manipulated variable, as sketched in Fig. 1.7.

4. *Feedforward control.* The basic idea is shown in Fig. 1.8. The disturbance is detected as it enters the process and an appropriate change is made in the manipulated variable such that the controlled variable is held constant. Thus we begin to take corrective action as soon as a disturbance entering the system is detected instead of waiting (as we do with feedback control) for the disturbance to propagate all the way through the process before a correction is made.

5. *Stability.* A process is said to be unstable if its output becomes larger and larger (either positively or negatively) as time increases. Examples are shown in Fig. 1.9. No real system really does this, of course, because some constraint will be met; for example, a control valve will completely shut or completely open, or a safety valve will "pop." A linear process is right at the limit of

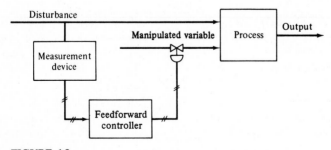

FIGURE 1.8
Feedforward control.

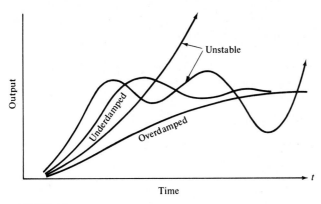

FIGURE 1.9
Stability.

stability if it oscillates, even when undisturbed, and the amplitude of the oscillations does not decay.

Most processes are *openloop stable*, i.e., stable with no controllers on the system. One important and very interesting exception that we will study in some detail is the exothermic chemical reactor which can be openloop unstable. All real processes can be made *closedloop unstable* (unstable when a feedback controller is in the system) if the controller gain is made large enough. Thus stability is of vital concern in feedback control systems.

The *performance* of a control system (its ability to control the process tightly) usually increases as we increase the controller gain. However, we get closer and closer to being closedloop unstable. Therefore the *robustness* of the control system (its tolerance to changes in process parameters) decreases: a small change will make the system unstable. Thus there is always a trade-off between robustness and performance in control system design.

1.6 LAWS AND LANGUAGES OF PROCESS CONTROL

1.6.1 Process Control Laws

There are several fundamental laws that have been developed in the process control field as a result of many years of experience. Some of these may sound similar to some of the laws attributed to Parkinson, but the process control laws are not intended to be humorous.

(1) FIRST LAW. The simplest control system that will do the job is the best.

Complex elegant control systems look great on paper but soon end up on "manual" in an industrial environment. Bigger is definitely not better in control system design.

(2) SECOND LAW. You must understand the process before you can control it.

No degree of sophistication in the control system (be it adaptive control, Kalman filters, expert systems, etc.) will work if you do not know how your process works. Many people have tried to use complex controllers to overcome ignorance about the process fundamentals, and they have failed! Learn how the process works before you start designing its control system.

1.6.2 Languages of Process Control

As you will see, several different approaches are used in this book to analyze the dynamics of systems. Direct solution of the differential equations to give functions of time is a "time domain" technique. Use of Laplace transforms to characterize the dynamics of systems is a "Laplace domain" technique. Frequency response methods provide another approach to the problem.

All of these methods are useful because each has its advantages and disadvantages. They yield exactly the same results when applied to the same problem. These various approaches are similar to the use of different languages by people around the world. A table in English is described by the word "TABLE." In Russian a table is described by the word "СТОЛ." In Chinese a table is "桌子." In German it is "der Tisch." But in any language a table is still a table.

In the study of process dynamics and control we will use several languages.

English = time domain (differential equations, yielding exponential time function solutions)

Russian = Laplace domain (transfer functions)

Chinese = frequency domain (frequency response Bode and Nyquist plots)

Greek = state variables (matrix methods applies to differential equations)

German = z domain (sampled-data systems)

You will find the languages are not difficult to learn because the vocabulary that is required is quite small. Only 8 to 10 "words" must be learned in each language. Thus it is fairly easy to translate back and forth between the languages.

We will use "English" to solve some simple problems. But we will find that more complex problems are easier to understand and solve using "Russian." As problems get even more complex and realistic, the use of "Chinese" is required. So we study in this book a number of very useful and practical process control languages.

I have chosen the five languages listed above simply because I have had some exposure to all of them over the years. Let me assure you that no political or nationalistic motives are involved. If you would prefer French, Spanish, Italian, Japanese, and Swahili, please feel free to make the appropriate substitutions! My purpose in using the language metaphor is to try to break some of the psychological barriers that students have to such things as Laplace transforms and frequency response. It is a pedagogical gimmick that I have used for over two decades and have found it to be very effective with students.

PART
I

MATHEMATICAL MODELS OF CHEMICAL ENGINEERING SYSTEMS

In the next two chapters we will develop dynamic mathematical models for several important chemical engineering systems. The examples should illustrate the basic approach to the problem of mathematical modeling.

Mathematical modeling is very much an art. It takes experience, practice, and brain power to be a good mathematical modeler. You will see a few models developed in these chapters. You should be able to apply the same approaches to your own process when the need arises. Just remember to always go back to basics: mass, energy, and momentum balances applied in their time-varying form.

2

FUNDAMENTALS

2.1 INTRODUCTION

2.1.1 Uses of Mathematical Models

Without doubt, the most important result of developing a mathematical model of a chemical engineering system is the understanding that is gained of what really makes the process "tick." This insight enables you to strip away from the problem the many extraneous "confusion factors" and to get to the core of the system. You can see more clearly the cause-and-effect relationships between the variables.

Mathematical models can be useful in all phases of chemical engineering, from research and development to plant operations, and even in business and economic studies.

1. Research and development: determining chemical kinetic mechanisms and parameters from laboratory or pilot-plant reaction data; exploring the effects of different operating conditions for optimization and control studies; aiding in scale-up calculations.
2. Design: exploring the sizing and arrangement of processing equipment for dynamic performance; studying the interactions of various parts of the process, particularly when material recycle or heat integration is used; evaluating alternative process and control structures and strategies; simulating start-up, shutdown, and emergency situations and procedures.

3. Plant operation: troubleshooting control and processing problems; aiding in start-up and operator training; studying the effects of and the requirements for expansion (bottleneck-removal) projects; optimizing plant operation. It is usually much cheaper, safer, and faster to conduct the kinds of studies listed above on a mathematical model than experimentally on an operating unit. This is not to say that plant tests are not needed. As we will discuss later, they are a vital part of confirming the validity of the model and of verifying important ideas and recommendations that evolve from the model studies.

2.1.2 Scope of Coverage

We will discuss in this book only deterministic systems that can be described by ordinary or partial differential equations. Most of the emphasis will be on *lumped* systems (with one independent variable, time, described by ordinary differential equations). Both English and SI units will be used. You need to be familiar with both.

2.1.3 Principles of Formulation

A. BASIS. The bases for mathematical models are the fundamental physical and chemical laws, such as the laws of conservation of mass, energy, and momentum. To study dynamics we will use them in their general form with time derivatives included.

B. ASSUMPTIONS. Probably the most vital role that the engineer plays in modeling is in exercising his engineering judgment as to what assumptions can be validly made. Obviously an extremely rigorous model that includes every phenomenon down to microscopic detail would be so complex that it would take a long time to develop and might be impractical to solve, even on the latest supercomputers. An engineering compromise between a rigorous description and getting an answer that is good enough is always required. This has been called "optimum sloppiness." It involves making as many simplifying assumptions as are reasonable without "throwing out the baby with the bath water." In practice, this optimum usually corresponds to a model which is as complex as the available computing facilities will permit. More and more this is a personal computer.

The development of a model that incorporates the basic phenomena occurring in the process requires a lot of skill, ingenuity, and practice. It is an area where the creativity and innovativeness of the engineer is a key element in the success of the process.

The assumptions that are made should be carefully considered and listed. They impose limitations on the model that should always be kept in mind when evaluating its predicted results.

C. MATHEMATICAL CONSISTENCY OF MODEL. Once all the equations of the mathematical model have been written, it is usually a good idea, particularly with big, complex systems of equations, to make sure that the number of variables equals the number of equations. The so-called "degrees of freedom" of the system must be zero in order to obtain a solution. If this is not true, the system is underspecified or overspecified and something is wrong with the formulation of the problem. This kind of consistency check may seem trivial, but I can testify from sad experience that it can save many hours of frustration, confusion, and wasted computer time.

Checking to see that the units of all terms in all equations are consistent is perhaps another trivial and obvious step, but one that is often forgotten. It is essential to be particularly careful of the time units of parameters in dynamic models. Any units can be used (seconds, minutes, hours, etc.), but they cannot be mixed. We will use "minutes" in most of our examples, but it should be remembered that many parameters are commonly on other time bases and need to be converted appropriately, e.g., overall heat transfer coefficients in Btu/h °F ft^2 or velocity in m/s. Dynamic simulation results are frequently in error because the engineer has forgotten a factor of "60" somewhere in the equations.

D. SOLUTION OF THE MODEL EQUATIONS. We will concern ourselves in detail with this aspect of the model in Part II. However, the available solution techniques and tools must be kept in mind as a mathematical model is developed. An equation without any way to solve it is not worth much.

E. VERIFICATION. An important but often neglected part of developing a mathematical model is proving that the model describes the real-world situation. At the design stage this sometimes cannot be done because the plant has not yet been built. However, even in this situation there are usually either similar existing plants or a pilot plant from which some experimental dynamic data can be obtained.

The design of experiments to test the validity of a dynamic model can sometimes be a real challenge and should be carefully thought out. We will talk about dynamic testing techniques, such as pulse testing, in Chap. 14.

2.2 FUNDAMENTAL LAWS

In this section, some fundamental laws of physics and chemistry are reviewed in their general time-dependent form, and their application to some simple chemical systems is illustrated.

2.2.1 Continuity Equations

A. TOTAL CONTINUITY EQUATION (MASS BALANCE). The principle of the conservation of mass when applied to a dynamic system says

$$\begin{bmatrix} \text{Mass flow} \\ \text{into system} \end{bmatrix} - \begin{bmatrix} \text{mass flow} \\ \text{out of system} \end{bmatrix} = \begin{bmatrix} \text{time rate of change} \\ \text{of mass inside system} \end{bmatrix} \qquad (2.1)$$

The units of this equation are mass per time. Only *one* total continuity equation can be written for one system.

The normal steadystate design equation that we are accustomed to using says that "what goes in, comes out." The dynamic version of this says the same thing with the addition of the world "eventually."

The right-hand side of Eq. (2.1) will be either a partial derivative $\partial/\partial t$ or an ordinary derivative d/dt of the mass inside the system with respect to the independent variable t.

Example 2.1 Consider the tank of perfectly mixed liquid shown in Fig. 2.1 into which flows a liquid stream at a volumetric rate of F_0 (ft^3/min or m^3/min) and with a density of ρ_0 (lb$_m$/ft^3 or kg/m^3). The volumetric holdup of liquid in the tank is V (ft^3 or m^3), and its density is ρ. The volumetric flow rate from the tank is F, and the density of the outflowing stream is the same as that of the tank's contents.

The system for which we want to write a total continuity equation is all the liquid phase in the tank. We call this a macroscopic system, as opposed to a microscopic system, since it is of definite and finite size. The mass balance is around the whole tank, not just a small, differential element inside the tank.

$$F_0 \rho_0 - F\rho = \text{time rate of change of } \rho V \qquad (2.2)$$

The units of this equation are lb$_m$/min or kg/min.

$$\left(\frac{\text{ft}^3}{\text{min}}\right)\left(\frac{\text{lb}_m}{\text{ft}^3}\right) - \left(\frac{\text{ft}^3}{\text{min}}\right)\left(\frac{\text{lb}_m}{\text{ft}^3}\right) = \frac{(\text{ft}^3)(\text{lb}_m/\text{ft}^3)}{\text{min}}$$

Since the liquid is perfectly mixed, the density is the same everywhere in the tank; it does not vary with radial or axial position; i.e., there are no spatial gradients in density in the tank. This is why we can use a macroscopic system. It also means that there is only one independent variable, t.

Since ρ and V are functions only of t, an ordinary derivative is used in Eq. (2.2).

$$\frac{d(\rho V)}{dt} = F_0 \rho_0 - F\rho \qquad (2.3)$$

Example 2.2. Fluid is flowing through a constant-diameter cylindrical pipe sketched in Fig. 2.2. The flow is turbulent and therefore we can assume plug-flow conditions, i.e., each "slice" of liquid flows down the pipe as a unit. There are no radial gradients in velocity or any other properties. However, axial gradients can exist.

Density and velocity can change as the fluid flows along the axial or z direction. There are now two independent variables: time t and position z. Density and

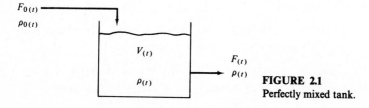

FIGURE 2.1
Perfectly mixed tank.

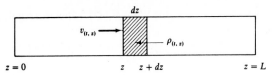

FIGURE 2.2
Flow through a pipe.

velocity are functions of both t and z: $\rho_{(t, z)}$ and $v_{(t, z)}$. We want to apply the total continuity equation [Eq. (2.1)] to a system that consists of a small slice. The system is now a "microscopic" one. The differential element is located at an arbitrary spot z down the pipe. It is dz thick and has an area equal to the cross-sectional area of the pipe A (ft^2 or m^2).

Time rate of change of mass inside system:

$$\frac{\partial(A\rho \; dz)}{\partial t} \tag{2.4}$$

$A \; dz$ is the volume of the system; ρ is the density. The units of this equation are lb$_m$/min or kg/min.

Mass flowing into system through boundary at z:

$$vA\rho \tag{2.5}$$

Notice that the units are still lb$_m$/min $=$ (ft/min)(ft^2)(lb$_m$/ft^3).

Mass flowing out of the system through boundary at $z + dz$:

$$vA\rho + \frac{\partial(vA\rho)}{\partial z} \; dz \tag{2.6}$$

The above expression for the flow at $z + dz$ may be thought of as a Taylor series expansion of a function $f_{(z)}$ around z. The value of the function at a spot dz away from z is

$$f_{(z+dz)} = f_{(z)} + \left(\frac{\partial f}{\partial z}\right)_{(z)} dz + \left(\frac{\partial^2 f}{\partial z^2}\right)_{(z)} \frac{(dz)^2}{2!} + \cdots \tag{2.7}$$

If the dz is small, the series can be truncated after the first derivative term. Letting $f_{(z)} = vA\rho$ gives Eq. (2.6).

Substituting these terms into Eq. (2.1) gives

$$\frac{\partial(A\rho \; dz)}{\partial t} = vA\rho - \left[vA\rho + \frac{\partial(vA\rho)}{\partial z} \; dz\right]$$

Canceling out the dz terms and assuming A is constant yield

$$\frac{\partial \rho}{\partial t} + \frac{\partial(v\rho)}{\partial z} = 0 \tag{2.8}$$

B. COMPONENT CONTINUITY EQUATIONS (COMPONENT BALANCES).
Unlike mass, chemical components are *not* conserved. If a reaction occurs inside a system, the number of moles of an individual component will increase if it is a

product of the reaction or decrease if it is a reactant. Therefore the component continuity equation of the jth chemical species of the system says

$$\begin{bmatrix} \text{Flow of moles of } j\text{th} \\ \text{component into system} \end{bmatrix} - \begin{bmatrix} \text{flow of moles of } j\text{th} \\ \text{component out of system} \end{bmatrix}$$

$$+ \begin{bmatrix} \text{rate of formation of moles of } j\text{th} \\ \text{component from chemical reactions} \end{bmatrix}$$

$$= \begin{bmatrix} \text{time rate of change of moles of } j\text{th} \\ \text{component inside system} \end{bmatrix} \quad (2.9)$$

The units of this equation are moles of component j per unit time.

The flows in and out can be both convective (due to bulk flow) and molecular (due to diffusion). We can write one component continuity equation for *each* component in the system. If there are NC components, there are NC component continuity equations for any one system. However, the *one* total mass balance and these NC component balances are not all independent, since the sum of all the moles times their respective molecular weights equals the total mass. Therefore a given system has only NC independent continuity equations. We usually use the total mass balance and $NC - 1$ component balances. For example, in a binary (two-component) system, there would be one total mass balance and one component balance.

> **Example 2.3.** Consider the same tank of perfectly mixed liquid that we used in Example 2.1 except that a chemical reaction takes place in the liquid in the tank. The system is now a CSTR (continuous stirred-tank reactor) as shown in Fig. 2.3. Component A reacts irreversibly and at a specific reaction rate k to form product, component B.
>
> $$A \xrightarrow{\;k\;} B$$
>
> Let the concentration of component A in the inflowing feed stream be C_{A0} (moles of A per unit volume) and in the reactor C_A. Assuming a simple first-order reaction, the rate of consumption of reactant A per unit volume will be directly proportional to the instantaneous concentration of A in the tank. Filling in the terms in Eq. (2.9) for a component balance on reactant A,
>
> $$\text{Flow of A into system} = F_0 C_{A0}$$
>
> $$\text{Flow of A out of system} = F C_A$$
>
> $$\text{Rate of formation of A from reaction} = -V k C_A$$

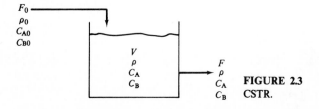

FIGURE 2.3 CSTR.

The minus sign comes from the fact that A is being consumed, not produced. The units of all these terms must be the same: moles of A per unit time. Therefore the VkC_A term must have these units, for example $(ft^3)(min^{-1})(moles of A/ft^3)$. Thus the units of k in this system are min^{-1}.

$$\text{Time rate of change of A inside tank} = \frac{d(VC_A)}{dt}$$

Combining all of the above gives

$$\frac{d(VC_A)}{dt} = F_0 C_{A0} - FC_A - VkC_A \qquad (2.10)$$

We have used an ordinary derivative since t is the only independent variable in this lumped system. The units of this component continuity equation are moles of A per unit time. The left-hand side of the equation is the dynamic term. The first two terms on the right-hand side are the convective terms. The last term is the generation term.

Since the system is binary (components A and B), we could write another component continuity equation for component B. Let C_B be the concentration of B in moles of B per unit volume.

$$\frac{d(VC_B)}{dt} = F_0 C_{B0} - FC_B + VkC_A$$

Note the plus sign before the generation term since B is being produced by the reaction. Alternatively we could use the total continuity equation [Eq. (2.3)] since C_A, C_B, and ρ are uniquely related by

$$M_A C_A + M_B C_B = \rho \qquad (2.11)$$

where M_A and M_B are the molecular weights of components A and B, respectively.

Example 2.4. Suppose we have the same macroscopic system as above except that now consecutive reactions occur. Reactant A goes to B at a specific reaction rate k_1, but B can react at a specific reaction rate k_2 to form a third component C.

$$A \xrightarrow{k_1} B \xrightarrow{k_2} C$$

Assuming first-order reactions, the component continuity equations for components A, B, and C are

$$\frac{d(VC_A)}{dt} = F_0 C_{A0} - FC_A - Vk_1 C_A$$

$$\frac{d(VC_B)}{dt} = F_0 C_{B0} - FC_B + Vk_1 C_A - Vk_2 C_B \qquad (2.12)$$

$$\frac{d(VC_C)}{dt} = F_0 C_{C0} - FC_C + Vk_2 C_B$$

The component concentrations are related to the density

$$\sum_{j=A}^{c} M_j C_j = \rho \tag{2.13}$$

Three component balances could be used or we could use two of the component balances and a total mass balance.

Example 2.5. Instead of fluid flowing down a pipe as in Example 2.2, suppose the pipe is a tubular reactor in which the same reaction A $\xrightarrow{k}$ B of Example 2.3 takes place. As a slice of material moves down the length of the reactor the concentration of reactant C_A decreases as A is consumed. Density ρ, velocity v, and concentration C_A can all vary with time and axial position z. We still assume plug-flow conditions so that there are no radial gradients in velocity, density, or concentration.

The concentration of A fed to the inlet of the reactor at $z = 0$ is defined as

$$C_{A(t,\,0)} = C_{A0(t)} \tag{2.14}$$

The concentration of A in the reactor effluent at $z = L$ is defined as

$$C_{A(t,\,L)} = C_{AL(t)} \tag{2.15}$$

We now want to apply the component continuity equation for reactant A to a small differential slice of width dz, as shown in Fig. 2.4. The inflow terms can be split into two types: bulk flow and diffusion. Diffusion can occur because of the concentration gradient in the axial direction. It is usually much less important than bulk flow in most practical systems, but we include it here to see what it contributes to the model. We will say that the diffusive flux of A, N_A (moles of A per unit time per unit area), is given by a Fick's law type of relationship

$$N_A = -\mathfrak{D}_A \frac{\partial C_A}{\partial z} \tag{2.16}$$

where $\mathfrak{D}_A$ is a diffusion coefficient due to both diffusion and turbulence in the fluid flow (so-called "eddy diffusivity"). $\mathfrak{D}_A$ has units of length^2 per unit time.

The terms in the general component continuity equation [Eq. (2.9)] are:
Molar flow of A into boundary at z (bulk flow and diffusion)

$$= vAC_A + AN_A \qquad \text{(moles of A/s)}$$

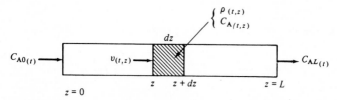

FIGURE 2.4
Tubular reactor.

Molar flow of A leaving system at boundary $z + dz$

$$= (vAC_A + AN_A) + \frac{\partial(vAC_A + AN_A)}{\partial z} dz$$

Rate of formation of A inside system $= -kC_A A\, dz$

Time rate of change of A inside system $= \dfrac{\partial(A\, dz\, C_A)}{\partial t}$

Substituting into Eq. (2.9) gives

$$\frac{\partial(A\, dz\, C_A)}{\partial t} = (vAC_A + AN_A) - \left(vAC_A + AN_A + \frac{\partial(vAC_A + AN_A)}{\partial z} dz \right) - kC_A A\, dz$$

$$\frac{\partial C_A}{\partial t} + \frac{\partial(vC_A + N_A)}{\partial z} + kC_A = 0$$

Substituting Eq. (2.16) for N_A,

$$\frac{\partial C_A}{\partial t} + \frac{\partial(vC_A)}{\partial z} + kC_A = \frac{\partial}{\partial z}\left(\mathcal{D}_A \frac{\partial C_A}{\partial z} \right) \tag{2.17}$$

The units of the equation are moles A per volume per time.

2.2.2 Energy Equation

The first law of thermodynamics puts forward the principle of conservation of energy. Written for a general "open" system (where flow of material in and out of the system can occur) it is

$$\begin{bmatrix} \text{Flow of internal, kinetic, and} \\ \text{potential energy into system} \\ \text{by convection or diffusion} \end{bmatrix} - \begin{bmatrix} \text{flow of internal, kinetic, and} \\ \text{potential energy out of system} \\ \text{by convection or diffusion} \end{bmatrix}$$

$$+ \begin{bmatrix} \text{heat added to system by} \\ \text{conduction, radiation, and} \\ \text{reaction} \end{bmatrix} - \begin{bmatrix} \text{work done by system on} \\ \text{surroundings (shaft work and} \\ \text{PV work)} \end{bmatrix}$$

$$= \begin{bmatrix} \text{time rate of change of internal, kinetic,} \\ \text{and potential energy inside system} \end{bmatrix} \tag{2.18}$$

Example 2.6. The CSTR system of Example 2.3 will be considered again, this time with a cooling coil inside the tank that can remove the exothermic heat of reaction λ (Btu/lb · mol of A reacted or cal/g · mol of A reacted). We use the normal convention that λ is negative for an exothermic reaction and positive for an endothermic reaction. The rate of heat generation (energy per time) due to reaction is the rate of consumption of A times λ.

$$Q_G = -\lambda V C_A k \tag{2.19}$$

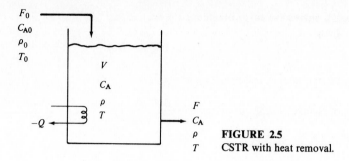

FIGURE 2.5
CSTR with heat removal.

The rate of heat removal from the reaction mass to the cooling coil is $-Q$ (energy per time). The temperature of the feed stream is T_0 and the temperature in the reactor is T (°R or K). Writing Eq. (2.18) for this system,

$$F_0 \rho_0 (U_0 + K_0 + \phi_0) - F\rho(U + K + \phi) + (Q_G + Q)$$

$$- (W + FP - F_0 P_0) = \frac{d}{dt}[(U + K + \phi)V\rho] \quad (2.20)$$

where U = internal energy (energy per unit mass)
K = kinetic energy (energy per unit mass)
ϕ = potential energy (energy per unit mass)
W = shaft work done by system (energy per time)
P = pressure of system
P_0 = pressure of feed stream

Note that all the terms in Eq. (2.20) must have the same units (energy per time) so the FP terms must use the appropriate conversion factor (778 ft·lb$_f$/Btu in English engineering units).

In the system shown in Fig. 2.5 there is no shaft work, so $W = 0$. If the inlet and outlet flow velocities are not very high, the kinetic-energy term is negligible. If the elevations of the inlet and outlet flows are about the same, the potential-energy term is small. Thus Eq. (2.20) reduces to

$$\frac{d(\rho V U)}{dt} = F_0 \rho_0 U_0 - F\rho U + Q_G + Q - F\rho \frac{P}{\rho} + F_0 \rho_0 \frac{P_0}{\rho_0}$$

$$= F_0 \rho_0 (U_0 + P_0 \bar{V}_0) - F\rho(U + P\bar{V}) + Q_G + Q \quad (2.21)$$

where $\bar{V}$ is the specific volume (ft³/lb$_m$ or m³/kg), the reciprocal of the density. Enthalpy, H or h, is defined:

$$H \text{ or } h \equiv U + P\bar{V} \quad (2.22)$$

We will use h for the enthalpy of a liquid stream and H for the enthalpy of a vapor stream. Thus, for the CSTR, Eq. (2.21) becomes

$$\frac{d(\rho V U)}{dt} = F_0 \rho_0 h_0 - F\rho h + Q - \lambda V k C_A \quad (2.23)$$

For liquids the $P\bar{V}$ term is negligible compared to the U term, and we use the time rate of change of the enthalpy of the system instead of the internal energy of the system.

$$\frac{d(\rho Vh)}{dt} = F_0 \rho_0 h_0 - F\rho h + Q - \lambda VkC_A \qquad (2.24)$$

The enthalpies are functions of composition, temperature, and pressure, but primarily temperature. From thermodynamics, the heat capacities at constant pressure, C_p, and at constant volume, C_v, are

$$C_p = \left(\frac{\partial H}{\partial T}\right)_p \qquad C_v = \left(\frac{\partial U}{\partial T}\right)_v \qquad (2.25)$$

To illustrate that the energy is primarily influenced by temperature, let us simplify the problem by assuming that the liquid enthalpy can be expressed as a product of absolute temperature and an average heat capacity C_p (Btu/lb$_m$°R or cal/g K) that is constant.

$$h = C_p T$$

We will also assume that the densities of all the liquid streams are constant. With these simplifications Eq. (2.24) becomes

$$\rho C_p \frac{d(VT)}{dt} = \rho C_p(F_0 T_0 - FT) + Q - \lambda VkC_A \qquad (2.26)$$

Example 2.7. To show what form the energy equation takes for a two-phase system, consider the CSTR process shown in Fig. 2.6. Both a liquid product stream F and a vapor product stream F_v (volumetric flow) are withdrawn from the vessel. The pressure in the reactor is P. Vapor and liquid volumes are V_v and V. The density and temperature of the vapor phase are ρ_v and T_v. The mole fraction of A in the vapor is y. If the phases are in thermal equilibrium, the vapor and liquid temperatures are equal ($T = T_v$). If the phases are in phase equilibrium, the liquid and vapor compositions are related by Raoult's law, a relative volatility relationship or some other vapor-liquid equilibrium relationship (see Sec. 2.2.6). The enthalpy of the vapor phase H (Btu/lb$_m$ or cal/g) is a function of composition y, temperature T_v, and pressure P. Neglecting kinetic-energy and potential-energy terms and the work term,

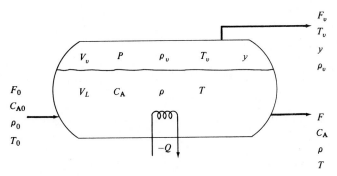

FIGURE 2.6
Two-phase CSTR with heat removal.

and replacing internal energies with enthalpies in the time derivative, the energy equation of the system (the vapor and liquid contents of the tank) becomes

$$\frac{d(\rho_v V_v H + \rho V_L h)}{dt} = F_0 \rho_0 h_0 - F\rho h - F_v \rho_v H + Q - \lambda V k C_A \qquad (2.27)$$

In order to express this equation explicitly in terms of temperature, let us again use a very simple form for h ($h = C_p T$) and an equally simple form for H.

$$H = C_p T + \lambda_v \qquad (2.28)$$

where λ_v is an average heat of vaporization of the mixture. In a more rigorous model λ_v could be a function of temperature T_v, composition y, and pressure P. Equation (2.27) becomes

$$\frac{d[\rho_v V_v(C_p T + \lambda_v) + \rho V_L C_p T]}{dt} = F_0 \rho_0 C_p T_0 - F\rho C_p T$$

$$- F_v \rho_v (C_p T + \lambda_v) + Q - \lambda V k C_A \qquad (2.29)$$

Example 2.8. To illustrate the application of the energy equation to a microscopic system, let us return to the plug-flow tubular reactor and now keep track of temperature changes as the fluid flows down the pipe. We will again assume no radial gradients in velocity, concentration, or temperature (a very poor assumption in some strongly exothermic systems if the pipe diameter is not kept small). Suppose that the reactor has a cooling jacket around it as shown in Fig. 2.7. Heat can be transferred from the process fluid reactants and products at temperature T to the metal wall of the reactor at temperature T_M. The heat is subsequently transferred to the cooling water. For a complete description of the system we would need energy equations for the process fluid, the metal wall, and the cooling water. Here we will concern ourselves only with the process energy equation.

Looking at a little slice of the process fluid as our system, we can derive each of the terms of Eq. (2.18). Potential-energy and kinetic-energy terms are assumed negligible, and there is no work term. The simplified forms of the internal energy and enthalpy are assumed. Diffusive flow is assumed negligible compared to bulk flow. We will include the possibility for conduction of heat axially along the reactor due to molecular or turbulent conduction.

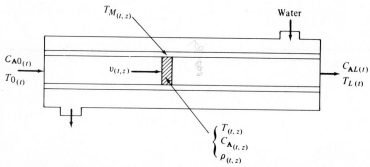

FIGURE 2.7
Jacketed tubular reactor.

Flow of energy (enthalpy) into boundary at z due to bulk flow:

$$v A \rho C_p T \quad \text{with English engineering units of} \quad \frac{\text{ft}}{\text{min}} \, \text{ft}^2 \, \frac{\text{lb}_\text{m}}{\text{ft}^3} \, \frac{\text{Btu}}{\text{lb}_\text{m} \, {}^\circ\text{R}} \, {}^\circ\text{R} = \text{Btu/min}$$

Flow of energy (enthalpy) out of boundary at $z + dz$:

$$v A \rho C_p T + \frac{\partial (v A \rho C_p T)}{\partial z} \, dz$$

$$\text{Heat generated by chemical reaction} = -A \, dz \, k C_A \lambda$$

$$\text{Heat transferred to metal wall} = -h_T (\pi D \, dz)(T - T_M)$$

where h_T = heat transfer film coefficient, Btu/min ft^2 °R
$\quad D$ = diameter of pipe, ft

$$\text{Heat conduction into boundary at } z = q_z A$$

where q_z is a heat flux in the z direction due to conduction. We will use Fourier's law to express q_z in terms of a temperature driving force:

$$q_z = -k_T \frac{\partial T}{\partial z} \tag{2.30}$$

where k_T is an effective thermal conductivity with English engineering units of Btu/ft min °R.

$$\text{Heat conduction out of boundary at } z + dz = q_z A + \frac{\partial (q_z A)}{\partial z} \, dz$$

$$\text{Rate of change of internal energy (enthalpy) of the system} = \frac{\partial (\rho A \, dz \, C_p T)}{\partial t}$$

Combining all the above gives

$$\frac{\partial (\rho C_p T)}{\partial t} + \frac{\partial (v \rho C_p T)}{\partial z} + \lambda k C_A + \frac{4 h_T}{D} (T - T_M) = \frac{\partial [k_T (\partial T / \partial z)]}{\partial z} \tag{2.31}$$

2.2.3   Equations of Motion

As any high school student knows, Newton's second law of motion says that force is equal to mass times acceleration for a system with constant mass M.

$$F = \frac{Ma}{g_c} \tag{2.32}$$

where F = force, lb$_\text{f}$
$\quad M$ = mass, lb$_\text{m}$
$\quad a$ = acceleration, ft/s^2
$\quad g_c$ = conversion constant needed when English engineering units are used to keep units consistent = 32.2 lb$_\text{m}$ ft/lb$_\text{f}$ s^2

This is the basic relationship that is used in writing the equations of motion for a system. In a slightly more general form, where mass can vary with time,

$$\frac{1}{g_c} \frac{d(Mv_i)}{dt} = \sum_{j=1}^{N} F_{ji} \qquad (2.33)$$

where v_i = velocity in the i direction, ft/s
$\quad F_{ji} = j$th force acting in the i direction

Equation (2.33) says that the time rate of change of momentum in the i direction (mass times velocity in the i direction) is equal to the net sum of the forces pushing in the i direction. It can be thought of as a dynamic force balance. Or more eloquently it is called the *conservation of momentum.*

In the real world there are three directions: x, y, and z. Thus, three force balances can be written for any system. Therefore, each system has three equations of motion (plus one total mass balance, one energy equation, and $NC - 1$ component balances).

Instead of writing three equations of motion, it is often more convenient (and always more elegant) to write the three equations as one vector equation. We will not use the vector form in this book since all our examples will be simple one-dimensional force balances. The field of fluid mechanics makes extensive use of the conservation of momentum.

Example 2.9. The gravity-flow tank system described in Chap. 1 provides a simple example of the application of the equations of motion to a macroscopic system. Referring to Fig. 1.1, let the length of the exit line be L (ft) and its cross-sectional area be A_p (ft^2). The vertical, cylindrical tank has a cross-sectional area of A_T (ft^2).

The part of this process that is described by a force balance is the liquid flowing through the pipe. It will have a mass equal to the volume of the pipe $(A_p L)$ times the density of the liquid ρ. This mass of liquid will have a velocity v (ft/s) equal to the volumetric flow divided by the cross-sectional area of the pipe. Remember we have assumed plug-flow conditions and incompressible liquid, and therefore all the liquid is moving at the same velocity, more or less like a solid rod. If the flow is turbulent, this is not a bad assumption.

$$M = A_p L\rho$$

$$v = \frac{F}{A_p} \qquad (2.34)$$

The amount of liquid in the pipe will not change with time, but if we want to change the rate of outflow, the velocity of the liquid must be changed. And to change the velocity or the momentum of the liquid we must exert a force on the liquid.

The direction of interest in this problem is the horizontal, since the pipe is assumed to be horizontal. The force pushing on the liquid at the left end of the pipe is the hydraulic pressure force of the liquid in the tank.

$$\text{Hydraulic force} = A_p \rho h \frac{g}{g_c} \qquad (2.35)$$

The units of this force are (in English engineering units):

$$\text{ft}^2 \, \frac{\text{lb}_m}{\text{ft}^3} \, \text{ft} \, \frac{32.2 \, \text{ft/s}^2}{32.2 \, \text{lb}_m \, \text{ft/lb}_f \, \text{s}^2} = \text{lb}_f$$

where g is the acceleration due to gravity and is 32.2 ft/s^2 if the tank is at sea level. The static pressures in the tank and at the end of the pipe are the same, so we do not have to include them.

The only force pushing in the opposite direction from right to left and opposing the flow is the frictional force due to the viscosity of the liquid. If the flow is turbulent, the frictional force will be proportional to the square of the velocity and the length of the pipe.

$$\text{Frictional force} = K_F L v^2 \qquad (2.36)$$

Substituting these forces into Eq. (2.33), we get

$$\frac{1}{g_c} \frac{d(A_p L \rho v)}{dt} = A_p \rho h \frac{g}{g_c} - K_F L v^2$$

$$\frac{dv}{dt} = \frac{g}{L} h - \frac{K_F g_c}{\rho A_p} v^2$$

(2.37)

The sign of the frictional force is negative because it acts in the direction opposite the flow. We have defined left to right as the positive direction.

Example 2.10. Probably the best contemporary example of a variable-mass system would be the equations of motion for a space rocket whose mass decreases as fuel is consumed. However, to stick with chemical engineering systems, let us consider the problem sketched in Fig. 2.8. Petroleum pipelines are sometimes used for transferring several products from one location to another on a batch basis, i.e., one product at a time. To reduce product contamination at the end of a batch transfer, a leather ball or "pig" that just fits the pipe is inserted in one end of the line. Inert gas is introduced behind the pig to push it through the line, thus purging the line of whatever liquid is in it.

To write a force balance on the liquid still in the pipe as it is pushed out, we must take into account the changing mass of material. Assume the pig is weightless and frictionless compared with the liquid in the line. Let z be the axial position of the pig at any time. The liquid is incompressible (density ρ) and flows in plug flow. It exerts a frictional force proportional to the square of its velocity and to the length of pipe still containing liquid.

$$\text{Frictional force} = K_F (L - z) v^2 \qquad (2.38)$$

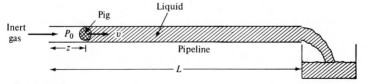

FIGURE 2.8
Pipeline and pig.

The cross-sectional area of the pipe is A_p. The mass of fluid in the pipe is $(L - z)A_p\rho$.

The pressure P_0 (lb_f/ft^2 gauge) of inert gas behind the pig is essentially constant all the way down the pipeline. The tank into which the liquid dumps is at atmospheric pressure. The pipeline is horizontal. A force balance in the horizontal z direction yields

$$\frac{1}{g_c} \frac{d}{dt} [\rho A_p v(L - z)] = P_0 A_p - K_F(L - z)v^2 \tag{2.39}$$

Substituting that $v = dz/dt$ we get

$$\frac{d}{dt}\left[(L - z)\frac{dz}{dt}\right] = \frac{P_0 g_c}{\rho} - \frac{g_c K_F}{\rho A_p}(L - z)\left(\frac{dz}{dt}\right)^2 \tag{2.40}$$

Example 2.11. As an example of a force balance for a microscopic system, let us look at the classic problem of the laminar flow of an incompressible, newtonian liquid in a cylindrical pipe. By "newtonian" we mean that its shear force (resistance that adjacent layers of fluid exhibit to flowing past each other) is proportional to the shear rate or the velocity gradient.

$$\tau_{rz} = -\frac{\mu}{g_c} \frac{\partial v_z}{\partial r} \tag{2.41}$$

where τ_{rz} = shear rate (shear force per unit area) acting in the z direction and perpendicular to the r axis, lb_f/ft^2

v_z = velocity in the z direction, ft/s

$\dfrac{\partial v_z}{\partial r}$ = velocity gradient of v_z in the r direction

μ = viscosity of fluid, lb_m/ft s

In many industries viscosity is reported in centipoise or poise. The conversion factor is 6.72×10^{-4} (lb_m/ft s)/centipoise.

We will pick as our system a small, doughnut-shaped element, half of which is shown in Fig. 2.9. Since the fluid is incompressible there is no radial flow of fluid, or $v_r = 0$. The system is symmetrical with respect to the angular coordinate (around the circumference of the pipe), and therefore we need consider only the two dimensions r and z. The forces in the z direction acting on the element are

Forces acting left to right:

$$\text{Shear force on face at } r = \tau_{rz}(2\pi r \; dz) \qquad \text{with units of} \quad \frac{lb_f}{ft^2} \; ft^2 \tag{2.42}$$

$$\text{Pressure force on face at } z = (2\pi r \; dr)P \tag{2.43}$$

Forces acting right to left:

$$\text{Shear force on face at } r + dr = 2\pi r \; dz \; \tau_{rz} + \frac{\partial}{\partial r}(2\pi r \; dz \; \tau_{rz}) \; dr$$

$$\text{Pressure force on face at } z + dz = 2\pi r \; dr \; P + \frac{\partial}{\partial z}(2\pi r \; dr \; P) \; dz$$

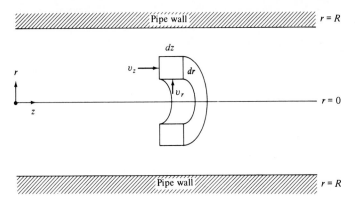

FIGURE 2.9
Laminar flow in a pipe.

The rate of change of momentum of the system is

$$\frac{1}{g_c} \frac{\partial}{\partial t} (2\pi r \, dz \, dr \, \rho v_z)$$

Combining all the above gives

$$\frac{r\rho}{g_c} \frac{\partial v_z}{\partial t} + \frac{\partial}{\partial r} (r\tau_{rz}) + r \frac{\partial P}{\partial z} = 0 \qquad (2.44)$$

The $\partial P/\partial z$ term, or the pressure drop per foot of pipe, will be constant if the fluid is incompressible. Let us call it $\Delta P/L$. Substituting it and Eq. (2.41) into Eq. (2.44) gives

$$\frac{\partial v_z}{\partial t} = \frac{\mu}{\rho r} \frac{\partial}{\partial r} \left(r \frac{\partial v_z}{\partial r} \right) - g_c \frac{\Delta P}{\rho L} \qquad (2.45)$$

2.2.4 Transport Equations

We have already used in the examples most of the laws governing the transfer of energy, mass, and momentum. These transport laws all have the form of a flux (rate of transfer per unit area) being proportional to a driving force (a gradient in temperature, concentration, or velocity). The proportionality constant is a physical property of the system (like thermal conductivity, diffusivity, or viscosity).

For transport on a molecular level, the laws bear the familiar names of Fourier, Fick, and Newton.

Transfer relationships of a more macroscopic overall form are also used; for example, film coefficients and overall coefficients in heat transfer. Here the difference in the bulk properties between two locations is the driving force. The proportionality constant is an overall transfer coefficient. Table 2.1 summarizes some to the various relationships used in developing models.

TABLE 2.1
Transport laws

Quantity	Heat	Mass	Momentum
Flux	q	N_A	τ_{rz}
	Molecular transport		
Driving force	$\dfrac{\partial T}{\partial z}$	$\dfrac{\partial C_A}{\partial z}$	$\dfrac{\partial v_z}{\partial r}$
Law	Fourier's	Fick's	Newton's
Property	Thermal conductivity	Diffusivity	Viscosity
	k_T	$\mathcal{D}_A$	μ
	Overall transport		
Driving force	ΔT	ΔC_A†	ΔP
Relationship	$q = h_T\,\Delta T$	$N_A = k_L\,\Delta C_A$	‡

† Driving forces in terms of partial pressures and mole fractions are also commonly used.

‡ The most common problem, determining pressure drops through pipes, uses friction factor correlations, $f = (g_c\, D\, \Delta P/L)/2\rho v^2$.

2.2.5 Equations of State

To write mathematical models we need equations that tell us how the physical properties, primarily density and enthalpy, change with temperature, pressure, and composition.

$$\text{Liquid density} = \rho_L = f_{(P,\, T,\, x_i)}$$
$$\text{Vapor density} = \rho_V = f_{(P,\, T,\, y_i)}$$
$$\text{Liquid enthalpy} = h = f_{(P,\, T,\, x_i)} \tag{2.46}$$
$$\text{Vapor enthalpy} = H = f_{(P,\, T,\, y_i)}$$

Occasionally these relationships have to be fairly complex to describe the system accurately. But in many cases simplification can be made without sacrificing much overall accuracy. We have already used some simple enthalpy equations in the examples of energy balances.

$$h = C_p T$$
$$H = C_p T + \lambda_v \tag{2.47}$$

The next level of complexity would be to make the C_p's functions of temperature:

$$h = \int_{T_0}^{T} C_{p(T)}\, dT \tag{2.48}$$

A polynomial in T is often used for C_p.

$$C_{p(T)} = A_1 + A_2 T \tag{2.49}$$

Then Eq. (2.48) becomes

$$h = \left[A_1 T + A_2 \frac{T^2}{2} \right]_{T_0}^{T} = A_1(T - T_0) + \frac{A_2}{2}(T^2 - T_0^2) \tag{2.50}$$

Of course, with mixtures of components the total enthalpy is needed. If heat-of-mixing effects are negligible, the pure-component enthalpies can be averaged:

$$h = \frac{\sum\limits_{j=1}^{NC} x_j h_j M_j}{\sum\limits_{j=1}^{NC} x_j M_j} \tag{2.51}$$

where x_j = mole fraction of jth component
M_j = molecular weight of jth component
h_j = pure-component enthalpy of jth component, energy per unit mass

The denominator of Eq. (2.51) is the average molecular weight of the mixture.

Liquid densities can be assumed constant in many systems unless large changes in composition and temperature occur. Vapor densities usually cannot be considered invariant and some sort of PVT relationship is almost always required. The simplest and most often used is the perfect-gas law:

$$PV = nRT \tag{2.52}$$

where P = absolute pressure (lb_f/ft^2 or kilopascals)
V = volume (ft^3 or m^3)
n = number of moles ($lb \cdot mol$ or $kg \cdot mol$)
R = constant = 1545 $lb_f \, ft/lb \cdot mol \, °R$ or 8.314 kPa $m^3/kg \cdot mol \, K$
T = absolute temperature ($°R$ or K)

Rearranging to get an equation for density ρ_v (lb_m/ft^3 or kg/m^3) of a perfect gas with a molecular weight M, we get

$$\rho_v = \frac{nM}{V} = \frac{MP}{RT} \tag{2.53}$$

2.2.6 Equilibrium

The second law of thermodynamics is the basis for the equations that tell us the conditions of a system when equilibrium conditions prevail.

A. CHEMICAL EQUILIBRIUM. Equilibrium occurs in a reacting system when

$$\sum_{j=1}^{NC} v_j \mu_j = 0 \tag{2.54}$$

where v_j = stoichiometric coefficient of the jth component with reactants having a negative sign and products a positive sign
μ_j = chemical potential of jth component

The usual way to work with this equation is in terms of an equilibrium constant for a reaction. For example, consider a reversible gas-phase reaction of A to form B at a specific rate k_1 and B reacting back to A at a specific reaction rate k_2. The stoichiometry of the reaction is such that v_a moles of A react to form v_b moles of B.

$$v_a A \underset{k_2}{\overset{k_1}{\rightleftharpoons}} v_b B \tag{2.55}$$

Equation (2.54) says equilibrium will occur when

$$v_b \mu_B - v_a \mu_A = 0 \tag{2.56}$$

The chemical potentials for a perfect-gas mixture can be written

$$\mu_j = \mu_j^0 + RT \ln \mathscr{P}_j \tag{2.57}$$

where μ_j^0 = standard chemical potential (or Gibbs free energy per mole) of the jth
            component, which is a function of temperature only
   $\mathscr{P}_j$ = partial pressure of the jth component
   R = perfect-gas law constant
   T = absolute temperature

Substituting into Eq. (2.56),

$$v_b(\mu_B^0 + RT \ln \mathscr{P}_B) - v_a(\mu_A^0 + RT \ln \mathscr{P}_A) = 0$$

$$RT \ln (\mathscr{P}_B)^{v_b} - RT \ln (\mathscr{P}_A)^{v_a} = v_a \mu_A^0 - v_b \mu_B^0$$

$$\ln \left(\frac{\mathscr{P}_B^{v_b}}{\mathscr{P}_A^{v_a}} \right) = \frac{v_a \mu_A^0 - v_b \mu_B^0}{RT} \tag{2.58}$$

The right-hand side of this equation is a function of temperature only. The term in parenthesis on the left-hand side is defined as the equilibrium constant K_p, and it tells us the equilibrium ratios of products and reactants.

$$K_p \equiv \frac{\mathscr{P}_B^{v_b}}{\mathscr{P}_A^{v_a}} \tag{2.59}$$

B. PHASE EQUILIBRIUM. Equilibrium between two phases occurs when the chemical potential of each component is the same in the two phases:

$$\mu_j^I = \mu_j^{II} \tag{2.60}$$

where μ_j^I = chemical potential of the jth component in phase I
   μ_j^{II} = chemical potential of the jth component in phase II

Since the vast majority of chemical engineering systems involve liquid and vapor phases, many vapor-liquid equilibrium relationships are used. They range from the very simple to the very complex. Some of the most commonly used relationships are listed below. More detailed treatments are presented in many thermodynamics texts. Some of the basic concepts are introduced by Luyben and

Wenzel in *Chemical Process Analysis: Mass and Energy Balances*, Chaps. 6 and 7, Prentice-Hall, 1988.

Basically we need a relationship that permits us to calculate the vapor composition if we know the liquid composition, or vice versa. The most common problem is a *bubblepoint* calculation: calculate the temperature T and vapor composition y_j, given the pressure P and the liquid composition x_j. This usually involves a trial-and-error, iterative solution because the equations can be solved explicitly only in the simplest cases. Sometimes we have bubblepoint calculations that start from known values of x_j and T and want to find P and y_j. This is frequently easier than when pressure is known because the bubblepoint calculation is usually noniterative.

Dewpoint calculations must be made when we know the composition of the vapor y_j and P (or T) and want to find the liquid composition x_j and T (or P). *Flash* calculations must be made when we know neither x_j nor y_j and must combine phase equilibrium relationships, component balance equations, and an energy balance to solve for all the unknowns.

We will assume ideal vapor-phase behavior in our examples, i.e., the partial pressure of the jth component in the vapor is equal to the total pressure P times the mole fraction of the jth component in the vapor y_j (Dalton's law):

$$\mathscr{P}_j = P y_j \tag{2.61}$$

Corrections may be required at high pressures.

In the liquid phase several approaches are widely used.

1. Raoult's law. Liquids that obey Raoult's are called *ideal*.

$$P = \sum_{j=1}^{NC} x_j P_j^S \tag{2.62}$$

$$y_j = \frac{x_j P_j^S}{P} \tag{2.63}$$

where P_j^S is the vapor pressure of pure component j. Vapor pressures are functions of temperature only. This dependence is often described by

$$\ln P_j^S = \frac{A_j}{T} + B_j \tag{2.64}$$

2. Relative volatility. The relative volatility α_{ij} of component i to component j is defined:

$$\alpha_{ij} = \frac{y_i/x_i}{y_j/x_j} \tag{2.65}$$

Relative volatilities are fairly constant in a number of systems. They are convenient so they are frequently used.

In a binary system the relative volatility α of the more volatile component compared with the less volatile component is

$$\alpha = \frac{y/x}{(1 - y)/(1 - x)}$$

Rearranging,

$$y = \frac{\alpha x}{1 + (\alpha - 1)x} \tag{2.66}$$

3. *K* values. Equilibrium vaporization ratios or *K* values are widely used, particularly in the petroleum industry.

$$K_j = \frac{y_j}{x_j} \tag{2.67}$$

The *K*'s are functions of temperature and composition, and to a lesser extent, pressure.

4. Activity coefficients. For nonideal liquids, Raoult's law must be modified to account for the nonideality in the liquid phase. The "fudge factors" used are called *activity coefficients*.

$$P = \sum_{j=1}^{NC} x_j P_j^S \gamma_j \tag{2.68}$$

where γ_j is the activity coefficient for the *j*th component. The activity coefficient is equal to 1 if the component is ideal. The γ's are functions of composition and temperature

2.2.7 Chemical Kinetics

We will be modeling many chemical reactors, and we must be familiar with the basic relationships and terminology used in describing the kinetics (rate of reaction) of chemical reactions. For more details, consult one of the several excellent texts in this field.

A. ARRHENIUS TEMPERATURE DEPENDENCE. The effect of temperature on the specific reaction rate k is usually found to be exponential:

$$k = \alpha e^{-E/RT} \tag{2.69}$$

where k = specific reaction rate
  α = preexponential factor
  E = activation energy; shows the temperature dependence of the reaction rate, i.e., the bigger E, the faster the increase in k with increasing temperature (Btu/lb · mol or cal/g · mol)
  T = absolute temperature
  R = perfect-gas constant = 1.99 Btu/lb · mol °R or 1.99 cal/g · mol K

This exponential temperature dependence represents one of the most severe non-linearities in chemical engineering systems. Keep in mind that the "apparent" temperature dependence of a reaction may not be exponential if the reaction is mass-transfer limited, not chemical-rate limited. If both zones are encountered in the operation of the reactor, the mathematical model must obviously include both reaction-rate and mass-transfer effects.

B. LAW OF MASS ACTION. Using the conventional notation, we will define an overall reaction rate $\mathcal{R}$ as the rate of change of moles of any component per volume due to chemical reaction divided by that component's stoichiometric coefficient.

$$\mathcal{R} = \frac{1}{v_j V} \left(\frac{dn_j}{dt}\right)_R \qquad (2.70)$$

The stoichiometric coefficients v_j are positive for products of the reaction and negative for reactants. Note that $\mathcal{R}$ is an intensive property and can be applied to systems of any size.

For example, assume we are dealing with an irreversible reaction in which components A and B react to form components C and D.

$$v_a A + v_b B \xrightarrow{\quad k \quad} v_c C + v_d D$$

Then

$$\mathcal{R} = \frac{1}{-v_a V}\left(\frac{dn_A}{dt}\right)_R = \frac{1}{-v_b V}\left(\frac{dn_B}{dt}\right)_R = \frac{1}{v_c V}\left(\frac{dn_C}{dt}\right)_R = \frac{1}{v_d V}\left(\frac{dn_D}{dt}\right)_R \qquad (2.71)$$

The law of mass action says that the overall reaction rate $\mathcal{R}$ will vary with temperature (since k is temperature-dependent) and with the concentration of reactants raised to some powers.

$$\mathcal{R} = k_{(T)}(C_A)^a (C_B)^b \qquad (2.72)$$

where C_A = concentration of component A
$\quad C_B$ = concentration of component B

The constants a and b are not, in general, equal to the stoichiometric coefficients v_a and v_b. The reaction is said to be first-order in A if $a = 1$. It is second-order in A if $a = 2$. The constants a and b can be fractional numbers.

As indicated earlier, the units of the specific reaction rate k depend on the order of the reaction. This is because the overall reaction rate $\mathcal{R}$ always has the same units (moles per unit time per unit volume). For a first-order reaction of A reacting to form B, the overall reaction rate $\mathcal{R}$, written for component A, would have units of moles of A/min ft^3.

$$\mathcal{R} = k C_A$$

If C_A has units of moles of A/ft^3, k must have units of min^{-1}.

If the overall reaction rate for the system above is second-order in A,

$$\mathcal{R} = kC_A^2$$

$\mathcal{R}$ still has units of moles of A/min ft³. Therefore k must have units of ft³/min mol A.

Consider the reaction A + B → C. If the overall reaction rate is first-order in both A and B,

$$\mathcal{R} = kC_A C_B$$

$\mathcal{R}$ still has units of moles of A/min ft³. Therefore k must have units of ft³/min mol B.

PROBLEMS

2.1. Write the component continuity equations describing the CSTR of Example 2.3 with:
   (*a*) Simultaneous reactions (first-order, isothermal)

$$A \xrightarrow{k_1} B \qquad A \xrightarrow{k_2} C$$

   (*b*) Reversible (first-order, isothermal)

$$A \underset{k_2}{\overset{k_1}{\rightleftharpoons}} B$$

2.2. Write the component continuity equations for a tubular reactor as in Example 2.5 with consecutive reactions occurring:

$$A \xrightarrow{k_1} B \xrightarrow{k_2} C$$

2.3. Write the component continuity equations for a perfectly mixed batch reactor (no inflow or outflow) with first-order isothermal reactions:
   (*a*) Consecutive
   (*b*) Simultaneous
   (*c*) Reversible

2.4. Write the energy equation for the CSTR of Example 2.6 in which consecutive first-order reactions occur with exothermic heats of reaction λ_1 and λ_2.

$$A \xrightarrow[\lambda_1]{k_1} B \xrightarrow[\lambda_2]{k_2} C$$

2.5. Charlie Brown and Snoopy are sledding down a hill that is inclined θ degrees from horizontal. The total weight of Charlie, Snoopy, and the sled is M. The sled is essentially frictionless but the air resistance of the sledders is proportional to the square of their velocity. Write the equations describing their position x, relative to the top of the hill ($x = 0$). Charlie likes to "belly flop," so their initial velocity at the top of the hill is v_0.

What would happen if Snoopy jumped off the sled halfway down the hill without changing the air resistance?

2.6. An automatic bale tosser on the back of a farmer's hay baler must throw a 60-pound bale of hay 20 feet back into a wagon. If the bale leaves the tosser with a velocity v_r, in

a direction $\theta = 45°$ above the horizontal, what must v_r be? If the tosser must acceler-
ate the bale from a dead start to v_r in 6 feet, how much force must be exerted?
What value of θ would minimize this acceleration force?

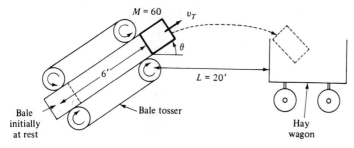

FIGURE P2.6

2.7. A mixture of two immiscible liquids is fed into a decanter. The heavier liquid α settles
to the bottom of the tank. The lighter liquid β forms a layer on the top. The two
interfaces are detected by floats and are controlled by manipulating the two flows F_α
and F_β.

$$F_\alpha = K_\alpha h_\alpha$$
$$F_\beta = K_\beta(h_\alpha + h_\beta)$$

The controllers increase or decrease the flows as the levels rise or fall.
The total feed rate is F_0. The weight fraction of liquid α in the feed is x_α. The
two densities ρ_α and ρ_β are constant.
Write the equations describing the dynamic behavior of this system.

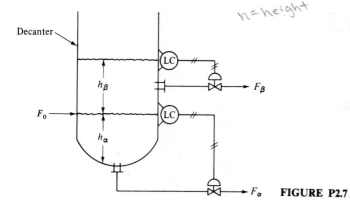

FIGURE P2.7

CHAPTER
3

EXAMPLES OF MATHEMATICAL MODELS OF CHEMICAL ENGINEERING SYSTEMS

3.1 INTRODUCTION

Even if you were only half awake when you read the preceding chapter, you should have recognized that the equations developed in the examples constituted parts of mathematical models. This chapter is devoted to more complete examples. We will start with simple systems and progress to more realistic and complex processes. The most complex example will be a nonideal, nonequimolal-overflow, multicomponent distillation column with a very large number of equations needed for a rigorous description of the system.

It would be impossible to include in this book mathematical models for all types of chemical engineering systems. The examples cover a number of very commonly encountered pieces of equipment: tanks, reactors of several types, and distillation columns (both continuous and batch). I hope that these specific examples (or case studies) of mathematical modeling will give you a good grasp of strategies and procedures so that you can apply them to your specific problem. Remember, just go back to basics when faced with a new situation. Use the dynamic mass and energy balances that apply to your system.

In each case we will set up all the equations required to describe the system. We will delay any discussion of solving these equations until Part II. Our purpose at this stage is to translate the important phenomena occurring in the physical process into quantitative, mathematical equations.

3.2 SERIES OF ISOTHERMAL, CONSTANT-HOLDUP CSTRs

The system is sketched in Fig. 3.1 and is a simple extension of the CSTR considered in Example 2.3. Product B is produced and reactant A is consumed in each of the three perfectly mixed reactors by a first-order reaction occurring in the liquid. For the moment let us assume that the temperatures and holdups (volumes) of the three tanks can be different, but both temperatures and the liquid volumes are assumed to be constant (isothermal and constant holdup). Density is assumed constant throughout the system, which is a binary mixture of A and B.

With these assumptions in mind, we are ready to formulate our model. If the volume and density of each tank are constant, the total mass in each tank is constant. Thus the total continuity equation for the first reactor is

$$\frac{d(\rho V_1)}{dt} = \rho F_0 - \rho F_1 = 0 \tag{3.1}$$

or $F_1 = F_0$.

Likewise total mass balances on tanks 2 and 3 give

$$F_3 = F_2 = F_1 = F_0 \equiv F \tag{3.2}$$

where F is defined as the throughput (m^3/min).

We want to keep track of the amounts of reactant A and product B in each tank, so component continuity equations are needed. However, since the system is binary and we know the total mass of material in each tank, only one component continuity equation is required. Either B or A can be used. If we arbitrarily choose A, the equations describing the dynamic changes in the amounts of

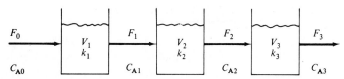

FIGURE 3.1
Series of CSTRs.

reactant A in each tank are (with units of kg · mol of A/min)

$$V_1 \frac{dC_{A1}}{dt} = F(C_{A0} - C_{A1}) - V_1 k_1 C_{A1}$$

$$V_2 \frac{dC_{A2}}{dt} = F(C_{A1} - C_{A2}) - V_2 k_2 C_{A2} \tag{3.3}$$

$$V_3 \frac{dC_{A3}}{dt} = F(C_{A2} - C_{A3}) - V_3 k_3 C_{A3}$$

The specific reaction rates k_n are given by the Arrhenius equation

$$k_n = \alpha e^{-E/RT_n} \qquad n = 1, 2, 3 \tag{3.4}$$

If the temperatures in the reactors are different, the k's are different. The n refers to the stage number.

The volumes V_n can be pulled out of the time derivatives because they are constant (see Sec. 3.3). The flows are all equal to F but can vary with time. An energy equation is not required because we have assumed isothermal operation. Any heat addition or heat removal required to keep the reactors at constant temperatures could be calculated from a steadystate energy balance (zero time derivatives of temperature).

The three first-order nonlinear ordinary differential equations given in Eqs. (3.3) are the mathematical model of the system. The parameters that must be known are V_1, V_2, V_3, k_1, k_2, and k_3. The variables that must be specified before these equations can be solved are F and C_{A0}. "Specified" does *not* mean that they must be constant. They can be time-varying, but they must be known or given functions of time. They are the *forcing functions.*

The initial conditions of the three concentrations (their values at time equal zero) must also be known.

Let us now check the degrees of freedom of the system. There are three equations and, with the parameters and forcing functions specified, there are only three unknowns or dependent variables: C_{A1}, C_{A2}, and C_{A3}. Consequently a solution should be possible, as we will demonstrate in Chap. 5.

We will use this simple system in many subsequent parts of this book. When we use it for controller design and stability analysis, we will use an even simpler version. If the throughput F is constant and the holdups and temperatures are the same in all three tanks, Eqs. (3.3) become

$$\frac{dC_{A1}}{dt} + \left(k + \frac{1}{\tau} \right) C_{A1} = \frac{1}{\tau} C_{A0}$$

$$\frac{dC_{A2}}{dt} + \left(k + \frac{1}{\tau} \right) C_{A2} = \frac{1}{\tau} C_{A1} \tag{3.5}$$

$$\frac{dC_{A3}}{dt} + \left(k + \frac{1}{\tau} \right) C_{A3} = \frac{1}{\tau} C_{A2}$$

where $\tau = V/F$ with units of minutes.

There is only one forcing function or input variable, C_{A0}.

3.3 CSTRs WITH VARIABLE HOLDUPS

If the previous example is modified slightly to permit the volumes in each reactor to vary with time, both total and component continuity equations are required for each reactor. To show the effects of higher-order kinetics, assume the reaction is now nth-order in reactant A.

Reactor 1:

$$\frac{dV_1}{dt} = F_0 - F_1$$

$$\frac{d}{dt}(V_1 C_{A1}) = F_0 C_{A0} - F_1 C_{A1} - V_1 k_1 (C_{A1})^n \tag{3.6}$$

Reactor 2:

$$\frac{dV_2}{dt} = F_1 - F_2$$

$$\frac{d}{dt}(V_2 C_{A2}) = F_1 C_{A1} - F_2 C_{A2} - V_2 k_2 (C_{A2})^n \tag{3.7}$$

Reactor 3:

$$\frac{dV_3}{dt} = F_2 - F_3$$

$$\frac{d}{dt}(V_3 C_{A3}) = F_2 C_{A2} - F_3 C_{A3} - V_3 k_3 (C_{A3})^n \tag{3.8}$$

Our mathematical model now contains six first-order nonlinear ordinary differential equations. Parameters that must be known are k_1, k_2, k_3, and n. Initial conditions for all the dependent variables that are to be integrated must be given: C_{A1}, C_{A2}, C_{A3}, V_1, V_2, and V_3. The forcing functions $C_{A0(t)}$ and $F_{0(t)}$ must also be given.

Let us now check the degrees of freedom of this system. There are six equations. *But* there are nine unknowns: C_{A1}, C_{A2}, C_{A3}, V_1, V_2, V_3, F_1, F_2, and F_3. Clearly this system is not sufficiently specified and a solution could not be obtained.

What have we missed in our modeling? A good plant operator could take one look at the system and see what the problem is. We have not specified how the flows out of the tanks are to be set. Physically there would probably be control valves in the outlet lines to regulate the flows. How are these control valves to be set? A common configuration is to have the level in the tank controlled by the outflow, i.e., a level controller opens the control valve on the exit line to increase the outflow if the level in the tank increases. Thus there must be a relationship between tank holdup and flow.

$$F_1 = f_{(V_1)} \qquad F_2 = f_{(V_2)} \qquad F_3 = f_{(V_3)} \tag{3.9}$$

The f functions will describe the level controller and the control valve. These three equations reduce the degrees of freedom to zero.

It might be worth noting that we could have considered the flow from the third tank F_3 as the forcing function. Then the level in tank 3 would probably be maintained by the flow into the tank, F_2. The level in tank 2 would be controlled by F_1, and tank 1 level by F_0. We would still have three equations.

The reactors shown in Fig. 3.1 would operate at atmospheric pressure if they were open to the atmosphere as sketched. If the reactors are not vented and if no inert blanketing is assumed, they would run at the bubblepoint pressure for the specified temperature and varying composition. Therefore the pressures could be different in each reactor, and they would vary with time, even though temperatures are assumed constant, as the C_A's change.

3.4   TWO HEATED TANKS

As our next fairly simple system let us consider a process in which two energy balances are needed to model the system. The flow rate F of oil passing through two perfectly mixed tanks in series is constant at 90 ft^3/min. The density ρ of the oil is constant at 40 lb$_m$/ft^3, and its heat capacity C_p is 0.6 Btu/lb$_m$°F. The volume of the first tank V_1 is constant at 450 ft^3, and the volume of the second tank V_2 is constant at 90 ft^3. The temperature of the oil entering the first tank is T_0 and is 150°F at the initial steadystate. The temperatures in the two tanks are T_1 and T_2. They are both equal to 250°F at the initial steadystate. A heating coil in the first tank uses steam to heat the oil. Let Q_1 be the heat addition rate in the first tank.

There is one energy balance for each tank, and each will be similar to Eq. (2.26) except there is no reaction involved in this process.

Energy balance for tank 1:

$$\frac{d(\rho C_p V_1 T_1)}{dt} = \rho C_p (F_0 T_0 - F_1 T_1) + Q_1 \tag{3.10}$$

Energy balance for tank 2:

$$\frac{d(\rho C_p V_2 T_2)}{dt} = \rho C_p (F_1 T_1 - F_2 T_2) \tag{3.11}$$

Since the throughput is constant $F_0 = F_1 = F_2 = F$. Since volumes, densities, and heat capacities are all constant, Eqs. (3.10) and (3.11) can be simplified.

$$\rho C_p V_1 \frac{dT_1}{dt} = \rho C_p F(T_0 - T_1) + Q_1 \tag{3.12}$$

$$\rho C_p V_2 \frac{dT_2}{dt} = \rho C_p F(T_1 - T_2) \tag{3.13}$$

Let's check the degrees of freedom of this system. The parameter values that are known are ρ, C_p, V_1, V_2, and F. The heat input to the first tank Q_1 would be set by the position of the control valve in the steam line. In later chapters we will use this example and have a temperature controller send a signal to the steam valve to position it. Thus we are left with two dependent variables, T_1 and T_2, and we have two equations. So the system is correctly specified.

3.5 GAS-PHASE, PRESSURIZED CSTR

Suppose a mixture of gases is fed into the reactor sketched in Fig. 3.2. The reactor is filled with reacting gases which are perfectly mixed. A reversible reaction occurs:

$$2A \underset{k_2}{\overset{k_1}{\rightleftharpoons}} B$$

The forward reaction is 1.5th-order in A; the reverse reaction is first-order in B. Note that the stoichiometric coefficient for A and the order of the reaction are not the same. The mole fraction of reactant A in the reactor is y. The pressure inside the vessel is P (absolute). Both P and y can vary with time. The volume of the reactor V is constant.

We will assume an isothermal system, so the temperature T is constant. Perfect gases are also assumed. The feed stream has a density ρ_0 and a mole fraction y_0 of reactant A. Its volumetric flow rate is F_0.

The flow out of the reactor passes through a restriction (control valve) into another vessel which is held at a constant pressure P_D (absolute). The outflow will vary with the pressure and the composition of the reactor. Flows through control valves are discussed in more detail in Part III; here let us use the formula

$$F = C_v \sqrt{\frac{P - P_D}{\rho}} \tag{3.14}$$

C_v is the valve-sizing coefficient. Density varies with pressure and composition.

$$\rho = \frac{MP}{RT} = [yM_A + (1 - y)M_B]\frac{P}{RT} \tag{3.15}$$

where M = average molecular weight
$\quad M_A$ = molecular weight of reactant A
$\quad M_B$ = molecular weight of product B

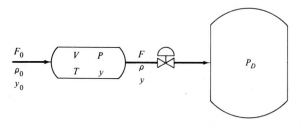

FIGURE 3.2
Gas-phase CSTR.

The concentration of reactant in the reactor is

$$C_A = \frac{Py}{RT} \qquad (3.16)$$

with units of moles of A per unit volume. The overall reaction rate for the forward reaction is

$$\mathcal{R}_F = k_1(C_A)^{1.5} = \overset{\text{Consumed}}{-\frac{1}{2V}\left(\frac{dn_A}{dt}\right)_R} = \overset{\text{Formed}}{\frac{1}{V}\left(\frac{dn_B}{dt}\right)_R}$$

The overall reaction rate for the reverse reaction is

$$\mathcal{R}_R = k_2\,C_B = \frac{1}{2V}\left(\frac{dn_A}{dt}\right)_R = -\frac{1}{V}\left(\frac{dn_B}{dt}\right)_R$$

With these fundamental relationships pinned down, we are ready to write the total and component continuity equations.

Total continuity:

$$V\frac{d\rho}{dt} = \rho_0 F_0 - \rho F \qquad (3.17)$$

Component A continuity:

$$V\frac{dC_A}{dt} = F_0 C_{A0} - FC_A - 2Vk_1(C_A)^{1.5} + 2Vk_2 C_B \qquad (3.18)$$

The 2 in the reaction terms comes from the stoichiometric coefficient of A.

There are five equations [Eqs. (3.14) through (3.18)] that make up the mathematical model of this system. The parameters that must be known are V, C_v, k_1, k_2, R, M_A, and M_B. The forcing functions (or inputs) could be P_D, ρ_0, F_0, and C_{A0}. This leaves five unknowns (dependent variables): C_A, ρ, P, F, and y.

3.6   NONISOTHERMAL CSTR

In the reactors studied so far, we have shown the effects of variable holdups, variable densities, and higher-order kinetics on the total and component continuity equations. Energy equations were not needed because we assumed isothermal operations. Let us now consider a system in which temperature can change with time. An irreversible, exothermic reaction is carried out in a single perfectly mixed CSTR as shown in Fig. 3.3.

$$A \xrightarrow{\ k\ } B$$

The reaction is nth-order in reactant A and has a heat of reaction λ (Btu/lb · mol of A reacted). Negligible heat losses and constant densities are assumed.

To remove the heat of reaction, a cooling jacket surrounds the reactor. Cooling water is added to the jacket at a volumetric flow rate F_J and with an

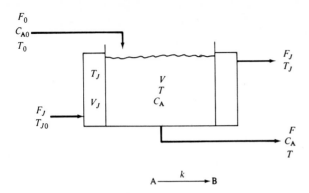

FIGURE 3.3
Nonisothermal CSTR.

inlet temperature of T_{J0}. The volume of water in the jacket V_J is constant. The mass of the metal walls is assumed negligible so the "thermal inertia" of the metal need not be considered. This is often a fairly good assumption because the heat capacity of steel is only about 0.1 Btu/lb$_m$°F, which is an order of magnitude less than that of water.

A. PERFECTLY MIXED COOLING JACKET. We assume that the temperature everywhere in the jacket is T_J. The heat transfer between the process at temperature T and the cooling water at temperature T_J is described by an overall heat transfer coefficient.

$$Q = UA_H(T - T_J) \tag{3.19}$$

where Q = heat transfer rate
  U = overall heat transfer coefficient
  A_H = heat transfer area

In general the heat transfer area could vary with the holdup in the reactor if some area was not completely covered with reaction mass liquid at all times. The equations describing the system are:

Reactor total continuity:

$$\frac{dV}{dt} = F_0 - F$$

Reactor component A continuity:

$$\frac{d(VC_A)}{dt} = F_0 C_{A0} - FC_A - Vk(C_A)^n$$

Reactor energy equation:

$$\rho \frac{d(Vh)}{dt} = \rho(F_0 h_0 - Fh) - \lambda Vk(C_A)^n - UA_H(T - T_J) \tag{3.20}$$

Jacket energy equation:

$$\rho_J V_J \frac{dh_J}{dt} = F_J \rho_J (h_{J0} - h_J) + U A_H (T - T_J) \tag{3.21}$$

where ρ_J = density of cooling water
h = enthalpy of process liquid
h_J = enthalpy of cooling water

The assumption of constant densities makes $C_p = C_v$ and permits us to use enthalpies in the time derivatives to replace internal energies.

A hydraulic relationship between reactor holdup and the flow out of the reactor is also needed. A level controller is assumed to change the outflow as the volume in the tank rises or falls: the higher the volume, the larger the outflow. The outflow is shut off completely when the volume drops to a minimum value $V_{\min}$.

$$F = K_V (V - V_{\min}) \tag{3.22}$$

The level controller is a proportional-only feedback controller.

Finally, we need enthalpy data to relate the h's to compositions and temperatures. Let us assume the simple forms

$$h = C_p T \quad \text{and} \quad h_J = C_J T_J \tag{3.23}$$

where C_p = heat capacity of the process liquid
C_J = heat capacity of the cooling water

Using Eqs. (3.23) and the Arrhenius relationship for k, the five equations that describe the process are

$$\frac{dV}{dt} = F_0 - F \tag{3.24}$$

$$\frac{d(V C_A)}{dt} = F_0 C_{A0} - F C_A - V(C_A)^n \alpha e^{-E/RT} \tag{3.25}$$

$$\rho C_p \frac{d(VT)}{dt} = \rho C_p (F_0 T_0 - FT) - \lambda V(C_A)^n \alpha e^{-E/RT} - U A_H (T - T_J) \tag{3.26}$$

$$\rho_J V_J C_J \frac{dT_J}{dt} = F_J \rho_J C_J (T_{J0} - T_J) + U A_H (T - T_J) \tag{3.27}$$

$$F = K_V (V - V_{\min}) \tag{3.28}$$

Checking the degrees of freedom, we see that there are five equations and five unknowns: V, F, C_A, T, and T_J. We must have initial conditions for these five dependent variables. The forcing functions are T_0, F_0, C_{A0}, and F_J.

The parameters that must be known are n, α, E, R, ρ, C_p, U, A_H, ρ_J, V_J, C_J, T_{J0}, K_V, and $V_{\min}$. If the heat transfer area varies with the reactor holdup it would be included as another variable, but we would also have another equation; the relationship between area and holdup. If the reactor is a flat-bottomed vertical cylinder with diameter D and if the jacket is only around the outside, not around the bottom

$$A_H = \frac{4}{D} V \tag{3.29}$$

We have assumed the overall heat transfer coefficient U is constant. It may be a function of the coolant flow rate F_J or the composition of the reaction mass, giving one more variable but also one more equation.

B. PLUG FLOW COOLING JACKET. In the model derived above, the cooling water inside the jacket was assumed to be perfectly mixed. In many jacketed vessels this is not a particularly good assumption. If the water flow rate is high enough so that the water temperature does not change much as it goes through the jacket, the mixing pattern makes little difference. However, if the water temperature rise is significant and if the flow is more like plug flow than a perfect mix (this would certainly be the case if a cooling coil is used inside the reactor instead of a jacket), then an average jacket temperature T_{JA} may be used.

$$T_{JA} = \frac{T_{J0} + T_{Jexit}}{2} \tag{3.30}$$

where T_{Jexit} is the outlet cooling-water temperature.

The average temperature is used in the heat transfer equation and to represent the enthalpy of jacket material. Equation (3.27) becomes

$$\rho_J V_J C_J \frac{dT_{JA}}{dt} = F_J \rho_J C_J (T_{J0} - T_{Jexit}) + U A_H (T - T_{JA}) \tag{3.31}$$

Equation (3.31) is integrated to obtain T_{JA} at each instant in time, and Eq. (3.30) is used to calculate T_{Jexit}, also as a function of time.

C. LUMPED JACKET MODEL. Another alternative is to break up the jacket volume into a number of perfectly mixed "lumps" as shown in Fig. 3.4.

An energy equation is needed for each lump. Assuming four lumps of equal volume and heat transfer area, we get four energy equations for the jacket:

$$\tfrac{1}{4}\rho_J V_J C_J \frac{dT_{J1}}{dt} = F_J \rho_J C_J (T_{J0} - T_{J1}) + \tfrac{1}{4} U A_H (T - T_{J1})$$

$$\tfrac{1}{4}\rho_J V_J C_J \frac{dT_{J2}}{dt} = F_J \rho_J C_J (T_{J1} - T_{J2}) + \tfrac{1}{4} U A_H (T - T_{J2})$$

$$\tag{3.32}$$

$$\tfrac{1}{4}\rho_J V_J C_J \frac{dT_{J3}}{dt} = F_J \rho_J C_J (T_{J2} - T_{J3}) + \tfrac{1}{4} U A_H (T - T_{J3})$$

$$\tfrac{1}{4}\rho_J V_J C_J \frac{dT_{J4}}{dt} = F_J \rho_J C_J (T_{J3} - T_{J4}) + \tfrac{1}{4} U A_H (T - T_{J4})$$

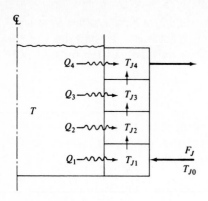

FIGURE 3.4
Lumped jacket model.

D. SIGNIFICANT METAL WALL CAPACITANCE. In some reactors, particularly high-pressure vessels or smaller-scale equipment, the mass of the metal walls and its effects on the thermal dynamics must be considered. To be rigorous, the energy equation for the wall should be a partial differential equation in time and radial position. A less rigorous but frequently used approximation is to "lump" the mass of the metal and assume the metal is all at one temperature T_M. This assumption is a fairly good one when the wall is not too thick and the thermal conductivity of the metal is large.

Then effective inside and outside film coefficients h_i and h_o are used as shown in Fig. 3.5.

The three energy equations for the process are:

$$\rho C_p \frac{d(VT)}{dt} = \rho C_p (F_0 T_0 - FT) - \lambda V(C_A)^n \alpha e^{-E/RT} - h_i A_i(T - T_M)$$

$$\rho_M V_M C_M \frac{dT_M}{dt} = h_i A_i(T - T_M) - h_o A_o(T_M - T_J) \tag{3.33}$$

$$\rho_J V_J C_J \frac{dT_J}{dt} = F_J \rho_J C_J(T_{J0} - T_J) + h_o A_o(T_M - T_J)$$

where h_i = inside heat transfer film coefficient
h_o = outside heat transfer film coefficient

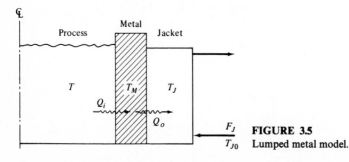

FIGURE 3.5
Lumped metal model.

ρ_M = density of metal wall
C_M = heat capacity of metal wall
V_M = volume of metal wall
A_i = inside heat transfer area
A_o = outside heat transfer area

3.7 SINGLE-COMPONENT VAPORIZER

Boiling systems represent some of the most interesting and important operations in chemical engineering processing and are among the most difficult to model. To describe these systems rigorously, conservation equations must be written for both the vapor and liquid phases. The basic problem is finding the rate of vaporization of material from the liquid phase into the vapor phase. The equations used to describe the boiling rate should be physically reasonable and mathematically convenient for solution.

Consider the vaporizer sketched in Fig. 3.6. Liquefied petroleum gas (LPG) is fed into a pressurized tank to hold the liquid level in the tank. We will assume that LPG is a pure component: propane. Vaporization of mixtures of components is discussed in Sec. 3.8.

The liquid in the tank is assumed perfectly mixed. Heat is added at a rate Q to hold the desired pressure in the tank by vaporizing the liquid at a rate W_v (mass per time). Heat losses and the mass of the tank walls are assumed negligible. Gas is drawn off the top of the tank at a volumetric flow rate F_v. F_v is the forcing function or load disturbance.

A. STEADYSTATE MODEL. The simplest model would neglect the dynamics of both vapor and liquid phases and relate the gas rate F_v to the heat input by

$$\rho_v F_v (H_v - h_0) = Q \qquad (3.34)$$

where H_v = enthalpy of vapor leaving tank (Btu/lb$_m$ or cal/g)
h_0 = enthalpy of liquid feed (Btu/lb$_m$ or cal/g)

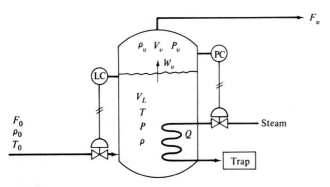

FIGURE 3.6
LPG vaporizer.

B. LIQUID-PHASE DYNAMICS MODEL. A somewhat more realistic model is obtained if we assume that the volume of the vapor phase is small enough to make its dynamics negligible. If only a few moles of liquid have to be vaporized to change the pressure in the vapor phase, we can assume that this pressure is always equal to the vapor pressure of the liquid at any temperature ($P = P_v$ and $W_v = \rho_v F_v$). An energy equation for the liquid phase gives the temperature (as a function of time), and the vapor-pressure relationship gives the pressure in the vaporizer at that temperature.

A total continuity equation for the liquid phase is also needed, plus the two controller equations relating pressure to heat input and liquid level to feed flow rate F_0. These feedback controller relationships will be expressed here simply as functions. In later parts of this book we will discuss these functions in detail.

$$Q = f_{1(P)} \qquad F_0 = f_{2(V_L)} \tag{3.35}$$

An equation of state for the vapor is needed to be able to calculate density ρ_v from the pressure or temperature. Knowing any one property (T, P, or ρ_v) pins down all the other properties since there is only one component, and two phases are present in the tank. The perfect-gas law is used.

The liquid is assumed incompressible so that $C_v = C_p$ and its internal energy is $C_p T$. The enthalpy of the vapor leaving the vaporizer is assumed to be of the simple form: $C_p T + \lambda_v$.

Total continuity:

$$\rho \frac{dV_L}{dt} = \rho_0 F_0 - \rho_v F_v \tag{3.36}$$

Energy:

$$C_p \rho \frac{d(V_L T)}{dt} = \rho_0 C_p F_0 T_0 - \rho_v F_v (C_p T + \lambda_v) + Q \tag{3.37}$$

State:

$$\rho_v = \frac{MP}{RT} \tag{3.38}$$

Vapor pressure:

$$\ln P = \frac{A}{T} + B \tag{3.39}$$

Equations (3.35) to (3.39) give us six equations. Unknowns are Q, F_0, P, V_L, ρ_v, and T.

C. LIQUID AND VAPOR DYNAMICS MODEL. If the dynamics of the vapor phase cannot be neglected (if we have a large volume of vapor), total continuity and energy equations must be written for the gas in the tank. The vapor leaving the tank, $\rho_v F_v$, is no longer equal, dynamically, to the rate of vaporization W_v.

The key problem now is to find a simple and reasonable expression for the boiling rate W_v. I have found in a number of simulations that a "mass-transfer" type of equation can be conveniently employed. This kind of relationship also makes physical sense. Liquid boils because, at some temperature (and composition if more than one component is present), it exerts a vapor pressure P greater than the pressure P_v in the vapor phase above it. The driving force is this pressure differential

$$W_v = K_{MT}(P - P_v) \tag{3.40}$$

where K_{MT} is the pseudo mass transfer coefficient. Naturally at equilibrium (*not* steadystate) $P = P_v$. If we assume that the liquid and vapor are in equilibrium, we are saying that K_{MT} is very large. When the equations are solved on a computer, several values of K_{MT} can be used to test the effects of nonequilibrium conditions.

The equations describing the system are:

Liquid phase

Total continuity:

$$\rho \frac{dV_L}{dt} = \rho_0 F_0 - W_v \tag{3.41}$$

Energy:

$$\rho \frac{d(V_L U_L)}{dt} = \rho_0 F_0 h_0 - W_v H_L + Q \tag{3.42}$$

Vapor pressure:

$$P = e^{A/T + B} \tag{3.43}$$

Vapor phase

Total continuity:

$$\frac{d(V_v \rho_v)}{dt} = W_v - \rho_v F_v \tag{3.44}$$

Energy:

$$\frac{d(V_v \rho_v U_v)}{dt} = W_v H_L - \rho_v F_v H_v \tag{3.45}$$

State:

$$\rho_v = \frac{M P_v}{R T_v} \tag{3.46}$$

where U_L = internal energy of liquid at temperature T
H_L = enthalpy of vapor boiling off liquid
U_v = internal energy of vapor at temperature T_v
H_v = enthalpy of vapor phase

Thermal-property data are needed to relate the enthalpies to temperatures. We would then have 10 variables: Q, F_0, V_L, W_v, T, V_v, ρ_v, T_v, P, and P_v. Counting Eqs. (3.35) and (3.40) to (3.46) we see there are only nine equations. Something is missing. A moment's reflection should generate the other relationship, a physical constraint: $V_L + V_v = $ total volume of tank.

D. THERMAL EQUILIBRIUM MODEL. The previous case yields a model that is about as rigorous as one can reasonably expect. A final model, not quite as rigorous but usually quite adequate, is one in which thermal equilibrium between liquid and vapor is assumed to hold at all times. More simply, the vapor and liquid temperatures are assumed equal to each other: $T = T_v$. This eliminates the need for an energy balance for the vapor phase. It probably works pretty well because the sensible heat of the vapor is usually small compared with latent-heat effects.

If the simple enthalpy relationships can be used, Eq. (3.42) becomes

$$\rho C_p \frac{d(V_L T)}{dt} = \rho_0 F_0 C_p T_0 - W_v(C_p T + \lambda_v) + Q \tag{3.47}$$

The simpler models discussed above (such as cases A and B) are usually good enough for continuous-flow systems where the changes in liquid and vapor holdups and temperatures are not very large. Batch systems may require the more rigorous models (cases C and D) because of the big variations of most variables.

3.8 MULTICOMPONENT FLASH DRUM

Let us look now at vapor-liquid systems with more than one component. A liquid stream at high temperature and pressure is "flashed" into a drum, i.e., its pressure is reduced as it flows through a restriction (valve) at the inlet of the drum. This sudden expansion is irreversible and occurs at constant enthalpy. If it were a reversible expansion, entropy (not enthalpy) would be conserved. If the drum pressure is lower than the bubblepoint pressure of the feed at the feed temperature, some of the liquid feed will vaporize.

Gas is drawn off the top of the drum through a control valve whose stem position is set by a pressure controller (Fig. 3.7). Liquid comes off the bottom of the tank on level control.

The pressure P_0 before the pressure letdown valve is high enough to prevent any vaporization of feed at its temperature T_0 and composition x_{0j} (mole fraction jth component). The forcing functions in this system are the feed temperature T_0, feed rate F, and feed composition x_{0j}. Adiabatic conditions (no heat losses) are assumed. The density of the liquid in the tank, ρ_L, is assumed to be a known function of temperature and composition.

$$\rho_L = f_{(x_j, T)} \tag{3.48}$$

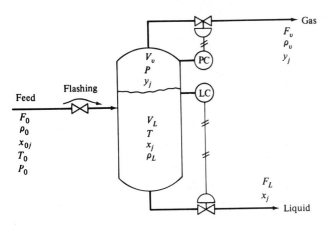

FIGURE 3.7
Flash drum.

The density ρ_v of the vapor in the drum is a known function of temperature T, composition y_j and pressure P. If the perfect-gas law can be used,

$$\rho_v = \frac{M_v^{\mathrm{av}} P}{RT} \tag{3.49}$$

where M_v^{av} is the average molecular weight of the gas.

$$M_v^{\mathrm{av}} = \sum_{j=1}^{NC} M_j y_j \tag{3.50}$$

where M_j is the molecular weight of the jth component.

A. STEADYSTATE MODEL. The simplest model of this system is one that neglects dynamics completely. Pressure is assumed constant, and the steadystate total and component continuity equations and a steadystate energy balance are used. Vapor and liquid phases are assumed to be in equilibrium.

Total continuity:

$$\rho_0 F_0 = \rho_v F_v + \rho_L F_L \tag{3.51}$$

Component continuity:

$$\frac{\rho_0 F_0}{M_0^{\mathrm{av}}} x_{0j} = \frac{\rho_v F_v}{M_v^{\mathrm{av}}} y_j + \frac{\rho_L F_L}{M_L^{\mathrm{av}}} x_j \tag{3.52}$$

Vapor-liquid equilibrium:

$$y_j = f_{(x_j, \, T, \, P)} \tag{3.53}$$

Energy equation:

$$h_0 \rho_0 F_0 = H \rho_v F_v + h \rho_L F_L \tag{3.54}$$

Thermal properties:

$$h_0 = f_{(x_{0j}, T_0)} \qquad h = f_{(x_j, T)} \qquad H = f_{(y_j, T, P)} \tag{3.55}$$

The average molecular weights M^{av} are calculated from the mole fractions in the appropriate stream [see Eq. (3.50)]. The number of variables in the system is $9 + 2(NC - 1)$: ρ_v, F_v, M_v^{av}, y_1, y_2, ..., y_{NC-1}, ρ_L, F_L, M_L^{av}, x_1, x_2, ..., x_{NC-1}, T, h, and H. Pressure P and all the feed properties are given. There are $NC - 1$ component balances [Eq. (3.52)].

There are a total of NC equilibrium equations. We can say that there are NC equations like Eq. (3.53). This may bother some of you. Since the sum of the y's has to add up to 1, you may feel that there are only $NC - 1$ equations for the y's. But even if you think about it this way, there is still one more equation: The sum of the partial pressures has to add up to the total pressure. Thus, whatever way you want to look at it, there are NC phase equilibrium equations.

	Equation	Number of equations
Total continuity	(3.51)	1
Energy	(3.54)	1
Component continuity	(3.52)	$NC - 1$
Vapor-liquid equilibrium	(3.53)	NC
Densities of vapor and liquid	(3.48) and (3.49)	2
Thermal properties for liquid and vapor streams	(3.55)	2
Average molecular weights	(3.50)	2
		$\overline{2NC + 7}$

The system is specified by the algebraic equations listed above. This is just a traditional steadystate "equilibrium-flash" calculation.

B. RIGOROUS MODEL. Dynamics can be included in a number of ways, with varying degrees of rigor, by using models similar to those in Sec. 3.7. Let us merely indicate how a rigorous model, like case C of Sec. 3.7, could be developed. Figure 3.8 shows the system schematically.

An equilibrium-flash calculation (using the same equations as in case A above) is made at each point in time to find the vapor and liquid flow rates and properties immediately after the pressure letdown valve (the variables with the primes: F_v', F_L', y_j', x_j', ... shown in Fig. 3.8). These two streams are then fed into the vapor and liquid phases. The equations describing the two phases will be similar to Eqs. (3.40) to (3.42) and (3.44) to (3.46) with the addition of (1) a multicomponent vapor-liquid equilibrium equation to calculate P_L and (2) $NC - 1$ component continuity equations for *each* phase. Controller equations relating V_L to F_L and P_v to F_v complete the model.

$$F_L = f_{(V_L)} \qquad F_v = f_{(P_v)} \tag{3.56}$$

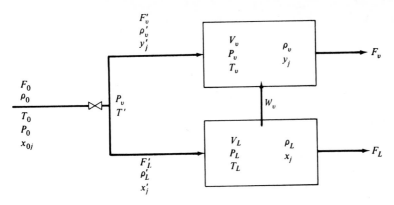

FIGURE 3.8
Dynamic flash drum.

C. PRACTICAL MODEL. A more workable dynamic model can be developed if we ignore the dynamics of the vapor phase (as we did in case B of Sec. 3.7). The vapor is assumed to be always in equilibrium with the liquid. The conservation equations are written for the liquid phase only.

Total continuity:

$$\frac{d(V_L \rho_L)}{dt} = \rho_0 F_0 - \rho_v F_v - \rho_L F_L \qquad (3.57)$$

Component continuity:

$$\frac{d\left(\dfrac{V_L \rho_L x_j}{M_L^{av}}\right)}{dt} = \frac{\rho_0 F_0}{M_0^{av}} x_{0j} - \frac{\rho_v F_v}{M_v^{av}} y_j - \frac{\rho_L F_L}{M_L^{av}} x_j \qquad (3.58)$$

Energy:

$$\frac{d(V_L \rho_L h)}{dt} = \rho_0 F_0 h_0 - \rho_v F_v H - \rho_L F_L h \qquad (3.59)$$

The NC vapor-liquid equilibrium equations [Eqs. (3.53)], the three enthalpy relationships [Eqs. (3.55)], the two density equations [Eqs. (3.48) and (3.49)], the two molecular-weight equations [Eq. (3.50)], and the feedback controller equations [Eqs. (3.56)] are all needed. The total number of equations is $2NC + 9$, which equals the total number of variables: P_v, V_L, ρ_v, F_v, M_v^{av}, y_1, y_2, ..., y_{NC-1}, ρ_L, F_L, M_L^{av}, x_1, x_2, ..., x_{NC-1}, T, h, and H.

Keep in mind that all the feed properties, or forcing functions, are given: F_0, ρ_0, h_0, x_{0j}, and M_0^{av}.

3.9 BATCH REACTOR

Batch processes offer some of the most interesting and challenging problems in modeling and control because of their inherent dynamic nature. Although most

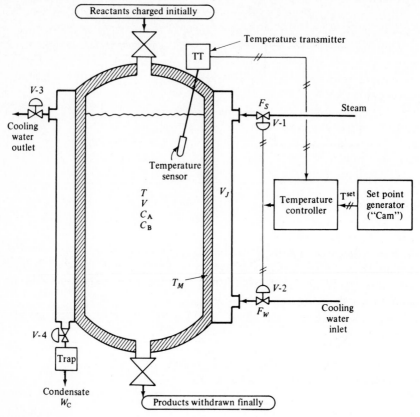

FIGURE 3.9
Batch reactor.

large-scale chemical engineering processes have traditionally been operated in a continuous fashion, many batch processes are still used in the production of smaller-volume specialty chemicals and pharmaceuticals. The batch chemical reactor has inherent kinetic advantages over continuous reactors for some reactions (primarily those with slow rate constants). The wide use of digital process control computers has permitted automation and optimization of batch processes and made them more efficient and less labor intensive.

Let us consider the batch reactor sketched in Fig. 3.9. Reactant is charged into the vessel. Steam is fed into the jacket to bring the reaction mass up to a desired temperature. Then cooling water must be added to the jacket to remove the exothermic heat of reaction and to make the reactor temperature follow the prescribed temperature-time curve. This temperature profile is fed into the temperature controller as a setpoint signal. The setpoint varies with time.

First-order consecutive reactions take place in the reactor as time proceeds.

$$A \xrightarrow{k_1} B \xrightarrow{k_2} C$$

The product that we want to make is component B. If we let the reaction go on too long, too much of B will react to form undesired C; that is, the yield will be low. If we stop the reaction too early, too little A will have reacted; i.e., the conversion and yield will be low. Therefore there is an optimum batch time when we should stop the reaction. This is often done by quenching it, i.e., cooling it down quickly.

There may also be an optimum temperature profile. If the temperature-dependences of the specific reaction rates k_1 and k_2 are the same (if their activation energies are equal), the reaction should be run at the highest possible temperature to minimize the batch time. This maximum temperature would be a limit imposed by some constraint: maximum working temperature or pressure of the equipment, further undesirable degradation or polymerization of products or reactants at very high temperatures, etc.

If k_1 is more temperature-dependent than k_2, we again want to run at the highest possible temperature to favor the reaction to B. In both cases we must be sure to stop the reaction at the right time so that the maximum amount of B is recovered.

If k_1 is less temperature-dependent that k_2, the optimum temperature profile is one that starts off at a high temperature to get the first reaction going but then drops to prevent the loss of too much B. Figure 3.10 sketches typical optimum temperature and concentration profiles. Also shown in Fig. 3.10 as the dashed line is an example of an actual temperature that could be achieved in a real reactor. The reaction mass must be heated up to T_{max}. We will use the optimum temperature profile as the setpoint signal.

With this background, let us now derive a mathematical model for this process. We will assume that the density of the reaction liquid is constant. The

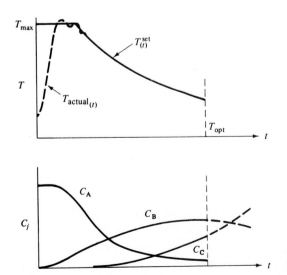

FIGURE 3.10
Batch profiles.

total continuity equation for the reaction mass, after the reactants have been charged and the batch cycle begun, is

$$\frac{d(\rho V)}{dt} = 0 - 0 \tag{3.60}$$

There is no inflow and no outflow. Since ρ is constant, $dV/dt = 0$. Therefore the volume of liquid in the reactor is constant.

Component continuity for A:

$$V \frac{dC_A}{dt} = -V k_1 C_A \tag{3.61}$$

Component continuity for B:

$$V \frac{dC_B}{dt} = V k_1 C_A - V k_2 C_B \tag{3.62}$$

Kinetic equations:

$$k_1 = \alpha_1 e^{-E_1/RT} \qquad k_2 = \alpha_2 e^{-E_2/RT} \tag{3.63}$$

Using a lumped model for the reactor metal wall and the simple enthalpy equation $h = C_p T$, the energy equations for the reaction liquid and the metal wall are:

Energy equation for process:

$$\rho V C_p \frac{dT}{dt} = -\lambda_1 V k_1 C_A - \lambda_2 V k_2 C_B - h_i A_i (T - T_M) \tag{3.64}$$

Energy equation for metal wall:

$$\rho_M V_M C_M \frac{dT_M}{dt} = h_o A_o (T_J - T_M) - h_i A_i (T_M - T) \tag{3.65}$$

where λ_1 and λ_2 are the exothermic heats of reaction for the two reactions.

Notice that when the reactor is heated with steam, T_J is bigger than T_M and T_M is bigger than T. When cooling with water, the temperature differentials have the opposite sign. Keep in mind also that the outside film coefficient h_o is usually significantly different for condensing steam and flowing cooling water.

This switching from heating to cooling is a pretty tricky operation, particularly if one is trying to heat up to T_{max} as fast as possible but cannot permit any overshoot. A commonly used system is shown in Fig. 3.9. The temperature controller keeps the steam valve (V-1) open and the cooling water valve (V-2) shut during the heat-up. This is accomplished by using *split-ranged valves*, discussed later in Part III. Also during the heat-up, the cooling-water outlet valve (V-3) is kept closed and the condensate valve (V-4) is kept open.

When cooling is required, the temperature controller shuts the steam valve and opens the cooling-water valve just enough to make the reactor temperature follow the setpoint. Valve V-3 must be opened and valve V-4 must be shut whenever cooling water is added.

We will study in detail the simulation and control of this system later in this book. Here let us simply say that there is a known relationship between the error signal E (or the temperature setpoint minus the reactor temperature) and the volumetric flow rates of steam F_s and cooling water F_w.

$$F_s = f_{1(E)} \qquad F_w = f_{2(E)} \tag{3.66}$$

To describe what is going on in the jacket we may need two different sets of equations, depending on the stage: heating or cooling. We may even need to consider a third stage: filling the jacket with cooling water. If the cooling-water flow rate is high and/or the jacket volume is small, the time to fill the jacket may be neglected.

A. HEATING PHASE. During heating, a total continuity equation and an energy equation for the steam vapor may be needed, plus an equation of state for the steam.

Total continuity:

$$V_J \frac{d\rho_J}{dt} = F_s \rho_s - W_c \tag{3.67}$$

where ρ_J = density of steam vapor in the jacket
V_J = volume of the jacket
ρ_s = density of incoming steam
W_c = rate of condensation of steam (mass per time)

The liquid condensate is assumed to be immediately drawn off through a steam trap.

Energy equation for steam vapor:

$$V_J \frac{d(U_J \rho_J)}{dt} = F_s \rho_s H_s - h_o A_o(T_J - T_M) - W_c h_c \tag{3.68}$$

where U_J = internal energy of the steam in the jacket
H_s = enthalpy of incoming steam
h_c = enthalpy of liquid condensate

The internal energy changes (sensible-heat effects) can usually be neglected compared with the latent-heat effects. Thus a simple algebraic steadystate energy equation can be used

$$W_c = \frac{h_o A_o(T_J - T_M)}{H_s - h_c} \tag{3.69}$$

The equations of state for steam (or the steam tables) can be used to calculate temperature T_J and pressure P_J from density ρ_J. For example, if the perfect-gas law and a simple vapor-pressure equation can be used,

$$\rho_J = \frac{M}{RT_J} \exp\left(\frac{A_w}{T_J} + B_w\right) \tag{3.70}$$

where M = molecular weight of steam = 18
A_w and B_w = vapor-pressure constants for water

Equation (3.70) can be solved (iteratively) for T_J if ρ_J is known [from Eq. (3.67)]. Once T_J is known, P_J can be calculated from the vapor-pressure equation. It is usually necessary to know P_J in order to calculate the flow rate of steam through the inlet valve since the rate depends on the pressure drop over the valve (unless the flow through the valve is "critical").

If the mass of the metal surrounding the jacket is significant, an energy equation is required for it. We will assume it negligible.

In most jacketed reactors or steam-heated reboilers the volume occupied by the steam is quite small compared to the volumetric flow rate of the steam vapor. Therefore the dynamic response of the jacket is usually very fast, and simple algebraic mass and energy balances can often be used. Steam flow rate is set equal to condensate flow rate, which is calculated by iteratively solving the heat-transfer relationship ($Q = UA \, \Delta T$) and the valve flow equation for the pressure in the jacket and the condensate flow rate.

B. COOLING PHASE. During the period when cooling water is flowing through the jacket, only one energy equation for the jacket is required if we assume the jacket is perfectly mixed.

$$\rho_J V_J C_J \frac{dT_J}{dt} = F_w C_J \rho_J (T_{J0} - T_J) + h_o A_o (T_M - T_J) \tag{3.71}$$

where T_J = temperature of cooling water in jacket
ρ_J = density of water
C_J = heat capacity of water
T_{J0} = inlet cooling-water temperature

Checking the degrees of freedom of the system during the heating stage, we have seven variables (C_A, C_B, T, T_M, T_J, ρ_J, and W_c) and seven equations [Eqs. (3.61), (3.62), (3.64), (3.65), (3.67), (3.69), and (3.70)]. During the cooling stage we use Eq. (3.71) instead of Eqs. (3.67), (3.69), and (3.70), but we have only T_J instead of T_J, ρ_J, and W_c.

3.10 REACTOR WITH MASS TRANSFER

As indicated in our earlier discussions about kinetics in Chap. 2, chemical reactors sometimes have mass-transfer limitations as well as chemical reaction-rate limitations. Mass transfer can become limiting when components must be moved

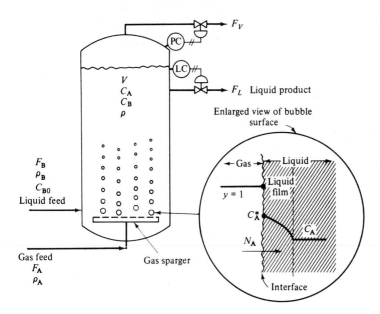

FIGURE 3.11
Gas-liquid bubble reactor.

from one phase into another phase, before or after reaction. As an example of the phenomenon, let us consider the gas-liquid bubble reactor sketched in Fig. 3.11.

Reactant A is fed as a gas through a distributor into the bottom of the liquid-filled reactor. A chemical reaction occurs between A and B in the liquid phase to form a liquid product C. Reactant A must dissolve into the liquid before it can react.

$$A + B \xrightarrow{\;k_1\;} C$$

If this rate of mass transfer of the gas A to the liquid is slow, the concentration of A in the liquid will be low since it is used up by the reaction as fast as it arrives. Thus the reactor is mass-transfer limited.

If the rate of mass transfer of the gas to the liquid is fast, the reactant A concentration will build up to some value as dictated by the steadystate reaction conditions and the equilibrium solubility of A in the liquid. The reactor is chemical-rate limited.

Notice that in the mass-transfer-limited region increasing or reducing the concentration of reactant B will make little difference in the reaction rate (or the reactor productivity) because the concentration of A in the liquid is so small. Likewise, increasing the reactor temperature will not give an exponential increase in reaction rate. The reaction rate may actually decrease with increasing temperature because of a decrease in the equilibrium solubility of A at the gas-liquid interface.

Let us try to describe some of these phenomena quantitatively. For simplicity, we will assume isothermal, constant-holdup, constant-pressure, and constant density conditions and a perfectly mixed liquid phase. The gas feed bubbles are assumed to be pure component A, which gives a constant equilibrium concentration of A at the gas-liquid interface of C_A^* (which would change if pressure and temperature were not constant). The total mass-transfer area of the bubbles is A_{MT} and could depend on the gas feed rate F_A. A constant-mass-transfer coefficient k_L (with units of length per time) is used to give the flux of A into the liquid through the liquid film as a function of the driving force.

$$N_A = k_L(C_A^* - C_A) \tag{3.72}$$

Mass transfer is usually limited by diffusion through the stagnant liquid film because of the low liquid diffusivities.

We will assume the vapor-phase dynamics are very fast and that any unreacted gas is vented off the top of the reactor.

$$F_V = F_A - \frac{A_{MT} N_A M_A}{\rho_A} \tag{3.73}$$

Component continuity for A:

$$V \frac{dC_A}{dt} = A_{MT} N_A - F_L C_A - V k C_A C_B \tag{3.74}$$

Component continuity for B:

$$V \frac{dC_B}{dt} = F_B C_{B0} - F_L C_B - V k C_A C_B \tag{3.75}$$

Total continuity:

$$\frac{d(\rho V)}{dt} = 0 = F_B \rho_B + M_A N_A A_{MT} - F_L \rho \tag{3.76}$$

Equations (3.72) through (3.76) give us five equations. Variables are N_A, C_A, C_B, F_V, and F_L. Forcing functions are F_A, F_B, and C_{B0}.

3.11 IDEAL BINARY DISTILLATION COLUMN

Next to the ubiquitous CSTR, the distillation column is probably the most popular and important process studied in the chemical engineering literature. Distillation is used in many chemical processes for separating feed streams and for purification of final and intermediate product streams.

Most columns handle multicomponent feeds. But many can be approximated by binary or pseudobinary mixtures. For this example, however, we will make several additional assumptions and idealizations that are sometimes valid but more frequently are only crude approximations.

The purpose of studying this simplified case first is to reduce the problem to its most elementary form so that the basic structure of the equations can be clearly seen. In the next example, a more realistic system will be modeled.

We will assume a binary system (two components) with constant relative volatility throughout the column and theoretical (100 percent efficient) trays, i.e., the vapor leaving the tray is in equilibrium with the liquid on the tray. This means the simple vapor-liquid equilibrium relationship can be used

$$y_n = \frac{\alpha x_n}{1 + (\alpha - 1)x_n} \tag{3.77}$$

where x_n = liquid composition on the nth tray (mole fraction more volatile component)

y_n = vapor composition on the nth tray (mole fraction more volatile component)

α = relative volatility

A single feed stream is fed as saturated liquid (at its bubblepoint) onto the feed tray N_F. See Fig. 3.12. Feed flow rate is F (mol/min) and composition is z (mole fraction more volatile component). The overhead vapor is totally condensed in a condenser and flows into the reflux drum, whose holdup of liquid is M_D (moles). The contents of the drum is assumed to be perfectly mixed with composition x_D. The liquid in the drum is at its bubblepoint. Reflux is pumped back to the top tray (N_T) of the column at a rate R. Overhead distillate product is removed at a rate D.

We will neglect any delay time (deadtime) in the vapor line from the top of the column to the reflux drum and in the reflux line back to the top tray (in industrial-scale columns this is usually a good assumption, but not in small-scale laboratory columns). Notice that y_{NT} is not equal, dynamically, to x_D. The two are equal only at steadystate.

At the base of the column, liquid bottoms product is removed at a rate B and with a composition x_B. Vapor boilup is generated in a thermosiphon reboiler at a rate V. Liquid circulates from the bottom of the column through the tubes in the vertical tube-in-shell reboiler because of the smaller density of the vapor-liquid mixture in the reboiler tubes. We will assume that the liquids in the reboiler and in the base of the column are perfectly mixed together and have the same composition x_B and total holdup M_B (moles). The circulation rates through well-designed thermosiphon reboilers are quite high, so this assumption is usually a good one. The composition of the vapor leaving the base of the column and entering tray 1 is y_B. It is in equilibrium with the liquid with composition x_B.

The column contains a total of N_T theoretical trays. The liquid holdup on each tray including the downcomer is M_n. The liquid on each tray is assumed to be perfectly mixed with composition x_n. The holdup of the vapor is assumed to be negligible throughout the system. Although the vapor volume is large, the number of moles is usually small because the vapor density is so much smaller than the liquid density. This assumption breaks down, of course, in high-pressure columns.

A further assumption we will make is that of equimolal overflow. If the molar heats of vaporization of the two components are about the same, whenever one mole of vapor condenses, it vaporizes a mole of liquid. Heat losses up the column and temperature changes from tray to tray (sensible-heat effects) are assumed negligible. These assumptions mean that the vapor and liquid rates through the stripping and rectifying sections will be constant under steadystate conditions. The "operating lines" on the familiar McCabe-Thiele diagram are straight lines.

However, we are interested here in dynamic conditions. The assumptions above, including negligible vapor holdup, mean that the vapor rate through all

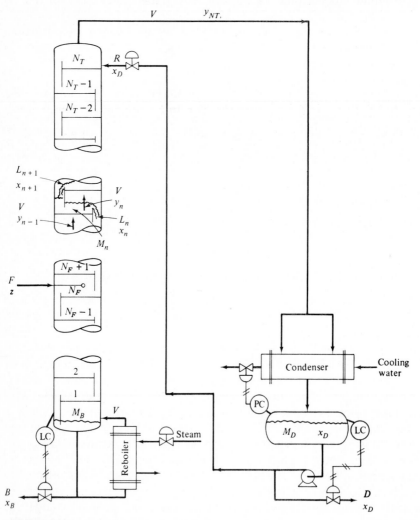

FIGURE 3.12
Binary distillation column.

trays of the column is the same, dynamically as well as at steadystate.

$$V = V_1 = V_2 = V_3 = \cdots = V_{NT}$$

Remember these V's are not necessarily constant with time. The vapor boilup can be manipulated dynamically. The mathematical effect of assuming equimolal overflow is that we do not need an energy equation for each tray. This is quite a significant simplification.

The liquid rates throughout the column will *not* be the same dynamically. They will depend on the fluid mechanics of the tray. Often a simple Francis weir formula relationship is used to relate the liquid holdup on the tray (M_n) to the liquid flow rate leaving the tray (L_n).

$$F_L = 3.33 L_w (h_{ow})^{1.5} \tag{3.78}$$

where F_L = liquid flow rate over weir (ft^3/s)
$\quad L_w$ = length of weir (ft)
$\quad h_{ow}$ = height of liquid over weir (ft)

More rigorous relationships can be obtained from the detailed tray hydraulic equations to include the effects of vapor rate, densities, compositions, etc. We will assume a simple functional relationship between liquid holdup and liquid rate.

$$M_n = f_{(L_n)} \tag{3.79}$$

Finally, we will neglect the dynamics of the condenser and the reboiler. In commercial-scale columns, the dynamic response of these heat exchangers is usually much faster than the response of the column itself. In some systems, however, the dynamics of this peripheral equipment are important and must be included in the model.

With all these assumptions in mind, we are ready to write the equations describing the system. Adopting the usual convention, our total continuity equations are written in terms of moles per unit time. This is kosher because no chemical reaction is assumed to occur in the column.

Condenser and Reflux Drum

Total continuity:

$$\frac{dM_D}{dt} = V - R - D \tag{3.80}$$

Component continuity (more volatile component):

$$\frac{d(M_D x_D)}{dt} = V y_{NT} - (R + D) x_D \tag{3.81}$$

Top Tray ($n = N_T$)

Total continuity:

$$\frac{dM_{NT}}{dt} = R - L_{NT} \tag{3.82}$$

Component continuity:

$$\frac{d(M_{NT} x_{NT})}{dt} = Rx_D - L_{NT} x_{NT} + V y_{NT-1} - V y_{NT} \tag{3.83}$$

Next to Top Tray $(n = N_T - 1)$

Total continuity:

$$\frac{dM_{NT-1}}{dt} = L_{NT} - L_{NT-1} \tag{3.84}$$

Component continuity:

$$\frac{d(M_{NT-1} x_{NT-1})}{dt} = L_{NT} x_{NT} - L_{NT-1} x_{NT-1} + V y_{NT-2} - V y_{NT-1} \tag{3.85}$$

nth Tray

Total continuity:

$$\frac{dM_n}{dt} = L_{n+1} - L_n \tag{3.86}$$

Component continuity:

$$\frac{d(M_n x_n)}{dt} = L_{n+1} x_{n+1} - L_n x_n + V y_{n-1} - V y_n \tag{3.87}$$

Feed Tray $(n = N_F)$

Total continuity:

$$\frac{dM_{NF}}{dt} = L_{NF+1} - L_{NF} + F \tag{3.88}$$

Component continuity:

$$\frac{d(M_{NF} x_{NF})}{dt} = L_{NF+1} x_{NF+1} - L_{NF} x_{NF} + V y_{NF-1} - V y_{NF} + Fz \tag{3.89}$$

First Tray $(n = 1)$

Total continuity:

$$\frac{dM_1}{dt} = L_2 - L_1 \tag{3.90}$$

Component continuity:

$$\frac{d(M_1 x_1)}{dt} = L_2 x_2 - L_1 x_1 + V y_B - V y_1 \tag{3.91}$$

Reboiler and Column Base

Total continuity:

$$\frac{dM_B}{dt} = L_1 - V - B \tag{3.92}$$

Component continuity:

$$\frac{d(M_B x_B)}{dt} = L_1 x_1 - V y_B - B x_B \tag{3.93}$$

Each tray and the column base have equilibrium equations [Eq. (3.77)]. Each tray also has a hydraulic equation [Eq. (3.79)]. We also need two equations representing the level controllers on the column base and reflux drum shown in Fig. 3.12.

$$D = f_{1(M_D)} \qquad B = f_{2(M_B)} \tag{3.94}$$

Let us now examine the degrees of freedom of the system. The feed rate F and composition z are assumed to be given.

Number of variables:

Tray compositions (x_n and y_n)	$= 2N_T$
Tray liquid flows (L_n)	$= N_T$
Tray liquid holdups (M_n)	$= N_T$
Reflux drum composition (x_D)	$= 1$
Reflux drum flows (R and D)	$= 2$
Reflux drum holdup (M_D)	$= 1$
Base compositions (x_B and y_B)	$= 2$
Base flows (V and B)	$= 2$
Base holdup (M_B)	$= 1$
	$4N_T + 9$

Number of equations:

		Equation number
Tray component continuity	$= N_T$	(3.87)
Tray total continuity	$= N_T$	(3.86)
Equilibrium (trays plus base)	$= N_T + 1$	(3.77)
Hydraulic	$= N_T$	(3.79)
Level controllers	$= 2$	(3.94)
Reflux drum component continuity	$= 1$	(3.81)
Reflux drum total continuity	$= 1$	(3.80)
Base component continuity	$= 1$	(3.93)
Base total continuity	$= 1$	(3.92)
	$4N_T + 7$	

Therefore the system is underspecified by two equations. From a control engineering viewpoint this means that there are only *two* variables that can be controlled (can be fixed). The two variables that must somehow be specified are reflux flow R and vapor boilup V (or heat input to the reboiler). They can be held constant (an openloop system) or they can be changed by two controllers to try to hold some other two variables constant. In a digital simulation of this column in Part II we will assume that two feedback controllers adjust R and V to control overhead and bottoms compositions x_D and x_B.

$$R = f_{1(x_D)} \qquad V = f_{2(x_B)} \tag{3.95}$$

3.12 MULTICOMPONENT NONIDEAL DISTILLATION COLUMN

As a more realistic distillation example, let us now develop a mathematical model for a multicomponent, nonideal column with NC components, nonequimolal overflow, and inefficient trays. The assumptions that we will make are:

1. Liquid on the tray is perfectly mixed and incompressible.
2. Tray vapor holdups are negligible. no ace
3. Dynamics of the condenser and the reboiler will be neglected. p.52
4. Vapor and liquid are in thermal equilibrium (same temperature) but not in phase equilibrium. A Murphree vapor-phase efficiency will be used to describe the departure from equilibrium.

$$E_{nj} = \frac{y_{nj} - y_{n-1,j}^T}{y_{nj}^* - y_{n-1,j}^T} \tag{3.96}$$

where y_{nj}^* = composition of vapor in phase equilibrium with liquid on nth tray with composition x_{nj}

y_{nj} = actual composition of vapor leaving nth tray

$y_{n-1,j}^T$ = actual composition of vapor entering nth tray

E_{nj} = Murphree vapor efficiency for jth component on nth tray

Multiple feeds, both liquid and vapor, and sidestream drawoffs, both liquid and vapor, are permitted. A general nth tray is sketched in Fig. 3.13. Nomenclature is summarized in Table 3.1. The equations describing this tray are:

Total continuity (one per tray):

$$\frac{dM_n}{dt} = L_{n+1} + F_n^L + F_{n-1}^V + V_{n-1} - V_n - L_n - S_n^L - S_n^V \tag{3.97}$$

Component continuity equations ($NC - 1$ per tray):

$$\frac{d(M_n x_{nj})}{dt} = L_{n+1} x_{n+1,j} + F_n^L x_{nj}^F + F_{n-1}^V y_{n-1,j}^F + V_{n-1} y_{n-1,j}$$

$$- V_n y_{nj} - L_n x_{nj} - S_n^L x_{nj} - S_n^V y_{nj} \tag{3.98}$$

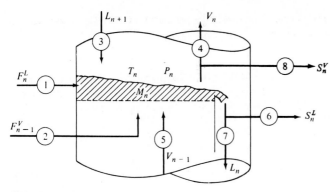

FIGURE 3.13
nth tray of multicomponent column.

Energy equation (one per tray):

$$\frac{d(M_n h_n)}{dt} = L_{n+1}h_{n+1} + F_n^L h_n^F + F_{n-1}^V H_{n-1}^F + V_{n-1}H_{n-1}$$

$$- V_n H_n - L_n h_n - S_n^L h_n - S_n^V H_n \qquad (3.99)$$

where the enthalpies have units of energy per mole.

Phase equilibrium (NC per tray):

$$y_{nj}^* = f_{(x_{nj},\, P_n,\, T_n)} \qquad (3.100)$$

An appropriate vapor-liquid equilibrium relationship, as discussed in Sec. 2.2.6, must be used to find y_{nj}^*. Then Eq. (3.96) can be used to calculate the y_{nj} for the inefficient tray. The $y_{n-1,j}^T$ would be calculated from the two vapors entering the tray: F_{n-1}^V and V_{n-1}.

Additional equations include physical property relationships to get densities and enthalpies, a vapor hydraulic equation to calculate vapor flow rates from known tray pressure drops, and a liquid hydraulic relationship to get liquid flow

TABLE 3.1
Streams on nth tray

Number	Flow rate	Composition	Temperature
1	F_n^L	x_{nj}^F	T_n^F
2	F_{n-1}^V	$y_{n-1,j}^F$	T_{n-1}^F
3	L_{n+1}	$x_{n+1,j}$	T_{n+1}
4	V_n	y_{nj}	T_n
5	V_{n-1}	$y_{n-1,j}$	T_{n-1}
6	S_n^L	x_{nj}	T_n
7	L_n	x_{nj}	T_n
8	S_n^V	y_{nj}	T_n

rates over the weirs from known tray holdups. We will defer any discussion of the very real practical problems of solving this large number of equations until Part II.

If we listed all the variables in this system and subtracted all the equations describing it and all the parameters that are fixed (all feeds), we would find that the degrees of freedom would be equal to the number of sidestreams plus two. Thus if we have no sidestreams, there are only two degrees of freedom in this multicomponent system. This is the same number that we found in the simple binary column. Typically we would want to control the amount of heavy key impurity in the distillate $x_{D, HK}$ and the amount of light key impurity in the bottoms $x_{B, LK}$.

3.13 BATCH DISTILLATION WITH HOLDUP

Batch distillation is frequently used for small-volume products. One column can be used to separate a multicomponent mixture instead of requiring $NC - 1$ continuous columns. The energy consumption in batch distillation is usually higher than in continuous, but with small-volume, high-value products energy costs seldom dominate the economics.

Figure 3.14 shows a typical batch distillation column. Fresh feed is charged into the still pot and heated until it begins to boil. The vapor works its way up the column and is condensed in the condenser. The condensate liquid runs into

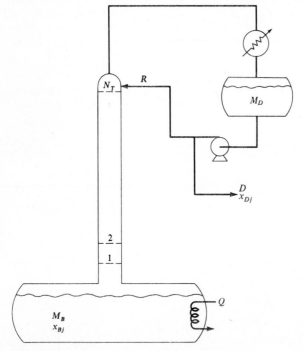

FIGURE 3.14
Batch distillation.

the reflux drum. When a liquid level has been established in the drum, reflux is pumped back to the top tray in the column.

The column is run on total reflux until the overhead distillate composition of the lightest component (component 1) x_{D1} reaches its specification purity. Then a distillate product, which is the lightest component, is withdrawn at some rate. Eventually the amount of component 1 in the still pot gets very low and the x_{D1} purity of the distillate drops. There is a period of time when the distillate contains too little of component 1 to be used for that product and also too little of component 2 to be used for the next heavier product. Therefore a "slop" cut must be withdrawn until x_{D2} builds up to its specification. Then a second product is withdrawn. Thus multiple products can be made from a single column.

The optimum design and operation of batch distillation columns are very interesting problems. The process can run at varying pressures and reflux ratios during each of the product and slop cuts. Optimum design of the columns (diameter and number of trays) and optimum operation can be important in reducing batch times, which results in higher capacity and/or improved product quality (less time at high temperatures reduces thermal degradation).

Theoretical trays, equimolal overflow, and constant relative volatilities are assumed. The total amount of material charged to the column is M_{B0} (moles). This material can be fresh feed with composition z_j or a mixture of fresh feed and the slop cuts. The composition in the still pot at the beginning of the batch is x_{B0j}. The composition in the still pot at any point in time is x_{Bj}. The instantaneous holdup in the still pot is M_B. Tray liquid holdup and reflux drum holdup are assumed constant. The vapor boilup rate is constant at V (moles per hour). The reflux drum, column trays, and still pot are all initially filled with material of composition x_{B0j}.

The equations describing the batch distillation of a multicomponent mixture are given below.

Still pot:

$$\frac{dM_B}{dt} = -D \tag{3.101}$$

$$\frac{d[M_B x_{Bj}]}{dt} = Rx_{1j} - Vy_{Bj} \tag{3.102}$$

$$y_{Bj} = \frac{\alpha_j x_{Bj}}{\sum\limits_{k=1}^{NC} \alpha_k x_{Bk}} \tag{3.103}$$

Tray n:

$$M_n \frac{dx_{nj}}{dt} = R[x_{n+1,j} - x_{nj}] + V[y_{n-1,j} - y_{nj}] \tag{3.104}$$

$$y_{nj} = \frac{\alpha_j x_{nj}}{\sum\limits_{k=1}^{NC} \alpha_k x_{nk}} \tag{3.105}$$

Tray N_T (top tray):

$$M_{NT} \frac{dx_{NT,j}}{dt} = R[x_{Dj} - x_{NT,j}] + V[y_{NT-1,j} - y_{NT,j}] \qquad (3.106)$$

$$y_{NT,j} = \frac{\alpha_j x_{NT,j}}{\sum_{k=1}^{NC} \alpha_k x_{NT,k}} \qquad (3.107)$$

Reflux drum:

$$M_D \frac{dx_{Dj}}{dt} = V y_{NT,j} - [R + D] x_{Dj} \qquad (3.108)$$

$$R = V - D \qquad (3.109)$$

3.14   pH SYSTEMS

The control of pH is a very important problem in many processes, particularly in effluent wastewater treatment. The development and solution of mathematical models of these systems is, therefore, a vital part of chemical engineering dynamic modeling.

3.14.1   Equilibrium-Constant Models

The traditional approach is to keep track of the amounts of the various chemical species in the system. At each point in time, the hydrogen ion concentration is calculated by solving a set of simultaneous nonlinear algebraic equations that result from the chemical equilibrium relationships for each dissociation reaction.

For example, suppose we have a typical wastewater pH control system. Several inlet feed streams with different chemical species, titration curves, and pH levels are fed into a perfectly mixed tank. If the feed streams are acidic, some source of OH^- ions is used to bring the pH up to the specification of seven. A slurry of $CaCO_3$ and/or caustic $(NaOH)$ are usually used.

The equilibrium-constant method uses a dynamic model that keeps track of all chemical species. Suppose, for example, that we have three dissociating acids in the system. Let the concentration of acid HA at some point in time be C_A. This concentration includes the part that is dissociated, plus the part that is not dissociated. The same quantity for acid HB is C_B and for acid C is C_C. These three acids come into the system in the feed streams.

$$HA \rightarrow H^+ + A^-$$

$$HB \rightarrow H^+ + B^-$$

$$HC \rightarrow H^+ + C^-$$

These dissociation reactions are reversible and have different forward and reverse rate constants. The equilibrium relationships for these three reactions are expressed in terms of the equilibrium constants K_A, K_B, and K_C.

$$K_A = \frac{[H^+][A^-]}{[HA]} \tag{3.110}$$

$$K_B = \frac{[H^+][B^-]}{[HB]} \tag{3.111}$$

$$K_C = \frac{[H^+][C^-]}{[HC]} \tag{3.112}$$

To solve for the concentration of hydrogen ion $[H^+]$ at each point in time, these three nonlinear algebraic equations must be solved simultaneously. Let

$$x = \text{fraction of HA dissociated}$$

$$y = \text{fraction of HB dissociated}$$

$$z = \text{fraction of HC dissociated}$$

Then

$$\text{Concentration of } A^- = x$$

$$\text{Concentration of } B^- = y$$

$$\text{Concentration of } C^- = z$$

$$\text{Concentration of undissociated HA} = C_A - x \tag{3.113}$$

$$\text{Concentration of undissociated HB} = C_B - y$$

$$\text{Concentration of undissociated HC} = C_C - z$$

$$\text{Concentration of } H^+ = x + y + z$$

These concentrations are substituted in Eqs. (3.110) to (3.112), giving three highly nonlinear algebraic equations in three unknowns: x, y, and z.

These nonlinear equations must be solved simultaneously at each point in time. Usually an iterative method is used and sometimes convergence problems occur. The complexity grows as the number of chemical species increases.

This modeling approach requires that the chemical species must be known and their equilibrium constants must be known. In many actual plant situations, this data is not available.

3.14.2   Titration-Curve Method

The information that is available in many chemical plants is a titration curve for each stream to be neutralized. The method outlined below can be used in this situation. It involves only a simple iterative calculation for one unknown at each point in time.

Let us assume that titration curves for the feed streams are known. These can be the typical sharp curves for strong acids or the gradual curves for weak acids, with or without buffering. The dynamic model keeps track of the amount of each stream that is in the tank at any point in time. Let C_n be the concentration of the nth stream in the tank, F_n be the flow rate of that stream into the tank, and F_{out} be the total flow rate of material leaving the tank.

If the volume of the liquid in the tank is constant, the outflow is the sum of all the inflows. The flow rates of caustic and lime slurry are usually negligible. For three feed streams

$$F_{out} = F_1 + F_2 + F_3 \qquad (3.114)$$

The dynamic component balance for the nth stream is

$$V \frac{dC_n}{dt} = F_n - C_n F_{out} \qquad (3.115)$$

where V = volume of the tank.

The dynamic balance for the OH^- ion in the system is

$$V \frac{dC_{OH}}{dt} = F_{OH} + R_{dis} - F_{out} C_{OH} \qquad (3.116)$$

where C_{OH} = concentration of OH^- ions in the system
F_{OH} = total flow rate of OH^- ion into the system in the caustic and lime slurry streams
R_{dis} = rate of OH^- ion generation due to the dissolving of the solid $CaCO_3$ particles

The rate of dissolution can be related to the particle size and the OH^- concentration.

$$R_{dis} = k_1 X_s \frac{k_2 - C_{OH}}{\tau} \qquad (3.117)$$

where k_1, k_2, and τ are constants determined from the dissolution rate data for solid $CaCO_3$ and X_s is the solid $CaCO_3$ concentration at any point in time.

The steps in the titration-curve method are:

1. At each point in time, all the C_n's and C_{OH} are known.
2. Guess a value for pH in the tank.
3. Use the titration curve for each stream to determine the amount of OH^- ion required for that stream to bring it up to the guess value of pH.
4. Check to see if the total amount of OH^- actually present (from C_{OH}) is equal to the total amount required for all streams.
5. Reguess pH if step 4 does not balance.

The method involves a simple iteration on only one variable, pH. Simple interval-halving convergence (see Chap. 4) can be used very effectively. The titration curves can be easily converted into simple functions to include in the computer program. For example, straight-line sections can be used to interpolate between data points.

This method has been applied with good success to a number of pH processes by Schnelle (Schnelle and Luyben, *Proceedings of ISA 88*, Houston, October 1988).

PROBLEMS

3.1. A fluid of constant density ρ is pumped into a cone-shaped tank of total volume $H\pi R^2/3$. The flow out of the bottom of the tank is proportional to the square root of the height h of liquid in the tank. Derive the equations describing the system.

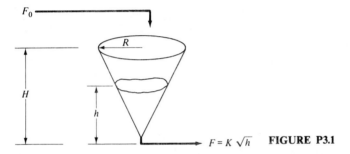

$F = K\sqrt{h}$ **FIGURE P3.1**

3.2. A perfect gas with molecular weight M flows at a mass flow rate W_0 into a cylinder through a restriction. The flow rate is proportional to the square root of the pressure drop over the restriction:

$$W_0 = K_0\sqrt{P_0 - P}$$

where P is the pressure in the cylinder and P_0 is the constant upstream pressure. The system is isothermal. Inside the cylinder, a piston is forced to the right as the pressure P builds up. A spring resists the movement of the piston with a force that is proportional to the axial displacement x of the piston.

$$F_s = K_s x$$

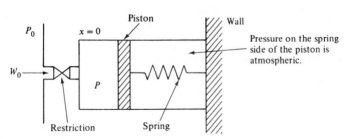

FIGURE P3.2

The piston is initially at $x = 0$ when the pressure in the cylinder is zero. The cross-sectional area of the cylinder is A. Assume the piston has negligible mass and friction.

(a) Derive the equations describing the system.

(b) What will the steadystate piston displacement be?

3.3. A perfectly mixed, isothermal CSTR has an outlet weir. The flow rate over the weir is proportional to the height of liquid over the weir, h_{ow}, to the 1.5 power. The weir height is h_w. The cross-sectional area of the tank is A. Assume constant density.

A first-order reaction takes place in the tank:

$$A \xrightarrow{\quad k \quad} B$$

Derive the equations describing the system.

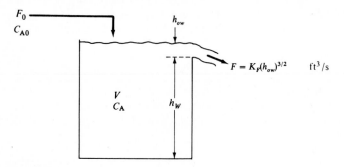

FIGURE P3.3

3.4. In order to ensure an adequate supply for the upcoming set-to with the Hatfields, Grandpa McCoy has begun to process a new batch of his famous Liquid Lightning moonshine. He begins by pumping the mash at a constant rate F_0 into an empty tank. In this tank the ethanol undergoes a first-order reaction to form a product that is the source of the high potency of McCoy's Liquid Lightning. Assuming that the concentration of ethanol in the feed, C_0, is constant and that the operation is isothermal, derive the equations that describe how the concentration C of ethanol in the tank and the volume V of liquid in the tank vary with time. Assume perfect mixing and constant density.

3.5. A rotating-metal-drum heat exchanger is half submerged in a cool stream, with its other half in a hot stream. The drum rotates at a constant angular velocity ω

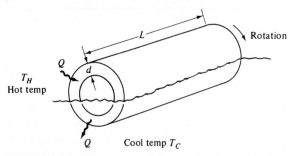

FIGURE P3.5

(radians per minute). Assume T_H and T_C are constant along their respective sections of the circumference. The drum length is L, thickness d, and radius R. Heat transfer coefficients in the heating and cooling zones are constant (U_H and U_C). Heat capacity C_p and density of the metal drum are constant. Neglect radial temperature gradients and assume steadystate operation.

(a) Write the equations describing the system.

(b) What are the appropriate boundary conditions?

3.6. Consider the system that has two stirred chemical reactors separated by a plug-flow deadtime of D seconds. Assume constant holdups (V_1 and V_2), constant throughput ~~no change~~ (F), constant density, isothermal operation at temperatures T_1 and T_2, and first- ~~in a tanks~~ order kinetics with simultaneous reactions:

$$A \xrightarrow{k_1} B \qquad A \xrightarrow{k_2} C$$

No reaction occurs in the plug-flow section.

Write the equations describing the system.

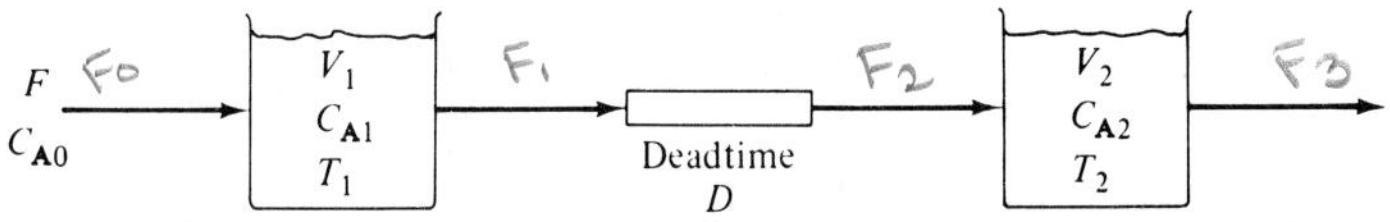

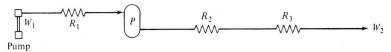

FIGURE P3.6

3.7. Consider the isothermal hydraulic system sketched below. A slightly compressible polymer liquid is pumped by a constant-speed, positive displacement pump so that the mass flow rate W_1 is constant. Liquid density is given by

$$\rho = \rho_0 + \beta(P - P_0)$$

where ρ_0, β, and P_0 are constants, ρ is the density, and P is the pressure.

Liquid is pumped through three resistances where the pressure drop is proportional to the square of the mass flow: $\Delta P = RW^2$. A surge tank of volume V is located between R_1 and R_2 and is liquid full. The pressure downstream of R_3 is atmospheric.

(a) Derive the differential equation that gives the pressure P in the tank as a function of time and W_1.

(b) Find the steadystate value of tank pressure P.

FIGURE P3.7

3.8. Develop the equations describing an "inverted" batch distillation column. This system has a large reflux drum into which the feed is charged. This material is fed to the top of the distillation column (which acts like a stripper). Vapor is generated in a reboiler in the base. Heavy material is withdrawn from the bottom of the column.

Derive a mathematical model of this batch distillation system for the case where the tray holdups cannot be neglected.

3.9. An ice cube is dropped into a hot, perfectly mixed, insulated cup of coffee. Develop the equations describing the dynamics of the system. List all assumptions and define all terms.

3.10. An isothermal, irreversible reaction

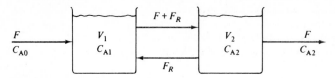

$$A \xrightarrow{\ k\ } B$$

takes place in the liquid phase in a constant-volume reactor. The mixing is *not* perfect. Observation of flow patterns indicates that a two-tank system with back mixing, as shown in the sketch below, should approximate the imperfect mixing.

Assuming F and F_R are constant, write the equations describing the system.

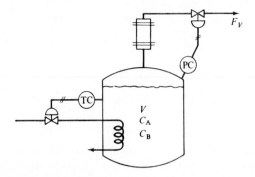

FIGURE P3.10

3.11. The liquid in a jacketed, nonisothermal CSTR is stirred by an agitator whose mass is significant compared with the reaction mass. The mass of the reactor wall and the mass of the jacket wall are also significant. Write the energy equations for the system. Neglect radial temperature gradients in the agitator, reactor wall, and jacket wall.

3.12. The reaction $3A \rightarrow 2B + C$ is carried out in an isothermal semibatch reactor. Product B is the desired product. Product C is a very volatile by-product that must be vented off to prevent a pressure buildup in the reactor. Gaseous C is vented off through a condenser to force any A and B back into the reactor to prevent loss of reactant and product.

Assume F_V is pure C. The reaction is first-order in C_A. The relative volatilities of A and C to B are $\alpha_{AB} = 1.2$ and $\alpha_{CB} = 10$. Assume perfect gases and constant pressure. Write the equations describing the system. List all assumptions.

FIGURE P3.12

3.13. Write the equations describing a simple version of the petroleum industry's important catalytic cracking operation. There are two vessels as shown in Fig. P3.13. Component A is fed to the reactor where it reacts to form product B while depositing component C on the solid fluidized catalyst.

$$A \rightarrow B + 0.1C$$

Spent catalyst is circulated to the regenerator where air is added to burn off C.

$$C + O \rightarrow P$$

Combustion products are vented overhead, and regenerated catalyst is returned to the reactor. Heat is added to or removed from the regenerator at a rate Q.

Your dynamic mathematical model should be based on the following assumptions:

(1) The perfect-gas law is obeyed in both vessels.
(2) Constant pressure is maintained in both vessels.
(3) Catalyst holdups in the reactor and in the regenerator are constant.
(4) Heat capacities of reactants and products are equal and constant in each vessel. Catalyst heat capacity is also constant.
(5) Complete mixing occurs in each vessel.

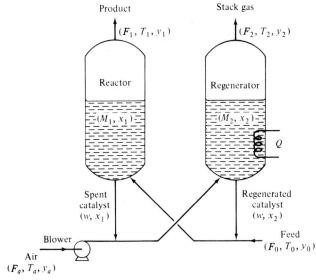

Reactor reaction: $A \xrightarrow{k_1} B + \frac{1}{10} C \downarrow$

Regenerator reaction: $C + O \xrightarrow{k_2} P$

FIGURE P3.13

3.14. Flooded condensers and flooded reboilers are sometimes used on distillation columns. In the sketch below, a liquid level is held in the condenser, covering some of the tubes. Thus a variable amount of heat transfer area is available to condense the vapor. Column pressure can be controlled by changing the distillate (or reflux) drawoff rate.

Write the equations describing the dynamics of the condenser.

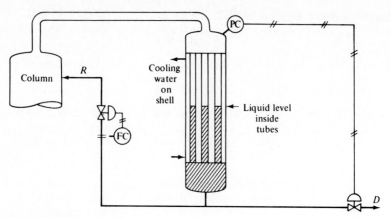

FIGURE P3.14

3.15. When cooling jackets and internal cooling coils do not give enough heat transfer area, a circulating cooling system is sometimes used. Process fluid from the reactor is pumped through an external heat exchanger and back into the reactor. Cooling water is added to the shell side of the heat exchanger at a rate F_w as set by the temperature controller. The circulation rate through the heat exchanger is constant. Assume that the shell side of the exchanger can be represented by two perfectly mixed "lumps" in series and that the process fluid flows countercurrent to the water flow, also through two perfectly mixed stages.

The reaction is irreversible and first-order in reactant A:

$$A \xrightarrow{k} B$$

The contents of the tank are perfectly mixed. Neglect reactor and heat-exchanger metal.

Derive a dynamic mathematical model of this system.

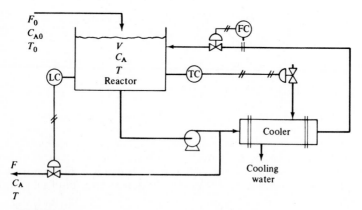

FIGURE P3.15

3.16. A semibatch reactor is run at constant temperature by varying the rate of addition of one of the reactants, A. The irreversible, exothermic reaction is first order in reactants A and B.

$$A + B \xrightarrow{k} C$$

The tank is initially filled to its 40 percent level with pure reactant B at a concentration C_{B0}. Maximum cooling-water flow is begun, and reactant A is slowly added to the perfectly stirred vessel.

Write the equations describing the system. Without solving the equations, try to sketch the profiles of F_A, C_A, and C_B with time during the batch cycle.

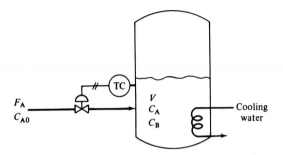

FIGURE P3.16

3.17. Develop a mathematical model for the three-column train of distillation columns sketched below. The feed to the first column is 400 kg·mol/h and contains four components (1, 2, 3, and 4), each at 25 mol %. Most of the lightest component is removed in the distillate of the first column, most of the next lightest in the second column distillate and the final column separates the final two heavy components. Assume constant relative volatilities throughout the system: α_1, α_2, and α_3. The condensers are total condensers and the reboilers are partial. Trays, column bases, and reflux drums are perfectly mixed. Distillate flow rates are set by reflux drum

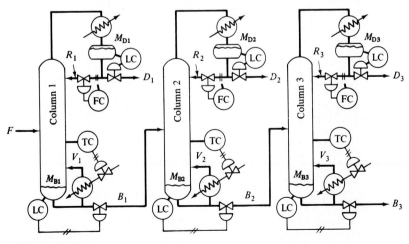

FIGURE P3.17

level controllers. Reflux flows are fixed. Steam flows to the reboilers are set by temperature controllers. Assume equimolal overflow, negligible vapor holdup, and negligible condenser and reboiler dynamics. Use a linear liquid hydraulic relationship

$$L_n = \bar{L}_n + \frac{M_n - \bar{M}_n}{\beta}$$

where $\bar{L}_n$ and $\bar{M}_n$ are the initial steadystate liquid rate and holdup and β is a constant with units of seconds.

3.18. The rate of pulp lay-down F on a paper machine is controlled by controlling both the pressure P and the height of slurry h in a feeder drum with cross-sectional area A. F is proportional to the square root of the pressure at the exit slit. The air vent rate G is proportional to the square root of the air pressure in the box P. Feedback controllers set the inflow rates of air G_0 and slurry F_0 to hold P and h. The system is isothermal.

Derive a dynamic mathematical model describing the system.

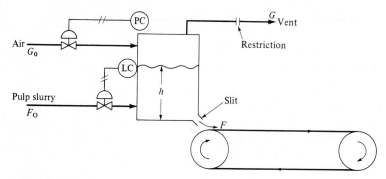

FIGURE P3.18

3.19. A wax filtration plant has six filters that operate in parallel, feeding from one common feed tank. Each filter can handle 1000 gpm when running, but the filters must be taken off-line every six hours for a cleaning procedure that takes ten minutes. The operating schedule calls for one filter to be cleaned every hour.

How many gallons a day can the plant handle? If the flow rate into the feed tank is held constant at this average flow rate, sketch how the liquid level in the feed tank varies over a typical three-hour period.

3.20. Alkylation is used in many petroleum refineries to react unsaturated butylenes with isobutane to form high octane iso-octane (alkylate). The reaction is carried out in a two liquid-phase system: sulfuric acid/hydrocarbon.

The butylene feed stream is split and fed into each of a series of perfectly mixed tanks (usually in one large vessel). This stepwise addition of butylene and the large excess of isobutane that is used both help to prevent undesirable reaction of butylene molecules with each other to form high-boiling, low octane polymers. Low temperature (40°F) also favors the desired $iC_4/C_4^=$ reaction.

The reaction is exothermic. One method of heat removal that is often used is *autorefrigeration*: the heat of vaporization of the boiling hydrocarbon liquid soaks up the heat of reaction.

The two liquid phases are completely mixed in the agitated sections, but in the last section the two phases are allowed to separate so that the acid can be recycled and the hydrocarbon phase sent off to a distillation column for separation.

Derive a dynamic mathematical model of the reactor.

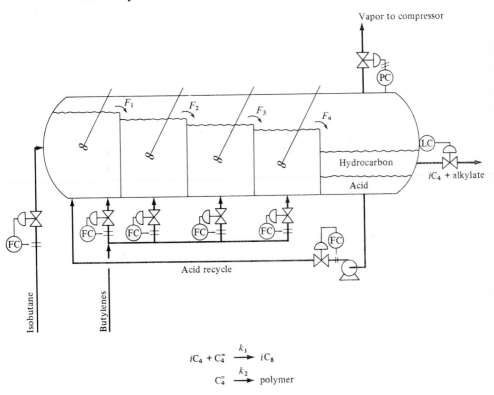

$$iC_4 + C_4^= \xrightarrow{k_1} iC_8$$

$$C_4^= \xrightarrow{k_2} polymer$$

FIGURE P3.20

3.21. Benzene is nitrated in an isothermal CSTR in three sequential irreversible reactions:

$$\text{Benzene} + HNO_3 \xrightarrow{k_1} \text{nitrobenzene} + H_2O$$

$$\text{Nitrobenzene} + HNO_3 \xrightarrow{k_2} \text{dinitrobenzene} + H_2O$$

$$\text{Dinitrobenzene} + HNO_3 \xrightarrow{k_3} \text{trinitrobenzene} + H_2O$$

Assuming each reaction is linearly dependent on the concentrations of each reactant, derive a dynamic mathematical model of the system. There are two feed streams, one pure benzene and one concentrated nitric acid (98 wt %). Assume constant densities and complete miscibility.

PART

II

COMPUTER
SIMULATION

In the next two chapters we will study computer simulation techniques for solving some of the systems of equations we generated in the two preceding chapters. A number of useful numerical methods are discussed in Chap. 4, including numerical integration of ordinary differential equations. Several examples are given in Chap. 5, starting with some simple systems and evolving into more realistic and complex processes to illustrate how to handle large numbers of equations.

Only digital simulation solutions for ordinary differential equations are presented. To present anything more than a very superficial treatment of simulation techniques for partial differential equations would require more space than is available in this book. This subject is covered in several texts. In many practical problems, distributed systems are often broken up into a number of "lumps" which can then be handled by ordinary differential equations.

Our discussions will be limited to only the most important and useful aspects of simulation. The techniques presented will be quite simple and unsophisticated, but I have found them to work just as well for most real systems as those that are more mathematically elegant. They also have the added virtues of being easy to understand and easy to program.

Some of the simple linear equations that we will simulate can, of course, be solved analytically by the methods covered in Part III to obtain general solutions. The nonlinear equations cannot, in general, be solved analytically, and computer simulation is usually required to get a solution. Keep in mind, however, that you must give the computer specific numerical values for parameters, initial conditions, and forcing functions. And you will get out of the computer specific numerical values for the solution. You cannot get a general solution in terms of arbitrary, unspecified inputs, parameters, etc., as you can with an analytic solution.

A working knowledge of FORTRAN 77 digital programming language is assumed and all programs are written in FORTRAN. Those who prefer other languages will find the conversion fairly easy since the programs are simple translations of the equations into source code. All programs can be run on any type of computer, personal computers or mainframes. The big multicomponent distillation column simulations require a lot of number crunching so are usually run on a mainframe. Most of the other programs can be conveniently run on personal computers.

CHAPTER
4

NUMERICAL
METHODS

4.1 INTRODUCTION

Digital simulation is a powerful tool for solving the equations describing chemical engineering systems. The principal difficulties are two: (1) solution of simultaneous nonlinear algebraic equations (usually done by some iterative method), and (2) numerical integration of ordinary differential equations (using discrete finite-difference equations to approximate continuous differential equations).

The accuracy and the numerical stability of these approximating equations must be kept in mind. Both accuracy and stability are affected by the finite-difference equation (or *integration algorithm*) employed. Many algorithms have been proposed in the literature. Some work better (i.e., faster and therefore cheaper for a specified degree of accuracy) on some problems than others. Unfortunately there is no one algorithm that works best for all problems. However, as we will discuss in more detail later, the simple first-order explicit Euler algorithm is the best for a large number of engineering applications.

Over the years a number of digital simulation packages have been developed. In theory, these simulation languages relieve the engineer of knowing anything about numerical integration. They automatically monitor errors and stability and adjust the integration interval or step size to stay within some accuracy criterion. In theory, these packages make it easier for the engineer to set up and solve problems.

In practice, however, these simulation languages have limited utility. In their push for generality, they usually have become inefficient. The computer execution time for a realistic engineering problem when run on one of these simulation packages is usually significantly longer than when run on a FORTRAN, BASIC, or PASCAL program written for the specific problem.

Proponents of these packages argue, however, that the setup and programming time is reduced by using simulation languages. This may be true for the engineer who doesn't know *any* programming and uses the computer only very occasionally and only for dynamic simulations. But almost all high school and certainly all engineering graduates know some computer programming language. So using a simulation package requires the engineer to learn a new language and a new system. Since some language is already known and since the simple, easily programmed numerical techniques work well, it has been my experience that it is much better for the engineer to develop a specific program for the problem at hand. Not only is it more computationally efficient, but it guarantees that the engineer knows what is in the program and what the assumptions and techniques are. This makes debugging when it doesn't work and modifying it to handle new situations much easier.

On the other hand, the use of special subroutines for doing specific calculations is highly recommended. The book by Franks (*Modeling and Simulation in Chemical Engineering*, John Wiley and Son, Inc., 1972) contains a number of useful subroutines. And of course there are usually extensive libraries of subroutines available at most locations such as the IMSL subroutines. These can be called very conveniently from a user's program.

4.2 COMPUTER PROGRAMMING

A comprehensive discussion of computer programming is beyond the scope of this book. I assume that you know some computer programming language. All the examples will use FORTRAN since it is the most widely used by practicing chemical engineers.

However, it might be useful to make a few comments and give you a few tips about programming. These thoughts are not coming from a computer scientist who is interested in generating the most efficient and elegant code, but from an engineer who is interested in solving problems.

Many people get all excited about including extensive comment statements in their code. Some go so far as to say that you should have two lines of comments for every one line of code. In my view this is ridiculous! If you use symbols in your program that are the same as you use in the equations describing the system, the code should be easy to follow. Some comment statements to point out the various sections of the program are fine.

For example, in distillation simulations the distillate and bottoms composition should be called "XD(J)" and "XB(J)" in the program. The tray compositions should be called "X(N,J)," where N is the tray number starting from the bottom and J is the component number. Many computer scientists put all the compositions into one variable "X(N,J)" and index it so that the distillate is X(1,J), the top tray is X(2,J), etc. This gives a more compact program but makes it much more difficult to understand the code.

Another important practical problem is debugging. Almost always, you will have some mistake in your program, either in coding or in logic. Always, repeat *always*, put loop counters in any iterative loops. This is illustrated in the bubble-point calculation programs given later in this chapter. If the program is not running, don't just throw up your hands and quit. It takes a fair amount of tenacity to get all the kinks out of some simulations. Maybe that's why the Dutch are pretty good at it.

When your program is not working correctly, the easiest thing to do is to put into the program print statements at each step in the calculations so that you can figure out what you have done wrong or where the coding or logic error is located.

One of the most frustrating coding errors, and one that is the toughest to find, is when you overfill an array. This usually just wipes out your program at some spot where there is nothing wrong with the program, and many FORTRAN compilers will not give a diagnostic warning of this problem. Be very careful to check that all dimensioned variables have been adequately dimensioned.

Subroutine arguments and COMMON statements can also be troublesome. Always make sure that the calls to a subroutine use the correct sequence of arguments and that all COMMON statements in all subroutines are exactly the same as that in the main program.

For the experienced programmer, the above comments are trivial, but they may save the beginner hours of frustration. The fun part of using the computer is getting useful results from the program once it is working correctly, not in coding and debugging.

4.3 ITERATIVE CONVERGENCE METHODS

One of the most common problems in digital simulation is the solution of simultaneous nonlinear algebraic equations. If these equations contain transcendental functions, analytical solutions are impossible. Therefore, an iterative trial-and-error procedure of some sort must be devised. If there is only one unknown, a value for the solution is guessed. It is plugged into the equation or equations to see if it satisfies them. If not, a new guess is made and the whole process is repeated until the iteration converges (we hope) to the right value.

The key problem is to find a method for making the new guess that converges rapidly to the correct answer. There are a host of techniques. Unfortunately there is no best method for all equations. Some methods that converge very rapidly for some equations will diverge for other equations; i.e., the series of new guesses will oscillate around the correct solution with ever-increasing deviations. This is one kind of numerical instability.

We will discuss only a few of the simplest and most useful methods. Fortunately, in dynamic simulations, we start out from some converged initial steady-state. At each instant in time, variables have changed very little from the values they had a short time before. Thus we always are close to the correct solution.

For this reason, the simple convergence methods are usually quite adequate for dynamic simulations.

The problem is best understood by considering an example. One of the most common iterative calculations is a vapor-liquid equilibrium bubblepoint calculation.

> **Example 4.1.** We are given the pressure P and the liquid composition x. We want to find the bubblepoint temperature and the vapor composition as discussed in Sec. 2.2.6. For simplicity let us assume a binary system of components 1 and 2. Component 1 is the more volatile, and the mole fraction of component 1 in the liquid is x and in the vapor is y. Let us assume also that the system is ideal: Raoult's and Dalton's laws apply.
>
> The partial pressures of the two components ($\mathcal{P}_1$ and $\mathcal{P}_2$) in the liquid and vapor phases are:

In liquid:
$$\mathcal{P}_1 = xP_1^s \qquad \mathcal{P}_2 = (1 - x)P_2^s \tag{4.1}$$

In vapor:
$$\mathcal{P}_1 = yP \qquad \mathcal{P}_2 = (1 - y)P \tag{4.2}$$

where $P_j^s = $ vapor pressure of pure component j which is a function of only temperature

$$\ln P_1^s = \frac{A_1}{T} + B_1 \qquad \ln P_2^s = \frac{A_2}{T} + B_2 \tag{4.3}$$

Equating partial pressures in liquid and vapor phases gives

$$P = xP_1^s + (1 - x)P_2^s \tag{4.4}$$

$$y = \frac{xP_1^s}{P} \tag{4.5}$$

Our convergence problem is to find the value of temperature T that will satisfy Eq. (4.4). The procedure is as follows:

1. Guess a temperature T.
2. Calculate the vapor pressures of components 1 and 2 from Eq. (4.3).
3. Calculate a total pressure P^{calc} using Eq. (4.4).

$$P^{\text{calc}} = xP_{1(T)}^s + (1 - x)P_{2(T)}^s \tag{4.6}$$

4. Compare P^{calc} with the actual total pressure given, P. If it is sufficiently close to P (perhaps using a relative convergence criterion of 10^{-6}), the guess T is correct. The vapor composition can then be calculated from Eq. (4.5).
5. If P^{calc} is greater than P, the guessed temperature was too high and we must make another guess of T that is lower. If P^{calc} is too low, we must guess a higher T.

Now let us discuss several ways of making a new guess for the example above.

4.3.1 Interval Halving

This technique is quite simple and easy to visualize and program. It is not very rapid in converging to the correct solution, but it is rock-bottom stable (it won't blow up on you numerically). It works well in dynamic simulations because the step size can be adjusted to correspond approximately to the rate at which the variable is changing with time during the integration time step.

Figure 4.1 sketches the interval-halving procedure graphically. An initial guess of temperature T_0 is made. P^{calc} is calculated from Eq. (4.6). Then P^{calc} is compared to P. A fixed increment in temperature ΔT is added to or subtracted from the temperature guess, depending on whether P^{calc} is greater or less than P.

We keep moving in the correct direction at this fixed step size until there is a change in the sign of the term $(P - P^{calc})$. This means we have crossed over the correct value of T. Then we back up halfway, i.e., we halve the increment ΔT. With each successive iteration we again halve ΔT, always moving either up or down in temperature.

Table 4.1 gives a little main program and a FORTRAN subroutine that uses interval halving to perform this bubblepoint calculation. Known values of x and P and an initial guess of temperature (T_0) are supplied as arguments of the subroutine. When the subroutine returns to the main program it supplies the correct value of T and the calculated vapor composition y. The vapor-pressure constants (A1 and B1 for component 1 and A2 and B2 for component 2) are calculated in the main program and transferred into the subroutine through the COMMON statement. The specific chemical system used in the main program is benzene-toluene at atmospheric pressure ($P = 760$ mmHg) with the mole fraction of benzene in the liquid phase (x) set at 0.50. Ideal VLE is assumed.

The initial guess of T can be either above or below the correct value. The program takes fixed steps DT equal to one degree until it crosses the correct value of T. The logic variables "FLAGM" and "FLAGP" are used to tell us when the solution has been crossed. Then interval halving is begun.

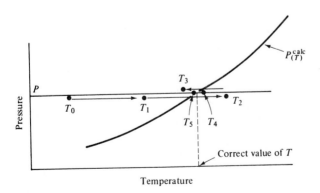

FIGURE 4.1
Interval-halving convergence.

TABLE 4.1

Example of iterative bubblepoint calculation using "interval-halving" algorithm

```
C
C  MAIN PROGRAM SETS DIFFERENT VALUES FOR
C      INITIAL GUESS OF TEMPERATURE
C      AND DIFFERENT INITIAL STEP SIZES
C  SPECIFIC CHEMICAL SYSTEM IS BENZENE/TOLUENE
C      AT 760 MM HG PRESSURE
C  PROGRAM MAIN
C
      COMMON A1,B1,A2,B2
C  CALCULATION OF VAPOR PRESSURE CONSTANTS FOR
C      BENZENE AND TOLUENE
C      DATA GIVEN AT 100 AND 125 DEGREES CELCIUS
      A1=LOG(2600./1360.)/((1./(125.+273.)) - (1./(100.+273.)))
      B1=LOG(2600.) - A1/(125.+273.)
      A2=LOG(1140./550.)/((1./(125.+273.)) - (1./(100.+273.)))
      B2=LOG(1140.) - A2/(125.+273.)
C   SET LIQUID COMPOSITION AND PRESSURE
      X=0.5
      P=760.
      WRITE(6,3)X,P
   3 FORMAT(' X = ',F8.5,'   P = ',F8.2,/)
      T0=80.
      DO 100 NT=1,3
      DT0=2.
      DO 50 NDT=1,4
      WRITE(6,1) T0,DT0
   1 FORMAT(' INITIAL TEMP GUESS = ',F7.2,'   INITIAL DT = ',F7.2)
      T=T0
      DT=DT0
      CALL BUBPT(X,T,DT,P,Y,LOOP)
      WRITE(6,2) T,Y,LOOP
   2 FORMAT('        T = ',F7.2,'   Y = ',F7.5,'   LOOP = ',I3,/)
  50 DT0=DT0*2.
 100 T0=T0+20.
      STOP
      END
C
C
      SUBROUTINE BUBPT(X,T,DT,P,Y,LOOP)
      COMMON A1,B1,A2,B2
      LOOP=0
      FLAGM=-1.
      FLAGP=-1.
C  TEMPERATURE ITERATION LOOP
 100 LOOP=LOOP+1
      IF(LOOP.GT.100)THEN
        WRITE(6,1)
   1    FORMAT(' LOOP IN BUBPT SUBROUTINE')
      . STOP
        ENDIF
      PS1=EXP(A1/(T+273.) + B1)
      PS2=EXP(A2/(T+273.) + B2)
      PCALC=X*PS1 + (1.-X)*PS2
```

TABLE 4.1 (*continued*)

```
C  TEST FOR CONVERGENCE
      IF(ABS(P-PCALC). LT. P/10000.)GO TO 50
      IF(P-PCALC) 20,20,30
C  TEMPERATURE GUESS WAS TOO HIGH
   20 IF(FLAGM.GT.0.)DT=DT/2.
      T=T-DT
      FLAGP=1.
      GO TO 100
C  TEMPERATURE GUESS WAS TOO LOW
   30 IF(FLAGP.GT.0.)DT=DT/2.
      T=T+DT
      FLAGM=1.
      GO TO 100
   50 Y=X*PS1/P
      RETURN
      END
```

Results of bubblepoint calculations

X = 0.50000 P = 760.00

INITIAL TEMP GUESS = 80.00 INITIAL DT = 2.00
    T = 92.20 Y = 0.71770 LOOP = 16

INITIAL TEMP GUESS = 80.00 INITIAL DT = 4.00
    T = 92.20 Y = 0.71770 LOOP = 14

INITIAL TEMP GUESS = 80.00 INITIAL DT = 8.00
    T = 92.20 Y = 0.71770 LOOP = 13

INITIAL TEMP GUESS = 80.00 INITIAL DT = 16.00
    T = 92.20 Y = 0.71770 LOOP = 13

INITIAL TEMP GUESS = 100.00 INITIAL DT = 2.00
    T = 92.20 Y = 0.71770 LOOP = 13

INITIAL TEMP GUESS = 100.00 INITIAL DT = 4.00
    T = 92.20 Y = 0.71770 LOOP = 12

INITIAL TEMP GUESS = 100.00 INITIAL DT = 8.00
    T = 92.20 Y = 0.71770 LOOP = 12

INITIAL TEMP GUESS = 100.00 INITIAL DT = 16.00
    T = 92.20 Y = 0.71770 LOOP = 13

INITIAL TEMP GUESS = 120.00 INITIAL DT = 2.00
    T = 92.20 Y = 0.71770 LOOP = 23

INITIAL TEMP GUESS = 120.00 INITIAL DT = 4.00
    T = 92.20 Y = 0.71770 LOOP = 17

INITIAL TEMP GUESS = 120.00 INITIAL DT = 8.00
    T = 92.20 Y = 0.71770 LOOP = 15

INITIAL TEMP GUESS = 120.00 INITIAL DT = 16.00
    T = 92.20 Y = 0.71770 LOOP = 14

Clearly, the number of iterations to converge depends on how far the initial guess is from the correct value and the size of the initial step. Table 4.1 gives results for several initial guesses of temperature (T0) and several step sizes (DT0). The interval-halving algorithm takes 10 to 20 iterations to converge to the correct temperature.

Note the presence of a loop counter. LOOP is the number of times a new guess has been made. If the iteration procedure diverges, the test for LOOP greater than 100 will stop the program with an appropriate explanation of where the problem is.

Interval halving can also be used when more than one unknown must be found. For example, suppose there are two unknowns. Two interval-halving loops could be used, one inside the other. With a fixed value of the outside variable, the inside loop is converged first to find the inside variable. Then the outside variable is changed, and the inside loop is reconverged. This procedure is repeated until both unknown variables are found that satisfy all the required equations.

Clearly this double-loop-iteration procedure can be very slow. However, for some simple problems it is quite effective.

4.3.2 Newton-Raphson Method

This method is probably the most popular convergence method. It is somewhat more complicated since it requires the evaluation of a derivative. It also can lead to stability problems if the initial guess is poor and if the function is highly nonlinear.

Newton-Raphson amounts to using the slope of the function curve to extrapolate to the correct value. Using the bubblepoint problem as a specific example, let us define the function $f_{(T)}$:

$$f_{(T)} = P_{(T)}^{\text{calc}} - P \tag{4.7}$$

We want to find the value of T that makes $f_{(T)}$ equal to zero; i.e., we want to find the root of $f_{(T)}$. We guess a value of temperature T_0. Then we evaluate the function at T_0, $f_{(T_0)}$. Next we evaluate the slope of the function at T_0, $f'_{(T_0)} = (df/dT)_{(T_0)}$. Then from the geometry shown in Fig. 4.2 we can see that

$$f'_{(T_0)} = \left(\frac{df}{dT}\right)_{(T_0)} = \frac{-f_{(T_0)}}{T_1 - T_0} \tag{4.8}$$

Solving for T_1 gives

$$T_1 = T_0 - \frac{f_{(T_0)}}{f'_{(T_0)}} \tag{4.9}$$

T_1 in Eq. (4.9) is the new guess of temperature. If the curve $f_{(T)}$ were a straight line, we would converge to the correct solution in just one iteration.

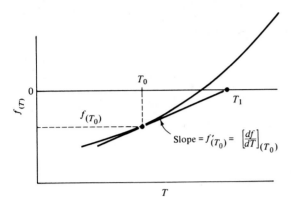

FIGURE 4.2
Graphical representation of
Newton-Raphson convergence.

Generalizing Eq. (4.9), we get the recursive iteration algorithm:

$$T_{n+1} = T_n - \frac{f_n}{f'_n} \tag{4.10}$$

where T_{n+1} = new guess of temperature
T_n = old guess of temperature
f_n = value of $f_{(T)}$ at $T = T_n$
f'_n = value of the derivative of f, df/dT, at $T = T_n$

The technique requires the evaluation of f', the derivative of the function $f_{(T)}$ with respect to temperature. In our bubblepoint example this can be obtained analytically.

$$f_{(T)} = P^{calc}_{(T)} - P = x\, e^{(A_1/T + B_1)} + (1 - x)\, e^{(A_2/T + B_2)}$$

$$f' = \frac{df}{dT} = -\frac{xA_1}{T^2}\, e^{(A_1/T + B_1)} - \frac{(1 - x)A_2}{T^2}\, e^{(A_2/T + B_2)}$$

$$= \frac{-xA_1 P^s_1 - (1 - x)A_2 P^s_2}{T^2} \tag{4.11}$$

If the function were so complex that an analytical derivative could not be obtained explicitly, an approximate derivative would have to be calculated numerically: make a small change in temperature ΔT, evaluate f at $T + \Delta T$ and use the approximation

$$f' = \frac{f_{(T+\Delta T)} - f_{(T)}}{\Delta T} \tag{4.12}$$

A digital computer program using Eqs. (4.10) and (4.11) is given in Table 4.2. The problem is the same bubblepoint calculation for benzene-toluene performed by interval-halving in Table 4.1, and the same initial guesses are made of temperature. The results given in Table 4.2 show that the Newton-Raphson algorithm is much more effective in this problem than interval halving: only 4 to 5

TABLE 4.2

Example of iterative bubblepoint calculation using "Newton-Raphson" algorithm

```
C
C  MAIN PROGRAM SETS DIFFERENT VALUES FOR INITIAL
C     GUESS OF TEMPERATURE
C  SPECIFIC CHEMICAL SYSTEM IS BENZENE/TOLUENE
C     AT 760 MM HG PRESSURE
C
      PROGRAM MAIN
      COMMON A1,B1,A2,B2
C  CALCULATION OF VAPOR PRESSURE CONSTANTS
C     FOR BENZENE AND TOLUENE
C     DATA GIVEN AT 100 AND 125 DEGREES CELCIUS
      A1=LOG(2600./1360.)/((1./(125.+273.)) - (1./(100.+273.)))
      B1=LOG(2600.) - A1/(125.+273.)
      A2=LOG(1140./550.)/((1./(125.+273.)) - (1./(100.+273.)))
      B2=LOG(1140.) - A2/(125.+273.)
C  SET LIQUID COMPOSITION AND PRESSURE
      X=0.5
      P=760.
      WRITE(6,3)X,P
    3 FORMAT(' X = ',F8.5,'   P = ',F8.2,/)
      T0=80.
      DO 100 NT=1,3
      WRITE(6,1) T0
    1 FORMAT(' INITIAL TEMP GUESS = ',F7.2)
      T=T0
      CALL BUBPT(X,T,DT,P,Y,LOOP)
      WRITE(6,2) T,Y,LOOP
    2 FORMAT('        T = ',F7.2,'    Y = ',F7.5,'   LOOP = ',I3,/)
  100 T0=T0+20.
      STOP
      END
```

iterations are required with Newton-Raphson, compared to 10 to 20 with interval-halving.

If the function is not as smooth and/or if the initial guess is not as close to the solution, the Newton-Raphson method can diverge instead of converge. Functions that are not monotonic are particularly troublesome, since the derivative approaching zero makes Eq. (4.10) blow up. Thus Newton-Raphson is a very efficient algorithm but one that can give convergence problems. Sometimes these difficulties can be overcome by constraining the size of the change permitted to be made in the new guess.

Newton-Raphson can be fairly easily extended to iteration problems involving more than one variable. For example, suppose we have two functions $f_{1(x_1, x_2)}$ and $f_{2(x_1, x_2)}$ that depend on two variables x_1 and x_2. We want to find the values of x_1 and x_2 that satisfy the two equations

$$f_{1(x_1, x_2)} = 0 \quad \text{and} \quad f_{2(x_1, x_2)} = 0$$

TABLE 4.2 (*continued*)

```
C
    SUBROUTINE BUBPT(X,T,DT,P,Y,LOOP)
    COMMON A1,B1,A2,B2
    LOOP=0
C TEMPERATURE ITERATION LOOP
  100 LOOP=LOOP+1
    IF(LOOP.GT.100)THEN
      WRITE(6,1)
    1   FORMAT(' LOOP IN BUBPT SUBROUTINE')
      STOP
      ENDIF
    PS1=EXP(A1/(T+273.) + B1)
    PS2=EXP(A2/(T+273.) + B2)
    PCALC=X*PS1 + (1.-X)*PS2
C TEST FOR CONVERGENCE
    IF(ABS(P-PCALC). LT. P/10000.)GO TO 50
    F=PCALC-P
    DF= - (X*PS1*A1 + (1.-X)*PS2*A2)/(T+273.)**2
    T=T-F/DF
    GO TO 100
  50 Y=X*PS1/P
    RETURN
    END
```

Results

X = 0.50000   P =   760.00

INITIAL TEMP GUESS =   80.00
     T =   92.19   Y = 0.71765   LOOP =   4

INITIAL TEMP GUESS =   100.00
     T =   92.19   Y = 0.71765   LOOP =   4

INITIAL TEMP GUESS =   120.00
     T =   92.19   Y = 0.71765   LOOP =   5

Expanding each of these functions around the point (x_{1n}, x_{2n}) in a Taylor series and truncating after the first derivative terms give

$$f_{1(x_{1,n+1}, x_{2,n+1})} = f_{1(x_{1n}, x_{2n})} + \left(\frac{\partial f_1}{\partial x_1}\right)_{(x_{1n}, x_{2n})} (x_{1,n+1} - x_{1n})$$

$$+ \left(\frac{\partial f_1}{\partial x_2}\right)_{(x_{1n}, x_{2n})} (x_{2,n+1} - x_{2n}) \quad (4.13)$$

$$f_{2(x_{1,n+1}, x_{2,n+1})} = f_{2(x_{1n}, x_{2n})} + \left(\frac{\partial f_2}{\partial x_1}\right)_{(x_{1n}, x_{2n})} (x_{1,n+1} - x_{1n})$$

$$+ \left(\frac{\partial f_2}{\partial x_2}\right)_{(x_{1n}, x_{2n})} (x_{2,n+1} - x_{2n}) \quad (4.14)$$

Setting $f_{1(x_1, n+1, x_2, n+1)}$ and $f_{2(x_1, n+1, x_2, n+1)}$ equal to zero and solving for the new guesses $x_{1, n+1}$ and $x_{2, n+1}$ give

$$x_{1, n+1} = x_{1n} + \frac{f_2(\partial f_1/\partial x_2) - f_1(\partial f_2/\partial x_2)}{(\partial f_1/\partial x_1)(\partial f_2/\partial x_2) - (\partial f_1/\partial x_2)(\partial f_2/\partial x_1)} \qquad (4.15)$$

$$x_{2, n+1} = x_{2n} + \frac{f_2(\partial f_1/\partial x_1) - f_1(\partial f_2/\partial x_1)}{(\partial f_1/\partial x_2)(\partial f_2/\partial x_1) - (\partial f_1/\partial x_1)(\partial f_2/\partial x_2)} \qquad (4.16)$$

All the partial derivatives and the functions are evaluated at x_{1n} and x_{2n}.

Equations (4.15) and (4.16) give the iteration algorithm for reguessing the two new values each time through the loop. Four partial derivatives must be calculated, either analytically or numerically, at each iteration step.

Equations (4.13) and (4.14) can be written more compactly in matrix form. We set the left sides of the equations equal to zero and call the changes in the guessed variables $\Delta x = x_{n+1} - x_n$.

$$\begin{bmatrix} 0 \\ 0 \end{bmatrix} = \begin{bmatrix} f_1 \\ f_2 \end{bmatrix} + \begin{bmatrix} \dfrac{\partial f_1}{\partial x_1} & \dfrac{\partial f_1}{\partial x_2} \\ \dfrac{\partial f_2}{\partial x_1} & \dfrac{\partial f_2}{\partial x_2} \end{bmatrix} \begin{bmatrix} \Delta x_1 \\ \Delta x_2 \end{bmatrix} \qquad (4.17)$$

All the functions and the partial derivatives are evaluated at x_{1n} and x_{2n}. The 2×2 matrix of partial derivatives is called the *jacobian* matrix.

$$\underline{J} = \begin{bmatrix} \dfrac{\partial f_1}{\partial x_1} & \dfrac{\partial f_1}{\partial x_2} \\ \dfrac{\partial f_2}{\partial x_1} & \dfrac{\partial f_2}{\partial x_2} \end{bmatrix} \qquad (4.18)$$

All of the terms in this matrix are just constants that are calculated at each iteration.

The Δx's can be calculated by solving the matrix equation

$$\begin{bmatrix} \Delta x_1 \\ \Delta x_2 \end{bmatrix} = \underline{J}^{-1} \begin{bmatrix} f_1 \\ f_2 \end{bmatrix} \qquad (4.19)$$

where $\underline{J}^{-1}$ is the inverse of the $\underline{J}$ matrix. We will discuss matrices in more detail in Chap. 15, but this notation is presented here to indicate how easy it is to extend the Newton-Raphson method to more than one variable. If there are three unknowns, the jacobian matrix contains nine partial derivative terms. If there are N unknowns, the jacobian matrix contains N^2 terms.

4.3.3 False Position

This convergence technique is a combination of Newton-Raphson and interval halving. An initial guess T_0 is made, and the function $f_{(T_0)}$ is evaluated. A step is taken in the correct direction to a new temperature T_1 and $f_{(T_1)}$ is evaluated. If

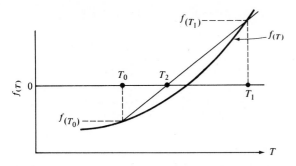

FIGURE 4.3
False-position convergence.

$f_{(T_1)}$ has the same sign as $f_{(T_0)}$, the solution has not been crossed and another step is taken (redefining T_0 as the old T_1). Stepping is continued until some temperature T_1 is reached where $f_{(T_1)}$ differs in sign from $f_{(T_0)}$. As shown in Fig. 4.3, a new guess for temperature T_2 can be found from the geometry. From similar triangles

$$\frac{T_2 - T_0}{f_{(T_0)}} = \frac{T_1 - T_2}{f_{(T_1)}}$$

Rearranging,

$$T_2 = T_1 - \frac{f_{(T_1)}(T_1 - T_0)}{f_{(T_1)} - f_{(T_0)}} \tag{4.20}$$

Generalizing, we get the recursive algorithm:

$$T_{n+1} = T_n - \frac{f_{(T_n)}(T_n - T_{n-1})}{f_{(T_n)} - f_{(T_{n-1})}} \tag{4.21}$$

Equation (4.21) is similar to Eq. (4.10), the Newton-Raphson algorithm. The derivative is approximated numerically in Eq. (4.21).

4.3.4 Explicit Convergence Methods

For some systems of equations it is possible to guess a value of a variable x_{guess}, and then use one of the equations to solve explicitly for a new calculated value of the same variable, x_{calc}. Then the calculated value and the original guess are compared and a new guess is made.

The new guess can be simply the calculated value (this is called *successive substitution*). Convergence may be very slow because of (1) a very slow rate of approach of x_{calc} to x_{guess}, or (2) an oscillation of x_{calc} back and forth around x_{guess}. The loop can even diverge.

Therefore a convergence factor β can be used to speed up or slow down the rate at which x_{guess} is permitted to change from iteration to iteration.

$$(x_{guess})_{new} = (x_{guess})_{old} + \beta[x_{calc} - (x_{guess})_{old}] \tag{4.22}$$

Note that letting $\beta = 1$ corresponds to successive substitution. This method is illustrated in the following example.

Example 4.2. A countercurrent heat exchanger is an important example of a system described by equations that are usually solved iteratively. Figure 4.4 shows the system. The problem is to find the steadystate outlet temperatures of the oil, T_{H2}, and cooling water, T_{C2}, and the heat transfer rate Q, given the inlet temperatures, flow rates, and heat transfer coefficient and area. The steadystate equations for heat transfer are

$$Q = UA(\Delta T)_{LM} = (120)(879)(\Delta T)_{LM} \tag{4.23}$$

$$Q = (70,000)(0.5)(250 - T_{H2}) \tag{4.24}$$

$$Q = (170.5)(60)(8.33)(T_{C2} - 80) \tag{4.25}$$

$$(\Delta T)_{LM} = \frac{(250 - T_{C2}) - (T_{H2} - 80)}{\ln\left(\dfrac{250 - T_{C2}}{T_{H2} - 80}\right)} \tag{4.26}$$

We have four equations and four variables: Q, $(\Delta T)_{LM}$, T_{H2}, and T_{C2}. The iterative procedure is:

1. Guess a value for the oil outlet temperature T_{H2}^{guess} (which must be greater than 80°F, for physical reasons).
2. Calculate Q_1 from Eq. (4.24).
3. Calculate T_{C2} from Eq. (4.25), using Q_1.
4. Calculate the log-mean-temperature driving force $(\Delta T)_{LM}$ from Eq. (4.26).
5. Calculate a new heat transfer rate Q_2 from Eq. (4.23).
6. Substitute the value of Q_2 into Eq. (4.24) and calculate a T_{H2}^{calc}.
7. Compare T_{H2}^{guess} and T_{H2}^{calc}.
8. Reguess T_{H2}^{guess} using Eq. (4.22).

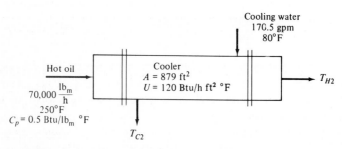

FIGURE 4.4
Countercurrent heat exchanger.

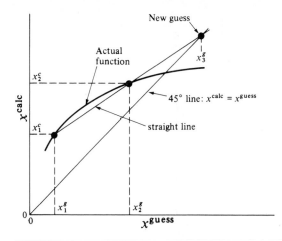

FIGURE 4.5
Wegstein method.

4.3.5 Wegstein

An alternative to using the "proportional" correction factor discussed in the previous section is to use two sets of "guessed" and "calculated" values to construct a straight line relationship and extrapolate this line to the solution. Figure 4.5 shows the method graphically.

Two values of x^{guess} are made: x_1^g and x_2^g. The corresponding two values of x^{calc} are calculated: x_1^c and x_2^c. The equation of a straight line joining these two points is

$$x^{calc} = mx^{guess} + b \tag{4.27}$$

where

$$m = \frac{x_2^c - x_1^c}{x_2^g - x_1^g} \tag{4.28}$$

$$b = \frac{x_1^g x_2^c - x_1^c x_2^g}{x_2^g - x_1^g} \tag{4.29}$$

Now, to calculate a new guess x_3^g, we want to find where this straight line intersects the 45° line (where $x^{guess} = x^{calc} = x_3^g$).

$$x_3^g = mx_3^g + b$$

Solving for x_3^g gives

$$x_3^g = \frac{x_1^g x_2^c - x_1^c x_2^g}{x_1^g - x_2^g + x_2^c - x_1^c} \tag{4.30}$$

4.3.6 Muller Method

The Newton-Raphson method uses a linear equation (straight line) to estimate the solution. The Muller method uses a quadratic equation to estimate the solu-

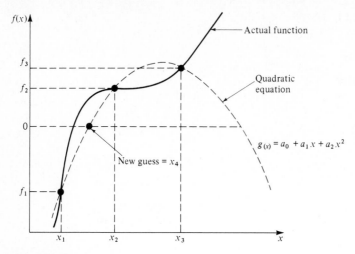

FIGURE 4.6
Muller method.

tion. The idea is illustrated in Fig. 4.6. Three values of the unknown x variable are guessed: x_1, x_2, and x_3. The function is evaluated at these three values of x, giving f_1, f_2, and f_3. A quadratic curve is drawn through these three points.

$$g_{(x)} = a_0 + a_1 x + a_2 x^2 \tag{4.31}$$

Then the equation $g_{(x)} = 0$ is solved for the two roots, and the appropriate one is selected for the next guess of x (x_4). In Fig. 4.6 the appropriate root is the one between x_1 and x_2.

The Muller method converges more quickly than the Newton-Raphson method when the functions have more curvature. However, it is more complex to program and more susceptible to numerical divergence problems.

The iterative algorithm is (Wang and Henke, *Hydro. Proc.*, Vol. 45, 1966, page 155)

$$x_n = x_{n-1} + (x_{n-1} - x_{n-2}) d_n \tag{4.32}$$

for $n = 4, 5, 6, \dots$ until $|f_{(x_k)}| < \varepsilon$, where

$$d_3 = \frac{x_3 - x_2}{x_2 - x_1} \tag{4.33}$$

$$d_n = \frac{-2f_{n-1}(1 + d_{n-1})}{b \pm \sqrt{b^2 - 4f_{n-1} d_{n-1}(1 + d_{n-1})c}} \tag{4.34}$$

$$b = f_{n-3}(d_{n-1})^2 - f_{n-2}(1 + d_{n-1})^2 + f_{n-1}(1 + 2 d_{n-1}) \tag{4.35}$$

$$c = f_{n-3} d_{n-1} - f_{n-2}(1 + d_{n-1}) + f_{n-1} \tag{4.36}$$

4.4 NUMERICAL INTEGRATION
OF ORDINARY DIFFERENTIAL EQUATIONS

As discussed in the introduction to this chapter, the solution of ordinary differential equations (ODEs) on a digital computer involves numerical integration. We will present several of the simplest and most popular numerical-integration algorithms. In Sec. 4.4.1 we will discuss *explicit methods* and in Sec. 4.4.2 we will briefly describe *implicit* algorithms. The differences between the two types and their advantages and disadvantages will be discussed.

The problems of *accuracy, numerical stability,* and *speed* of any numerical integration method must be kept in mind when solving ODEs. If the numerical integration step size is not kept small enough, two things can happen: (1) the calculated solution is not accurate enough, or (2) the calculations may "blow up," indicating numerical instability (values of variables swinging wildly from step to step and variables calculated to have unreal values; e.g., mole fractions becoming negative or greater than one). Of course as the step size is made smaller, the computer time it takes to solve the problem increases. Computer time also increases as the number of differential equations increases. For some algorithms the increase is linear with the number (N) of ODEs. In other algorithms the increase in computer time is proportional to a higher power of N. All of these aspects must be balanced in selecting an algorithm.

Let me make my own personal preference clear from the outset. I have solved literally hundreds of systems of ODEs for chemical engineering systems over my 30 years of experience, and I have found only one or two situations where the plain old simple-minded first-order Euler algorithm was not the best choice for the problem. We will show some comparisons of different types of algorithms on different problems in this chapter and the next.

We need to study the numerical integration of only first-order ODEs. Any higher-order equations, say with Nth-order derivatives, can be reduced to N first-order ODEs. For example, suppose we have a third-order ODE:

$$\frac{d^3x}{dt^3} + a_2 \frac{d^2x}{dt^2} + a_1 \frac{dx}{dt} + a_0 x = b_1 m \qquad (4.37)$$

If we define the new variables

$$x_1 = x \qquad x_2 = \frac{dx}{dt} \qquad x_3 = \frac{d^2x}{dt^2} \qquad (4.38)$$

Eq. (4.37) becomes

$$\frac{dx_3}{dt} + a_2 x_3 + a_1 x_2 + a_0 x_1 = b_1 m$$

Thus we have three first-order ODEs to solve:

$$\frac{dx_1}{dt} = x_2 \tag{4.39}$$

$$\frac{dx_2}{dt} = x_3 \tag{4.40}$$

$$\frac{dx_3}{dt} = -a_2 x_3 - a_1 x_2 - a_0 x_1 + b_1 m \tag{4.41}$$

4.4.1 Explicit Numerical Integration Algorithms

Explicit algorithms involve explicit calculation of derivatives and stepping out in time, with no iteration. Two popular methods that are self-starting and easy to use are described below: Euler and fourth-order Runge-Kutta. There are literally hundreds of other algorithms. Many are quite complex, difficult to program and debug and often quite inefficient for realistic practical chemical engineering problems.

A. EULER ALGORITHM. The simplest possible numerical-integration scheme (and the most useful) is Euler (pronounced "oiler"), illustrated in Fig. 4.7. Assume we wish to solve the ODE

$$\frac{dx}{dt} = f_{(x,\, t)} \tag{4.42}$$

where $f_{(x,\, t)}$ is, in general, a nonlinear function. We need to know where we are starting from, i.e., we need a known initial condition for x. Usually this is at time equal zero.

$$x_{(0)} = x_0 \quad \text{at} \quad t = 0 \tag{4.43}$$

Now if we move forward in time by a small step Δt to $t = t_1 = \Delta t$, we can get an estimate of the new value of x at $t = \Delta t$, $x_{(\Delta t)}$, from a linear extrapolation using the initial time rate of change of x (the derivative of x at $t = 0$). The new value of x (x_1) is approximately equal to the old value of x (x_0) plus the product of the

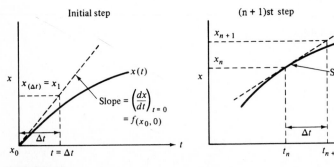

FIGURE 4.7
Graphical representation of Euler method.

derivative of x times the step size.

$$x_{(\Delta t)} = x_{(0)} + \left(\frac{dx}{dt}\right)_{t=0} \Delta t \tag{4.44}$$

$$x_1 = x_0 + f_{(x_0,\, 0)} \Delta t \tag{4.45}$$

If the step size (the integration interval) is small enough, this estimate of x_1 will be very close to the correct value.

To step out another Δt to $t = t_2 = 2\,\Delta t$, we estimate $x_{(2\,\Delta t)} \equiv x_2$ from

$$x_{(2\,\Delta t)} = x_{(\Delta t)} + \left(\frac{dx}{dt}\right)_{t=\Delta t} \Delta t$$

$$x_2 = x_1 + f_{(x_1,\, t_1)} \Delta t \tag{4.46}$$

Generalizing to the $(n + 1)$st step in time,

$$x_{n+1} = x_n + f_{(x_n,\, t_n)} \Delta t = x_n + \left(\frac{dx}{dt}\right)_{(x_n,\, t_n)} \Delta t \tag{4.47}$$

$$t_{n+1} = t_n + \Delta t \tag{4.48}$$

Euler integration is extremely simple to program, as will be illustrated in Example 4.3. This simplicity is retained, even as the number of ODEs increases and as the derivative functions become more complex and nonlinear.

If we have two simultaneous, coupled ODEs to numerically integrate

$$\frac{dx_1}{dt} = f_{1(x_1,\, x_2,\, t)} \qquad \frac{dx_2}{dt} = f_{2(x_1,\, x_2,\, t)} \tag{4.49}$$

the Euler integration algorithms would be

$$x_{1,\, n+1} = x_{1n} + \Delta t\, f_{1(x_{1n},\, x_{2n},\, t_n)} \tag{4.50}$$

$$x_{2,\, n+1} = x_{2n} + \Delta t\, f_{2(x_{1n},\, x_{2n},\, t_n)} \tag{4.51}$$

Notice that only one derivative evaluation is required per ODE at each point in time. If we had a set of N ordinary differential equations, we would have N equations like Eqs. (4.50) and (4.51).

Example 4.3. Suppose we have a system that is described by the ODE

$$\tau \frac{dx}{dt} = 1 - x \tag{4.52}$$

with $x = 0$ at $t = 0$. For the moment let the parameter τ (the time constant) be equal to 1. A FORTRAN digital computer program using Euler is given in Table 4.3, together with output results. An integration step size (DELTA) of 0.05 is used. Figure 4.8 compares the computed values of x for different step sizes. The analytical solution (see Chap. 6) of Eq. (4.52) is also shown, to indicate the accuracy of the method.

$$x_{(t)} = 1 - e^{-t} \tag{4.53}$$

TABLE 4.3
First-order explicit Euler integration

```
C
C    ODE IS    XDOT = 1 - X
C    INITIAL CONDITIONS:   X = 0  AT  T = 0
C
     X=0.
     T=0.
     DELTA=0.05
     WRITE(6,1)
   1 FORMAT(' TIME    X      XDOT')
 100 XDOT=1. - X
     WRITE(6,2)T,X,XDOT
   2 FORMAT(3X,3F8.4)
     X=X+XDOT*DELTA
     T=T+DELTA
     IF(T.LE.2.)GO TO 100
     STOP
     END
```

Results

TIME	X	XDOT	TIME	X	XDOT
0.0000	0.0000	1.0000	1.0500	0.6594	0.3406
0.0500	0.0500	0.9500	1.1000	0.6765	0.3235
0.1000	0.0975	0.9025	1.1500	0.6926	0.3074
0.1500	0.1426	0.8574	1.2000	0.7080	0.2920
0.2000	0.1855	0.8145	1.2500	0.7226	0.2774
0.2500	0.2262	0.7738	1.3000	0.7365	0.2635
0.3000	0.2649	0.7351	1.3500	0.7497	0.2503
0.3500	0.3017	0.6983	1.4000	0.7622	0.2378
0.4000	0.3366	0.6634	1.4500	0.7741	0.2259
0.4500	0.3698	0.6302	1.5000	0.7854	0.2146
0.5000	0.4013	0.5987	1.5500	0.7961	0.2039
0.5500	0.4312	0.5688	1.6000	0.8063	0.1937
0.6000	0.4596	0.5404	1.6500	0.8160	0.1840
0.6500	0.4867	0.5133	1.7000	0.8252	0.1748
0.7000	0.5123	0.4877	1.7500	0.8339	0.1661
0.7500	0.5367	0.4633	1.8000	0.8422	0.1578
0.8000	0.5599	0.4401	1.8500	0.8501	0.1499
0.8500	0.5819	0.4181	1.9000	0.8576	0.1424
0.9000	0.6028	0.3972	1.9500	0.8647	0.1353
0.9500	0.6226	0.3774	2.0000	0.8715	0.1285
1.0000	0.6415	0.3585			

Fairly small steps must be taken (<0.1) if an accurate dynamic curve of $x_{(t)}$ is desired. Fairly large steps can be taken, but the solution is not accurate. However, if the step size is made bigger than 2, the solution goes numerically unstable. The physical system or process is *not* unstable. What is unstable is the numerical integration algorithm at this step size for the ODE given in Eq. (4.52). The Euler algorithm has the property that if the steps are made small enough to achieve reasonable accuracy (four or five significant figures), the solution is stable.

These step sizes scale directly with the time constant τ. If τ were 10, we could take steps that were 10 times bigger. So the maximum stable step size for the Euler integration is twice the time constant.

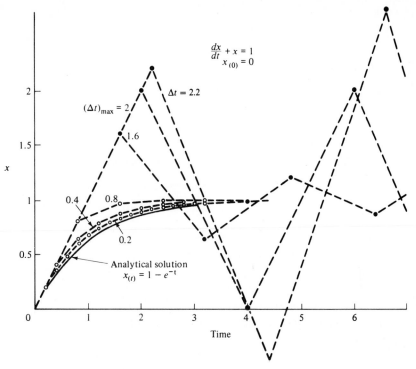

FIGURE 4.8
Effect of integration step size with Euler integration.

B. RUNGE-KUTTA (FOURTH-ORDER). The fourth-order Runge-Kutta algorithm is widely used in chemical engineering. For the ODE given in Eq. (4.42) the Runge-Kutta algorithm is

$$k_1 = \Delta t \, f_{(x_n, \, t_n)}$$
$$k_2 = \Delta t \, f_{(x_n + \frac{1}{2}k_1, \, t_n + \frac{1}{2} \Delta t)}$$
$$k_3 = \Delta t \, f_{(x_n + \frac{1}{2}k_2, \, t_n + \frac{1}{2} \Delta t)} \qquad (4.54)$$
$$k_4 = \Delta t \, f_{(x_n + k_3, \, t_n + \Delta t)}$$
$$x_{n+1} = x_n + \frac{1}{6}(k_1 + 2k_2 + 2k_3 + k_4)$$

For numerically integrating two first-order ODEs

$$\frac{dx_1}{dt} = f_{1(x_1, \, x_2, \, t)}$$

$$\frac{dx_2}{dt} = f_{2(x_1, \, x_2, \, t)}$$

with fourth-order Runge-Kutta, four k's are evaluated for each ODE.

$$k_{11} = \Delta t\, f_{1(x_{n1},\, x_{n2},\, t_n)}$$

$$k_{12} = \Delta t\, f_{2(x_{n1},\, x_{n2},\, t_n)}$$

$$k_{21} = \Delta t\, f_{1(x_{n1}+\frac{1}{2}k_{11},\, x_{n2}+\frac{1}{2}k_{12},\, t_n+\frac{1}{2}\Delta t)}$$

$$k_{22} = \Delta t\, f_{2(x_{n1}+\frac{1}{2}k_{11},\, x_{n2}+\frac{1}{2}k_{12},\, t_n+\frac{1}{2}\Delta t)}$$

$$k_{31} = \Delta t\, f_{1(x_{n1}+\frac{1}{2}k_{21},\, x_{n2}+\frac{1}{2}k_{22},\, t_n+\frac{1}{2}\Delta t)}$$

$$k_{32} = \Delta t\, f_{2(x_{n1}+\frac{1}{2}k_{21},\, x_{n2}+\frac{1}{2}k_{22},\, t_n+\frac{1}{2}\Delta t)}$$

$$k_{41} = \Delta t\, f_{1(x_{n1}+k_{31},\, x_{n2}+k_{32},\, t_n+\Delta t)}$$

$$k_{42} = \Delta t\, f_{2(x_{n1}+k_{31},\, x_{n2}+k_{32},\, t_n+\Delta t)}$$

Then the new values of x_1 and x_2 are calculated:

$$x_{n+1,\,1} = x_{n1} + \tfrac{1}{6}(k_{11} + 2k_{21} + 2k_{31} + k_{41}) \tag{4.55}$$

$$x_{n+1,\,2} = x_{n2} + \tfrac{1}{6}(k_{12} + 2k_{22} + 2k_{32} + k_{42}) \tag{4.56}$$

Table 4.4 gives a computer program and results for the ODE of Eq. (4.52) using fourth-order Runge-Kutta with a step size of 0.2. Figure 4.9 shows the effect of the integration step size on the computed values of x. Three-significant-figure accuracy is obtained for $\Delta t = 0.8$. The maximum stable Δt is 2.7.

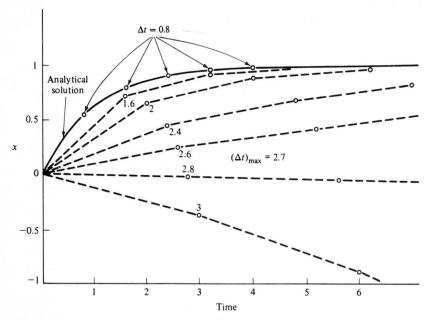

FIGURE 4.9
Effect of integration step size with fourth-order Runge-Kutta integration.

TABLE 4.4

Runge-Kutta integration algorithm

```
C
C     ODE IS    XDOT = 1 - X
C     INITIAL CONDITIONS:    X = 0   AT  T = 0
C
      REAL K1,K2,K3,K4
C USE FUNCTION STATEMENT FOR DERIVATIVE EVALUATION
      DF(XX)=DELTA*(1.-XX)
      DELTA=0.2
      T=0.
      X=0.
      WRITE(6,1)
    1 FORMAT('     TIME    X       K1      K2      K3      K4')
C EVALUATE FOUR K VALUES
  100 K1=DF(X)
      K2=DF(X+K1/2.)
      K3=DF(X+K2/2.)
      K4=DF(X+K3)
      WRITE(6,2) T,X,K1,K2,K3,K4
    2 FORMAT(3X,6F8.4)
C INTEGRATE ALA RUNGE-KUTTA
      X=X+(K1+2.*K2+2.*K3+K4)/6.
      T=T+DELTA
      IF(T.LE.2.01)GO TO 100
      STOP
      END
```

Results

TIME	X	K1	K2	K3	K4
0.0000	0.0000	0.2000	0.1800	0.1820	0.1636
0.2000	0.1813	0.1637	0.1474	0.1490	0.1339
0.4000	0.3297	0.1341	0.1207	0.1220	0.1097
0.6000	0.4512	0.1098	0.0988	0.0999	0.0898
0.8000	0.5507	0.0899	0.0809	0.0818	0.0735
1.0000	0.6321	0.0736	0.0662	0.0670	0.0602
1.2000	0.6988	0.0602	0.0542	0.0548	0.0493
1.4000	0.7534	0.0493	0.0444	0.0449	0.0403
1.6000	0.7981	0.0404	0.0363	0.0367	0.0330
1.8000	0.8347	0.0331	0.0298	0.0301	0.0270
2.0000	0.8647	0.0271	0.0244	0.0246	0.0221

Notice that four derivative evaluations are required per ODE at each time step. Thus the computer time required to run Euler with a step size of 0.05 would be about the same as the time required to run Runge-Kutta with a step size of 0.2.

We can draw some very important conclusions about the two algorithms from the numerical results obtained in the sample example considered above:

1. If an accurate integration is required, the fourth-order Runge-Kutta is superior to Euler. For the same computing time (with the step size used in Runge-Kutta four times that used in Euler) the Runge-Kutta is more accurate.

2. If accuracy is not required for the particular ODE being integrated, Euler is superior to Runge-Kutta. Euler is stable for step sizes that are almost as large

as those for which Runge-Kutta is stable. Since Runge-Kutta requires four derivative evaluations compared with only one for Euler, the Euler algorithm will run almost four times as fast.

You may wonder why we would ever be satisfied with anything less than a very accurate integration. The ODEs that make up the mathematical models of most practical chemical engineering systems usually represent a mixture of fast dynamics and slow dynamics. For example, in a distillation column the liquid flow or hydraulic dynamic response occurs fairly rapidly, of the order of a few seconds per tray. The composition dynamics, the rate of change of liquid mole fractions on the trays, are usually much slower—minutes or even hours for columns with many trays. Systems with this mixture of fast and slow ODEs are called *stiff* systems.

If accurate integration is specified for all the ODEs, the fast ones will require a small step size, much smaller than would be required for the slow ODEs. Therefore it is often quite acceptable to sacrifice accuracy on the fast ODEs and run at a step size for which the fast ODEs are still stable and the slow ODEs are quite accurate. This is illustrated in Fig. 4.10. Since the process is often dominated by the slow ODEs, the inaccuracy of the rapidly changing variables has little effect on the accuracy of the slowly changing variables.

Therefore my experience has been that, for most of the complex systems that chemical engineers have to deal with, a simple Euler integration is just as good as, if not better than, the more complex fourth-order Runge-Kutta.

One final practical tip about numerical integration. Many digital simulation experts advocate the use of variable step-size algorithms. The notion is that small steps must be taken while the process is changing rapidly, but big steps can be taken when variables are changing slowly. In terms of accuracy this is true. If, however, the fastest ODE is running at its numerical stability limit, the step size cannot be increased no matter how slowly other variables are changing. And of course this constant checking and readjusting of step size chews up additional computer time.

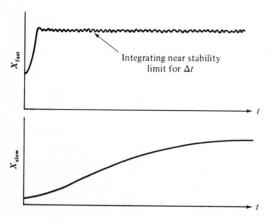

FIGURE 4.10
Responses of fast and slow ODEs.

Therefore my recommended technique is to start with a very small step size to get the program debugged and running. Then make a few quick empirical tests in which you keep doubling the step size. Check the accuracy of successive runs and continue increasing the step size until you find the biggest Δt that gives sufficiently accurate answers and is stable.

4.4.2   Implicit Methods

The explicit methods considered in the previous section involved derivative evaluations, followed by explicit calculation of new values for variables at the next point in time. As the name implies, implicit integration methods use algorithms that result in implicit equations that must be solved for the new values at the next time step. A single-ODE example illustrates the idea.

The first-order *implicit* Euler algorithm is

$$x_{n+1} = x_n + \left(\frac{dx}{dt}\right)_{(x_{n+1},\, t_{n+1})} \Delta t \tag{4.57}$$

Compare this carefully with the *explicit* algorithm given in Eq. (4.47). The derivative is evaluated at the next step in time where we do not know the variable x_{n+1}. Thus, the unknown x_{n+1} appears on both sides of the equation. Consider the simple ODE

$$\frac{dx}{dt} = 1 - x$$

The implicit Euler algorithm is

$$x_{n+1} = x_n + \left(\frac{dx}{dt}\right)_{(x_{n+1},\, t_{n+1})} \Delta t$$
$$x_{n+1} = x_n + (1 - x_{n+1}) \, \Delta t \tag{4.58}$$

Solving for the unknown x_{n+1} gives

$$x_{n+1} = \frac{x_n + \Delta t}{1 + \Delta t} \tag{4.59}$$

The main advantage of the implicit algorithms is that they do not become numerically unstable. Very large step sizes can be taken without having to worry about the instability problems that plague the explicit methods. Thus, the implicit methods are very useful for stiff systems.

Equation (4.58) can be solved easily for the unknown x_{n+1}. However, suppose we have a large number (N) of ODEs to numerically integrate. In general, all of the N derivatives depend on all the variables, so we end up with N simultaneous (usually nonlinear) algebraic equations that must be solved at each point in time for the N unknown values of the variables at the next time step.

Consider the set of N linear ODEs

$$\frac{d}{dt}\begin{bmatrix} x_1 \\ x_2 \\ x_3 \\ \cdots \\ x_N \end{bmatrix} = \begin{bmatrix} a_{11} & a_{12} & a_{13} & \cdots & a_{1N} \\ a_{21} & a_{22} & \cdots & \cdots & \cdots \\ a_{31} & \cdots & \cdots & \cdots & \cdots \\ \cdots & \cdots & \cdots & \cdots & \cdots \\ a_{N1} & \cdots & \cdots & \cdots & a_{NN} \end{bmatrix}\begin{bmatrix} x_1 \\ x_2 \\ x_3 \\ \cdots \\ x_N \end{bmatrix} \qquad (4.60)$$

$$\frac{d\underline{x}}{dt} = \underline{A}\underline{x}$$

where $\underline{x}$ is a vector of variables and $\underline{A}$ is a square matrix of constants. Applying the implicit Euler algorithm to these equations gives

$$\underline{x}_{n+1} = \underline{x}_n + [\underline{A}\underline{x}_{n+1}\ \Delta t] \qquad (4.61)$$

Solving for $\underline{x}_{n+1}$ gives

$$\underline{x}_{n+1} = [\underline{I} - \underline{A}\ \Delta t]^{-1}\underline{x}_n \qquad (4.62)$$

where the -1 superscript on the term in brackets means "matrix inverse." We will deal with matrices and vectors in more detail in Chap. 15 when we study multivariable systems. For the moment the point that I am trying to make is that the use of implicit algorithms involves calculating the inverse of a matrix (usually at each point in time because the ODEs are usually not linear). This calculation can require a fair amount of computer time, particularly as the number of ODEs increases. Many chemical engineering models contain hundreds of ODEs. This means that the inverses of very large matrices must be calculated.

Thus the implicit methods become slower and slower as the number of ODEs increases, despite the fact that large step sizes can be taken. Therefore plain old *explicit* Euler turns out to run faster than the implicit methods on many realistically large problems, unless the stiffness of the system is very, very severe. We will talk more about this in Chap. 5.

The implicit methods are much more complicated to program and to debug. Fortunately a number of fairly easy-to-use packages are available: DGEAR, LSODE, etc. The review paper by Byrne and Hindmarsh (*Journal of Computational Physics*, Vol. 70, No. 1, May 1987) gives a good summary of the history and selection of stiff-system implicit integration algorithms.

PROBLEMS

4.1. Write a BUBPT subroutine that uses false-position convergence.

4.2. Compare convergence times, using interval halving, Newton-Raphson, and false position, for an ideal, four-component, vapor-liquid equilibrium system. The pure component vapor pressures are:

Component	Vapor pressure of pure component P_j^s (psia)	
	at 150°F	at 200°F
1	25	200
2	14.7	60
3	4	14.7
4	0.5	5

Calculate the correct temperature and vapor compositions for a liquid at 75 psia with a composition $x_1 = 0.10$, $x_2 = 0.54$, $x_3 = 0.30$, and $x_4 = 0.06$.

4.3. The design of ejectors requires trial and error to find the "motive" pressure P_m that, with a fixed motive flow rate of gas W_m, will suck a design flow rate W_s of suction gas from a suction pressure of P_s and discharge against a higher pressure P_D.

The motive gas is at 300°F and has a molecular weight of 60. Its flow rate is 5000 lb$_m$/h. The suction gas is at 400°F and 150 psia and has a molecular weight of 50. A suction flow of 7000 lb$_m$/h must be ejected into a discharge pressure of 160 psia.

Assume perfect gases and frictionless, reversible, adiabatic operation of the jet; i.e., the expansions and contractions into and out of the throat are isentropic. The ratio of C_p to C_v for all gases is 1.2 and C_p heat capacities are constant and equal to 0.6 Btu/lb$_m$°R.

Find the motive pressure P_m required and the areas in the throat on the motive and suction sides, A_m and A_s.

4.4. Find the optimum liquid concentration of the propane-isobutane mixture in an auto-refrigerated alkylation reactor. The exothermic heat Q_R (10^6 Btu/h) of the alkylation reaction is removed by vaporization of the liquid in the reactor. The vapor is compressed, condensed, and flashed back into the reactor through a pressure letdown valve. The reactor must operate at 50°F, and the compressed vapors must be condensed at 110°F.

Find the liquid mole fraction x of propane that minimizes the compressor horsepower requirements for a given Q_R. Assume the compressor adiabatic efficiency is 100 percent.

CHAPTER
5

SIMULATION EXAMPLES

Now that we understand some of the numerical-analysis tools, let us illustrate their application to some chemical engineering systems. We will start with simple examples and work our way up to more realistic systems that involve many simultaneous ordinary differential and nonlinear algebraic equations.

In all the programs presented the emphasis is not on programming or computational efficiency but on easy translation of the equations and the solution logic into a workable and understandable FORTRAN program.

5.1 GRAVITY-FLOW TANK

The gravity-flow tank that we considered in Chap. 1 and later in Example 2.9 makes a nice simple system to start our simulation examples. The force balance on the outlet line gave us the nonlinear ODE

$$\frac{dv}{dt} = \frac{g}{L} h - \frac{K_F g_c}{\rho A_p} v^2 \tag{5.1}$$

To describe the system completely a total continuity equation on the liquid in the tank is also needed.

$$A_T \frac{dh}{dt} = F_0 - F \tag{5.2}$$

We have to pick a specific numerical case to solve these two coupled ordinary differential equations. Equation (5.1) is nonlinear because of the v^2 term. Physical dimensions, parameter values, and steadystate flow rate and liquid height are given in Table 5.1.

116

TABLE 5.1
Gravity-flow tank data

Pipe:
ID = 3 ft   Area = 7.06 ft^2 Length = 3000 ft
Tank:
ID = 12 ft   Area = 113 ft^2 Height = 7 ft
Steadystate values:
$\bar{F}$ = 35.1 ft^3/s (15,700 gpm)
$\bar{h}$ = 4.72 ft
$\bar{v}$ = 4.97 ft/s
Parameters:
Reynolds number = 1,380,000
Friction factor = 0.0123
K_F = 2.81 × 10^{-2} lb$_f$/(ft/s)^2 ft

Using the relationship $F = vA_p$ and substituting the numerical values of parameters into Eqs. (5.1) and (5.2) give

$$\frac{dv}{dt} = 0.0107h - 0.00205v^2 \tag{5.3}$$

$$\frac{dh}{dt} = 0.311 - 0.0624v \tag{5.4}$$

Table 5.2 gives a FORTRAN program that numerically integrates the two ODEs describing this system for two different initial conditions of flow and liquid level in the tank: (1) when the initial flow rate is 50 percent of the design rate, and (2) when the initial flow rate is 67 percent of the design flow rate. At time equal zero, the flow rate into the tank is increased to the maximum design flow rate of 35.1 ft^3/s.

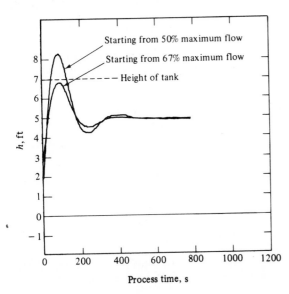

FIGURE 5.1
Gravity-flow tank.

TABLE 5.2

Gravity-flow tank simulation

```
C        (TIME IS IN SECONDS)
C
C  TWO CASES ARE RUN. AT TIME EQUAL ZERO THE FEED FLOW RATE IS
C      INCREASED TO 100% OF THE DESIGN FLOW RATE (15,700 GPM).
C      CASE NO.1 STARTS FROM 67% OF DESIGN FLOW RATE
C      CASE NO.2 STARTS FROM 50% OF DESIGN FLOW RATE
   DATA V1,H1/3.40,2.05/
   DATA V2,H2/2.50,1.2/
   DO 200 N=1,2
   IF(N.EQ.1)THEN
      V=V1
      H=H1
   ELSE
      V=V2
      H=H2
   ENDIF
   TIME=0.
   DELTA=1.
   TPRINT=0.
   WRITE(6,1)V,H
 1 FORMAT (' INITIAL CONDITIONS:  V = ',F6.3,'  H = ',F8.3)
   WRITE(6,2)
 2 FORMAT('  TIME     V       H')
C
C EVALUATE DERIVATIVES
C
100 VDOT=0.0107*H-0.00205*V**2
   HDOT=0.311-0.0624*V
   IF(TIME.LT.TPRINT)GO TO 10
   WRITE(6,3)TIME,V,H
 3 FORMAT(3X,3F8.2)
   TPRINT=TPRINT+20.
C
C INTEGRATE USING EULER
```

The explicit first-order Euler algorithm is used. The variables that we are solving for as functions of time are V and H. The right-hand sides of Eqs. (5.3) and (5.4) are the derivative functions. These are called VDOT and HDOT in the program. At the nth step in time

$$(\text{VDOT})_n = 0.0107(H)_n - 0.00205[(V)_n]^2 \tag{5.5}$$

$$(\text{HDOT})_n = 0.311 - 0.0624(V)_n \tag{5.6}$$

The new values of H and V at the $(n + 1)$st step are calculated from the Euler algorithm with a step size of DELTA.

$$(H)_{n+1} = (H)_n + \text{DELTA}(\text{HDOT})_n \tag{5.7}$$

$$(V)_{n+1} = (V)_n + \text{DELTA}(\text{VDOT})_n \tag{5.8}$$

Results are plotted in Fig. 5.1. Notice that the tank can overflow if the inflow rate is changed from 50 to 100 percent of design. So even though the level

TABLE 5.2 (*continued*)

```
C
  10 V=V+VDOT*DELTA
     H=H+HDOT*DELTA
     TIME=TIME+DELTA
     IF(TIME.LE.200.)GO TO 100
 200 CONTINUE
     STOP
     END
```

Results

INITIAL CONDITIONS: V = 3.400 H = 2.050

TIME	V	H	
0.00	3.40	2.05	
20.00	3.55	3.98	
40.00	3.99	5.52	
60.00	4.54	6.44	
80.00	5.02	6.69	Liquid level near top of tank.
100.00	5.33	6.45	
120.00	5.46	5.92	
140.00	5.44	5.32	
160.00	5.33	4.81	
180.00	5.19	4.46	
200.00	5.05	4.29	

INITIAL CONDITIONS: V = 2.500 H = 1.200

TIME	V	H	
0.00	2.50	1.20	
20.00	2.79	4.19	
40.00	3.54	6.51	
60.00	4.43	7.78	Liquid level exceeds tank height!
80.00	5.19	8.00	"
100.00	5.64	7.44	"
120.00	5.79	6.50	
140.00	5.72	5.52	
160.00	5.51	4.72	
180.00	5.27	4.20	
200.00	5.05	3.98	

in the tank at the design flow rate would be lower than the height of the tank (7 ft), the dynamic change in the height exceeds 7 ft. This is due to the inertia of the mass of liquid in the pipe.

5.2 THREE CSTRs IN SERIES

The equations describing the series of three isothermal CSTRs were developed in Sec. 3.2.

$$\frac{dC_{A1}}{dt} = \frac{1}{\tau}(C_{A0} - C_{A1}) - kC_{A1} \tag{5.9}$$

$$\frac{dC_{A2}}{dt} = \frac{1}{\tau}(C_{A1} - C_{A2}) - kC_{A2} \tag{5.10}$$

$$\frac{dC_{A3}}{dt} = \frac{1}{\tau}(C_{A2} - C_{A3}) - kC_{A3} \tag{5.11}$$

TABLE 5.3

Three-isothermal CSTR (openloop)

```
C
C    DISTURBANCE IS A STEP CHANGE IN FEED CONCENTRATION AT TIME
C        EQUAL ZERO FROM 0.8 TO 1.8 KG-MOLES OF A/CUBIC METER.
C    TIME IS IN MINUTES.
     REAL K
C
C    INITIAL CONDITIONS:
     TIME=0.
     CA1=0.4
     CA2=0.2
     CA3=0.1
     CA0=1.8
C PARAMETER VALUES:
     TAU=2.
     K=0.5
     DELTA=0.1
     TPRINT=0.
     WRITE(6,1)
  1 FORMAT('    TIME    CA1    CA2    CA3')
C EVALUATE DERIVATIVES
 100 CA1DOT=(CA0-CA1)/TAU-K*CA1
     CA2DOT=(CA1-CA2)/TAU-K*CA2
     CA3DOT=(CA2-CA3)/TAU-K*CA3
     IF(TIME.LT.TPRINT)GO TO 10
     WRITE(6,2)TIME,CA1,CA2,CA3
  2 FORMAT(3X,4F8.3)
     TPRINT=TPRINT+0.1
 10 CA1=CA1+CA1DOT*DELTA
     CA2=CA2+CA2DOT*DELTA
     CA3=CA3+CA3DOT*DELTA
     TIME=TIME+DELTA
     IF(TIME.LE.3.)GO TO 100
     STOP
     END
```

The initial conditions are $C_{A1(0)} = 0.4$ kg·mol of component A/m³, $C_{A2(0)} = 0.2$ kg·mol of component A/m³, and $C_{A3(0)} = 0.1$ kg·mol of component A/m³. The forcing function is C_{A0}. We will assume that at time zero C_{A0} is set at 1.8 kg·mol of A/m³ and held constant. The parameter τ is set equal to 2 min and the value of k is 0.5 min⁻¹.

The right-hand sides of the ODEs [Eqs. (5.9) to 5.11)] are the functions $f_{(x, t)}$ discussed in Sec. 4.4.1. Let us call these derivatives CA1DOT, CA2DOT, and CA3DOT. At the nth step in time,

$$(CA1DOT)_n = \frac{1}{\tau}\left[(C_{A0})_n - (C_{A1})_n\right] - k(C_{A1})_n \tag{5.12}$$

$$(CA2DOT)_n = \frac{1}{\tau}\left[(C_{A1})_n - (C_{A2})_n\right] - k(C_{A2})_n \tag{5.13}$$

$$(CA3DOT)_n = \frac{1}{\tau}\left[(C_{A2})_n - (C_{A3})_n\right] - k(C_{A3})_n \tag{5.14}$$

TABLE 5.3 (*continued*)

Results

TIME	CA1	CA2	CA3
0.000	0.400	0.200	0.100
0.100	0.450	0.200	0.100
0.200	0.495	0.203	0.100
0.300	0.536	0.207	0.100
0.400	0.572	0.213	0.100
0.500	0.605	0.220	0.101
0.600	0.634	0.229	0.102
0.700	0.661	0.237	0.103
0.800	0.685	0.247	0.105
0.900	0.706	0.256	0.107
1.000	0.726	0.266	0.109
1.100	0.743	0.276	0.111
1.200	0.759	0.285	0.114
1.300	0.773	0.295	0.117
1.400	0.786	0.304	0.120
1.500	0.797	0.313	0.123
1.600	0.807	0.321	0.126
1.700	0.817	0.330	0.130
1.800	0.825	0.337	0.133
1.900	0.832	0.345	0.137
2.000	0.839	0.352	0.140
2.100	0.845	0.359	0.144
2.200	0.851	0.365	0.147
2.300	0.856	0.371	0.151
2.400	0.860	0.377	0.154
2.500	0.864	0.382	0.158
2.600	0.868	0.387	0.161
2.700	0.871	0.392	0.164
2.800	0.874	0.396	0.168
2.900	0.876	0.400	0.171

Then to step to the next point in time, using Euler integration with a step size DELTA,

$$(CA1)_{n+1} = (CA1)_n + DELTA(CA1DOT)_n \tag{5.15}$$

$$(CA2)_{n+1} = (CA2)_n + DELTA(CA2DOT)_n \tag{5.16}$$

$$(CA3)_{n+1} = (CA3)_n + DELTA(CA3DOT)_n \tag{5.17}$$

Converting these equations into a FORTRAN program is simple, as shown in Table 5.3. Results show that the concentrations in the system increase gradually with time after the step increase in the inlet concentration C_{A0}.

Now let us make life a little more interesting. The system considered above is an "openloop" system, i.e., no feedback control is used. If we add a feedback controller, we have a "closedloop" system. The controller looks at the product concentration leaving the third tank C_{A3} and makes adjustments in the inlet concentration to the first reactor C_{A0} in order to keep C_{A3} near its desired setpoint value C_{A3}^{set}. The variable C_{AD} is a disturbance concentration and the variable C_{AM} is a manipulated concentration that is changed by the controller. We assume that

$$C_{A0} = C_{AM} + C_{AD} \tag{5.18}$$

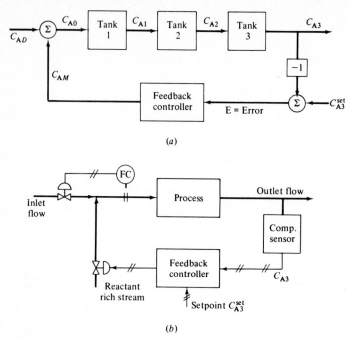

FIGURE 5.2
Closedloop three-CSTR process. (*a*) Idealized system; (*b*) actual system.

This is an idealization of the real physical system in which the control signal from the controller would actually move the position of a control valve that would bleed a stream with a high concentration of reactant A into the feed stream to the process. See Fig. 5.2.

The feedback controller has proportional and integral action. It changes C_{AM} based on the magnitude of the error (the difference between the setpoint and C_{A3}) and the integral of this error.

$$C_{AM} = 0.8 + K_c\left(E + \frac{1}{\tau_I}\int E_{(t)}\,dt\right) \tag{5.19}$$

where $E = C_{A3}^{set} - C_{A3}$

K_c = feedback controller gain (dimensionless)

τ_I = feedback controller integral time constant or reset time (minutes)

The 0.8 term in Eq. (5.19) is the *bias* value of the controller, i.e., the value of C_{AM} at time equal zero. Numerical values of $K_c = 30$ and $\tau_I = 5$ min are used in the program given in Table 5.4.

To simulate Eq. (5.19) we need to numerically integrate the error to get the integral term.

TABLE 5.4
Three-isothermal CSTR (closedloop)

```
C          ( WITH PROPORTIONAL FEEDBACK CONTROLLER)
C
C   DISTURBANCE (CAD) IS A STEP CHANGE AT TIME EQUAL ZERO
C          FROM 0 TO 0.2.
C          TIME IS IN MINUTES.
      REAL K,KC
C
C     INITIAL CONDITIONS:
      TIME=0.
      CA1=0.4
      CA2=0.2
      CA3=0.1
      CA3SET=0.1
      ERINT=0.
C DISTURBANCE
      CAD=0.2
C PARAMETER VALUES:
      TAU=2.
      K=0.5
      KC=30.
      TAUI=5.
      DELTA=0.01
      TPRINT=0.
      WRITE(6,1)
    1 FORMAT('    TIME    CA1    CA2    CA3    CAM')
  100 CONTINUE
C FEEDBACK CONTROLLER
      E=CA3SET-CA3
      CAM=0.8+KC*(E+ERINT/TAUI)
      CA0=CAM+CAD
C EVALUATE DERIVATIVES
      CA1DOT=(CA0-CA1)/TAU-K*CA1
      CA2DOT=(CA1-CA2)/TAU-K*CA2
      CA3DOT=(CA2-CA3)/TAU-K*CA3
      IF(TIME.LT.TPRINT)GO TO 10
      WRITE(6,2)TIME,CA1,CA2,CA3,CAM
    2 FORMAT(3X,5F8.4)
      TPRINT=TPRINT+0.5
   10 CA1=CA1+CA1DOT*DELTA
      CA2=CA2+CA2DOT*DELTA
      CA3=CA3+CA3DOT*DELTA
      ERINT=ERINT+E*DELTA
      TIME=TIME+DELTA
      IF(TIME.LE.10.)GO TO 100
      STOP
      END
```

Results

TIME	CA1	CA2	CA3	CAM	TIME	CA1	CA2	CA3	CAM
0.0000	0.4000	0.2000	0.1000	0.8000	4.5000	0.3952	0.1961	0.1012	0.6536
0.5100	0.4394	0.2046	0.1004	0.7888	5.0000	0.4112	0.1984	0.1001	0.6830
1.0100	0.4564	0.2127	0.1020	0.7373	5.5000	0.4235	0.2026	0.1002	0.6810
1.5100	0.4529	0.2188	0.1044	0.6549	6.0000	0.4275	0.2068	0.1011	0.6521
2.0100	0.4339	0.2200	0.1066	0.5728	6.5001	0.4227	0.2092	0.1023	0.6110
2.5100	0.4090	0.2162	0.1076	0.5206	7.0001	0.4119	0.2090	0.1032	0.5748
3.0100	0.3882	0.2093	0.1071	0.5134	7.5001	0.3999	0.2065	0.1035	0.5566
3.5100	0.3788	0.2021	0.1053	0.5470	8.0001	0.3915	0.2029	0.1030	0.5605
4.0100	0.3823	0.1972	0.1031	0.6023	8.5001	0.3891	0.1997	0.1021	0.5818

123

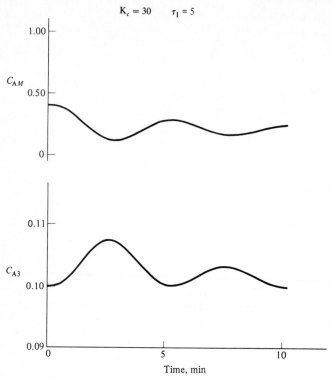

FIGURE 5.3
Plotted results for the three-CSTR example.

We define ERINT as the integral $\int E \, dt$. The derivative of ERINT is just the error E, since

$$\frac{d}{dt}\left(\int E_{(t)} \, dt\right) = E_{(t)} \tag{5.20}$$

Therefore we have an additional ODE that comes from the controller. This must be solved at the same time as the three ODEs describing the process. Using Euler with a DELTA integration step gives

$$(\text{ERINT})_{n+1} = (\text{ERINT})_n + \text{DELTA}(E)_n \tag{5.21}$$

Figure 5.3 shows results for a step change in the disturbance C_{AD} of 0.2 at time equal zero. An integration step size of 0.1 min is used. We will return to this simple system later in this book to discuss the selection of values for K_c and τ_I, that is, how we *tune* the controller.

5.3 NONISOTHERMAL CSTR

The jacketed exothermic CSTR discussed in Sec. 3.6 provides a good example of the simulation of very nonlinear ODEs. Both flow rates and holdups will be variable. A proportional level controller manipulates the liquid leaving the tank, F, as a linear function of the volume in the tank.

$$F = 40 - 10(48 - V) \qquad (5.22)$$

A second controller manipulates the flow rate of cooling water to the jacket, F_J, in direct proportion to the temperature in the reactor.

$$F_J = 49.9 - K_c(600 - T) \qquad (5.23)$$

Constant holdup and perfect mixing are assumed in the cooling jacket. Disturbances in inlet feed flow rate F_0 and feed concentration C_{A0} are step changes at time equal zero.

The ODEs describing the system are

$$\frac{dV}{dt} = F_0 - F \qquad (5.24)$$

$$\frac{d(VC_A)}{dt} = F_0 C_{A0} - FC_A - VkC_A \qquad (5.25)$$

$$\frac{d(VT)}{dt} = F_0 T_0 - FT - \frac{\lambda VkC_A}{\rho C_p} - \frac{UA_H}{\rho C_p}(T - T_J) \qquad (5.26)$$

$$\frac{dT_J}{dt} = \frac{F_J(T_{J0} - T_J)}{V_J} + \frac{UA_H}{\rho_J V_J C_J}(T - T_J) \qquad (5.27)$$

The algebraic equations describing the system are Equations (5.22) and (5.23) and the following:

$$k = \alpha e^{-E/R\bar{T}} \qquad (5.28)$$

Table 5.5 gives values of parameters and steadystate conditions. The variables with overscores or "bars" over them are steadystate values. Note that the time basis used in this problem is hours. Table 5.6 gives a FORTRAN program that simulates this system using Euler integration. The right-hand sides of the

TABLE 5.5
Nonisothermal CSTR parameter values

Steadystate values:

$\bar{F} = 40$ ft^3/h	$\bar{V} = 48$ ft^3
$\bar{C}_{A0} = 0.50$ lb·mol A/ft^3	$\bar{C}_A = 0.245$ lb·mol A/ft^3
$\bar{T} = 600°$R	$\bar{T}_J = 594.6°$R
$\bar{F}_J = 49.9$ ft^3/h	$\bar{T}_0 = 530°$R

Parameter values:

$V_J = 3.85$ ft^3	$\alpha = 7.08 \times 10^{10}$ h^{-1}
$E = 30{,}000$ Btu/lb·mol	$R = 1.99$ Btu/lb·mol °R
$U = 150$ Btu/h ft^2 °R	$A_H = 250$ ft^2
$T_{J0} = 530$ °R	$\lambda = -30{,}000$ Btu/lb·mol
$C_p = 0.75$ Btu/lb$_m$ °R	$C_J = 1.0$ Btu/lb$_m$ °R
$\rho = 50$ lb$_m$/ft^3	$\rho_J = 62.3$ lb$_m$/ft^3
$K_c = 4$ (ft^3/h)/°R	$T^{\text{set}} = 600°$R

TABLE 5.6
Nonisothermal CSTR

```
C
C    DISTURBANCE IS STEP CHANGE IN FEED COMPOSITION AT TIME ZERO
C             FROM 0.50 TO 0.55.
     REAL K,KC
C INITIAL CONDITIONS
     CA=0.245
     T=600.
     TJ=594.59
     V=48.
     TIME=0.
     VC=V*CA
     VT=V*T
C PARAMETER VALUES
     TJ0=530.
     F0=40.
     T0=530.
     CAO=0.5
     KC=4.
     DELTA=0.01
     TPRINT=0.
C DISTURBANCE
     CA0=0.55
     WRITE(6,1)
   1 FORMAT('    TIME   CA   T    V    F    TJ    FJ')
 100 CONTINUE
C   FEEDBACK CONTROLLERS
     FJ=49.9-KC*(600.-T)
     F=40.-10.*(48.-V)
C   REACTION RATE
     K=7.08E10*EXP(-30000./(1.99*T))
     Q=150.*250.*(T-TJ)
C EVALUATE ALL DERIVATIVES
     VDOT=F0-F
     VCDOT=F0*CA0-F*CA-V*K*CA
     VTDOT=F0*T0-F*T+(30000.*V*K*CA-Q)/(0.75*50.)
     TJDOT=FJ*(TJ0-TJ)/3.85 +Q/240.
     IF(TIME.LT.TPRINT)GO TO 10
     WRITE(6,2)TIME,CA,T,V,F,TJ,FJ
   2 FORMAT(1X,7F8.3)
     TPRINT=TPRINT+0.2
  10 V=V+VDOT*DELTA
     VC=VC+VCDOT*DELTA
     VT=VT+VTDOT*DELTA
     TJ=TJ+TJDOT*DELTA
     TIME=TIME+DELTA
     CA=VC/V
     T=VT/V
     IF(TIME.LT.4.1)GO TO 100
     STOP
     END
```

Results

TIME	CA	T	V	F	TJ	FJ
0.000	0.245	600.000	48.000	40.000	594.590	49.900
0.200	0.252	600.370	48.000	40.000	594.822	51.379
0.410	0.256	601.138	48.000	40.000	595.222	54.453

TABLE 5.6 (*continued*)

0.610	0.257	601.810	48.000	40.000	595.572	57.139
0.810	0.258	602.267	48.000	40.000	595.810	58.967
1.010	0.257	602.499	48.000	40.000	595.932	59.896
1.210	0.256	602.565	48.000	40.000	595.968	60.161
1.410	0.256	602.542	48.000	40.000	595.957	60.070
1.610	0.256	602.490	48.000	40.000	595.930	59.859
1.810	0.256	602.442	48.000	40.000	595.906	59.668
2.010	0.256	602.411	48.000	40.000	595.889	59.544
2.210	0.256	602.397	48.000	40.000	595.882	59.488
2.410	0.256	602.394	48.000	40.000	595.881	59.477
2.610	0.256	602.397	48.000	40.000	595.882	59.487
2.810	0.256	602.401	48.000	40.000	595.884	59.503
3.010	0.256	602.404	48.000	40.000	595.886	59.516
3.210	0.256	602.406	48.000	40.000	595.887	59.525
3.410	0.256	602.407	48.000	40.000	595.887	59.529
3.610	0.256	602.407	48.000	40.000	595.887	59.529
3.810	0.256	602.407	48.000	40.000	595.887	59.529
4.010	0.256	602.407	48.000	40.000	595.887	59.529

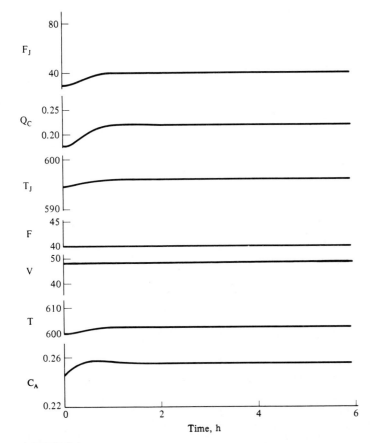

FIGURE 5.4
Nonisothermal CSTR: $+10\%\ \Delta C_{A0}$.

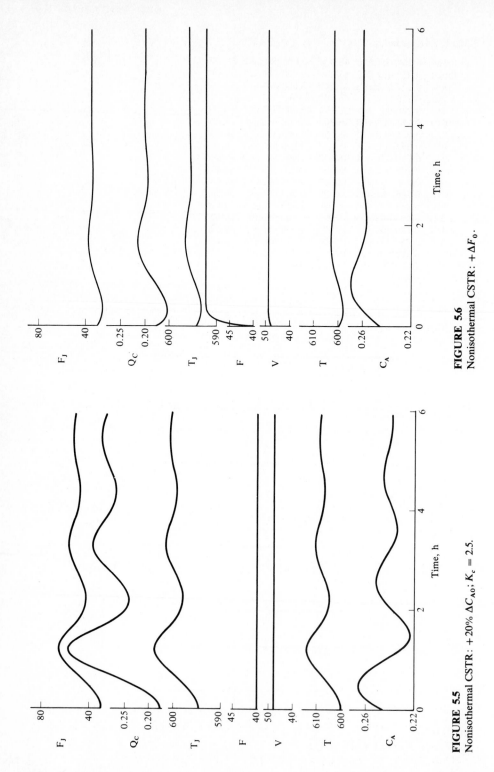

FIGURE 5.5
Nonisothermal CSTR: $+20\%\ \Delta C_{AO}$; $K_c = 2.5$.

FIGURE 5.6
Nonisothermal CSTR: $+\Delta F_0$.

ODEs are defined as VDOT, VCDOT, VTDOT, and TJDOT. At each point in time, integration gives values for V, VC, VT, and TJ. Then C_A and T are found by dividing VC and VT by V.

Figures 5.4 through 5.6 give results for disturbances in feed composition and feed flow rate. Note that in Fig. 5.5 the controller gain has been decreased to 2.5 and a larger feed composition disturbance has been made. The response is quite oscillatory. We will discuss the tuning of temperature controllers in this type of system in much more detail later in this book.

5.4 BINARY DISTILLATION COLUMN

The digital simulation of a distillation column is fairly straightforward. The main complication is the large number of ODEs and algebraic equations that must be solved. We will illustrate the procedure first with the simplified binary distillation column for which we developed the equations in Chap. 3 (Sec. 3.11). Equimolal overflow, constant relative volatility, and theoretical plates have been assumed. There are two ODEs per tray (a total continuity equation and a light component continuity equation) and two algebraic equations per tray (a vapor-liquid phase equilibrium relationship and a liquid-hydraulic relationship).

$$\frac{dM_n}{dt} = L_{n+1} - L_n \tag{5.29}$$

$$\frac{d(M_n x_n)}{dt} = L_{n+1} x_{n+1} + V y_{n-1} - L_n x_n - V y_n \tag{5.30}$$

$$y_n = \frac{\alpha x_n}{1 + (\alpha - 1)x_n} \tag{5.31}$$

$$L_n = \bar{L}_n + \frac{M_n - \bar{M}_n}{\beta} \tag{5.32}$$

Equation (5.32) is a simple linear relationship between the liquid holdup on a tray, M_n, and the liquid flow rate leaving the tray, L_n. The parameter β is the hydraulic time constant, typically 3 to 6 seconds per tray.

Since there are many trays and most are described by Eqs. (5.29) through (5.32), it is logical to use "dimensioned" variables and to evaluate derivatives and integrate using FORTRAN "DO" loops. It also makes sense to use a SUBROUTINE or FUNCTION to find y_n, given x_n, because the same equation is used over and over again.

At each instant in time we know all holdups M_n and all liquid compositions x_n. Our simulation logic is:

1. Calculate vapor compositions on all trays from Eq. (5.31).
2. Calculate all liquid flow rates from Eq. (5.32).
3. Evaluate all derivatives. These are the right-hand sides of Eqs. (5.29) and (5.30) applied to all trays. These derivatives are called MDOT(N) and MXDOT(N) in the program given in Table 5.7.
4. Integrate with Euler all ODEs and start again at step 1 above.

TABLE 5.7

Binary distillation column dynamics

```
C
C    ASSUMPTIONS: CONSTANT RELATIVE VOLATILITY, EQUIMOLAL
C        OVERFLOW, THEORETICAL TRAYS, SIMPLE LIQUID TRAY
C        HYDRAULICS
C    FEEDBACK CONTROLLERS MANIPULATE R AND V TO CONTROL XD AND XB
C    DISTURBANCE IS A FEED COMPOSITION CHANGE FROM 0.50 TO 0.55
C        AT TIME EQUAL ZERO
C
     DIMENSION X(20),Y(20),L(20),LO(20),M(20)
     DIMENSION MX(20),MDOT(20),MXDOT(20)
     REAL L,LO,M,MD,MB,MX,MDOT,MXDOT,MDO,MO,MBO,KCD,KCB
C USE A FUNCTION STATEMENT FOR VLE
     EQUIL(XX)=ALPHA*XX/(1.+(ALPHA-1.)*XX)
C INITIAL CONDITIONS AND PARAMETER VALUES
     DATA NT,NF,MDO,MBO,MO,RO,VO,F,BETA,ALPHA/20,10,100.,100.,
    + 10.,128.01,178.01,100.,0,1,2./
     DATA XB,X,XD/.02,.035,.05719,.08885,.1318,.18622,.24951,
    + .31618,.37948,.43391,.47688,.51526,.56295,.61896,.68052,
    + .74345,.80319,.85603,.89995,.93458,.96079,.98/
     DATA KCD,KCB,TAUD,TAUB,DELTA,TIME,TPRINT,ERINTD,ERINTB/
    + 1000.,1000.,5.,1.25, .005,4*0./
C DISTURBANCE
     Z=0.55
     WRITE(6,1) Z,F
   1 FORMAT(7X,'Z = ',F10.5,'   F = ',F10.2)
C    INITIAL CONDITIONS
     DO 3 N=1,NT
     M(N)=MO
     MX(N)=M(N)*X(N)
     LO(N)=RO+F
     IF(N.GT.NF) LO(N)=RO
   3 CONTINUE
     WRITE(6,2)
   2 FORMAT(6X,' TIME    XB     X10     XD      R       V')
C TRAY LIQUID HYDRAULICS AND VLE
 100 DO 20 N=1,NT
     Y(N)=EQUIL(X(N))
     L(N)=LO(N)+(M(N)-MO)/BETA
  20 CONTINUE
     YB=EQUIL(XB)
C    TWO PI FEEDBACK CONTROLLERS
     ERRB=.02-XB
     ERRD=.98-XD
     V=VO-KCB *(ERRB+ERINTB/TAUB)
     R=RO+KCD *(ERRD+ERINTD/TAUD)
C PERFECT LEVEL CONTROLLERS IN REFLUX DRUM AND COLUMN BASE
     D=V-R
     B=L(1)-V
     IF(R.LT.0.) GO TO 500
     IF(V.LT.0.) GO TO 500
     IF(D.LT.0.) GO TO 500
     IF(B.LT.0.) GO TO 500
C EVALUATE DERIVATIVES
     XBDOT=(L(1)*X(1)-V*YB-B*XB)/MBO
     MDOT(1)=L(2)-L(1)
     MXDOT(1)=V*(YB-Y(1))+ L(2)*X(2)- L(1)*X(1)
     DO 30 N=2,NF-1
```

TABLE 5.7 (*continued*)

```
     MDOT(N)= L(N+1)- L(N)
  30 MXDOT(N)=V*(Y(N-1)-Y(N))+ L(N+1)*X(N+1)- L(N)*X(N)
C  FEED PLATE
     MDOT(NF)= L(NF+1)- L(NF)+F
     MXDOT(NF)=V*(Y(NF-1)-Y(NF))+ L(NF+1)*X(NF+1)- L(NF)*X(NF)+F*Z
     DO 40 N=NF+1,NT-1
     MDOT(N)= L(N+1)- L(N)
  40 MXDOT(N)=V*(Y(N-1)-Y(N))+ L(N+1)*X(N+1)- L(N)*X(N)
     MDOT(NT)=R-L(NT)
     MXDOT(NT)=V*(Y(NT-1)-Y(NT))+R*XD-L(NT)*X(NT)
     XDDOT=V*(Y(NT)-XD)/MDO
     IF(TIME.LT.TPRINT) GO TO 50
     WRITE(6,41)TIME,XB,X(10),XD,R,V
  41 FORMAT(7X,F5.1,3F9.5,2F9.2)
     TPRINT=TPRINT+.5
  50 CONTINUE
C  EULER INTEGRATION
     TIME=TIME+DELTA
     XB=XB+DELTA*XBDOT
     DO 60 N=1,NT
     M(N)=M(N)+MDOT(N)*DELTA
     MX(N)=MX(N)+MXDOT(N)*DELTA
     X(N)=MX(N)/M(N)
     IF(X(N).LT.0.) GO TO 500
     IF(X(N).GT.1.) GO TO 500
  60 CONTINUE
     XD=XD+XDDOT*DELTA
     ERINTD=ERINTD+ERRD*DELTA
     ERINTB=ERINTB+ERRB*DELTA
     IF(TIME.LE.10.)GO TO 100
     STOP
 500 WRITE(6,501)
 501 FORMAT(' LEVEL TOO LOW OR COMPOSITION UNREAL')
     STOP
     END
```

Results

Z =   0.55000 F =   100.00

TIME	XB	X10	XD	R	V
0.0	0.02000	0.47688	0.98000	128.01	178.01
0.5	0.02014	0.51325	0.98000	128.01	178.16
1.0	0.02108	0.52434	0.98010	127.90	179.32
1.5	0.02218	0.53030	0.98035	127.64	181.08
2.0	0.02276	0.53229	0.98061	127.33	182.67
2.5	0.02268	0.53141	0.98076	127.11	183.69
3.0	0.02212	0.52879	0.98077	127.02	184.10
3.5	0.02132	0.52560	0.98065	127.06	183.99
4.0	0.02051	0.52282	0.98048	127.18	183.55
4.5	0.01987	0.52108	0.98030	127.32	182.98
5.0	0.01950	0.52056	0.98019	127.41	182.47
5.5	0.01939	0.52104	0.98014	127.44	182.13
6.0	0.01950	0.52207	0.98016	127.41	182.01
6.5	0.01972	0.52318	0.98022	127.33	182.07
7.0	0.01995	0.52399	0.98029	127.24	182.24
7.5	0.02012	0.52433	0.98034	127.15	182.43
8.0	0.02019	0.52423	0.98036	127.09	182.56
8.5	0.02016	0.52382	0.98035	127.07	182.60
9.0	0.02007	0.52332	0.98032	127.06	182.57
9.5	0.01997	0.52290	0.98028	127.07	182.47

It is very important to note that all derivatives are evaluated using the current values of all variables *before* integrating any of the ODEs. A fairly common mistake is to evaluate one derivative and to integrate that ODE before going on to the next ODE. This procedure is *not* correct and will lead to inaccurate answers.

We will assume constant holdups in the reflux drum M_D and in the column base M_B. Proportional-integral feedback controllers at both ends of the column will change the reflux flow rate R and the vapor boilup V to control overhead composition x_D and bottoms composition x_B at setpoint values of 0.98 and 0.02 respectively.

Table 5.7 gives the program, the initial conditions, and the printed output results for a step change in feed composition from 0.50 to 0.55 at time equal zero.

5.5 MULTICOMPONENT DISTILLATION COLUMN

The extension of the simple ideal binary system considered in the preceding section to a nonideal multicomponent column is not difficult. The only changes that have to be made to the basic structure of the solution algorithm are:

1. More ordinary differential equations must be added per tray. We need one per component per tray. But this is easily programmed using doubly dimensioned variables $X(N, J)$, where N is the tray number and J is the component number.
2. One energy balance per tray must be included if equimolal overflow cannot be assumed.
3. An appropriate multicomponent bubblepoint subroutine must be used. This may be a little more complex because of nonidealities, but as far as the main program is concerned, the bubblepoint subroutine is provided with known liquid compositions and a known pressure, and its job is to calculate the temperature and vapor compositions.

The general model was developed in Sec. 3.12. Table 5.8 gives a fairly general program for continuous multicomponent distillation.

The specific column simulated is assumed to have the following equipment configurations and conditions:

1. There is one feed plate onto which vapor feed and liquid feed are introduced.
2. Pressure is constant and known on each tray. It varies linearly up the column from P_B in the base to P_D at the top (psia).
3. Coolant and steam dynamics are negligible in the condenser and reboiler.
4. Vapor and liquid products D_V and D_L are taken off the reflux drum and are in equilibrium. Dynamics of the vapor space in the reflux drum and throughout the column are negligible.

TABLE 5.8
Multicomponent distillation dynamics

```
C
C       INPUTS R, QR AND DV ARE FIXED
C
      REAL MW,LO,MVB,MVD,MWA,MV,LV,M,L,MB,MD
      CHARACTER*6 NAME
      COMMON NC,MW(5),DENS(5),C1(5),C2(5),C3(5),BPT(5),AVP(5),
     + BVP(5)
      DIMENSION LV(50),L(50),P(50),XF(5),YF(5),DXD(5),YAV(5),
     + YY(5),HL(50),HV(50),V(50),DM(50),DXM(50,5),XM(50,5),DXB(5)
      DIMENSION NAME(5),T(50),XB(5),X(50,5),Y(50,5),LO(50)
     + ,XD(5),YB(5),YD(5),XX(5),MV(50),M(50)
C...................
C     READ COLUMN DATA
      READ(5,1) NT,NF,NC,WHS,WHR,DS,DR,WLS,WLR,MVB,MVD
    1 FORMAT (3I6,10F6.2)
      WRITE(6,300)
  300 FORMAT(' NT NF  NC  WHS   WHR   DS    DR   WLS   WLR   MVB   MVD
     + ')
      WRITE(6,13) NT,NF,NC,WHS,WHR,DS,DR,WLS,WLR,MVB,MVD
   13 FORMAT(1X,3I3,1X,8F6.2)
C...................
C     READ PHYSICAL PROPERTY DATA
      WRITE(6,301)
  301 FORMAT(' NAME   MW   DENS   HVAP   BPT   HCAPV   HCAPL   VP1
     +  T1   VP2    T2')
      DO 5 J=1,NC
      READ(5,6) NAME(J),MW(J),DENS(J),HVAP,BPT(J),HCAPV,HCAPL,VP1,T1,VP2
     + ,T2
    6 FORMAT (A6,10F6.2)
      WRITE(6,7) NAME(J),MW(J),DENS(J),HVAP,BPT(J),HCAPV,HCAPL,VP1,T1,
     + VP2,T2
    7 FORMAT(1X,A6,4F7.2,2F7.3,4F7.1)
      AVP(J)=(T1+460.)*(T2+460.)*ALOG(VP2/VP1)/(T1-T2)
      BVP(J)=ALOG(VP2)-AVP(J)/(T2+460.)
      C2(J) = HCAPV*MW(J)
      C3(J) = HCAPL*MW(J)
    5 C1(J)=HVAP*MW(J)+(C3(J)-C2(J))*BPT(J)
C...................
C     READ FEED
      READ(5,10) TFL,FL,(XF(J),J=1,NC)
      READ(5,10) TFV,FV,(YF(J),J=1,NC)
   10 FORMAT (12F6.2)
      WRITE(6,306)
  306 FORMAT(4X,'FL      TFL      XF1     XF2     XF3     XF4
     +   XF5')
      WRITE(6,308) FL,TFL,(XF(J),J=1,NC)
  308 FORMAT(1X,2F10.2,5E10.2)
      WRITE(6,307)
  307 FORMAT(4X,'FV      TFV      YF1     YF2     YF3     YF4
     +   YF5')
      WRITE(6,308) FV,TFV,(YF(J),J=1,NC)
      CALL ENTH(TFL,XF,YF,HLF,HVF)
    8 FORMAT(1X,2F8.2,10E10.2)
C...................
C     READ CONDITIONS
      READ (5,10) PD,PB,QR,R,DV,EFF
      WRITE(6,304)
```

TABLE 5.8 (*continued*)

```
 304 FORMAT(6X,'PD     PB     QR     R     DV     EFF')
     WRITE(6,9)PD,PB,QR,R,DV,EFF
   9 FORMAT(1X,10F8.2)
C.....................
C    READ INITIAL CONDITIONS
     WRITE(6,305)
 305 FORMAT(4X,'N    TEMP    L     X1     X2     X3     X4
    +     X5')
     READ(5,18)TB,(XB(J),J=1,NC)
  18 FORMAT(F10.2,5E10.3)
     BLANK=0.
     WRITE(6,11)TB,BLANK,(XB(J),J=1,NC)
  11 FORMAT(5X,2F8.2,5E10.3)
     DO 15 N=1,NT
     READ (5,17 )T(N),LO(N),(X(N,J),J=1,NC)
  17 FORMAT(2F5.1,5E10.3)
  15 WRITE(6,12)N,T(N),LO(N),(X(N,J),J=1,NC)
  12 FORMAT(1X,I3,1X,2F8.2,5E10.3)
     READ (5,18 )TD,(XD(J),J=1,NC)
     WRITE(6,11 )TD,R,(XD(J),J=1,NC)
C.....................
C    CALCULATE INITIAL HOLDUPS
     CALL MWDENS(TB,XB,MWA,DENSA)
     MB = MVB*DENSA/MWA
     DO 20 N=1,NF
     DO 21 J=1,NC
  21 XX(J) = X(N,J)
     CALL MWDENS(T(N),XX,MWA,DENSA)
     LV(N) = LO(N) * MWA/DENSA
     L(N) = LO(N)
     HFOW = (LV(N)/(999.*WLS))**.66667
     MV(N) = (HFOW+WHS/12.)*3.1416*DS*DS/(4.*144.)
  20 M(N) = MV(N)*DENSA/MWA
     DO 25 N=NF+1,NT
     DO 26 J=1,NC
  26 XX(J) = X(N,J)
     CALL MWDENS(T(N),XX,MWA,DENSA)
     LV(N) = LO(N) * MWA/DENSA
     L(N) = LO(N)
     HFOW = (LV(N)/(999.*WLR))**.66667
     MV(N) = (HFOW+WHR/12.)*3.1416*DR*DR/(4.*144.)
  25 M(N) = MV(N) * DENSA/MWA
     DO 30 N=1,NT
     DO 31 J=1,NC
     XM(N,J) = M(N)*X(N,J)
  31 CONTINUE
  30 CONTINUE
     CALL MWDENS(TD,XD,MWA,DENSA)
     MD=MVD*DENSA/MWA
C    CALCULATE PRESSURE PROFILE
     DO 35 N=1,NT
  35 P(N)=(PB-(N*(PB-PD)))/NT)
     DELTA=.0001
     WRITE(6,37)DELTA
  37 FORMAT(1X,' DELTA = ',F8.5)
     TIME = 0.
     TPRINT = 0.
C INITIAL GUESS OF V(5) FOR FIRST EFFICIENCY CALC.
     V(5)=822.
C.....................
```

TABLE 5.8 (*continued*)

```
C   MAIN LOOP FOR EACH TIME STEP
C....................
 100 CONTINUE
     CALL BUBPT (TB,XB,YB,PB)
     CALL ENTH (TB,XB,YB,HLB,HVB)
     DO 105 J=1,NC
 105 XX(J)=X(1,J)
     CALL BUBPT(T(1),XX,YY,P(1))
     DO 106 J=1,NC
     Y(1,J)=YB(J)+EFF*(YY(J)-YB(J))
 106 YY(J)=Y(1,J)
     CALL ENTH(T(1),XX,YY,HL(1),HV(1))
     DO 110 N = 2,NF
     DO 111 J = 1,NC
 111 XX(J)=X(N,J)
     CALL BUBPT(T(N),XX,YY,P(N))
     DO 112 J = 1,NC
     Y(N,J) =(YY(J) -Y(N-1,J))*EFF+Y(N-1,J)
 112 YY(J)=Y(N,J)
     CALL ENTH (T(N),XX,YY,HL(N),HV(N))
 110 CONTINUE
     DO 113 J=1,NC
 113 XX(J)=X(NF+1,J)
     CALL BUBPT(T(NF+1),XX,YY,P(NF+1))
     DO 114 J=1,NC
     YAV(J)=(YF(J)*FV+Y(NF,J)*V(NF))/(V(NF)+FV)
     Y(NF+1,J)=(YY(J)-YAV(J))*EFF+YAV(J)
 114 YY(J)=Y(NF+1,J)
     CALL ENTH(T(NF+1),XX,YY,HL(NF+1),HV(NF+1))
     DO 115 N=NF+2,NT
     DO 116 J=1,NC
 116 XX(J)=X(N,J)
     CALL BUBPT(T(N),XX,YY,P(N))
     DO  117 J=1,NC
     Y(N,J) =(YY(J) -Y(N-1,J))*EFF+Y(N-1,J)
 117 YY(J)=Y(N,J)
     CALL ENTH (T(N),XX,YY,HL(N),HV(N))
 115 CONTINUE
     CALL BUBPT(TD,XD,YD,PD)
     CALL ENTH(TD,XD,YD,HLD,HVD)
C....................
C   CALCULATE VAPOR RATES
C....................
     VB = (QR*1000000.-L(1)*(HLB-HL(1)))/(HVB-HLB)
     B = L(1)-VB
     IF (B .LT. 0.) STOP
     V(1) = (HL(2)*L(2)+HVB*VB-HL(1)*L(1))/HV(1)
     DO 120 N = 2,NF-1
     V(N) = (HL(N+1)*L(N+1)+HV(N-1)*V(N-1)-HL(N)*L(N))/HV(N)
 120 CONTINUE
     V(NF) = (HL(NF+1)*L(NF+1)+HV(NF-1)*V(NF-1)-HL(NF)*L(NF)+HLF*FL)/HV
    + (NF)
     V(NF+1) = (HL(NF+2)*L(NF+2)+HV(NF)*V(NF)+HVF*FV-HL(NF+1)*L(NF+1)
    + )/HV(NF+1)
     DO 130 N = NF+2,NT-1
 130 V(N) = (HL(N+1)*L(N+1)+HV(N-1)*V(N-1)-HL(N)*L(N))/HV(N)
     V(NT) = (HLD*R+HV(NT-1)*V(NT-1)-HL(NT)*L(NT))/HV(NT)
     DL=V(NT)-DV-R
C....................
```

TABLE 5.8 (*continued*)

```
C     EVALUATE DERIVATIVES
C...................
      DM(1) = L(2) + VB-V(1)-L(1)
      DO 140 N = 2,NF-1
  140 DM(N) = L(N+1) + V(N-1)-L(N)-V(N)
      DM(NF) = L(NF+1) + FL + V(NF-1) - L(NF) - V(NF)
      DM(NF+1) = L(NF+2) + FV + V(NF) - L(NF+1) - V(NF+1)
      DO 150 N = NF+2,NT-1
  150 DM(N) = L(N+1) + V(N-1)-L(N)-V(N)
      DM(NT)= R + V(NT-1) - L(NT) - V(NT)
      DO 160 J=1,NC
      DXB(J) = (X(1,J)*L(1)-YB(J)*VB-XB(J)*B)/MB
      DXM(1,J) = X(2,J)*L(2)+YB(J)*VB-X(1,J)*L(1)-Y(1,J)*V(1)
      DO 165 N=2,NF-1
  165 DXM(N,J)=X(N+1,J)*L(N+1)+Y(N-1,J)*V(N-1)-X(N,J)*L(N)-V(N)*Y(N,J)
      DXM(NF,J)=X(NF+1,J)*L(NF+1)+Y(NF-1,J)*V(NF-1)-X(NF,J)*L(NF)-V(NF)
     +  *Y(NF,J)+FL*XF(J)
      DXM(NF+1,J) = X(NF+2,J)*L(NF+2)+Y(NF,J)*V(NF)-X(NF+1,J)*L(NF+1)
     +  -V(NF+1)*Y(NF+1,J)+FV*YF(J)
      DO 170 N = NF+2,NT-1
  170 DXM(N,J) = X(N+1,J)*L(N+1)+Y(N-1,J)*V(N-1)-X(N,J)*L(N)-V(N)*Y(N,J)
      DXM(NT,J) = XD(J)*R+Y(NT-1,J)*V(NT-1)-X(NT,J)*L(NT)-Y(NT,J)*V(NT)
      DXD(J)=(V(NT)*Y(NT,J)-DV*YD(J)-(R+DL)*XD(J))/MD
  160 CONTINUE
      IF (TIME.GT..0011) GO TO 400
      IF (TIME .LT. TPRINT) GO TO 210
      WRITE(6,201)
  201 FORMAT (5X,'TIME   T       X1      X2      X3      X4
     1   X5    L')
      WRITE(6,202) TIME,TB,(XB(J),J=1,NC),B
  202 FORMAT(1X,F5.4,3X,F7.2,5F10.6,F7.1)
      DO 203 N=1,NT
  203 WRITE(6,204)  N,T(N),(X(N,J),J=1,NC),L(N)
  204 FORMAT(3X,I3,3X,F7.2,5F10.6,F7.1)
      WRITE(6,205) TD,(XD(J),J=1,NC),R
  205 FORMAT (9X,F7.2,5F10.6,F7.1)
      WRITE(6,206) (YD(J),J=1,NC),DL
  206 FORMAT(16X,5F10.6,F7.1)
      TPRINT = TPRINT+.0005
C...................
C     INTEGRATION ALA EULER
C...................
  210 TIME = TIME + DELTA
      DO 215 N = 1,NT
  215 M(N) = M(N) + DM(N) * DELTA
      DO 220 J = 1,NC
      XB(J) = XB(J) + DXB(J) * DELTA
      IF (XB(J) .LT. 0.) XB(J) = 0.0
      IF (XB(J) .GT. 1.) XB(J) = 1.
      DO 225 N = 1,NT
      XM(N,J) = XM(N,J) + DXM(N,J)*DELTA
      X(N,J) = XM(N,J)/M(N)
      IF (X(N,J).GT.1.) X(N,J) = 1.
      IF (X(N,J) .LT. 0.) X(N,J) = 0.0
  225 CONTINUE
      XD(J)=XD(J)+DXD(J)*DELTA
      IF(XD(J).LT.0.) XD(J)=0.
      IF(XD(J).GT.1.) XD(J)=1.
  220 CONTINUE
```

TABLE 5.8 (*continued*)

```
C   CALC NEW LIQUID RATES
      DO 270 N=1,NF
      DO 271 J=1,NC
 271 XX(J)=X(N,J)
      CALL HYDRAU(M(N),T(N),XX,L(N),WHS,WLS,DS)
 270 CONTINUE
      DO 273 N=NF+1,NT
      DO 275 J=1,NC
 275 XX(J)=X(N,J)
      CALL HYDRAU(M(N),T(N),XX,L(N),WHR,WLR,DR)
 273 CONTINUE
      GO TO 100
 400 STOP
      END
C
      SUBROUTINE HYDRAU(M,T,X,L,WH,WL,DCOL)
      REAL M,L,MW,MWA
      COMMON NC,MW(5),DENS(5),C1(5),C2(5),C3(5),BPT(5),AVP(5),
     + BVP(5)
      DIMENSION X(5)
      CALL MWDENS(T,X,MWA,DENSA)
      CONST=183.2*M*MWA/(DENSA*DCOL*DCOL)-WH/12.
      IF(CONST.LE.0.) GO TO 10
      L=DENSA*WL* 999.*((183.2*M*MWA/(DENSA*DCOL*DCOL)-
WH/12.)**1.5)/MWA
      RETURN
  10 L=0.
      RETURN
      END
C
      SUBROUTINE ENTH(T,X,Y,HL,HV)
      COMMON NC,MW(5),DENS(5),C1(5),C2(5),C3(5),BPT(5),AVP(5),
     + BVP(5)
      DIMENSION X(5),Y(5)
      HL=0.0
      HV=0.0
      DO 1 J=1,NC
      HL=HL+X(J)*C3(J)*T
      HV=HV+Y(J)*(C1(J)+C2(J)*T)
  1 CONTINUE
      RETURN
      END
C
      SUBROUTINE MWDENS(T,X,MWA,DENSA)
      COMMON NC,MW(5),DENS(5),C1(5),C2(5),C3(5),BPT(5),AVP(5),
     + BVP(5)
      DIMENSION X(5)
      REAL MW,MWA
      DENSA=0.0
      MWA=0.
      DO 1 J=1,NC
      MWA=X(J)*MW(J)+MWA
  1 DENSA=X(J)*DENS(J) +DENSA
      RETURN
      END
C
      SUBROUTINE BUBPT(T,X,Y,P)
      COMMON NC,MW(5),DENS(5),C1(5),C2(5),C3(5),BPT(5),AVP(5),
     + BVP(5)
```

TABLE 5.8 (*continued*)

```
   DIMENSION X(5),Y(5),PS(5)
   LOOP=0
10 LOOP=LOOP+1
   IF(LOOP.GT.50) GO TO 30
   SUMY=0.0
   DO 15 J=1,NC
   PS(J)=EXP(BVP(J)+AVP(J)/(T+460.))
   Y(J)=PS(J)*X(J) /P
15 SUMY=SUMY+Y(J)
   IF(ABS(SUMY-1.).LT..00001)RETURN
   F=SUMY*P-P
   FSLOPE=0.
   TSQ=(T+460.)**2
   DO 20 J=1,NC
20 FSLOPE=FSLOPE-AVP(J)*X(J)*PS(J)/TSQ
   T=T-F/FSLOPE
   GO TO 10
30 WRITE(6,21)
21 FORMAT(1X,'TEMP LOOP')
   STOP
   END
```

Results

NT	NF	NC	WHS	WHR	DS	DR	WLS	WLR	MVB	MVD
15	5	5	0.75	1.25	72.00	72.00	48.00	48.00	10.00	10.00

NAME	MW	DENS	HVAP	BPT	HCAPV	HCAPL	VP1	T1	VP2	T2
LLK	30.00	40.00	100.00	10.00	0.200	0.600	14.7	10.0	50.0	30.0
LK	50.00	40.00	90.00	90.00	0.400	0.600	14.7	90.0	500.0	200.0
INTER	90.00	60.00	70.00	150.00	0.300	0.500	14.7	150.0	150.0	200.0
HK	130.00	70.00	80.00	210.00	0.300	0.400	14.7	210.0	150.0	300.0
HHK	300.00	90.00	80.00	360.00	0.300	0.400	14.7	360.0	150.0	420.0

FL	TFL	XF1	XF2	XF3	XF4	XF5
800.00	120.00	0.50E-01	0.60E+00	0.10E-01	0.30E+00	0.40E-01

FV	TFV	YF1	YF2	YF3	YF4	YF5
200.00	120.00	0.40E+00	0.53E+00	0.20E-01	0.50E-01	0.00E+00

PD	PB	QR	R	DV	EFF
19.70	21.20	5.00	400.00	200.00	0.50

N	TEMP	L	X1	X2	X3	X4	X5
	201.58	0.00	0.000E+00	0.725E-02	0.488E-01	0.836E+00	0.108E+00
1	154.90	740.10	0.999E-11	0.110E+00	0.240E+00	0.607E+00	0.433E-01
2	132.60	814.40	0.156E-08	0.286E+00	0.202E+00	0.473E+00	0.393E-01
3	120.20	892.00	0.182E-06	0.457E+00	0.131E+00	0.376E+00	0.359E-01
4	114.00	960.10	0.133E-04	0.572E+00	0.803E-01	0.314E+00	0.333E-01
5	108.40	986.00	0.760E-03	0.634E+00	0.496E-01	0.284E+00	0.325E-01
6	101.20	320.00	0.112E-02	0.818E+00	0.866E-01	0.942E-01	0.174E-05
7	98.20	381.90	0.129E-02	0.910E+00	0.446E-01	0.440E-01	0.776E-06
8	96.90	409.60	0.136E-02	0.953E+00	0.238E-01	0.218E-01	0.371E-06
9	96.20	423.70	0.140E-02	0.975E+00	0.128E-01	0.110E-01	0.181E-06
10	95.80	431.20	0.142E-02	0.986E+00	0.694E-02	0.563E-02	0.893E-07
11	95.50	435.20	0.143E-02	0.992E+00	0.374E-02	0.286E-02	0.440E-07
12	95.30	437.50	0.144E-02	0.995E+00	0.199E-02	0.145E-02	0.216E-07
13	95.10	438.70	0.144E-02	0.997E+00	0.104E-02	0.718E-03	0.104E-07
14	94.90	439.50	0.145E-02	0.998E+00	0.519E-03	0.342E-03	0.484E-08
15	94.20	438.60	0.175E-02	0.998E+00	0.236E-03	0.149E-03	0.205E-08
	77.26	400.00	0.174E-01	0.982E+00	0.824E-04	0.493E-04	0.659E-09

DELTA = 0.00010

TABLE 5.8 (*continued*)

TIME	T	X1	X2	X3	X4	X5	L
0.000	201.58	0.000000	0.007254	0.048850	0.836300	0.107600	298.1
1	154.92	0.000000	0.109600	0.240300	0.606900	0.043260	740.1
2	132.64	0.000000	0.285700	0.202100	0.472900	0.039310	814.4
3	120.23	0.000000	0.457000	0.131300	0.375900	0.035890	892.0
4	114.05	0.000013	0.572100	0.080310	0.314200	0.033340	960.1
5	108.42	0.000760	0.633600	0.049590	0.283600	0.032460	986.0
6	101.18	0.001121	0.818000	0.086620	0.094240	0.000002	320.0
7	98.23	0.001288	0.910200	0.044560	0.043990	0.000001	381.9
8	96.91	0.001362	0.953000	0.023800	0.021840	0.000000	409.6
9	96.20	0.001399	0.974700	0.012840	0.011050	0.000000	423.7
10	95.77	0.001419	0.986000	0.006940	0.005625	0.000000	431.2
11	95.49	0.001430	0.992000	0.003737	0.002863	0.000000	435.2
12	95.28	0.001436	0.995100	0.001991	0.001446	0.000000	437.5
13	95.10	0.001440	0.996800	0.001039	0.000718	0.000000	438.7
14	94.93	0.001450	0.997700	0.000519	0.000342	0.000000	439.5
15	94.24	0.001749	0.997900	0.000236	0.000149	0.000000	438.6
	77.26	0.017450	0.982400	0.000082	0.000049	0.000000	400.0
		0.556223	0.443775	0.000001	0.000000	0.000000	500.8

TIME	T	X1	X2	X3	X4	X5	L
0.0005	201.61	0.000000	0.007233	0.048763	0.836398	0.107614	295.5
1	154.94	0.000000	0.109514	0.240196	0.607077	0.043266	738.4
2	132.65	0.000000	0.285653	0.202034	0.473022	0.039314	812.7
3	120.23	0.000000	0.456967	0.131257	0.375952	0.035891	890.2
4	114.05	0.000014	0.572098	0.080287	0.314233	0.033341	958.2
5	108.43	0.000756	0.633541	0.049565	0.283677	0.032468	984.2
6	101.18	0.001120	0.818013	0.086583	0.094269	0.000002	318.7
7	98.24	0.001285	0.910208	0.044540	0.044002	0.000001	380.4
8	96.92	0.001356	0.953008	0.023790	0.021846	0.000000	408.1
9	96.21	0.001397	0.974704	0.012835	0.011053	0.000000	422.1
10	95.78	0.001415	0.986008	0.006937	0.005627	0.000000	429.6
11	95.50	0.001425	0.992002	0.003735	0.002864	0.000000	433.6
12	95.28	0.001431	0.995107	0.001990	0.001446	0.000000	435.9
13	95.11	0.001436	0.996805	0.001039	0.000718	0.000000	437.1
14	94.94	0.001447	0.997704	0.000519	0.000342	0.000000	437.9
15	94.25	0.001745	0.997901	0.000236	0.000149	0.000000	437.3
	77.25	0.017462	0.982389	0.000082	0.000049	0.000000	400.0
		0.556381	0.443620	0.000001	0.000000	0.000000	505.0

TIME	T	X1	X2	X3	X4	X5	L
0.001	201.64	0.000000	0.007221	0.048687	0.836477	0.107628	295.5
1	154.96	0.000000	0.109428	0.240103	0.607246	0.043272	738.2
2	132.65	0.000000	0.285587	0.201986	0.473139	0.039318	812.5
3	120.24	0.000000	0.456921	0.131229	0.376010	0.035892	890.1
4	114.06	0.000012	0.572070	0.080270	0.314280	0.033343	958.2
5	108.43	0.000758	0.633487	0.049547	0.283738	0.032475	984.3
6	101.18	0.001117	0.818018	0.086555	0.094297	0.000002	318.7
7	98.24	0.001286	0.910208	0.044525	0.044014	0.000001	380.4
8	96.92	0.001358	0.953008	0.023782	0.021852	0.000000	408.1
9	96.21	0.001395	0.974706	0.012831	0.011056	0.000000	422.1
10	95.78	0.001415	0.986011	0.006935	0.005628	0.000000	429.6
11	95.49	0.001426	0.991998	0.003734	0.002865	0.000000	433.6
12	95.28	0.001433	0.995107	0.001990	0.001447	0.000000	435.9
13	95.11	0.001437	0.996806	0.001038	0.000718	0.000000	437.1
14	94.94	0.001446	0.997706	0.000519	0.000343	0.000000	438.0
15	94.25	0.001743	0.997900	0.000236	0.000149	0.000000	437.5
	77.25	0.017456	0.982396	0.000082	0.000049	0.000000	400.0
		0.556307	0.443694	0.000001	0.000000	0.000000	504.9

5. Liquid hydraulics are calculated from the Francis weir formula.

6. Volumetric liquid holdups in the reflux drum and column base are held perfectly constant by changing the flow rates of bottoms product B and liquid distillate product D_L.

7. Dynamic changes in internal energies on the trays are much faster than the composition or total holdup changes, so the energy equation on each tray [Eq. (3.99)] is just algebraic.

8. Reflux R and heat input to the reboiler Q_R are the manipulated variables. In the program given in Table 5.8, they are simply held constant, thus giving the openloop response of the column. If the closedloop response is desired, the program can be easily changed to use R to hold a temperature or a composition in the top of the column and to use Q_R to hold a temperature or a composition in the bottom of the column. There are two degrees of freedom, so two variables can be specified.

The program in Table 5.8 is very similar in structure to that for the simple binary case. The steps taken are:

1. Input data on the column size, components, physical properties, feeds, and initial conditions (liquid compositions, liquid flow rates, and initial guesses of temperatures on all trays).

2. Calculate initial tray holdups and the pressure profile.

3. Calculate the temperatures and vapor compositions from the vapor-liquid equilibrium data, using the subroutine BUBPT. Raoult's law is used in the example, but nonideality can be included by adding activity coefficient equations. Newton-Raphson convergence is used.

4. Calculate liquid and vapor enthalpies, using subroutine ENTH.

5. Calculate vapor flow rates on all trays, starting in the column base, using the algebraic form of the energy equations.

6. Evaluate all derivatives of the component continuity equations for all NC components on all NT trays plus the reflux drum and the column base.

7. Integrate all ODEs (using Euler, what else!).

8. Calculate new total liquid holdups from the sum of the component holdups. Then calculate the new liquid mole fractions from the component holdups and the total holdups.

9. Calculate new liquid flow rates from the new total holdups for all trays, using subroutine HYDRAU.

10. Go back to step 3 and repeat for the next step in time.

Table 5.9 gives a list of terms for the input and output variables.

TABLE 5.9
Nomenclature for multicomponent distillation program

NT	Total number of trays
NF	Feed tray (counting from the bottom)
NC	Total number of components
WHS,WLS,DS	Weir height and length and column diameter in stripping section (in)
WHR,WLR,DR	Weir height and length and column diameter in rectifying section (in)
MVB,MVD	Volumetric holdup in column base and in reflux drum (ft^3)
MW	Molecular weight (lb$_m$/lb·mol)
HVAP	Heat of vaporization at normal boiling point (Btu/lb$_m$)
HCAPV	Heat capacity of vapor (Btu/lb$_m$ °F)
HCAPL	Heat capacity of liquid (Btu/lb$_m$ °F)
VP1	Vapor pressure (psia) at temperature T_1 (°F)
VP2	Vapor pressure (psia) at temperature T_2 (°F)
TFL,FL,XF	Liquid feed temperature (°F), flow rate (lb·mol/h) and composition (m.f.)
TFV,FV,YF	Vapor feed temperature (°F), flow rate (lb·mol/h) and composition (m.f.)
PD,PB	Pressures in the top and base of the column (psia)
QR	Reboiler heat input (10^6 Btu/h)
R	Reflux flow rate (lb·mol/h)
DV,DL	Vapor and liquid distillate product flow rates (lb·mol/h)
EFF	Murphree vapor-phase tray efficiency
T,LO,X,	Initial conditions of temperature (°F), liquid flow rate (lb·mol/h) and liquid compositions for all trays and all components (mole fractions)
TD,TB,XD,XB	Initial conditions of temperature and compositions in reflux drum and column base
V	Vapor flow rates (lb·mol/h)
M	Molar liquid holdup on tray (lb·mol)
MV	Volumetric liquid holdup on tray (ft^3)

5.6 VARIABLE PRESSURE DISTILLATION

Pressures were assumed constant in the distillation column considered in Sec. 5.5. In many distillation columns this is a good assumption. However, there are quite a few columns in which this assumption is not valid. In vacuum columns, pressure changes can be significant. This is also true in columns whose pressures can vary greatly because of heat integration schemes (their pressures must rise or fall to provide the changing temperature difference driving force in the condenser/reboiler as throughputs and compositions vary).

There are several ways to account for variable pressures. If the total pressure of the column changes but not the pressure drop through the trays (the normal situation in heat-integrated columns, particularly with valve trays whose pressure drops are fairly constant), an approximate variable-pressure model can be used.

5.6.1 Approximate Variable-Pressure Model

A total molar balance is written for the entire vapor volume in the column, reflux drum, and overhead piping (V_{tot}). The molar flow rates into this lumped vapor volume are the vapor boilup in the reboiler and any vapor in the feed stream. The molar flow rates out of this volume are the rate of condensation in the condenser and the vapor product from the reflux drum. An average temperature for the entire column is used.

$$\frac{V_{\text{tot}}}{R T_{\text{av}}} \frac{dP_D}{dt} = V_B + F_V - D_V - L_c \qquad (5.33)$$

The condensation rate in the condenser L_c changes as the pressure in the condenser P_D varies since the condensing temperature depends on pressure. Thus L_c depends on column pressure, overhead vapor composition, and the temperature of the coolant in the condenser. Equation (5.33) assumes ideal gas behavior, which is usually adequate in these low-pressure columns where pressure changes are significant.

This approximate approach is admittedly crude, but I have used it quite effectively for several distillation simulations. At each point in time the pressure P_D at the top of the column is calculated from Eq. (5.33), and new pressures on all the trays are calculated using a constant pressure drop per tray.

5.6.2 Rigorous Variable-Pressure Model

For vacuum columns, where both absolute pressure and tray pressure drops vary significantly, a rigorous vapor-hydraulic model may have to be used. The modeling and simulation are easy. The numerical integration is quite difficult. This is because the ODEs become very, very stiff when vapor hydraulics are included in the model.

Instead of using an algebraic form of the energy balance, the energy-balance ODE is integrated along with the rest of the ODEs (component continuity equations). As given in Chap. 3 [Eqs. (3.97) to (3.99)], integration of these ODEs gives values for all the liquid compositions (x_{nj}), the total liquid holdup (M_n), and the liquid enthalpy (h_n) on each tray at each point in time.

Knowing x_{nj} and h_n we can go into the physical property data and calculate the temperature T_n. Note that this is the reverse of the normal procedure where we calculate enthalpy from known temperature and known composition.

Now using temperature and liquid compositions, we can do a bubblepoint calculation to determine the pressure on the tray P_n and the vapor composition y_n. Note that this bubblepoint calculation is usually not iterative since we know the temperature.

Finally we can now calculate the vapor flow rate through the tray from the pressure drop through the tray ($P_{n-1} - P_n$) and the liquid height on the tray, which we can get from the weir height h_w and the height of liquid over the weir h_{ow}. The total pressure drop is the sum of the "dry hole" pressure drop plus the hydraulic pressure of the liquid.

$$P_{n-1} - P_n = \rho_{L, n}(h_{w, n} + h_{ow, n}) + K_{DH} \rho_{V, n-1}(v_n)^2 \qquad (5.34)$$

TABLE 5.10
Design parameters for variable-pressure toluene/o-xylene column

Feed flow rate	= 18 kg·mol/min
Feed composition	= 0.33 mole fraction toluene
Feed temperature	= 95°C (liquid feed)
Distillate composition	= 0.9955 mole fraction toluene
Distillate flow rate	= 5.94 kg·mol/min
Bottoms composition	= 0.001 mole fraction toluene
Bottoms flow rate	= 12.06 kg·mol/min
Pressure reflux drum	= 90 mmHg
Pressure drop (average)	= 5 mmHg per tray
Reflux ratio	= 2.22
Temperature reflux drum	= 49°C
Temperature base	= 106°C
Number of trays	= 30
Feed tray (from bottom)	= 14
Reboiler heat input	= 0.15×10^6 kcal/min
Diameter	= 3.962 meter
Weir height	= 0.0612 meter
Weir length	= 3.78 meter
Dry hole pressure drop coefficient	= 0.134 mmHg/(kg·mol/m³)/(m/s)²
Holdup in reflux drum	= 65 kg·mol
Holdup in base	= 90 kg·mol

where $\rho_{L,n}$ and $\rho_{V,n}$ = liquid and vapor densities on the nth tray

$\qquad v_n$ = vapor velocity

All the terms in Eq. (5.34) must have consistent units (psi, mmHg, Pascals, atmospheres, or bar).

To illustrate the use of a vapor-hydraulic model, let us consider a vacuum distillation column in which we are separating a binary mixture of toluene and o-xylene at 90 mmHg reflux drum pressure. The base pressure at design is 245 mmHg with a heat input of 150,000 kcal/min. Table 5.10 gives the parameter values and steadystate operating conditions. The column has 30 trays and is 13 feet in diameter. Theoretical trays and ideal (Raoult's law) VLE are assumed in the program given in Table 5.11.

The BUBPT2 subroutine calculates pressure from given liquid composition and temperature. The ENTH2 subroutine calculates the temperature from given liquid enthalpy and composition.

The stiff-ODE, implicit numerical-integration-algorithm LSODE is used to integrate the ODEs. The computer time to run out 60 minutes of process time was 160 seconds on a Cyber 850 using LSODE. The computer time using first-order explicit Euler for the same problem was 1400 seconds. Because of the extremely stiff system of ODEs, a very small step size had to be taken in Euler (0.00025 min). This example illustrates that there are some chemical engineering processes (but not many) where the implicit algorithm is better than Euler.

TABLE 5.11
Variable-pressure column simulation

```
C
C         TOLUENE/O-XYLENE SEPARATION AT 90 MM HG
C         RIGOROUS VAPOR-HYDRAULIC MODEL (VAPOR RATES
C            CALCULATED FROM PRESSURE DROP THROUGH TRAYS)
C      USING LSODE IMPLICIT STIFF INTEGRATOR PROGRAM
C       ASSUMPTIONS:
C          IDEAL VLE, THEORETICAL TRAYS
      EXTERNAL FEX,JEX
      COMMON A1,A2,B1,B2
      COMMON F,P,PB,B,V,L,VB,Z,HF,NT,PD,R,QR,YB,Y,YD,TB,T,TD,B0,MB0
      COMMON HLB,HL,D
      DIMENSION X(30),Y(30),T(30),P(30),XM(30),HM(30),L(30),V(30)
    + ,M(30),MDOT(30),XMDOT(30),HMDOT(30),HL(30),HV(30),HTOT(30)
      DIMENSION RWORK(9704),IWORK(114),YY(94),YYDOT(94)
      REAL L,MB,M,MD,MDOT,MB0,MBDOT
C CALCULATE VAPOR PRESSURE CONSTANTS
      A1=(273.+51.9)*(273.+89.5)*ALOG(400./100.)/(51.9-89.5)
      B1=ALOG(400.)-A1/(273.+89.5)
      A2=(273.+81.3)*(273.+121.7)*ALOG(400./100.)/(81.3-121.7)
      B2=ALOG(400.)-A2/(273.+121.7)
      NT=30
      NF=14
C FEED LIQUID ENTHALPY
      F=18.
      Z=0.33
      TF=95.
      CALL ENTH1(Z,Z,TF,HF,DUM)
C CALCULATE INITIAL PRESSURE PROFILE AND INITIAL ENTHALPIES
      PD=90.
      READ(7,80)XB,X,XD,M,HLB,HL,B0,D,MB0,MD,TB,T,TD,QR
      REWIND 7
      XMB=XB*MB0
      MB=MB0
      HLMB=HLB*MB0
      DO 10 N=1,30
      XM(N)=M(N)*X(N)
   10 HM(N)=M(N)*HL(N)
   12 FORMAT(' TIME     N      X      Y      T      P      V
     +L     M')
      D=5.94
      QR=150000.
      WRITE(6,999)QR
  999 FORMAT(' QR = ',F10.1)
C SET PARAMETERS FOR LSODE
      DO 20 I=1,114
   20 IWORK(I)=0.
      DO 21 I=1,9704
   21 RWORK(I)=0.
      NEQ=94
      TIME=0.
      TOUT=1.
      ITOL=1
      RTOL=1.E-4
      ATOL=1.E-6
      ITASK=1
      ISTATE=1
      IOPT=0
```

TABLE 5.11 (*continued*)

```
      LRW=9704
      LIW=114
      MF=22
C****************************************************
C   MAIN INTEGRATION LOOP
C****************************************************
 100 CONTINUE
      YY(1)=MB
      YY(2)=XMB
      YY(3)=HLMB
      DO 33 N=1,30
      YY(3+N)=M(N)
      YY(33+N)=XM(N)
  33 YY(63+N)=HM(N)
      YY(94)=XD
      CALL LSODE(FEX,NEQ,YY,TIME,TOUT,ITOL,RTOL,ATOL,ITASK,ISTATE,
     + IOPT,RWORK,LRW,IWORK,LIW,JEX,MF)
      MB=YY(1)
      XMB=YY(2)
      XB=XMB/MB
      DO 40 N=1,30
      M(N)=YY(N+3)
      XM(N)=YY(N+33)
  40 X(N)=XM(N)/M(N)
      XD=YY(94)
      NN=0
      WRITE(6,12)
      WRITE(8,12)
      WRITE(6,41)TIME,NN,XB,YB,TB,PB,VB,B,MB
      WRITE(8,41)TIME,NN,XB,YB,TB,PB,VB,B,MB
  41 FORMAT(1X,F6.3,I3,2X,2F9.5,5F8.1)
      DO 45 N=1,30
      WRITE(8,46)N,X(N),Y(N),T(N),P(N),V(N),L(N),M(N)
  45 WRITE(6,46)N,X(N),Y(N),T(N),P(N),V(N),L(N),M(N)
  46 FORMAT(7X,I3,2X,2F9.5,5F8.1)
      WRITE(6,47)XD,YD,TD,PD,D,R,MD
      WRITE(8,47)XD,YD,TD,PD,D,R,MD
  47 FORMAT(12X,2F9.5,5F8.1)
      IF(ISTATE.LT.0)GO TO 81
      TOUT=TOUT+1.
      IF(TOUT.LT.2.1)GO TO 100
  80 FORMAT(6E12.5)
      STOP
  81 WRITE(6,82)
  82 FORMAT(' ISTATE ERROR IN SLODE')
      STOP
      END
C
      SUBROUTINE FEX(NEQ,TIME,YY,YYDOT)
      COMMON A1,A2,B1,B2
      COMMON F,P,PB,B,V,L,VB,Z,HF,NT,PD,R,QR,YB,Y,YD,TB,T,TD,B0,MB0
      COMMON HLB,HL,D
      DIMENSION X(30),Y(30),T(30),P(30),XM(30),HM(30),L(30),V(30)
     + ,M(30),MDOT(30),XMDOT(30),HMDOT(30),HL(30),HV(30),HTOT(30)
      DIMENSION RWORK(9704),IWORK(114),YY(94),YYDOT(94)
      REAL L,MB,M,MD,MDOT,MB0,MBDOT
      MB=YY(1)
      XMB=YY(2)
```

TABLE 5.11 (*continued*)

```
      HLMB=YY(3)
      DO 10 N=1,30
      M(N)=YY(N+3)
      XM(N)=YY(N+33)
  10  HM(N)=YY(N+63)
      XD=YY(94)
      XB=XMB/MB
      HLB=HLMB/MB
C  CALCULATE LIQUID FLOW RATES FROM HOLDUPS
      DO 15 N=1,30
      HL(N)=HM(N)/M(N)
      X(N)=XM(N)/M(N)
  15  CALL LIQHYD(M(N),X(N),L(N),HTOT(N))
C  CALCULATE TEMPERATURE FROM ENTHALPY AND COMPOSITION
      CALL ENTH2(XB,HLB,TB)
      DO 70 N=1,30
  70  CALL ENTH2(X(N),HL(N),T(N))
C  CALCULATE PRESSURE FROM TEMPERATURE AND COMPOSITION
      CALL BUBPT2(XB,TB,PB,YB)
      DO 75 N=1,30
  75  CALL BUBPT2(X(N),T(N),P(N),Y(N))
      CALL BUBPT1(XD,PD,TD,YD)
C  CALCULATE VAPOR FLOW RATES FROM PRESSURE DROPS
      CALL VAPHYD(PB,P(1),X(1),TB,HTOT(1),VB)
      DO 20 N=1,29
  20  CALL VAPHYD(P(N),P(N+1),X(N+1),T(N),HTOT(N+1),V(N))
      V(30)=13.*SQRT((P(30)-PD)/4.)
C  CALCULATE VAPOR ENTHALPIES
      CALL ENTH1(XB,YB,TB,DUM,HVB)
      DO 85 N=1,30
  85  CALL ENTH1(X(N),Y(N),T(N),DUM,HV(N))
C  CALCULATE REFLUX ENTHALPY
      CALL ENTH1(XD,YD,TD,HLD,HVD)
C  REFLUX DRUM AND BASE LEVEL CONTROLLERS
      B=B0*MB/MB0
      R=V(30)-D
      IF(B.LE.0.)B=0.
      IF(R.LE.0.)R=0.
C  EVALUATE DERIVATIVES
      MBDOT=L(1)-B-VB
      XMBDOT=L(1)*X(1)-B*XB-VB*YB
      HLMBD=L(1)*HL(1)+QR-B*HLB-VB*HVB
      MDOT(1)=L(2)+VB-L(1)-V(1)
      XMDOT(1)=L(2)*X(2)+VB*YB-L(1)*X(1)-V(1)*Y(1)
      HMDOT(1)=L(2)*HL(2)+VB*HVB-L(1)*HL(1)-V(1)*HV(1)
      DO 30 N=2,13
      MDOT(N)=L(N+1)+V(N-1)-L(N)-V(N)
      HMDOT(N)=L(N+1)*HL(N+1)+V(N-1)*HV(N-1)-L(N)*HL(N)-V(N)*HV(N)
  30  XMDOT(N)=L(N+1)*X(N+1)+V(N-1)*Y(N-1)-L(N)*X(N)-V(N)*Y(N)
      MDOT(14)=L(15)+V(13)+F-L(14)-V(14)
      HMDOT(14)=L(15)*HL(15)+V(13)*HV(13)+F*HF
     +   -L(14)*HL(14)-V(14)*HV(14)
      XMDOT(14)=L(15)*X(15)+V(13)*Y(13)+F*Z-L(14)*X(14)-V(14)*Y(14)
      DO 40 N=15,29
      MDOT(N)=L(N+1)+V(N-1)-L(N)-V(N)
      HMDOT(N)=L(N+1)*HL(N+1)+V(N-1)*HV(N-1)-L(N)*HL(N)-V(N)*HV(N)
  40  XMDOT(N)=L(N+1)*X(N+1)+V(N-1)*Y(N-1)-L(N)*X(N)-V(N)*Y(N)
      MDOT(30)=R+V(29)-L(30)-V(30)
```

TABLE 5.11 (*continued*)

```
    HMDOT(30)=R*HLD+V(29)*HV(29)-L(30)*HL(30)-V(30)*HV(30)
    XMDOT(30)=R*XD+V(29)*Y(29)-L(30)*X(30)-V(30)*Y(30)
    XDDOT=V(30)*Y(30)-(R+D)*XD
    YYDOT(1)=MBDOT
    YYDOT(2)=XMBDOT
    YYDOT(3)=HLMBD
    DO 33 N=1,30
    YYDOT(3+N)=MDOT(N)
    YYDOT(33+N)=XMDOT(N)
 33 YYDOT(63+N)=HMDOT(N)
    YYDOT(94)=XDDOT
    RETURN
    END
C
    SUBROUTINE JEX(NEQ,TIME,YY,ML,MU,NRPD)
    COMMON A1,A2,B1,B2
    COMMON F,P,PB,B,V,L,VB,Z,HF,NT,PD,R,QR,YB,Y,YD,TB,T,TD,B0,MB0
    COMMON HLB,HL,D
    DIMENSION X(30),Y(30),T(30),P(30),XM(30),HM(30),L(30),V(30)
   + ,M(30),MDOT(30),XMDOT(30),HMDOT(30),HL(30),HV(30),HTOT(30)
    DIMENSION RWORK(9704),IWORK(114),YY(94),YYDOT(94)
    REAL L,MB,M,MD,MDOT,MB0,MBDOT
    RETURN
    END
C
    SUBROUTINE ENTH1(X,Y,T,HL,HV)
    REAL M1,M2
    DATA M1,M2,C1,C2/92.1,106.2,0.5,0.5/
    DATA DH1,DH2,BP1,BP2/86.8,82.9,110.8,144./
    X1=X
    X2=1.-X
    HL=X1*M1*C1*(T-BP1)+X2*M2*C2*(T-BP2)
    Y1=Y
    Y2=1.-Y
    HV=Y1*M1*(C1*(T-BP1)+DH1)+Y2*M2*(C2*(T-BP2)+DH2)
    RETURN
    END
C
    SUBROUTINE ENTH2(X,H,T)
    REAL M1,M2
    DATA M1,M2,C1,C2/92.1,106.2,0.5,0.5/
    DATA DH1,DH2,BP1,BP2/86.8,82.9,110.8,144./
    X1=X
    X2=1.-X
    T=(H+X1*M1*C1*BP1+X2*M2*C2*BP2)/(X1*M1*C1+X2*M2*C2)
    RETURN
    END
C
    SUBROUTINE BUBPT1(X,P,T,Y)
    COMMON A1,A2,B1,B2
    LOOP=0
  1 LOOP=LOOP+1
    IF(LOOP.GT.50)THEN
    WRITE(6,2)
  2 FORMAT(' BUBPT LOOP')
    STOP
    ENDIF
    PS1=EXP(A1/(T+273.)+B1)
```

TABLE 5.11 (*continued*)

```
      PS2=EXP(A2/(T+273.)+B2)
      PCALC= X*PS1+(1.-X)*PS2
      IF(ABS((PCALC-P)/P).LT..00001) GO TO 10
      F=PCALC-P
      DF=(-A1*PS1*X-A2*PS2*(1.-X))/((273.+T)**2)
      T=T-F/DF
      GO TO 1
   10 Y=PS1*X/P
      RETURN
      END
C
      SUBROUTINE BUBPT2(X,T,P,Y)
      COMMON A1,A2,B1,B2
      PS1=EXP(A1/(T+273.)+B1)
      PS2=EXP(A2/(T+273.)+B2)
      P= X*PS1+(1.-X)*PS2
      Y=PS1*X/P
      RETURN
      END
C
      SUBROUTINE LIQHYD(M,X,L,HTOT)
C DIMENSIONS ARE IN METERS, FLOW IS IN KG-MOLE/MIN
C      HOLDUP IS IN KG-MOLE
C FRANCIS WEIR FORMULA USED
      REAL M,L,M1,M2
      DATA M1,M2,SPGR,AREA/92.1,106.2,0.85,12.33/
      AVMW=X*M1+(1.-X)*M2
      VOL=M*AVMW/SPGR/1000.
      HTOT=VOL/12.33
C 1.25 INCH WEIR HEIGHT
      HOW=HTOT-0.0306
      IF(HOW.LE.0.) THEN
      L=0.
      RETURN
      ENDIF
      L= 3.33*3.78*3.281*60.*28.32*SPGR*((3.281*HOW)**1.5)/AVMW
      RETURN
      END
C
      SUBROUTINE VAPHYD(P1,P2,X,T,HTOT,V)
C "KDH" IS DRY-HOLE PRESSURE DROP COEFFICIENT; HOLE AREA = 1.233
C      PERFECT GAS CONSTANT IS IN "MM HG CU.M/K KG-MOLE
C PRESSURES ARE ALL IN "MM HG"
C VELOCITY IS IN "M/SEC"
C V IS IN "KG-MOLE/MIN"
      REAL KDH
      DATA SPGR,KDH/.85,.134/
      DENVAP=P2/(62.36*(T+273.))
      DPLIQ=HTOT*SPGR*73.06
      DPVAP=P1-P2-DPLIQ
      IF(DPVAP.LE.0.)THEN
      V=0.
      RETURN
      ENDIF
      VEL=SQRT(DPVAP/DENVAP/KDH)
      V=VEL*60.*DENVAP*1.233
      RETURN
      END
```

TABLE 5.11 (*continued*)

Initial conditions

```
.58379E-03  .11891E-02  .21718E-02  .37639E-02  .63341E-02  .10457E-01
.17000E-01  .27200E-01  .42666E-01  .65139E-01  .95843E-01  .13443E+00
.17812E+00  .22206E+00  .26125E+00  .26106E+00  .26204E+00  .26541E+00
.27368E+00  .29164E+00  .32778E+00  .39443E+00  .50184E+00  .64212E+00
.78094E+00  .88386E+00  .94433E+00  .97492E+00  .98917E+00  .99553E+00
.99831E+00  .99950E+00  .71894E+01  .71850E+01  .71806E+01  .71765E+01
.71727E+01  .71695E+01  .71674E+01  .71673E+01  .71701E+01  .71772E+01
.71897E+01  .72074E+01  .72279E+01  .72479E+01  .53878E+01  .53908E+01
.53953E+01  .54027E+01  .54161E+01  .54413E+01  .54878E+01  .55678E+01
.56837E+01  .58136E+01  .59223E+01  .59935E+01  .60334E+01  .60548E+01
.60669E+01  .60749E+01 -.19939E+04 -.20286E+04 -.20645E+04 -.21016E+04
-.21404E+04 -.21814E+04 -.22253E+04 -.22729E+04 -.23250E+04 -.23814E+04
-.24408E+04 -.24997E+04 -.25544E+04 -.26029E+04 -.26459E+04 -.26816E+04
-.27183E+04 -.27558E+04 -.27938E+04 -.28309E+04 -.28633E+04 -.28802E+04
-.28612E+04 -.27916E+04 -.26947E+04 -.26197E+04 -.25902E+04 -.25987E+04
-.26306E+04 -.26760E+04 -.27295E+04  .12179E+02  .59400E+01  .89721E+02
.65000E+02  .10643E+03  .10576E+03  .10505E+03  .10429E+03  .10347E+03
.10256E+03  .10151E+03  .10025E+03  .98730E+02  .96869E+02  .94648E+02
.92129E+02  .89477E+02  .86905E+02  .84590E+02  .83899E+02  .83146E+02
.82285E+02  .81226E+02  .79805E+02  .77758E+02  .74779E+02  .70791E+02
.66318E+02  .62296E+02  .59274E+02  .57142E+02  .55541E+02  .54183E+02
.52901E+02  .51609E+02  .49367E+02  .15000E+06
```

Results

TIME	N	X	Y	T	P	V	L	M
1.000	0	.00058	.00162	106.4	245.5	16.9	12.1	89.6
	1	.00119	.00331	105.8	240.3	16.9	29.0	7.2
	2	.00217	.00605	105.0	235.1	16.8	28.9	7.2
	3	.00377	.01047	104.3	229.8	16.8	28.9	7.2
	4	.00634	.01759	103.5	224.6	16.7	28.8	7.2
	5	.01046	.02891	102.6	219.4	16.6	28.8	7.2
	6	.01701	.04662	101.5	214.1	16.5	28.7	7.2
	7	.02721	.07354	100.3	208.9	16.4	28.6	7.2
	8	.04268	.11282	98.7	203.6	16.3	28.5	7.2
	9	.06516	.16670	96.9	198.4	16.2	28.4	7.2
	10	.09587	.23467	94.6	193.2	16.1	28.3	7.2
	11	.13446	.31182	92.1	188.0	16.0	28.2	7.2
	12	.17815	.38954	89.5	182.7	16.0	28.1	7.2
	13	.22208	.45890	86.9	177.5	15.9	28.0	7.2
	14	.26127	.51442	84.6	172.3	17.3	28.0	7.2
	15	.26108	.51479	83.9	168.1	17.3	11.4	5.4
	16	.26205	.51672	83.1	163.8	17.3	11.4	5.4
	17	.26543	.52184	82.3	159.5	17.4	11.4	5.4
	18	.27371	.53326	81.2	155.2	17.4	11.4	5.4
	19	.29167	.55651	79.8	150.8	17.4	11.5	5.4
	20	.32782	.59958	77.8	146.4	17.5	11.5	5.4
	21	.39449	.66915	74.8	142.0	17.7	11.6	5.5
	22	.50191	.76054	70.8	137.5	17.9	11.7	5.6
	23	.64219	.85199	66.3	133.0	18.3	12.0	5.7
	24	.78099	.92078	62.3	128.4	18.6	12.3	5.8
	25	.88389	.96172	59.3	123.6	18.8	12.7	5.9
	26	.94434	.98261	57.1	118.8	19.0	12.9	6.0
	27	.97493	.99239	55.5	113.9	19.1	13.0	6.0
	28	.98917	.99676	54.2	108.9	19.1	13.1	6.1
	29	.99553	.99868	52.9	103.9	19.2	13.2	6.1
	30	.99831	.99950	51.6	98.7	19.2	13.2	6.1
		.99950	.99986	49.4	90.0	5.9	13.2	65.0

TABLE 5.11 (*continued*)

TIME	N	X	Y	T	P	V	L	M
2.000	0	.00058	.00162	106.4	245.5	16.9	12.1	89.5
	1	.00119	.00331	105.8	240.3	16.9	29.0	7.2
	2	.00217	.00605	105.0	235.1	16.8	28.9	7.2
	3	.00376	.01047	104.3	229.8	16.8	28.9	7.2
	4	.00633	.01759	103.5	224.6	16.7	28.8	7.2
	5	.01046	.02891	102.6	219.4	16.6	28.8	7.2
	6	.01700	.04662	101.5	214.1	16.5	28.7	7.2
	7	.02720	.07354	100.3	208.9	16.4	28.6	7.2
	8	.04267	.11282	98.7	203.6	16.3	28.5	7.2
	9	.06514	.16670	96.9	198.4	16.2	28.4	7.2
	10	.09584	.23467	94.6	193.2	16.1	28.3	7.2
	11	.13443	.31182	92.1	188.0	16.0	28.2	7.2
	12	.17812	.38954	89.5	182.7	16.0	28.1	7.2
	13	.22206	.45890	86.9	177.5	15.9	28.0	7.2
	14	.26125	.51442	84.6	172.3	17.3	28.0	7.2
	15	.26106	.51479	83.9	168.1	17.3	11.4	5.4
	16	.26204	.51672	83.1	163.8	17.3	11.4	5.4
	17	.26541	.52184	82.3	159.5	17.4	11.4	5.4
	18	.27368	.53326	81.2	155.2	17.4	11.4	5.4
	19	.29164	.55651	79.8	150.8	17.4	11.5	5.4
	20	.32778	.59958	77.8	146.4	17.5	11.5	5.4
	21	.39443	.66915	74.8	142.0	17.7	11.6	5.5
	22	.50184	.76054	70.8	137.5	17.9	11.7	5.6
	23	.64212	.85199	66.3	133.0	18.3	12.0	5.7
	24	.78094	.92078	62.3	128.4	18.6	12.3	5.8
	25	.88386	.96172	59.3	123.6	18.8	12.7	5.9
	26	.94433	.98261	57.1	118.8	19.0	12.9	6.0
	27	.97492	.99239	55.5	113.9	19.1	13.0	6.0
	28	.98917	.99676	54.2	108.9	19.1	13.1	6.1
	29	.99553	.99868	52.9	103.9	19.2	13.2	6.1
	30	.99831	.99950	51.6	98.7	19.2	13.2	6.1
		.99950	.99986	49.4	90.0	5.9	13.2	65.0

5.7 BATCH REACTOR

Let us consider the batch reactor modeled in Sec. 3.9 (Fig. 3.9). Steam is initially fed into the jacket to heat up the system to temperatures at which the consecutive reactions begin. Then cooling water must be used in the jacket to remove the exothermic heats of the reactions.

The output signal of the temperature controller goes to two split-ranged valves, a steam valve and a water valve. The instrumentation is all pneumatic, so the controller output pressure P_c goes from 3 to 15 psig. The valves will be adjusted so that the steam valve is wide open when the controller output pressure P_c is at 15 psig and is closed at $P_c = 9$ psig (i.e., half the full range of the controller output). The water valve will be closed at $P_c = 9$ psig and wide open at $P_c = 3$ psig. The reason for hooking up the valves in this manner is to have the correct fail-safe action in the event of an instrument air failure. The steam valve takes air pressure to open it and therefore it will fail closed. We call this an "air-to-open" (AO) valve. On the other hand, the water valve takes air pressure to close it and therefore it will fail open. This is an "air-to-close" (AC) valve. If an emergency occurs, we want to remove the source of heat (steam) and go to full cooling.

Controller output range (psig)	3		9	15
Steam valve fraction open X_s:			0 (closed)	1 (open)
Water valve fraction open X_w:	1 (open)		0 (closed)	

The equations for the reaction liquid inside the tank and the vessel metal are

$$\frac{dC_A}{dt} = -k_1 C_A \tag{5.35}$$

$$\frac{dC_B}{dt} = k_1 C_A - k_2 C_B \tag{5.36}$$

$$\frac{dT}{dt} = \frac{-\lambda_1}{\rho C_p} k_1 C_A - \frac{-\lambda_2}{\rho C_p} k_2 C_B - \frac{Q_M}{V \rho C_p} \tag{5.37}$$

$$Q_M = h_i A_i(T - T_M) \tag{5.38}$$

$$\frac{dT_M}{dt} = \frac{Q_M - Q_J}{\rho_M C_M V_M} \tag{5.39}$$

The equations for the jacket are different for the three phases of the batch cycle.

A. With steam in the jacket (35 psia supply pressure steam):

$$V_J \frac{d\rho_s}{dt} = w_s - w_c \tag{5.40}$$

$$\rho_s = \frac{M P_J}{R(T_J + 460)} \tag{5.41}$$

$$P_J = \exp\left(\frac{A_{vp}}{T_J + 460} + B_{vp}\right) \tag{5.42}$$

$$w_s = C_{Vs} X_s \sqrt{35 - P_J} \tag{5.43}$$

$$Q_J = -h_{os} A_{os}(T_J - T_M) \tag{5.44}$$

$$w_c = -\frac{Q_J}{H_s - h_c} \tag{5.45}$$

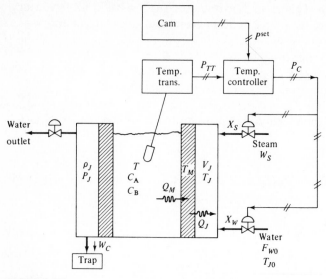

FIGURE 5.7
Batch reactor.

B. During filling with water (20 psig water header pressure):

$$A_o = \left(\frac{A_o}{V_J}\right)_{total} V_J \tag{5.46}$$

$$\frac{dV_J}{dt} = F_{WO} \tag{5.47}$$

$$\frac{d(V_J T_J)}{dt} = F_{WO} T_{JO} + \frac{Q_J}{\rho_J C_J} \tag{5.48}$$

$$Q_J = h_{ow} A_o (T_M - T_J) \tag{5.49}$$

$$F_{WO} = C_{Vw} X_w \sqrt{20} \tag{5.50}$$

C. When the jacket is full of water:

$$\frac{dT_J}{dt} = \frac{F_{WO}}{V_J}(T_{JO} - T_J) + \frac{Q_J}{C_J V_J \rho_J} \tag{5.51}$$

The system is sketched in Fig. 5.7, and numerical values of parameters are given in Table 5.12. The digital program is given in Table 5.13. Plotted results are shown in Fig. 5.8.

The temperature transmitter has a range of 50 to 250°F, so its output pneumatic pressure signal goes from 3 psig at 50°F to 15 psig at 250°F.

$$P_{TT} = 3 + (T - 50)\frac{12}{200} \tag{5.52}$$

TABLE 5.12
Parameters for batch reactor

α_1	729.55 min^{-1}	V_J	18.83 ft^3
α_2	6567.6 min^{-1}	C_{Vw}	100 gpm/psi$^{0.5}$
E_1	15,000 Btu/lb·mol	T_{J0}	80°F
E_2	20,000 Btu/lb·mol	A_i	56.5 ft^2
A_{vp}	$-8744.4°$R	λ_1	$-40,000$ Btu/lb·mol
B_{vp}	15.70	λ_2	$-50,000$ Btu/lb·mol
C_{A0}	0.80 lb·mol A/ft^3	C_p	1 Btu/lb$_m$ °F
T_0	80°F	V	42.5 ft^3
K_c	10 psi/psi	ρ	50 lb$_m$/ft^3
C_{Vs}	112 lb$_m$/min psi$^{0.5}$	C_M	0.12 Btu/lb$_m$ °F
h_{os}	1000 Btu/h °F ft^2	V_M	9.42 ft^3
h_{ow}	400 Btu/h °F ft^2	ρ_M	512 lb$_m$/ft^3
h_i	160 Btu/h °F ft^2	ρ_J	62.3 lb$_m$/ft^3
A_0	56.5 ft^2	C_J	1 Btu/lb$_m$ °F
$H_s - h_c$	939 Btu/lb$_m$		

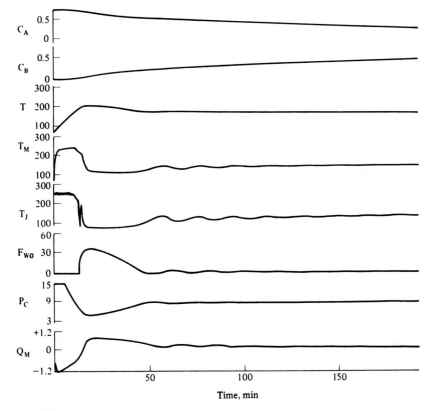

FIGURE 5.8
Plotted results for batch reactor.

TABLE 5.13

Batch reactor simulation

```
      REAL K1,K2,KC
      ALPHA1=729.5488
      ALPHA2=6567.587
C VAPOR PRESSURE CONSTANTS FOR STEAM
      AVP=-8744.4
      BVP=15.70036
C USE SMALL DELTA DURING STEAM PERIOD
      DELTA=.002
      KC=2.
      TPRINT=1.
C INITIAL CONDITIONS
      CA=0.8
      CB=0.
      TIME=0.
      T=80.
      TM=80.
      TJ=259.
      PJ=34.4
      DENS=18.*PJ*144./(1545.*(TJ+460.))
      PSET=12.6
      START=1.
      FWO=0.
      FULL=-1.
      VJ=0.
      VJTJ=0.
      TFLAG=0.
      CAM=-1.
      RAMP =.005
      WRITE(6,54)
   54 FORMAT(' TIME   CA   CB   XS   XW   T    TM   TJ   FWO
     +  QJ   QM')
C MAIN LOOP FOR EACH STEP IN TIME
  100 K1=ALPHA1*EXP(-15000./(1.99*(T+460.)))
      K2=ALPHA2*EXP(-20000./((T+460.)*1.99))
C   TRANSMITTER
      PTT=3.+(T-50.)*12./200.
C  CONTROLLER
      PC=7.+KC*(PSET-PTT)
      IF(PC.GT.15.) PC=15.
      IF(PC.LT.3.) PC=3.
C   VALVES
      XS=(PC-9.)/6.
      XW=(9.-PC)/6.
      IF(XS.GT.1.) XS=1.
      IF(XS.LT.0.) XS=0.
      IF(XW.GT.1.) XW=1.
      IF(XW.LT.0.) XW=0.
C   TEST FOR STEAM
      IF(START.LT.0.) GO TO 20
      IF(PJ.GE.35.) GO TO 40
      WS=XS*112.*SQRT(35.-PJ)
      GO TO 41
   40 WS=0.
   41 CONTINUE
      QJ=-1000.*56.5*(TJ-TM)/60.
      WC=-QJ/939.
      DENDOT=(WS-WC)/18.83
      DENS=DENS+DELTA*DENDOT
C ITERATIVE LOOP TO CALCULATE STEAM TEMPERATURE
```

TABLE 5.13 (*continued*)

```
C     AND PRESSURE FROM KNOWN DENSITY USING
C     INTERVAL HALVING
      FLAGP=-1.
      FLAGM=-1.
      DTJ=1.
      LOOP=0
   15 PJ=EXP(BVP+AVP/(TJ+460.))
      LOOP=LOOP+1
      IF(LOOP.GT.50)  GO TO 70
      DCALC=18.*PJ*144./(1545.*(TJ+460.))
      IF(ABS(DENS-DCALC).LT. .0011) GO TO 50
      IF(DENS.GT.DCALC) GO TO 17
      IF(FLAGM.LT.0.) GO TO 16
      DTJ=DTJ/2.
   16 TJ=TJ-DTJ
      FLAGP=1.
      GO TO 15
   70 WRITE(6,71)
   71 FORMAT(1X,'STEAM TEMP LOOP')
      STOP
   17 IF(FLAGP.LT.0.) GO TO 18
      DTJ=DTJ/2.
   18 TJ=TJ+DTJ
      FLAGM=1.
      GO TO 15
   20 FWO=100.*SQRT(20.)*8.33*XW/62.3
      WS=0.
      WC=0.
      DENS=0.
      PJ=0.
      XS=0.
C   TEST FOR JACKET FILLING
      IF(FULL.GT.0.) GO TO 30
      AO=VJ*56.5/18.83
      VJ=VJ+DELTA*FWO
      IF(VJ.GE. 18.83) FULL=1.
      QJ=400.*AO*(TM-TJ)/60.
      VJTJ=VJTJ+DELTA*(FWO*80.+QJ)
      IF(VJ.LE.0.) GO TO 25
      TJ=VJTJ/VJ
      GO TO 50
   25 TJ=80.
      GO TO 50
C   FULL JACKET
   30 QJ=400.*56.5*(TM-TJ)/60.
      TPRINT=2.
C USE BIGGER DELTA ONCE JACKET IS FULL OF WATER
      DELTA=.05
C EVALUATE DERIVATIVES
      TJDOT=FWO*(80.-TJ)/18.83+QJ/(18.83*62.3)
      TJ=TJ+DELTA*TJDOT
   50 CADOT=-K1*CA
      CBDOT=K1*CA-K2*CB
      QM=160.*56.5*(T-TM)/60.
      TDOT=(K1*CA*40000.+K2*CB*50000.)/50.-QM/(42.4*50.)
      TMDOT=(QM-QJ)/(512.*.12*9.42)
C   INTEGRATION
      TIME=TIME+DELTA
      CA=CA+CADOT*DELTA
```

TABLE 5.13 (*continued*)

```
    CB=CB+CBDOT*DELTA
    T=T+TDOT*DELTA
    TM=TM+TMDOT*DELTA
    IF(T.GT.300.) STOP
    IF(T.GT.200.) START=-1.
    IF(T.GT.200.) CAM=1.
    IF(CAM.GT.0.) PSET=PSET-DELTA*RAMP
    IF(TIME.GT.100.) GO TO 56
    IF(TIME.LT.TFLAG) GO TO 100
    WRITE(6,55)TIME,CA,CB,XS,XW,T,TM,TJ,FWO,QJ,QM
 55 FORMAT(1X,F5.1,2F7.3,2F5.2,4F6.1,2E10.2)
    TFLAG=TFLAG+TPRINT
    GO TO 100
 56 STOP
    END
```

Results

TIME	CA	CB	XS	XW	T	TM	TJ	FWO	QJ	QM
0.0	0.800	0.000	1.00	0.00	80.0	80.6	251.0	0.0	-0.17E+06	0.00E+00
1.0	0.799	0.001	1.00	0.00	86.3	211.9	265.0	0.0	-0.43E+05	-0.19E+05
2.0	0.799	0.001	1.00	0.00	96.4	234.2	261.0	0.0	-0.27E+05	-0.21E+05
3.0	0.798	0.002	1.00	0.00	106.7	239.0	260.0	0.0	-0.22E+05	-0.20E+05
4.0	0.797	0.003	1.00	0.00	116.7	240.8	262.0	0.0	-0.22E+05	-0.19E+05
5.0	0.796	0.005	1.00	0.00	126.3	242.6	261.0	0.0	-0.19E+05	-0.18E+05
6.0	0.794	0.006	1.00	0.00	135.7	243.8	265.0	0.0	-0.21E+05	-0.16E+05
7.0	0.792	0.008	0.97	0.00	144.7	244.9	260.0	0.0	-0.16E+05	-0.15E+05
8.0	0.789	0.011	0.80	0.00	153.5	245.9	262.0	0.0	-0.12E+05	-0.14E+05
9.0	0.787	0.014	0.62	0.00	162.1	246.7	263.0	0.0	-0.11E+05	-0.13E+05
10.0	0.783	0.017	0.46	0.00	170.6	247.3	259.0	0.0	-0.12E+05	-0.12E+05
11.0	0.779	0.021	0.29	0.00	178.9	248.0	259.0	0.0	-0.10E+05	-0.10E+05
12.0	0.775	0.026	0.12	0.00	187.3	248.8	259.0	0.0	-0.96E+04	-0.93E+04
13.0	0.769	0.031	0.00	0.04	195.6	243.4	244.0	0.0	-0.54E+03	-0.72E+04
14.0	0.763	0.037	0.00	0.20	203.2	232.6	212.7	11.9	0.17E+04	-0.44E+04
15.0	0.756	0.044	0.00	0.34	210.4	222.9	203.2	20.1	0.51E+04	-0.20E+04
17.0	0.741	0.059	0.00	0.53	219.7	159.1	98.5	31.8	0.23E+05	0.89E+04
19.0	0.724	0.076	0.00	0.59	222.4	132.9	88.5	35.3	0.17E+05	0.13E+05
21.0	0.707	0.092	0.00	0.61	223.0	127.1	86.9	36.2	0.15E+05	0.14E+05
23.0	0.691	0.108	0.00	0.61	222.9	125.8	86.6	36.3	0.15E+05	0.15E+05
25.0	0.675	0.123	0.00	0.60	222.5	125.5	86.5	36.0	0.15E+05	0.15E+05
27.0	0.659	0.138	0.00	0.59	221.8	125.3	86.6	35.4	0.15E+05	0.15E+05
29.0	0.644	0.152	0.00	0.58	220.9	125.1	86.7	34.5	0.14E+05	0.14E+05
31.0	0.630	0.166	0.00	0.55	219.6	124.9	86.9	33.2	0.14E+05	0.14E+05
33.0	0.616	0.178	0.00	0.53	218.0	124.7	87.1	31.5	0.14E+05	0.14E+05
35.0	0.603	0.191	0.00	0.49	216.1	124.4	87.5	29.5	0.14E+05	0.14E+05
37.0	0.591	0.202	0.00	0.45	214.0	124.1	87.9	27.1	0.14E+05	0.14E+05
39.0	0.579	0.213	0.00	0.41	211.5	123.8	88.5	24.4	0.13E+05	0.13E+05
41.0	0.568	0.223	0.00	0.36	208.9	123.6	89.3	21.4	0.13E+05	0.13E+05
43.0	0.558	0.232	0.00	0.30	206.1	123.4	90.4	18.2	0.12E+05	0.12E+05
45.0	0.549	0.241	0.00	0.25	203.0	123.5	91.9	14.7	0.12E+05	0.12E+05
47.0	0.540	0.249	0.00	0.19	199.9	123.9	94.0	11.2	0.11E+05	0.11E+05
49.0	0.531	0.257	0.00	0.13	196.8	124.9	97.2	7.7	0.10E+05	0.11E+05
51.0	0.523	0.264	0.00	0.07	193.9	126.8	101.9	4.4	0.94E+04	0.10E+05
53.0	0.516	0.270	0.00	0.03	191.3	130.2	109.1	1.5	0.80E+04	0.92E+04
55.0	0.509	0.277	0.00	0.00	189.3	135.8	119.6	0.0	0.62E+04	0.81E+04
57.0	0.503	0.282	0.00	0.00	188.3	142.5	129.0	0.0	0.51E+04	0.69E+04
59.0	0.496	0.288	0.00	0.00	188.1	148.5	137.0	0.0	0.43E+04	0.60E+04
61.0	0.490	0.294	0.00	0.00	188.7	153.9	143.9	0.0	0.38E+04	0.53E+04

TABLE 5.13 (*continued*)

```
63.0  0.483  0.300 0.00 0.01 189.9 158.6 148.9   0.7  0.36E+04  0.47E+04
65.0  0.477  0.305 0.00 0.05 191.4 159.1 143.9   2.8  0.57E+04  0.48E+04
67.0  0.470  0.311 0.00 0.07 192.5 153.8 132.9   4.3  0.78E+04  0.58E+04
69.0  0.464  0.317 0.00 0.08 192.6 146.8 123.7   4.7  0.87E+04  0.69E+04
71.0  0.457  0.322 0.00 0.07 191.8 141.9 119.4   3.9  0.85E+04  0.75E+04
73.0  0.451  0.328 0.00 0.04 190.5 140.1 119.9   2.5  0.77E+04  0.76E+04
75.0  0.445  0.333 0.00 0.02 189.1 141.3 124.3   1.1  0.65E+04  0.72E+04
77.0  0.439  0.338 0.00 0.00 188.0 145.2 131.8   0.0  0.51E+04  0.65E+04
79.0  0.434  0.343 0.00 0.00 187.7 150.4 139.5   0.0  0.41E+04  0.56E+04
81.0  0.428  0.348 0.00 0.00 188.0 155.4 145.7   0.3  0.36E+04  0.49E+04
83.0  0.422  0.352 0.00 0.02 188.8 158.1 146.3   1.5  0.44E+04  0.46E+04
85.0  0.417  0.357 0.00 0.04 189.7 156.6 140.8   2.7  0.59E+04  0.50E+04
87.0  0.411  0.362 0.00 0.06 190.0 152.2 133.4   3.3  0.71E+04  0.57E+04
89.0  0.406  0.366 0.00 0.05 189.6 147.8 128.3   3.1  0.73E+04  0.63E+04
91.0  0.401  0.371 0.00 0.04 188.7 145.3 127.1   2.2  0.69E+04  0.65E+04
93.0  0.395  0.375 0.00 0.02 187.6 145.5 129.7   1.1  0.60E+04  0.64E+04
95.0  0.390  0.379 0.00 0.00 186.8 148.0 135.3   0.3  0.48E+04  0.59E+04
97.0  0.385  0.383 0.00 0.00 186.4 152.3 142.2   0.0  0.38E+04  0.52E+04
99.0  0.381  0.387 0.00 0.01 186.6 156.6 147.2   0.4  0.35E+04  0.45E+04
```

A proportional feedback controller is used with its output biased at 7 psig (i.e., its output pressure is 7 psig when there is zero error).

$$P_c = 7 + K_c(P^{\text{set}} - P_{TT}) \tag{5.53}$$

The setpoint signal P^{set} comes from a pneumatic function generator. When the process temperature gets up to 200°F the P^{set} signal is ramped slowly downward to prevent too much loss of component B, as discussed in Sec. 3.9.

$$P^{\text{set}} = 12 - \text{RAMP}(t - t_{200}) \tag{5.54}$$

where RAMP = rate of P^{set} change with time, psi/min

t = batch time, min

t_{200} = time when process temperature T reaches 200°F

5.8 TERNARY BATCH DISTILLATION WITH HOLDUP

The model of a multicomponent batch distillation column was derived in Sec. 3.13. For a simulation example, let us consider a ternary mixture. Three products will be produced and two "slop" cuts may also be produced. Constant relative volatility, equimolal overflow, constant tray holdup, and ideal trays are assumed.

Table 5.14 gives a digital computer FORTRAN program for this three-component batch distillation dynamic simulation. The specific example is a column with 20 trays and relative volatilities of 9, 3, and 1. The vapor flow rate is constant at 100 mol/h.

The column starts up on total reflux (no distillate is withdrawn) until the distillate composition reaches the desired purity level. The time at total reflux is called TE in the program. Then distillate is withdrawn at a fixed rate of 40 mol/h.

TABLE 5.14
Ternary batch distillation with slop cuts

```
C
C        AFTER TOTAL REFLUX STARTUP, DISTILLATE FLOW RATE IS FIXED
C        ASSUMPTIONS: CONSTANT RELATIVE VOLATILITIES (TERNARY)
C                     EQUIMOLAL OVERFLOW, IDEAL TRAYS
C
      DIMENSION X(100,3),Y(100,3),Z(3),ALPHA(3),HBXB(3),XD(3),XB(3),
     + YB(3),DX(100,3),DHBXB(3),DXD(3),XX(3),YY(3),XS1(3),HXS1(3)
      DIMENSION XBO(3),XS2(3),HXS2(3)
      REAL KC
      DATA XD1SP,XD2SP,XB3SP,Z,ALPHA/.95,.95,.95,.3,.3,.4,9.,3.,1./
      DATA HBO,HD,HN,V,DELTA/400.,10.,1.,100.,.001/
      DATA NT,DFIX/20,40./
      WRITE(6,1) Z,ALPHA
    1 FORMAT(//,' FEED COMPOSITION = ',3F8.5,/,' ALPHA = ',3F8.3)
      DO 5 J=1,3
    5 XBO(J)=Z(J)
      WRITE(6,2)NT,HBO,DFIX
    2 FORMAT(//,' NT = ',I3,' HBO = ',F8.4,' DFIX = ',F8.4)
C INITIAL CONDITIONS
      HXD1=0.
      HXD2=0.
      HD1=0.
      HD2=0.
      XD1AV=1.
      XD2AV=1.
      TP2=0.
      HS2=0.
      TPRINT=0.
      TIME=0.
      HB=HBO-HD-NT*HN
      HS1=0.
      DO 10 J=1,3
      HXS1(J)=0.
      XS2(J)=0.
      HXS2(J)=0.
      XB(J)=XBO(J)
      HBXB(J)=HB*XBO(J)
      DO 6 N=1,NT
    6 X(N,J)=XBO(J)
   10 XD(J)=XBO(J)
      WRITE(6,19)
   19 FORMAT(' TIME   D    XB1   XB2   XB3   XD1   XD2   XD3
     +   XD1AV   XD2AV')
      FLAGTE=-1.
      FLAGP1=-1.
      FLAGP2=-1.
      FLAGS1=-1.
   20 CALL BUBPT(XB,YB,ALPHA)
      DO 25 N=1,NT
      DO 24 J=1,3
   24 XX(J)=X(N,J)
      CALL BUBPT(XX,YY,ALPHA)
      DO 26 J=1,3
   26 Y(N,J)=YY(J)
   25 CONTINUE
C TOTAL REFLUX UNTIL XD(1) REACHES XD1SP
      D=DFIX
      IF(FLAGTE.LT.0.) D=0.
```

TABLE 5.14 (*continued*)

```
     R=V-D
     DHB=-D
     DO 40 J=1,3
     DHBXB(J)=R*X(1,J)-V*YB(J)
     DX(1,J)=(V*YB(J)+R*X(2,J)-V*Y(1,J)-R*X(1,J))/HN
     DO 35 N=2,NT-1
  35 DX(N,J)=(V*Y(N-1,J)+R*X(N+1,J)-V*Y(N,J)-R*X(N,J))/HN
     DX(NT,J)=(V*Y(NT-1,J)+R*XD(J)-V*Y(NT,J)-R*X(NT,J))/HN
  40 DXD(J)=V*(Y(NT,J)-XD(J))/HD
     IF(TIME.LT.TPRINT) GO TO 50
     WRITE(6,41)TIME,D,XB,XD,XD1AV,XD2AV
  41 FORMAT(F5.2,F6.1,8F8.5)
     TPRINT=TPRINT+.1
  50 TIME=TIME+DELTA
     HB=HB+DHB*DELTA
     IF(HB.LE.0.)THEN
     WRITE(6,*) 'STILL POT EMPTY'
     STOP
     ENDIF
     DO 60 J=1,3
     HBXB(J)=HBXB(J)+DHBXB(J)*DELTA
     XB(J)=HBXB(J)/HB
     DO 55 N=1,NT
  55 X(N,J)=X(N,J)+DX(N,J)*DELTA
  60 XD(J)=XD(J)+DXD(J)*DELTA
     IF(XB(1).LT..00001)THEN
     XB(1)=0.
     XB(3)=1.-XB(2)
     ENDIF
     IF(FLAGTE.GT.0.)GO TO 65
C START P1 WITHDRAWAL; FLAGTE = 1 IF P1 PRODUCT IS BEING REMOVED
     IF(XD(1).GE.XD1SP ) THEN
        FLAGTE=1.
        TE=TIME
        WRITE(6,61)TE
  61    FORMAT(' TE = ',F8.4)
     GO TO 20
        ENDIF
  65 CONTINUE
     IF(FLAGP1.GT.0.)GO TO 70
     HD1=HD1+D*DELTA
     HXD1=HXD1+D*XD(1)*DELTA
     IF(D.GT.0.)XD1AV=HXD1/HD1
C START SLOP CUT NO. 1; FLAGP1 = 1 IF SLOP NO. 1 IS BEING REMOVED
     IF(XD1AV.LT.XD1SP)THEN
        FLAGP1=1.
        TP1=TIME
        WRITE(6,66)TP1
  66 FORMAT(' TP1 = ',F8.4)
        ENDIF
  70 IF(FLAGP1.LE.0.)GO TO 20
     IF(FLAGS1.GT.0.)GO TO 80
C START P2 WITHDRAWAL; FLAGS1 =1 IF P2 PRODUCT IS BEING REMOVED
     HS1=0.
     DO 73 J=1,3
     HXS1(J)=HXS1(J)+D*XD(J)*DELTA
  73 HS1=HS1+HXS1(J)
     DO 74 J=1,3
  74 XS1(J)=HXS1(J)/HS1
```

TABLE 5.14 (*continued*)

```
   IF(XD(2).GE.XD2SP ) THEN
     FLAGS1=1.
     TSLOP1=TIME
   WRITE(6,71)TSLOP1
71 FORMAT(' TSLOP1 = ',F8.4)
     ENDIF
80 IF(FLAGS1.LT.0.)GO TO 20
   IF(FLAGP2.GT.0.)GO TO 90
   HD2=HD2+D*DELTA
   HXD2=HXD2+D*XD(2)*DELTA
   IF(HD2.GT.0.) XD2AV=HXD2/HD2
   IF(XD2AV.LE.XD2SP)THEN
     FLAGP2=1.
     TP2=TIME
   WRITE(6,89)TP2
89 FORMAT(' TP2 = ',F8.4)
     ENDIF
C START SLOP CUT NO. 2; FLAGP2 = 1 IF SLOP NO. 2 IS BEING REMOVED
90 IF(FLAGP2.LT.0.)GO TO 95
   HS2=0.
   DO 93 J=1,3
   HXS2(J)=HXS2(J)+D*XD(J)*DELTA
93 HS2=HS2+HXS2(J)
   DO 94 J=1,3
94 XS2(J)=HXS2(J)/HS2
95 SUM=0.
   DO 91 N=1,NT
91 SUM=SUM+X(N,3)*HN
   SUM=SUM+XB(3)*HB
   XB3AV=SUM/(HB+NT*HN)
   IF(XB3AV.GE.XB3SP)GO TO 99
   GO TO 20
99 HS2TOT=HD+HS2
   IF(FLAGP2.LT.0.)THEN
   HTEST=HD+HD2
   XTEST2=(HXD2+HD*XD(2))/HTEST
   IF(XTEST2.GT.XD2SP)THEN
   HD2=HD2+HD
   HS2TOT=0.
   DO 96 J=1,3
96 XS2(J)=0.
   ENDIF
   ENDIF
   TF=TIME
   HB=HB+NT*HN
   CAP=(HD1+HB+HD2 )/(TF+.5)
   WRITE(6,98)HD1,HD2,HB,TE,TP1,TSLOP1,TP2,TF,CAP
98 FORMAT(' P1 = ',F6.2,' P2 = ',F6.2,' P3 = ',F6.2,/,
  +' TE = ',F5.2,' TP1 = ',F5.2,' TSLOP1 = ',F5.2,' TP2 = ',F5.2,
  +' TF = ',F5.2,' CAP = ',F6.2)
   WRITE(6,104)HS1,XS1
104 FORMAT(' S1 = ',F8.4,' XS1 = ',3F8.5)
   IF(HS2TOT.GT.0.)THEN
   DO 106 J=1,3
106 XS2(J)=(HS2*XS2(J)+HD*XD(J))/HS2TOT
   ENDIF
   WRITE(6,105)HS2TOT,XS2
105 FORMAT(' S2   = ',F8.4,' XS2 = ',3F8.5)
```

TABLE 5.14 (*continued*)

```
    STOP
    END
C****************************************************
    SUBROUTINE BUBPT(X,Y,ALPHA)
    DIMENSION X(3),Y(3),ALPHA(3)
    SUM=0.
    DO 10 J=1,3
 10 SUM=SUM+ALPHA(J)*X(J)
    DO 20 J=1,3
 20 Y(J)=ALPHA(J)*X(J)/SUM
    RETURN
    END
```

Results

```
 FEED COMPOSITION =   .30000   .30000   .40000
 ALPHA =    9.000   3.000   1.000
```

NT = 20 HBO = 400.0000 DFIX = 40.0000

TIME	D	XB1	XB2	XB3	XD1	XD2	XD3	XD1AV	XD
.00	.0	.30000	.30000	.40000	.30000	.30000	.40000	1.00000	1.00
.10	.0	.28982	.30201	.40817	.59870	.21791	.18339	1.00000	1.00
.20	.0	.27992	.30395	.41613	.78220	.13734	.08046	1.00000	1.00
.30	.0	.27025	.30631	.42343	.88711	.07884	.03406	1.00000	1.00
.40	.0	.26153	.31013	.42834	.94275	.04312	.01413	1.00000	1.00

TE = .4220

TIME	D	XB1	XB2	XB3	XD1	XD2	XD3	XD1AV	XD
.50	40.0	.25291	.31377	.43332	.97014	.02393	.00593	.96131	1.00
.60	40.0	.24406	.31719	.43874	.98366	.01382	.00252	.97043	1.00
.70	40.0	.23512	.32078	.44410	.99086	.00808	.00106	.97663	1.00
.80	40.0	.22616	.32446	.44938	.99477	.00479	.00045	.98097	1.00
.90	40.0	.21718	.32816	.45466	.99693	.00288	.00019	.98410	1.00
1.00	40.0	.20818	.33186	.45996	.99816	.00176	.00008	.98644	1.00
1.10	40.0	.19914	.33552	.46534	.99887	.00110	.00003	.98823	1.00
1.20	40.0	.19009	.33911	.47080	.99928	.00070	.00001	.98962	1.00
1.30	40.0	.18093	.34263	.47644	.99953	.00047	.00001	.99075	1.00
1.40	40.0	.17186	.34599	.48215	.99967	.00033	.00000	.99165	1.00
1.50	40.0	.16281	.34921	.48799	.99975	.00025	.00000	.99240	1.00
1.60	40.0	.15380	.35224	.49396	.99980	.00020	.00000	.99303	1.00
1.70	40.0	.14486	.35506	.50008	.99982	.00018	.00000	.99356	1.00
1.80	40.0	.13600	.35765	.50635	.99981	.00019	.00000	.99401	1.00
1.90	40.0	.12727	.35997	.51277	.99977	.00023	.00000	.99440	1.00
2.00	40.0	.11867	.36198	.51935	.99968	.00032	.00000	.99474	1.00
2.10	40.0	.11025	.36366	.52609	.99947	.00053	.00000	.99503	1.00
2.20	40.0	.10203	.36496	.53301	.99898	.00102	.00000	.99527	1.00
2.30	40.0	.09403	.36587	.54010	.99767	.00233	.00000	.99543	1.00
2.40	40.0	.08636	.36634	.54730	.99366	.00634	.00000	.99546	1.00
2.50	40.0	.07889	.36635	.55476	.97995	.02005	.00000	.99511	1.00
2.60	40.0	.07172	.36586	.56242	.93873	.06127	.00000	.99360	1.00
2.70	40.0	.06488	.36485	.57027	.86334	.13666	.00000	.98961	1.00
2.80	40.0	.05838	.36328	.57834	.78067	.21933	.00000	.98253	1.00
2.90	40.0	.05225	.36114	.58662	.70573	.29427	.00000	.97282	1.00
3.00	40.0	.04649	.35839	.59512	.63939	.36061	.00000	.96114	1.00

TP1 = 3.0860

TIME	D	XB1	XB2	XB3	XD1	XD2	XD3	XD1AV	XD
3.10	40.0	.04112	.35503	.60385	.58007	.41993	.00000	.94990	1.00
3.20	40.0	.03615	.35103	.61282	.52624	.47376	.00000	.94990	1.00
3.30	40.0	.03157	.34639	.62204	.47676	.52324	.00000	.94990	1.00
3.40	40.0	.02740	.34110	.63150	.43082	.56918	.00000	.94990	1.00
3.50	40.0	.02361	.33517	.64122	.38790	.61210	.00000	.94990	1.00
3.60	40.0	.02020	.32860	.65121	.34772	.65228	.00000	.94990	1.00
3.70	40.0	.01715	.32139	.66145	.31010	.68990	.00000	.94990	1.00

TABLE 5.14 (*continued*)

3.80	40.0	.01446	.31357	.67197	.27499	.72501	.00000	.94990	1.00
3.90	40.0	.01209	.30516	.68275	.24236	.75764	.00000	.94990	1.00
4.00	40.0	.01002	.29618	.69380	.21222	.78778	.00000	.94990	1.00
4.10	40.0	.00824	.28667	.70509	.18457	.81543	.00000	.94990	1.00
4.20	40.0	.00672	.27666	.71662	.15937	.84062	.00000	.94990	1.00
4.30	40.0	.00543	.26622	.72835	.13661	.86339	.00000	.94990	1.00
4.40	40.0	.00434	.25539	.74027	.11621	.88379	.00000	.94990	1.00
4.50	40.0	.00344	.24425	.75232	.09808	.90192	.00001	.94990	1.00
4.60	40.0	.00269	.23285	.76445	.08210	.91789	.00001	.94990	1.00
4.70	40.0	.00209	.22128	.77663	.06816	.93183	.00001	.94990	1.00
4.80	40.0	.00160	.20961	.78879	.05609	.94389	.00002	.94990	1.00

TSLOP1 =   4.8580

4.90	40.0	.00121	.19790	.80089	.04575	.95422	.00003	.94990	.95
5.00	40.0	.00091	.18621	.81288	.03697	.96297	.00006	.94990	.95
5.10	40.0	.00067	.17460	.82473	.02960	.97029	.00011	.94990	.96
5.20	40.0	.00049	.16310	.83641	.02346	.97629	.00025	.94990	.96
5.30	40.0	.00035	.15178	.84787	.01841	.98090	.00069	.94990	.96
5.40	40.0	.00025	.14068	.85908	.01430	.98344	.00226	.94990	.97
5.50	40.0	.00017	.12983	.87000	.01100	.98016	.00885	.94990	.97
5.60	40.0	.00012	.11929	.88059	.00837	.95491	.03672	.94990	.97
5.70	40.0	.00008	.10909	.89083	.00629	.88245	.11126	.94990	.96
5.80	40.0	.00005	.09928	.90067	.00466	.78866	.20667	.94990	.95

TP2 =   5.8130

5.90	40.0	.00003	.08988	.91008	.00342	.70398	.29260	.94990	.94
6.00	40.0	.00002	.08095	.91903	.00249	.63133	.36618	.94990	.94
6.10	40.0	.00001	.07250	.92749	.00179	.56837	.42983	.94990	.94
6.20	40.0	.00000	.06455	.93545	.00128	.51278	.48594	.94990	.94
6.30	40.0	.00000	.05713	.94287	.00086	.46280	.53633	.94990	.94
6.40	40.0	.00000	.05024	.94976	.00058	.41717	.58225	.94990	.94

P1 = 106.56 P2 =   38.24 P3 = 149.84
TE =   .42 TP1 =   3.09 TSLOP1 =   4.86 TP2 =   5.81 TF =   6.43 CAP =   42.
S1 =   70.9200 XS1 =   .25292 .74708 .00000
S2   =   34.5600 XS2 =   .00154 .52131 .47715

Three products (P1, P2, and P3) and two slop cuts (S1 and S2) are produced. The average composition of the products are 95 mole percent. The P1 product is mostly the lightest component (component 1). The P2 product is mostly intermediate component (number 2) with some impurities of both the light and the heavy components. The final product P3 is what is left in the still pot and on the trays. The times to produce the various products and slop cuts are given in the results shown in Table 5.14. The total time for the batch distillation in this example is 6.4 hours.

Note that the 70.92 moles of the first slop cut contain mostly light and intermediate component (25/75 mol %), while the 34.5 moles of second slop cut contain mostly intermediate and heavy components (52/48 mol %). Recycling these slop cuts back to the next batch cycle makes little thermodynamic sense, but that is the normal procedure in practice.

PROBLEMS

5.1. Simulate the nonisothermal CSTR of Sec. 5.3, using Euler and fourth-order Runge-Kutta, and compare maximum step sizes and computation times that give 0.1% accuracy.

5.2. Simulate the ideal binary distillation column of Sec. 5.4, using Euler and fourth-order Runge-Kutta and compare computation times.

5.3. The initial startup of an adiabatic, gas-phase packed tubular reactor makes a good example of how a distributed system can be lumped into a series of CSTRs in order to study the dynamic response. The reactor is a cylindrical vessel (3 feet ID by 20 feet long) packed with a metal packing. The packing occupies 5 percent of the total volume, provides 50 ft^2 of area per ft^3 of total volume, weighs 400 lb$_m$/ft^3 and has a heat capacity of 0.1 Btu/lb$_m$ °F. The heat transfer coefficient between the packing and the gas is 10 Btu/h ft^2 °F.

The reaction occurring is first order:

$$A \xrightarrow{\ k\ } B$$

A dilute mixture of reactant A in product B is fed into the reactor at y_0 mole fraction A and temperature $T_0 = 500$°F. The heat of reaction is $-30,000$ Btu/lb·mol A. The specific reaction rate is given by

$$k = 4 \times 10^2\, e^{-15,000/RT}$$

Assume perfect gases with molecular weights of 40 and heat capacities equal to 0.15 Btu/lb$_m$ °F.

The pressure at the inlet of the reactor is 100 psia. The pressure drop over the reactor is 5 psi at the design superficial velocity is 1 ft/s at inlet conditions.

Assume that this distributed system can be adequately modeled by a five-lump model of equal lengths. Inside each lump the gas temperature and the composition vary with time, as does the packing temperature.

The packing and gas in each section are initially at 500°F with no reactant in the system. At time zero, y_0 is raised to 0.10 mole fraction A. Simulate the system on a digital computer and find the dynamic changes in temperatures and concentrations in all the sections.

5.4 A 6-inch ID pipe, 300 feet long, connects two process units. The liquid flows through the pipe in essentially plug-flow conditions, so the pipe acts as a pure deadtime. This deadtime varies with the flow rate through the pipe. From time equals zero, the flow rate is 1000 gpm for 2 minutes. Then it drops to 500 gpm and holds constant for 3 minutes. Then it jumps to 2000 gpm for 2 minutes and finally returns to 1000 gpm. Liquid density is 50 lb$_m$/ft^3.

While these flow rate changes are occurring, the temperature of the fluid entering the pipe varies sinusoidally:

$$T_{in(t)} = 100 + 10 \sin(\omega t)$$

where T_{in} = inlet temperature, °F
ω = 30 radians per minute

Write a digital computer program that gives the dynamic changes in the temperature of the liquid leaving the pipe, $T_{out(t)}$, for this variable deadtime process.

Hint: The easiest way to handle deadtime in a digital simulation is to set up an array for the variable to be delayed. At each point in time you use the variable at the bottom of the array as the delayed variable. Then each value is moved down one position in the array and the current undelayed value is stuffed into the top of the array. For fixed step sizes and fixed deadtimes, this is easy to program. For variable step sizes and variable deadtimes, the programming is more complex.

PART
III

TIME-DOMAIN DYNAMICS AND CONTROL

In this section we will study the time-dependent behavior of some chemical engineering systems, both openloop (without control) and closedloop (with controllers included). Systems will be described by differential equations, and solutions will be in terms of time-dependent functions. Thus, our language for this part of the book will be "English." In the next part we will learn a little "Russian" in order to work in the Laplace domain where the notation is more simple than in "English." Then in Part V we will study some "Chinese" because of its ability to easily handle much more complex systems.

In the computer simulation studies of the two preceding chapters, the systems and their describing equations could be quite complex and nonlinear. In the remaining parts of this book only systems described by *linear* ordinary differential equations will be considered (linearity is defined in Chap. 6). The reason we are limited to linear systems is that practically all the analytical mathematical techniques currently available are applicable only to linear equations.

Since most chemical engineering systems are nonlinear, it might appear to be a waste of time to study methods that are limited to linear systems. However, linear techniques are of great practical importance, particularly for continuous processes, because the nonlinear equations describing most systems can be linearized around some steadystate operating condition. The resulting linear equations adequately describe the dynamic response of the system in some region around the steadystate conditions. The size of the region over which the linear model is valid will vary with the degree of nonlinearity of the process and the magnitude of the disturbances. In many processes the linear model can be successfully used to study dynamics and, more importantly, to design controllers.

Complex systems can usually be broken down into a number of simple elements. We must understand the dynamics of these simple systems before we tackle the more complex. We will start out looking at some simple uncontrolled processes in Chap. 6. We will examine the openloop dynamics or the response of the system to a disturbance, starting from an initial condition and with no feedback controllers.

Then in Chaps. 7 and 8 we will look at closedloop systems. Instrumentation hardware, controller types and performance, controller tuning, and various types of control systems structures will be discussed.

CHAPTER
6

TIME-DOMAIN DYNAMICS

Studying the dynamics of systems in the time domain involves direct solutions of differential equations. The computer simulation techniques of Part II are very general in the sense that they can give solutions to very complex nonlinear problems. However, they are also very specific in the sense that they provide a solution to only the particular numerical case fed into the computer.

The classical analytical techniques discussed in this chapter are limited to linear ordinary differential equations. But they yield general analytical solutions that apply for any values of parameters, initial conditions, and forcing functions.

We will start by briefly classifying and defining types of systems and types of disturbances. Then we will learn how to linearize nonlinear equations. It is assumed that you have had a course in differential equations, but we will review some of the most useful solution techniques for simple ordinary differential equations. Finally we will show how useful dynamic insights can sometimes be obtained from steadystate equations alone.

6.1 CLASSIFICATION AND DEFINITION

Processes and their dynamics can be classified in several ways:

1. Number of independent variables

   (a) *Lumped* if time is the only independent variable; described by ordinary differential equations.

(b) *Distributed* if time and spatial independent variables are required; described by partial differential equations.

2. Linearity

(a) *Linear* if all functions in the equations are linear functions (see Sec. 6.2)
(b) *Nonlinear* if not linear.

3. Stability

(a) *Stable* if "self-regulatory" so that variables converge to some steadystate when disturbed.
(b) *Unstable* if variables go to infinity (mathematically).

Most processes are *openloop* stable. However, the exothermic irreversible chemical reactor is a notable example of a process that can be openloop unstable. *All* real processes can be made *closedloop* unstable (unstable with a feedback controller in service) and therefore one of the principal objectives in feedback controller design is to avoid closedloop instability.

4. Order

If a system is described by one ordinary differential equation with derivatives of order N, the system is called Nth-order.

$$a_N \frac{d^N x}{dt^N} + a_{N-1} \frac{d^{N-1} x}{dt^{N-1}} + \cdots + a_1 \frac{dx}{dt} + a_0 x = f_{(t)} \qquad (6.1)$$

where a_i are constants and $f_{(t)}$ is the forcing function or disturbance. Two very important special cases are for $N = 1$ and $N = 2$.

First-order:

$$a_1 \frac{dx}{dt} + a_0 x = f_{(t)} \qquad (6.2)$$

Second-order:

$$a_2 \frac{d^2 x}{dt^2} + a_1 \frac{dx}{dt} + a_0 x = f_{(t)} \qquad (6.3)$$

The "standard" forms that we will usually employ for the above are

First-order:

$$\tau \frac{dx}{dt} + x = f_{(t)} \qquad (6.4)$$

Second-order:

$$\tau^2 \frac{d^2 x}{dt^2} + 2\tau\zeta \frac{dx}{dt} + x = f_{(t)} \qquad (6.5)$$

where τ = process time constant (either openloop or closedloop)
ζ = damping coefficient (either openloop or closedloop)

One of the most important parameters that we will use in the remaining sections of this book is the damping coefficient of the closedloop system. We typically tune a controller to give a closedloop system that has a damping coefficient of about 0.3.

Disturbances can also be classified and defined in several ways.

1. Shape (see Fig. 6.1)

(a) *Step.* Step disturbances are functions that change instantaneously from one level to another and are thereafter constant. If the size of the step is equal to unity, the disturbance is called the *unit step function* $u_{(t)}$ defined as

$$
\begin{aligned}
u_{(t)} &= 1 \qquad \text{for } t > 0 \\
u_{(t)} &= 0 \qquad \text{for } t \leq 0
\end{aligned}
\tag{6.6}
$$

The response of a system to a step disturbance is called the *step response* or the *transient response.*

(b) *Pulse.* A pulse is a function of arbitrary shape (but usually rectangular or triangular) that begins and ends at the same level. A rectangular pulse is simply the sum of one positive step function made at time zero and one negative step function made D minutes later. D is the delay time or deadtime.

$$
\text{Rectangular pulse of height 1 and width } D = u_{(t)} - u_{(t-D)} \tag{6.7}
$$

(c) *Impulse.* The impulse is defined as the Dirac delta function, an infinitely high pulse whose width is zero and whose area is unity. This kind of disturbance is, of course, a pure mathematical fiction, but we will find it a useful tool.

(d) *Ramp.* Ramp inputs are functions that change linearly with time.

$$
\text{Ramp function} = Kt \tag{6.8}
$$

where K is a constant. The classical example is the change in the setpoint to an antiaircraft gun as the airplane sweeps across the sky. Chemical engineering examples include batch reactor temperature or pressure setpoint changes with time.

(e) *Sinusoidal.* Pure periodic sine and cosine inputs seldom occur in real chemical engineering systems. However, the response of systems to this kind of forcing function (called the *frequency response* of the system) is of great practical importance, as we will show in our "Chinese" lessons in Part V and in multivariable control applications in Part VI.

2. Location of disturbance in feedback loop

Let us now consider a process with a feedback controller in service. This closedloop system can experience disturbances at two different spots in the feedback loop: load disturbances and setpoint disturbances.

1. Step

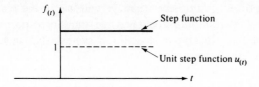

2. Pulses

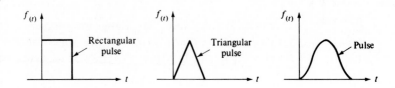

3. Impulse (Dirac delta function $\delta_{(t)}$)

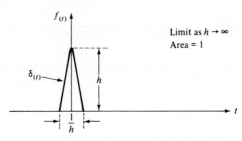

4. Ramp

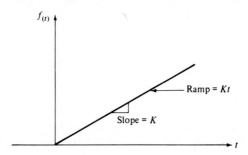

5. Sine wave

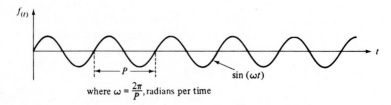

FIGURE 6.1
Disturbance shapes.

Most disturbances in chemical engineering systems are load disturbances, such as changes in throughput, feed composition, supply steam pressure, cooling water temperature, etc. The feedback controller's function when a load disturbance occurs is to return the controlled variable to its setpoint by suitable changes in the manipulated variable. The closedloop response to a load disturbance is called the *regulator response* or the *closedloop load response*.

Setpoint changes can also be made, particularly in batch processes or in changing from one operating condition to another in a continuous process. These setpoint changes also act as disturbances to the closedloop system. The function of the feedback controller is to drive the controlled variable to match the new setpoint. The closedloop response to a setpoint disturbance is called the *servo response* (from the early applications of feedback control in mechanical servomechanism tracking systems).

6.2 LINEARIZATION AND PERTURBATION VARIABLES

6.2.1 Linearization

As mentioned earlier, we must convert the rigorous nonlinear differential equations describing the system into linear differential equations if we are to be able to use the powerful linear mathematical techniques.

The first question to be answered is just what is a linear differential equation. Basically it is one that contains variables only to the first power in any one term of the equation. If square roots, squares, exponentials, products of variables, etc., appear in the equation, it is nonlinear:

Linear example

$$a_1 \frac{dx}{dt} + a_0 x = f_{(t)} \tag{6.9}$$

where a_0 and a_1 are constants or functions of time only, not of dependent variables or their derivatives.

Nonlinear examples

$$a_1 \frac{dx}{dt} + a_0 x^{0.5} = f_{(t)} \tag{6.10}$$

$$a_1 \frac{dx}{dt} + a_0(x)^2 = f_{(t)} \tag{6.11}$$

$$a_1 \frac{dx}{dt} + a_0 e^x = f_{(t)} \tag{6.12}$$

$$a_1 \frac{dx_1}{dt} + a_0 x_{1(t)} x_{2(t)} = f_{(t)} \tag{6.13}$$

where x_1 and x_2 are both dependent variables.

Mathematically, a linear differential equation is one for which the following two properties hold:

1. *If $x_{(t)}$ is a solution, then $cx_{(t)}$ is also a solution, where c is a constant.*
2. *If x_1 is a solution and x_2 is also a solution, then $x_1 + x_2$ is a solution.*

Linearization is quite straightforward. All we do is take the nonlinear functions, expand them in Taylor series expansions around the steadystate operating level, and neglect all terms after the first partial derivatives.

Let us assume we have a nonlinear function f of the process variables x_1 and x_2: $f_{(x_1, x_2)}$. For example, x_1 could be mole fraction or temperature or flow rate. We will denote the steadystate values of these variables by using an overscore:

$$\bar{x}_1 \equiv \text{steadystate value of } x_1$$

$$\bar{x}_2 \equiv \text{steadystate value of } x_2$$

Now we expand the function $f_{(x_1, x_2)}$ around its steadystate value $f_{(\bar{x}_1, \bar{x}_2)}$.

$$f_{(x_1, x_2)} = f_{(\bar{x}_1, \bar{x}_2)} + \left(\frac{\partial f}{\partial x_1}\right)_{(\bar{x}_1, \bar{x}_2)} (x_1 - \bar{x}_1)$$
$$+ \left(\frac{\partial f}{\partial x_2}\right)_{(\bar{x}_1, \bar{x}_2)} (x_2 - \bar{x}_2) + \left(\frac{\partial^2 f}{\partial x_1^2}\right)_{(\bar{x}_1, \bar{x}_2)} \frac{(x_1 - \bar{x}_1)^2}{2!} + \cdots \quad (6.14)$$

Linearization consists of truncating the series after the first partial derivatives.

$$f_{(x_1, x_2)} \simeq f_{(\bar{x}_1, \bar{x}_2)} + \left(\frac{\partial f}{\partial x_1}\right)_{(\bar{x}_1, \bar{x}_2)} (x_1 - \bar{x}_1) + \left(\frac{\partial f}{\partial x_2}\right)_{(\bar{x}_1, \bar{x}_2)} (x_2 - \bar{x}_2) \quad (6.15)$$

We are approximating the real function by a linear function. The process is sketched graphically in Fig. 6.2 for a function of a single variable. The method is best illustrated in some common examples.

Example 6.1. Consider the square-root dependence of flow out of a tank on the liquid height in the tank.

$$F_{(h)} = K\sqrt{h} \quad (6.16)$$

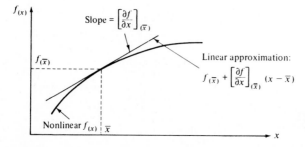

FIGURE 6.2
Linearization.

The Taylor series expansion around the steadystate value of h, which is $\bar{h}$ in our nomenclature, is

$$F_{(h)} = F_{(\bar{h})} + \left(\frac{\partial F}{\partial h}\right)_{(\bar{h})} (h - \bar{h}) + \left(\frac{\partial^2 F}{\partial h^2}\right)_{(\bar{h})} \frac{(h - \bar{h})^2}{2!} + \cdots$$

$$\simeq F_{(\bar{h})} + (\tfrac{1}{2} K h^{-1/2})_{(\bar{h})} (h - \bar{h})$$

$$F_{(h)} = K\sqrt{\bar{h}} + \frac{K}{2\sqrt{\bar{h}}} (h - \bar{h}) \tag{6.17}$$

Example 6.2. The Arrhenius temperature dependence of the specific reaction rate k is a highly nonlinear function that is linearized as follows:

$$k_{(T)} = \alpha e^{-E/RT}$$

$$k_{(T)} \simeq k_{(\bar{T})} + \left(\frac{\partial k}{\partial T}\right)_{(\bar{T})} (T - \bar{T}) \tag{6.18}$$

$$k_{(T)} = \bar{k} + \frac{E\bar{k}}{R\bar{T}^2} (T - \bar{T}) \tag{6.19}$$

where $\bar{k} \equiv k_{(\bar{T})}$

Example 6.3. The product of two dependent variables is a nonlinear function of the two variables:

$$f_{(C_A, F)} = C_A F \tag{6.20}$$

Linearizing:

$$f_{(C_A, F)} \simeq f_{(\bar{C}_A, \bar{F})} + \left(\frac{\partial f}{\partial C_A}\right)_{(\bar{C}_A, \bar{F})} (C_A - \bar{C}_A) + \left(\frac{\partial f}{\partial F}\right)_{(\bar{C}_A, \bar{F})} (F - \bar{F}) \tag{6.21}$$

$$C_{A(t)} F_{(t)} \simeq \bar{C}_A \bar{F} + \bar{F}(C_{A(t)} - \bar{C}_A) + \bar{C}_A(F_{(t)} - \bar{F}) \tag{6.22}$$

Notice that the linearization process converts the nonlinear function (the product of two dependent variables) into a linear function containing two terms.

Example 6.4. Consider the nonlinear ordinary differential equation for the gravity-flow tank of Example 2.9.

$$\frac{dv}{dt} = \left(\frac{g}{L}\right) h - \left(\frac{K_F g_c}{\rho A_p}\right) v^2 \tag{6.23}$$

Linearizing the v^2 term gives

$$v^2 = \bar{v}^2 + (2\bar{v})(v - \bar{v}) \tag{6.24}$$

Thus Eq. (6.23) becomes

$$\frac{dv}{dt} = \left(\frac{g}{L}\right) h - \left(\frac{2\bar{v} K_F g_c}{\rho A_p}\right) v + \left(\frac{\bar{v}^2 K_F g_c}{\rho A_p}\right) \tag{6.25}$$

This ODE is now linear. The terms in the parentheses are constants, which depend, of course, on the steadystate around which the system is linearized.

Example 6.5. The component continuity equation for an irreversible nth-order, nonisothermal reaction occurring in a constant-volume, variable-throughput CSTR is

$$V \frac{dC_A}{dt} = F_0 C_{A0} - F C_A - V(C_A)^n \alpha e^{-E/RT} \tag{6.26}$$

Linearization gives

$$V \frac{dC_A}{dt} = [\bar{F}_0 \bar{C}_{A0} + \bar{F}_0(C_{A0} - \bar{C}_{A0}) + \bar{C}_{A0}(F_0 - \bar{F}_0)]$$

$$- [\bar{F}\bar{C}_A + \bar{F}(C_A - \bar{C}_A) + \bar{C}_A(F - \bar{F})]$$

$$- V\left[\bar{k}\bar{C}_A^n + n\bar{k}\bar{C}_A^{n-1}(C_A - \bar{C}_A) + \frac{\bar{k}\bar{C}_A^n E}{R\bar{T}^2}(T - \bar{T})\right] \tag{6.27}$$

So far we have looked at examples where all the nonlinearity is in the derivative terms, i.e., the right-hand sides of the ODE. Quite often the model of a system will give an ODE which contains nonlinear terms inside the time derivative itself. For example, suppose the model of a nonlinear system is

$$\frac{d(h^3)}{dt} = K\sqrt{h} \tag{6.28}$$

The correct procedure for linearizing this type of equation is to rearrange it so that all the nonlinear functions appear only on the right-hand side of the ODE and then linearize in the normal way. For the example given in Eq. (6.28), we differentiate the h^3 term to get

$$3h^2 \frac{dh}{dt} = K\sqrt{h} \tag{6.29}$$

Then rearrangement gives

$$\frac{dh}{dt} = \frac{K}{3}(h)^{-1.5} \tag{6.30}$$

Now we are ready to linearize.

$$\frac{dh}{dt} = \frac{K}{3}(\bar{h})^{-1.5} + \frac{\partial}{\partial h}\left(\frac{K}{3}(h)^{-1.5}\right)_{(\bar{h})}(h - \bar{h}) \tag{6.31}$$

$$\frac{dh}{dt} = \frac{K}{3}(\bar{h})^{-1.5} + \left(\frac{-1.5K}{3}(\bar{h})^{-2.5}\right)(h - \bar{h}) \tag{6.32}$$

$$\frac{dh}{dt} = \frac{5K}{6}(\bar{h})^{-1.5} + \left(\frac{-K}{2}(\bar{h})^{-2.5}\right)h \tag{6.33}$$

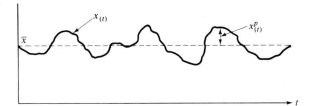

FIGURE 6.3
Perturbation variables.

This is a linear ODE with constant coefficients:

$$\frac{dh}{dt} = a_0 + a_1 h \tag{6.34}$$

6.2.2 Perturbation Variables

We will find it very useful in practically all the linear dynamics and control studies in the rest of the book to look at the changes of variables away from steadystate values instead of the absolute variables themselves. Why this is useful will become apparent in the discussion below.

Since the total variables are functions of time, $x_{(t)}$, their departures from the steadystate values $\bar{x}$ will also be functions of time, as sketched in Fig. 6.3. These departures from steadystate are called *perturbations* or *perturbation variables*. We will use, for the present, the symbol $x^p_{(t)}$. Thus the perturbation in x is defined:

$$x^p_{(t)} \equiv x_{(t)} - \bar{x} \tag{6.35}$$

The equations describing the linear system can now be expressed in terms of these perturbation variables. When this is done, two very useful results occur:

1. The terms in the ordinary differential equation with just constants in them drop out.
2. The initial conditions for the perturbation variables are all equal to zero if the starting point is the steadystate operating condition around which the equations have been linearized.

Both of the above greatly simplify the linearized equations. For example, if the perturbations in velocity and liquid height are used in Eq. (6.25), we get

$$\frac{d(\bar{v} + v^p_{(t)})}{dt} = \left(\frac{g}{L}\right)(\bar{h} + h^p_{(t)}) - \left(\frac{2\bar{v}K_F g_c}{\rho A_p}\right)(\bar{v} + v^p_{(t)}) + \left(\frac{\bar{v}^2 K_F g_c}{\rho A_p}\right) \tag{6.36}$$

Since $\bar{v}$ is a constant

$$\frac{dv^p_{(t)}}{dt} = \left(\frac{g}{L}\right)h^p_{(t)} - \left(\frac{2\bar{v}K_F g_c}{\rho A_p}\right)v^p_{(t)} + \left(\frac{g\bar{h}}{L} - \frac{\bar{v}^2 K_F g_c}{\rho A_p}\right) \tag{6.37}$$

Now consider Eq. (6.23) under steadystate conditions. At steadystate v will be equal to $\bar{v}$, a constant, and h will be equal to $\bar{h}$, another constant.

$$\frac{d\bar{v}}{dt} = 0 = \left(\frac{g}{L}\right)\bar{h} - \left(\frac{K_F g_c}{\rho A_p}\right)\bar{v}^2 \tag{6.38}$$

Therefore the last term in Eq. (6.37) is equal to zero. We end up with a linear ordinary differential equation with constant coefficients in terms of perturbation variables.

$$\frac{dv^p_{(t)}}{dt} = \left(\frac{g}{L}\right)h^p_{(t)} - \left(\frac{2\bar{v}K_F g_c}{\rho A_p}\right)v^p_{(t)} \tag{6.39}$$

In a similar way Eq. (6.27) can be written in terms of perturbations in C_A, C_{A0}, F_0, F, and T.

$$V\frac{d(\bar{C}_A + C^p_A)}{dt} = (\bar{F}_0)C^p_{A0} + (\bar{C}_{A0})F^p_0 - (\bar{F})C^p_A - (\bar{C}_A)F^p$$

$$- (Vnk\bar{C}_A^{n-1})C^p_A + \left(\frac{Vk\bar{C}_A^n E}{R\bar{T}^2}\right)T^p$$

$$+ [\bar{F}_0\bar{C}_{A0} - \bar{F}\bar{C}_A - Vk\bar{C}_A^n] \tag{6.40}$$

Application of Eq. (6.26) under steadystate conditions shows that the last term in Eq. (6.40) is just equal to zero. So we end up with a simple linear ODE in terms of perturbation variables.

$$V\frac{dC^p_A}{dt} = (\bar{F}_0)C^p_{A0} + (\bar{C}_{A0})F^p_0 - (\bar{F})C^p_A - (\bar{C}_A)F^p$$

$$- (Vnk\bar{C}_A^{n-1})C^p_A + \left(\frac{Vk\bar{C}_A^n E}{R\bar{T}^2}\right)T^p \tag{6.41}$$

Since we will be using perturbation variables most of the time, we will often not bother to use the superscript p. It will be understood that whenever we write the linearized equations for the system all variables will be perturbation variables. Thus Eqs. (6.39) and (6.41) can be written

$$\frac{dv}{dt} = \left(\frac{g}{L}\right)h - \left(\frac{2\bar{v}K_F g_c}{\rho A_p}\right)v \tag{6.42}$$

$$V\frac{dC_A}{dt} = (\bar{F}_0)C_{A0} + (\bar{C}_{A0})F_0 - (\bar{F})C_A - (\bar{C}_A)F$$

$$- (Vnk\bar{C}_A^{n-1})C_A + \left(\frac{Vk\bar{C}_A^n E}{R\bar{T}^2}\right)T \tag{6.43}$$

Note that the initial conditions of all these perturbation variables are zero since all variables start at the initial steadystate values. This will prove to simplify things significantly when we start using Laplace transforms in Part IV.

6.3 RESPONSES OF SIMPLE LINEAR SYSTEMS

6.3.1 First-Order Linear Ordinary Differential Equation

Consider the general first-order linear ODE

$$\frac{dx}{dt} + P_{(t)} x = Q_{(t)} \tag{6.44}$$

with a given value of x known at a fixed point in time: $x_{(t_0)} = x_0$. Usually this is an initial condition where $t_0 = 0$.

Multiply both sides of Eq. (6.44) by the integrating factor $\exp(\int P\, dt)$.

$$\frac{dx}{dt} \exp\left(\int P\, dt\right) + P_{(t)} x \exp\left(\int P\, dt\right) = Q_{(t)} \exp\left(\int P\, dt\right)$$

Combining the two terms on the left-hand side of the equation above gives

$$\frac{d}{dt}\left[x \exp\left(\int P\, dt\right) \right] = Q_{(t)} \exp\left(\int P\, dt\right)$$

Integrating yields

$$x \exp\left(\int P\, dt\right) = \int\left[Q_{(t)} \exp\left(\int P\, dt\right) \right] dt + c_1$$

where c_1 is a constant of integration and can be evaluated by using the boundary or initial condition. Therefore the general solution of Eq. (6.44) is

$$x = \exp\left(-\int P\, dt\right)\left\{\int\left[Q_{(t)} \exp\left(\int P\, dt\right) \right] dt + c_1\right\} \tag{6.45}$$

Example 6.6. An isothermal, constant-holdup, constant-throughput CSTR with a first-order irreversible reaction is described by a component continuity equation that is a first-order linear ODE:

$$\frac{dC_A}{dt} + \left(\frac{F}{V} + k\right)C_A = \left(\frac{F}{V}\right)C_{A0} \tag{6.46}$$

Let the concentrations C_{A0} and C_A be total values, not perturbations, for the present. The reactant concentration in the tank is initially zero.

Initial condition

$$C_{A(0)} = 0$$

At time equal zero a step change in feed concentration is made from zero to a constant value $\bar{C}_{A0}$.

Forcing function

$$C_{A0(t)} = \bar{C}_{A0}.$$

Comparing Eqs. (6.44) and (6.46),

$$x = C_A \qquad P = \frac{F}{V} + k \qquad Q = \frac{F\bar{C}_{A0}}{V}$$

Therefore

$$\exp\left(\int P \, dt\right) = e^{[(F/V)+k]t}$$

$$\int\left[Q_{(t)} \exp\left(\int P \, dt\right)\right] dt = \int\left(\frac{F\bar{C}_{A0}}{V}\right)e^{[(F/V)+k]t} \, dt$$

$$= \left(\frac{F\bar{C}_{A0}}{V}\right)\left(\frac{1}{F/V + k}\right)e^{[(F/V)+k]t} + c_1$$

The solution to Eq. (6.46) is, according to Eq. (6.45),

$$C_{A(t)} = e^{-[(F/V)+k]t}\left\{\left(\frac{F\bar{C}_{A0}}{V}\right)\left(\frac{1}{F/V + k}\right)e^{[(F/V)+k]t} + c_1\right\}$$

$$= \frac{F\bar{C}_{A0}}{F + kV} + c_1 e^{-[(F/V)+k]t} \tag{6.47}$$

The initial condition is now used to find the value of c_1.

$$C_{A(0)} = 0 = \frac{F\bar{C}_{A0}}{F + kV} + c_1(1)$$

Therefore the time-dependent response of C_A to the step disturbance in feed concentration is

$$C_{A(t)} = \frac{\bar{C}_{A0}}{1 + k\tau}\left[1 - e^{-[(1/\tau)+k]t}\right] \tag{6.48}$$

where $\tau \equiv V/F$ and is the residence time of the vessel.

The response is sketched in Fig. 6.4 and is the classical first-order exponential rise to the new steadystate.

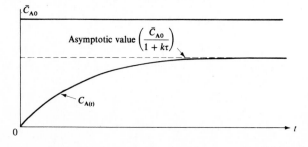

FIGURE 6.4
Step response of a first-order system.

The first thing you should always do when you get a solution is check to see if it is consistent with the initial conditions and if it is reasonable physically. At $t = 0$, Eq. (6.48) becomes

$$C_{A(t=0)} = \frac{\bar{C}_{A0}}{1 + k\tau} [1 - 1] = 0$$

so the initial condition is satisfied.

Does the solution make sense from a steadystate point of view? The new steadystate value of C_A that is approached asymptotically by the exponential function can be found from either the solution [Eq. (6.48)], letting time t go to infinity, or from the original ODE [Eq. (6.46)], setting the time derivative dC_A/dt equal to zero. Either method predicts that at the final steadystate

$$C_{A(t \to \infty)} \equiv \bar{C}_A = \frac{\bar{C}_{A0}}{1 + k\tau} \qquad (6.49)$$

Is this reasonable? It says that the consumption of reactant will be greater (the ratio of $\bar{C}_A$ to $\bar{C}_{A0}$ will be smaller) the bigger k and τ are. This certainly makes good chemical engineering sense. If k is zero (i.e., no reaction) the final steadystate value of $\bar{C}_A$ will be equal to the feed concentration $\bar{C}_{A0}$, as it should be. Note that $C_{A(t)}$ would not be dynamically equal to $\bar{C}_{A0}$; it would start at 0 and rise asymptotically up to its final steadystate value. Thus the predictions of the solution seem to check the real physical world.

The ratio of the change in the steadystate value of the output divided by the magnitude of the step change made in the input is called the *steadystate gain* of the process K_p.

$$K_p \equiv \frac{\bar{C}_A}{\bar{C}_{A0}} = \frac{1}{1 + k\tau} \qquad (6.50)$$

These steadystate gains will be extremely important in our dynamic studies and in controller design.

Does the solution make sense dynamically? The rate of rise will be determined by the magnitude of the $(k + 1/\tau)$ term in the exponential. The bigger this term, the faster the exponential term will decay to zero as time increases. The smaller this term, the slower the decay will be. Therefore the dynamics are set by $(k + 1/\tau)$.

The reciprocal of this term is called the *process openloop time constant* and we use the symbol τ_p. The bigger the time constant, the slower the dynamic response will be. The solution [Eq. (6.48)] predicts that a small value of k or a big value of τ will give a large process time constant. Again, this makes good physical sense. If there is no reaction, the time constant is just equal to $\tau = V/F$, the residence time.

Before we leave this example let us put Eq. (6.46) in the standard form

$$\tau_p \frac{dC_A}{dt} + C_A = K_p C_{A0} \qquad (6.51)$$

This will be the form in which we want to look at many systems of this type. Dividing by the term $(k + 1/\tau)$ does the trick.

$$\frac{1}{k + 1/\tau} \frac{dC_A}{dt} + C_A = \frac{1/\tau}{k + 1/\tau} C_{A0} = \frac{1}{k\tau + 1} C_{A0} \qquad (6.52)$$

$$\tau_p = \frac{1}{k + 1/\tau} = \text{process time constant with units of time}$$

$$K_p = \frac{1}{k\tau + 1} = \frac{\text{process steadystate gain with units of concentration in}}{\text{product stream divided by concentration in feed stream}}$$

Then the solution [Eq. (6.48)] becomes

$$C_{A(t)} = \bar{C}_{A0} K_p [1 - e^{-t/\tau_p}] \qquad (6.53)$$

In this example we have used total variables. If we convert Eq. (6.46) into perturbation variables, we get

$$\frac{d(\bar{C}_A + C_A^p)}{dt} + \left(\frac{F}{V} + k\right)(\bar{C}_A + C_A^p) = \left(\frac{F}{V}\right)(\bar{C}_{A0} + C_{A0}^p)$$

$$\frac{dC_A^p}{dt} + \left(\frac{F}{V} + k\right)C_A^p = \left(\frac{F}{V}\right)C_{A0}^p - \left\{\left(\frac{F}{V} + k\right)\bar{C}_A - \left(\frac{F}{V}\right)\bar{C}_{A0}\right\} \qquad (6.54)$$

The last term in the equation above is zero. Therefore Eqs. (6.54) and (6.46) are identical, except one is in terms of total variables and the other is in terms of perturbations. Whenever the original ODE is already linear, either total or perturbation variables can be used. Initial conditions will, of course, differ by the steadystate values of all variables.

Example 6.7. Suppose the feed concentration in the CSTR system considered above is ramped up with time:

$$C_{A0(t)} = Kt \qquad (6.55)$$

where K is a constant. C_A is initially zero.

Rearranging Eq. (6.51) gives

$$\frac{dC_A}{dt} + \frac{1}{\tau_p} C_A = \frac{K_p K}{\tau_p} t \qquad (6.56)$$

The solution, according to Eq. (6.45), is

$$C_{A(t)} = \exp\left(-\int \frac{1}{\tau_p} dt\right)\left\{\int\left[\frac{K_p K}{\tau_p} t \exp\left(\int \frac{1}{\tau_p} dt\right)\right] dt + c_1\right\}$$

$$= e^{-t/\tau_p}\left(\frac{K_p K}{\tau_p}\int t e^{t/\tau_p} dt + c_1\right) \qquad (6.57)$$

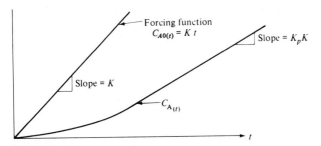

FIGURE 6.5
Ramp response of a first-order system.

The integral in Eq. (6.57) can be looked up in mathematics tables or can be found by integrating by parts.

Let

$$u = t \quad \text{and} \quad dv = e^{t/\tau_p} \, dt$$

Then

$$du = dt \quad \text{and} \quad v = \tau_p e^{t/\tau_p}$$

Since

$$\int u \, dv = uv - \int v \, du$$

$$\int t e^{t/\tau_p} \, dt = \tau_p t e^{t/\tau_p} - \int \tau_p e^{t/\tau_p} \, dt$$

$$= \tau_p t e^{t/\tau_p} - (\tau_p)^2 e^{t/\tau_p} \tag{6.58}$$

Therefore Eq. (6.57) becomes

$$C_{A(t)} = K_p K(t - \tau_p) + c_1 e^{-t/\tau_p} \tag{6.59}$$

Using the initial condition to find c_1,

$$C_{A(0)} = 0 = K_p K(-\tau_p) + c_1 \tag{6.60}$$

The final solution is

$$C_{A(t)} = K_p K \tau_p \left(\frac{t}{\tau_p} - 1 + e^{-t/\tau_p} \right) \tag{6.61}$$

The ramp response is sketched in Fig. 6.5.

It is frequently useful to be able to determine the time constant of a first-order system from experimental step response data. This is easy to do. When time is equal to τ_p in Eq. (6.53), the term $[1 - e^{-t/\tau_p}]$ becomes $[1 - e^{-1}] = 0.623$. This means that the output variable has changed 62.3 percent of the total change that it is going to make. Thus the time constant of a first-order system is simply the time it takes the step response to reach 62.3 percent of its new final steadystate value.

6.3.2   Second-Order Linear ODEs With Constant Coefficients

The first-order system considered in the previous section yields well-behaved exponential responses. Second-order systems can be much more exciting since they can give an oscillatory or *underdamped* response.

The first-order linear equation [Eq. (6.44)] could have a time-variable coefficient; that is, $P_{(t)}$ could be a function of time. We will consider only linear second-order ODEs that have constant coefficients (τ_p and ζ are constants).

$$\tau_p^2 \frac{d^2x}{dt^2} + 2\zeta\tau_p \frac{dx}{dt} + x = m_{(t)} \tag{6.62}$$

Analytical methods are available for linear ODEs with variable coefficients, but their solutions are usually messy infinite series. We will not consider them here.

The solution of a second-order ODE can be deduced from the solution of a first-order ODE. Equation (6.45) can be broken up into two parts:

$$x_{(t)} = \left\{ c_1 \exp\left(-\int P \, dt \right) \right\}$$

$$+ \left\{ \exp\left(-\int P \, dt \right) \int \left[Q_{(t)} \exp\left(\int P \, dt \right) \right] dt \right\} \equiv x_c + x_p \tag{6.63}$$

The variable x_c is called the *complementary solution*. It is the function that satisfies the original ODE with the forcing function $m_{(t)}$ set equal to zero (called the *homogeneous* differential equation):

$$\frac{dx}{dt} + P_{(t)} x = 0 \tag{6.64}$$

The variable x_p is called the *particular* solution. It is the function that satisfies the original ODE with a specified $m_{(t)}$. One of the most useful properties of linear ODEs is that the total solution is the sum of the complementary solution and the particular solution.

Now we are ready to extend the above ideas to the second-order ODE of Eq. (6.62). First we will obtain the complementary solution x_c by solving the homogeneous equation

$$\tau_p^2 \frac{d^2x}{dt^2} + 2\zeta\tau_p \frac{dx}{dt} + x = 0 \tag{6.65}$$

Then we will solve for the particular solution x_p and add the two to get the entire solution.

A. COMPLEMENTARY SOLUTION. Since the complementary solution of the first-order ODE is an exponential, it is reasonable to guess that the complementary solution of the second-order ODE will also be of exponential form. Let us guess that

$$x_c = ce^{st} \tag{6.66}$$

where c and s are constants. Differentiating x_c with respect to time twice gives

$$\frac{dx_c}{dt} = cse^{st} \quad \text{and} \quad \frac{d^2x_c}{dt^2} = cs^2e^{st}$$

Now we substitute the guessed solution and its derivatives into Eq. (6.65) to find the values of s that will make the assumed form [Eq. (6.66)] satisfy it.

$$\tau_p^2(cs^2e^{st}) + 2\zeta\tau_p(cse^{st}) + (ce^{st}) = 0$$

$$\boxed{\tau_p^2 s^2 + 2\zeta\tau_p s + 1 = 0} \tag{6.67}$$

The above equation is called the *characteristic equation* of the system. It is the system's most important dynamic feature. The values of s that satisfy Eq. (6.67) are called the *roots* of the characteristic equation (they are also called the *eigenvalues* of the system). Their values, as we will shortly show, will dictate if the system is fast or slow, stable or unstable, overdamped or underdamped. Dynamic analysis and controller design consists of finding out the values of the roots of the characteristic equation of the system and changing their values to give the desired response. The rest of this book is devoted to looking at roots of characteristic equations. They are an extremely important concept that you should fully understand.

Using the general solution for a quadratic equation, we can solve Eq. (6.67) for its two roots

$$s = \frac{-2\zeta\tau_p \pm \sqrt{(2\zeta\tau_p)^2 - 4\tau_p^2}}{2\tau_p^2} = -\frac{\zeta}{\tau_p} \pm \frac{\sqrt{\zeta^2 - 1}}{\tau_p} \tag{6.68}$$

There are two values of s that satisfy Eq. (6.67). Therefore there are two exponentials of the form given in Eq. (6.66) that are solutions of the original homogeneous ODE [Eq. (6.65)]. The sum of these solutions is also a solution since the ODE is linear. Therefore the complementary solution is (for $s_1 \neq s_2$)

$$x_c = c_1 e^{s_1 t} + c_2 e^{s_2 t} \tag{6.69}$$

where c_1 and c_2 are constants. The two roots s_1 and s_2 are

$$s_1 = -\frac{\zeta}{\tau_p} + \frac{\sqrt{\zeta^2 - 1}}{\tau_p} \tag{6.70}$$

$$s_2 = -\frac{\zeta}{\tau_p} - \frac{\sqrt{\zeta^2 - 1}}{\tau_p} \tag{6.71}$$

The shape of the solution curve depends strongly on the values of the physical parameter ζ, the damping coefficient. Let us now look at the various possibilities.

1. For $\zeta > 1$ (overdamped system). If the damping coefficient is greater than unity, the quantity inside the square root is positive. Then s_1 and s_2 will both be real numbers, and they will be different (called *distinct roots*).

Example 6.8. Consider the ODE

$$\frac{d^2x}{dt^2} + 5\frac{dx}{dt} + 6x = 0$$

$$\left(\frac{1}{\sqrt{6}}\right)^2 \frac{d^2x}{dt^2} + 2\left(\frac{1}{\sqrt{6}}\right)\left(\frac{5}{2\sqrt{6}}\right)\frac{dx}{dt} + x = 0 \tag{6.72}$$

Its characteristic equation can be written in several forms:

$$s^2 + 5s + 6 = 0 \tag{6.73}$$

$$(s + 3)(s + 2) = 0 \tag{6.74}$$

$$\left(\frac{1}{\sqrt{6}}\right)^2 s^2 + 2\left(\frac{1}{\sqrt{6}}\right)\left(\frac{5}{2\sqrt{6}}\right)s + 1 = 0 \tag{6.75}$$

All three are completely equivalent. The time constant and the damping coefficient for the system are

$$\tau_p = \frac{1}{\sqrt{6}} \qquad \zeta = \frac{5}{2\sqrt{6}}$$

The roots of the characteristic equation are obvious from Eq. (6.74), but the use of Eq. (6.68) gives

$$s = -\frac{\zeta}{\tau_p} \pm \frac{\sqrt{\zeta^2 - 1}}{\tau_p} = -\tfrac{5}{2} \pm \tfrac{1}{2}$$

$$s_1 = -2$$

$$s_2 = -3$$

The two roots are real. The complementary solution is

$$x_c = c_1 e^{-2t} + c_2 e^{-3t} \tag{6.76}$$

2. For $\zeta = 1$ (critically damped system). If the damping coefficient is equal to unity, the term inside the square root of Eq. (6.68) is zero. There is only one value of s that satisfies the characteristic equation.

$$s = -\frac{1}{\tau_p} \tag{6.77}$$

The two roots are the same and are called *repeated* roots. This is clearly seen if a value of $\zeta = 1$ is substituted into the characteristic equation [Eq. (6.67)]:

$$\tau_p^2 s^2 + 2\tau_p s + 1 = 0 = (\tau_p s + 1)(\tau_p s + 1) \tag{6.78}$$

The complementary solution with a repeated root is

$$x_c = (c_1 + c_2 t)e^{st} = (c_1 + c_2 t)e^{-t/\tau_p} \tag{6.79}$$

This is easily proved by substituting it into Eq. (6.65) with ζ set equal to unity.

Example 6.9. If two CSTRs like the one considered in Example 6.6 are run in series, two first-order ODEs describe the system:

$$\frac{dC_{A1}}{dt} + \left(\frac{1}{\tau_1} + k_1\right)C_{A1} = \left(\frac{1}{\tau_1}\right)C_{A0} \tag{6.80}$$

$$\frac{dC_{A2}}{dt} + \left(\frac{1}{\tau_2} + k_2\right)C_{A2} = \left(\frac{1}{\tau_2}\right)C_{A1} \tag{6.81}$$

Differentiating the second equation with respect to time and eliminating C_{A1} give a second-order ODE:

$$\frac{d^2C_{A2}}{dt^2} + \left(\frac{1}{\tau_1} + k_1 + \frac{1}{\tau_2} + k_2\right)\frac{dC_{A2}}{dt} + \left(\frac{1}{\tau_1} + k_1\right)\left(\frac{1}{\tau_2} + k_2\right)C_{A2} = \left(\frac{1}{\tau_1\tau_2}\right)C_{A0} \tag{6.82}$$

If temperatures and holdups are the same in both tanks, the specific reaction rates k and holdup times τ will be the same:

$$k_1 = k_2 \equiv k \qquad \tau_1 = \tau_2 \equiv \tau$$

The characteristic equation is

$$s^2 + 2\left(\frac{1}{\tau} + k\right)s + \left(\frac{1}{\tau} + k\right)^2 = 0$$

$$\left(s + \frac{1}{\tau} + k\right)\left(s + \frac{1}{\tau} + k\right) = 0 \tag{6.83}$$

The damping coefficient is unity and there is a real, repeated root:

$$s = -\left(\frac{1}{\tau} + k\right)$$

The complementary solution is

$$(C_{A2})_c = (c_1 + c_2 t)e^{-(k + 1/\tau)t} \tag{6.84}$$

3. For $\zeta < 1$ (underdamped system). Things begin to get interesting when the damping coefficient is less than unity. Now the term inside the square root in Eq.

(6.68) is negative, giving an imaginary number in the roots.

$$s = -\frac{\zeta}{\tau_p} \pm \frac{\sqrt{\zeta^2 - 1}}{\tau_p} = -\frac{\zeta}{\tau_p} \pm i \frac{\sqrt{1 - \zeta^2}}{\tau_p} \tag{6.85}$$

The roots are now complex numbers with real and imaginary parts.

$$s_1 = -\frac{\zeta}{\tau_p} + i \frac{\sqrt{1 - \zeta^2}}{\tau_p} \tag{6.86}$$

$$s_2 = -\frac{\zeta}{\tau_p} - i \frac{\sqrt{1 - \zeta^2}}{\tau_p} \tag{6.87}$$

To be more specific, they are *complex conjugates* since they have the same real parts and their imaginary parts differ only in sign. The complementary solution is

$$
\begin{aligned}
x_c &= c_1 e^{s_1 t} + c_2 e^{s_2 t} \\
&= c_1 \exp\left\{\left(-\frac{\zeta}{\tau_p} + i \frac{\sqrt{1 - \zeta^2}}{\tau_p}\right)t\right\} + c_2 \exp\left\{\left(-\frac{\zeta}{\tau_p} - i \frac{\sqrt{1 - \zeta^2}}{\tau_p}\right)t\right\} \\
&= e^{-\zeta t/\tau_p}\left\{c_1 \exp\left(+i \frac{\sqrt{1 - \zeta^2}}{\tau_p} t\right) + c_2 \exp\left(-i \frac{\sqrt{1 - \zeta^2}}{\tau_p} t\right)\right\}
\end{aligned} \tag{6.88}
$$

Now we use the relationships

$$e^{ix} = \cos x + i \sin x \tag{6.89}$$

$$\cos (-x) = \cos x \tag{6.90}$$

$$\sin (-x) = -\sin x \tag{6.91}$$

Substituting into Eq. (6.88) gives

$$
\begin{aligned}
x_c &= e^{-\zeta t/\tau_p}\left(c_1\left\{\cos\left(\frac{\sqrt{1 - \zeta^2}}{\tau_p} t\right) + i \sin\left(\frac{\sqrt{1 - \zeta^2}}{\tau_p} t\right)\right\}\right. \\
&\quad \left. + c_2\left\{\cos\left(\frac{\sqrt{1 - \zeta^2}}{\tau_p} t\right) - i \sin\left(\frac{\sqrt{1 - \zeta^2}}{\tau_p} t\right)\right\}\right) \\
&= e^{-\zeta t/\tau_p}\left\{(c_1 + c_2) \cos\left(\frac{\sqrt{1 - \zeta^2}}{\tau_p} t\right) + i(c_1 - c_2) \sin\left(\frac{\sqrt{1 - \zeta^2}}{\tau_p} t\right)\right\}
\end{aligned} \tag{6.92}
$$

The complementary solution consists of oscillating sinusoidal terms multiplied by an exponential. Thus the solution is oscillatory or underdamped for $\zeta < 1$. Note that as long as the damping coefficient is positive ($\zeta > 0$), the exponential term will decay to zero as time goes to infinity. Therefore the amplitude of the oscillations will decrease to zero. This is sketched in Fig. 6.6.

Since the solution x_c must be a real quantity, if we are describing a real physical system, the terms with the constants in Eq. (6.92) must all be real. So the term $c_1 + c_2$ and the term $i(c_1 - c_2)$ must both be real. This can be true only if c_1 and c_2 are complex conjugates, as proved below.

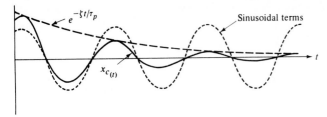

FIGURE 6.6
Complementary solution for $\zeta < 1$.

Let z be a complex number and $\bar{z}$ be its complex conjugate.

$$z = x + iy \qquad \text{and} \qquad \bar{z} = x - iy$$

Now look at the sum and the difference:

$$z + \bar{z} = (x + iy) + (x - iy) = 2x \qquad \text{a real number}$$

$$z - \bar{z} = (x + iy) - (x - iy) = 2yi \qquad \text{a pure imaginary number}$$

$$i(z - \bar{z}) = -2y \qquad \text{a real number}$$

So we have shown that to get real numbers for both $c_1 + c_2$ and $i(c_1 - c_2)$ the numbers c_1 and c_2 must be a complex conjugate pair. Let $c_1 = c^R + ic^I$ and $c_2 = c^R - ic^I$. Then the complementary solution becomes

$$x_{c(t)} = e^{-\zeta t/\tau_p}\left\{(2c^R)\cos\left(\frac{\sqrt{1-\zeta^2}}{\tau_p}t\right) - (2c^I)\sin\left(\frac{\sqrt{1-\zeta^2}}{\tau_p}t\right)\right\} \qquad (6.93)$$

Example 6.10. Consider the ODE

$$\frac{d^2x}{dt^2} + \frac{dx}{dt} + x = 0$$

Writing this in the standard form

$$(1)^2\frac{d^2x}{dt^2} + 2(1)(0.5)\frac{dx}{dt} + x = 0$$

we see that the time constant $\tau_p = 1$ and the damping coefficient $\zeta = 0.5$. The characteristic equation is

$$s^2 + s + 1 = 0$$

Its roots are

$$s = -\frac{\zeta}{\tau_p} \pm i\frac{\sqrt{1-\zeta^2}}{\tau_p}$$

$$= -\tfrac{1}{2} \pm i\sqrt{1 - (\tfrac{1}{2})^2} = -\tfrac{1}{2} \pm i\frac{\sqrt{3}}{2} \qquad (6.94)$$

The complementary solution is

$$x_{c(t)} = e^{-t/2}\left\{2c^R\cos\left(\frac{\sqrt{3}}{2}t\right) - 2c^I\cos\left(\frac{\sqrt{3}}{2}t\right)\right\} \qquad (6.95)$$

4. For $\zeta = 0$ (undamped system). The complementary solution is the same as Eq. (6.93) with the exponential term equal to unity. There is no decay of the sine and cosine terms and therefore the system will oscillate forever.

The result is obvious if we go back to Eq. (6.65) and set $\zeta = 0$.

$$\tau_p^2 \frac{d^2x}{dt^2} + x = 0 \tag{6.96}$$

You might remember from your physics that this is the differential equation that describes a harmonic oscillator. The solution is a sine wave with a frequency of $1/\tau_p$. We will discuss these kinds of functions in detail in Part V when we begin our "Chinese" lessons covering the frequency domain.

5. For $\zeta < 0$ (unstable system). If the damping coefficient is negative, the exponential term increases without bound as time becomes large. Thus the system is unstable.

This situation is extremely important. We have found the limit of stability of a second-order system. The roots of the characteristic equation are

$$s = -\frac{\zeta}{\tau_p} \pm i \frac{\sqrt{1 - \zeta^2}}{\tau_p}$$

If the real part of the root of the characteristic equation $(-\zeta/\tau_p)$ is a positive number, the system is unstable. So the stability requirement is:

> A system is stable if the real parts of all the roots of the characteristic equation are negative.

We will use this result extensively throughout the rest of the book. It is the foundation on which almost all controller designs are based.

B. PARTICULAR SOLUTION. Up to this point we have found only the complementary solution of the homogeneous equation

$$\tau_p^2 \frac{d^2x}{dt^2} + 2\zeta\tau_p \frac{dx}{dt} + x = 0$$

This corresponds to the solution for the unforced or undisturbed system. Now we must find the particular solutions for some specific forcing functions $m_{(t)}$. Then the total solution will be the sum of the complementary and particular solutions.

There are several methods for finding particular solutions. Laplace transform methods are probably the most convenient, and we will use them in Part IV. Here we will present the *method of undetermined coefficients*. It consists of assuming a particular solution that has the same form as the forcing function. It is illustrated in the examples below.

Example 6.11. The overdamped system of Example 6.8 is forced with a unit step function.

$$\frac{d^2x}{dt^2} + 5\frac{dx}{dt} + 6x = 1 \tag{6.97}$$

Initial conditions are

$$x_{(0)} = 0 \quad \text{and} \quad \left(\frac{dx}{dt}\right)_{(0)} = 0$$

The forcing function is a constant, so we assume that the particular solution is also a constant: $x_p = c_3$. Substituting into Eq. (6.97) gives

$$0 + 5(0) + 6c_3 = 1 \quad \Rightarrow \quad c_3 = \tfrac{1}{6} \tag{6.98}$$

Now the total solution is [using the complementary solution given in Eq. (6.76)]

$$x = x_c + x_p = c_1 e^{-2t} + c_2 e^{-3t} + \tfrac{1}{6} \tag{6.99}$$

The constants are evaluated from the initial conditions, using the total solution. A common mistake is to evaluate them using only the complementary solution.

$$x_{(0)} = 0 = c_1 + c_2 + \tfrac{1}{6}$$

$$\left(\frac{dx}{dt}\right)_{(0)} = 0 = (-2c_1 e^{-2t} - 3c_2 e^{-3t})_{(t=0)} = -2c_1 - 3c_2 = 0$$

Therefore

$$c_1 = -\tfrac{1}{2} \quad \text{and} \quad c_2 = \tfrac{1}{3}$$

The final total solution for the constant forcing function is

$$x_{(t)} = -\tfrac{1}{2} e^{-2t} + \tfrac{1}{3} e^{-3t} + \tfrac{1}{6} \tag{6.100}$$

Example 6.12. A general underdamped second-order system is forced by a unit step function:

$$\tau_p^2 \frac{d^2x}{dt^2} + 2\zeta\tau_p \frac{dx}{dt} + x = 1 \tag{6.101}$$

Initial conditions are

$$x_{(0)} = 0 \quad \text{and} \quad \left(\frac{dx}{dt}\right)_{(0)} = 0$$

Since the forcing function is a constant, the particular solution is assumed to be a constant, giving $x_p = 1$. The total solution is the sum of the particular and complementary solutions [see Eq. (6.93)].

$$x_{(t)} = 1 + e^{-\zeta t/\tau_p} \left\{ (2c^R) \cos\left(\frac{\sqrt{1-\zeta^2}}{\tau_p} t\right) - (2c^I) \sin\left(\frac{\sqrt{1-\zeta^2}}{\tau_p} t\right) \right\} \tag{6.102}$$

Using the initial conditions to evaluate constants,

$$x_{(0)} = 0 = 1 + [2c^R(1) - 2c^I(0)]$$

$$\frac{dx}{dt} = -\frac{\zeta}{\tau_p} e^{-\zeta t/\tau_p} \left\{ 2c^R \cos\left(\frac{\sqrt{1-\zeta^2}}{\tau_p} t\right) - 2c^I \sin\left(\frac{\sqrt{1-\zeta^2}}{\tau_p} t\right) \right\}$$

$$+ e^{-\zeta t/\tau_p} \left\{ -2c^R \frac{\sqrt{1-\zeta^2}}{\tau_p} \sin\left(\frac{\sqrt{1-\zeta^2}}{\tau_p} t\right) \right.$$

$$\left. - 2c^I \frac{\sqrt{1-\zeta^2}}{\tau_p} \cos\left(\frac{\sqrt{1-\zeta^2}}{\tau_p} t\right) \right\}$$

$$\left(\frac{dx}{dt}\right)_{(0)} = 0 = -\frac{\zeta}{\tau_p} (2c^R) + \left(-2c^I \frac{\sqrt{1-\zeta^2}}{\tau_p}\right)$$

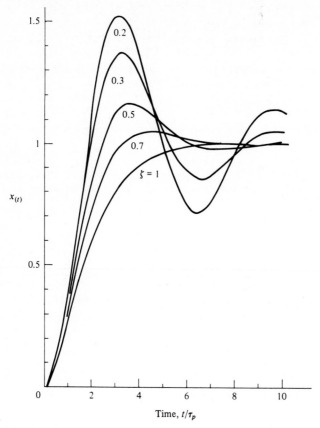

FIGURE 6.7
Step responses of a second-order underdamped system.

Solving for the constants gives

$$2c^R = -1 \quad \text{and} \quad 2c^I = \frac{\zeta}{\sqrt{1 - \zeta^2}}$$

The total solution is

$$x_{(t)} = 1 - e^{-\zeta t/\tau_p}\left\{\cos\left(\frac{\sqrt{1 - \zeta^2}}{\tau_p}t\right) + \frac{\zeta}{\sqrt{1 - \zeta^2}}\sin\left(\frac{\sqrt{1 - \zeta^2}}{\tau_p}t\right)\right\} \quad (6.103)$$

This step response is sketched in Fig. 6.7 for several values of the damping coefficient. Note that the amount of overshoot of the final steadystate value increases as the damping coefficient decreases. The system also becomes more oscillatory. In Chap. 7 we will tune feedback controllers so that we get a reasonable amount of overshoot by selecting a damping coefficient in the 0.3 to 0.5 range.

It is frequently useful to be able to calculate damping coefficients and time constants for second-order systems from experimental step response data.

Problem 6.11 gives some very useful relationships between these parameters (damping coefficient and time constant) and the shape of the response curve. There is a simple relationship between the "peak overshoot ratio" and the damping coefficient. Then the time constant can be calculated from the "rise time" and the damping coefficient. Refer to Prob. 6.11 for the definitions of these terms.

Example 6.13. The overdamped system of Example 6.8 is now forced with a ramp input:

$$\frac{d^2x}{dt^2} + 5\frac{dx}{dt} + 6x = t \tag{6.104}$$

Since the forcing function is the first term of a polynomial in t, we will assume that the particular solution is also a polynomial in t.

$$x_p = b_0 + b_1 t + b_2 t^2 + b_3 t^3 + \cdots \tag{6.105}$$

where the b_i are constants to be determined. Differentiating Eq. (6.105) twice gives

$$\frac{dx_p}{dt} = b_1 + 2b_2 t + 3b_3 t^2 + \cdots$$

$$\frac{d^2x_p}{dt^2} = 2b_2 + 6b_3 t + \cdots$$

Substituting into Eq. (6.104) gives

$$(2b_2 + 6b_3 t + \cdots) + 5(b_1 + 2b_2 t + 3b_3 t^2 + \cdots)$$
$$+ 6(b_0 + b_1 t + b_2 t^2 + b_3 t^3 + \cdots) = t$$

Now we rearrange the above to group together all terms with equal powers of t.

$$\cdots + t^3(6b_3 + \cdots) + t^2(6b_2 + 15b_3 + \cdots)$$
$$+ t(6b_3 + 10b_2 + 6b_1) + (2b_2 + 5b_1 + 6b_0) = t$$

Equating like powers of t on the left-hand and right-hand sides of this equation gives the simultaneous equations

$$6b_3 + \cdots = 0$$
$$6b_2 + 15b_3 + \cdots = 0$$
$$6b_3 + 10b_2 + 6b_1 = 1$$
$$2b_2 + 5b_1 + 6b_0 = 0$$

Solving simultaneously gives

$$b_0 = -\tfrac{5}{36} \qquad b_1 = \tfrac{1}{6} \qquad b_2 = b_3 = \cdots = 0$$

The particular solution is

$$x_p = -\tfrac{5}{36} + \tfrac{1}{6}t \tag{6.106}$$

The total solution is

$$x_{(t)} = -\tfrac{5}{36} + \tfrac{1}{6}t + c_1 e^{-2t} + c_2 e^{-3t} \tag{6.107}$$

If the initial conditions are

$$x_{(0)} = 0 \quad \text{and} \quad \left(\frac{dx}{dt}\right)_{(0)} = 0$$

the constants c_1 and c_2 can be evaluated:

$$x_{(0)} = 0 = -\tfrac{5}{36} + c_1 + c_2$$

$$\left(\frac{dx}{dt}\right)_{(0)} = 0 = \tfrac{1}{6} - 2c_1 - 3c_2$$

Solving simultaneously gives

$$c_1 = \tfrac{1}{4} \quad \text{and} \quad c_2 = -\tfrac{1}{9} \tag{6.108}$$

And the final solution is

$$x_{(t)} = -\tfrac{5}{36} + \tfrac{1}{6}t + \tfrac{1}{4}e^{-2t} - \tfrac{1}{9}e^{-3t} \tag{6.109}$$

6.3.3 *N*th-Order Linear ODEs With Constant Coefficients

The results obtained in the last two sections for simple first- and second-order systems can now be generalized to higher-order systems. Consider the *N*th-order ODE

$$a_N \frac{d^N x}{dt^N} + a_{N-1} \frac{d^{N-1}x}{dt^{N-1}} + \cdots + a_1 \frac{dx}{dt} + a_0 x = m_{(t)} \tag{6.110}$$

The solution of this equation is the sum of a particular solution x_p and a complementary solution x_c. The complementary solution is the sum of N exponential terms. The characteristic equation is an *N*th-order polynomial:

$$a_N s^N + a_{N-1} s^{N-1} + \cdots + a_1 s + a_0 = 0 \tag{6.111}$$

There are N roots s_k of the characteristic equation, some of which may be repeated (twice or more). Factoring Eq. (6.111) gives

$$(s - s_1)(s - s_2)(s - s_3) \cdots (s - s_{N-1})(s - s_N) = 0 \tag{6.112}$$

where the s_k are the roots (or *zeros*) of the polynomial. The complementary solution is (for all distinct roots, i.e., no repeated roots)

$$x_{c(t)} = c_1 e^{s_1 t} + c_2 e^{s_2 t} + \cdots + c_N e^{s_N t}$$

And therefore the total solution is

$$x_{(t)} = x_{p(t)} + \sum_{k=1}^{N} c_k e^{s_k t} \tag{6.113}$$

The roots of the characteristic equation can be real or complex. But if they are complex they must appear in complex conjugate pairs. The reason for this is illustrated for a second-order system with the characteristic equation

$$s^2 + a_1 s + a_0 = 0 \tag{6.114}$$

Let the two roots be s_1 and s_2.

$$(s - s_1)(s - s_2) = 0$$

$$s^2 + (-s_1 - s_2)s + s_1 s_2 = 0 \tag{6.115}$$

The coefficients a_0 and a_1 can then be expressed in terms of the roots.

$$a_0 = s_1 s_2 \quad \text{and} \quad a_1 = -(s_1 + s_2) \tag{6.116}$$

If Eq. (6.114) is the characteristic equation for a real physical system, the coefficients a_0 and a_1 must be real numbers. These are the coefficient that multiply the derivatives in the Nth-order differential equation. So they cannot be imaginary.

If the roots s_1 and s_2 are both real numbers, Eq. (6.116) shows that a_0 and a_1 are certainly both real. If the roots s_1 and s_2 are complex, the coefficients a_0 and a_1 must still be real and must also satisfy Eq. (6.116). Complex conjugates are the only complex numbers that give real numbers when they are multiplied together *and* when added together. To illustrate this, let z be a complex number: $z = x + iy$. Let $\bar{z}$ be the complex conjugate of z: $\bar{z} = x - iy$. Now

$$z\bar{z} = x^2 + y^2 \qquad \text{(a real number)}$$

$$z + \bar{z} = 2x \qquad \text{(a real number)}$$

Therefore the roots s_1 and s_2 must be a complex conjugate pair if they are complex. This is exactly what we found in Eq. (6.85) in the previous section.

For a third-order system with three roots s_1, s_2, and s_3, the roots could all be real: $s_1 = \alpha_1$, $s_2 = \alpha_2$, and $s_3 = \alpha_3$. Or there could be one real root and two complex conjugate roots:

$$s_1 = \alpha_1 \tag{6.117}$$

$$s_2 = \alpha_2 + i\omega_2 \tag{6.118}$$

$$s_3 = \alpha_2 - i\omega_2 \tag{6.119}$$

where α_k = real part of $s_k \equiv \text{Re}\,[s_k]$
ω_k = imaginary part of $s_k \equiv \text{Im}\,[s_k]$

These are the only two possibilities. We cannot have three complex roots.

The complementary solution would be either (for distinct roots)

$$x_c = c_1 e^{s_1 t} + c_2 e^{s_2 t} + c_3 e^{s_3 t} \tag{6.120}$$

or

$$x_c = c_1 e^{\alpha_1 t} + e^{\alpha_2 t}[(c_2 + c_3) \cos(\omega_2 t) + i(c_2 - c_3) \sin(\omega_2 t)] \tag{6.121}$$

where the constants c_2 and c_3 must also be complex conjugates in the latter equation, as discussed in the previous section.

If some of the roots are repeated (not distinct) the complementary solution will contain exponential terms that are multiplied by various powers of t. For example, if α_1 is a repeated root of order 2, the characteristic equation would be

$$(s - \alpha_1)^2(s - s_3)(s - s_4) \cdots (s - s_N) = 0$$

and the resulting complementary solution is

$$x_c = (c_1 + c_2 t)e^{\alpha_1 t} + \sum_{k=3}^{N} c_k e^{s_k t} \tag{6.122}$$

If α_1 is a repeated root of order 3, the characteristic equation would be

$$(s - \alpha_1)^3(s - s_4) \cdots (s - s_N) = 0$$

and the resulting complementary solution is

$$x_c = (c_1 + c_2 t + c_3 t^2)e^{\alpha_1 t} + \sum_{k=4}^{N} c_k e^{s_k t} \tag{6.123}$$

The stability of the system is dictated by the values of the real parts α_k of the roots. The system is stable if the real parts of *all* roots are negative, since the exponential terms go to zero as time goes to infinity. If the real part of *any* one of the roots is positive, the system is unstable.

The roots of the characteristic equation can be very conveniently plotted in a two-dimensional figure (Fig. 6.8) called the "s plane." The ordinate is the imaginary part ω of the root s, and the abscissa is the real part α of the root s. The roots of Eqs. (6.117) to (6.119) are shown in Fig. 6.8. We will use these s-plane plots extensively in Part IV.

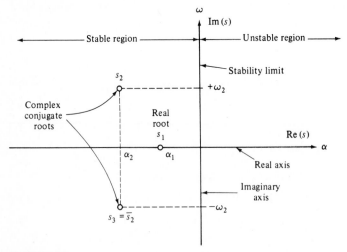

FIGURE 6.8
s-plane plot of the roots of the characteristic equation.

The stability criterion for an Nth-order system is:

> The system is stable if all the roots of its characteristic equation lie in the left half of the s plane.

6.4 STEADYSTATE TECHNIQUES

Sometimes useful information and insight can be obtained about the dynamics of a system from just the steadystate equations of the system. Van Heerden (*Ind. Eng. Chem.* Vol. 45, 1953, p. 1242) proposed the application of the following steadystate analysis to a continuous perfectly mixed chemical reactor. Consider a nonisothermal CSTR described by the two nonlinear ODEs

$$\frac{d(VC_A)}{dt} = F(C_{A0} - C_A) - VC_A \alpha e^{-E/RT} \tag{6.124}$$

$$\frac{d(VC_p \rho T)}{dt} = F\rho C_p(T_0 - T) - \lambda VC_A \alpha e^{-E/RT} - UA(T - T_J) \tag{6.125}$$

Inflow and outflow rates have been assumed equal in these equations.

Under steadystate conditions these two equations become

$$0 = \bar{F}(\bar{C}_{A0} - \bar{C}_A) - \bar{V}\bar{C}_A \alpha e^{-E/R\bar{T}} \tag{6.126}$$

$$0 = \bar{F}\rho C_p(\bar{T}_0 - \bar{T}) - \lambda \bar{V}\bar{C}_A \alpha e^{-E/R\bar{T}} + UA(\bar{T} - \bar{T}_J) \tag{6.127}$$

We want to find the steadystate value (or values) of temperature. Let us pick a trial value of steadystate temperature $\bar{T}'$. If it satisfies the two steadystate equations above, it is a bona fide steadystate temperature $\bar{T}$. Solving Eq. (6.126) for $\bar{C}_A$ gives

$$\bar{C}_A = \frac{\bar{F}\bar{C}_{A0}}{\bar{F} + \bar{V}\alpha e^{-E/R\bar{T}}} \tag{6.128}$$

If we assume a series of values of $\bar{T}'$, the above equation gives the corresponding values of $\bar{C}_{A(\bar{T}')}$. At low temperatures $\bar{C}_A$ will be essentially equal to $\bar{C}_{A0}$ since the reaction rate is very small. As temperature is increased, the reaction rate term becomes larger and larger. This causes $\bar{C}_A$ to approach zero. Most of the reactant (component A) is consumed in the reaction and there is little of it left in the reactor.

Now let us look at the second term on the right-hand side of Eq. (6.127). This is defined as the "heat generation" term Q_G.

$$Q_G = -\lambda \bar{V}\bar{C}_A \alpha e^{-E/R\bar{T}} \tag{6.129}$$

It is the rate at which the reaction is generating heat. Remember λ is negative if the reaction is exothermic. Figure 6.9a shows how Q_G varies with the assumed temperature $\bar{T}'$. Q_G is low at low temperatures because the reaction rate is low. Q_G begins to increase as temperature increases. But as the "$\alpha e^{-E/R\bar{T}}$" part of the

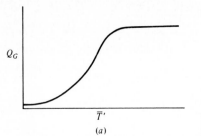

(a)

FIGURE 6.9a
Heat generation.

term continues to increase, the "C_A" part is decreasing toward zero. The result is that the Q_G term flattens out at a maximum value. The rate of generation cannot increase beyond the value that corresponds to complete conversion of all the reactant A in the feed stream. So the largest that Q_G can become is $-\lambda \bar{F} \bar{C}_{A0}$.

Now let us group the other terms in Eq. (6.127) together into a "heat removal" term (note the change in signs).

$$Q_R = \bar{F} \rho C_p (\bar{T} - \bar{T}_0) + UA(\bar{T} - \bar{T}_J) \tag{6.130}$$

The first term is a sensible-heat term. If the feed temperature $\bar{T}_0$ is lower than the reactor temperature $\bar{T}$, the feed stream tends to cool the reactor. The second term is the rate of heat removal by heat transfer to the cooling jacket.

Figure 6.9b shows a plot of Q_R versus the assumed temperature $\bar{T}'$. If U, A, $\bar{F}$, C_p, $\bar{T}_0$, ρ, and $\bar{T}_J$ are all constant, Q_R is a linear function of temperature:

$$Q_R = (UA + \bar{F} C_p \rho)\bar{T} - (UA\bar{T}_J + \bar{F} C_p \rho \bar{T}_0) \tag{6.131}$$

Thus the slope of the straight line is $UA + \bar{F} C_p \rho$.

With these definitions, Eq. (6.127) becomes

$$Q_G - Q_R = 0 \tag{6.132}$$

Therefore a temperature at which the curves of Q_G and Q_R intersect is a steady-state solution to Eqs. (6.126) and (6.127). Figure 6.9c shows both curves plotted together with one intersection at $\bar{T}$.

What can be concluded about the dynamics of the system from this steady-state plot? Imagine that a small disturbance causes the temperature to increase slightly above its steadystate value $\bar{T}$. At the higher temperature the heat-

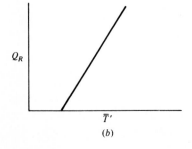

(b)

FIGURE 6.9b
Heat removal.

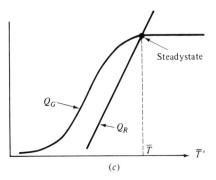

FIGURE 6.9c
Heat generation and heat removal.

removal rate is greater than the heat-generation rate, according to the curves in Fig. 6.9c. Therefore, the temperature will tend to decrease toward the steady-state $\bar{T}$.

If the disturbance in temperature had been in the other direction, resulting in a decrease in temperature, the heat-generation rate is now higher than the heat-removal rate. So the temperature would tend to be driven back up again. Thus the system is self-regulatory.

We can intuitively say that in order for a steadystate to be stable, the slope of the Q_R curve must be greater than the slope of the Q_G curve.

$$\left(\frac{dQ_R}{dT}\right)_{(\bar{T})} > \left(\frac{dQ_G}{dT}\right)_{(\bar{T})} \tag{6.133}$$

This is a necessary but not sufficient condition for stability, as we will see in more detail in Chap. 10.

Figure 6.9d shows the Q_G and Q_R for another reactor which has some very interesting features. Now there are *three* intersections of the curves. This means that there can be three different temperatures that are steadystates. For exactly the same feed conditions and parameter values, the reactor could settle out at three different temperatures: $\bar{T}_A$, $\bar{T}_B$, and $\bar{T}_C$. This phenomenon of *multiple steadystates* may be hard to believe, if you haven't thought about it before. But it is a real occurrence. It has been demonstrated experimentally in the laboratory.

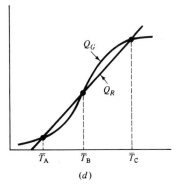

FIGURE 6.9d
Multiple steadystates.

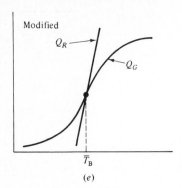

FIGURE 6.9e

Q_R curve modified to make $\bar{T}_B$ a stable steadystate.

In a linear system, the phenomenon of multiple steadystates cannot occur. It is the nonlinearity of the process, the exponential temperature dependence of the reaction rate, that can lead to more than one steadystate.

With multiple steadystates, the process outputs can be different with the same process inputs. The reverse of this can also occur. This interesting possibility, called *input multiplicity*, can occur in some nonlinear systems. In this situation we have the same process outputs, but with different process inputs. For example, we could have the same reactor temperature and concentration, but with different values of feed flow rate and cooling water flow rate.

The three steadystate temperatures shown in Fig. 6.9d correspond to three different steadystate compositions with low, medium, or high conversions (high, medium, or low concentrations of reactant).

The stability criterion [Eq. (6.133)] predicts that the steadystates at $\bar{T}_A$ and $\bar{T}_C$ would be stable but that the steadystate at $\bar{T}_B$ would be unstable. Any little temperature disturbance would cause the reaction to "quench" down to the low temperature $\bar{T}_A$ (with low conversion of reactant) or "run away" up to the high temperature $\bar{T}_C$ (with high conversion).

If we want to run the reactor at the steadystate temperature $\bar{T}_B$, the heat-removal curve must be modified by changing the parameters of the system (or by adding a feedback controller, as we will show in the next part of this book) to make the Q_R curve intersect the Q_G curve at $\bar{T}_B$ with a slope greater than $(dQ_G/dT)_{(\bar{T}_B)}$ as sketched in Fig. 6.9e.

PROBLEMS

6.1. Linearize the following nonlinear functions:

(a) $f_{(x)} = y_{(x)} = \dfrac{\alpha x}{1 + (\alpha - 1)x}$ where α is a constant

(b) $f_{(T)} = P^s_{(T)} = e^{A/T + B}$ where A and B are constants

(c) $f_{(v)} = U_{(v)} = K(v)^{0.8}$ where K is a constant

(d) $f_{(h)} = L_{(h)} = K(h)^{3/2}$ where K is a constant

6.2. Linearize the ODE describing the conical tank modeled in Prob. 3.1 and convert to perturbation variables.

6.3. Linearize the equations describing a variable-volume CSTR similar to the one considered in Sec. 3.3.

6.4. Solve the ODEs:

(a) $\dfrac{d^2x}{dt^2} + 5\dfrac{dx}{dt} + 4x = 2 \qquad x_{(0)} = 0, \left(\dfrac{dx}{dt}\right)_{(0)} = 1$

(b) $\dfrac{d^2x}{dt^2} + 2\dfrac{dx}{dt} + 2x = 1 \qquad x_{(0)} = 2, \left(\dfrac{dx}{dt}\right)_{(0)} = 0$

6.5. Show that the linearized system describing the gravity-flow tank of Example 6.4 is a second-order system. Solve for the damping coefficient and the time constant in terms of the parameters of the system.

6.6. Solve the second-order ODE describing the steadystate flow of an incompressible, newtonian liquid through a pipe:

$$\frac{d}{dr}\left(r\,\frac{d\bar{v}_z}{dr}\right) = \left(\frac{\Delta P\,g_c}{\mu L}\right)r$$

What are the boundary conditions?

6.7. Find the responses of general first- and second-order systems given below to the following forcing functions:

$$\tau_p\,\frac{dx}{dt} + x = m_{(t)}$$

$$\tau_p^2\,\frac{d^2x}{dt^2} + 2\tau_p\zeta\,\frac{dx}{dt} + x = m_{(t)} \qquad 0 < \zeta < 1$$

(a) $m_{(t)} = \delta_{(t)}$
(b) $m_{(t)} = \sin(\omega t)$

6.8. Solve for the unit step response of a general second-order system for:
(a) $\zeta = 1$
(b) $\zeta > 1$

6.9. A feedback controller is added to the CSTR of Example 6.6. The inlet concentration C_{A0} is now changed by the controller to hold C_A near its setpoint value C_A^{set}.

$$C_{A0} = C_{AM} + C_{AD}$$

where C_{AD} is a disturbance composition. The controller has proportional and integral action:

$$C_{AM} = \bar{C}_{AM} + K_c\left(E + \frac{1}{\tau_I}\int E\,dt\right)$$

where K_c and τ_I are constants
  $\bar{C}_{AM}$ = steadystate value of C_{AM}
  $E = C_A^{\text{set}} - C_A$

Derive the second-order equation describing the closedloop process in terms of perturbation variables. Show that the damping coefficient is

$$\zeta = \frac{1 + k\tau + K_c}{2\sqrt{K_c \tau/\tau_I}}$$

What value of K_c will give critical damping? At what value of K_c will the system become unstable?

6.10. Combine the three first-order ODEs describing the three-CSTR system of Sec. 3.2 into one third-order ODE in terms of C_{A3}. Then solve for the response of C_{A3} to a unit step change in C_{A0}, assuming all k's and τ's are identical.

6.11. Consider the second-order underdamped system

$$\tau_p^2 \frac{d^2x}{dt^2} + 2\tau_p \zeta \frac{dx}{dt} + x = K_p m_{(t)}$$

where K_p is the process steadystate gain and $m_{(t)}$ is the forcing function. The unit step response of such a system can be characterized by rise time t_R, peak time t_P, settling time t_S, and peak overshoot ratio POR. The values of t_R and t_P are defined in the sketch below. The value of t_S is the time it takes the exponential portion of the response to decay to a given fraction F of the final steadystate value of x, x_{SS}. The POR is defined:

$$\text{POR} \equiv \frac{x_{(t_p)} - x_{SS}}{x_{SS}}$$

Show that

(a) $\dfrac{x_{(t)}}{x_{SS}} = 1 - \dfrac{e^{-\zeta t/\tau_p}}{\sqrt{1 - \zeta^2}} \sin\left(\dfrac{\sqrt{1 - \zeta^2}}{\tau_p} t + \phi\right)$ where $\phi \equiv \cos^{-1} \zeta$

(b) $\dfrac{t_R}{\tau_p} = \dfrac{\pi - \phi}{\sin \phi}$

(c) $\dfrac{t_S}{\tau_p} = \dfrac{\ln \left[1/(F \sin \phi)\right]}{\cos \phi}$

(d) $\text{POR} = e^{-\pi \cos \phi}$

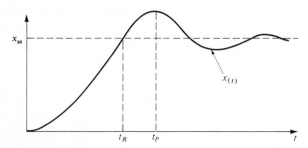

FIGURE P6.11

6.12. (a) Linearize the two ODEs given below that describe a nonisothermal CSTR with constant volume. The input variables are T_0, T_J, C_{A0}, and F.

$$V \frac{dC_A}{dt} = F(C_{A0} - C_A) - VkC_A$$

$$V\rho C_p \frac{dT}{dt} = FC_p\rho(T_0 - T) - \lambda VkC_A - UA(T - T_J)$$

where $k = \alpha e^{-E/RT}$

(b) Convert to perturbation variables and arrange in the form

$$\frac{dC_A}{dt} = a_{11}C_A + a_{12}T + a_{13}C_{A0} + a_{14}T_0 + a_{15}F + a_{16}T_J$$

$$\frac{dT}{dt} = a_{21}C_A + a_{22}T + a_{23}C_{A0} + a_{24}T_0 + a_{25}F + a_{26}T_J$$

(c) Combine the two linear ODEs above into one second-order ODE and find the roots of the characteristic equation in terms of the a_{ij} coefficients.

6.13. The flow rate F of a manipulated stream through a control valve with equal-percentage trim is given by the following equation:

$$F = C_v \alpha^{x-1}$$

where F is the flow in gallons per minute and C_v and α are constants set by the valve size and type. The control-valve stem position x (fraction of wide open) is set by the output signal CO of an analog electronic feedback controller whose signal range is 4 to 20 milliamperes. The valve cannot be moved instantaneously. It is approximately a first-order system:

$$\tau_V \frac{dx}{dt} + x = \frac{CO - 4}{16}$$

The effect of the flow rate of the manipulated variable on the process temperature T is given by

$$\tau_p \frac{dT}{dt} + T = K_p F$$

Derive one linear ordinary differential equation that gives the dynamic dependence of process temperature on controller output signal CO.

6.14. Solve the ODE derived in Prob. 3.4 to show that the concentration C in Grandpa McCoy's batch of Liquid Lightning is

$$C_{(t)} = \frac{C_0(1 - e^{-kt})}{kt}$$

6.15. Suicide Sam slipped his 2000 lb_m hot rod into neutral as he came over the crest of a mountain at 55 mph. In front of him the constant downgrade dropped 2000 feet in 5 miles, and the local acceleration of gravity was 31.0 ft/s².

Sam maintained a constant 55 mph speed by riding his brakes until they heated up to 600°F and burned up. The brakes weighed 40 lb_m and had a heat capacity of 0.1 Btu/lb_m °F. At the crest of the hill they were at 60°F.

Heat was lost from the brakes to the air, as the brakes heated up, at a rate proportional to the temperature difference between the brake temperature and the air temperature. The proportionality constant was 30 Btu/h °F.

Assume that the car was frictionless and encountered negligible air resistance.

(a) At what distance down the hill did Sam's brakes burn up?

(b) What speed did his car attain by the time it reached the bottom of the hill?

6.16. A farmer fills his silo with chopped corn. The entire corn plant (leaves, stem, and ear) is cut up into small pieces and blown into the top of the cylindrical silo at a rate W_0. This is similar to a fed-batch chemical reaction system.

The diameter of the silo is D and its height is H. The density of the chopped corn in the silo varies with the depth of the bed. The density ρ at a point that has z feet of material above it is

$$\rho_{(z)} = \rho_0 + \beta z$$

where ρ_0 and β are constants.

(a) Write the equations that describe the system and show how the height of the bed $h_{(t)}$ varies as a function of time.

(b) What is the total weight of corn fodder that can be stored in the silo?

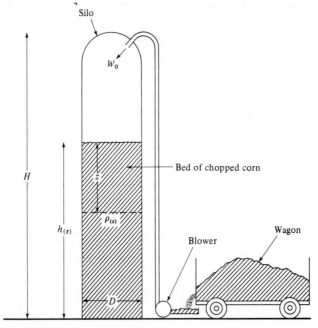

FIGURE P6.16

6.17. Two consecutive, first-order reactions take place in a perfectly mixed, isothermal batch reactor.

$$A \xrightarrow{\ k_1\ } B \xrightarrow{\ k_2\ } C$$

Assuming constant density, solve analytically for the dynamic changes in the concentrations of components A and B in the situation where $k_1 = k_2$. The initial concentration of A at the beginning of the batch cycle is C_{A0}. There is initially no B or C in the reactor.

What is the maximum concentration of component B that can be produced and at what point in time does it occur?

6.18. The same reactions considered in Prob. 6.17 are now carried out in a single, perfectly mixed, isothermal continuous reactor. Flow rates, volume and densities are constant.
 (a) Derive a mathematical model describing the system.
 (b) Solve for the dynamic change in the concentration of component A, C_A, if the concentration of A in the feed stream is constant at C_{A0} and the initial concentrations of A, B, and C at time equal zero are $C_{A(0)} = C_{A0}$ and $C_{B(0)} = C_{C(0)} = 0$.
 (c) In the situation where $k_1 = k_2$, find the value of holdup time ($\tau = V/F$) that maximizes the steadystate ratio of C_B/C_{A0}. Compare this ratio with the maximum found in Prob. 6.17.

6.19. The same consecutive reactions considered in Prob. 6.18 are now carried out in *two* perfectly mixed continuous reactors. Flow rates and densities are constant. The volumes of the two tanks (V) are the same and constant. The reactors operate at the same constant temperature.
 (a) Derive a mathematical model describing the system.
 (b) If $k_1 = k_2$, find the value of the holdup time ($\tau = V/F$) that maximizes the steadystate ratio of concentration of component B in the product to the concentration of reactant A in the feed.

6.20. A vertical, cylindrical tank is filled with well water at 65°F. The tank is insulated at the top and bottom but is exposed on its vertical sides to cold 10°F night air. The diameter of the tank is 2 feet and its height is 3 feet. The overall heat transfer coefficient is 20 Btu/h °F ft². Neglect the metal wall of the tank and assume that the water in the tank is perfectly mixed.
 (a) Calculate how many minutes it will be until the first crystal of ice is formed.
 (b) How long will it take to completely freeze the water in the tank? The heat of fusion of water is 144 Btu/lb$_m$.

6.21. An isothermal, first-order, liquid-phase, reversible reaction is carried out in a constant-volume, perfectly mixed continuous reactor.

$$A \underset{k_2}{\overset{k_1}{\rightleftharpoons}} B$$

The concentration of product B is zero in the feed and in the reactor is C_B. Feed rate is F.
 (a) Derive a mathematical model describing the dynamic behavior of the system.
 (b) Derive the steadystate relationship between C_A and C_{A0}. Show that the conversion of A and the yield of B decrease as k_2 increases.
 (c) Assuming that the reactor is at this steadystate concentration and that a step change is made in C_{A0} to $(C_{A0} + \Delta C_{A0})$, find the analytical solution that gives the dynamic response of $C_{A(t)}$.

6.22. An isothermal, first-order, liquid-phase, irreversible reaction is conducted in a constant volume batch reactor.

$$A \xrightarrow{k} B$$

The initial concentration of reactant A at the beginning of the batch is C_{A0}. The specific reaction rate k decreases with time because of catalyst degradation: $k = k_0 e^{-\beta t}$.

(a) Solve for $C_{A(t)}$.

(b) Show that in the limit as $\beta \to 0$, $C_{A(t)} = C_{A0} e^{-k_0 t}$

(c) Show that in the limit as $\beta \to \infty$, $C_{A(t)} = C_{A0}$

6.23. There are 3460 pounds of water in the jacket of a reactor that are initially at 145°F. At time equal zero, 70°F cooling water is added to the jacket at a constant rate of 416 pounds per minute. The holdup of water in the jacket is constant since the jacket is completely filled with water and excess water is removed from the system on pressure control as cold water is added. Water in the jacket can be assumed to be perfectly mixed.

(a) How many minutes does it take the jacket water to reach 99°F if no heat is transferred into the jacket?

(b) Suppose a constant 362,000 Btu/h of heat is transferred into the jacket from the reactor, starting at time equal zero when the jacket is at 145°F. How long will it take the jacket water to reach 99°F if the cold water addition rate is constant at 416 pounds per minute?

6.24. Hay dries, after being cut, at a rate which is proportional to the amount of moisture it contains. During a hot (90°F) July summer day, this proportionality constant is 0.30 h⁻¹. Hay cannot be baled until it has dried down to no more than 5 wt % moisture. Higher moisture levels will cause heating and mold formation, making it unsuitable for horses.

The effective drying hours are from 11:00 a.m. to 5:00 p.m. If hay cannot be baled by 5:00 p.m. it must stay in the field overnight and picks up moisture from the dew. It picks up 25 percent of the moisture that is lost during the previous day.

If the hay is cut at 11:00 a.m. Monday morning and contains 40 wt % moisture at the moment of cutting, when can it be baled?

6.25. Process liquid is continuously fed into a perfectly mixed tank in which it is heated by a steam coil. Feed rate F is 50,000 lb_m/h of material with a constant density ρ of 50 lb_m/ft³ and heat capacity C_p of 0.5 Btu/lb_m °F. Holdup in the tank V is constant at 4000 lb_m. Inlet feed temperature T_0 is 80°F.

Steam is added at a rate S lb_m/h that heats the process liquid up to temperature T. At the initial steadystate, T is 190°F. The latent heat of vaporization λ_s of the steam is 900 Btu/lb_m.

(a) Derive a mathematical model of the system and prove that process temperature is described dynamically by the ODE

$$\tau \frac{dT}{dt} + T = K_1 T_0 + K_2 S$$

where $\tau = V/F$ $\qquad K_1 = 1$ $\qquad K_2 = \lambda_s/C_p F$

(b) Solve for the steadystate value of steam flow $\bar{S}$.

(c) Suppose a proportional feedback controller is used to adjust steam flow rate,

$$S = \bar{S} + K_c(190 - T)$$

Solve analytically for the dynamic change in $T_{(t)}$ for a step change inlet feed temperature from 80°F down to 50°F. What will the final values of T and S be at the new steadystate for a K_c of 100 lb_m/h/°F?

CHAPTER
7

CONVENTIONAL
CONTROL
SYSTEMS
AND
HARDWARE

In this chapter we will study control equipment, controller performance, controller tuning, and general control-systems design concepts. Some of the questions that we will explore are how do we decide what kind of control valve to use; what type of sensor can be used and what are some of the pitfalls that you should be aware of that can give faulty signals; what type of controller should we select for a given application; and how do we "tune" the controller.

First we will look briefly at some of the control hardware that is currently used in process control systems: transmitters, control valves, controllers, etc. Then we will discuss the performance of conventional controllers and present empirical tuning techniques. Finally we will talk about some important design concepts and heuristics that are useful in specifying the structure of a control system for a process.

7.1 CONTROL INSTRUMENTATION

Some familiarity with control hardware and software is required before we can discuss selection and tuning. We are not concerned with the details of how the various mechanical, pneumatic, hydraulic, electronic, and computing devices are constructed. The nitty-gritty details can be obtained from the instrumentation and process-control-computer vendors. We need to know only how they basically work and what they ....osed to do. Pictures of some typical hardware are given in the Appendix.

There has been a real revolution in instrumentation hardware in the last several decades. Twenty years ago, most control hardware was mechanical and pneumatic (using instrument air pressure to drive gadgets and for control signals). Tubing had to be run back and forth between the process equipment and the control room. Signals were recorded on strip-chart paper recorders.

Today most new control systems use "distributed control" hardware: microprocessors that serve several control loops simultaneously. Information is displayed on CRTs (cathode ray tubes). Most signals are transmitted in analog electronic form (usually current signals).

Despite all these changes in hardware, the basic concepts of control system structure and control algorithms (types of controllers) remain essentially the same as they were thirty years ago. It is now easier to implement control structures; we just reprogram a computer. But the process control engineers job is the same: come up with a control system that will give good, stable, robust control.

As we preliminarily discussed in Chap 1, the basic feedback control loop consists of a sensor to detect the process variable; a transmitter to convert the sensor signal into an equivalent "signal" (an air-pressure signal in pneumatic systems or a current signal in analog electronic systems); a controller that compares this process signal with a desired setpoint value and produces an appropriate controller output signal; and a final control element that changes the manipulated variable. Usually the final control element is an air-operated control valve that opens and closes to change the flow rate of the manipulated stream. See Fig. 7.1.

The sensor, transmitter, and control valve are physically located on the process equipment ("in the field"). The controller is usually located on a panel or in a computer in a control room that is some distance from the process equipment. Wires connect the two locations, carrying current signals from transmitters to the controller and from the controller to the final control element.

The control hardware used in chemical and petroleum plants is either analog (pneumatic or electronic) or digital. The analog systems use air-pressure signals (3 to 15 psig) or current/voltage signals (4 to 20 milliamperes, 10 to 50 milliamperes or 0 to 10 volts dc). They are powered by instrument air supplies (25 psig air) or 24 volt dc electrical power. Pneumatic systems send air-pressure signals through small tubing. Analog electronic systems use wires.

Since most valves are still actuated by air pressure, current signals are usually converted into an air pressure. An "*I* to *P*" transducer (current to pressure) is used to convert 4 to 20 mA signals into 3 to 15 psig signals.

Also located in the control room is the manual-automatic switching hardware (or software). During start-up or under abnormal conditions, the plant operator may want to be able to set the position of the control valve himself instead of having the controller position it. A switch is usually provided on the control panel or in the computer system as sketched in Fig. 7.2. In the "manual" position the operator can stroke the valve by changing a knob (a pressure regula-

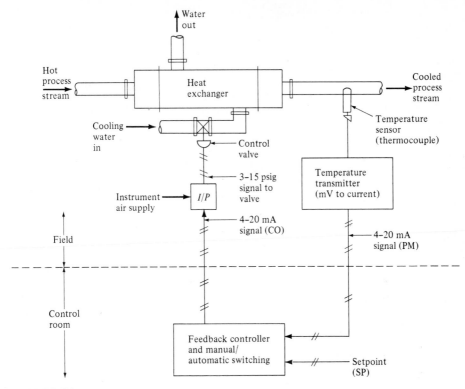

FIGURE 7.1
Feedback control loop.

tor in a pneumatic system or a potentiometer in an analog electronic system). In the "automatic" position the controller output goes directly to the valve.

Each controller must provide the following:

1. Indicate the value of the controlled variable: the signal from the transmitter.
2. Indicate the value of the signal being sent to the valve: the controller output.
3. Indicate the setpoint.
4. Have a manual/automatic switch.
5. Have a knob to set the setpoint when the controller is on automatic.
6. Have a knob to set the signal to the valve when the controller is on manual.

All controllers, be they 30-year-old pneumatic controllers or modern distributed microprocessor-based controllers, have these features.

7.1.1 Sensors

Let's start from the beginning of the control loop, at the sensor. Instruments for on-line measurement of many properties have been developed. The most

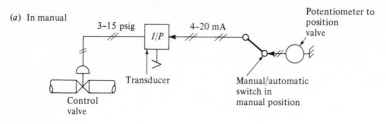

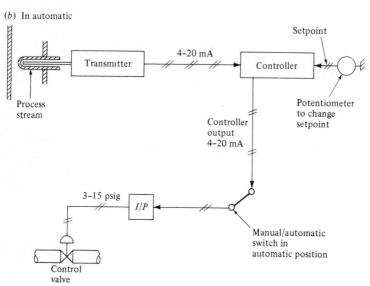

FIGURE 7.2
Manual/automatic switching.

important variables are flow rate, temperature, pressure, and level. Devices for measuring other properties, such as pH, density, viscosity, infrared and ultraviolet absorption, and refractive index are available. Direct measurement of chemical composition by means of on-line gas chromatographs is quite widespread. They pose interesting control problems because of their intermittent operation (a composition signal is generated only every few minutes). We will study the analysis of these discontinuous, "sampled-data" systems in Part VII.

We will briefly discuss below some of the common sensing elements. Details of their operation, construction, relative merits, and costs are given in several handbooks: B. G. Liptak *Instrument Engineers Handbook*, Chilton, 1970; R. L. Moore *Measurement Fundamentals*, Instrument Society of America, 1982).

A. FLOW. Orifice plates are by far the most common type of flow-rate sensor. The pressure drop across the orifice varies with the square of the flow in turbulent flow, so measuring the differential pressure gives a signal that can be related to flow rate. Normally orifice plates are designed to give pressure drops in the range of 20 to 200 inches of water. Turbine meters are also widely used. They are more expensive but give more accurate flow measurement. Other types of flow meters include sonic flow meters, magnetic flow meters, rotameters, vortex-shedding devices, and pitot tubes. In gas recycle systems where the pressure drop through the flow meter can mean a significant amount of compressor work, low-pressure drop flow meters are used, like the last two mentioned above.

When a flow sensor is installed for accurate accounting measurements of the absolute flow rate, many precautions must be taken, such as providing a long section of straight pipe before the orifice plate. For control purposes, however, one may not need to know the absolute value of the flow but only the changes in flow rate. Therefore pressure drops over pieces of equipment, around elbows or over sections of pipe can sometimes be used to get a rough indication of flow rate changes.

The signals from flow rate measurements are usually noisy (fluctuate around the actual value) because of the turbulent flow. These signals often need to be filtered to smooth out the signal sent to the controller.

B. TEMPERATURE. Thermocouples are the most commonly used temperature sensing devices. The two dissimilar wires produce a millivolt signal that varies with the "hot-junction" temperature. Iron-constantan thermocouples are commonly used over the 0 to 1300°F temperature range.

Filled-bulb temperature sensors are also widely used. An inert gas is enclosed in a constant-volume system. Changes in process temperature cause the pressure exerted by the gas to change. Resistance thermometers are used where accurate temperature or differential-temperature measurement is required. They use the principle that the electrical resistance of wire changes with temperature.

The dynamic response of most sensors is usually much faster than the dynamics of the process itself. Temperature sensors are a notable and sometimes troublesome exception. The time constant of a thermocouple and a heavy thermowell can be 30 seconds or more. If the thermowell is coated with polymer or other goo, the response time can be several minutes. This can significantly degrade control performance.

C. PRESSURE AND DIFFERENTIAL PRESSURE. Bourdon tubes, bellows, and diaphragms are used to sense pressure and differential pressure. For example, in a mechanical system the process pressure force is balanced by the movement of a spring. The position of the spring can be related to the process pressure.

D. LEVEL. Liquid levels are detected in a variety of ways. The three most common are

1. Following the position of a float that is lighter than the fluid (as in a bathroom toilet).

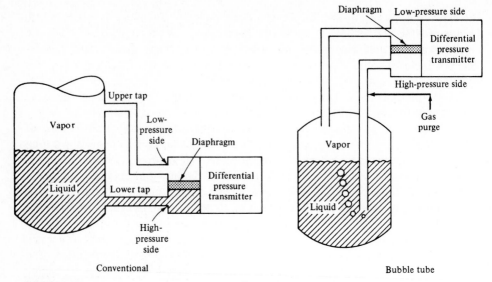

FIGURE 7.3
Differential-pressure level measurement.

2. Measuring the apparent weight of a heavy cylinder as it is buoyed up more or less by the liquid (these are called *displacement meters*).

3. Measuring the difference in static pressure between two fixed elevations, one in the vapor above the liquid and the other under the liquid surface. As sketched in Fig. 7.3, the differential pressure between the two level taps is directly related to the liquid level in the vessel.

In the last scheme the process liquid and vapor are normally piped directly to the differential-pressure measuring device (ΔP transmitter). Some care has to be taken to account for or to prevent condensation of vapor in the connecting line (called "impulse line") from the top level tap. If the line fills up with liquid, the differential pressure will be zero even though the liquid level is all the way up to the top level tap. You will think that the level is low, but the level is actually at or above the top level tap. If safety problems can occur because of a high level, a second level sensor should be used to *independently* detect high level. Keeping the vapor impulse line hot or purging it with a small vapor flow can sometimes keep it from filling with liquid. Purging it with a small liquid flow works, too, because you know that the line is always filled with liquid, so the "zero" (the ΔP at which the transmitter puts out its 4 mA signal) can be adjusted appropriately to indicate the correct level.

Because of plugging or corrosion problems, it is sometimes necessary to keep the process fluid out of the ΔP transmitter. This is accomplished by mechanical diaphragm seals or by purges (introducing a small amount of liquid or gas into the connecting lines which flows back into the process).

If it is difficult to provide a level tap in the base of the vessel (for mechanical design reasons, for example in a glass-lined vessel); a bubble tube can be suspended from the top of the vessel down under the liquid surface, as shown in Fig. 7.3. A small gas purge through the tube gives a pressure on the high-pressure side of the ΔP transmitter that is the same as the static pressure at the base of the bubble tube. This type of level measurement can give incorrect level readings when the pressure in the vessel is increasing rapidly because the liquid can back up in the dip-tube if the gas purge flow rate is not large enough to compensate for the pressure increase.

For very hard-to-handle process fluids nuclear radiation gauges are used to detect interfaces and levels.

As you can tell from the above discussion, it is very easy to be fooled by a differential pressure measurement of level. As one who has been bitten many times by these problems, I highly recommend redundant sensors and judicious skepticism about the validity of instrument readings.

7.1.2 Transmitters

The transmitter is the interface between the process and its control system. The job of the transmitter is to convert the sensor signal (millivolts, mechanical movement, pressure differential, etc.) into a control signal (4 to 20 mA, for example).

Consider the pressure transmitter shown in Fig. 7.4a. Let us assume that this particular transmitter is set up so that its output current signal varies from 4 to 20 mA as the process pressure in the vessel varies from 100 to 1000 kPa gauge. This is called the *range* of the transmitter. The *span* of the transmitter is 900 kPa. The *zero* of the transmitter is 100 kPa gauge. The transmitter has two adjustment knobs in it somewhere that can be changed to modify the span and/or the zero. Thus, if we shifted the zero up to 200 kPa gauge, the range of the transmitter would now be 200 to 1100 kPa gauge. Its span is still 900 kPa.

The dynamic response of most transmitters is usually much faster than the process and the control valves. Consequently we can normally consider the transmitter as a simple "gain" (a step change in the input to the transmitter gives an instantaneous step change in the output). The gain of the pressure transmitter considered above would be

$$\frac{20 \text{ mA} - 4 \text{ mA}}{1000 \text{ kPa} - 100 \text{ kPa}} = \frac{16 \text{ mA}}{900 \text{ kPa}} \tag{7.1}$$

Thus the transmitter is just a "transducer" that converts the process variable into an equivalent control signal.

Figure 7.4b shows a temperature transmitter which accepts thermocouple input signals and is set up so that its current output goes from 4 to 20 mA as the process temperature varies from 50 to 250°F. The range of the temperature transmitter is 50 to 250°F, its span is 200°F, and its zero is 50°F. The gain of the temperature transmitter is

$$\frac{20 \text{ mA} - 4 \text{ mA}}{250°F - 50°F} = \frac{16 \text{ mA}}{200°F} \tag{7.2}$$

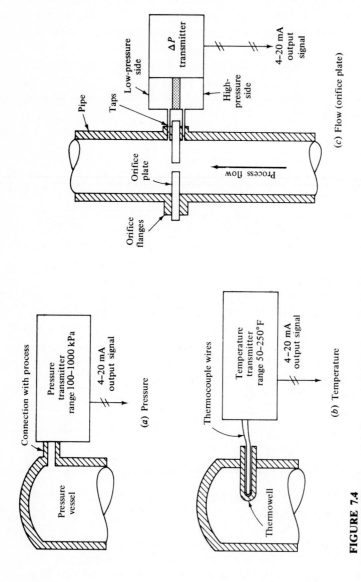

FIGURE 7.4

Typical transmitters. (a) Pressure; (b) temperature; (c) flow (orifice plate).

As noted in the previous section, the dynamics of the thermowell-thermocouple sensor are often not negligible and should be included in the dynamic analysis.

Figure 7.4c shows a ΔP transmitter used with an orifice plate as a flow transmitter. The pressure drop over the orifice plate (the sensor) is converted into a control signal. Suppose the orifice plate is sized to give a pressure drop of 100 in H_2O at a process flow rate of 2000 kg/h. The ΔP transmitter converts inches of H_2O into milliamperes, and its gain is 16 mA/100 in H_2O. However, we really want flow rate, not orifice-plate pressure drop. Since ΔP is proportional to the square of the flow rate, there is a nonlinear relationship between flow rate F and the transmitter output signal:

$$PM = 4 + 16\left(\frac{F}{2000}\right)^2 \tag{7.3}$$

where PM = transmitter output signal, mA
$\quad\quad F$ = flow rate in kg/h

Dropping the flow by a factor of two cuts the ΔP signal by a factor of four. For system analysis we usually linearize Eq. (7.3) around the steadystate value of flow rate, $\bar{F}$.

$$PM = \frac{32\bar{F}}{(F_{max})^2} F \tag{7.4}$$

where PM and F = perturbations from steadystate
$\quad\quad \bar{F}$ = steadystate flow rate, kg/h
$\quad\quad F_{max}$ = maximum full-scale flow rate = 2000 kg/h in this example

7.1.3 Control Valves

The interface with the process at the other end of the control loop is made by the final control element. In a vast majority of chemical engineering processes the final control element is an automatic control valve which throttles the flow of a manipulated variable. Most control valves consist of a plug on the end of a stem that opens or closes an orifice opening as the stem is raised or lowered. As sketched in Fig. 7.5, the stem is attached to a diaphragm that is driven by changing air pressure above the diaphragm. The force of the air pressure is opposed by a spring.

There are several aspects of control valves: their *action, characteristics,* and *size.*

A. ACTION OF THE VALVE. Valves are designed to either fail in the wide-open position or completely shut. Which action is appropriate depends on the effect of the manipulated variable on the process. For example, if the valve is handling steam or fuel, you would want the flow to be cut off in an emergency, i.e., you want the valve to fail shut. If the valve is handling cooling water to a reactor, you want the flow to go to a maximum in an emergency, i.e., you want the valve to fail wide open.

The valve shown in Fig. 7.5 is closed when the stem is completely down and wide open when the stem is at the top of its stroke. Since increasing air pressure closes the valve, this valve is an "air-to-close" (AC) valve. If the air-pressure signal dropped to zero because of some failure (for example, suppose the instrument-air supply line was cut), this valve would fail wide open since the spring would push the valve open. The valve can be made "air-to-open" (AO) by reversing the action of the plug to close the opening in the up position or by reversing the locations of the spring and air pressure (put the air pressure under the diaphragm).

Thus we use either AO or AC valves, and the decision as to which to use depends on whether we want the valve to fail shut or wide open.

B. SIZE. Sizing of control valves is one of the more controversial subjects in process control. The flow rate through a control valve depends on the size of the valve, the pressure drop over the valve, the stem position, and the fluid properties. The design equation for liquids (nonflashing) is

$$F = C_v f_{(x)} \sqrt{\frac{\Delta P_v}{\text{sp gr}}} \qquad (7.5)$$

where F = flow rate, gpm
$\quad C_v$ = valve size coefficient
$\quad x$ = valve stem position (fraction of wide open)
$\quad f(x)$ = fraction of the total flow area of the valve. (The curve of $f_{(x)}$ versus x is called the "inherent characteristics" of the valve. We will discuss this later.)
$\quad$ sp gr = specific gravity (relative to water)
$\quad \Delta P_v$ = pressure drop over the valve, psi

More detailed equations are available in publications of the control-valve manufacturers (for example, the Masonielan Handbook for Control Valve Sizing, Dresser Industries, 6th edition, 1977) that handle flows of gases, flashing liquids, and critical flows with either English or SI units.

The sizing of control valves is a good example of the engineering trade-off that must be made in designing a plant. Consider the process sketched in Fig. 7.6. Suppose the flow rate at design conditions is 100 gpm, the pressure in the feed tank is atmospheric, the pressure drop over the heat exchanger (ΔP_H) at the design flow rate is 40 psi, and the pressure in the final tank, P_2, is 150 psig. Let us assume that we will have the control valve half open ($f_{(x)} = 0.5$) at the design flow. The specific gravity of the liquid is 1.

The process engineer's job is to size both the centrifugal pump and the control valve. The bigger the control valve, the less pressure drop it will take. This means a lower-head pump can be used and energy costs will be lower

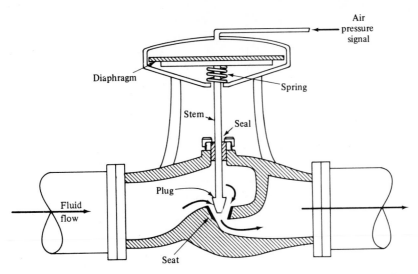

FIGURE 7.5
Typical air-operated control valve.

because the power consumed by the motor driving the pump will be less. So the process engineer, knowing little about control, wants to design a system that has a low pressure drop across the control valve. From a steadystate standpoint, this makes perfect sense.

However, the process engineer goes to talk with the control engineer, and the control engineer wants to take a lot of pressure drop over the valve. Why? Basically it is a question of "rangeability": the larger the pressure drop, the larger the size of the changes that can be made in the flow rate (in *both* directions: increase and decrease). Let's examine two different designs to show why it is desirable from a dynamic point of view to take more pressure drop over the control valve.

In case 1 we will size the valve so that it takes 20 psi pressure drop at design flow when it is half open. This means that the pump must produce a

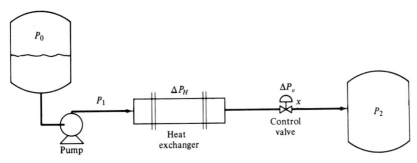

FIGURE 7.6

differential head of $150 + 40 + 20 = 210$ psi at design. In case 2 we will size the valve so that it takes 80 psi pressure drop at design. Now a higher-head pump will be needed: $150 + 40 + 80 = 270$ psi.

Using Eq. (7.5), both control valves can be sized.

Case 1:

$$F = C_v f_{(x)} \sqrt{\frac{\Delta P_v}{\text{sp gr}}}$$

$$100 = C_{v1}(0.5)\sqrt{20} \quad \Rightarrow \quad C_{v1} = 44.72$$

when the design valve pressure drop is 20 psi

Case 2:

$$100 = C_{v2}(0.5)\sqrt{80} \quad \Rightarrow \quad C_{v2} = 22.36$$

when the design valve pressure drop is 80 psi

Naturally the control valve in case 2 is smaller than that in case 1.

Now let's see what happens in the two cases when we open the control valve all the way: $f_{(x)} = 1$. Certainly the flow rate will increase, but how much? From a control point of view, we may want to be able to increase the flow substantially. Let's call this unknown flow F_{max}.

The higher flow rate will increase the pressure drop over the heat exchanger as the square of the flow rate.

$$\Delta P_H = 40\left(\frac{F_{max}}{F_{des}}\right)^2 = 40\left(\frac{F_{max}}{100}\right)^2 \tag{7.6}$$

The higher flow rate might also reduce the head that the centrifugal pump produces if we are out on the pump curve where head is dropping rapidly with throughput. For simplicity, let us assume that the pump curve is flat. This means that the total pressure drop across the heat exchanger and the control valve is constant. Therefore, the pressure drop over the control valve must decrease as the the pressure drop over the heat exchanger increases.

$$\Delta P_v = \Delta P_{Total} - \Delta P_H \tag{7.7}$$

Plugging in the numbers for the two cases yields the following results.

Case 1 (20 psi design):

$$\Delta P_{Total} = 60 \text{ psi} \qquad C_{v1} = 44.72$$

$$F_{max} = (44.72)(1.0)\sqrt{60 - 40\left(\frac{F_{max}}{100}\right)^2} \tag{7.8}$$

This equation can be solved for F_{max}: 115 gpm. So the maximum flow through the valve is only 15 percent more than design if a 20 psi pressure drop over the valve is used at design flow rate.

Case 2 (80 psi design):

$$\Delta P_{Total} = 120 \text{ psi} \qquad C_v = 22.36$$

$$F_{max} = (23.36)(1.0)\sqrt{120 - 40\left(\frac{F_{max}}{100}\right)^2} \tag{7.9}$$

Solving for F_{max} yields 141 gpm. So the maximum flow through this valve, which has been designed for a higher pressure drop, is over 40 percent more than design.

We can see from the results above that the valve that has been designed for the larger pressure drop can produce more of an increase in the flow rate at its maximum capacity.

Now let's see what happens when we want to reduce the flow. Control valves don't work too well when they are less than about 10 percent open. They can become mechanically unstable, shutting off completely and then popping partially open. The resulting fluctuations in flow are undesirable. Therefore, we want to design for a minimum valve opening of 10 percent. Let's see what the minimum flow rates will be in the two cases considered above when the two valves are pinched down so that $f_{(x)} = 0.1$.

In this case the lower flow rate will mean a decrease in the pressure drop over the heat exchanger and therefore an *increase* in the pressure drop over the control valve.

Case 1 (20 psi design):

$$F_{min} = (0.1)(44.72)\sqrt{60 - 40\left(\frac{F_{min}}{100}\right)^2} \tag{7.10}$$

Solving gives $F_{min} = 33.3$ gpm.

Case 2 (80 psi design):

$$F_{min} = (0.1)(22.36)\sqrt{120 - 40\left(\frac{F_{min}}{100}\right)^2} \tag{7.11}$$

This F_{min} is 24.2 gpm.

These results show that the minimum flow rate is lower for the valve that was designed for a larger pressure drop. So not only can we increase the flow more, but we also can reduce it more. Thus the *turndown* (the ratio of F_{max} to F_{min}) of the big ΔP valve is larger.

$$\text{Turndown ratio for 20-psi design valve} = \frac{115}{33.3} = 3.46$$

$$\text{Turndown ratio for 80-psi design valve} = \frac{141}{24.2} = 5.83$$

We have demonstrated why the control engineer wants more pressure drop over the valve.

So how do we resolve this conflict between the process engineer wanting low pressure drop and the control engineer wanting large pressure drop? A commonly used heuristic recommends that the pressure drop over the control valve at design should be 50 percent of the total pressure drop through the system. Although widely used, this procedure makes little sense to me. A more logical design procedure is outlined below.

In some situations it is very important to be able to increase the flow rate above the design conditions (for example, the cooling water to an exothermic reactor may have to be doubled or tripled to handle dynamic upsets). In other cases this is not as important (for example, the feed flow rate to a unit). Therefore it is logical to base the design of the control valve and the pump on having a process that can attain both the maximum and the minimum flow conditions. The design flow conditions are only used to get the pressure drop over the heat exchanger (or fixed resistance part of the process).

The designer must specify the maximum flow rate that is required under the worst conditions *and* the minimum flow rate that is required. Then the valve flow equations for the maximum and minimum conditions give two equations and two unknowns: the pressure head of the centrifugal pump ΔP_P and the control valve size C_v.

Example 7.1. Suppose we want to design a control valve for admitting cooling water to a cooling coil in an exothermic chemical reactor. The normal flow rate is 50 gpm. To prevent reactor runaways, the valve must be able to provide three times the design flow rate. Because the sales forecast could be overly optimistic, a minimum flow rate of 50 percent of the design flow rate must be achievable. The pressure drop through the cooling coil is 10 psi at the design flow rate of 50 gpm. The cooling water is to be pumped from an atmospheric tank. The water leaving the coil runs into a pipe in which the pressure is constant at 2 psig. Size the control valve and the pump.

The pressure drop through the coil depends on the flow rate F:

$$\Delta P_c = 10 \left(\frac{F}{50} \right)^2 \tag{7.12}$$

The pressure drop over the control valve is the total pressure drop available (which we don't know yet) minus the pressure drop over the coil.

$$\Delta P_v = \Delta P_T - 10 \left(\frac{F}{50} \right)^2 \tag{7.13}$$

Now we write one equation for the maximum flow conditions and one for the minimum.

At the maximum conditions:

$$150 = C_v (1.0) \sqrt{\Delta P_T - 10 \left(\frac{150}{50} \right)^2} \tag{7.14}$$

At the minimum conditions:

$$25 = C_v (0.1) \sqrt{\Delta P_T - 10 \left(\frac{25}{50}\right)^2} \tag{7.15}$$

Solving simultaneously for the two unknowns yields the control valve size (C_v = 21.3) and the pump head ($\Delta P_p = \Delta P_T + 2 = 139.2 + 2 = 141.2$ psi).

At the design conditions (50 gpm), the valve fraction open (f_{des}) will be given by

$$50 = 21.3 f_{des} \sqrt{139.2 - 10} \quad \Rightarrow \quad f_{des} = 0.206 \tag{7.16}$$

The control valve/pump sizing procedure proposed above is not without its limitations. The two design equations for the maximum and minimum conditions in general terms are:

$$F_{max} = C_v \sqrt{\Delta P_T - (\Delta P_H)_{des} \left(\frac{F_{max}}{F_{des}}\right)^2} \tag{7.17}$$

$$F_{min} = f_{min} C_v \sqrt{\Delta P_T - (\Delta P_H)_{des} \left(\frac{F_{min}}{F_{des}}\right)^2} \tag{7.18}$$

where ΔP_T = total pressure drop through the system at design flow rates
$(\Delta P_H)_{des}$ = pressure drop through the fixed resistances in the system at design flow
f_{min} = minimum valve opening
F_{des} = flow rate at design

A flat pump curve is assumed in the above derivation. Solving these two equations for ΔP_T gives:

$$\frac{\Delta P_T}{(\Delta P_H)_{des}} = \frac{\left\{\dfrac{(F_{max})^2 - (F_{min})^2}{(F_{des})^2}\right\}}{1 - \left(\dfrac{f_{min} F_{max}}{F_{min}}\right)^2} \tag{7.19}$$

It is clear from Eq. (7.19) that as the second term in the denominator approaches unity, the required pressure drop goes to infinity! So there is a limit to the achievable rangeability of a system.

Let us define this term as the rangeability index of the system, $\mathcal{R}$.

$$\mathcal{R} \equiv \frac{f_{min} F_{max}}{F_{min}} \tag{7.20}$$

The parameters on the right side of Eq. (7.20) *must* be chosen such that $\mathcal{R}$ is less than unity.

This can be illustrated, using the numbers from Example 7.1. If the minimum flow rate is reduced from 50 percent of design (where ΔP_T was 139.2 psi) to 40 percent, the new ΔP_T becomes 202 psi. If F_{min} is reduced further to 35 percent of design, ΔP_T is 335 psi. In the limit as F_{min} goes to 30 percent of design, the rangeability index becomes

$$\mathcal{R} \equiv \frac{f_{min} \, F_{max}}{F_{min}} = \frac{(0.1)(150)}{15} = 1$$

and the total pressure drop available goes to infinity.

The value of f_{min} can be reduced below 0.1 if a large turndown ratio is required. This is accomplished by using two control valves in parallel, one large and one small, in a split-range configuration. The small valve opens first and then the large valve opens as the signal to the two valves changes over its full range.

Characteristics. By changing the shape of the plug and the seat in the valve, different relationships between stem position and flow area can be attained. The common flow *characteristics* used are *linear trim* valves and *equal-percentage trim* valves as shown in Fig. 7.7. The term "equal percentage" comes from the slope of the $f_{(x)}$ curve being a constant fraction of f.

If constant pressure drop over the valve is assumed and if the stem position is 50 percent open, a linear-trim valve gives 50 percent of the maximum flow and an equal-percentage-trim valve gives only 15 percent of the maximum flow. The equations for these valves are:

Linear:

$$f_{(x)} = x \tag{7.21}$$

Equal percentage:

$$f_{(x)} = \alpha^{x-1} \tag{7.22}$$

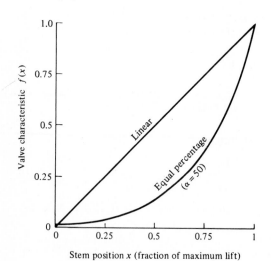

Stem position x (fraction of maximum lift)

3 psig ⟵ air-to-open valve ⟶ 15 psig
15 psig ⟵ air-to-close valve ⟶ 3 psig

FIGURE 7.7
Control-valve characteristics.

where α is a constant (20 to 50) that depends on the valve design. A value of 50 is used in Fig. 7.7.

The basic reason for using different control-valve trims is to keep the stability of the control loop fairly constant over a wide range of flows. Linear-trim valves are used, for example, when the pressure drop over the control valve is fairly constant and a linear relationship exists between the controlled variable and the flow rate of the manipulated variable. Consider the flow of steam from a constant-pressure supply header. The steam flows into the shell side of a heat exchanger. A process liquid stream flows through the tube side and is heated by the steam. There is a linear relationship between the process outlet temperature and steam flow (with constant process flow rate and inlet temperature) since every pound of steam provides a certain amount of heat.

Equal-percentage valves are often used when the pressure drop available over the control valve is not constant. This occurs when there are other pieces of equipment in the system that act as fixed resistances. The pressure drops over these parts of the process vary as the square of the flow rate. We saw this in the examples discussing control valve sizing.

At low flow rates, most of the pressure drop is taken over the control valve since the pressure drop over the rest of the process equipment is low. At high flow rates, the pressure drop over the control valve is low. In this situation the equal-percentage trim tends to give a more linear relationship between flow and control-valve position than does linear trim. Figure 7.8 shows the *installed characteristics* of linear and equal-percentage valves for different ratios of the pressure drop over the fixed resistance (ΔP_H for the heat exchanger example) over the pressure drop over the valve at design conditions. The larger this ratio, the more nonlinear are the installed characteristics of a linear valve.

The *inherent characteristics* are those that relate flow to valve position in the situation where the pressure drop over the control valve is constant. These are the $(\Delta P_H / \Delta P_v) = 0$ curves in Fig. 7.8. *Installed characteristics* are those that result from the variation in the pressure drop over the valve.

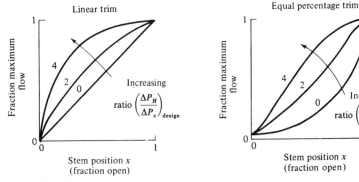

FIGURE 7.8
Control-valve performance in a system ("installed characteristics").

In conventional valves the air-pressure signal to the diaphragm comes from an *I/P* transducer in analog electronic systems. "Valve positioners" are often used to improve control, particularly for large valves and with dirty or gooky fluids which can make the valve stick. A sticky valve can cause a control loop to oscillate; the controller output signal changes but the valve position doesn't do anything until the pressure force gets high enough to move the valve. Then, of course, the valve moves too far and the controller must reverse the direction of change of its output, and the same thing occurs in the other direction. So the loop will fluctuate around its setpoint even with no other disturbances.

Valve positioners are little feedback controllers that sense the actual position of the stem, compare it with the desired position as given by the signal from the controller and adjust the air pressure on the diaphragm to drive the stem to its correct position. Valve positioners can also be used to make valves open and close over various ranges (split-range valves).

Control valves are usually fairly fast compared with the process. With large valves (greater than 4 inches) it may take 20 to 40 seconds for the valve to move full stroke.

7.1.4 Analog and Digital Controllers

The part of the control loop that we will spend most of our time with in this book is the controller. The job of the controller is to compare the process signal from the transmitter with the setpoint signal and to send out an appropriate signal to the control valve. We will go into more detail about the performance of the controller in Sec. 7.2. In this section we will describe what kind of action standard commercial controllers take when they see an error.

Analog controllers use continuous electronic or pneumatic signals. The controllers see transmitter signals continuously, and control valves are changed continuously.

Digital computer controllers are discontinuous in operation, looking at a number of loops sequentially. Each individual loop is only looked at every sampling period. The analog signals from transmitters must be sent through analog-to-digital (A/D) converters to get the information into the computer in a form that it can use. After the computer performs its calculations (control algorithm) it sends out a signal which must pass through a digital-to-analog (D/A) converter and a "hold" that sends a continuous signal to the control valve. We will study these sampled-data systems in detail in Chaps. 18 through 20.

There are three basic types of controllers that are commonly used for continuous feedback control. The details of construction of the analog devices and the programming of the digital devices vary from one manufacturer to the next, but their basic functions are essentially the same.

A. PROPORTIONAL ACTION. A proportional-only feedback controller changes its output signal, CO, in direct proportion to the error signal, E, which is the difference between the setpoint, SP, and the process measurement signal, PM,

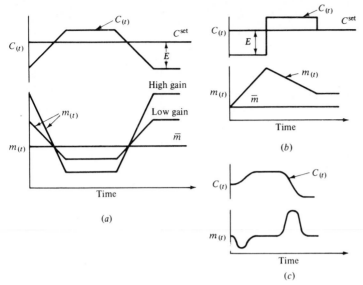

FIGURE 7.9
Action of a feedback controller. (a) Proportional; (b) integral; (c) ideal derivative.

coming from the transmitter.

$$CO = bias \pm K_c(SP - PM) \qquad (7.23)$$

The bias signal is a constant and is the value of the controller output when there is no error. The "K_c" is called the *controller gain*. The larger the gain, the more the controller output will change for a given error. For example, if the gain is 1, an error of 10 percent of scale (1.6 mA in an analog electronic 4 to 20 mA system) will change the controller output by 10 percent of scale. Figure 7.9a sketches the action of a proportional controller for given error signals, E.

Many instrument manufacturers use an alternative term, *proportional band* (PB), instead of gain. The two are related by

$$PB = \frac{100}{K_c} \qquad (7.24)$$

The higher or "wider" the proportional band, the lower the gain and vice versa. The term proportional band refers to the range over which the error must change to drive the controller output over its full range. Thus a wide PB is a low gain, and a narrow PB is a high gain.

The gain on the controller can be made either positive or negative by setting a switch in an analog controller or specifying the desired sign in a digital controller. A positive gain results in the controller output decreasing when the process measurement increases. This "increase-decrease" action is called a *reverse-acting* controller. For a negative gain the controller output increases when the process measurement increases, and this is called a *direct-acting* controller. The correct sign depends on the action of the transmitter (which is usually

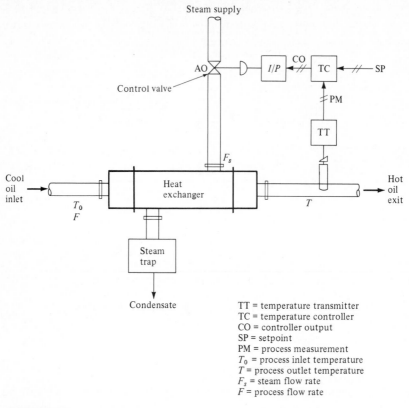

FIGURE 7.10
Heat exchanger.

direct), the action of the valve (air-to-open to air-to-close), and the effect of the manipulated variable on the controlled variable.

For example, suppose we are controlling the process outlet temperature of a heat exchanger as sketched in Fig. 7.10. A control valve on the steam to the shell side of the heat exchanger is positioned by a temperature controller. To decide what action the controller should have we first look at the valve. Since this valve puts steam into the process, we would want it to fail shut. Therefore we choose an air-to-open (AO) control valve.

Next we look at the temperature transmitter. It is direct acting (when the process temperature goes up, the transmitter output signal, PM, goes up). Now if PM increases, we want to have less steam. This means that the controller output must decrease since the valve is AO. Thus the controller must be reverse-acting and have a positive gain.

If we were cooling instead of heating, we would want the coolant flow to increase when the temperature increased. But the controller action would still be reverse because the control valve would be an air-to-close valve, since we want it to fail wide open.

As a final example, suppose we are controlling the base level in a distillation column with the bottoms product flow rate. The valve would be AO because we want it to fail shut (we don't want to lose base level in an emergency). The level transmitter signal increases if the level increases. If the level goes up, we want the bottoms flow rate to increase. Therefore the base level controller should be "increase-increase" (direct acting).

One of the most important items to check in setting up a feedback control loop on the plant is that the action of the controller is correct.

B. INTEGRAL ACTION (RESET). Proportional action moves the control valve in direct proportion to the magnitude of the error. Integral action moves the control valve based on the time integral of the error, as sketched in Fig. 7.9b.

$$CO = bias + \frac{1}{\tau_I} \int E_{(t)} \, dt \qquad (7.25)$$

where τ_I is the integral time or the reset time with units of minutes.

If there is no error, the controller output does not move. As the error goes positive or negative, the integral of the error drives the controller output either up or down, depending on the action (reverse or direct) of the controller.

Most controllers are calibrated in *minutes* (or *minutes/repeat*, a term that comes from the test of putting into the controller a fixed error and seeing how long it takes the integral action to ramp up the controller output to produce the same change that a proportional controller would make when its gain is 1; the integral *repeats* the action of the proportional controller).

The basic purpose of integral action is to drive the process back to its setpoint when it has been disturbed. A proportional controller will *not* usually return the controlled variable to the setpoint when a load or setpoint disturbance occurs. This permanent error (SP − PM) is called *steadystate error* or *offset*. Integral action reduces the offset to zero.

Integral action degrades the dynamic response of a control loop. We will demonstrate this quantitatively in Chap. 10. It makes the control loop more oscillatory and moves it toward instability. But integral action is usually needed if it is desirable to have zero offset. This is another example of an engineering trade-off that must be made between dynamic performance and steadystate performance.

C. DERIVATIVE ACTION. The purpose of derivative action (also called *rate* or *preact*) is to anticipate where the process is heading by looking at the time rate of change of the controlled variable (its derivative). If we were able to take the derivative of the error signal (which we cannot do perfectly, as we will explain more fully in Chap. 10), we would have ideal derivative action.

$$CO = bias + \tau_D \frac{dE}{dt} \qquad (7.26)$$

where τ_D is the derivative time (minutes).

In theory, derivative action should always improve dynamic response, and it does in many loops. In others, however, the problem of noisy signals (fluctuating process-measurement signals) makes the use of derivative action undesirable.

D. COMMERCIAL CONTROLLERS. The three actions described above are used individually or combined in commercial controllers. Probably 60 percent of all controllers are PI (proportional-integral), 20 percent are PID (proportional-integral-derivative) and 20 percent are P-only (proportional). We will discuss the reasons for selecting one type over another in Sec. 7.2.

7.1.5. Computing and Logic Devices

A host of gadgets and software are available to perform a variety of computations and logical operations with control signals. For example, adders, multipliers, dividers, low selectors, high selectors, high limiters, low limiters, and square-root extractors can all be implemented in both analog and computer systems. They are widely used in *ratio* control, in *computed variable* control, in *feedforward* control, and in *override* control. These will be discussed in the next chapter.

In addition to the basic control loops, all processes have instrumentation that (1) sounds alarms to alert the operator to any abnormal or unsafe condition, and (2) shuts down the process if unsafe conditions are detected or equipment fails. For example, if a compressor motor overloads and the electrical control system on the motor shuts down the motor, the rest of the process will usually have to be shut down immediately. This type of instrumentation is called an "interlock." It either shuts a control valve completely or drives the control valve wide open. Other examples of conditions that can "interlock" a process down include failure of a feed or reflux pump, detection of high pressure or temperature in a vessel, and indication of high or low liquid level in a tank or column base. Interlocks are usually achieved by pressure, mechanical, or electrical switches. They can be included in the computer software in a computer control system, but they are usually "hard-wired" for reliability and redundancy.

7.2 PERFORMANCE OF FEEDBACK CONTROLLERS

7.2.1 Specifications for Closedloop Response

There are a number of criteria by which the desired performance of a closedloop system can be specified in the time domain. For example, we could specify that the closedloop system be critically damped so that there is no overshoot or oscillation. We must then select the type of controller and set its tuning constants so that it will give, when coupled with the process, the desired closedloop response. Naturally the control specification must be physically attainable. We cannot make a Boeing 747 jumbo jet airplane behave like an F-15 fighter. We cannot violate constraints on the manipulated variable (the control valve can only go wide open or completely shut), and we cannot require a physically unrealizable controller (more about the mathematics of this in Chap. 10).

There are a number of time-domain specifications. A few of the most frequently used dynamic specifications are listed below (see also Prob. 6.11). The traditional test input signal is a step change in setpoint.

1. Closedloop damping coefficient (as discussed in Chap. 6)
2. Overshoot: the magnitude by which the controlled variable swings past the setpoint
3. Rise time (speed of response): the time it takes the process to come up to the new setpoint
4. Decay ratio: the ratio of maximum amplitudes of successive oscillations
5. Settling time: the time it takes the amplitude of the oscillations to decay to some fraction (0.05) of the change in setpoint
6. Integral of the squared error:

$$\text{ISE} = \int_0^\infty (E_{(t)})^2 \, dt$$

Notice that the first five of these assume an underdamped closedloop system, i.e., one that has some oscillatory nature.

My personal preference is to design for a closedloop damping coefficient of 0.3 to 0.5. As we will see throughout the rest of this book, this criterion is easy to use and reliable. Criterion like ISE can be used for any type of disturbance, setpoint, or load. Some "experts" (remember an "expert" is one who is seldom in doubt, but *often* in error) recommend different tuning parameters for the two types of disturbances. This makes little sense to me. What you want is a reasonable compromise between performance (tight control; small closedloop time constants) and robustness (not too sensitive to changes in process parameters). This compromise is achieved by using a closedloop damping coefficient of 0.3 to 0.5 since it keeps the real parts of the roots of the closedloop characteristic equation a reasonable distance from the imaginary axis, the point where the system becomes unstable (see Chap. 6). The closedloop damping coefficient specification is independent of the type of input disturbance.

The steadystate error is another time-domain specification. It is not a dynamic specification, but it is an important performance criterion. In many loops (but not all) a steadystate error of zero is desired, i.e., the value of the controlled variable should eventually level out at the setpoint.

7.2.2 Load Performance

The job of most control loops in a chemical process is one of regulation or load rejection, i.e., holding the controlled variable at its setpoint in the face of load disturbances. Let us look at the effects of load changes when the standard types of controllers are used.

We will use a simple heat-exchanger process (Fig. 7.10) in which an oil stream is heated with steam. The process outlet temperature T is controlled by manipulating steam flow rate F_s to the shell side of the heat exchanger. The oil flow rate F and the inlet oil temperature T_0 are load disturbances. The signal from the temperature transmitter (TT) is the process measurement signal, PM. The setpoint signal is SP. The output signal, CO, from the temperature controller (TC) goes through an I/P transducer to the steam control valve. The valve is AO because we want it to fail closed.

A. ON-OFF CONTROL. The simplest controller would be an on-off controller like the thermostat in your home heating system. The manipulated variable is either at maximum flow or at zero flow. The on-off controller is a proportional controller with a very high gain and gives "bang-bang" control action. This type of control is seldom used in a continuous process because of the cycling nature of the response, surging flows, and wear on control valves.

In the heat-exchanger example the controlled variable T cycles as shown in Fig. 7.11a. When a load disturbance in inlet temperature (a step decrease in T_0) occurs, both the period and the average value of the controlled variable T change. You have observed this in your heating system. When the outside temperature is colder, the furnace runs longer and more frequently, and the room temperature is lower on average. This is one of the reasons why you feel colder inside on a cold day than on a warm day for the same setting of the thermostat.

The system is really unstable in the classic linear sense. The nonlinear bounds or constraints on the manipulated variable (control-valve position) keep it in a "limit cycle."

B. PROPORTIONAL CONTROLLER. The output of a proportional controller changes only if the error signal changes. Since a load change requires a new control-valve position, the controller must end up with a new error signal. This means that a proportional controller usually gives a *steadystate error* or *offset*. This is an inherent limitation of P controllers and why integral action is usually added.

As shown in Fig. 7.11b for the heat-exchanger example, a decrease in process inlet temperature T_0 requires more steam. Therefore the error must increase to open the steam valve more. The magnitude of the offset depends on the size of the load disturbance and on the controller gain. The bigger the gain, the smaller the offset. As the gain is made bigger, however, the process becomes underdamped and eventually, at still higher gains, the loop will go unstable, acting like an on-off controller.

Steadystate error is not always undesirable. In many level control loops the absolute level is unimportant as long as the tank does not run dry or overflow. Thus a proportional controller is often the best type for level control. We will discuss this in more detail in Sec. 7.3.

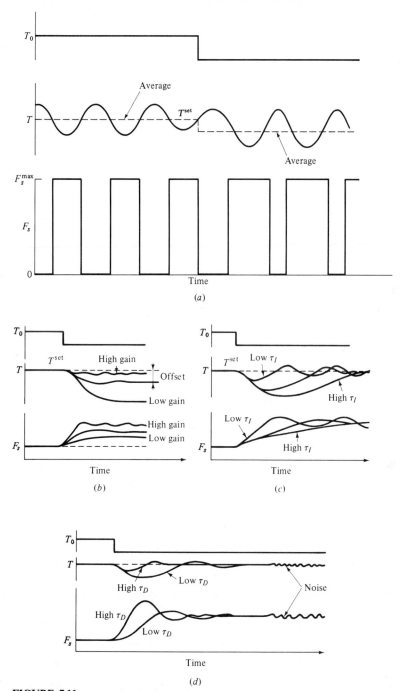

FIGURE 7.11
Controller load performance. (a) On-off controller; (b) proportional (P); (c) proportional-integral;
(d) proportional-integral-derivative (PID).

(a) Derivative on error

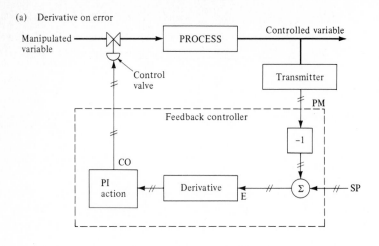

(b) Derivative on process measurement

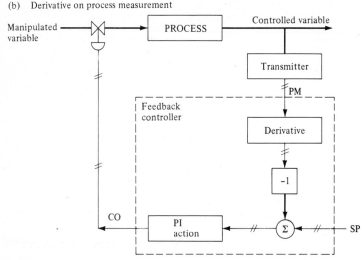

FIGURE 7.12
Derivative action.

C. PROPORTIONAL-INTEGRAL (PI) CONTROLLER. Most control loops use PI controllers. The integral action eliminates steadystate error in T (see Fig. 7.11c). The smaller the integral time τ_I, the faster the error is reduced. But the system becomes more underdamped as τ_I is reduced. If it is made too small, the loop becomes unstable.

D. PROPORTIONAL-INTEGRAL-DERIVATIVE (PID) CONTROLLER. PID controllers are used in loops where signals are not noisy and where tight dynamic response is important. The derivative action helps to compensate for lags in the loop. Temperature controllers in reactors are usually PID. The controller senses the rate of movement away from the setpoint and starts moving the control valve earlier than with only PI action (see Fig. 7.11*d*).

Derivative action can be used on either the error signal (SP − PM) or just the process measurement (PM). If it is on the error signal, step changes in setpoint will produce large bumps in the control valve. Therefore, in most process control applications, the derivative action is applied only to the PM signal as it enters the controller. The P and I action is then applied to the difference between the setpoint and the output signal from the derivative unit (see Fig. 7.12).

7.3 CONTROLLER TUNING

There are a variety of feedback controller tuning methods. Probably 80 percent of all loops are tuned experimentally by an instrument mechanic, and 75 percent of the time the mechanic can guess approximately what the settings will be by drawing on experience with similar loops. We will discuss a few of the time-domain methods below. In subsequent chapters we will present other techniques for finding controller constants in the Laplace and frequency domains.

7.3.1 Rules of Thumb

The common types of control loops are level, flow, temperature, and pressure. The type of controller and the settings used for any one type are sometimes pretty much the same from one application to another. For example, most flow control loops use PI controllers with wide proportional band and fast integral action.

Some heuristics are given below. They are not to be taken as gospel. They merely indicate common practice and they work in most applications.

A. FLOW LOOPS. PI controllers are used in most flow loops. A wide proportional band setting (PB = 150) or low gain is used to reduce the effect of the noisy flow signal due to flow turbulence. A low value of integral or reset time (τ_I = 0.1 minute per repeat) is used to get fast, snappy setpoint tracking.

The dynamics of the process are usually very fast. The sensor sees the change in flow almost immediately. The control valve dynamics are the slowest element in the loop. So a small reset time can be used.

There is one notable exception to fast PI flow control: flow control of condensate-throttled reboilers. As sketched in Fig. 7.13, the flow rate of vapor to a reboiler is sometimes controlled by manipulating the liquid condensate valve. Since the vapor flow depends on the rate of condensation, vapor flow can only be varied by changing the area for heat transfer in the reboiler. This is accomplished by raising or lowering the liquid level in this "flooded" reboiler. Changing the liquid level takes some time. Typical time constants are 3 to 6 minutes. Therefore, this flow control loop would have much different controller tuning constants than suggested in the rule-of-thumb cited above. Some derivative action may even be used in the loop to give faster flow control.

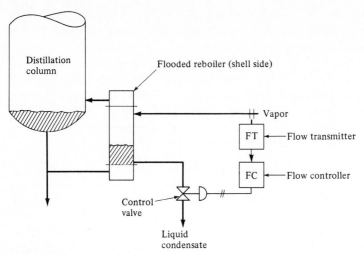

FIGURE 7.13
Condensate-throttling flow control.

B. LEVEL LOOPS. Most liquid levels represent material inventory used as surge capacity. In these cases it is relatively unimportant where the level is, as long as it is between some maximum and minimum levels. Therefore, proportional controllers are often used on level loops to give smooth changes in flow rates and to filter out fluctuations in flow rates to downstream units.

One of the most common errors in laying out a control structure for a plant with multiple units in series is the use of PI level controllers. If P controllers are used, the process flows rise or fall slowly down the train of units with no overshoot of flow rates. Liquid levels rise if flows increase and fall if flows decrease. Levels are *not* maintained at setpoints. See Fig. 7.14.

If PI level controllers are used, the integral action forces the level back to its setpoint. In fact if the level controller is doing a "perfect" job, the level is held right at its setpoint. This means that any change in the flow rate into the surge tank will immediately change the flow rate out of the tank. We might as well not even use a tank: just run the inlet pipe right into the outlet pipe! Thus, this is an example of where tight control is *not* desirable. We want the flow rate out of the tank to increase gradually when the inflow increases so as to not upset the downstream units.

Suppose the flow rate F_0 increases to the first tank in Fig. 7.14. The level h_1 in the first tank will start to increase. The level controller will start to increase F_1. When F_1 has increased to the point that it is equal to F_0, the level will stop changing since the tank is just an integrator. Now, if we use a P level controller, nothing else will happen. The level will remain at the higher level and the entering and exiting flows will be equal.

If, however, we use a PI level controller, the controller will continue to increase the outflow *beyond* the value of the inflow in order to drive the level

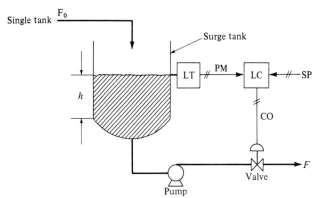

Single tank

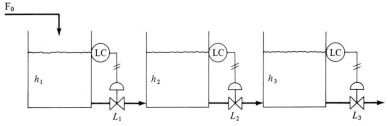

Units in series:

FIGURE 7.14
P versus PI level control.

back down to its setpoint. So an inherent problem with PI level controllers is that they amplify flow rate changes of this type. The change in the flow rate out of the tank is actually larger (for a period of time) than the change in the flow rate into the tank. This amplification gets worse as it works its way down through the series of units. What started out at the beginning as a small disturbance can result in large fluctuations by the time it reaches the last unit in the train.

There are, of course, many situations where it is desired to control level tightly, for example, in a reactor where control of residence time is important.

The tuning of proportional level controllers is a trivial job. For example, we could set the bias value at 50 percent of full scale, the setpoint at 50 percent of full scale, and the proportional band at 50. This means that the control valve will be half open when the tank is half full, wide open when the tank is 75 percent full, and completely shut when the tank is 25 percent full. Changing the proportional band to 100 would mean that the tank would be completely full to have the valve wide open and completely empty to have the valve shut.

C. PRESSURE LOOPS. Pressure loops vary from very tight, fast loops (almost like flow control) to slow averaging loops (almost like level control). An example of a fast pressure loop is the case of a valve throttling the flow of vapor from a vessel, as shown in Fig. 7.15a. The valve has a direct handle on pressure, and

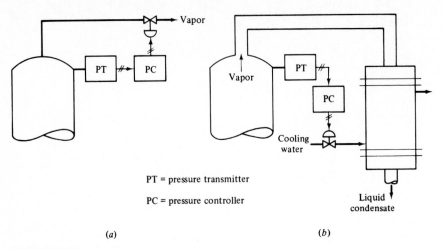

PT = pressure transmitter

PC = pressure controller

(a)

(b)

FIGURE 7.15
Pressure control. (a) Fast pressure loop; (b) slow pressure loop.

tight control can be achieved. An example of a slower pressure loop is shown in Fig. 7.15b. Pressure is held by throttling the water flow to a condenser. The water changes the ΔT driving force for condensation in the condenser. Therefore, the heat transfer dynamics and the lag of the water flowing through the shell side of the condenser are introduced into the pressure control loop.

D. TEMPERATURE LOOPS. Temperature control loops are usually moderately slow because of the sensor lags and the process heat transfer lags. PID controllers are often used. Proportional band settings are fairly low, depending on temperature transmitter spans and control-valve sizes. The reset time is of the same order as the process time constant; i.e., the faster the process, the smaller τ_I can be set. Derivative time is set something like one-fourth the process time constant, depending on the noise in the transmitter signal. We will quantify these tuning numbers later in the book.

7.3.2 On-Line Trial and Error

To tune a controller on line, a good instrument mechanic follows a procedure something like the following:

1. With the controller on manual, take all the integral and derivative action out of the controller, i.e., set τ_I at maximum minutes per repeat and τ_D at minimum minutes.
2. Set the PB at a high value, perhaps 200.
3. Put the controller on automatic.

4. Make a small setpoint or load change and observe the response of the controlled variable. The gain is low so the response will be sluggish.

5. Reduce the PB by a factor of 2 (double the gain) and make another small change in setpoint or load.

6. Keep reducing PB, repeating step 5, until the loop becomes very underdamped and oscillatory. The gain at which this occurs is called the *ultimate gain*.

7. Back-off on the PB to twice this ultimate value.

8. Now start bringing in integral action by reducing τ_I by factors of 2, making small disturbances at each value of τ_I to see the effect.

9. Find the value of τ_I that makes the loop very underdamped and set τ_I at twice this value.

10. Start bringing in derivative action by increasing τ_D. Load changes should be used to disturb the system and the derivative should act on the process measurement signal. Find the value of τ_D that gives the tightest control without amplifying the noise in the process measurement signal.

11. Reduce the PB again by steps of 10 percent until the desired specification on damping coefficient or overshoot is satisfied.

It should be noted that there are some loops where these procedures do *not* work. Systems that exhibit "conditional stability" are the classic example. These processes are unstable at high values of controller gain *and* are also unstable at low values of controller gain, but are stable over some intermediate range of gains. We will discuss some of these in Chap. 10.

7.3.3 Ziegler-Nichols Method

The Ziegler-Nichols (ZN) controller settings (J. G. Ziegler and N. B. Nichols, *Trans. ASME*, Vol. 64, 1942, p. 759) are pseudostandards in the control field. They are easy to find and to use and give reasonable performance on some loops. The ZN settings are benchmarks against which the performance of other controller settings are compared in many studies. They give reasonable first guesses of settings. There are many loops where the ZN settings are *not* very good. They tend to be too underdamped for most process control applications. Some on-line tuning can improve control significantly. But the ZN settings are useful as a place to start.

The ZN method consists of first finding the ultimate gain K_u, the value of gain at which the loop is at the limit of stability with a proportional-only feedback controller. The period of the resulting oscillation is called the *ultimate period*, P_u (minutes per cycle). The ZN settings are then calculated from K_u and P_u by the formulas given in Table 7.1 for the three types of controllers. Notice that a lower gain is used when integration is included in the controller (PI) and that the addition of derivative permits a higher gain and faster reset.

TABLE 7.1
Ziegler-Nichols settings

	P	PI	PID
K_c	$\dfrac{K_u}{2}$	$\dfrac{K_u}{2.2}$	$\dfrac{K_u}{1.7}$
τ_I (minutes)	....	$\dfrac{P_u}{1.2}$	$\dfrac{P_u}{2}$
τ_D (minutes)	....	....	$\dfrac{P_u}{8}$

The isothermal three-CSTR process of Sec 5.2 has, as we will prove in Chap. 10, an ultimate gain of 64 and an ultimate period of 3.63 minutes. The ZN settings for this system are given in Table 7.2. The response of the closedloop system to a step load disturbance in C_{AD} is shown in Fig. 7.16 with P, PI, and PID controllers and the ZN settings.

These results show several interesting things:

1. There is a steadystate error in the controlled variable C_{A3} when a P controller is used. This offset results because there is no integral term to drive the error to zero.

2. The ZN settings for all the controllers result in a fairly underdamped system: the responses show significant oscillation. The closedloop damping coefficient of this system is about 0.1 to 0.2. By way of contrast, the curve marked "PI_{+2}" (which will be explained in Chap. 13) gives a closedloop damping coefficient of about 0.5. The control is less tight (more deviation in the controlled variable) but there is a more gradual change in the manipulated variable and the response is not as oscillatory.

   In many process control applications, this kind of response is more desirable than snappy response that calls for rapid and large changes in the manipulated variable. For example, in the control of a tray temperature in a distillation column we want tight temperature control, but we do not want rapid or large changes in the heat input to the reboiler because the column may flood during one of the transients. If the surge in vapor rate is too rapid, we could even mechanically damage the trays in the column. So we must

TABLE 7.2
ZN settings for 3-CSTR process

	P	PI	PID
K_c	32	29.1	37.6
τ_I (minutes)	...	3.03	1.82
τ_D (minutes)	...	......	0.453

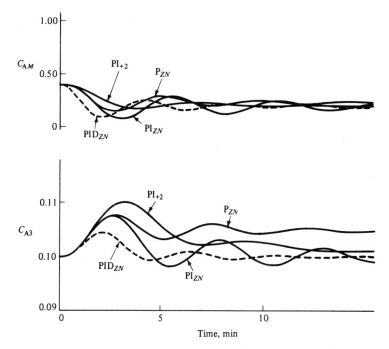

FIGURE 7.16
Three-CSTR example with Ziegler-Nichols settings.

sacrifice some performance (tight control) for smoother and less violent changes in the manipulated variable. The tuning procedure that gives the PI_{+2} curve achieves this looser control.

3. Control is improved when the PID controller is used. There is less deviation in the controlled variable because the manipulated variable C_{AM} changes more quickly. As discussed above, if rapid and large changes in the manipulated variable cannot be tolerated, derivative action cannot be used to improve the control performance.

There are many other tuning methods. One of the most simple uses the step response of the process to determine steadystate gain, time constant, and dead-time (see Chap. 14 for more details). Then controller tuning constants can be calculated from these values. Smith and Corripio (*Principles and Practice of Automatic Process Control*, 1985, p. 216, John Wiley & Sons, New York) give a clear discussion of this method. I prefer the ultimate gain and ultimate frequency approach because of the very significant problem of nonlinearity that exists in most chemical engineering systems. Step testing drives the process away from its initial steadystate and is, therefore, much more sensitive to nonlinearity than is the closedloop ultimate-gain method in which the process is held in a region near the initial steadystate. We will discuss this more in Chap. 14.

PROBLEMS

7.1. (a) Calculate the gain of an orifice plate and differential-pressure transmitter for flow rates from 10 to 90 percent of full scale.

(b) Calculate the gain of linear and equal-percentage valves over the same range, assuming constant pressure drop over the valve.

(c) Calculate the total loop gain of the valve and the sensor-transmitter system over this range.

7.2. The temperature of a CSTR is controlled by an electronic (4 to 20 mA) feedback control system containing (1) a 100 to 200°F temperature transmitter, (2) a PI controller with integral time set at 3 minutes and proportional band at 25, and (3) a control valve with linear trim, air-to-open action, and a $C_v = 4$ through which cooling water flows. The pressure drop across the valve is a constant 25 psi. If the steadystate controller output is 12 mA, how much cooling water is going through the valve? If a sudden disturbance increases reactor temperature by 5°F, what will be the immediate effect on the controller output signal and the water flow rate?

7.3. Simulate the three-CSTR system on a digital computer with an on-off feedback controller. Assume the manipulated variable C_{AM} is limited to ± 1 mol of A/ft^3 around the steadystate value. Find the period of oscillation and the average value of C_{A3} for values of the load variable C_{AD} of 0.6 and 1.

7.4. Two ways to control the outlet temperature of a heat-exchanger cooler are sketched below. Comment on the relative merits of these two systems from the standpoints of both control and heat-exchanger design.

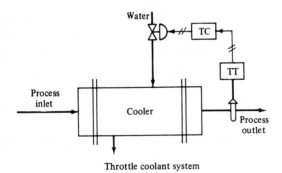

Throttle coolant system

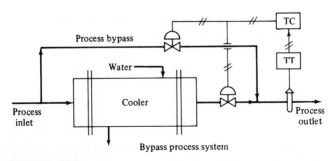

Bypass process system

FIGURE P7.4

7.5. Specify the following items for the bypass cooler system of Prob. 7.4:
(a) The action of the valves (AO or AC) and kind of trim.
(b) The action and type of controller.

7.6. Assume that the bypass cooling system of Prob. 7.4 is designed so that the total process flow of 50,000 lb_m/h (heat capacity of 0.5 Btu/lb_m °F) is split under normal conditions, 25 percent going around the bypass and 75 percent going through the cooler. Process inlet and outlet temperatures under these conditions are 250 and 150°F. Inlet and outlet water temperatures are 80 and 120°F. Process side pressure drop through the exchanger is 10 psi. The control valves have linear trim and are designed to be half open at design rates with a 10 psi drop over the valve in series with the cooler. Liquid density is constant at 62.3 lb_m/ft^3.

What will the valve positions be if the total process flow is reduced to 25 percent of design and the process outlet temperature is held at 150°F?

7.7. A liquid (sp gr = 1) is pumped through a heat exchanger and a control valve at a design rate of 200 gpm. The exchanger pressure drop is 30 psi at design throughput. Make plots of flow rate versus valve position x for linear and equal-percentage ($\alpha = 50$) control valves. Both valves are set at $f_{(x)} = 0.5$ at design rate. The total pressure drop over the entire system is constant. The pressure drop over the control valve at design rate is:
(a) 10 psi
(b) 30 psi
(c) 120 psi

7.8. Process designers sometimes like to use "dephlegmators" or partial condensers mounted directly in the top of the distillation column when the overhead product is taken off as a vapor. They are particularly popular for corrosive, toxic, or hard-to-handle chemicals since they eliminate a separate condenser shell, a reflux drum, and a reflux pump. Comment on the relative controllability of the two process systems sketched below.

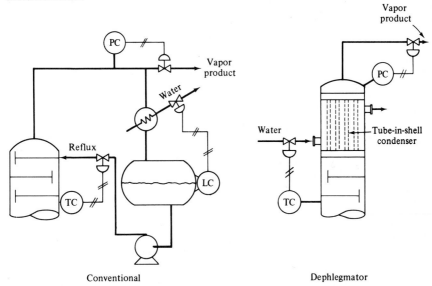

Conventional   Dephlegmator

FIGURE P7.8

7.9. Compare quantitatively by digital simulation the dynamic performance of the three coolers sketched below with countercurrent flow, cocurrent flow, and circulating water systems. Assume the tube and shell sides can each be represented by four perfectly mixed lumps.

Process design conditions are:
> Flow rate = 50,000 lb_m/h
> Inlet temperature = 250°F
> Outlet temperature = 130°F
> Heat capacity = 0.5 Btu/lb_m °F

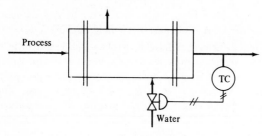

Countercurrent

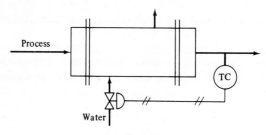

Cocurrent

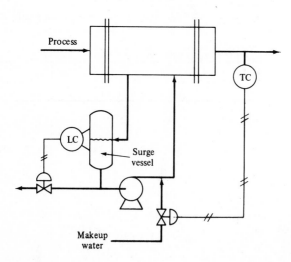

FIGURE P7.9

Cooling-water design conditions are:

    A. *Countercurrent*
        Inlet temperature = 80°F
        Outlet temperature = 130°F
    B. *Cocurrent*
        Inlet temperature = 80°F
        Outlet temperature = 125°F
    C. *Circulating system*
        Inlet temperature to cooler = 120°F
        Outlet temperature from cooler = 125°F
        Makeup water temperature to system = 80°F

Neglect the tube and shell metal. Tune PI controllers experimentally for each system. Find the outlet temperature deviations for a 25 percent step increase in process flow rate.

7.10. The overhead vapor from a depropanizer distillation column is totally condensed in a water-cooled condenser at 120°F and 227 psig. The vapor is 95 mol % propane and 5 mol % isobutane. Its design flow rate is 25,500 lb_m/h and average latent heat of vaporization is 125 Btu/lb_m.

    Cooling water inlet and outlet temperatures are 80 and 105°F, respectively. The condenser heat transfer area is 1000 ft². The cooling water pressure drop through the condenser at design rate is 5 psi. A linear-trim control valve is installed in the cooling water line. The pressure drop over the valve is 30 psi at design with the valve half open.

    The process pressure is measured by an electronic (4–20 mA) pressure transmitter whose range is 100–300 psig. An analog electronic proportional controller with a gain of 3 is used to control process pressure by manipulating cooling water flow. The electronic signal from the controller (CO) is converted into a pneumatic signal in the I/P transducer.

(a) Calculate the cooling water flow rate (gpm) at design conditions.
(b) Calculate the size coefficient (C_v) of the control valve.
(c) Specify the action of the control valve and the controller.
(d) What are the values of the signals PM, CO, SP, and PV at design conditions?

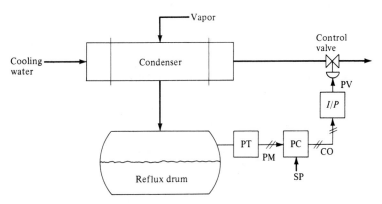

FIGURE P7.10

(e) Suppose the process pressure jumps 10 psi. How much will the cooling water flow rate increase? Give values for PM, CO, and PV at this higher pressure. Assume that the total pressure drop over the condenser and control valve is constant.

7.11. A circulating chilled-water system is used to cool an oil stream from 90 to 70°F in a tube-in-shell heat exchanger. The temperature of the chilled water entering the process heat exchanger is maintained constant at 50°F by pumping the chilled water through a refrigerated cooler located upstream of the process heat exchanger.

The design chilled-water rate for normal conditions is 1000 gpm, with chilled water leaving the process heat exchanger at 60°F. Chilled-water pressure drop through the process heat exchanger is 15 psi at 1000 gpm. Chilled-water pressure drop through the refrigerated cooler is 15 psi at 1000 gpm. The heat transfer area of the process heat exchanger is 1143 ft².

The temperature transmitter on the process oil stream leaving the heat exchanger has a range of 50–150°F. The range of the orifice-differential pressure flow transmitter on the chilled water is 0–1500 gpm. All instrumentation is electronic (4 to 20 mA). Assume the chilled-water pump is centrifugal with a flat pump curve.

(a) Design the chilled-water control valve so that it is 25 percent open at the 1000 gpm design rate and can pass a maximum flow of 1500 gpm. Assume linear trim is used.

(b) Give values of the signals from the temperature transmitter, temperature controller, and chilled-water flow transmitter when the chilled-water flow is 1000 gpm.

(c) What is the pressure drop over the chilled-water valve when it is wide open?

(d) What are the pressure drop and fraction open of the chilled-water control valve when the chilled-water flow rate is reduced to 500 gpm? What is the chilled-water flow transmitter output at this rate?

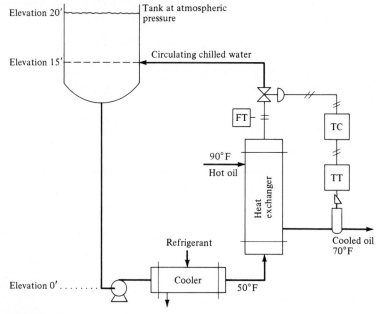

FIGURE P7.11

(e) If electric power costs 2.5 cents per kilowatthour, what are the annual pumping costs for the chilled-water pump at the design 1000 gpm rate? What horsepower motor is required to drive the chilled-water pump? (1 hp = 550 ft·lb$_f$/s = 746 watts).

7.12. Tray 4 temperature on the Lehigh distillation column is controlled by a pneumatic PI controller with a 2-minute reset time and a 50 percent proportional band. Temperature controller output (CO$_T$) adjusts the setpoint of a steam flow controller (reset time 0.1 min and proportional band 100 percent). Column base level is controlled by a pneumatic proportional-only controller setting bottoms product withdrawal rate.

Transmitter ranges are:

Temperature tray 4	60–120°C
Steam flow	0–4.2 lb$_m$/min (orifice/ΔP transmitter)
Bottoms flow	0–1 gpm (orifice/ΔP transmitter)
Base level	0–20 in H$_2$O

Steadystate operating conditions are:

$$\text{Tray 4 temperature} = 83°C$$
$$\text{Base level} = 55\% \text{ full}$$
$$\text{Steam flow} = 3.5 \text{ lb}_m/\text{min}$$
$$\text{Bottom flow} = 0.6 \text{ gpm}$$

Pressure drop over the control valve on the bottoms product is a constant at 30 psi. This control valve has linear trim and a C_v of 0.5. The formula for steam flow through a control valve (when the upstream pressure P_s in psia is greater than twice the downstream pressure) is

$$W = \frac{3C_v}{2} P_s X$$

where W = steam flow rate (lb$_m$/h)
  $C_v = 4$
  X = valve fraction open (linear trim)

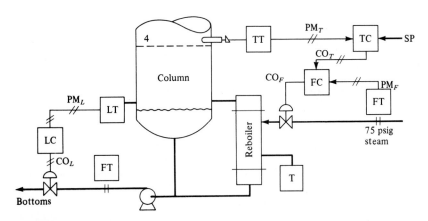

FIGURE P7.12

(a) Calculate the control signals from the base level transmitter, temperature transmitter, steam flow transmitter, bottoms flow transmitter, temperature controller, steam flow controller, and base level controller.

(b) What is the instantaneous effect of a $+5°C$ step change in tray 4 temperature on the control signals and flow rates?

7.13. A reactor is cooled by a circulating jacket water system. A double cascade reactor temperature control to jacket temperature control to makeup cooling water flow control is employed.

   Instrumentation details are given below (electronic, 4–20 mA):
      Reactor temperature transmitter range: 50–250°F
      Circulating jacket water temperature transmitter range: 50–150°F
      Makeup cooling water flow transmitter range: 0–250 gpm (orifice plate + differential pressure transmitter)
      Control valve: linear trim, constant 35 psi pressure drop
   Normal operating conditions are:
      Reactor temperature = 140°F
      Circulating water temperature = 106°
      Makeup water flow rate = 63 gpm
      Control valve is 25 percent open

(a) Specify the action and size of the makeup cooling water control valve.

(b) Calculate the milliampere control signals from all transmitters and controllers at normal operating conditions.

(c) Specify whether each controller is reverse- or direct-acting.

(d) Calculate the instantaneous values of all control signals if reactor temperature increases suddenly 10°F.

Proportional band settings of the reactor temperature controller, circulating jacket water temperature controller, and cooling water flow controller are 20, 67, and 200, respectively.

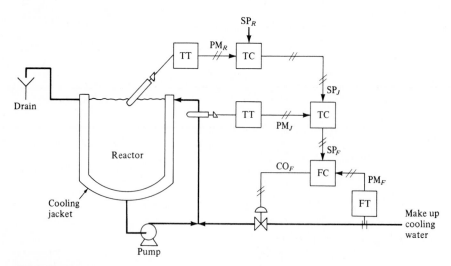

FIGURE P7.13

7.14. Three vertical cylindrical tanks (10 feet high, 10 feet diameter) are used in a process. Two tanks are process tanks and are level controlled by manipulating outflows using proportional-only level controllers (PB = 100). Level transmitter spans are 10 feet. Control valves are linear, 50 percent open at the normal liquid rate of 1000 gpm, air-to-open, constant pressure drop. These two process tanks are 50 percent full at the normal liquid rate of 1000 gpm.

The third tank is a surge tank whose level is uncontrolled. Liquid is pumped from this tank to the first process vessel, on to second tank in series, and then back to the surge tank. If the surge tank is half full when 1000 gpm of liquid is circulated, how full will the surge tank be, at the new steadystate, when circulating rate around the system is cut to 500 gpm?

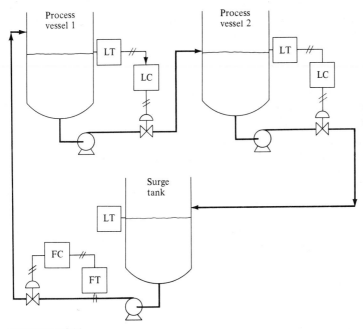

FIGURE P7.14

7.15. Liquid (sp gr = 1) is pumped from a tank at atmospheric pressure through a heat exchanger and a control valve into a process vessel held at 100 psig pressure. The system is designed for a maximum flow rate of 400 gpm. At this maximum flow rate the pressure drop across the heat exchanger is 50 psi.

A centrifugal pump is used with a performance curve that can be approximated by the relationship

$$\Delta P_p = 198.33 - 1.458 \times 10^{-4} F^2$$

where ΔP_p = pump head in psi
F = flow rate in gpm

The control valve has linear trim.

(a) Calculate the fraction that the control valve is open when the throughput is reduced to 200 gpm by pinching down on the control valve.

(b) An orifice-plate/differential pressure transmitter is used for flow measurement. If the maximum full-scale flow reading is 400 gpm, what will the output signal from the electronic flow transmitter be when the flow rate is reduced to 150 gpm?

7.16. Design liquid level control systems for the base of a distillation column and for the vaporizer shown below. Steam flow to the vaporizer is held constant and cannot be used to control level. Liquid feed to the vaporizer can come from the column and/or from the surge tank. Liquid from the column can go to the vaporizer and/or to the surge tank.

Since the liquid must be cooled if it is sent to the surge tank and then reheated in the vaporizer, there is an energy cost penalty associated with sending more material to the surge tank than is absolutely necessary. Your level control system should therefore hold both levels and also minimize the amount of material sent to the surge tank. (*Hint:* One way to accomplish this is to make sure that the valves in the lines to and from the surge tank cannot be opened simultaneously.)

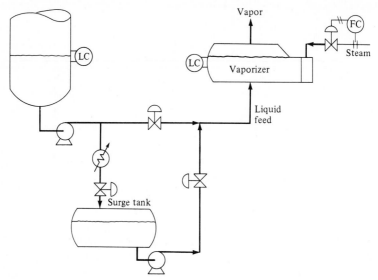

FIGURE P7.16

7.17. A chemical reactor is cooled by a circulating oil system as sketched in Fig. P7.17. Oil is circulated through a water-cooled heat exchanger and through control valve V_1. A portion of the oil stream can be bypassed around the heat exchanger through control valve V_2. The system is to be designed so that at design conditions:

- The oil flow rate through the heat exchanger is 50 gpm (sp gr = 1) with a 10 psi pressure drop across the heat exchanger and with the V_1 control valve 25 percent open.
- The oil flow rate through the bypass is 100 gpm with the V_2 control valve 50 percent open.

Both control valves have linear trim. The circulating pump has a flat pump curve. A maximum oil flow rate through the heat exchanger of 100 gpm is required.

(a) Specify the action of the two control valves and the two temperature controllers.

(b) Calculate the size (C_v's) of the two control valves and the design pressure drops over the two valves.

(c) How much oil will circulate through the bypass valve if it is wide open and the valve in the heat exchanger loop is shut?

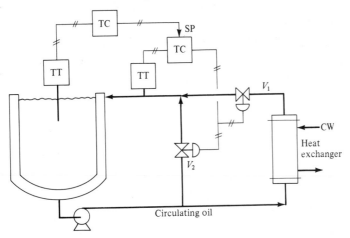

FIGURE P7.17

7.18. The formula for the flow of saturated steam through a control valve is

$$W = 2.1 C_v f_{(x)} \sqrt{(P_1 + P_2)(P_1 - P_2)}$$

where $W = \text{lb}_m/\text{h}$ steam
$P_1 = $ upstream pressure, psia
$P_2 = $ downstream pressure, psia

The temperature of the steam-cooled reactor shown below is 285°F. The heat that must be transferred from the reactor into the steam generation system is 25×10^6 Btu/h. The overall heat transfer coefficient for the cooling coils is 300 Btu/h ft² °F. The steam discharges into a 25-psia steam header. The enthalpy difference

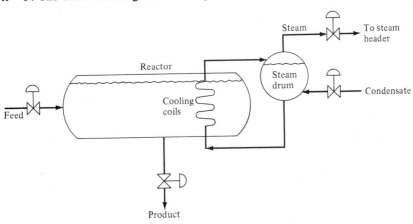

FIGURE P7.18

between saturated steam and liquid condensate is 1000 Btu/lb$_m$. The vapor pressure of water can be approximated over this range of pressure by a straight line.

$$T \ (°F) = 195 + 1.8P \ (psia)$$

Design two systems, one where the steam drum pressure is 40 psia at design and another where it is 30 psia.

(a) Calculate the area of the cooling coils for each case.
(b) Calculate the C_v value for the steam valve in each case, assuming that the valve is half open at design conditions: $f_{(x)} = 0.5$.
(c) What is the maximum heat removal capacity of the system for each case?

7.19. Cooling water is pumped through the jacket of a reactor. The pump and the control valve must be designed so that:

(a) The normal cooling water flow rate is 250 gpm.
(b) The maximum emergency rate is 500 gpm.
(c) The valve cannot be less than 10 percent open when the flow rate is 100 gpm.

Pressure drop through the jacket is 10 psi at design. The pump curve has a linear slope of -0.1 psi/gpm.

Calculate the C_v value of the control valve, the pump head at design rate, the size of the motor required to drive the pump, the fraction that the valve is open at design, and the pressure drop over the valve at design rate.

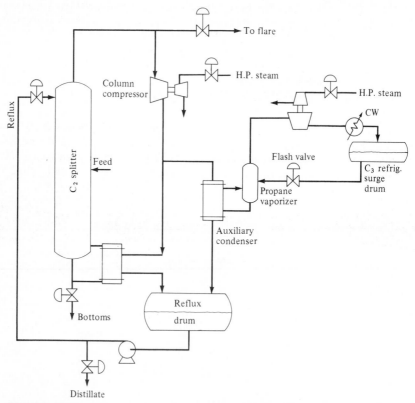

FIGURE P7.20

7.20. A C_2 splitter column uses vapor recompression. Because of the low temperature required to stay below the critical temperatures of ethylene and ethane, the auxiliary condenser must be cooled by a propane refrigeration system.

(a) Specify the action of all control valves.

(b) Sketch a control concept diagram which accomplishes the following objectives:

(i) Level in the propane vaporizer is controlled by the liquid propane flow from the refrigeration surge drum.

(ii) Column pressure is controlled by adjusting the speed of the column compressor through a steam-flow-control–speed-control–pressure-control cascade system.

(iii) Reflux is flow controlled. Reflux drum level sets distillate flow. Base level sets bottoms flow.

(iv) Column tray 10 temperature is controlled by adjusting the pressure in the propane vaporizer, which is controlled by refrigeration compressor speed.

(v) High column pressure opens the valve to the flare.

(c) How effective do you think the column temperature control will be? Suggest an improved control system which still achieves minimum energy consumption in the two compressors.

7.21. Hot oil from the base of a distillation column is used to reboil two other distillation columns that operate at lower temperatures. The design flow rates through reboilers 1 and 2 are 100 and 150 gpm, respectively. At these flow rates, the pressure drops through the reboilers are 20 and 30 psi. The hot oil pump has a flat pump curve.

Size the two control valves and the pump so that:

(a) Maximum flow rates through each reboiler can be at least twice design.

(b) At minimum turndown rates where only half the design flow rates are required, the control valves are no less than 10 percent open.

What is the fraction of valve opening for each valve at design rates?

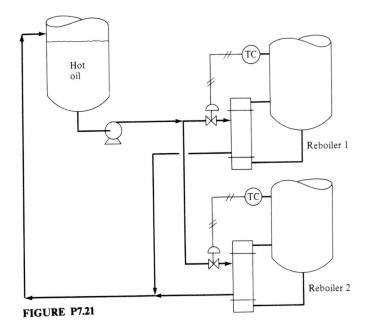

FIGURE P7.21

7.22. A reactor is cooled by circulating liquid through a heat exchanger that produces low-pressure (10 psig) steam. This steam is then split between a compressor and a turbine. The portion that goes through the turbine drives the compressor. The portion that goes through the compressor is used by 50 psig steam users. 100 psig steam can also be used in the turbine to provide power required beyond that available in the 10 psig steam.

Sketch a control concept diagram that includes all valve actions and the following control strategies:

(a) Reactor temperature is controlled by changing the setpoint of the turbine speed controller.

(b) Turbine speed is controlled by two split-range valves, one on the 10 psig inlet to the turbine and the other on the 100 psig steam that can also be used to drive the turbine. Your instrumentation system should be designed so that the valve on the 10 psig steam is wide open before any 100 psig steam is used.

(c) Liquid circulation from the reactor to the heat exchanger is flow-controlled.

(d) Condensate level in the condensate drum is controlled by manipulating BFW (boiler feed water).

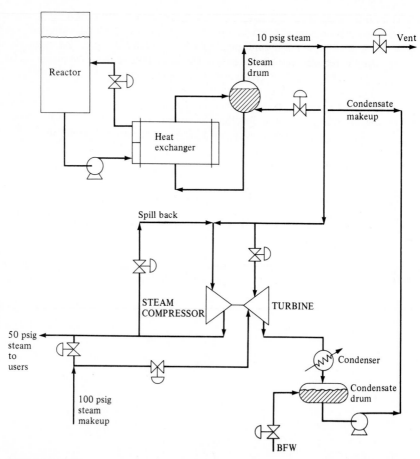

FIGURE P7.22

(e) Condensate makeup to the steam drum is ratioed to the 10 psig steam flow rate from the steam drum. This ratio is then reset by the steam drum level controller.

(f) Pressure in the 50 psig steam header is controlled by adding 100 psig steam.

(g) A high-pressure controller opens the vent valve on the 10 psig header when the pressure in the 10 psig header is too high.

(h) Compressor surge is prevented by using a low flow controller that opens the valve in the spill-back line from compressor discharge to compressor suction.

7.23. Water is pumped from an atmospheric tank, through a heat exchanger and a control valve, into a pressurized vessel. The operating pressure in the vessel can vary from 200 to 300 psig, but is 250 psig at design. Design flow rate is 100 gpm with a 20 psi pressure drop through the heat exchanger. Maximum flow rate is 150 gpm. Minimum flow rate is 25 gpm. A centrifugal pump is used which has a straight-line pump curve with a slope of -0.1 psi/gpm.

Design the control valve and pump so that both the maximum and minimum flow rates can be handled with the valve never less than 10 percent open.

7.24. Reactant liquid is pumped into a batch reactor at a variable rate. The reactor pressure also varies during the batch cycle. Specify the control valve size and the centrifugal pump head required. Assume a flat pump curve.

The initial flow rate into the reactor is 20 gpm (sp gr $= 1$). It is decreased linearly with time down to 5 gpm at 5 hours into the batch cycle. The initial reactor pressure is 50 psig. It increases linearly with time up to 350 psig at 5 hours. The reactant liquid comes from a tank at atmospheric pressure.

7.25. Water is pumped from an atmospheric tank into a vessel at 50 psig through a heat exchanger. There is a bypass around the heat exchanger. The pump has a flat curve. The heat exchanger pressure drop is 30 psi with 200 gpm of flow through it.

Size the pump and the two control valves so that:

(a) 200 gpm can be bypassed.

(b) Flow through the heat exchanger can be varied from 75 to 300 gpm.

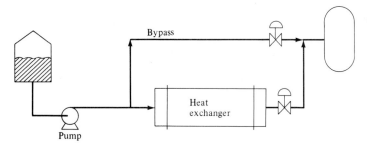

FIGURE P7.25

7.26. An engineer from Catastrophic Chemical Company has designed a system in which a positive-displacement pump is used to pump water from an atmospheric tank into a pressurized tank operating at 150 psig. A control valve is installed between the pump discharge and the pressurized tank.

With the pump running at a constant speed and stroke length, 350 gpm of water is pumped when the control valve is wide open and the pump discharge pressure is 200 psig.

If the control valve is pinched back to 50 percent open, what will be the flow rate of water and the pump discharge pressure?

7.27. Hot oil from a tank at 400°F is pumped through a heat exchanger to vaporize a liquid boiling at 200°F. A control valve is used to set the flow rate of oil through the loop. Assume the pump has a flat pump curve. The pressure drop over the control valve is 30 psi and the pressure drop over the heat exchanger is 35 psi under the normal design conditions given below:

> Heat transferred in heat exchanger = 17×10^6 Btu/h
> Hot oil inlet temperature = 400°F
> Hot oil exit temperature = 350°F
> Fraction valve open = 0.8

The hot oil gives off sensible heat only (heat capacity = 0.5 Btu/lb$_m$ °F, density = 4.58 lb$_m$/gal). The heat transfer area in the exchanger is 652 ft^2. Assume the temperature on the tube side of the heat exchanger stays constant at 200°F and the inlet hot oil temperature stays constant at 400°F. A log mean temperature difference must be used.

Assuming the heat transfer coefficient does not change with flow rate, what will the valve opening be when the heat transfer rate in the heat exchanger is half the normal design value?

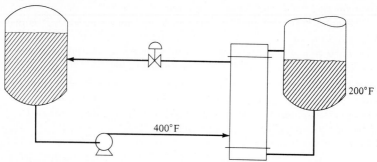

FIGURE P7.27

7.28. A control valve/pump system proposed by Connell (*Chemical Engineering*, September 28, 1987, p. 123) consists of a centrifugal pump, several heat exchangers, a furnace, an orifice, and a control valve. Liquid is pumped through this circuit and up into a column that operates at 20 psig. Because the line running up the column is full of liquid, there is a hydraulic pressure differential between the base of the column and the point of entry into the column of 15 psi.

The pump suction pressure is constant at 10 psig. The design flow rate is 500 gpm. At this flow rate the pressure drop over the flow orifice is 2 psi, through the piping is 30 psi, over three heat exchangers is 32 psi, and over the furnace is 60 psi. Assume a flat pump curve and a specific gravity of 1.

Connell recommends that a control valve be used that takes a 76 psi pressure drop at design flow rate. The system should be able to increase flow to 120 percent of design.

(*a*) Calculate the pressure drop over the valve at the maximum flow rate.
(*b*) Calculate pump discharge pressure and the control valve C_v.
(*c*) Calculate the fraction that the valve is open at design.
(*d*) If turndown is limited to a valve opening of 10 percent, what is the minimum flow rate?

CHAPTER
8

ADVANCED
CONTROL
SYSTEMS

In the previous chapter we discussed the elements of a conventional single-input–single-output (SISO) feedback control loop. This configuration forms the backbone of almost all process control structures.

However, over the years a number of slightly more complex structures have been developed that can, in some cases, significantly improve the performance of a control system. These structures include ratio control, cascade control, override control, etc. We will devote much of this chapter to these subjects.

Also covered in this chapter will be some guidelines for developing an appropriate control system structure for a single unit and for a group of units that form a plant. Several realistic examples will be presented.

Finally, a brief discussion is given of a new type of control algorithm called *dynamic matrix control*. This is a time-domain method that uses a model of the process to calculate future changes in the manipulated variable such that an objective function is minimized. It is basically a least-squares solution.

8.1 RATIO CONTROL

As the name implies, ratio control involves keeping constant the ratio of two or more flow rates. The flow rate of the "wild" or uncontrolled stream is measured and the flow rate of the manipulated stream is changed to keep the two streams at a constant ratio with each other. Common examples include (1) holding a constant reflux ratio on a distillation column, (2) keeping stoichiometric amounts of two reactants being fed into a reactor, and (3) purging off a fixed percentage of the feed stream to a unit.

Ratio control is achieved by two alternative schemes, shown in Fig. 8.1. In the scheme shown at the top of the figure, the two flow rates are measured and their ratio is computed (by the divider). This computed ratio signal is fed into a conventional PI controller as the process measurement signal. The setpoint of the ratio controller is the desired ratio. The output of the controller goes to the valve

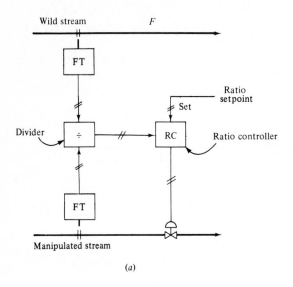

(a)

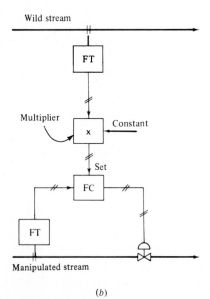

(b)

FIGURE 8.1
Ratio control. (a) Ratio compute; (b) flow set.

on the manipulated variable stream that changes its flow rate in the correct direction to hold the ratio of the two flows constant. This computed ratio signal can also be used to trigger an alarm or an interlock.

In the scheme shown at the bottom of Fig. 8.1, the wild flow is measured and this flow signal is multiplied by a constant, which is the desired ratio. The output of the multiplier is the setpoint of a remote-set flow controller on the manipulated variable.

If orifice plates are used as flow sensors, the signals from the differential-pressure transmitters are really the squares of the flow rates. Some instrument engineers prefer to put in square-root extractors and convert everything to linear flow signals.

Ratio control is often part of a *feedforward* control structure that we will discuss in Sec. 8.6.

8.2 CASCADE CONTROL

One of the most useful concepts in advanced control is cascade control. A cascade control structure has two feedback controllers with the output of the primary (or master) controller changing the setpoint of the secondary (or slave) controller. The output of the secondary goes to the valve, as shown in Fig. 8.2a.

There are two purposes for cascade control: (1) to eliminate the effects of some disturbances, and (2) to improve the dynamic performance of the control loop.

To illustrate the disturbance rejection effect, consider the distillation column reboiler shown in Fig. 8.2a. Suppose the steam supply pressure increases. The pressure drop over the control valve will be larger, so the steam flow rate will increase. With the single-loop temperature controller, no correction will be made until the higher steam flow rate increases the vapor boilup and the higher vapor rate begins to raise the temperature on tray 5. Thus the whole system is disturbed by a supply-steam pressure change.

With the cascade control system, the steam flow controller will immediately see the increase in steam flow and will pinch back on the steam valve to return the steam flow rate to its setpoint. Thus the reboiler and the column are only slightly affected by the steam supply-pressure disturbance.

Figure 8.2b shows another common system where cascade control is used. The reactor temperature controller is the primary controller; the jacket temperature controller is the secondary controller. The reactor temperature control is isolated by the cascade system from disturbances in cooling-water inlet temperature and supply pressure.

This system also is a good illustration of the improvement in dynamic performance that cascade control can provide in some systems. As we will show quantitatively in Chap. 11, the closedloop time constant of the reactor temperature will be smaller when the cascade system is used than when reactor temperature sets the cooling water makeup valve directly. Therefore performance has been improved by using cascade control.

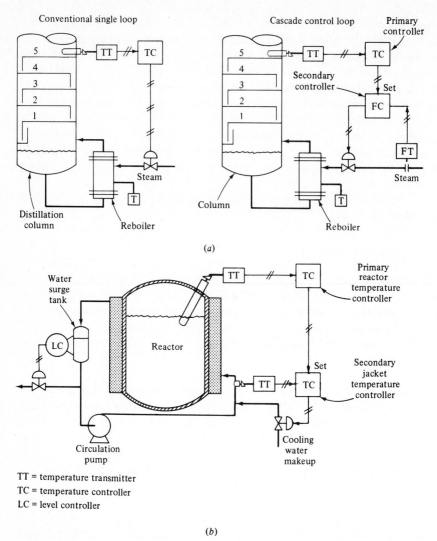

FIGURE 8.2
Conventional versus cascade control. (a) Distillation-column-reboiler temperature control; (b) CSTR temperature control.

We will also talk in Chap. 11 about the two types of cascade control: series cascade and parallel cascade. The two examples discussed above are both series cascade systems because the manipulated variable affects the secondary controlled variable, which then affects the primary variable. In a parallel cascade system the manipulated variable affects *both* the primary and the secondary controlled variables directly. Thus the two processes are basically different and result in different dynamic characteristics. We will quantify these ideas later.

8.3 COMPUTED VARIABLE CONTROL

One of the most logical and earliest extensions of conventional control was the idea of controlling the variable that was of real interest by computing its value from other measurements.

For example, suppose we want to control the mass flow rate of a gas. Controlling the pressure drop over the orifice plate gives only an approximate mass flow rate because gas density varies with temperature and pressure in the line. By measuring temperature, pressure, and orifice-plate pressure drop, and feeding these signals into a mass-flow-rate computer, the mass flow rate can be controlled as sketched in Fig. 8.3a.

Another example is sketched in Fig. 8.3b. A hot oil stream is used to reboil a distillation column. Controlling the flow rate of the hot oil does not guarantee a fixed heat input because the inlet oil temperature can vary and the ΔT requirements in the reboiler can change. The heat input Q can be computed from the flow rate and the inlet and outlet temperatures, and this Q can then be controlled.

As a final example, consider the problem of controlling the temperature in a distillation column where significant pressure changes occur. We really want to measure and control composition, but temperature is used to infer composition because temperature measurements are much more reliable and inexpensive than composition measurements.

In a binary system, composition depends only on pressure and temperature:

$$x = f_{(T, P)} \tag{8.1}$$

Thus changes in composition depend on changes in temperature and pressure.

$$\Delta x = \left(\frac{\partial x}{\partial P}\right)_T \Delta P + \left(\frac{\partial x}{\partial T}\right)_P \Delta T \tag{8.2}$$

where x = mole fraction of the more volatile component in the liquid.

The partial derivatives are usually assumed to be constants that are evaluated at the steadystate operating level from the vapor-liquid equilibrium data. Thus, pressure and temperature on a tray can be measured, as shown in Fig. 8.3c, and a composition signal or pressure-compensated temperature signal generated and controlled.

$$\Delta T^{PC} = K_1 \Delta P - K_2 \Delta T \tag{8.3}$$

where T^{PC} = pressure-compensated temperature signal
K_1 and K_2 = constants

Thirty years ago these computed variables were calculated using pneumatic devices. Today they are much more easily done in the digital control computer. Much more complex types of computed variables can now be calculated. Several variables of a process can be measured and all the other variables can be calculated from a rigorous model of the process. For example, the nearness to flooding in distillation columns can be calculated from heat input, feed flow rate, and

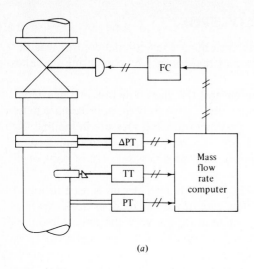

(a)

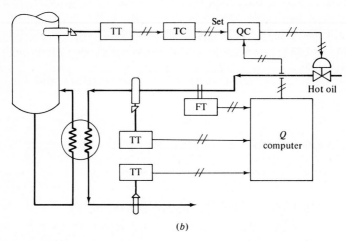

(b)

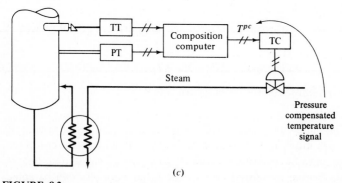

(c)

FIGURE 8.3
Computed variable control. (*a*) Mass flow rate; (*b*) heat input; (*c*) composition (pressure-compensated temperature).

temperature and pressure data. Another application is the calculation of product purities in a distillation column from measurements of several tray temperatures and flow rates by the use of mass and energy balances, physical property data, and vapor-liquid equilibrium information.

The computer makes these "rigorous estimators" feasible. It opens up a number of new possibilities in the control field. The limitation in applying these more powerful methods is the scarcity of engineers who understand both control and chemical engineering processes well enough to apply them effectively. Hopefully, this book will help a little to ease this shortage.

8.4 OVERRIDE CONTROL

There are situations where the control loop should be aware of more than just one controlled variable. This is particularly true in highly automated plants where the operator cannot be expected to make all the decisions that must be made under abnormal conditions. This includes the startup and shutdown of the process.

Override control (or "selective control" as it is sometimes called) is a form of multivariable control in which a *manipulated* variable can be set at any point in time by one of a number of different *controlled* variables.

8.4.1 Basic System

The idea is best explained with an example. Suppose the base level in a distillation column is normally held by bottoms product withdrawal as shown in Fig. 8.4a. A temperature in the stripping section is held by steam to the reboiler. Situations can arise where the base level continues to drop even with the bottoms flow at zero (vapor boilup is greater than the liquid rate from tray 1). If no corrective action is taken, the reboiler may boil dry (which could foul the tubes) and the bottoms pump could lose suction.

If an operator saw this problem developing, he would switch the temperature loop into "manual" and cut back on the steam flow. The control system in Fig. 8.4a will perform this "override" control automatically. The low selector (LS) sends to the steam valve the lower of the two signals. If the steam valve is air-to-open, the valve will be pinched back by either high temperature (through the reverse-acting temperature controller) or by low base level (through the low-base-level override controller).

In level control applications, this override controller can be a simple fixed-gain relay which acts like a proportional controller. The gain of the controller shown in Fig. 8.4a is five. It would be "zeroed" so that as the level transmitter dropped from 20 to 0 percent of full scale, the output of the relay would drop from 100 to 0 percent of scale. This means that under normal conditions when the level is above 20 percent, the output of the relay will be at 100 percent. This will be higher than the signal from the temperature controller, so the low selector

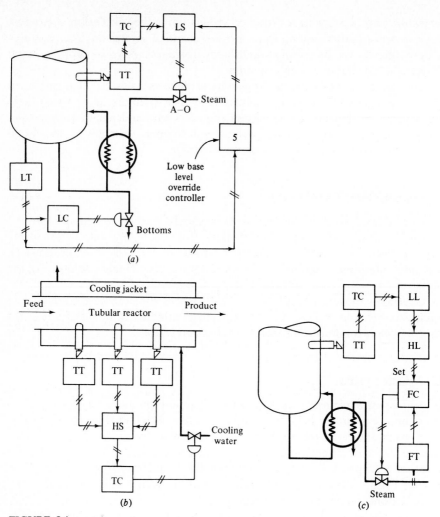

FIGURE 8.4
Selective control loops.

will pass the temperature controller output signal to the valve. However, when the base level drops below 20 percent and continues to fall toward 0 percent, the signal from the relay will drop and at some point it will become lower than the temperature controller output. At this point the temperature controller is overridden by the low-base-level override controller.

There may be other variables that might also take over control of the steam valve. If the pressure in the column gets too high, we might want to pinch the steam valve. If the temperature in the base gets too high, we might want to do the same. So there could be a number of inputs to the low selector from various override controllers. The lowest signal will be the one that goes to the valve.

In temperature and pressure override applications the override controller usually must be a PI controller, not a P controller as used in the level override controller. This is because the typical change in the transmitter signal over which we want to take override action in these applications (high pressure, high temperature, etc.) is only a small part of the total transmitter span. A very high-gain P controller would have to be used to achieve the override control action, and the override control loop would probably be closedloop unstable at this high gain. Therefore a PI controller must be used with a lower gain and a reasonably fast reset time to achieve as tight control as possible.

Figure 8.4b shows another type of *selective control* system. The signals from the three temperature transmitters located at various positions along a tubular reactor are fed into a high selector. The highest temperature is sent to the temperature controller whose output manipulates cooling water. Thus, this system controls the peak temperature in the reactor, wherever it is located.

Another very common example of this type of system is in controlling two feed streams to a reactor where an excess of one of the reactants could move the composition in the reactor into a region where an explosion could occur. Therefore, it is vital that the flow rate of this reactant be less than some critical amount, relative to the other flow. Multiple, redundant flow measurements would be used, and the highest flow signal would be used for control. In addition, if the differences between the flow measurements exceeded some reasonable quantity, the whole system would be "interlocked down" until the cause of the discrepancy was found.

Thus, override and selective controls are widely used to handle safety problems and constraint problems. High and low limits on controller outputs, as illustrated in Fig. 8.4c, are also widely used to limit the amount of change permitted.

8.4.2 Reset Windup

When a controller with integral action (PI or PID) sees an error signal for a long period of time, it integrates the error until it reaches a maximum (usually 100 percent of scale) or a minimum (usually 0 percent). This is called *reset windup.*

A sustained error signal can occur for a number of reasons, but the use of override control is one major cause. If the main controller has integral action, it will windup when the override controller has control of the valve. And if the override controller is a PI controller, it will windup when the normal controller is setting the valve. So this reset windup problem must be recognized and solved.

This is accomplished in a number of different ways, depending on the controller hardware and software used. In pneumatic controllers, reset windup can be prevented by using *external reset feedback* (feeding back the signal of the control valve to the reset chamber of the controller instead of the controller output). This lets the controller integrate the error when its output is going to the valve, but breaks the integration loop when the override controller is setting the valve. Similar strategies are used in analog electronics. In computer control systems, the integration action is turned off when the controller does not have control of the valve.

8.5 NONLINEAR AND ADAPTIVE CONTROL

8.5.1 Nonlinear Control

Since many of our chemical engineering processes are nonlinear, it would seem intuitively advantageous to use nonlinear controllers in some systems. The idea is to modify the controller action and/or settings in some way to compensate for the nonlinearity of the process.

For example, we could use a variable gain controller in which the gain K_c varies with the magnitude of the error.

$$K_c = K_{c0}(1 + b|E|) \tag{8.4}$$

where K_{c0} = controller gain with zero error
$|E|$ = absolute magnitude of the error
b = adjustable constant

This would permit us to use a low value of gain so that the system is stable near the setpoint over a broad range of operating levels with changing process gains. When the process is disturbed away from the setpoint, the gain will become larger. The system may even be closedloop unstable at some point. But the instability is in the direction of driving the loop rapidly back toward the stable setpoint region.

Another advantage of this kind of nonlinear controller is that the low gain at the setpoint reduces the effects of noise.

The parameter b can be different for positive and negative errors if the nonlinearity of the process is different for increasing or decreasing changes. For example, in distillation columns a change in a manipulated variable that moves product compositions in the direction of *higher* purity has less of an effect than a change in the direction toward *lower* purity. Thus higher controller gains can be used as product purities rise and lower gains can be used when purities fall.

Another type of nonlinear control can be achieved by using nonlinear transformations of the controlled variables. For example, in chemical reactor control the rate of reaction can be controller instead of the temperature. The two are, of course, related through the exponential temperature relationship. In high-purity distillation columns, a transformation of the type shown below can sometimes be useful to "linearize" the composition signal and produce improved control while still using a conventional linear controller.

$$(x_D)_{TR} = \frac{1 - x_D}{1 - x_D^{set}} \tag{8.5}$$

$$(x_B)_{TR} = \frac{x_B}{x_B^{set}} \tag{8.6}$$

where the subscript TR indicates transformed variables.

8.5.2 Adaptive Control

Adaptive control has been an active area of research for many years. The full-blown ideal adaptive controller continuously identifies (on-line) the parameters of the process as they change, and retunes the controller appropriately. Unfortunately, this on-line adaptation is fairly complex and has some pitfalls that can lead to poor performance. Also, it takes considerable time for the on-line identification to be achieved, which means that the plant may have already changed to a different condition. These are some of the reasons why on-line adaptive controllers are not widely used in the chemical industry.

However, the main reason for the lack of wide application of on-line adaptive control is the lack of economic incentive. On-line identification is rarely required because it is usually possible to predict with off-line tests how the controller must be retuned as conditions vary. The dynamics of the process are determined at different operating conditions, and appropriate controller settings are determined for all the different conditions. Then, when the process moves from one operating region to another, the controller settings are automatically changed. This is called *openloop-adaptive control* or *gain scheduling.*

These openloop-adaptive controllers are really just another form of nonlinear control. They have been quite successfully used in many industrial processes, particularly in batch processes where operating conditions can vary widely.

The one notable case where on-line adaptive control *has* been widely used is in pH control. The wide variations in titration curves as changes in buffering occur makes pH control ideal for on-line adaptive control methods.

Several instrument vendors have developed commercial on-line adaptive controllers. Difficulties have been reported in two situations. First, when they are applied in a multivariable environment, the interaction among control loops can cause the adaptation to fail. Second, when few disturbances are occurring, the adaptive controller has little to work with and its performance may degrade drastically.

Seborg, Edgar, and Shah (*AIChE Journal*, 1986, Vol. 32, p. 881) give a survey of adaptive control strategies in process control.

8.6 VALVE-POSITION CONTROL

Shinskey [*Chem. Eng. Prog.*, Vol. 72(5), 1976, p. 73, and *Chem. Eng. Prog.*, Vol. 74(5), 1978, p. 43] proposed the use of a type of control configuration that he called *valve position control.* This strategy provides a very simple and effective method for achieving "optimizing control." The basic idea is illustrated by several important applications.

Since relative volatilities increase in most distillation systems as pressure decreases, the optimum operation would be to minimize the pressure at all times. One way to do this is to just completely open the control valve on the cooling water. The pressure would then float up and down as cooling water temperatures changed.

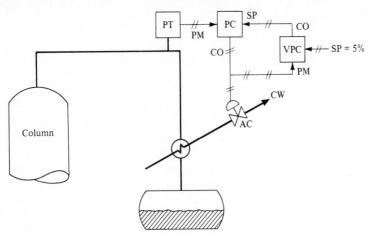

FIGURE 8.5
Floating pressure control (VPC).

However, if there is a sudden drop in cooling water temperature (as can occur during a thunder shower or "blue norther"), the pressure in the column can fall rapidly. This can cause flashing of the liquid on the trays, will upset the composition and level controls on the column, and could even cause the column to flood.

To prevent this rapid drop, Shinskey developed a "floating-pressure" control system, sketched in Fig. 8.5. A conventional PI pressure controller is used. The output of the pressure controller goes to the cooling water valve, which is AC so that it will fail open. The pressure controller output is also sent to another controller, the "valve-position controller" (VPC). This controller looks at the signal to the valve, compares it with the VPC setpoint signal, and sends out a signal which is the setpoint of the pressure controller. Since the valve is AC, the setpoint of the VPC is about 5 percent of scale so as to keep the cooling water valve almost wide open.

The VPC scheme is a different type of cascade control system. The primary control is the position of the valve. The secondary control is the column pressure. The pressure controller is PI and tuned fairly tightly so that it can prevent the sudden drops in pressure. Its setpoint is slowly changed by the VPC to drive the cooling water valve nearly wide open. A slow-acting, integral-only controller should be used in the VPC.

Figure 8.6 shows another example of the application of VPC to optimize a process. We want to control the temperature of a reactor. The reactor is cooled by both cooling water flowing through a jacket surrounding the reactor and by condensing vapor that boils off the reactor in a heat exchanger that is cooled by a refrigerant. This form of cooling is called *autorefrigeration*.

From an energy-cost perspective, we would like to use cooling water and not refrigerant because water is much cheaper. However, the dynamic response of

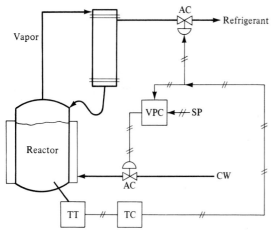

FIGURE 8.6
Use of VPC to minimize energy cost.

the temperature to a change in cooling water may be much slower than to a change in refrigerant flow. This is because the change in water flow must change the jacket temperature, which then changes the metal wall temperature, which then begins to change the reaction-mass temperature. Changes in refrigerant flow quickly raise or lower the pressure in the condenser and change the amount of vaporization in the reactor, which is reflected in reactor temperature almost immediately.

So, from a control point of view, we would like to use refrigerant to control temperature. Much tighter control could be achieved as compared to using cooling water. The VPC approach handles this optimization problem very nicely. Simply control temperature with refrigerant, but send the signal that is going to the refrigerant valve (the temperature controller output) into a valve-position controller which will slowly move the cooling water valve to keep the refrigerant valve nearly closed. Since the refrigerant valve is AC, the setpoint signal to the VPC will be about 5 to 10 percent of full scale.

Note that in the floating-pressure application, there was only one manipulated variable (cooling-water flow) and one primary controlled variable (valve position). In the reactor temperature-control application, there are two manipulated variables and two controlled variables (temperature and refrigerant valve position).

8.7 FEEDFORWARD CONTROL CONCEPTS

Up to this point we have used only feedback controllers. An error must be detected in a controlled variable before the feedback controller can take action to change the manipulated variable. So disturbances must upset the system before the feedback controller can do anything.

It seems very reasonable that if we could detect a disturbance entering a process, we should begin to correct for it *before* it upsets the process. This is the basic idea of feedforward control. If we can measure the disturbance, we can send

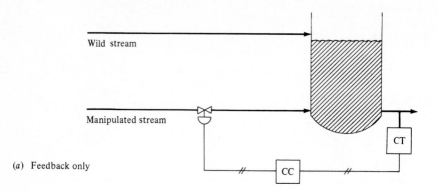

(a) Feedback only

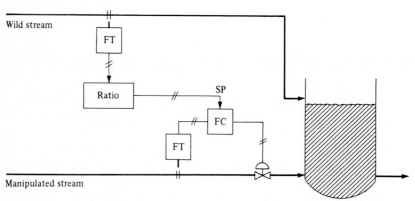

(b) Feedforward for flow rate

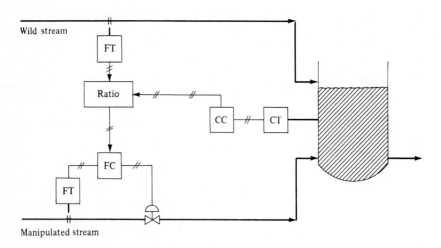

(c) Feedforward/feedback

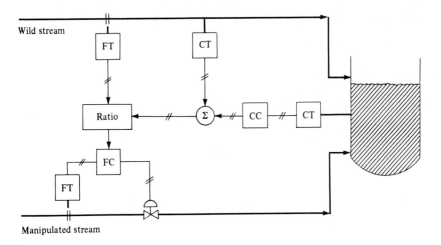

Wild stream

Manipulated stream

(d)   Feedforward for both flow and composition

FIGURE 8.7
Feedforward control.

this signal through a feedforward control algorithm that makes appropriate changes in the manipulated variable so as to keep the controlled variable near its desired value.

We do not yet have all the tools to deal quantitatively with feedforward controller design. When our Russian lessons have been learned (Laplace transforms), we will come back to this subject in Chap. 11.

However, we can describe the basic structure of several feedforward control systems. Figure 8.7 shows a blending system with one stream which acts as a disturbance; both its flow rate and its composition can change. In Fig. 8.7a the conventional feedback controller senses the controlled composition of the total blended stream and changes the flow rate of a manipulated flow. In Fig. 8.7b the manipulated flow is simply ratioed to the wild flow. This provides feedforward control for flow rate changes. Note that the disturbance must be measured to implement feedforward control.

In Fig. 8.7c the ratio of the two flows is changed by the output of a composition controller. This system is a combination of feedforward and feedback control. Finally in Fig. 8.7d a feedforward system is shown that measures both the flow rate and the composition of the disturbance stream and changes the flow rate of the manipulated variable appropriately. The feedback controller can also change the ratio. Note that *two* composition measurements are required, one measuring the disturbance and one measuring the controlled stream.

We will go into the details of how these feedforward controllers are calculated in Chap. 11.

8.8   CONTROL SYSTEM DESIGN CONCEPTS

Having learned a little about hardware and about several strategies used in control, we are now ready to talk about some basic concepts for designing a control system. At this point the discussion will be completely qualitative. In later chapters we will quantify most of the statements and recommendations made in this section. Our purpose here is to provide a broad overview of how to go about finding an effective control structure *and* designing an easily controlled process.

A consideration of dynamics should be factored into the design of a plant at an early stage, preferably during pilot-plant design and operation. It is often easy and inexpensive in the early stages of a project to design a piece of process equipment so that it is easy to control. If the plant is designed with little or no consideration of dynamics, it may take an elaborate control system to try to make the most of a poor situation.

For example, it is important to have large enough holdups in surge vessels, reflux drums, column bases, etc., to provide effective damping of disturbances (a much-used rule of thumb is 5 to 10 minutes). A sufficient excess of heat transfer area must be available in reboilers, condensers, cooling jackets, etc., to be able to handle the dynamic changes and upsets during operation. The same is true of flow rates of manipulated variables. Measurements and sensors should be located so that they can be used for effective control.

8.8.1   General Guidelines

Some guidelines and recommendations are discussed below, together with a few examples of their application. The books by Buckley (*Techniques of Process Control*, Wiley, 1964) and Shinskey (*Process-Control Systems*, McGraw-Hill, 1967) are highly recommended for additional coverage of this important topic.

1. Keep the control system as simple as possible. Everyone involved in the process, from the operators up to the plant manager, should be able to understand the system. Use as few pieces of control hardware as possible. Every additional gadget that is included in the system is one more item that can fail or drift. The instrument salesperson will never tell you this, of course.
2. Use feedforward control to compensate for large, frequent, and measurable disturbances.
3. Use override control to operate at or to avoid constraints.
4. Avoid lags and deadtimes in feedback loops. Control is improved by keeping the lags and deadtimes inside the loop as small as possible. This means that sensors should be located close to where the manipulated variable enters the process.

> **Example 8.1.** Consider the two blending systems shown in Fig. 8.8. The flow rate or composition of stream 1 is the disturbance. The flow rate of stream 2 is the manipulated variable. In scheme *A* the sensor is located after the tank and therefore the

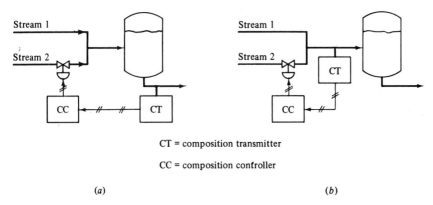

CT = composition transmitter

CC = composition controller

(a)   (b)

FIGURE 8.8
Blending systems. (a) With tank inside loop; (b) with tank outside loop.

dynamic lag of the tank is included in the feedback control loop. In scheme B the sensor is located at the inlet of the tank. The process lag is now very small since the tank is not inside the loop. Control performance of scheme B, in terms of speed of response and load rejection, would be better than the performance of scheme A. In addition, the tank now acts as a filter to average out any fluctuations in composition.

Example 8.2. The location of the best temperature-control tray in a distillation column is a popular subject in the process-control literature. Ideally, the best location for controlling distillate composition x_D with reflux flow by using a tray temperature would be at the top of the column for a binary system. See Fig. 8.9a. This is desirable dynamically because it keeps the measurement lags as small as possible. It is also desirable from a steadystate standpoint because it keeps the distillate composition constant at steadystate in a constant pressure, binary system. Holding a temperature on a tray farther down in the column does not guarantee that x_D will be constant, particularly when feed composition changes occur.

However, in many applications the temperature profile is quite flat (very little temperature change per tray) near the top of the column if the distillate product is of reasonable purity. The sensitivity of the temperature sensor may become limiting, but it is more probable that the limiting factor will be pressure changes swamping the effects of composition. In addition, if the system is not binary but has some lighter-than-light key components, these components will be at their highest concentration near the top of the column. In this case, the optimum temperature to hold constant is *not* at the top of the column, even from a steadystate standpoint.

For these reasons an intermediate tray is selected down the column where the temperature profile begins to break. Pressure compensation of the temperature signal should be used if column pressure or pressure drop varies significantly.

If bottoms composition is to be controlled by vapor boilup, the control tray should be located as close to the base of the column as possible in a binary system. In multicomponent systems with heavy components in the feed which have their highest concentration in the base of the column, the optimum control tray moves up in the column.

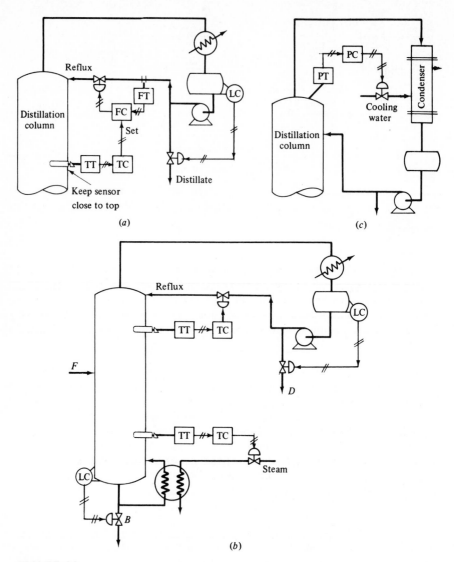

FIGURE 8.9
(a) Temperature-control-tray location; (b) interaction; (c) pressure control.

5. Use proportional-only level controls in surge tanks and column bases to smooth out disturbances.

6. Eliminate minor disturbances by using cascade control systems where possible.

7. Avoid control-loop interaction if possible, but if not, make sure the controllers are tuned to make the entire system stable. Up to this point we have discussed tuning only single-input–single-output (SISO) control loops. Many

chemical engineering systems are multivariable and inherently interacting, i.e., one control loop affects other control loops.

The classic example of an interacting system is a distillation column in which two compositions or two temperatures are controlled. As shown in Fig. 8.9b, the upper temperature sets reflux and the lower temperature sets heat input. Interaction occurs because both manipulated variables affect both controlled variables.

A common way to avoid interaction is to tune one loop very tight and the other loop loose. The performance of the slow loop is thus sacrificed. We will discuss other approaches to this problem in Part VI.

8. Check the control system for potential dynamic problems during abnormal conditions or at operating conditions that are not the same as the design. The ability of the control system to work well over a range of conditions is called *flexibility*. Startup and shutdown situations should also be studied. Operation at low throughputs can also be a problem. Process gains and time constants can change drastically at low flow rate, and controller retuning may be required. Installation of dual control valves (one big and one little) may be required.

   Rangeability problems can also be caused by seasonal variations in cooling-water temperature. Consider the distillation-column pressure control system shown in Fig. 8.9c. During the summer, cooling-water temperatures may be as high as 90°F and require a large flow rate and a big control valve. During the winter, the cooling-water temperatures may drop to 50°F, requiring much less water. The big valve may be almost on its seat, and poor pressure control may result. In addition, the water outlet temperature may get quite high under these low-flow conditions, presenting corrosion problems. In fact, if the process vapor temperature entering the condenser is above 212°F, the cooling water may even start to boil! Ambient effects can be even more severe in air-cooled condensers.

9. Avoid saturation of a manipulated variable. A good example of saturation is the level control of a reflux drum in a distillation column that has a very high reflux ratio. Suppose the reflux ratio (R/D) is 20, as shown in Fig. 8.10. Scheme A uses distillate flow rate D to control reflux drum level. If the vapor boilup dropped only 5 percent, the distillate flow would go to zero. Any bigger drop in vapor boilup would cause the drum to run dry (unless a low-level override controller were used to pinch back on the reflux valve). Scheme B is preferable for this high reflux-ratio case.

10. Avoid "nesting" control loops. Control loops are nested if the operation of the external loop depends on the operation of the internal loop. Figure 8.11 illustrates a nested loop. A *vapor* sidestream is drawn off a column to hold the column base level, and a temperature higher up in the column is held by heat input to the reboiler. The base liquid level is only affected by the liquid stream entering and the vapor boiled off, and therefore it is not directly influenced by the amount of vapor sidestream withdrawn. Thus the base level

Scheme *A*

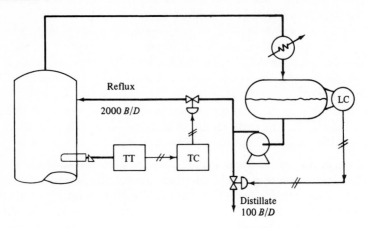

Scheme *B*

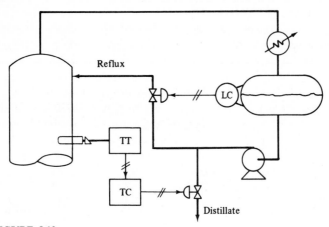

FIGURE 8.10
High-reflux-ratio column.

cannot be held by the vapor sidestream unless the temperature control loop is in operation. Then the change in the net vapor sent up the column will affect the temperature, and the vapor boilup will be changed by the temperature controller. This finally has an effect on the base level.

If the temperature controller is on "manual," the level loop cannot work. In this process it probably would be better to reverse the pairing of the loops: control temperature with vapor sidestream and control base level with heat input. Notice that if the sidestream were removed as a liquid, the control system would not be nested. Sometimes, of course, nested loops cannot be avoided. Notice that the recommended Scheme *B* in Fig. 8.10 is just such a nested system. Distillate has no direct effect on tray temperature. It is only

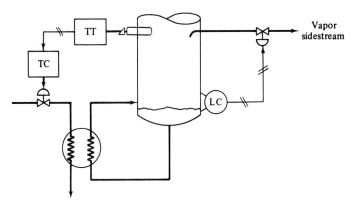

FIGURE 8.11
Nested control loops.

through the level loop and its changes in reflux that the temperature is affected.

8.8.2   Trade-Offs Between Steadystate Design and Control

The design of a chemical engineering system always involves a number of trade-offs. We have already discussed the conflicts between the process engineer and the control engineer in the question of control valve sizing. There are many other such conflicts between what would be the optimum from only a steadystate standpoint and what is needed to handle the dynamics of the process.

A. LIQUID HOLDUPS. The most common and most important trade-off is that of specifying holdup volumes in tanks, column bases, reflux drums, etc. From a steadystate standpoint, these volumes should be kept as small as possible because this will minimize capital investment. The more holdup that is needed in the base of a distillation column, the taller the column must be. In addition, if the material in the base of the column is heat-sensitive, it is very desirable to keep the holdup in the base as small as possible in order to reduce the time that the material is at the high base temperature. Large holdups also increase the potential pollution and safety risks if hazardous or toxic material is being handled.

So all of these considerations suggest small liquid holdups. However, from a dynamics and control point of view we want enough holdup to be able to ride through disturbances without losing the base level and having to shut down the column, and to be able to change flow rates and compositions slowly to downstream processes.

Over the years some heuristics have been developed that work pretty well in most systems. Holdup times (based on *total* flow in and out of the surge volume) of about 5 to 10 minutes seem to work well. If the column has a fired

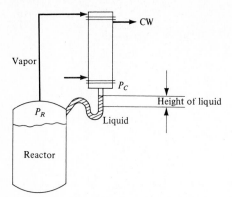

FIGURE 8.12
Gravity-flow condensers.

reboiler, the base holdup should be made larger. If a downstream unit is particularly sensitive to changes, the holdup volumes in the upstream equipment should be doubled or tripled.

B. GRAVITY FLOW CONDENSERS. Another trade-off example is a gravity-flow condenser design. As shown in Fig. 8.12, a condenser is located above a reactor (or distillation column). Vapor from the reactor flows up into the condenser because it is at a lower pressure than the reactor ($P_R > P_C$). After the vapor is condensed, the liquid must flow back into the reactor against this positive pressure gradient. This is accomplished by building up a leg of liquid in the reflux return line to a height sufficient to overcome the pressure gradient, and any frictional pressure drop in the liquid line.

Under normal design conditions, the liquid height may be several feet. But since the pressure drop through the vapor line increases as the square of the flow rate, the liquid height must increase by a factor of four if the vapor rate doubles. Thus the control engineer pushes the design engineer to mount the condenser much higher in the air than he would like because this increases the cost of the equipment.

C. DISTILLATION COLUMN DESIGN. Control engineers usually would like to have a few extra trays added to the column and the column diameter made bigger in order to make the control of the column easier. The additional trays and larger cross-sectional area give a cushion which permits the column to ride through disturbances. Of course the design engineer wants to keep the size of the column at the steadystate optimum because the additional trays and larger diameter raise capital investment.

8.8.3 Plant-Wide Control

We have discussed setting up a single feedback controller and establishing a control strategy for one unit operation: a reactor, a column, etc. The next level of complexity is to look at an entire operating plant which is made up of many unit operations connected in series and parallel, with recycle of material and energy among the various parts of the plant. This is one of the most challenging jobs of the process control engineer.

A. BUCKLEY PROCEDURE. Buckley was one of the pioneers in this aspect of control. He developed a procedure that is still widely used today. His plant-wide design methodology consists of the following steps:

1. Lay out a logical control scheme to handle all the liquid levels and pressure loops throughout the plant so that the flows from one unit to the next are as smooth as possible. Buckley called these the *material-balance* loops. If the feed rate is set into the front of the process, the material-balance loops should be set up in the direction of flow; i.e., the flow out of each unit is set by a liquid level or pressure in the unit. If the product flow rate out of the plant is set, the material-balance loops should be in the direction opposite flow; i.e., the flow into each unit is set by a liquid level or pressure in the unit.
2. Then design the composition control loops for each unit operation. Buckley called these the *product-quality* loops. Determine the closedloop time constants of these product-quality loops.
3. Size the holdup volumes so that the closedloop time constants of the material-balance loops are a factor of ten bigger than the closedloop time constants of the product-quality loops. This breaks the interaction between the two types of loops.

B. EIGENSTRUCTURE. The plant-wide control problem is still a very active area of research. In recent years there has been more emphasis on finding control structures for individual unit operations and for entire plants that are inherently simple, self-regulatory, and self-optimizing. This approach has been called a search for the *eigenstructure* of the plant.

To illustrate the concept, consider a single distillation column with distillate and bottoms products. To produce these products while using the minimum amount of energy, the compositions of both products should be controlled at their specifications. Figure 8.13a shows a "dual composition" control system. The disadvantages of this structure are (1) two composition analyzers are required, (2) the instrumentation is more complex, and (3) there may be dynamic interaction problems since the two loops are interacting. This system may be difficult to design and to tune.

If the only disturbances were feed flow rate changes, we could simply ratio the reflux flow rate to the feed rate and control the composition of only one end of the column (or even one temperature in the column). However, changes in feed composition may require changes in reflux and vapor boilup for the same feed flow rate.

The eigenstructure concept is to see if we can find a simple control strategy for this column. One way to do this is to see how the various manipulated vari-

(*a*) Dual composition control

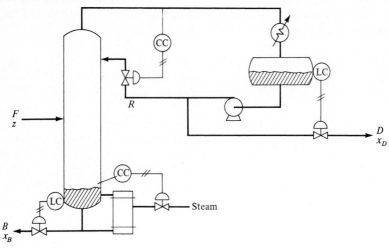

(*b*) Changes in *R*, *V*, and *RR*
(for x_D and x_B constant)

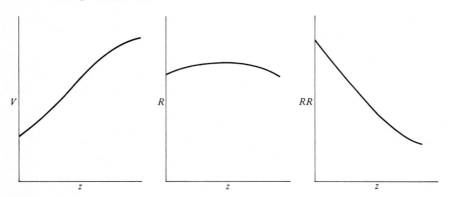

FIGURE 8.13

ables must change as feed composition changes while holding products at the specified purities. Figure 8.13*b* sketches some typical curves of vapor boilup, reflux, and reflux ratio. For this example we can see that the reflux flow rate really doesn't change much as feed composition changes. Therefore, the eigenstructure would be to simply ratio reflux flow to feed flow and control only one end of the column by manipulating heat input. A small amount of energy would be wasted, but the control would be much tighter and simpler and less susceptible to upsets. Figure 8.14 shows the eigenstructure configuration.

Note that the approach was to explore the effects of various load disturbances, to find the optimum operating conditions when these disturbances

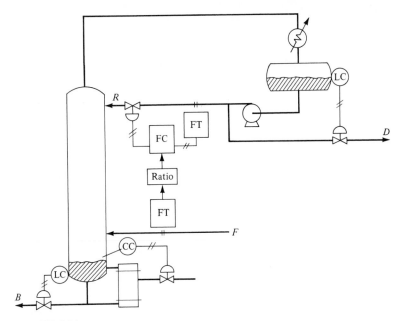

FIGURE 8.14

occurred, and to see if a simple control structure could be found which easily held the unit near these optimum conditions.

These same notions can be extended to an entire plant in which several unit operations are connected together. The HDA process for hydrodealkylation of toluene to form benzene is a good example of where an eigenstructure can be found that provides a more easily and simply controlled plant. See Fig. 8.15. Assuming that the toluene feed rate to the unit is fixed, this plant has 22 valves that must be set. There are 11 inventory loops (levels and pressures), so they require 11 valves. One possible conventional control structure is shown in Fig. 8.15.

Douglas and coworkers (Fisher, W. R., Doherty, M. F., Douglas, J. M., "The Interface between Design and Control. 3. Selecting a Set of Controlled Variables," *I&EC Research*, 1988, Vol. 27, p. 611) have studied this process from a steadystate point of view and determined that the process can be held very close to its optimum for a variety of expected load disturbances (cooling-water temperature, hydrogen feed purity and throughput) by using the following strategy:

1. Fix the flow of recycle gas through the compressor at its maximum value.
2. Hold a constant heat input flow rate in the stabilizer.
3. Eliminate the reflux entirely in the recycle column (make it a stripper).
4. Maintain a constant hydrogen-to-aromatics ratio in the reactor inlet by adjusting hydrogen fresh feed.

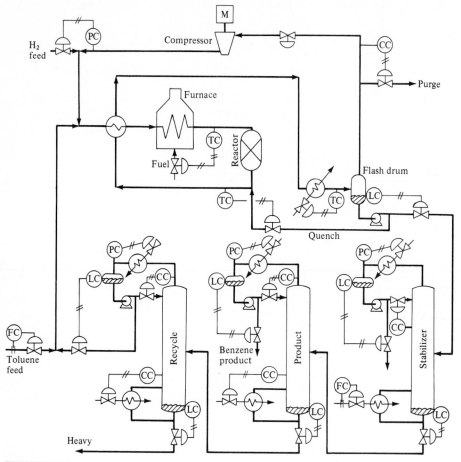

FIGURE 8.15

5. Hold the recycle toluene flow rate constant by adjusting fuel to the furnace.

6. Hold the temperature of the cooling water leaving the partial condenser constant.

Notice that this is not the control strategy shown in Fig. 8.15. Item 5 above makes little sense from a dynamic point of view. The same is true about item 4 if the hydrogen fresh feed is significantly larger than the purge stream. If item 4 is implemented, the pressure in the reactor would have to be controlled by the purge stream. This configuration would probably give poor results because it is better to hold the pressure in the system with the big stream, i.e., the hydrogen fresh feed.

C. MAKEUP CONTROL. Another aspect of plant-wide control is the problem of "makeup" control strategy. When a fresh feed stream is brought into a plant it

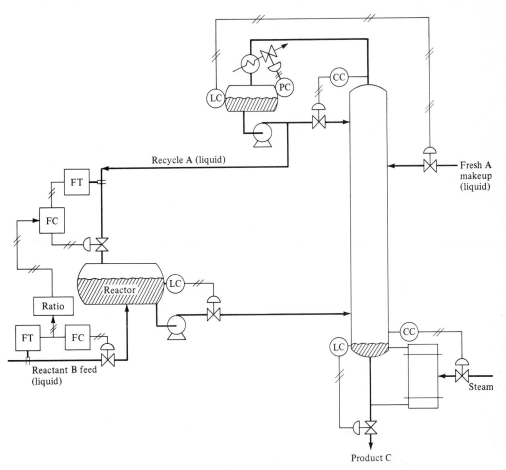

FIGURE 8.16
Liquid makeup.

can enter in a number of locations. The best place to add it depends upon a number of factors including impurities it might contain and its phase (is it liquid or vapor).

Figure 8.16 shows a process in which two reactants (A and B) are fed into a process and react to form product C. An excess of A is used to avoid undesirable side reactions. Reactant B is essentially totally consumed in the reactor.

The reactor effluent is separated in a distillation column. The overhead is mostly excess reactant A which is recycled back to the reactor. The bottoms from the column is mostly product C. The reaction occurs in the liquid phase so the reactor feed streams are liquid. Reactant B is added directly to the reactor on flow control. The flow rate of the recycle stream is ratioed to the flow rate of the B feed stream. The composition of A in the column base sets heat input. The composition of C in the column overhead sets reflux.

The makeup of fresh A is added into the column as a liquid feed. This might be done so that any heavy impurities in this feed stream will go out the bottom of the column and not go into the reactor where they could cause catalyst poisoning. The makeup of fresh A might be fed into the column because it contains both A and C, and it is desirable to let the column purify the stream so that it is richer in A.

Note that the makeup of liquid A to the column is set by the level controller on the reflux drum of the column. This may seem like an unusual arrangement, and it certainly involves nested loops (the reflux drum level is not directly affected by makeup feed, but only through the bottoms composition loop changing heat input). If the makeup could be added directly into the reflux drum, there would be no nesting of control loops.

Figure 8.17 shows a similar process, this time with a vapor-phase reaction, vapor streams fed to the reactor, and a vapor makeup stream of A. A compressor is used to recycle gas from the top of the column back into the reactor. Now the

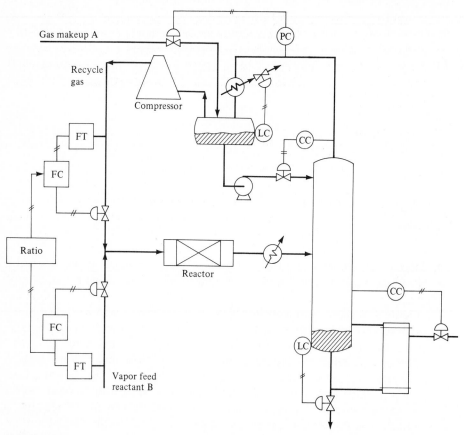

FIGURE 8.17
Vapor makeup.

makeup is added at the compressor suction (or into the reflux drum to make sure that no liquid is carried into the compressor). This makeup rate is set by the pressure controller on the column. The makeup gas is added to hold the desired pressure in the column and in the reactor. Note the difference in control structure between the vapor-filled system in Fig. 8.17 and the liquid-filled system in Fig. 8.16.

8.9 DYNAMIC MATRIX CONTROL

The last decade has seen the development a several control concepts that are based on using a model of the process as part of the controller. Most of these methods use Laplace or z-transform representations of the process, which we are not yet ready to handle. After our Russian lessons have been completed, we will discuss some of these.

There is one method that is based on a time-domain model. It was developed at Shell Oil Company (C. R. Cutler and B. L. Ramaker, "Dynamic Matrix Control: A Computer Control Algorithm," paper presented at the 86th National AIChE Meeting, 1979) and is called *dynamic matrix control* (DMC). Several other methods have also been proposed that are quite similar. The basic idea is to use a time-domain step-response model of the process to calculate the future changes in the manipulated variable that will minimize some performance index. Much of the explanation of DMC given in this section follows the development presented by C. C. Yu in his Ph.D. thesis (Lehigh University, 1987).

8.9.1 Review of Least Squares

The DMC method uses the same statistical mathematics that are used in a standard least-squares procedure for determining the best values of parameters of an equation to fit a number of data points. In the DMC approach, we would like to have NP future output responses match some "optimum" trajectory by finding the "best" values of NC future changes in the manipulated variables. This is exactly the concept of a least-squares problem of fitting NP data points with an equation with NC coefficients. This is a valid least-squares problem as long as NP is greater than NC.

Suppose we have NP data points that give values of a measured variable x with known values of two variables a_1 and a_2 (we will generalize to NC variables later).

Data points		
x values	**a_1 values**	**a_2 values**
x_1	a_{11}	a_{12}
x_2	a_{21}	a_{22}
$\vdots$	$\vdots$	$\vdots$
x_{NP}	$a_{NP,1}$	$a_{NP,2}$

Suppose we want to find the values of two parameters (m_1 and m_2) in an equation

$$\bar{x}_{(a_1, a_2)} = (m_1)a_1 + (m_2)a_2 \tag{8.7}$$

that fit the data points "best." The value of x calculated from the equation is $\bar{x}$.

The performance index is the sum of the squares of the differences between the NP actual data points (x_i) and the values calculated from the equation ($\bar{x}_i$).

$$\bar{x}_i = m_1 a_{i1} + m_2 a_{i2} \tag{8.8}$$

The performance index is

$$J = \sum_{i=1}^{NP} (x_i - \bar{x}_i)^2 \tag{8.9}$$

At each of the NP data points, the values of x, a_1, and a_2 are known. The job is to find the values of m_1 and m_2 that do the best job of making Eq. (8.7) fit the data.

As you may recall, the procedure for solving this problem is to substitute Eq. (8.8) into Eq. (8.9) and to take partial derivatives with respect to the two unknown parameters m_1 and m_2. Then the two partial derivative equations are set equal to zero. This gives two equations in two unknowns.

$$J = \sum_{i=1}^{NP} (x_i - \bar{x}_i)^2 = \sum_{i=1}^{NP} (x_i - m_1 a_{i1} - m_2 a_{i2})^2 \tag{8.10}$$

$$\frac{\partial J}{\partial m_1} = 2 \sum_{i=1}^{NP} (x_i - m_1 a_{i1} - m_2 a_{i2})(-a_{i1}) = 0 \tag{8.11}$$

$$\frac{\partial J}{\partial m_2} = 2 \sum_{i=1}^{NP} (x_i - m_1 a_{i1} - m_2 a_{i2})(-a_{i2}) = 0 \tag{8.12}$$

Rearranging Eqs. (8.11) and (8.12) gives

$$m_1 \sum_{i=1}^{NP} (a_{i1})^2 + m_2 \sum_{i=1}^{NP} (a_{i1}a_{i2}) = \sum_{i=1}^{NP} (x_i a_{i1}) \tag{8.13}$$

$$m_1 \sum_{i=1}^{NP} (a_{i1}a_{i2}) + m_2 \sum_{i=1}^{NP} (a_{i2})^2 = \sum_{i=1}^{NP} (x_i a_{i2}) \tag{8.14}$$

These two equations can be compactly written in matrix form

$$\underline{A}^T \underline{A} \underline{m} = \underline{A}^T \underline{x} \tag{8.15}$$

where

$$\underline{A} \equiv \begin{bmatrix} a_{11} & a_{12} \\ a_{21} & a_{22} \\ \vdots & \vdots \\ a_{NP,\,1} & a_{NP,\,2} \end{bmatrix} \tag{8.16}$$

$$m \equiv \begin{bmatrix} m_1 \\ m_2 \end{bmatrix} \qquad x \equiv \begin{bmatrix} x_1 \\ x_2 \\ \vdots \\ x_{NP} \end{bmatrix} \qquad (8.17)$$

Equation (8.15) can be solved for the unknown parameters m_1 and m_2.

$$m = [A^T A]^{-1} A^T x \qquad (8.18)$$

We can easily extend Eq. (8.8) to contain NC values of the unknown parameters m and NC variables (a_i)

$$\bar{x}_i = m_1 a_{i1} + m_2 a_{i2} + m_3 a_{i3} + \cdots + m_{NC} a_{i, NC} = \sum_{k=1}^{NC} m_k a_{ik} \qquad (8.19)$$

by simply redefining the m vector of unknown parameters and the A matrix of data points:

$$m = \begin{bmatrix} m_1 \\ m_2 \\ \vdots \\ m_{NC} \end{bmatrix} \qquad A = \begin{bmatrix} a_{11} & a_{12} & \cdots & a_{1, NC} \\ a_{21} & a_{22} & \cdots & \cdots \\ \cdots\cdots\cdots\cdots\cdots\cdots\cdots\cdots \\ a_{NP, 1} & a_{NP, 2} & & a_{NP, NC} \end{bmatrix} \qquad (8.20)$$

Equations (8.15) and (8.18) still apply.

The performance index can be changed to include another term (there is no sense in doing this in the parameter estimation problem, but as we will see in the next section, there is a good reason for doing it in the DMC problem). If the magnitudes of the m_i parameter values are included in the J performance index

$$J = \sum_{i=1}^{NP} (x_i - \bar{x}_i)^2 + f^2 \sum_{k=1}^{NC} (m_k)^2$$

where f = weighting factor on the parameter values

$$J = \sum_{i=1}^{NP} \left[x_i - \sum_{k=1}^{NC} m_k a_{ik} \right]^2 + f^2 \sum_{k=1}^{NC} (m_k)^2 \qquad (8.21)$$

The partial derivative of J with respect to the m_k is

$$\frac{\partial J}{\partial m_k} = 2 \sum_{i=1}^{NP} \left[\left\{ x_i - \sum_{k=1}^{NC} m_k a_{ik} \right\}(-a_{ik}) \right] + 2f^2 m_k = 0 \qquad (8.22)$$

There are k of these equations. The first one is

$$m_1 \left\{ \sum_{i=1}^{NP} (a_{i1})^2 + f^2 \right\} + m_2 \sum_{i=1}^{NP} (a_{i1} a_{i2}) + \cdots + m_{NC} \sum_{i=1}^{NP} (a_{i1} a_{i, NC}) = \sum_{i=1}^{NP} (x_i a_{i1}) \qquad (8.23)$$

These k equations can be written in matrix form:

$$[A^T A + f^2 I] m = A^T x \qquad (8.24)$$

where $\underline{I}$ is the identity matrix (see Chap. 15). Solving for the $\underline{m}$ vector gives

$$\underline{m} = [\underline{A}^T\underline{A} + f^2\underline{I}]^{-1}\underline{A}^T\underline{x} \tag{8.25}$$

We will use this solution in the DMC algorithm discussed in the next section.

8.9.2 Step-Response Models

Dynamic matrix control uses time-domain step-response models (called *convolution models*). As sketched in Fig. 8.18, the response (x) of a process to a unit step change in the input ($\Delta m_1 = 1$) made at time equal zero can be described by the values of x at discrete points in time (the b_i's shown on the figure). At $t = nT_s$, the value of x is b_{nT_s}. If Δm_1 is not equal to one, the value of x at $t = nT_s$ is $b_{nT_s}\,\Delta m_1$. The complete response can be described using a finite number (NP) values of b_i coefficients. NP is typically chosen such that the response has reached 90 to 95 percent of its final value.

It should be noted that we have chosen a slightly unusual nomenclature for the changes in the manipulated variable, the Δm_i. Usually the value of the variable x at zero time would be called x_0, the value at $t = T_s$ would be called x_1, etc. We have chosen to call the value of Δm at zero time Δm_1 instead of the more conventional Δm_0. This is simply to make the matrix notation to be used below more convenient (everything starts with the index of 1).

Now suppose there are two step changes of the input, Δm_1 occurring at $t = 0$ and Δm_2 occurring at $t = T_s$. Using the principal of superposition (the total

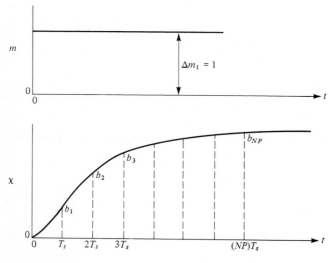

FIGURE 8.18
Step response.

output is the sum of the effects of the individual inputs), we can calculate x at each discrete point in time.

$$\text{At } t = 0: \qquad x_0 = 0$$

$$\text{At } t = T_s: \qquad x_1 = b_1 \, \Delta m_1$$

$$\text{At } t = 2T_s: \qquad x_2 = b_2 \, \Delta m_1 + b_1 \, \Delta m_2$$

$$\text{At } t = 3T_s: \qquad x_3 = b_3 \, \Delta m_1 + b_2 \, \Delta m_2$$

Now if changes occur in the input over four steps in time, the output for the next six steps would be:

$$x_1 = b_1 \, \Delta m_1$$

$$x_2 = b_2 \, \Delta m_1 + b_1 \, \Delta m_2$$

$$x_3 = b_3 \, \Delta m_1 + b_2 \, \Delta m_2 + b_1 \, \Delta m_3$$

$$x_4 = b_4 \, \Delta m_1 + b_3 \, \Delta m_2 + b_2 \, \Delta m_3 + b_1 \, \Delta m_4$$

$$x_5 = b_5 \, \Delta m_1 + b_4 \, \Delta m_2 + b_3 \, \Delta m_3 + b_2 \, \Delta m_4$$

$$x_6 = b_6 \, \Delta m_1 + b_5 \, \Delta m_2 + b_4 \, \Delta m_3 + b_3 \, \Delta m_4$$

The above equations can be generalized to give NP values of the output for NC changes in the inputs by using matrix notation:

$$\begin{bmatrix} x_1 \\ x_2 \\ x_3 \\ \cdots \\ x_{NP} \end{bmatrix} = \begin{bmatrix} b_1 & 0 & \cdots & 0 \\ b_2 & b_1 & 0 & \cdots \\ \cdots\cdots\cdots\cdots\cdots\cdots\cdots \\ \cdots\cdots\cdots\cdots\cdots\cdots\cdots \\ b_{NP} & b_{NP-1} & \cdots & b_{NP+1-NC} \end{bmatrix} \begin{bmatrix} \Delta m_1 \\ \Delta m_2 \\ \Delta m_3 \\ \cdots \\ \Delta m_{NC} \end{bmatrix} \qquad (8.26)$$

$$\underline{x} = \underline{B} \, \underline{\Delta m} \qquad (8.27)$$

where $\underline{B}$ is the matrix defined in Eq. (8.26) which has NP rows and NC columns. Now let us define a new $NP \times NC$ matrix $\underline{A}$ and set it equal to matrix $\underline{B}$ above.

$$\underline{A} = \begin{bmatrix} a_{11} & a_{12} & \cdots & a_{1,NC} \\ a_{21} & a_{22} & \cdots & \cdots \\ \cdots\cdots\cdots\cdots\cdots\cdots\cdots\cdots \\ a_{NP,1} & a_{NP,2} & \cdots & a_{NP,NC} \end{bmatrix} \qquad (8.28)$$

Notice that the elements of $\underline{A}(a_{ik})$ are related to the elements of $\underline{B}(b_i)$ through the following relationship (with $b_i = 0$ for $i \le 0$):

$$a_{ik} = b_{i+1-k} \qquad (8.29)$$

So the ith value of the output can be written:

$$x_i = \sum_{k=1}^{NC} b_{i+1-k} \, \Delta m_k = \sum_{k=1}^{NC} a_{ik} \, \Delta m_k \qquad (8.30)$$

Equation (8.30) describes how Δm_k affects the ith output x_i using the step-response coefficient b_{i+1-k}. Note that the sum of the indices in Δm and b is always $i+1$. The summation gives the effects of all the NC terms using the principle of superposition.

From a control point of view, we would like to distinguish between changes of the manipulated variable in the past and in the future. If there have been NP changes in the manipulated variable during the previous NP steps and if no other changes were made, the output would change in the future because of the "old" changes in the input. Let us call these old changes in the input $(\Delta m)^{old}$ and the response to these old changes we will call the "openloop" response x_{OL}. Using the step-response model to predict this openloop response, we will use the notation $\tilde{x}_{OL,\,i}$ for the predicted value of the output at the ith step in the future.

$$\tilde{x}_{OL,\,i} = \sum_{k=0}^{-NP+1} b_{i+1-k} \, (\Delta m_k)^{old} \qquad (8.31)$$

However, at the current sampling instance (the 0th) we can measure the actual output of the process. Let us call this x_0^{meas}. If our model were perfect and, more importantly, if no load disturbances occurred, the predicted and the measured values would be equal. However, this is usually not the case. So we will use the difference between the two to give a better prediction into the future. The predicted value at the current sampling time is $\tilde{x}_{OL,\,0}$ and is given by Eq. (8.31).

$$\tilde{x}_{OL,\,0} = \sum_{k=0}^{-NP+1} b_{1-k} \, (\Delta m_k)^{old} \qquad (8.32)$$

The difference between the actual measured present value x_0^{meas} and predicted present value $\tilde{x}_{OL,\,0}$ is added to the predicted value at the ith sampling period to give a better prediction of the openloop response.

$$\tilde{x}_{OL,\,i} = \sum_{k=0}^{-NP+1} b_{i+1-k} \, (\Delta m_k)^{old} + x_0^{meas} - \tilde{x}_{OL,\,0}$$

$$= \sum_{k=0}^{-NP+1} b_{i+1-k} \, (\Delta m_k)^{old} + x_0^{meas} - \sum_{k=0}^{-NP+1} b_{1-k} \, (\Delta m_k)^{old}$$

$$\tilde{x}_{OL,\,i} = x_0^{meas} + \sum_{k=0}^{-NP+1} [b_{i+1-k} - b_{1-k}] \, (\Delta m_k)^{old} \qquad (8.33)$$

Equation (8.33) gives the prediction of the openloop response we will use in the DMC calculations.

Now the real response will be the "openloop" response plus the effects of the future changes in the manipulated variables: $(\Delta m_k)^{new}$. We call this the "closedloop" response and use the symbol $x_{CL, i}$ for the value at the ith step into the future. Equation (8.30) can be used to predict the output when NC changes in the manipulated variables have been made in the future.

$$x_{CL, i} = x_{OL, i} + \sum_{k=1}^{NC} a_{ik} (\Delta m_k)^{new} \tag{8.34}$$

8.9.3   DMC Algorithm

The whole idea of DMC is to find the "best" values of these future changes in the manipulated variables (the future $(\Delta m)^{new}$'s) such that a performance index J is minimized. The performance index is the sum of two terms:

1. The squares of the errors $(x^{set} - x_{CL, i})$ summed over NP time intervals so as to cover 90 to 95 percent of the system response.
2. The squares of the manipulated variables summed over NC time intervals. If NC is small, only a few moves of the manipulated variables can be used. NC is typically set at 50 percent of NP. The reason for adding the manipulated variables into the performance index is to prevent large swings in the manipulated inputs. The bigger the weighting factor f, the smaller the changes in the Δm's. Increasing f increases the damping coefficient of the closedloop system.

$$J = \sum_{i=1}^{NP} [x^{set} - x_{CL, i}]^2 + f^2 \sum_{i=1}^{NC} [(\Delta m_k)^{new}]^2 \tag{8.35}$$

$$J = \sum_{i=1}^{NP} \left[x^{set} - x_{OL, i} - \sum_{k=1}^{NC} a_{ik} (\Delta m_k)^{new} \right]^2 + f^2 \sum_{i=1}^{NC} [(\Delta m_k)^{new}]^2 \tag{8.36}$$

Now compare Eqs. (8.21) and (8.36). They are identical if we let $x_i = x^{set} - x_{OL, i}$ and let $\underline{m} = (\Delta m)^{new}$. The $\underline{A}$ matrix is defined in Eqs. (8.26) and (8.28). So the solution is

$$(\Delta \underline{m})^{new} = [\underline{A}^T \underline{A} + f^2 \underline{I}]^{-1} \underline{A}^T \underline{x} \tag{8.37}$$

where the vector $(\Delta \underline{m})^{new}$ is the NC values of the future changes in the manipulated variables that minimize the performance index.

The DMC algorithm has the following steps at each point in time:

1. Calculate the NP values of $x_{OL, i}$ from Eq. (8.33). They depend only on the past values of the manipulated variables and the present measured value of the controlled variable x^{meas}.
2. Calculate the NC values of the future changes in the manipulated variables from Eq. (8.37) using the "dynamic matrix" $\underline{A}$ given in Eq. (8.28).
3. Implement the first change $(\Delta m_1)^{new}$.
4. At the next sampling period, measure the controlled variable to get a new value of x^{meas} and repeat the steps above.

Note that the $[\underline{A}^T\underline{A} + f^2\underline{I}]^{-1}\underline{A}^T$ term in Eq. (8.37) is just a matrix of constant elements which do not change. Therefore it only has to be calculated once, not at every time step.

The f factor is a detuning parameter that prevents large changes in the manipulated variables. The larger the value of f, the less underdamped and the more robust the closedloop system will be.

The DMC discussed in this chapter is for a SISO system. We will say more about DMC in Chap. 17 since this methodology is fairly easily extended to multivariable systems, which is where its real potential usefulness occurs.

PROBLEMS

8.1. The suction pressure of an air compressor is controlled by manipulating an air stream from an off-site process. An override system is to be used in conjunction with the basic loop to prevent overpressuring or underpressuring the compressor suction during upsets. Valve actions are indicated on the sketch below.

The pressure transmitter span is 0 to 20 psig. The pressure controller setpoint is 10 psig. If the pressure gets above 15 psig the vent valve is to start opening and is to be wide open at 20 psig. If the pressure drops below 5 psig the recycle valve is to start opening and is to be wide open at 0 psig.

Specify the range and action of the override control elements required to achieve this control strategy.

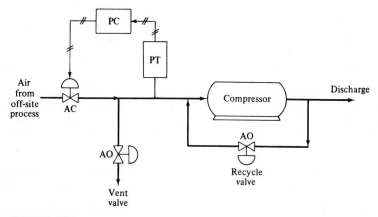

FIGURE P8.1

8.2. Design an override control system that will prevent the liquid level in a reflux drum from dropping below 5 percent of the level transmitter span by pinching the reflux control valve. The system must also prevent the liquid level from rising above 90 percent of the level transmitter span by opening the reflux control valve. Normal level control is achieved by manipulating distillate flow over the middle 50 percent of the level transmitter span using a proportional level controller (proportional band setting is 50).

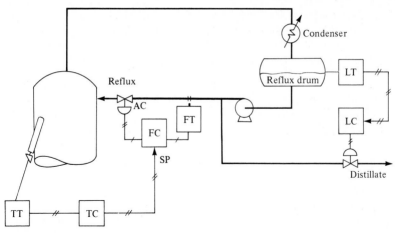

FIGURE P8.2

8.3. Design an override control system for the chilled-water loop considered in Prob. 7.11. The flow rate of chilled water is not supposed to drop below 500 gpm. Your override control circuit should open the chilled-water control valve if chilled-water flow gets below 500 gpm, overriding the temperature controller.

8.4. Vapor feed to an adiabatic tubular reactor is heated to about 700°F in a furnace. The reaction is endothermic. The exit temperature of gas leaving the reactor is to be controlled at 600°F.

   Draw an instrumentation and control diagram that accomplishes the following objectives:
(a) Feed is flow-controlled.
(b) Fuel gas is flow-controlled and ratioed to feed rate.
(c) The fuel to feed ratio is adjusted by a furnace exit temperature controller.
(d) The setpoint of the furnace exit temperature controller is adjusted by a reactor exit temperature controller.
(e) Furnace exit temperature is not to exceed 750°F.
(f) High furnace stack-gas temperature should pinch the fuel gas control valve.

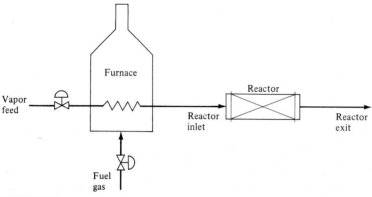

FIGURE P8.4

8.5. Sketch control system diagrams for the following systems:

(a) Temperature in a reactor is controlled by manipulating cooling water to a cooling coil. Pressure in the reactor is controlled by admitting a gas feed into the reactor. A high temperature override pinches reactant feed gas.

(b) Reflux drum level is controlled by a reflux flow rate back to a distillation column. Distillate flow is manipulated to maintain a specified reflux ratio (reflux/distillate). This specified reflux ratio can be changed by a composition controller in the top of the column.

8.6. Sketch a control concept diagram for the distillation column shown below. The objectives of your control system are:

(a) Reflux is flow-controlled, ratioed to feed rate, and overridden by low reflux drum level.

(b) Steam is flow-controlled, with the flow controller setpoint coming from a temperature controller which controls a tray temperature in the stripping section of the column. Low base level or high column pressure pinch the steam valve.

(c) Base level is controlled by bottom product flow rate.

(d) Reflux drum level is controlled by distillate product flow rate.

(e) Column pressure is controlled by changing the setpoint of a speed controller on the compressor turbine. The speed controller output sets a flow controller on the high-pressure steam to the turbine.

(f) A minimum flow controller ("antisurge") sets the valve in the compressor bypass line to prevent the flow rate through the compressor from dropping below some minimum flow rate.

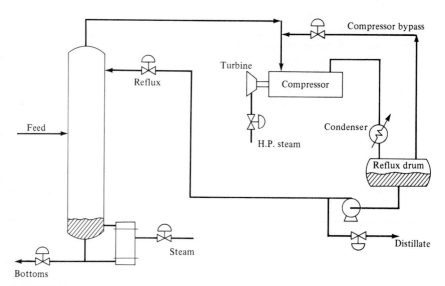

FIGURE P8.6

8.7. A distillation column operates with vapor recompression.

(a) Specify the action of all control valves.

(b) Sketch a control concept diagram with the following loops:

(i) Reflux and bottoms are flow-controlled and overridden by low reflux drum level and low base level, respectively.

(ii) Reflux drum level is controlled by distillate flow.

(iii) Steam to the turbine driving the compressor is flow-controlled, reset by a speed controller, which is reset by a column pressure controller.

(iv) Column base level is controlled by the valve under the condenser which floods or exposes tubes in the condenser to vary the heat transfer area.

(v) High reflux drum pressure overrides the steam valve on the compressor turbine.

(c) At design conditions, the total flow through the compressor is 120,000 lb_m/h, the flow through the hot side of the reboiler is 80,000 lb_m/h, the pressure drop through the condenser is 2 psi, the pressure drop through the hot side of the reboiler is 10 psi, and the control valve below the condenser is 25 percent open. Density of the liquid is 5 lb_m/gal.

(i) What is the C_v value of the control valve?

(ii) What is the flow rate through the hot side of the reboiler when the control valve is wide open? Assume that the pressure in the reflux drum is constant and that the column pressure control will adjust compressor speed to keep the total flow through the compressor constant.

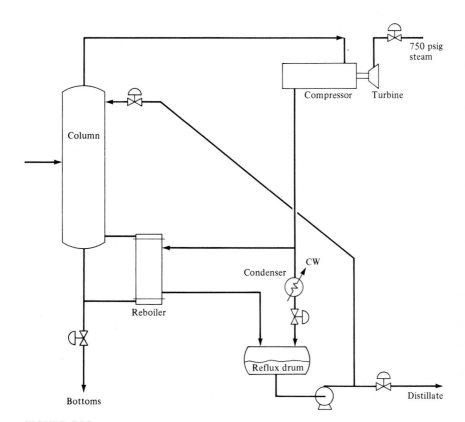

FIGURE P8.7

8.8. Sketch a control concept diagram for a chemical reactor that is cooled by generating steam (see Prob. 7.18).

(a) Steam drum pressure is controlled by the valve in the steam exit line.

(b) Condensate flow is ratioed to steam flow.

(c) Steam drum liquid level is controlled by adjusting the condensate to steam ratio.

(d) Feed is flow-controlled.

(e) Reactor liquid level is controlled by product withdrawal.

(f) Reactor temperature is controlled by resetting the setpoint of the steam pressure controller.

(g) The override controls are as follows:

(i) High reactor temperature pinches reactor feed valve.

(ii) Low steam drum level pinches reactor feed.

8.9. A chemical plant has a four-header steam system. A boiler generates 900 psig steam which is let down through turbines to a 150 psig header and to a 25 psig header. There are several consumers at each pressure level. There are also other producers of 25 psig steam and 10 psig steam. Sketch a control system that:

(a) Controls pressure in the 900 psig header by fuel firing rate to the boiler.

(b) Controls pressure in the 150 psig header by valve A.

(c) Controls pressure in the 25 psig header by opening valve B if pressure is low and opening valve C if the pressure is high.

(d) Controls pressure in the 10 psig header by opening valve C if pressure is low and opening valve D if pressure is high.

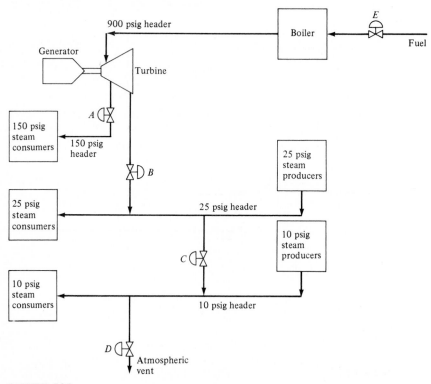

FIGURE P8.9

8.10. Sketch a control system for the two-column heat-integrated distillation system shown below.

(a) Reflux drum levels are controlled by distillate flows on each column.

(b) Reflux flow is ratioed to distillate flow on each column.

(c) Column pressure drop is controlled on the first column by manipulating steam flow to the auxiliary reboiler.

(d) Temperature in the second column is controlled by steam flow to its reboiler.

(e) Base levels are controlled by bottoms flow rates on each column.

(f) High or low base level in the first column overrides both steam valves.

(g) High or low base level in the second column overrides the steam valve on its reboiler.

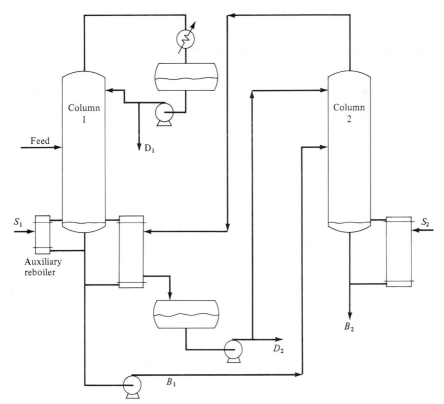

FIGURE P8.10

8.11. The distillation column sketched in Fig. P8.11 has an intermediate reboiler and a vapor sidestream. Sketch a control concept diagram showing the following control objectives:

(a) Column top pressure is controlled by a vent/bleed system, i.e., inert gas is added if pressure is low and gas from the reflux drum is vented if pressure is high.

(b) Reflux drum level is controlled by distillate flow.

(c) Reflux is ratioed to feed rate.

(d) Low-pressure steam flow rate to the intermediate reboiler is ratioed to feed rate.

(e) Vapor sidestream flow rate is set by a temperature controller holding tray 10 temperature.

(f) Base level is controlled by high-pressure steam flow rate to the base reboiler.

(g) Bottoms purge flow is flow controlled.

(h) Low base level overrides the low-pressure steam flow to the intermediate reboiler.

(i) High column pressure overrides both steam valves.

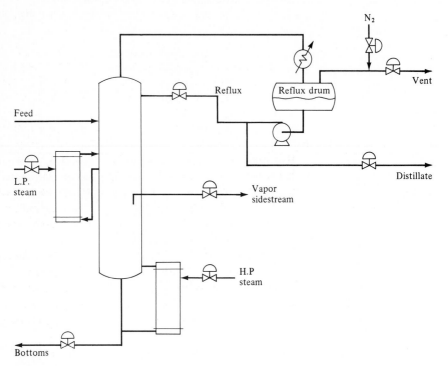

FIGURE P8.11

8.12. Hot oil is heated in a furnace in three parallel passes. The oil is used as a heat source in four parallel process heat exchangers. Draw a control concept diagram that achieves the following objectives:

(a) The temperature of each process stream leaving the four process heat exchangers is controlled by manipulating the hot oil flow rate through each exchanger.

(b) Furnace hot oil exit temperature is controlled by manipulating fuel flow rate.

(c) A valve position controller is used to reset the setpoint of the furnace exit temperature controller such that the control valve that is the most open of the four control valves on the hot oil streams flowing through the process heat exchangers is 80 percent open.

(d) The flow rates of hot oil through the three parallel passes in the furnace are maintained equal to each other.

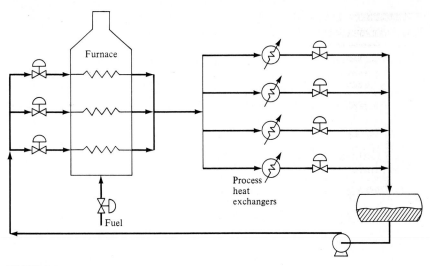

FIGURE P8.12

8.13. Terephthalic acid (TPA) is produced by air oxidation of paraxylene at 200 psig. The reaction is exothermic. Nitrogen plus excess oxygen leave the top of the reactor.

$$\text{Xylene} + O_2 \rightarrow \text{TPA} + \text{water}$$

Liquid products (TPA and water) are removed from the bottom of the reactor. Heat is removed by circulating liquid from the base of the reactor through a water-cooled heat exchanger. The air compressor is driven by a steam turbine.

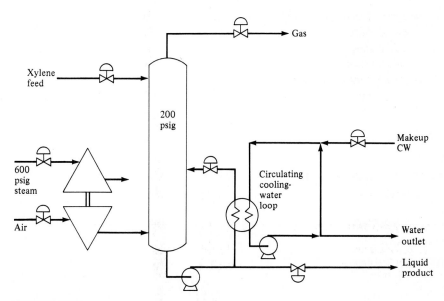

FIGURE P8.13

Specify the action of all control valves and sketch a control system that will achieve the following objectives:

(a) Xylene feed is flow-controlled.

(b) Air is flow-controlled by adjusting the speed of the turbine.

(c) Reactor temperature is controlled through a cascade system. Circulating water temperature is controlled by makeup cooling water. The setpoint of this temperature controller is set by the reactor temperature controller. The circulation rate of process liquid through the cooler is flow-controlled.

(d) Reactor pressure is controlled by the gas leaving the reactor.

(e) Base liquid level is controlled by liquid product flow.

(f) Air flow rate is adjusted by a composition controller that holds 2 percent oxygen in the gas leaving the reactor.

(g) High reactor pressure overrides steam flow to the air compressor turbine.

8.14. Overhead vapor from a distillation column passes through a partial condenser. The uncondensed portion is fed into a vapor-phase reactor. The condensed portion is used for reflux in the distillation column.

The vapor fed to the reactor can also come from a vaporizer which is fed from a surge tank. To conserve energy, it is desirable to feed the reactor with vapor directly from the column instead of from the vaporizer. The only time that the vaporizer should be used is when there is not enough vapor produced by the column.

Sketch a control concept diagram showing:

(a) Total vapor flow rate to the reactor is flow controlled by valve V-1.

(b) Reflux drum pressure is controlled by valve V-2.

(c) Vaporizer pressure is controlled by valve V-3.

(d) Vaporizer liquid level is controlled by valve V-4.

(e) Column reflux is flow controlled by valve V-5.

(f) Reflux drum level is controlled by valve V-6.

(g) High reflux drum level opens valve V-7.

(h) High vaporizer pressure overrides valve V-2.

(i) High reflux drum pressure overrides valve V-6.

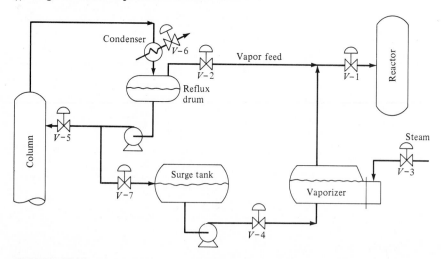

FIGURE P8.14

8.15. Two distillation columns are heat-integrated as shown in the sketch below. The first column has an auxiliary condenser to take any excess vapor that the second column does not need. The second column has an auxiliary reboiler that provides additional heat if required.

Prepare a control concept diagram that includes the following control objectives:

(a) Base levels are controlled by bottoms flows.

(b) Reflux drum levels are controlled by distillate flows.

(c) Reflux flows are flow-controlled.

(d) The pressure in the first column is controlled by vapor flow rate to the auxiliary condenser. A low pressure override pinches the vapor valve to the second column reboiler.

(e) The pressure in the second column is controlled by manipulating cooling water to the condenser.

(f) A temperature in the stripping section of the first column is controlled by manipulating steam to the reboiler.

(g) A temperature in the stripping section of the second column is controlled by manipulating the vapor to the reboiler of the second column that comes from the first column and by manipulating the steam to the auxiliary reboiler. A

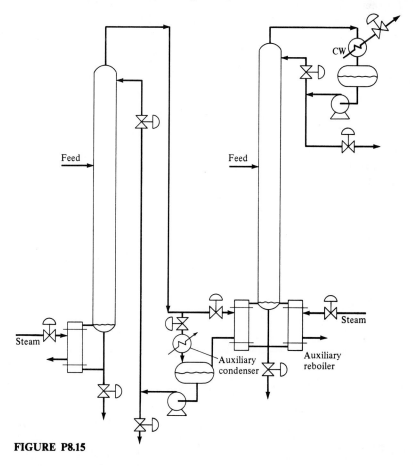

Feed

Feed

CW

Steam

Steam

Auxiliary
condenser

Auxiliary
reboiler

FIGURE P8.15

split-range system is used so that steam to the auxiliary reboiler is only used when insufficient heat is available from the vapor from the first column.

(*h*) High column pressures in both columns pinch reboiler steam.

8.16. The sketch below shows a distillation column that is heat-integrated with an evaporator. Draw a control concept diagram which accomplishes the following objectives:

(*a*) In the evaporator, temperature is controlled by steam, level by liquid product, and pressure by auxiliary cooling or vapor to the reboiler. Level in the condensate receiver is controlled by condensate.

(*b*) In the column, reflux is flow-controlled, reflux drum level is controlled by distillate, base level by bottoms, pressure by vent vapor, and temperature by steam to

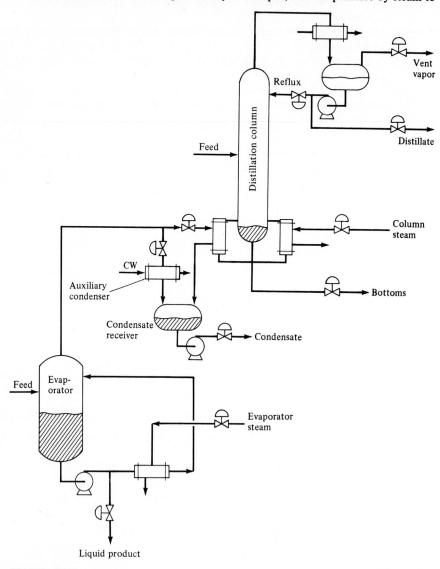

FIGURE P8.16

the auxiliary reboiler or vapor from the evaporator.

(c) A high column pressure override controller pinches steam to both the evaporator and the auxiliary reboiler.

8.17. Sketch a control scheme for the cryogenic stripper shown below that is used for removing small amounts of propane from natural gas.

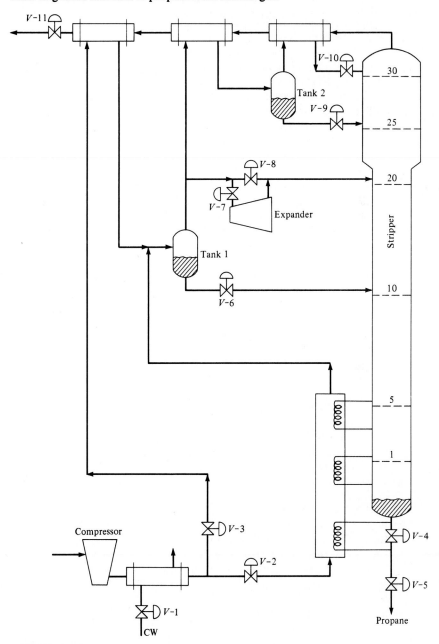

FIGURE P8.17

(a) Cooling-water valve V-1 is manipulated to control the gas temperature leaving the cooler.

(b) Valve V-2 controls a temperature on tray 15 in the stripper.

(c) Valve V-3 controls the total flow rate of gas into the compressor.

(d) Valve V-4 controls the temperature of the propane bottoms product leaving the unit.

(e) Valve V-5 controls the column base level.

(f) Valve V-6 controls the liquid level in tank 1.

(g) Valve V-7 controls the speed of the expander turbine.

(h) A "valve position controller" is used to keep valve V-7 nearly wide open by adjusting the setpoint of the expander speed controller.

(i) A pressure controller opens valve V-8 if the pressure in tank 1 gets too high.

(j) Valve V-9 controls tank 2 level.

(k) Valve V-10 controls tank 2 pressure.

(l) Valve V-11 controls the pressure in the stripper.

(m) If the pressure in the stripper gets too high, an override controller pinches both valves V-2 and V-3.

8.18. A distillation column is used to separate two close-boiling components that have a relative volatility close to one. The reflux ratio is quite high (15) and many trays are required (150). To control the compositions of both products the flow rates of the product streams (distillate D and bottoms B) are manipulated. Gas chromatographs are used to measure the product compositions. Base level is controlled by steam flow rate to the reboiler and reflux drum level is controlled by reflux flow rate.

(a) Design an override control system that will use bottoms flow rate to control base level if a high limit is reached on the steam flow.

(b) Design an override control system that will use very low composition transmitter output signals to detect if either of the chromatographs has failed or is out of service. The control system should switch the structure of the control loops as follows:

   (i) If the distillate composition signal is not available, reflux should be held constant at given flow rate and reflux drum level should be held by distillate flow rate.

   (ii) If the bottoms composition signal is not available, steam should be held constant and base level should be held by bottoms flow rate.

8.19. A furnace control system consists of the following loops:

- Temperature of the process stream leaving the furnace changes the setpoint of a flow controller on fuel.
- Air flow is ratioed to fuel flow, with the ratio changed by a stack-gas excess-air controller.

Show how some simple first-order lags and selectors can be used to produce a control system in which

- The air flow *leads* the fuel flow when an increase in fuel flow occurs.
- The air flow *lags* the fuel flow when a decrease in fuel flow occurs.

    Hint: Sketch the responses for both positive and negative step changes in the input to a circuit which consists of a first-order lag with unity gain and a low-selector. The input signal goes in parallel to the lag and to the low-selector. The output of the lag goes to the other input of the low-selector.

PART
IV

LAPLACE-DOMAIN DYNAMICS AND CONTROL

Well, now it is finally time for our Russian lessons! We have explored dynamics and control in the "English" time domain, using differential equations and finding exponential solutions. We saw that the important parameters are the time constant, the steadystate gain, and the damping coefficient of the system. If the process has no controllers, the system is openloop, and we look at openloop time constants and openloop damping coefficients. If controllers are included in the analysis, the system is a closedloop one. We tune the controller to achieve certain desired closedloop time constants and closedloop damping coefficients. It is important that you keep track of what system you are looking at, openloop or closedloop. We do not want to compare apples and oranges in our studies.

In the next three chapters we will develop methods of analysis of dynamic systems, both openloop and closedloop, that use Laplace transforms. This form of representation of systems is much more compact and convenient than using time-domain representation. The Laplace-domain description of a process is a "transfer function." This is a relationship between the input to a system and the output of the system. Transfer functions contain all the steadystate and dynamic information about a process in a very compact form.

So we will find it extremely useful to learn a little "Russian." Now don't get too concerned about having to learn an extensive Russian vocabulary. As you will see, there are only about ten "words" that you have to learn in Russian: ten transfer functions can describe almost all chemical engineering processes.

Laplace transformation can *only* be applied to *linear* ordinary differential equations. So for most of the rest of the book, we will be dealing with linear systems.

CHAPTER
9

LAPLACE-DOMAIN DYNAMICS

The use of Laplace transformations yields some very useful simplifications in notation and computation. Laplace-transforming the linear ordinary differential equations describing our processes in terms of the independent variable t converts them into algebraic equations in the Laplace transform variable s. This provides a very convenient representation of system dynamics.

Most of you have probably been exposed to Laplace transforms in a mathematics course, but we will lead off this chapter with a brief review of some of the most important relationships. Then we will derive the Laplace transformations of commonly encountered functions. Next we will develop the idea of transfer functions by observing what happens to the differential equations describing a process when they are Laplace-transformed. Finally, we will apply these techniques to some chemical engineering systems.

9.1 LAPLACE-TRANSFORMATION FUNDAMENTALS

9.1.1 Definition

The Laplace transformation of a function of time $f_{(t)}$ consists of "operating on" the function by multiplying it by e^{-st} and integrating with respect to time t from 0 to infinity. The operation of Laplace transforming will be indicated by the notation

$$\mathcal{L}[f_{(t)}] \equiv \int_0^\infty f_{(t)} e^{-st} \, dt \tag{9.1}$$

where $\mathcal{L}$ = Laplace transform operator
s = Laplace transform variable

In integrating between the definite limits of 0 and infinity we "integrate out" the time variable t and are left with a new quantity that is a function of s. We will use the notation

$$\mathcal{L}[f_{(t)}] \equiv F_{(s)} \tag{9.2}$$

The variable s is a complex number.

Thus Laplace transformation converts functions from the time domain (where t is the independent variable) into the Laplace domain (where s is the independent variable). The advantages of using this transformation will become clear later in this chapter.

9.1.2 Linearity Property

One of the most important properties of Laplace transformation is that it is linear.

$$\mathcal{L}[f_{1(t)} + f_{2(t)}] = \mathcal{L}[f_{1(t)}] + \mathcal{L}[f_{2(t)}] \tag{9.3}$$

This property is easily proved:

$$\mathcal{L}[f_{1(t)} + f_{2(t)}] = \int_0^\infty [f_{1(t)} + f_{2(t)}] e^{-st} \, dt$$

$$= \int_0^\infty f_{1(t)} e^{-st} \, dt + \int_0^\infty f_{2(t)} e^{-st} \, dt$$

$$= \mathcal{L}[f_{1(t)}] + \mathcal{L}[f_{2(t)}] = F_{1(s)} + F_{2(s)} \tag{9.4}$$

9.2 LAPLACE TRANSFORMATION OF IMPORTANT FUNCTIONS

Let us now apply the definition of the Laplace transformation to some important time functions: steps, ramps, exponential, sines, etc.

9.2.1 Step Function

Consider the function

$$f_{(t)} = K u_{(t)} \tag{9.5}$$

where K is a constant and $u_{(t)}$ is the unit step function defined in Sec. 6.1 as

$$u_{(t)} = 1 \qquad t > 0$$
$$= 0 \qquad t \leq 0 \qquad (9.6)$$

Note that the step function is just a constant (for time greater than zero).
Laplace-transforming this function gives

$$\mathcal{L}[Ku_{(t)}] \equiv \int_0^\infty [Ku_{(t)}]e^{-st} \, dt = K \int_0^\infty e^{-st} \, dt$$

since $u_{(t)}$ is just equal to unity over the range of integration.

$$\mathcal{L}[Ku_{(t)}] = \left[-\frac{K}{s} e^{-st} \right]_{t=0}^{t=\infty} = -\frac{K}{s}[0 - 1] = \frac{K}{s} \qquad (9.7)$$

Therefore the Laplace transformation of a step function (or a constant) of magnitude K is simply $K(1/s)$.

9.2.2 Ramp

The ramp function is one that changes continuously with time at a constant rate K.

$$f_{(t)} = Kt \qquad (9.8)$$

Then the Laplace transformation is

$$\mathcal{L}[Kt] \equiv \int_0^\infty [Kt]e^{-st} \, dt$$

Integrating by parts, we let

$$u = t \qquad \text{and} \qquad dv = e^{-st} \, dt$$

Then

$$du = dt \qquad \text{and} \qquad v = -\frac{1}{s} e^{-st}$$

Since

$$\int_0^\infty u \, dv = [uv]_0^\infty - \int_0^\infty v \, du \qquad (9.9)$$

$$K \int_0^\infty t e^{-st} \, dt = \left[-\frac{Kt}{s} e^{-st} \right]_{t=0}^{t=\infty} + \int_0^\infty \frac{K}{s} e^{-st} \, dt$$

$$= [0 - 0] - \left[\frac{K}{s^2} e^{-st} \right]_{t=0}^{t=\infty} = K\left(\frac{1}{s^2} \right)$$

Therefore the Laplace transformation of a ramp function is

$$\mathcal{L}[Kt] = K\left(\frac{1}{s^2}\right) \tag{9.10}$$

9.2.3 Sine

$$f_{(t)} = \sin(\omega t) \tag{9.11}$$

where ω = frequency (radians per time)

$$\mathcal{L}[\sin(\omega t)] \equiv \int_0^\infty [\sin(\omega t)]e^{-st}\, dt$$

Using

$$\sin(\omega t) = \frac{e^{i\omega t} - e^{-i\omega t}}{2i} \tag{9.12}$$

$$\mathcal{L}[\sin(\omega t)] \equiv \int_0^\infty \left[\frac{e^{i\omega t} - e^{-i\omega t}}{2i}\right] e^{-st}\, dt$$

$$= \int_0^\infty \left[\frac{e^{-(s-i\omega)t} - e^{-(s+i\omega)t}}{2i}\right] dt$$

$$= \frac{1}{2i}\left[\frac{-e^{-(s-i\omega)t}}{s-i\omega} + \frac{e^{-(s+i\omega)t}}{s+i\omega}\right]_{t=0}^{t=\infty} = \frac{1}{2i}\left[\frac{1}{s-i\omega} - \frac{1}{s+i\omega}\right]$$

Therefore

$$\mathcal{L}[\sin(\omega t)] = \frac{\omega}{s^2 + \omega^2} \tag{9.13}$$

9.2.4 Exponential

Since we found in Chap. 6 that the responses of linear systems are a series of exponential terms, the Laplace transformation of the exponential function is the most important of any of the functions.

$$f_{(t)} = e^{-at} \tag{9.14}$$

$$\mathcal{L}[e^{-at}] \equiv \int_0^\infty [e^{-at}]e^{-st}\, dt = \int_0^\infty [e^{-(s+a)t}]\, dt$$

$$= \left[\frac{-1}{s+a}e^{-(s+a)t}\right]_{t=0}^{t=\infty} = \frac{1}{s+a}$$

Therefore

$$\boxed{\mathcal{L}[e^{-at}] = \frac{1}{s+a}} \tag{9.15}$$

We will use this function repeatedly throughout the rest of this book.

9.2.5 Exponential Multiplied By Time

If you remember from Chap. 6, repeated roots of the characteristic equation yielded time functions that contained an exponential multiplied by time.

$$f_{(t)} = te^{-at} \tag{9.16}$$

$$\mathcal{L}[te^{-at}] \equiv \int_0^\infty [te^{-at}]e^{-st}\,dt = \int_0^\infty [te^{-(s+a)t}]\,dt$$

Integrating by parts

$$u = t \qquad dv = e^{-(s+a)t}\,dt$$

$$du = dt \qquad v = -\frac{1}{s+a}e^{-(s+a)t}$$

$$\int_0^\infty [te^{-(s+a)t}]\,dt = \left[\frac{-te^{-(s+a)t}}{s+a}\right]_{t=0}^{t=\infty} + \int_0^\infty \frac{e^{-(s+a)t}}{s+a}\,dt$$

$$= [0-0] - \left[\frac{1}{(s+a)^2}e^{-(s+a)t}\right]_{t=0}^{t=\infty}$$

Therefore

$$\mathcal{L}[te^{-at}] = \frac{1}{(s+a)^2} \tag{9.17}$$

Equation (9.17) can be generalized for a repeated root of nth order to give

$$\mathcal{L}[t^n e^{-at}] = \frac{n!}{(s+a)^{n+1}} \tag{9.18}$$

9.2.6 Impulse (Dirac Delta Function $\delta_{(t)}$)

The impulse function is an infinitely high spike that has zero width and an area of one (see Fig. 9.1*a*). It is a function that cannot occur in any real system, but it is a useful mathematical function that will be used in several spots in this book.

One way to define $\delta_{(t)}$ is to call it the derivative of the unit step function, as sketched in Fig. 9.1*b*.

$$\delta_{(t)} = \frac{du_{(t)}}{dt} \tag{9.19}$$

Now the unit step function can be expressed as a limit of the first-order exponential step response as the time constant goes to zero.

$$u_{(t)} = \lim_{\tau \to 0} [1 - e^{-t/\tau}] \tag{9.20}$$

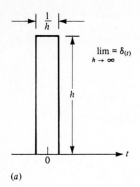

(a)

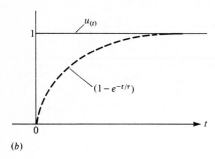

(b)

FIGURE 9.1
$\delta_{(t)}$ function.

Now

$$\mathcal{L}[\delta_{(t)}] = \mathcal{L}\left[\frac{d}{dt}\left\{\lim_{\tau \to 0}(1 - e^{-t/\tau})\right\}\right] = \lim_{\tau \to 0}\mathcal{L}\left[\frac{1}{\tau}e^{-t/\tau}\right]$$

$$= \lim_{\tau \to 0}\left[\frac{1/\tau}{s + 1/\tau}\right] = \lim_{\tau \to 0}\left[\frac{1}{\tau s + 1}\right] = 1$$

Therefore

$$\mathcal{L}[\delta_{(t)}] = 1 \tag{9.21}$$

9.3 INVERSION OF LAPLACE TRANSFORMS

After transforming equations into the Laplace domain and solving for output variables as functions of s, we sometimes want to transform back into the time domain. This operation is called *inversion* or *inverse Laplace transformation*. We are translating from Russian into English. We will use the notation.

$$\mathcal{L}^{-1}[F_{(s)}] = f_{(t)} \tag{9.22}$$

There are several ways to invert functions of s into functions of t. Since s is a complex number, a contour integration in the complex s plane can be used.

$$f_{(t)} = \frac{1}{2\pi i} \int_{\alpha - i\omega}^{\alpha + i\omega} e^{st} F_{(s)} \, ds \tag{9.23}$$

Another method is simply to look up the function in mathematics tables.

The most common inversion method is called *partial fractions expansion*. The function to be inverted, $F_{(s)}$, is merely rearranged into a series of simple functions:

$$F_{(s)} = F_{1(s)} + F_{2(s)} + \cdots + F_{N(s)} \tag{9.24}$$

Then each term is inverted (usually by inspection because they are simple). The total time-dependent function is the sum of the individual time-dependent functions [see Eq. (7.3)].

$$f_{(t)} = \mathcal{L}^{-1}[F_{1(s)}] + \mathcal{L}^{-1}[F_{2(s)}] + \cdots + \mathcal{L}^{-1}[F_{N(s)}]$$
$$= f_{1(t)} + f_{2(t)} + \cdots + f_{N(t)} \tag{9.25}$$

As we will shortly find out, the $F_{(s)}$'s normally appear as ratios of polynomials in s.

$$F_{(s)} = \frac{Z_{(s)}}{P_{(s)}} \tag{9.26}$$

where $Z_{(s)} = $ Mth-order polynomial in s
$P_{(s)} = $ Nth-order polynomial in s

Factoring the denominator into its roots (or *zeros*) gives

$$F_{(s)} = \frac{Z}{(s - p_1)(s - p_2)(s - p_3) \cdots (s - p_N)} \tag{9.27}$$

where the p_i are the roots of the polynomial $P_{(s)}$, which may be distinct or repeated.

If all the p_i are different (i.e., distinct roots), we can express $F_{(s)}$ as a sum of N terms:

$$F_{(s)} = \frac{A}{s - p_1} + \frac{B}{s - p_2} + \frac{C}{s - p_3} + \cdots + \frac{W}{s - p_N} \tag{9.28}$$

The numerators of each of the terms in Equation (9.28) can be evaluated as shown below and then each term is inverted.

$$A = \lim_{s \to p_1} \ [(s - p_1)F_{(s)}]$$

$$B = \lim_{s \to p_2} \ [(s - p_2)F_{(s)}] \tag{9.29}$$

$$\cdots \cdots \cdots \cdots \cdots \cdots$$

$$W = \lim_{s \to p_N} \ [(s - p_N)F_{(s)}]$$

Example 9.1. Given the $F_{(s)}$ below, find its inverse $f_{(t)}$ by partial fractions expansion.

$$F_{(s)} = \frac{K_p \bar{C}_{A0}}{s(\tau_p s + 1)}$$

$$F_{(s)} = \frac{K_p \bar{C}_{A0}/\tau_p}{s(s + 1/\tau_p)} = \frac{A}{s} + \frac{B}{s + 1/\tau_p}$$

(9.30)

The roots of the denominator are 0 and $-1/\tau_p$.

$$A = \lim_{s \to 0} [s F_{(s)}] = \lim_{s \to 0} \left[s \frac{K_p \bar{C}_{A0}}{s(\tau_p s + 1)} \right] = K_p \bar{C}_{A0}$$

$$B = \lim_{s \to -1/\tau_p} \left[\left(s + \frac{1}{\tau_p} \right) F_{(s)} \right] = \lim_{s \to -1/\tau_p} \left[\frac{K_p \bar{C}_{A0}/\tau_p}{s} \right] = -K_p \bar{C}_{A0}$$

Therefore

$$F_{(s)} = K_p \bar{C}_{A0} \left[\frac{1}{s} - \frac{1}{s + 1/\tau_p} \right]$$

(9.31)

The two simple functions in Eq. (9.31) can be inverted by using Eqs. (9.7) and (9.15).

$$f_{(t)} = K_p \bar{C}_{A0} [1 - e^{-t/\tau_p}]$$

(9.32)

If there are some repeated roots in the denominator of Eq. (9.27), we must expand $F_{(s)}$ as a sum of N terms:

$$F_{(s)} = \frac{Z}{(s - p_1)^2 (s - p_3)(s - p_4) \cdots (s - p_N)}$$

(9.33)

$$F_{(s)} = \frac{A}{(s - p_1)^2} + \frac{B}{s - p_1} + \frac{C}{s - p_3} + \cdots + \frac{W}{s - p_N}$$

(9.34)

The above is for a repeated root of order 2. If the root is repeated three times (of order 3) the expansion would be

$$F_{(s)} = \frac{Z}{(s - p_1)^3 (s - p_4)(s - p_5) \cdots (s - p_N)}$$

(9.35)

$$F_{(s)} = \frac{A}{(s - p_1)^3} + \frac{B}{(s - p_1)^2} + \frac{C}{s - p_1} + \cdots + \frac{W}{s - p_N}$$

(9.36)

The numerators of the terms in Eq. (9.34) are found from the relationships given below. These are easily proved by merely carrying out the indicated operations on Eq. (9.34).

$$A = \lim_{s \to p_1} [(s - p_1)^2 F_{(s)}]$$

$$B = \lim_{s \to p_1} \left\{ \frac{d}{ds} [(s - p_1)^2 F_{(s)}] \right\} \tag{9.37}$$

$$C = \lim_{s \to p_3} [(s - p_3) F_{(s)}]$$

To find the C numerator in Eq. (9.36) a second derivative with respect to s would have to be taken. Generalizing to the mth term A_m of an Nth-order root at p_1,

$$A_m = \lim_{s \to p_1} \left\{ \frac{d^{m-1}}{ds^{m-1}} [(s - p_1)^N F_{(s)}] \right\} \frac{1}{(m-1)!} \tag{9.38}$$

Example 9.2. Given the $F_{(s)}$ below, find its inverse.

$$F_{(s)} = \frac{K_p}{s(\tau_p s + 1)^2} = \frac{K_p/\tau_p^2}{s(s + 1/\tau_p)^2}$$

$$= \frac{A}{s} + \frac{B}{(s + 1/\tau_p)^2} + \frac{C}{s + 1/\tau_p}$$

$$A = \lim_{s \to 0} \frac{K_p}{(\tau_p s + 1)^2} = K_p$$

$$B = \lim_{s \to -1/\tau_p} \frac{K_p/\tau_p^2}{s} = -\frac{K_p}{\tau_p}$$

$$C = \lim_{s \to -1/\tau_p} \left[\frac{d}{ds} \left(\frac{K_p/\tau_p^2}{s} \right) \right] = \lim_{s \to -1/\tau_p} \frac{-K_p/\tau_p^2}{s^2} = -K_p$$

Therefore

$$F_{(s)} = K_p \left[\frac{1}{s} - \frac{1/\tau_p}{(s + 1/\tau_p)^2} - \frac{1}{s + 1/\tau_p} \right] \tag{9.39}$$

Inverting term by term yields

$$f_{(t)} = K_p \left(1 - \frac{t}{\tau_p} e^{-t/\tau_p} - e^{-t/\tau_p} \right) \tag{9.40}$$

9.4 TRANSFER FUNCTIONS

Our primary use of Laplace transformations in process control involves representing the dynamics of the process in terms of "transfer functions." These are output-input relationships and are obtained by Laplace-transforming algebraic and differential equations. In the discussion below, the output variable of the process is $x_{(t)}$. The input variable or the forcing function is $m_{(t)}$.

9.4.1   Multiplication by a Constant

Consider the algebraic equation

$$x_{(t)} = Km_{(t)} \tag{9.41}$$

Laplace-transforming both sides of the equation gives

$$\int_0^\infty x_{(t)} e^{-st} \, dt = K \int_0^\infty m_{(t)} e^{-st} \, dt \tag{9.42}$$

$$X_{(s)} = KM_{(s)}$$

where $X_{(s)}$ and $M_{(s)}$ are the Laplace transforms of $x_{(t)}$ and $m_{(t)}$. Note that $m_{(t)}$ is an arbitrary function of time. We have not specified at this point the exact form of the input. Comparing Eqs. (9.41) and (9.42) shows that the input and output variables are related in the Laplace domain in exactly the same way as they are related in the time domain. Thus the English and Russian words describing this situation are the same.

Equation (9.42) can be put into transfer-function form by finding the output-input ratio:

$$\frac{X_{(s)}}{M_{(s)}} = K \tag{9.43}$$

For any input $M_{(s)}$ the output $X_{(s)}$ is found by simply multiplying $M_{(s)}$ by the constant K. Thus the transfer function relating $X_{(s)}$ and $M_{(s)}$ is a constant or a "gain." We can represent this in block-diagram form as shown in Fig. 9.2.

9.4.2   Differentiation With Respect to Time

Consider what happens when the time derivative of a function $x_{(t)}$ is Laplace-transformed.

$$\mathcal{L}\left[\frac{dx}{dt}\right] = \int_0^\infty \left(\frac{dx}{dt}\right) e^{-st} \, dt \tag{9.44}$$

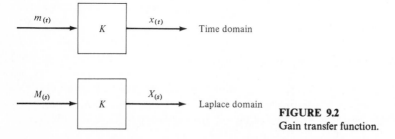

FIGURE 9.2
Gain transfer function.

Integrating by parts gives

$$u = e^{-st} \qquad\qquad dv = \frac{dx}{dt}\, dt$$

$$du = -se^{-st}\, dt \qquad v = x$$

Therefore

$$\int_0^\infty \left(\frac{dx}{dt}\right) e^{-st}\, dt = [xe^{-st}]_{t=0}^{t=\infty} + \int_0^\infty sxe^{-st}\, dt$$

$$= 0 - x_{(t=0)} + s\int_0^\infty x_{(t)} e^{-st}\, dt$$

The integral is, by definition, just the Laplace transformation of $x_{(t)}$, which we call $X_{(s)}$.

$$\mathcal{L}\left[\frac{dx}{dt}\right] = sX_{(s)} - x_{(t=0)} \tag{9.45}$$

The result is the most useful of all the Laplace transformations. It says that the operation of differentiation in the time domain is replaced by multiplication by s in the Laplace domain, minus an initial condition. This is where perturbation variables become so useful. If the initial condition is the steadystate operating level, all the initial conditions like $x_{(t=0)}$ are equal to zero. Then simple multiplication by s is equivalent to differentiation. An ideal derivative unit or a perfect differentiator can be represented in block-diagram form as shown in Fig. 9.3.

The same procedure, applied to a second-order derivative, gives the following:

$$\mathcal{L}\left[\frac{d^2x}{dt^2}\right] = s^2 X_{(s)} - sx_{(t=0)} - \left(\frac{dx}{dt}\right)_{t=0} \tag{9.46}$$

Thus differentiation twice is equivalent to multiplying twice by s, if all initial conditions are zero. The block diagram is shown in Fig. 9.3.

The above can be generalized to an Nth-order derivative with respect to time. In going from the time domain into the Laplace domain, $d^N x/dt^N$ is replaced by s^N. Therefore an Nth-order differential equation becomes an Nth-order algebraic equation.

$$a_N \frac{d^N x}{dt^N} + a_{N-1} \frac{d^{N-1}x}{dt^{N-1}} + \cdots + a_1 \frac{dx}{dt} + a_0 x = m_{(t)} \tag{9.47}$$

$$a_N s^N X_{(s)} + a_{N-1} s^{N-1} X_{(s)} + \cdots + a_1 s X_{(s)} + a_0 X_{(s)} = M_{(s)} \tag{9.48}$$

$$(a_N s^N + a_{N-1} s^{N-1} + \cdots + a_1 s + a_0) X_{(s)} = M_{(s)} \tag{9.49}$$

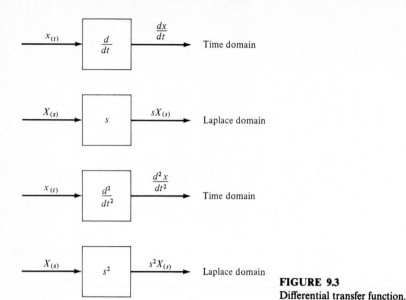

FIGURE 9.3
Differential transfer function.

Notice that the polynomial in Eq. (9.49) looks exactly like the characteristic equation discussed in Chap. 6. We will return to this not-accidental similarity in the next section.

9.4.3 Integration

Laplace-transforming the integral of a function $x_{(t)}$ gives

$$\mathcal{L}\left[\int x_{(t)}\, dt\right] = \int_0^\infty \left(\int x_{(t)}\, dt\right) e^{-st}\, dt$$

Integrating by parts,

$$u = \int x\, dt \qquad dv = e^{-st}\, dt$$

$$du = x\, dt \qquad v = -\frac{1}{s} e^{-st}$$

$$\int_0^\infty \left(\int x_{(t)}\, dt\right) e^{-st}\, dt = \left[-\frac{1}{s} e^{-st} \int x\, dt\right]_{t=0}^{t=\infty} + \frac{1}{s} \int_0^\infty x_{(t)} e^{-st}\, dt$$

Therefore

$$\mathcal{L}\left[\int x_{(t)}\, dt\right] = \frac{1}{s} X_{(s)} + \frac{1}{s}\left(\int x\, dt\right)_{(t=0)} \tag{9.50}$$

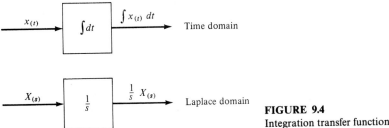

Time domain

Laplace domain

FIGURE 9.4
Integration transfer function.

The operation of integration is equivalent to division by s in the Laplace domain, using zero initial conditions. Thus, integration is the inverse of differentiation. Figure 9.4 gives a block-diagram representation.

The $1/s$ is an operator or a transfer function showing what operation is performed on the input signal. This is a completely different idea than the simple Laplace transformation of a function. Remember, the Laplace transform of the unit step function was also equal to $1/s$. But this is the Laplace transformation of a function. The $1/s$ operator discussed above is a transfer function, not a function.

9.4.4 Deadtime

Delay time, transportation lag, or deadtime is frequently encountered in chemical engineering systems since we did not earn our reputation as underpaid plumbers for nothing!

Suppose a process stream is flowing through a pipe in essentially plug flow and that it takes D minutes for any individual element of fluid to flow from the entrance to the exit of the pipe. Then the pipe represents a deadtime element.

If a certain dynamic variable $f_{(t)}$, such as temperature or composition, enters the front end of the pipe, it will emerge from the other end D minutes later with exactly the same shape, as shown in Fig. 9.5.

Let us see what happens when we Laplace-transform a function $f_{(t-D)}$ that has been delayed by a deadtime. Laplace transformation is defined in Eq. (9.1).

$$\mathcal{L}[f_{(t)}] \equiv \int_0^\infty f_{(t)} e^{-st}\, dt = F_{(s)} \tag{9.51}$$

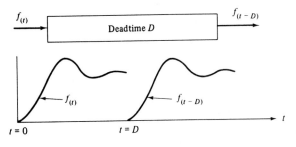

FIGURE 9.5
Effect of a dead-time element.

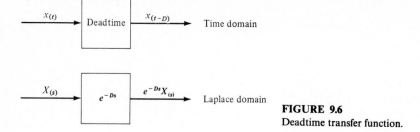

FIGURE 9.6
Deadtime transfer function.

The variable t in the above equation is just a "dummy variable" of integration. It is integrated out, leaving a function of only s. Thus we can write Eq. (9.51) in a completely equivalent mathematical form:

$$F_{(s)} = \int_0^\infty f_{(y)} e^{-sy} \, dy \qquad (9.52)$$

where y is now the dummy variable of integration. Now let $y = t - D$.

$$F_{(s)} = \int_0^\infty f_{(t-D)} e^{-s(t-D)} \, d(t-D) = e^{Ds} \int_0^\infty f_{(t-D)} e^{-st} \, dt$$

$$F_{(s)} = e^{Ds} \mathcal{L}[f_{(t-D)}] \qquad (9.53)$$

Therefore

$$\mathcal{L}[f_{(t-D)}] = e^{-Ds} F_{(s)} \qquad (9.54)$$

Thus time delay or deadtime in the time domain is equivalent to multiplication by e^{-Ds} in the Laplace domain.

If the input into the deadtime element is $x_{in(t)}$ and the output of the deadtime element is $x_{out(t)}$, then x_{in} and x_{out} are related by

$$x_{out(t)} = x_{in(t-D)}$$

and in the Laplace domain,

$$X_{out(s)} = e^{-Ds} X_{in(s)} \qquad (9.55)$$

Thus the transfer function between output and input variables for a pure deadtime process is e^{-Ds}, as sketched in Fig. 9.6.

9.5 EXAMPLES

Now we are ready to apply all these Laplace-transformation techniques to some typical chemical engineering processes.

Example 9.3. Consider the isothermal CSTR of Example 6.6. The equation describing the system in terms of perturbation variables is

$$\frac{dC_A}{dt} + \left(\frac{1}{\tau} + k\right)C_{A(t)} = \frac{1}{\tau}C_{A0(t)} \tag{9.56}$$

where k and τ are constants. The initial condition is $C_{A(0)} = 0$. We will not specify what $C_{A0(t)}$ is for the moment but will just leave it as an arbitrary function of time. Laplace-transforming each term in Eq. (9.56) gives

$$sC_{A(s)} - C_{A(t=0)} + \left(\frac{1}{\tau} + k\right)C_{A(s)} = \frac{1}{\tau}C_{A0(s)} \tag{9.57}$$

The second term drops out because of the initial condition. Grouping like terms in $C_{A(s)}$ gives

$$\left(s + \frac{1}{\tau} + k\right)C_{A(s)} = \frac{1}{\tau}C_{A0(s)}$$

Thus the ratio of the output to the input (the "transfer function" $G_{(s)}$) is

$$G_{(s)} \equiv \frac{C_{A(s)}}{C_{A0(s)}} = \frac{1/\tau}{s + k + 1/\tau} \tag{9.58}$$

The denominator of the transfer function is exactly the same as the polynomial in s that was called the characteristic equation in Chap. 6. The roots of the denominator of the transfer function are called the *poles* of the transfer function. These are the values of s at which $G_{(s)}$ goes to infinity.

> The roots of the characteristic equation
> are equal to the poles of the transfer function.

This relationship between the poles of the transfer function and the roots of characteristic equation is an extremely important and useful one.

The transfer function given in Eq. (9.58) has one pole with a value of $-(k + 1/\tau)$. Rearranging Eq. (9.58) into the standard form of Eq. (6.34) gives

$$G_{(s)} = \frac{\left(\dfrac{1}{1 + k\tau}\right)}{\left(\dfrac{1}{k + 1/\tau}\right)s + 1} = \frac{K_p}{\tau_p s + 1} \tag{9.59}$$

where K_p is the process steadystate gain and τ_p is the process time constant. The pole of the transfer function is the reciprocal of the time constant.

This particular type of transfer function is called a *first-order lag*. It tells us how the input C_{A0} affects the output C_A, both dynamically and at steadystate. The form of the transfer function (polynomial of degree one in the denominator, i.e., one pole), and the numerical values of the parameters (steadystate gain and time constant) give a complete picture of the system in a very compact and usable form. The transfer function is property of the system only and is applicable for any input.

We can determine the dynamics and the steadystate characteristics of the system without having to pick any specific forcing function.

If the same input as used in Example 6.6 is imposed on the system, we should be able to use Laplace transforms to find the response of C_A to a step change of magnitude $\bar{C}_{A0}$.

$$C_{A0(t)} = \bar{C}_{A0}\, u_{(t)} \tag{9.60}$$

We will Laplace transform $C_{A0(t)}$, substitute into the system transfer function, solve for $C_{A(s)}$, and invert back into the time domain to find $C_{A(t)}$.

$$\mathcal{L}[C_{A0(t)}] = C_{A0(s)} = \bar{C}_{A0}\,\frac{1}{s} \tag{9.61}$$

$$C_{A(s)} = G_{(s)}\,C_{A0(s)} = \left[\frac{K_p}{\tau_p s + 1}\right]\left[\bar{C}_{A0}\,\frac{1}{s}\right] = \frac{K_p \bar{C}_{A0}}{s(\tau_p s + 1)} \tag{9.62}$$

Using partial fractions expansion to invert (see Example 9.1) gives

$$C_{A(t)} = K_p \bar{C}_{A0}[1 - e^{-t/\tau_p}]$$

This is exactly the solution obtained in Example 6.6 [Eq. (6.53)].

Example 9.4. The ODE of Example 6.8 with an arbitrary forcing function $m_{(t)}$ is

$$\frac{d^2x}{dt^2} + 5\frac{dx}{dt} + 6x = m_{(t)} \tag{9.63}$$

with the initial conditions

$$x_{(0)} = \left(\frac{dx}{dt}\right)_{(0)} = 0 \tag{9.64}$$

Laplace transforming gives

$$s^2 X_{(s)} + 5s X_{(s)} + 6X_{(s)} = M_{(s)}$$

$$X_{(s)}(s^3 + 5s + 6) = M_{(s)}$$

The process transfer function $G_{(s)}$ is

$$\frac{X_{(s)}}{M_{(s)}} = G_{(s)} = \frac{1}{s^2 + 5s + 6} = \frac{1}{(s + 2)(s + 3)} \tag{9.65}$$

Notice that the denominator of the transfer function is again the same polynomial in s as appeared in the characteristic equation of the system [Eq. (6.73)]. The poles of the transfer function are located at $s = -2$ and $s = -3$. So the poles of the transfer function are the roots of the characteristic equation.

If $m_{(t)}$ is a ramp input as in Example 6.13,

$$\mathcal{L}[m_{(t)}] = \mathcal{L}[t] = \frac{1}{s^2} \tag{9.66}$$

$$X_{(s)} = G_{(s)} M_{(s)} = \left(\frac{1}{s^2 + 5s + 6}\right)\left(\frac{1}{s^2}\right) = \frac{1}{s^2(s + 2)(s + 3)} \tag{9.67}$$

Partial fractions expansion gives

$$X_{(s)} = \frac{A}{s^2} + \frac{B}{s} + \frac{C}{s+2} + \frac{D}{s+3}$$

$$A = \lim_{s \to 0} [s^2 X_{(s)}] = \lim_{s \to 0} \left[\frac{1}{(s+2)(s+3)} \right] = \frac{1}{6}$$

$$B = \lim_{s \to 0} \left[\frac{d}{ds} (s^2 X_{(s)}) \right] = \lim_{s \to 0} \left[\frac{d}{ds} \left(\frac{1}{s^2 + 5s + 6} \right) \right]$$

$$= \lim_{s \to 0} \left[\frac{-(2s+5)}{(s^2 + 5s + 6)^2} \right] = -\frac{5}{36}$$

$$C = \lim_{s \to -2} [(s+2)X_{(s)}] = \lim_{s \to -2} \left[\frac{1}{s^2(s+3)} \right] = \frac{1}{4}$$

$$D = \lim_{s \to -3} [(s+3)X_{(s)}] = \lim_{s \to -3} \left[\frac{1}{s^2(s+2)} \right] = -\frac{1}{9}$$

Therefore

$$X_{(s)} = \frac{\frac{1}{6}}{s^2} - \frac{\frac{5}{36}}{s} + \frac{\frac{1}{4}}{s+2} - \frac{\frac{1}{9}}{s+3} \tag{9.68}$$

Inverting into the time domain gives the same solution as Eq. (6.109).

$$x_{(t)} = \tfrac{1}{6}t - \tfrac{5}{36} + \tfrac{1}{4}e^{-2t} - \tfrac{1}{9}e^{-3t} \tag{9.69}$$

Example 9.5. The isothermal three CSTR system is described by the three linear ODEs

$$\frac{dC_{A1}}{dt} + \left(k_1 + \frac{1}{\tau_1} \right) C_{A1} = \frac{1}{\tau_1} C_{A0}$$

$$\frac{dC_{A2}}{dt} + \left(k_2 + \frac{1}{\tau_2} \right) C_{A2} = \frac{1}{\tau_2} C_{A1} \tag{9.70}$$

$$\frac{dC_{A3}}{dt} + \left(k_3 + \frac{1}{\tau_3} \right) C_{A3} = \frac{1}{\tau_3} C_{A2}$$

The variables can be either total or perturbation variables since the equations are linear (all k's and τ's are constant). Let us use perturbation variables, and therefore the initial conditions for all variables are zero.

$$C_{A1(0)} = C_{A2(0)} = C_{A3(0)} = 0 \tag{9.71}$$

Laplace transforming gives

$$\left(s + k_1 + \frac{1}{\tau_1}\right)C_{A1(s)} = \frac{1}{\tau_1}\,C_{A0(s)}$$

$$\left(s + k_2 + \frac{1}{\tau_2}\right)C_{A2(s)} = \frac{1}{\tau_2}\,C_{A1(s)} \qquad (9.72)$$

$$\left(s + k_3 + \frac{1}{\tau_3}\right)C_{A3(s)} = \frac{1}{\tau_3}\,C_{A2(s)}$$

These can be rearranged to put them in terms of transfer functions for each tank.

$$G_{1(s)} \equiv \frac{C_{A1(s)}}{C_{A0(s)}} = \frac{1/\tau_1}{s + k_1 + 1/\tau_1}$$

$$G_{2(s)} \equiv \frac{C_{A2(s)}}{C_{A1(s)}} = \frac{1/\tau_2}{s + k_2 + 1/\tau_2} \qquad (9.73)$$

$$G_{3(s)} \equiv \frac{C_{A3(s)}}{C_{A2(s)}} = \frac{1/\tau_3}{s + k_3 + 1/\tau_3}$$

If we are interested in the total system and want only the effect of the input C_{A0} on the output C_{A3}, the three equations can be combined to eliminate C_{A1} and C_{A2}.

$$C_{A3(s)} = G_3\,C_{A2(s)} = G_3(G_2\,C_{A1(s)}) = G_3\,G_2(G_1\,C_{A0(s)}) \qquad (9.74)$$

The overall transfer function $G_{(s)}$ is

$$G_{(s)} \equiv \frac{C_{A3(s)}}{C_{A0(s)}} = G_{1(s)}\,G_{2(s)}\,G_{3(s)} \qquad (9.75)$$

The above demonstrates one very important and useful property of transfer functions. The total effect of a number of transfer functions connected in series is just the product of all the individual transfer functions. Figure 9.7 shows this in block-diagram form. The overall transfer function is a third-order lag with three poles.

$$G_{(s)} = \frac{1/\tau_1\tau_2\tau_3}{(s + k_1 + 1/\tau_1)(s + k_2 + 1/\tau_2)(s + k_3 + 1/\tau_3)} \qquad (9.76)$$

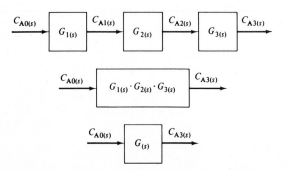

FIGURE 9.7
Transfer functions in series.

Further rearrangement puts the above in the standard form with time constants τ_{pi} and a steadystate gain K_p.

$$G = \cfrac{\cfrac{1}{1 + k_1 \tau_1} \cfrac{1}{1 + k_2 \tau_2} \cfrac{1}{1 + k_3 \tau_3}}{\left(\cfrac{\tau_1}{1 + k_1 \tau_1} s + 1\right)\left(\cfrac{\tau_2}{1 + k_2 \tau_2} s + 1\right)\left(\cfrac{\tau_3}{1 + k_3 \tau_3} s + 1\right)} \qquad (9.77)$$

$$G_{(s)} = \frac{K_p}{(\tau_{p1} s + 1)(\tau_{p2} s + 1)(\tau_{p3} s + 1)}$$

Let us assume a unit step change in the feed concentration C_{A0} and solve for the response of C_{A3}. We will take the case where all the τ_{pi}'s are the same, giving a repeated root of order 3 (a third-order pole at $s = -1/\tau_p$).

$$C_{A0(t)} = u_{(t)} \qquad \Rightarrow \qquad C_{A0(s)} = \frac{1}{s}$$

$$C_{A3(s)} = G_{(s)} C_{A0(s)} = \frac{K_p}{(\tau_p s + 1)^3} \frac{1}{s} = \frac{K_p/\tau_p^3}{s(s + 1/\tau_p)^3} \qquad (9.78)$$

Applying partial fractions expansion,

$$C_{A3(s)} = \frac{A}{s} + \frac{B}{(s + 1/\tau_p)^3} + \frac{C}{(s + 1/\tau_p)^2} + \frac{D}{s + 1/\tau_p} \qquad (9.79)$$

$$A = \lim_{s \to 0} \left(\frac{K_p/\tau_p^3}{(s + 1/\tau_p)^3}\right) = K_p$$

$$B = \lim_{s \to -1/\tau_p} \left(\frac{K_p/\tau_p^3}{s}\right) = -\frac{K_p}{\tau_p^2}$$

$$C = \lim_{s \to -1/\tau_p} \left[\frac{d}{ds}\left(\frac{K_p/\tau_p^3}{s}\right)\right] = \lim_{s \to -1/\tau_p} \left[-\frac{K_p/\tau_p^3}{s^2}\right] = -\frac{K_p}{\tau_p}$$

$$D = \lim_{s \to -1/\tau_p} \left(\frac{1}{2!}\frac{d^2}{ds^2}\left(\frac{K_p/\tau_p^3}{s}\right)\right) = \lim_{s \to -1/\tau_p} \left[\frac{1}{2}\frac{2K_p/\tau_p^3}{s^3}\right] = -K_p$$

Inverting Eq. (9.79), with the use of Eq. (9.18), yields

$$C_{A3(t)} = K_p\left[1 - \frac{1}{2}\left(\frac{t}{\tau_p}\right)^2 e^{-t/\tau_p} - \frac{t}{\tau_p} e^{-t/\tau_p} - e^{-t/\tau_p}\right] \qquad (9.80)$$

Example 9.6. The nonisothermal CSTR modeled in Sec. 3.6 can be linearized (see Prob. 6.12) to give two linear ODEs in terms of perturbation variables.

$$\frac{dC_A}{dt} = a_{11} C_A + a_{12} T + a_{13} C_{A0} + a_{15} F$$

$$\frac{dT}{dt} = a_{21} C_A + a_{22} T + a_{24} T_0 + a_{25} F + a_{26} T_J \qquad (9.81)$$

where

$$a_{11} = -\frac{\bar{F}}{V} - \bar{k} \qquad a_{12} = \frac{-\bar{C}_A E \bar{k}}{R \bar{T}^2}$$

$$a_{13} = \frac{\bar{F}}{V} \qquad a_{15} = \frac{\bar{C}_{A0} - \bar{C}_A}{V}$$

$$a_{21} = \frac{-\lambda \bar{k}}{\rho C_p} \qquad a_{22} = \frac{-\lambda \bar{k} E \bar{C}_A}{\rho C_p R \bar{T}^2} - \frac{\bar{F}}{V} - \frac{UA}{V \rho C_p} \qquad (9.82)$$

$$a_{24} = \frac{\bar{F}}{V} \qquad a_{25} = \frac{\bar{T}_0 - \bar{T}}{V}$$

$$a_{26} = \frac{UA}{V \rho C_p}$$

The variables C_{A0}, T_0, F, and T_J are all considered inputs. The output variables are C_A and T. Therefore eight different transfer functions are required to completely

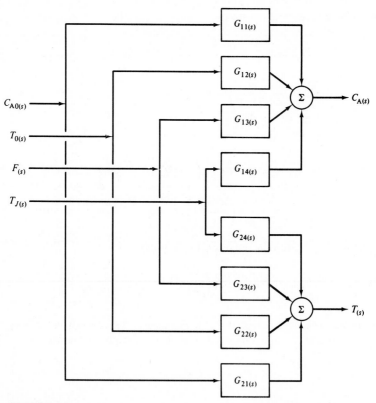

FIGURE 9.8
Block diagram of a multivariable linearized nonisothermal CSTR system.

describe the system. This multivariable aspect is the usual situation in most chemical engineering systems.

$$C_{A(s)} = G_{11(s)} C_{A0(s)} + G_{12(s)} T_{0(s)} + G_{13(s)} F_{(s)} + G_{14(s)} T_{J(s)}$$

$$T_{(s)} = G_{21(s)} C_{A0(s)} + G_{22(s)} T_{0(s)} + G_{23(s)} F_{(s)} + G_{24(s)} T_{J(s)}$$

(9.83)

The G_{ij}'s are, in general, functions of s and are the transfer functions relating inputs and outputs. Since the system is linear, the output is the sum of the effects of each individual input. This is called the principle of *superposition*.

To find these transfer functions, Eqs. (9.81) are Laplace-transformed and solved simultaneously.

$$sC_A = a_{11} C_A + a_{12} T + a_{13} C_{A0} + a_{15} F$$

$$sT = a_{21} C_A + a_{22} T + a_{24} T_0 + a_{25} F + a_{26} T_J$$

$$(s - a_{11})C_A = a_{12} T + a_{13} C_{A0} + a_{15} F$$

$$(s - a_{22})T = a_{21} C_A + a_{24} T_0 + a_{25} F + a_{26} T_J$$

Combining,

$$(s - a_{11})C_A = a_{12}\left[\frac{a_{21} C_A + a_{24} T_0 + a_{25} F + a_{26} T_J}{s - a_{22}}\right] + a_{13} C_{A0} + a_{15} F$$

$$\left(s - a_{11} - \frac{a_{12} a_{21}}{s - a_{22}}\right)C_A = \left(\frac{a_{12} a_{24}}{s - a_{22}}\right)T_0 + \left(\frac{a_{12} a_{25}}{s - a_{22}} + a_{15}\right)F$$

$$+ \left(\frac{a_{12} a_{26}}{s - a_{22}}\right)T_J + a_{13} C_{A0}$$

Finally,

$$C_{A(s)} = \left[\frac{a_{13}(s - a_{22})}{s^2 - (a_{11} + a_{22})s + a_{11} a_{22} - a_{12} a_{21}}\right] C_{A0(s)}$$

$$+ \left[\frac{a_{12} a_{24}}{s^2 - (a_{11} + a_{22})s + a_{11} a_{22} - a_{12} a_{21}}\right] T_{0(s)}$$

$$+ \left[\frac{a_{12} a_{25} + a_{15}(s - a_{22})}{s^2 - (a_{11} + a_{22})s + a_{11} a_{22} - a_{12} a_{21}}\right] F_{(s)}$$

$$+ \left[\frac{a_{12} a_{26}}{s^2 - (a_{11} + a_{22})s + a_{11} a_{22} - a_{12} a_{21}}\right] T_{J(s)}$$

(9.84)

$$T_{(s)} = \left[\frac{a_{13} a_{21}}{s^2 - (a_{11} + a_{22})s + a_{11} a_{22} - a_{12} a_{21}}\right] C_{A0(s)}$$

$$+ \left[\frac{a_{24}(s - a_{11})}{s^2 - (a_{11} + a_{22})s + a_{11} a_{22} - a_{12} a_{21}}\right] T_{0(s)}$$

$$+ \left[\frac{a_{15} a_{21} + a_{25}(s - a_{11})}{s^2 - (a_{11} + a_{22})s + a_{11} a_{22} - a_{12} a_{21}}\right] F_{(s)}$$

$$+ \left[\frac{a_{26}(s - a_{11})}{s^2 - (a_{11} + a_{22})s + a_{11} a_{22} - a_{12} a_{21}}\right] T_{J(s)}$$

(9.85)

The system is shown in block-diagram form in Fig. 9.8.

Notice that the G's are ratios of polynomials in s. The $s - a_{11}$ and $s - a_{22}$ terms in the numerators are called *first-order leads*. Notice also that the denominators of all the G's are exactly the same.

Example 9.7. The two-heated-tank example modeled in Sec. 3.4 was described by two linear ODEs:

$$\rho C_p V_1 \frac{dT_1}{dt} = \rho C_p F(T_0 - T_1) + Q_1 \tag{9.86}$$

$$\rho C_p V_2 \frac{dT_2}{dt} = \rho C_p F(T_1 - T_2) \tag{9.87}$$

The numerical values of variables are:

$$F = 90 \text{ ft}^3/\text{min} \qquad \rho = 40 \text{ lb}_m/\text{ft}^3 \qquad C_p = 0.6 \text{ Btu/lb}_m \, °F$$

$$V_1 = 450 \text{ ft}^3 \qquad V_2 = 90 \text{ ft}^3$$

Plugging these into Eqs. (9.86) and (9.87) gives

$$(40)(0.6)(450) \frac{dT_1}{dt} = (40)(90)(0.6)(T_0 - T_1) + Q_1 \tag{9.88}$$

$$(40)(0.6)(90) \frac{dT_2}{dt} = (40)(90)(0.6)(T_1 - T_2) \tag{9.89}$$

$$5 \frac{dT_1}{dt} + T_1 = T_0 + \frac{Q_1}{2160} \tag{9.90}$$

$$\frac{dT_2}{dt} + T_2 = T_1 \tag{9.91}$$

Laplace-transforming gives

$$(5s + 1)T_{1(s)} = T_{0(s)} + \frac{1}{2160} Q_{1(s)}$$

$$(s + 1)T_{2(s)} = T_{1(s)}$$

Rearranging and combining to eliminate T_1 give the output variable T_2 as a function of the two input variables, T_0 and Q_1.

$$T_{2(s)} = \left[\frac{1}{(s + 1)(5s + 1)} \right] T_{0(s)} + \left[\frac{1/2160}{(s + 1)(5s + 1)} \right] Q_{1(s)} \tag{9.92}$$

The two terms in the brackets represent the transfer functions of this openloop process. In the next chapter we will look at this system again and will use a temperature controller to control T_2 by manipulating Q_1. The transfer function relating the controlled variable T_2 to the manipulated variable Q_1 is defined as $G_{M(s)}$. The transfer function relating the controlled variable T_2 to the load disturbance T_0 is defined as $G_{L(s)}$.

$$T_{2(s)} = G_{L(s)} T_{0(s)} + G_{M(s)} Q_{1(s)} \tag{9.93}$$

Both of these transfer functions are second-order lags with time constants of 1 minute and 5 minutes.

9.6 PROPERTIES OF TRANSFER FUNCTIONS

An Nth-order system is described by the linear ODE

$$a_N \frac{d^N x}{dt^N} + a_{N-1} \frac{d^{N-1} x}{dt^{N-1}} + \cdots + a_1 \frac{dx}{dt} + a_0 x$$

$$= b_M \frac{d^M m}{dt^M} + b_{M-1} \frac{d^{M-1} m}{dt^{M-1}} + \cdots + b_1 \frac{dm}{dt} + b_0 m \quad (9.94)$$

where a_i and b_i = constant coefficients
$\qquad\quad x$ = output
$\qquad\quad m$ = input or forcing function

9.6.1 Physical Realizability

For this equation to describe a real physical system the order of the right-hand side, M, cannot be greater than the order of the left-hand side, N. This criterion for physical realizability is

$$N \geq M \quad (9.95)$$

This requirement can be proved intuitively from the following reasoning. Take a case where $N = 0$ and $M = 1$.

$$a_0 x = b_1 \frac{dm}{dt} + b_0 m \quad (9.96)$$

This equation says that we have a process whose output x depends on the value of the input and the value of the derivative of the input. Therefore the process must be able to differentiate, perfectly, the input signal. But it is impossible for any real system to differentiate perfectly. This would require that a step change in the input produce an infinite spike in the output. This is physically impossible.

This example can be generalized to any case where $M \geq N$ to show that differentiation would be required. Therefore N must always be greater than or equal to M. Laplace-transforming Eq. (9.96) gives

$$\frac{X_{(s)}}{M_{(s)}} = \frac{b_1}{a_0} s + \frac{b_0}{a_0}$$

This is a first-order lead. It is physically unrealizable; i.e., a real device cannot be built that has exactly this transfer function.

Consider the case where $M = N = 1$.

$$a_1 \frac{dx}{dt} + a_0 x = b_1 \frac{dm}{dt} + b_0 m \quad (9.97)$$

TABLE 9.1
Common transfer functions

Terminology	$G_{(s)}$
Gain	K
Derivative	s
Integrator	$\dfrac{1}{s}$
First-order lag	$\dfrac{1}{\tau s + 1}$
First-order lead	$\tau s + 1$
Second-order lag	
$\quad$ Underdamped $\zeta < 1$	$\dfrac{1}{\tau^2 s^2 + 2\tau\zeta s + 1}$
$\quad$ Critically damped $\zeta = 1$	$\dfrac{1}{(\tau s + 1)^2}$
$\quad$ Overdamped $\zeta > 1$	$\dfrac{1}{(\tau_{p1} s + 1)(\tau_{p2} s + 1)}$
Deadtime	e^{-Ds}
Lead-lag	$\dfrac{\tau_z s + 1}{\tau_p s + 1}$

It appears that a derivative of the input is again required. But Eq. (9.97) can be rearranged, grouping the derivative terms together:

$$\frac{d}{dt}(a_1 x - b_1 m) = \frac{dz}{dt} = b_0 m - a_0 x \tag{9.98}$$

The right-hand side of this equation contains functions of time but no derivatives. This ODE can be integrated by evaluating the right-hand side (the derivative) at each point in time and integrating to get z at the new point in time. Then the new value of x is calculated from the known value of m: $x = (z + b_1 m)/a_1$. Differentiation is not required and this transfer function is physically realizable.

Remember, nature always integrates. It never differentiates!

Laplace-transforming Eq. (9.98) gives

$$\frac{X_{(s)}}{M_{(s)}} = \frac{b_1 s + b_0}{a_1 s + a_0}$$

This is called a *lead-lag* element and contains a first-order lag and a first-order lead. See Table 9.1 for some commonly used transfer function elements.

9.6.2 Poles and Zeros

Returning now to Eq. (9.94), let us Laplace-transform and solve for the ratio of output $X_{(s)}$ to input $M_{(s)}$, the system transfer function $G_{(s)}$.

$$G_{(s)} = \frac{X_{(s)}}{M_{(s)}} = \frac{b_M s^M + b_{M-1} s^{M-1} + \cdots + b_1 s + b_0}{a_N s^N + a_{N-1} s^{N-1} + \cdots + a_1 s + a_0} \tag{9.99}$$

The denominator is a polynomial in s that is the same as the characteristic equation of the system. Remember the characteristic equation is obtained from the homogeneous ODE, that is, considering the right-hand side of Eq. (9.94) equal to zero.

The roots of the denominator are called the *poles* of the transfer function. The roots of the numerator are called the *zeros* of the transfer function (these values of s make the transfer function equal zero). Factoring both numerator and denominator yields

$$G_{(s)} = \left(\frac{b_M}{a_N}\right) \frac{(s - z_1)(s - z_2) \cdots (s - z_M)}{(s - p_1)(s - p_2) \cdots (s - p_N)} \tag{9.100}$$

where z_i = zeros of the transfer function
 p_i = poles of the transfer function

As noted in Chap. 6, the roots of the characteristic equation, which are the poles of the transfer function, must be real or must occur as complex conjugate pairs. In addition, the real parts of all the poles must be negative for the system to be stable.

> A system is stable if all its poles lie in the left half of the s plane.

The locations of the zeros of the transfer function have *no* effect on the stability of the system! They certainly affect the dynamic response, but they do not affect stability.

9.6.3 Steadystate Gains

One final point should be made about transfer functions. The steadystate gain K_p for all the transfer functions derived in the examples was obtained by expressing the transfer function in terms of time constants instead of in terms of poles and zeros. For the general system of Eq. (9.91) this would be

$$G_{(s)} = K_p \frac{(\tau_{z1} s + 1)(\tau_{z2} s + 1) \cdots (\tau_{zM} s + 1)}{(\tau_{p1} s + 1)(\tau_{p2} s + 1) \cdots (\tau_{pN} s + 1)} \tag{9.101}$$

The steadystate gain is the ratio of output steadystate perturbation over the input perturbation.

$$K_p = \left(\frac{x_{\text{out}}^p}{x_{\text{in}}^p}\right)_{(t \to \infty)} = \frac{\bar{x}_{\text{out}}^p}{\bar{x}_{\text{in}}^p} \tag{9.102}$$

In terms of total variables,

$$K_p = \left(\frac{x_{\text{out}} - \bar{x}_{\text{out}}}{x_{\text{in}} - \bar{x}_{\text{in}}}\right)_{(t \to \infty)} = \frac{\Delta \bar{x}_{\text{out}}}{\Delta \bar{x}_{\text{in}}}$$

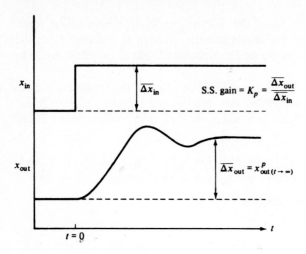

FIGURE 9.9
Steady-state gain.

Thus, for a step change in the input variable of $\Delta \bar{x}_{in}$, the steadystate gain is simply found by dividing the steadystate change in the output variable $\Delta \bar{x}_{out}$ by $\Delta \bar{x}_{in}$, as sketched in Fig. 9.9.

Instead of rearranging the transfer function to put it into the time-constant form, it is sometimes more convenient to find the steadystate gain by an alternative method that does not require factoring of polynomials. This consists of merely letting $s = 0$ in the transfer function.

$$K_p = \lim_{s \to 0} G_{(s)} \tag{9.103}$$

By definition, steadystate corresponds to the condition that all time derivatives are equal to zero. Since the variable s replaces d/dt in the Laplace domain, letting s go to zero is equivalent to the steadystate gain.

This can be proved more rigorously by using the *final-value theorem* of Laplace transforms:

$$\lim_{t \to \infty} [f_{(t)}] = \lim_{s \to 0} [sF_{(s)}] \tag{9.104}$$

Consider the arbitrary transfer function

$$G_{(s)} = \frac{X_{(s)}}{M_{(s)}}$$

If a unit step disturbance is used

$$M_{(s)} = \frac{1}{s}$$

This means that the output is

$$X_{(s)} = G_{(s)} \frac{1}{s}$$

The final steadystate value of the output will be equal to the steadystate gain since the magnitude of the input was 1.

$$K_p = \lim_{t \to \infty} [x_{(t)}] = \lim_{s \to 0} [sX_{(s)}] = \lim_{s \to 0} \left[sG_{(s)} \frac{1}{s} \right] = \lim_{s \to 0} [G_{(s)}]$$

For example, the steadystate gain for the transfer function given in Eq. (9.99) is

$$K_p = \lim_{s \to 0} \left[\frac{b_M s^M + b_{M-1} s^{M-1} + \cdots + b_1 s + b_0}{a_N s^N + a_{N-1} s^{N-1} + \cdots + a_1 s + a_0} \right] = \frac{b_0}{a_0} \qquad (9.105)$$

It is obvious that this must be the right value of gain since at steadystate Eq. (9.94) reduces to

$$a_0 \bar{x} = b_0 \bar{M} \qquad (9.106)$$

In the two-heated-tank process of Example 9.7, the two transfer functions were given in Eq. (9.92). The steadystate gain between the inlet temperature T_0 and the output T_1 is found to be $1°F/°F$ when s is set equal to zero. This says that a one degree change in the inlet temperature will raise the outlet temperature by one degree, which seems reasonable. The steadystate gain between T_2 and the heat input Q_1 is $1/2160°F/Btu/min$. You should be careful about the units of gains. Sometimes they have engineering units, as in this example. Other times dimensionless gains are used. We will discuss this in more detail in Chap. 10.

9.7 TRANSFER FUNCTIONS FOR FEEDBACK CONTROLLERS

As discussed in Chap. 7, the three common commercial feedback controllers are proportional (P), proportional-integral (PI) and proportional-integral-derivative (PID). The transfer functions for these devices are developed below.

The equation describing a proportional controller in the time domain is:

$$CO_{(t)} = \text{bias} \pm K_c(SP_{(t)} - PM_{(t)}) \qquad (9.107)$$

where CO = controller output signal sent to the control valve
bias = constant
SP = setpoint
PM = process measurement signal from the transmitter

Equation (9.107) is written in terms of total variables. If we are dealing with perturbation variables, we simply drop the bias term. Laplace transforming gives

$$CO_{(s)} = \pm K_c(SP_{(s)} - PM_{(s)}) = \pm K_c E_{(s)} \qquad (9.108)$$

where E = error signal = SP − PM

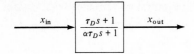

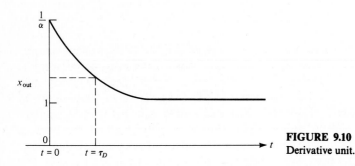

FIGURE 9.10
Derivative unit.

Rearranging to get the output over the input gives the transfer function $B_{(s)}$ for the controller.

$$\frac{CO_{(s)}}{E_{(s)}} = \pm K_c \equiv B_{(s)} \tag{9.109}$$

So the transfer function for a proportional controller is just a gain.

The equation describing a proportional-integral controller in the time domain is

$$CO_{(t)} = \text{bias} \pm K_c\left[E_{(t)} + \frac{1}{\tau_I}\int E_{(t)}\,dt\right] \tag{9.110}$$

where τ_I = reset time, minutes

Equation (9.110) is in terms of total variables. Converting to perturbation variables and Laplace-transforming give

$$CO_{(s)} = \pm K_c\left[E_{(s)} + \frac{1}{\tau_I s}E_{(s)}\right]$$

$$\frac{CO_{(s)}}{E_{(s)}} = B_{(s)} = \pm K_c\left[1 + \frac{1}{\tau_I s}\right] = \pm K_c\left(\frac{\tau_I s + 1}{\tau_I s}\right) \tag{9.111}$$

Thus the transfer function for a PI controller contains a first-order lead and an integrator. It is a function of s, having numerator and denominator polynomials of order one.

The transfer function of a "real" PID controller, as opposed to an "ideal" one, is the PI transfer function with a lead-lag element placed in series.

$$\frac{CO_{(s)}}{E_{(s)}} = B_{(s)} = \pm K_c \left(\frac{\tau_I s + 1}{\tau_I s} \right) \left(\frac{\tau_D s + 1}{\alpha \tau_D s + 1} \right) \tag{9.112}$$

where τ_D = derivative time constant, minutes
$\quad\;\; \alpha$ = a constant = 0.1 to 0.05 for many commercial controllers

The lead-lag unit is called a *derivative unit*, and its step response is sketched in Fig. 9.10. For a unit step change in the input, the output jumps to $1/\alpha$ and then decays at a rate that depends on τ_D. So the derivative unit approximates an ideal derivative. It is physically realizable since the order of its numerator polynomial is the same as the order of its denominator polynomial.

PROBLEMS

9.1. Prove that the Laplace transformation of the following functions are:

(a) $\mathcal{L}\left[\dfrac{d^2 f}{dt^2} \right] = s^2 F_{(s)} - s f_{(0)} - \left(\dfrac{df}{dt} \right)_{(t=0)}$

(b) $\mathcal{L}[\cos (\omega t)] = \dfrac{s}{s^2 + \omega^2}$

(c) $\mathcal{L}[e^{-at} \sin (\omega t)] = \dfrac{\omega}{(s + a)^2 + \omega^2}$

9.2. Find the Laplace transformation of a rectangular pulse of height H_p and duration T_p.

9.3. An isothermal perfectly mixed batch reactor has consecutive first-order reactions

$$A \xrightarrow{\;k_1\;} B \xrightarrow{\;k_2\;} C$$

The initial material charged to the vessel contains only A at a concentration C_{A0}. Use Laplace transform techniques to solve for the changes in C_A and C_B with time during the batch cycle for:
(a) $k_1 > k_2$.
(b) $k_1 = k_2$.

9.4. Two isothermal CSTRs are connected by a long pipe that acts like a pure deadtime of D minutes at the steadystate flow rates. Assume constant throughputs and holdups and a first-order irreversible reaction

$$A \xrightarrow{\;k\;} B$$

in each tank. Derive the transfer function relating the feed concentration to the first tank C_{A0} and the concentration of A in the stream leaving the second tank C_{A2}. Use inversion to find $C_{A2(t)}$ for a unit step disturbance in C_{A0}.

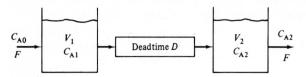

FIGURE P9.4

9.5. A general second-order system is described by the ODE

$$\tau_p^2 \frac{d^2x}{dt^2} + 2\tau_p \zeta \frac{dx}{dt} + x = K_p m_{(t)}$$

If $\zeta > 1$, show that the system transfer function has two first-order lags with time constants τ_{p1} and τ_{p2}. Express these time constants in terms of τ_p and ζ.

9.6. Use Laplace transform techniques to solve Example 6.7 where a ramp disturbance drives a first-order system.

9.7. Find the transfer function of an underdamped second-order system:

$$\tau_p^2 \frac{d^2x}{dt^2} + 2\tau_p \zeta \frac{dx}{dt} + x = K_p m_{(t)}$$

What are the poles of the transfer function? Solve for the response of $x_{(t)}$ to
(a) A unit step disturbance
(b) An impulse disturbance

9.8. Solve Prob. 6.6 using Laplace transforms. Note that this is a steadystate problem and that radial position r is the independent variable. You must therefore Laplace-transform with respect to r, not time t.

9.9. The imperfect mixing in a chemical reactor can be modeled by splitting the total volume into two perfectly mixed sections with circulation between them. Feed enters and leaves one section. The other section acts like a "side-capacity" element.

Assume holdups and flow rates are constant. The reaction is an irreversible, first-order consumption of reactant A. The system is isothermal. Solve for the transfer function relating C_{A0} and C_A. What are the zeros and poles of the transfer function? What is the steadystate gain?

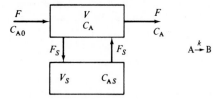

FIGURE P9.9

9.10 One way to determine the rate of change of a process variable is to measure the differential pressure $\Delta P = P_{out} - P_{in}$ over a device called a derivative unit that has a transfer function

$$\frac{P_{out(s)}}{P_{in(s)}} = \frac{\tau s + 1}{(\tau/6)s + 1}$$

(a) Derive the transfer function between ΔP and P_{in}.

(b) Show that the ΔP signal will be proportional to the rate of rise in P_{in}, after an initial transient period, when P_{in} is a ramp function.

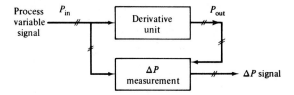

FIGURE P9.10

9.11. A convenient way to measure the density of a liquid is to pump it slowly through a vertical pipe and measure the differential pressure between the top and the bottom of the pipe. This differential head is directly related to the density of the liquid in the pipe if frictional pressure losses are negligible.

Suppose the density can change with time. What is the transfer function relating a perturbation in density to the differential-pressure measurement? Assume the fluid moves up the vertical column in plug flow at constant velocity.

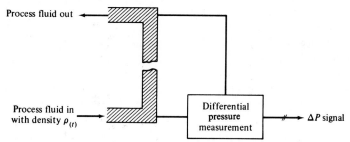

FIGURE P9.11

9.12. A thick-walled kettle of mass M_M, temperature T_M, and specific heat C_M is filled with a perfectly mixed process liquid of mass M, temperature T, and specific heat C. A heating fluid at temperature T_J is circulated in a jacket around the kettle wall. The heat transfer coefficient between the process fluid and the metal wall is U and between the metal outside wall and the heating fluid is U_M. Inside and outside heat transfer areas A are approximately the same. Neglecting any radial temperature gradients through the metal wall, show that the transfer function between T and T_J is two first-order lags.

$$G_{(s)} = \frac{K_p}{(\tau_{p1} s + 1)(\tau_{p2} s + 1)}$$

The value of the steadystate gain K_p is unity. Is this reasonable?

9.13. An ideal three-mode PID (proportional, integral, and derivative) feedback controller is described by the equation:

$$CO_{(t)} = \text{bias} + K_c\left[E_{(t)} + \frac{1}{\tau_I}\int E_{(t)}\,dt + \tau_D\frac{dE}{dt}\right]$$

Derive the transfer function between $CO_{(s)}$ and $E_{(s)}$. Is this transfer function physically realizable?

9.14. Show that the linearized nonisothermal CSTR of Example 9.6 can be stable only if

$$\frac{UA}{V\rho C_p} > \frac{-\lambda\bar{k}\bar{C}_A E}{\rho C_p R\bar{T}^2} - 2\frac{\bar{F}}{V} - \bar{k}$$

9.15. Write a digital computer program that numerically integrates the following ODE:

$$a_2\frac{d^2x}{dt^2} + a_1\frac{dx}{dt} + a_0 x = b_2\frac{d^2m}{dt^2} + b_1\frac{dm}{dt} + b_0 m$$

9.16. A deadtime element is basically a distributed system. One approximate way to get the dynamics of distributed systems is to lump them into a number of perfectly mixed sections. Prove that a series of N mixed tanks is equivalent to a pure deadtime as N goes to infinity.

   Hint: Keep the total volume of the system constant as more and more lumps are used.

9.17. A feedback controller is added to the three-CSTR system of Example 9.5. Now C_{A0} is changed by the feedback controller to keep C_{A3} at its setpoint, which is the steady-state value of C_{A3}. The error signal is therefore just $-C_{A3}$ (the perturbation in C_{A3}). Find the transfer function of this closedloop system between the disturbance C_{AD} and C_{A3}. List the values of poles, zeros, and steadystate gain when the feedback controller is:

(a) Proportional: $C_{A0} = C_{AD} + K_c(-C_{A3})$

(b) Proportional-integral: $C_{A0} = C_{AD} + K_c\left[-C_{A3} + \frac{1}{\tau_I}\int(-C_{A3})\,dt\right]$

Note the above equations are in terms of perturbation variables.

9.18. The partial condenser sketched in Fig. P9.18 is described by two ODEs:

$$\left(\frac{\text{Vol}}{RT}\right)\frac{dP}{dt} = F - V - \frac{Q_c}{\Delta H}$$

$$\frac{dM_R}{dt} = \frac{Q_c}{\Delta H} - L$$

where P = pressure
   Vol = volume of condenser
   M_R = liquid holdup
   F = vapor feed rate
   V = vapor product
   L = liquid product

(a) Draw a block diagram showing the transfer functions describing the openloop system.

(b) Draw a block diagram of the closedloop system if a proportional controller is used to manipulate Q_c to hold M_R and PI controller is used to manipulate V to hold P.

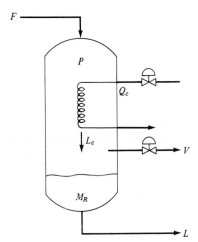

FIGURE P9.18

9.19. Show that a proportional-only level controller on a tank will give zero steadystate error for a step change in level setpoint.

9.20. Use Laplace transforms to prove mathematically that a P controller produces steadystate offset and that a PI controller does not. The disturbance is a step change in the load variable. The process openloop transfer functions, G_M and G_L, are both first-order lags with different gains but identical time constants.

9.21. Two 100-barrel tanks are available to use as surge volume to filter liquid flow rate disturbances in a petroleum refinery. Average throughput is 14,400 barrels per day. Should these tanks be piped up for parallel operation or for series operation? Assume proportional-only level controllers.

9.22. A perfectly mixed batch reactor, containing 7500 lb_m of liquid with a heat capacity of 1 Btu/lb_m °F, is surrounded by a cooling jacket that is filled with 2480 lb_m of perfectly mixed cooling water.

At the beginning of the batch cycle, both the reactor liquid and the jacket water are at 203°F. At this point in time, catalyst is added to the reactor and a reaction occurs which generates heat at a constant rate of 15,300 Btu/min. At this same moment in time, makeup cooling water at 68°F is fed into the jacket at a constant 832 lb_m/min flow rate.

The heat transfer area between the reactor and jacket is 140 ft². The overall heat transfer coefficient is 70 Btu/h °F ft². Mass of the metal walls can be neglected. Heat losses are negligible.

(a) Develop a mathematical model of the process.

(b) Use Laplace transforms to solve for the dynamic change in reactor temperature $T_{(t)}$.

(c) What is the peak reactor temperature and when does it occur?

(d) What is the final steadystate reactor temperature?

9.23. A milk tank on a dairy farm is equipped with a refrigeration compressor which removes q (Btu/min) of heat from the warm milk. The insulated, perfectly mixed tank is initially filled with V_0 (ft^3) of warm milk (99.5°F). The compressor is then turned on and begins to chill down the milk. At the same time fresh warm (99.5°F) milk is continuously added at a constant rate F (ft^3/min) through a milk pipeline from the milking parlor. The total volume after all cows have been milked is V_T (ft^3).

Derive the equation describing how the temperature T of milk in the tank varies with time. Solve for $T_{(t)}$. What is the temperature of the milk at the end of the milking? How long does it take to chill the milk down to 35°F? Parameter values are $F = 1$ ft^3/min, $\rho = 62.3$ lb$_m$/ft^3, $C_p = 1$ Btu/lb$_m$ °F, $V_0 = 5$ ft^3, $V_T = 100$ ft^3, $q = 300$ Btu/min.

9.24. The flow of air into the regenerator on a catalytic cracking unit is controlled by two control values. One is a large, slow-moving value that is located on the suction of the air blower. The other is a small, fast-acting valve that vents to the atmosphere.

The fail-safe condition is to not feed air into the regenerator. Therefore, the suction valve is air-to-open and the vent valve is air-to-close. What action should the flow controller have, direct or reverse?

The device with the following transfer function $G_{(s)}$ is installed in the control line to the vent valve.

$$G_{(s)} = \frac{\tau s}{\tau s + 1} = \frac{P_{V(s)}}{CO_{(s)}}$$

The purpose of this device is to cause the vent valve to respond quickly to changes in CO but to minimize the amount of air vented (since this wastes power) under steadystate conditions. What will be the dynamic response of the perturbation in P_V for a step change of 10 percent of full scale in CO? What is the new steadystate value of P_V?

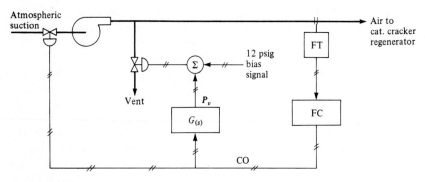

FIGURE P9.24

9.25. An openloop process has the transfer function:

$$G_M = \frac{s}{\tau s + 1}$$

Calculate the openloop response of this process to a unit step change in its input. What is the steadystate gain of this process?

9.26. A chemical reactor is cooled by both jacket cooling and autorefrigeration (boiling liquid in the reactor). Sketch a block diagram, using appropriate process and control system transfer functions, describing the system. Assume these transfer functions are known, either from fundamental mathematic models or from experimental dynamic testing.

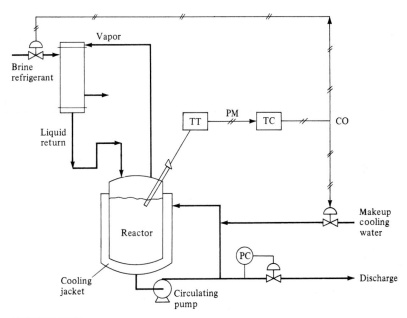

FIGURE P9.26

9.27. Solve the problem given below, which is part of a problem given in Levenspiel's *Chemical Reaction Engineering* (1962, John Wiley), using Laplace-transform techniques. Find analytical expressions for the number of Nelson's ships $N_{(t)}$ and the number of Villeneuve's ships as functions of time.

> The great naval battle, to be known to history as the battle of Trafalgar (1805), was soon to be joined. Admiral Villeneuve proudly surveyed his powerful fleet of 33 ships stately sailing in single file in the light breeze. The British fleet under Lord Nelson was now in sight, 27 ships strong. Estimating that it would still be two hours before the battle, Villeneuve popped open another bottle of burgundy and point by point reviewed his carefully thought-out battle strategy. As was the custom of naval battles at that time, the two fleets would sail in single file parallel to each other and in the same direction, firing their cannons madly. Now, by long experience in battles of this kind, it was a well-known fact that the rate of destruction of a fleet was proportional to the fire power of the opposing fleet. Considering his ships to be on a par, one for one, with the British, Villeneuve was confident of victory. Looking at his sundial,

Villeneuve sighed and cursed the light wind; he'd never get it over with in time for his favorite television western. "Oh well," he sighed "c'est la vie." He could see the headlines next morning. "British Fleet annihilated, Villeneuve's losses are ...." Villeneuve stopped short. How many ships would he lose? Villeneuve called over his chief bottle cork popper, Monsieur Dubois, and asked this question. What answer did he get?

CHAPTER
10

LAPLACE-DOMAIN ANALYSIS OF CONVENTIONAL FEEDBACK CONTROL SYSTEMS

Now that we have learned a little Russian, we are ready to see how useful it is in analyzing the dynamics and stability of systems. Laplace-domain methods provide a lot of insight into what is happening to the damping coefficients and time constants as we change the settings on the controller. The *root-locus* plots that we will use are similar in value to the graphical McCabe-Thiele diagram in binary distillation: they provide a nice picture in which the effects of parameters can be easily seen.

In this chapter we will demonstrate the significant computational and notational advantages of Laplace transforms. The techniques involve finding the transfer function of the openloop process, specifying the desired performance of the closedloop system (process plus controller) and finding the feedback controller transfer function that is required to do the job.

10.1 OPENLOOP AND CLOSEDLOOP SYSTEMS

10.1.1 Openloop Characteristic Equation

Consider the general openloop system sketched in Fig. 10.1a. The load variable $L_{(s)}$ enters through the openloop process transfer function $G_{L(s)}$. The manipulated

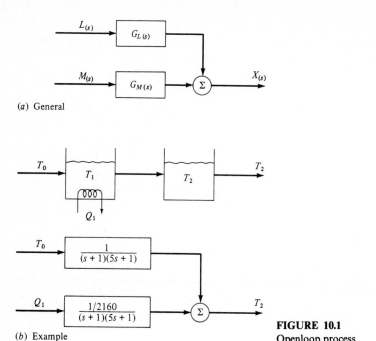

FIGURE 10.1
Openloop process.

(a) General

(b) Example

variable $M_{(s)}$ enters through the openloop process transfer function $G_{M(s)}$. The controlled variable $X_{(s)}$ is the sum of the effects of the manipulated variable and the load variable. Remember we are working with linear systems in the Laplace domain so superposition applies.

Figure 10.1b shows a specific example: the two-heated-tank process discussed in Example 9.7. The load variable is the inlet temperature T_0. The manipulated variable is the heat input to the first tank Q_1. The two transfer functions $G_{L(s)}$ and $G_{M(s)}$ were derived in Chap. 9.

The dynamics of this openloop system depend on the roots of the openloop characteristic equation, i.e., on the roots of the polynomials in the denominators of the openloop transfer functions. These are the poles of the openloop transfer functions. If all the roots lie in the left half of the s plane, the system is openloop stable. For the two-heated-tank example shown in Fig. 10.1b, the poles of the openloop transfer function are $s = -1$ and $s = -\frac{1}{5}$, so the system is openloop stable.

Note that the $G_{L(s)}$ transfer function for the two-heated-tank process has a steadystate gain that has units of °F/°F. The $G_{M(s)}$ transfer function has a steadystate gain that has units of °F/Btu/min.

10.1.2 Closedloop Characteristic Equation and Closedloop Transfer Functions

Now let us put a feedback controller on the process, as shown in Fig. 10.2a. The controlled variable is converted into a process measurement signal PM by the sensor/transmitter element $G_{T(s)}$. The feedback controller compares the PM signal to the desired setpoint signal SP, feeds the error signal E through a feedback-controller transfer function $B_{(s)}$ and sends out a controller output signal CO. The controller output signal changes the position of a control valve which changes the flow rate of the manipulated variable M.

Figure 10.2b gives a sketch of the feedback control system and a block diagram for the two-heated-tank process with a controller. Let us use an analog electronic system with 4 to 20 mA control signals. The temperature sensor has a range of 100°F, so the G_T transfer function (neglecting any dynamics in the temperature measurement) is

$$G_{T(s)} \equiv \frac{PM_{(s)}}{T_{2(s)}} = \frac{16 \text{ mA}}{100°F} \qquad (10.1)$$

The controller output signal CO goes to an I/P transducer that converts 4 to 20 mA into 3 to 15 psig air-pressure signal to drive the control valve through which steam is added to the heating coil. Let us assume that the valve has linear installed characteristics (see Chap. 7) and can pass enough steam to add 500,000 Btu/min to the liquid in the tank when the valve is wide open. Therefore, the transfer function between Q_1 and CO (lumping together the transfer function for the I/P transducer and the control valve) is

$$G_{V(s)} \equiv \frac{Q_{1(s)}}{CO_{(s)}} = \frac{500,000 \text{ Btu/min}}{16 \text{ mA}} \qquad (10.2)$$

Looking at the block diagram in Fig. 10.2a, we can see that the output $X_{(s)}$ is given by

$$X = G_{L(s)} L + G_{M(s)} M \qquad (10.3)$$

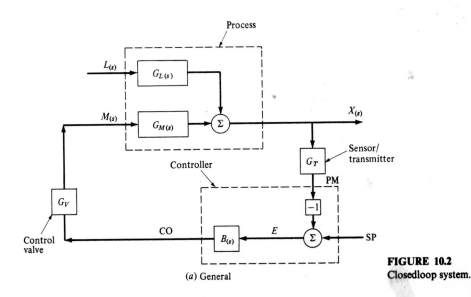

(a) General

FIGURE 10.2
Closedloop system.

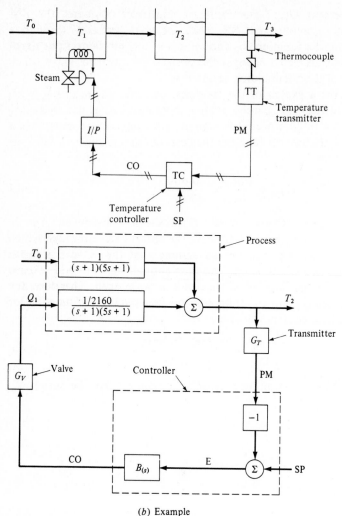

FIGURE 10.2
Closedloop system.

(b) Example

But in this closedloop system, $M_{(s)}$ is related to $X_{(s)}$:

$$M = G_{V(s)} \, CO = G_{V(s)} \, B_{(s)} \, E = G_{V(s)} \, B_{(s)} \, (SP - PM)$$

$$M = G_{V(s)} \, B_{(s)} \, (SP - G_{T(s)} \, X) \tag{10.4}$$

Combining Eqs. (10.3) and (10.4) gives

$$X = G_{L(s)} \, L + G_{M(s)} \, G_{V(s)} \, B_{(s)} \, (SP - G_{T(s)} \, X)$$

$$[1 + G_{M(s)} \, G_{V(s)} \, B_{(s)} \, G_{T(s)}] X = G_{L(s)} \, L + G_{M(s)} \, G_{V(s)} \, B_{(s)} \, SP$$

$$X_{(s)} = \left[\frac{G_{L(s)}}{1 + G_{M(s)} \, G_{V(s)} \, B_{(s)} \, G_{T(s)}} \right] L_{(s)}$$

$$+ \left[\frac{G_{M(s)} \, G_{V(s)} \, B_{(s)}}{1 + G_{M(s)} \, G_{V(s)} \, B_{(s)} \, G_{T(s)}} \right] SP_{(s)} \tag{10.5}$$

Equation 10.5 gives the transfer functions describing the *closedloop* system, so these are closedloop transfer functions. The two inputs are the load $L_{(s)}$ and the setpoint $SP_{(s)}$. The controlled variable is $X_{(s)}$. Note that the denominators of both of these closedloop transfer functions are identical.

Example 10.1. The closedloop transfer functions for the two-heated-tank process can be calculated from the openloop process transfer functions and the feedback controller transfer function. We will choose a proportional controller, so $B_{(s)} = K_c$. Note that the dimensions of the gain of the controller are mA/mA, that is, the gain is dimensionless. The controller looks at a milliampere signal (PM) and puts out a milliampere signal (CO).

$$G_{L(s)} = \frac{1\ °F/°F}{(s + 1)(5s + 1)}$$

$$G_{M(s)} = \frac{1/2160\ °F/Btu/min}{(s + 1)(5s + 1)}$$

$$G_{V(s)} = \frac{500{,}000\ Btu/min}{16\ mA}$$

$$G_{T(s)} = \frac{16\ mA}{100°F}$$

The closedloop transfer function for load changes is

$$\frac{T_2}{T_0} = \frac{G_{L(s)}}{1 + G_{M(s)}\,G_{V(s)}\,B_{(s)}\,G_{T(s)}}$$

$$= \frac{\dfrac{1\ °F/°F}{(s + 1)(5s + 1)}}{1 + \left(\dfrac{1/2160\ °F/Btu/min}{(s + 1)(5s + 1)}\right)\left(\dfrac{500{,}000\ Btu/min}{16\ mA}\right)(K_c)\left(\dfrac{16\ mA}{100°F}\right)}$$

$$= \frac{1\ °F/°F}{(s + 1)(5s + 1) + 500K_c/216} = \frac{1\ °F/°F}{5s^2 + 6s + 1 + 500K_c/216} \tag{10.6}$$

The closedloop transfer function for setpoint changes is

$$\frac{T}{SP} = \frac{G_{M(s)}\,G_{V(s)}\,B_{(s)}}{1 + G_{M(s)}\,G_{V(s)}\,B_{(s)}\,G_{T(s)}}$$

$$= \frac{\left(\dfrac{1/2160\ °F/Btu/min}{(s + 1)(5s + 1)}\right)\left(\dfrac{500{,}000\ Btu/min}{16\ mA}\right)(K_c)}{1 + \left(\dfrac{1/2160\ °F/Btu/min}{(s + 1)(5s + 1)}\right)\left(\dfrac{500{,}000\ Btu/min}{16\ mA}\right)(K_c)\left(\dfrac{16\ mA}{100°F}\right)}$$

$$= \frac{50{,}000K_c/216/16\ °F/mA}{5s^2 + 6s + 1 + 500K_c/216} \tag{10.7}$$

If we look at the closedloop transfer function between PM and SP, we must multiply the above by G_T.

$$\frac{\text{PM}}{\text{SP}} = \frac{500K_c/216 \text{ mA/mA}}{5s^2 + 6s + 1 + 500K_c/216} \tag{10.8}$$

Notice that the denominators of all of these closedloop transfer functions are identical. Notice also that the steadystate gain of the closedloop servo transfer function PM/SP is not unity; i.e., there is a steadystate offset. This is because of the proportional controller. We can calculate the PM/SP ratio at steadystate by letting s go to zero in Eq. (10.8).

$$\lim_{t \to \infty} \left(\frac{\text{PM}}{\text{SP}} \right) = \frac{500K_c/216}{1 + 500K_c/216} = \frac{1}{216/500K_c + 1} \tag{10.9}$$

Equation (10.9) shows that the bigger the gain, the smaller the offset.

Since the characteristic equation of any system (openloop or closedloop) is the denominator of the transfer function describing it, the closedloop characteristic equation for this system is

$$\boxed{1 + G_{M(s)} G_{V(s)} B_{(s)} G_{T(s)} = 0} \tag{10.10}$$

This equation shows that closedloop dynamics depend on the process openloop transfer functions (G_M, G_V, and G_T) and on the feedback controller transfer function (B). Equation (10.10) applies for simple single-input–single-output systems. We will derive closedloop characteristic equations for other systems in later chapters.

The first closedloop transfer function in Eq. (10.5) relates the controlled variable to the load variable. It is called the closedloop *regulator* transfer function. The second closedloop transfer function in Eq. (10.5) relates the controlled variable to the setpoint. It is called the closedloop *servo* transfer function.

Normally we design the feedback controller $B_{(s)}$ to give some desire closedloop performance. For example, we might specify a desired closedloop damping coefficient.

It is useful to consider the ideal situation. If we could design an ideal controller without any regard for physical realizability, what would the ideal closedloop regular and servo transfer functions be? Clearly, we would wish a load disturbance to have *no* effect on the controlled variable. So the ideal closedloop regulator transfer function is zero. For setpoint changes, we would like the controlled variable to track the setpoint perfectly at all times. So the ideal servo transfer function is unity.

A look at Eq. (10.5) will show that both of these could be achieved if we could simply make $B_{(s)}$ infinitely large. This would make the first term zero and the second term unity. However, as we will see in the next section, stability limitations prevent us from achieving this ideal situation.

Instead of considering the process, transmitter, and valve transfer functions separately, it is often convenient to combine them into just one transfer function.

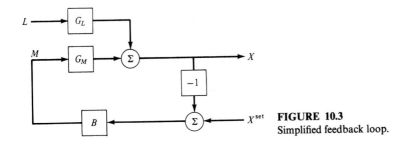

FIGURE 10.3
Simplified feedback loop.

Then the closedloop block diagram, shown in Fig. 10.3, becomes more simple. The equation describing this simplified closedloop system is

$$X_{(s)} = \left[\frac{G_{L(s)}}{1 + G_{M(s)}B_{(s)}}\right]L_{(s)} + \left[\frac{G_{M(s)}B_{(s)}}{1 + G_{M(s)}B_{(s)}}\right]X_{(s)}^{set} \qquad (10.11)$$

These are the equations that we will use in most cases because they are more convenient. Keep in mind that the $G_{M(s)}$ transfer function in Eq. (10.11) is a combination of the process, transmitter, and the valve transfer functions. The closed-loop characteristic equation is

$$\boxed{1 + G_{M(s)}B_{(s)} = 0} \qquad (10.12)$$

10.2 STABILITY

The most important dynamic aspect of any system is its stability. We learned in Chap. 6 that stability was dictated by the location of the roots of the character-istic equation of the system. In Chap. 9 we learned that the roots of the denomi-nator of the system transfer function, its poles, are exactly the same as the roots of the characteristic equation. Thus, for the system to be stable, the poles of the transfer function must lie in the left half of the s plane (LHP).

This stability requirement applies to any system, openloop or closedloop. The stability of an openloop process depends upon the location of the poles of its openloop transfer function. The stability of a closedloop process depends upon the location of the poles of its closedloop transfer function. These closedloop poles will naturally be different from the openloop poles.

Thus the criteria for openloop and closedloop stability are different. Most systems are openloop stable but can be either closedloop stable or unstable, depending on the values of the controller parameters. We will show that any real process can be made closedloop unstable by making the gain of the feedback controller high enough.

There are some processes that are openloop unstable. We will show that these systems can usually be made closedloop stable by the correct choice of the type of controller and its settings.

There are several methods for testing for stability in the Laplace domain. Some of the most useful are discussed below. Frequency-domain methods will be discussed in Chap. 13.

10.2.1 Routh Stability Criterion

The Routh method can be used to find out if there are any roots of a polynomial in the RHP. It can be applied to either closedloop or openloop systems by using the appropriate characteristic equation.

Assume the characteristic equation of interest is an Nth-order polynomial:

$$a_N s^N + a_{N-1} s^{N-1} + \cdots + a_1 s + a_0 = 0 \tag{10.13}$$

The Routh array is formed as given below:

$$\text{Routh array} = \begin{bmatrix} a_N & a_{N-2} & a_{N-4} & \cdots & a_0 \\ a_{N-1} & a_{N-3} & a_{N-5} & & \\ A_1 & A_2 & A_3 & & \\ B_1 & B_2 & B_3 & & \\ C_1 & C_2 & & & \\ \hdotsfor{5} \end{bmatrix} \tag{10.14}$$

where the A_i, B_i, ... are calculated from the equations

$$A_1 = \frac{a_{N-1} a_{N-2} - a_N a_{N-3}}{a_{N-1}}$$

$$A_2 = \frac{a_{N-1} a_{N-4} - a_N a_{N-5}}{a_{N-1}}$$

$$\cdots\cdots\cdots\cdots\cdots\cdots\cdots \tag{10.15}$$

$$B_1 = \frac{A_1 a_{N-3} - a_{N-1} A_2}{A_1}$$

$$\cdots\cdots\cdots\cdots\cdots\cdots$$

Then the first column of the array of Eq. (10.14) is examined. The number of sign changes of this first column is equal to the number of roots of the polynomial that are in the RHP.

$$\text{First column} = \begin{bmatrix} a_N \\ a_{N-1} \\ A_1 \\ B_1 \\ C_1 \\ \vdots \end{bmatrix} \tag{10.16}$$

Thus for the system to be stable there can be *no* sign changes in the first column of the Routh array.

Let us illustrate the application of the Routh stability criterion in some specific examples.

Example 10.2. Assume the characteristic equation of the system is

$$s^5 + 2s^4 + s^3 + 3s^2 + 4s + 5 = 0 \tag{10.17}$$

The Routh array is

$$
\begin{vmatrix}
1 & 1 & 4 \\
2 & 3 & 5 \\
\left(\dfrac{2-3}{2} = -\dfrac{1}{2}\right) & \left(\dfrac{8-5}{2} = \dfrac{3}{2}\right) & 0 \\
\left(\dfrac{-\frac{3}{2}-3}{-\frac{1}{2}} = 9\right) & \left(\dfrac{-\frac{5}{2}-0}{-\frac{1}{2}} = 5\right) & \\
\left(\dfrac{\frac{27}{2}+\frac{5}{2}}{9} = \dfrac{32}{18}\right) & 0 & \\
5 & & \\
0 & & \\
\end{vmatrix}
\tag{10.18}
$$

Examining the first column, we see that there are two sign changes. There must be two roots in the RHP. The system is unstable.

Example 10.3. Consider our old friend the three-CSTR process with a process transfer function

$$G_{M(s)} = \frac{\frac{1}{8}}{(s+1)^3} = \left(\frac{C_{A3}}{C_{A0}}\right)_{(s)} \tag{10.19}$$

We want to look at the stability of the closedloop system with a proportional controller: $B_{(s)} = K_c$. First, however, let us check the openloop stability of this system. The openloop characteristic equation is

$$s^3 + 3s^2 + 3s + 1 = 0 \tag{10.20}$$

The Routh array becomes

$$
\begin{vmatrix}
1 & 3 \\
3 & 1 \\
\left(\dfrac{9-1}{3} = \dfrac{8}{3}\right) & 0 \\
1 & \\
\end{vmatrix}
\tag{10.21}
$$

There are no sign changes in the first column, so the system is openloop stable. This finding should be no great shock since, from our simulations, we know the openloop system is stable. We also can see by inspection of Eq. (10.19) that the three poles of the openloop transfer function are located at -1 in the LHP, which is the stable region.

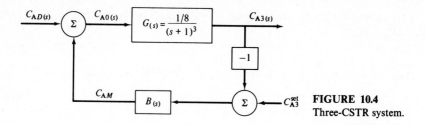

FIGURE 10.4
Three-CSTR system.

Now let us check for closedloop stability. The system is sketched in Fig. 10.4. The closedloop characteristic equation is

$$1 + G_{M(s)} B_{(s)} = 0 = 1 + K_c \frac{\frac{1}{8}}{(s+1)^3}$$

$$s^3 + 3s^2 + 3s + 1 + \frac{K_c}{8} = 0 \qquad (10.22)$$

The Routh array is

$$\begin{vmatrix} 1 & 3 \\ 3 & 1 + \dfrac{K_c}{8} \\ \dfrac{9 - (1 + K_c/8)}{3} & 0 \\ 1 + K_c/8 & \end{vmatrix} \qquad (10.23)$$

Looking at the first column we can see that there will be a sign change if the third term is negative.

$$\frac{8 - K_c/8}{3} < 0 \quad \Rightarrow \quad -8 + K_c/8 > 0 \quad \Rightarrow \quad K_c > 64$$

Therefore the system is closedloop stable for feedback controller gains less than 64 but closedloop unstable for gains greater than 64. The maximum stable value of K_c is what we defined in Chap. 7 as the ultimate gain K_u.

The Routh stability criterion is quite useful, but it has definite limitations. It cannot handle systems with deadtime. It tells if the system is stable or unstable but it gives no information about how stable or unstable the system is. That is, if the test tells us that the system is stable, we do not know how close to instability it is. Another limitation of the Routh method is the need to express the characteristic equation explicitly as a polynomial in s. This can become complex in high-order systems.

10.2.2 Direct Substitution For Stability Limit

The direct-substitution method is a simple and useful method for finding the values of parameters in the characteristic equation that put the system just at the limit of stability.

We know the system is stable if all the roots of the characteristic equation are in the LHP and unstable if any of the roots are in the RHP. Therefore the imaginary axis represents the stability boundary. On the imaginary axis s is equal to some pure imaginary number: $s = i\omega$.

The technique consists of substituting $i\omega$ for s in the characteristic equation and solving for the values of ω and other parameters (e.g., controller gain) that satisfy the resulting equations. The method is best understood by looking at the example below.

Example 10.4. Consider again the three-CSTR system. We have already developed its closedloop characteristic equation with a proportional controller [Equation (10.22)].

$$s^3 + 3s^2 + 3s + 1 + \frac{K_c}{8} = 0$$

Substituting $s = i\omega$ gives

$$-i\omega^3 - 3\omega^2 + 3i\omega + 1 + \frac{K_c}{8} = 0$$

$$\left(1 + \frac{K_c}{8} - 3\omega^2\right) + i(3\omega - \omega^3) = 0 + i0 \tag{10.24}$$

Equating the real and imaginary parts of the left- and right-hand sides of the above equation gives two equations:

$$1 + \frac{K_c}{8} - 3\omega^2 = 0 \qquad \text{and} \qquad 3\omega - \omega^3 = 0$$

Therefore

$$\omega^2 = 3 \qquad \Rightarrow \qquad \omega = \pm\sqrt{3} \tag{10.25}$$

$$\frac{K_c}{8} = 3\omega^2 - 1 = 3(3) - 1 \qquad \Rightarrow \qquad K_c = 64 \tag{10.26}$$

The value of the gain at the limit of stability is 64. This is the same value that the Routh stability criterion gave us. It is K_u. The ω at this limit is the value of the imaginary part of s when the roots lie right on the imaginary axis. Since the real part of s is zero, the system will show a sustained oscillation with this frequency ω_u, called the *ultimate frequency*, in radians per time. The period of the oscillation is exactly the same as the ultimate period P_u that we defined in Chap. 7 in the Ziegler-Nichols tuning method.

$$P_u = \frac{2\pi}{\omega_u} \tag{10.27}$$

Example 10.5. Suppose the closedloop characteristic equation for a system is

$$\tfrac{5}{2}s^3 + 8s^2 + \tfrac{13}{2}s + 1 + K_c = 0 \tag{10.28}$$

To find the ultimate gain K_u and ultimate frequency ω_u, we substitute $i\omega$ for s.

$$-\tfrac{5}{2}i\omega^3 - 8\omega^2 + \tfrac{13}{2}i\omega + 1 + K_c = 0$$

$$[1 + K_c - 8\omega^2] + i[-\tfrac{5}{2}\omega^3 + \tfrac{13}{2}\omega] = 0 + i0 \tag{10.29}$$

Solving the resulting two equations in two unknowns gives

$$\omega_u = \sqrt{\frac{13}{5}} \quad \text{and} \quad K_u = 19.8 \tag{10.30}$$

In Chap. 12 we will show that we can convert from the Laplace domain (Russian) into the frequency domain (Chinese) by merely substituting $i\omega$ for s in the transfer function of the process. This is similar to the direct substitution method, but keep in mind that these two operations are different. In one we use the transfer function. In the other we use the characteristic equation.

10.3 PERFORMANCE SPECIFICATIONS

In order to design feedback controllers, we must have some way to evaluate their effect on the performance of the closedloop system, both dynamically and at steadystate.

10.3.1 Steadystate Performance

The usual steadystate performance specification is zero steadystate error. We will show below that this steadystate performance depends on both the system (process and controller) and the type of disturbance. This is different from the question of stability of the system which, as we have previously shown, is only a function of the system (roots of the characteristic equation) and does not depend on the input.

The error signal in the Laplace domain, $E_{(s)}$, is defined as the difference between the setpoint $X_{(s)}^{\text{set}}$ and the process output $X_{(s)}$.

$$E_{(s)} = X_{(s)}^{\text{set}} - X_{(s)} \tag{10.31}$$

Assuming that there is a change in the setpoint $X_{(s)}^{\text{set}}$ but no change in the load disturbance ($L_{(s)} = 0$) and substituting for $X_{(s)}$ from Eq. (10.11) give

$$E_{(s)} = X_{(s)}^{\text{set}} - \left[\frac{G_{M(s)} B_{(s)}}{1 + G_{M(s)} B_{(s)}}\right] X_{(s)}^{\text{set}}$$

$$\frac{E_{(s)}}{X_{(s)}^{\text{set}}} = \frac{1}{1 + G_{M(s)} B_{(s)}} \tag{10.32}$$

To find the steadystate value of the error, we will use the final-value theorem from Chap. 9.

$$\bar{E} \equiv \lim_{t \to \infty} E_{(t)} = \lim_{s \to 0} [s E_{(s)}] \tag{10.33}$$

Now let us look at two types of setpoint inputs: a step and a ramp.

A. UNIT STEP INPUT. $X^{set}_{(s)} = 1/s$.

$$\bar{E} = \lim_{s \to 0} [sE_{(s)}] = \lim_{s \to 0} \left[s \frac{1}{1 + G_{M(s)} B_{(s)}} \frac{1}{s} \right] = \lim_{s \to 0} \left[\frac{1}{1 + G_{M(s)} B_{(s)}} \right]$$

If the steadystate error is to go to zero, the term $1/(1 + G_{M(s)} B_{(s)})$ must go to zero as s goes to zero. This means that the term $B_{(s)} G_{M(s)}$ must go to infinity as s goes to zero. Thus $B_{(s)} G_{M(s)}$ must contain a $1/s$ term, which is an integrator. If the process $G_{M(s)}$ does not contain integration, we must put it into the controller $B_{(s)}$. We add reset or integral action to eliminate steadystate error for step input changes in setpoint.

If we use a proportional controller, the steadystate error is

$$\bar{E} = \lim_{s \to 0} \left[\frac{1}{1 + G_{M(s)} B_{(s)}} \right] = \frac{1}{1 + (K_c z_1 z_2 \cdots z_M/p_1 p_2 \cdots p_N)} \qquad (10.34)$$

where z_i = zeros of $G_{M(s)}$
p_i = poles of $G_{M(s)}$

Thus the steadystate error is reduced by increasing K_c, the controller gain.

B. RAMP INPUT. $X^{set}_{(s)} = 1/s^2$.

$$\bar{E} = \lim_{s \to 0} \left[s \frac{1}{1 + G_{M(s)} B_{(s)}} \frac{1}{s^2} \right] = \lim_{s \to 0} \left[\frac{1}{s(1 + G_{M(s)} B_{(s)})} \right]$$

If the steadystate error is to go to zero, the term $1/s(1 + G_{M(s)} B_{(s)})$ must go to zero as s goes to zero. This requires that $B_{(s)} G_{M(s)}$ must contain a $1/s^2$ term. Double integration is needed to drive the steadystate error to zero for a ramp input (to make the output track the changing setpoint).

10.3.2 Dynamic Specifications

The dynamic performance of a system can be deduced by merely observing the location of the roots of the system characteristic equation in the s plane. The time-domain specifications of time constants and damping coefficients for a closedloop system can be used directly in the Laplace domain.

1. If all the roots lie in the LHP, the system is stable.
2. If all the roots lie on the negative real axis, we know the system is overdamped or critically damped (all real roots).
3. The farther out on the negative axis the roots lie, the faster will be the dynamics of the system (the smaller the time constants).
4. The roots that lie close to the imaginary axis will dominate the dynamic response since the ones farther out will die out quickly.
5. The farther any complex conjugate roots are from the real axis the more underdamped the system will be.

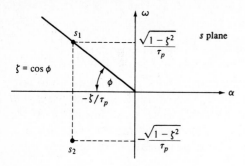

FIGURE 10.5
Dominant second-order root in the s plane.

There is a quantitative relationship between the location of roots in the s plane and the damping coefficient. Assume we have a second-order system or, if it is of higher order, assume it is dominated by the second-order roots closest to the imaginary axis. As shown in Fig. 10.5 the two roots are s_1 and s_2 and they are, of course, complex conjugates. From Eq. (6.68) the two roots are

$$s_1 = -\frac{\zeta}{\tau} + i\frac{\sqrt{1 - \zeta^2}}{\tau}$$

$$s_2 = -\frac{\zeta}{\tau} - i\frac{\sqrt{1 - \zeta^2}}{\tau}$$

The τ and ζ are the time constant and damping coefficient of the system. If the system is an openloop one, these are openloop time constant and openloop damping coefficient. If the system is a closedloop one, these are closedloop time constant and closedloop damping coefficient.

The hypotenuse of the triangle shown in Fig. 10.5 is the distance from the origin out to the root s_1.

$$\sqrt{\left(\frac{\sqrt{1 - \zeta^2}}{\tau}\right)^2 + \left(\frac{\zeta}{\tau}\right)^2} = \frac{1}{\tau} \tag{10.35}$$

The angle ϕ can be defined from the hypotenuse and the adjacent side of the triangle.

$$\cos\phi = \frac{\zeta/\tau}{1/\tau} = \zeta \tag{10.36}$$

Thus the location of a complex root can be converted directly to a damping coefficient and a time constant. The damping coefficient is equal to the cosine of the angle between the negative real axis and a radial line from the origin to the root. The time constant is equal to the reciprocal of the radial distance from the origin to the root.

Notice that lines of constant damping coefficient are radial lines in the s plane. Lines of constant time constant are circles.

10.4 ROOT LOCUS ANALYSIS

10.4.1 Definition

A *root locus* plot is a figure that shows how the roots of the closedloop characteristic equation vary as the gain of the feedback controller is varied from zero to infinity. The abscissa is the real part of the closedloop root; the ordinate is the imaginary part. Since we are plotting closedloop roots, the time constants and damping coefficients that we will pick off these root locus plots are all *closedloop time constants* and *closedloop damping coefficients*.

The examples below show the types of curves obtained and illustrate some important general principles.

Example 10.6. Let us start with the simplest of all processes, a first-order lag. We will choose a proportional controller. The system and controller transfer functions are

$$G_{M(s)} B_{(s)} = \left(\frac{K_0}{\tau_0 s + 1} \right) K_c \tag{10.37}$$

where K_0 = steadystate gain of the openloop process
$\quad \tau_0$ = time constant of the openloop process
$\quad K_c$ = controller gain

The closedloop characteristic equation is

$$1 + G_{M(s)} B_{(s)} = 0$$

$$1 + \frac{K_0 K_c}{\tau_0 s + 1} = 0 \tag{10.38}$$

$$\tau_0 s + 1 + K_0 K_c = 0$$

Solving for the closedloop root gives

$$s = -\frac{1 + K_0 K_c}{\tau_0} \tag{10.39}$$

There is one root and there will be only one curve in the s plane. Figure 10.6 gives the root locus plot. The curve starts at $s = -1/\tau_0$ when $K_c = 0$. The closedloop root moves out along the negative real axis as K_c is increased.

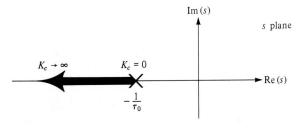

FIGURE 10.6
Root locus for first-order system.

For a first-order system, the closedloop root is always real, so the system can never be underdamped or oscillatory. The closedloop damping coefficient of this system is always greater than one. The larger the value of controller gain, the smaller the closedloop time constant because the root moves farther away from the origin (remember the time constant is the reciprocal of the distance from the root to the origin). If we wanted a closedloop time constant of $\frac{1}{10}\tau_0$ (the closedloop system is ten times as fast as the openloop system), we would set K_c equal to $9/K_0$. Equation (10.39) shows that at this value of gain the closedloop root is equal to $-10/\tau_0$.

This first-order system can never be closedloop unstable because the root always lies in the LHP. No real system is only first-order. There are always small lags in the process, in the control valve or in the instrumentation, that make all real systems of higher order than first.

Example 10.7. Now let's move up to a second-order system with a proportional controller.

$$G_{M(s)} = \frac{1}{(s+1)(5s+1)} \tag{10.40}$$

The closedloop characteristic equation is

$$1 + G_{M(s)} B_{(s)} = 0 = 1 + \frac{1}{(s+1)(5s+1)} K_c$$

$$5s^2 + 6s + 1 + K_c = 0$$

The quadratic formula gives the two closedloop roots:

$$s = \frac{-6 \pm \sqrt{(6)^2 - (4)(5)(1 + K_c)}}{(2)(5)}$$

$$s = -\tfrac{3}{5} \pm \tfrac{1}{5}\sqrt{4 - 5K_c} \tag{10.41}$$

The location of these roots for various values of K_c are shown in Fig. 10.7a.

When K_c is zero the closedloop roots are at $s = -\tfrac{1}{5}$ and $s = -1$. Notice that these values of s are the poles of the openloop transfer function. The root locus plot always starts at the poles of the openloop transfer function.

For K_c between zero and $\frac{4}{5}$, the two roots are real and lie on the negative real axis. The closedloop system is critically damped (the closedloop damping coefficient is 1) at $K_c = \frac{4}{5}$ since the roots are equal. For values of gain greater than $\frac{4}{5}$, the roots will be complex.

$$s = -\tfrac{3}{5} \pm i\tfrac{1}{5}\sqrt{5K_c - 4} \tag{10.42}$$

As the gain goes to infinity, the real parts of both roots are constant at $-\tfrac{3}{5}$ and the imaginary parts go to plus and minus infinity. Thus the system becomes increasingly underdamped. The closedloop damping coefficient goes to zero as the gain becomes infinite.

However, this second-order system never becomes closedloop unstable since the roots are always in the LHP.

Suppose we wanted to design this system for a closedloop damping coefficient of 0.707. Equation (10.36) tells us that

$$\phi = \arccos 0.707 = 45°$$

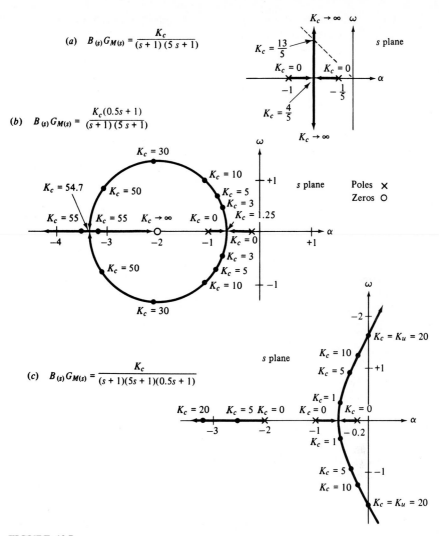

(a) $\quad B_{(s)}G_{M(s)} = \dfrac{K_c}{(s+1)(5s+1)}$

(b) $\quad B_{(s)}G_{M(s)} = \dfrac{K_c(0.5s+1)}{(s+1)(5s+1)}$

(c) $\quad B_{(s)}G_{M(s)} = \dfrac{K_c}{(s+1)(5s+1)(0.5s+1)}$

FIGURE 10.7
Root locus curves.

Therefore, we must find the value of gain on the root locus plot where it intersects a 45° line from the origin. At the point of intersection the real and imaginary parts of the roots must be equal. This occurs when $K_c = \frac{13}{5}$. The closedloop time constant τ_c of the system at this value of gain can be calculated from the reciprocal of the radial distance from the origin.

$$\tau_c = \frac{1}{\sqrt{(\frac{3}{5})^2 + (\frac{3}{5})^2}} = \frac{5}{3\sqrt{2}} \tag{10.43}$$

Example 10.8. Let us change the system transfer function from the above example by adding a lead or a zero.

$$G_{M(s)} B_{(s)} = \frac{K_c(\frac{1}{2}s + 1)}{(s + 1)(5s + 1)}$$

The closedloop characteristic equation becomes

$$1 + G_{M(s)} B_{(s)} = 1 + \frac{K_c(\frac{1}{2}s + 1)}{(s + 1)(5s + 1)}$$

$$5s^2 + \left(6 + \frac{K_c}{2}\right)s + K_c + 1 = 0 \tag{10.44}$$

The roots are

$$s = -\left(\frac{3}{5} + \frac{K_c}{20}\right) \pm \frac{1}{10} \sqrt{\frac{(K_c)^2}{4} - 14K_c + 16} \tag{10.45}$$

For low values of K_c the term inside the square root will be positive, since the $+16$ will dominate; the two closedloop roots are real and distinct. For very big values of gain, the K_c^2 term will dominate and the roots will again be real. For intermediate values of K_c the term inside the square root will be negative and the roots will be complex.

The range of K_c values that give complex roots can be found from the roots of

$$\frac{K_c^2}{4} - 14K_c + 16 = 0 \tag{10.46}$$

$$K_{c1} = 28 - 12\sqrt{5}$$
$$K_{c2} = 28 + 12\sqrt{5} \tag{10.47}$$

where K_{c1} = smaller value of K_c where the square-root term is zero
K_{c2} = larger value of K_c where the square-root term is zero

The root locus plot is shown in Fig. 10.7b.

Note that the effect of adding a zero or a lead is to pull the root locus toward a more stable region of the s plane. The root locus starts at the poles of the openloop transfer function. As the gain goes to infinity the two paths of the root locus go to minus infinity and to the zero of the transfer function at $s = -2$. We will find that this is true in general: the root locus plot ends at the zeros of the openloop transfer function.

The system is closedloop stable for all values of gain. The fastest responding system would be obtained with $K_c = K_{c2}$, where the two roots are equal and real.

Example 10.9. Now let us add a pole or a lag, instead of a zero, to the system of Example 10.7. The system is now third-order.

$$G_{M(s)} B_{(s)} = \frac{K_c}{(s + 1)(5s + 1)(\frac{1}{2}s + 1)} \tag{10.48}$$

The closedloop characteristic equation becomes

$$1 + G_{M(s)} B_{(s)} = 1 + \frac{K_c}{(s + 1)(5s + 1)(\frac{1}{2}s + 1)} \tag{10.49}$$

$$\tfrac{5}{2}s^3 + 8s^2 + \tfrac{13}{2}s + 1 + K_c = 0$$

We will discuss how to solve for the roots of this cubic equation in the next section. The root locus curves are sketched in Fig. 10.7c. There are three curves because there are three roots. The root locus plot starts at the three openloop poles of the transfer function: -1, -2, and $-\frac{1}{5}$.

The effect of adding a lag or a pole is to pull the root locus plot toward the unstable region. The two curves that start at $s = -\frac{1}{5}$ and $s = -1$ become complex conjugates and curve off into the RHP. Therefore this third-order system is closed-loop unstable if K_c is greater than $K_u = 20$. This was the same result that we obtained in Example 10.5.

The examples above have illustrated a very important point: the higher the order of the system, the worse the dynamic response of the closedloop system. The first-order system is never underdamped and cannot be made closedloop unstable for any value of gain. The second-order system becomes underdamped as gain is increased but never goes unstable. Third-order (and higher) systems can be made closedloop unstable.

One of the basic limitations of root locus techniques is that deadtime cannot be handled conveniently. The *first-order Pade* approximation of deadtime is frequently used, but it is often not very accurate.

$$e^{-Ds} \simeq \frac{1 - (\frac{1}{2}D)s}{1 + (\frac{1}{2}D)s} \tag{10.50}$$

10.4.2 Construction of Root Locus Curves

Root locus plots are easy to generate for first- and second-order systems since the roots can be found analytically as explicit functions of controller gain. For higher-order systems things become more difficult. Both numerical and graphical methods are available. Root-solving subroutines can be easily used on any computer to do the job. First, however, it is worthwhile to understand a few of the graphical techniques that have been developed to enable engineers to quickly sketch the curves. These are worth knowing so that you can check the numbers that are coming out of the computer. Some of the most useful are summarized below.

A. GRAPHICAL.

1. The root loci start ($K_c = 0$) at the poles of the system openloop transfer function $G_{M(s)} B_{(s)}$.

2. The root loci end ($K_c = $ infinity) at the zeros of $G_{M(s)} B_{(s)}$.

3. The number of loci is equal to the order of the system, i.e., the number of poles of $G_{M(s)} B_{(s)}$.
4. The complex parts of the curves always appear as complex conjugates.
5. The angle of the asymptotes of the loci (as $s \to \infty$) is equal to $\pm 180°/(N - M)$, where N = number of poles of $G_{M(s)} B_{(s)}$
   M = number of zeros of $G_{M(s)} B_{(s)}$

Rules 1 to 4 are fairly self-evident. Rule 5 comes from the fact that at a point on the root locus plot the complex number s must satisfy the equation:

$$1 + G_{M(s)} B_{(s)} = 0$$
$$G_{M(s)} B_{(s)} = -1 + i0 \tag{10.51}$$

Therefore, the argument of $G_{M(s)} B_{(s)}$ on a root locus must always be

$$\arg G_{M(s)} B_{(s)} = \arctan \frac{0}{-1} = \pm \pi \tag{10.52}$$

Now $G_{M(s)} B_{(s)}$ is a ratio of polynomials, Mth-order in the numerator and Nth-order in the denominator.

$$G_{M(s)} B_{(s)} = \frac{b_M s^M + b_{M-1} s^{M-1} + \cdots + b_1 s + b_0}{a_N s^N + a_{N-1} s^{N-1} + \cdots + a_1 s + a_0}$$

On the asymptotes, s gets very big, so only the s^N and s^M terms remain significant.

$$\lim_{s \to \infty} [G_{M(s)} B_{(s)}] = \frac{b_M s^M}{a_N s^N} = \frac{b_M/a_N}{s^{N-M}} \tag{10.53}$$

Putting s into polar form ($s = re^{i\theta}$) gives

$$\lim_{s \to \infty} [G_{M(s)} B_{(s)}] = \frac{b_M/a_N}{r^{N-M} e^{i\theta(N-M)}}$$

The angle or argument of $G_{M(s)} B_{(s)}$ is

$$\lim_{s \to \infty} [\arg G_{M(s)} B_{(s)}] = -(N - M)\theta$$

Equation (10.52) must still be satisfied on the asymptote, and therefore Q.E.D.

$$(N - M)\theta = \pm \pi$$

Applying rule 5 to a first-order process ($N = 1$ and $M = 0$) gives asymptotes that go off at $180°$ (see Example 10.6). Applying it to a second-order process ($N = 2$ and $M = 0$) gives asymptotes that go off at $90°$ (see Example 10.7). Example 10.8 has a second-order denominator ($N = 2$) but it also has a first-order numerator ($M = 1$). So this system has a "net order" ($N - M$) of 1, and the asymptotes go off at $180°$. Example 10.9 shows that the asymptotes go off at $60°$ since the order of the system is third.

B. DIGITAL COMPUTER NUMERICAL SOLUTION. With the wide availability of digital computers, root locus plots are easily obtained by using numerical techniques. Standard polynomial root-solving packages are readily accessible and easy to use. They are computationally inefficient for generating the root locus curves, but this is not worth worrying about unless you plan to generate a very large number of root locus curves.

The polynomial root-solving program POLRT was used in the examples below. It is from the IBM Scientific Subroutines. A listing of the source program is given in Table 10.1. POLRT is very fast, even on a personal computer.

TABLE 10.1
Root locus program for three-CSTR process

```
      DIMENSION XCOF(4),COF(4),ROOTR(3),ROOTI(3)
      REAL KC
      XCOF(4)=1.
      XCOF(3)=3.
      XCOF(2)=3.
      KC=0.
      DO 100 N=1,100
      XCOF(1)=1.+KC/8.
      CALL POLRT(XCOF,COF,3,ROOTR,ROOTI,IER)
      WRITE(6,1) KC,(ROOTR(I),ROOTI(I),I=1,3)
    1 FORMAT(1X,7F8.3)
  100 KC=KC+1.
      STOP
      END
C
C      ..................................................
C
C      SUBROUTINE POLRT
C
C      PURPOSE
C        COMPUTES THE REAL AND COMPLEX ROOTS OF A REAL POLYNOMIAL
C
C      USAGE
C        CALL POLRT(XCOF,COF,M,ROOTR,ROOTI,IER)
C
C      DESCRIPTION OF PARAMETERS
C        XCOF -VECTOR OF M+1 COEFFICIENTS OF THE POLYNOMIAL
C             ORDERED FROM SMALLEST TO LARGEST POWER
C        COF  -WORKING VECTOR OF LENGTH M+1
C        M    -ORDER OF POLYNOMIAL
C        ROOTR-RESULTANT VECTOR OF LENGTH M CONTAINING REAL ROOTS
C             OF THE POLYNOMIAL
C        ROOTI-RESULTANT VECTOR OF LENGTH M CONTAINING THE
C             CORRESPONDING IMAGINARY ROOTS OF THE POLYNOMIAL
C        IER  -ERROR CODE WHERE
C             IER 0  NO ERROR
C             IER 1  M LESS THAN ONE
C             IER 2  M GREATER THAN 36
C             IER 3  UNABLE TO DETERMINE ROOT WITH 500 INTERATIONS
C                 ON 5 STARTING VALUES
C             IER 4  HIGH ORDER COEFFICIENT IS ZERO
C
```

TABLE 10.1 (*continued*)

```
C     REMARKS
C        LIMITED TO 36TH ORDER POLYNOMIAL OR LESS.
C        FLOATING POINT OVERFLOW MAY OCCUR FOR HIGH ORDER
C     POLYNOMIALS BUT WILL NOT AFFECT THE ACCURACY OF THE RESULTS.
C
C     SUBROUTINES AND FUNCTION SUBPROGRAMS REQUIRED
C        NONE
C
C     METHOD
C        NEWTON-RAPHSON ITERATIVE TECHNIQUE.  THE FINAL ITERATIONS
C        ON EACH ROOT ARE PERFORMED USING THE ORIGINAL POLYNOMIAL
C        RATHER THAN THE REDUCED POLYNOMIAL TO AVOID ACCUMULATED
C        ERRORS IN THE REDUCED POLYNOMIAL.
C
C     ............................................................
C
      SUBROUTINE POLRT(XCOF,COF,M,ROOTR,ROOTI,IER)
      DIMENSION XCOF(1),COF(1),ROOTR(1),ROOTI(1)
      DOUBLE PRECISION XO,YO,X,Y,XPR,YPR,UX,UY,V,YT,XT,U,XT2,YT2,SUMSQ,
     1 DX,DY,TEMP,ALPHA
C
C     ............................................................
C
C     IF A DOUBLE PRECISION VERSION OF THIS ROUTINE IS DESIRED, THE
C     C IN COLUMN 1 SHOULD BE REMOVED FROM THE DOUBLE PRECISION
C     STATEMENT WHICH FOLLOWS.
C
C     DOUBLE PRECISION XCOF,COF,ROOTR,ROOTI
C
C     THE C MUST ALSO BE REMOVED FROM DOUBLE PRECISION STATEMENTS
C     APPEARING IN OTHER ROUTINES USED IN CONJUNCTION WITH THIS
C     ROUTINE.
C     THE DOUBLE PRECISION VERSION MAY BE MODIFIED BY CHANGING THE
C     CONSTANT IN STATEMENT 78 TO 1.0D-12 AND IN STATEMENT 122 TO
C     1.0D-10.  THIS WILL PROVIDE HIGHER PRECISION RESULTS AT THE
C     COST OF EXECUTION TIME
C
C     ............................................................
C
      IFIT=0
      N=M
      IER=0
      IF(XCOF(N+1))10,25,10
   10 IF(N) 15,15,32
C
C     SET ERROR CODE TO 1
C
   15 IER=1
   20 RETURN
C
C     SET ERROR CODE TO 4
C
   25 IER=4
      GO TO 20
C
C     SET ERROR CODE TO 2
C
   30 IER=2
```

TABLE 10.1 (*continued*)

```
      GO TO 20
   32 IF(N-36) 35,35,30
   35 NX=N
      NXX=N+1
      N2=1
      KJ1=  N+1
      DO 40 L=1,KJ1
      MT=KJ1-L+1
   40 COF(MT)=XCOF(L)
C
C      SET INITIAL VALUES
C
   45 XO=.00500101
      YO=0.01000101
C
C      ZERO INITIAL VALUE COUNTER
C
      IN=0
   50 X=XO
C
C      INCREMENT INITIAL VALUES AND COUNTER
C
      XO=-10.0*YO
      YO=-10.0*X
C
C      SET X AND Y TO CURRENT VALUE
C
      X=XO
      Y=YO
      IN=IN+1
      GO TO 59
   55 IFIT=1
      XPR=X
      YPR=Y
C
C      EVALUATE POLYNOMIAL AND DERIVATIVES
C
   59 ICT=0
   60 UX=0.0
      UY=0.0
      V= 0.0
      YT=0.0
      XT=1.0
      U=COF(N+1)
      IF(U) 65,130,65
   65 DO 70 I=1,N
      L= N-I+1
      TEMP=COF(L)
      XT2=X*XT-Y*YT
      YT2=X*YT+Y*XT
      U=U+TEMP*XT2
      V=V+TEMP*YT2
      FI=I
      UX=UX+FI*XT*TEMP
      UY=UY-FI*YT*TEMP
      XT=XT2
   70 YT=YT2
      SUMSQ=UX*UX+UY*UY
      IF(SUMSQ) 75,110,75
```

TABLE 10.1 (*continued*)

```
  75 DX=(V*UY-U*UX)/SUMSQ
     X=X+DX
     DY=-(U*UY+V*UX)/SUMSQ
     Y=Y+DY
  78 IF(DABS(DY)+DABS(DX)-1.0D-05) 100,80,80
C
C      STEP ITERATION COUNTER
C
  80 ICT=ICT+1
     IF(ICT-500) 60,85,85
  85 IF(IFIT)100,90,100
  90 IF(IN-5) 50,95,95
C
C      SET ERROR CODE TO 3
C
  95 IER=3
     GO TO 20
 100 DO 105 L=1,NXX
     MT=KJ1-L+1
     TEMP=XCOF(MT)
     XCOF(MT)=COF(L)
 105 COF(L)=TEMP
     ITEMP=N
     N=NX
     NX=ITEMP
     IF(IFIT) 120,55,120
 110 IF(IFIT) 115,50,115
 115 X=XPR
     Y=YPR
 120 IFIT=0
 122 IF(DABS(Y)-1.0D-4*DABS(X)) 135,125,125
 125 ALPHA=X+X
     SUMSQ=X*X+Y*Y
     N=N-2
     GO TO 140
 130 X=0.0
     NX=NX-1
     NXX=NXX-1
 135 Y=0.0
     SUMSQ=0.0
     ALPHA=X
     N=N-1
 140 COF(2)=COF(2)+ALPHA*COF(1)
 145 DO 150 L=2,N
 150 COF(L+1)=COF(L+1)+ALPHA*COF(L)-SUMSQ*COF(L-1)
 155 ROOTI(N2)=Y
     ROOTR(N2)=X
     N2=N2+1
     IF(SUMSQ) 160,165,160
 160 Y=-Y
     SUMSQ=0.0
     GO TO 155
 165 IF(N) 20,20,45
     END
```

In recent years a number of commercial programs have been developed that produce root locus plots (and provide other types of analysis tools). These software packages can speed up controller design. Some of the most popular include "CC," "CONSYD," and "MATRIX-X." We will refer to these packages again later in the book since they are also useful in the frequency and z domains, as well as for handling multivariable systems.

10.4.3   Examples

Example 10.10. Our three-CSTR system is an interesting one to explore via root locus. With the same process $G_{M(s)}$ as shown in Fig. 10.4 we will use different types of feedback controllers and different settings and see how the root loci change.

(a) *Proportional controller*

$$B_{(s)} = K_c$$

The closedloop characteristic equation is

$$1 + G_{M(s)} B_{(s)} = 1 + \frac{\frac{1}{8}K_c}{(s + 1)^3} = 0 \tag{10.54}$$

$$s^3 + 3s^2 + 3s + 1 + \tfrac{1}{8}K_c = 0$$

Figure 10.8a gives the root locus curves that were generated by solving for the three roots numerically on a digital computer at different values of gain K_c. Table 10.1 gives a FORTRAN program that calculates the roots of the closedloop characteristic equation [Eq. (10.54)] using the polynomial root-solving subroutine POLRT. It took 30 seconds on a personal computer to calculate the roots at 100 values of gain.

Notice that the construction rules are satisfied. The root loci start ($K_c = 0$) at the poles of the system openloop transfer function, $s = -1$. There are three loci. The angle of the asymptotes is $180/3 = 60°$.

Notice also that the two curves cross into the RHP when $K_c = 64$ and when s has the values of $\pm i\sqrt{3}$. This confirms our findings of Example 10.4.

A gain of 17 gives a closedloop damping coefficient of 0.316 and a dominant second-order closedloop time constant of 0.85 minutes. The third root is real and lies far out on the negative real axis at -2.3. Thus the largest first-order time constant is 0.43 minutes.

The Ziegler-Nichols recommended value of gain for this system ($K_c = 32$) is also shown on the plot. This setting gives a closedloop system that has a damping coefficient of only 0.15. This is more underdamped than most instrument engineers would normally like to see on a process loop. The ZN settings are often found to be more underdamped than desired in chemical engineering applications.

(b) *Proportional-integral controller*

$$B_{(s)} = K_c \frac{\tau_I s + 1}{\tau_I s} = K_c \frac{s + 1/\tau_I}{s}$$

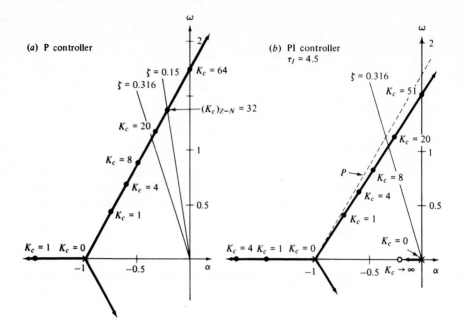

The closedloop characteristic equation is now

$$1 + G_{M(s)} B_{(s)} = 1 + \frac{\frac{1}{8}}{(s+1)^3} K_c \frac{s+1/\tau_I}{s} = 0$$

(10.55)

$$s^4 + 3s^3 + 3s^2 + (1 + \tfrac{1}{8}K_c)s + \frac{K_c}{8\tau_I} = 0$$

Figure 10.8*b,c* gives the root locus plots for two values of the reset time constant τ_I. With $\tau_I = 4.5$ in Fig. 10.8*b*, the ultimate gain is reduced to 51 (compared to the ultimate gain of 64 for a proportional-only controller), and the gain that gives a closedloop damping coefficient of 0.316 is 15. There is a first-order root on the real axis near the origin.

With $\tau_I = 3.03$, the recommended ZN value, the ultimate gain is 44 and the gain that gives a damping coefficient of 0.316 is 13. The ZN value of gain, $K_c = 29.1$, gives a damping coefficient of 0.09. These results are summarized in Table 10.2.

The construction rules are again satisfied. The root loci start at the poles $s = 0$ and $s = -1$. They end at the zero $s = -1/\tau_I$. There are four curves. The angle of the asymptotes is equal to 60° since $N = 4$ and $M = 1$ (i.e., the net order of the system is 3).

(c) *Proportional-integral-derivative controller*

$$B_{(s)} = K_c \frac{\tau_I s + 1}{\tau_I s} \frac{\tau_D s + 1}{(\tau_D/20)s + 1}$$

(10.56)

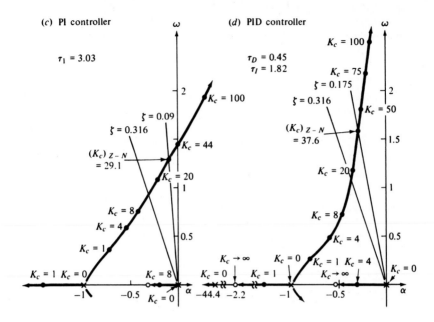

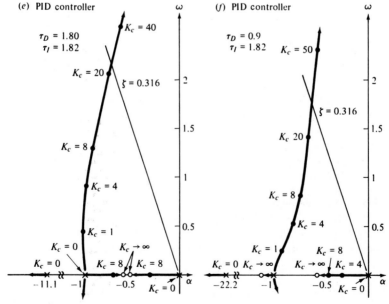

FIGURE 10.8
Root locus curves for a three-CSTR system.

TABLE 10.2
Root locus results for three-CSTR system

| | τ_D | τ_I | K_u | For $\zeta_c = 0.316$ | | |
				K_c	τ_{c2}	τ_{c1}
Proportional	...	...	64	17	0.85	0.43
Proportional-integral	...	3.03	44	13	1.03	3.8
	...	4.5	51	15	0.94	6.3
Proportional-integral-derivative	0.45	1.82	275	17	0.86	1.9
	0.9	1.82	280	30	0.53	1.9
	1.8	1.82	111	22	0.43	2.3

The closedloop characteristic equation is now

$$1 + G_{M(s)} B_{(s)} = 1 + \frac{\frac{1}{8}}{(s+1)^3} K_c \frac{s + 1/\tau_I}{s} \frac{\tau_D s + 1}{(\tau_D/20)s + 1} = 0 \qquad (10.57)$$

$$s^5 + \left(3 + \frac{20}{\tau_D}\right)s^4 + \left(1 + \frac{60}{\tau_D} + \frac{20K_c}{8}\right)s^3$$

$$+ \left[\frac{20}{\tau_D} + \frac{20K_c}{8}\left(\frac{1}{\tau_I} + \frac{1}{\tau_D}\right)\right]s + \frac{20K_c}{8\tau_I\tau_D} = 0 \qquad (10.58)$$

Figure 10.8d,e,f gives root locus curves for three values of derivative time τ_D with integral time τ_I constant at 1.82. There are now five loci. They start at the poles of

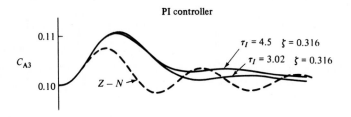

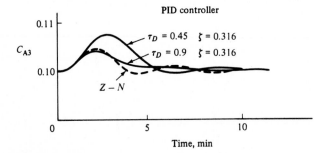

FIGURE 10.9
Load step response of a three-CSTR system with Ziegler-Nichols and $\zeta = 0.316$ controller settings.

the openloop transfer function $s = -1$, $s = 0$, and $s = -20/\tau_D$. They end at the zeros $s = -1/\tau_I$ and $s = -1/\tau_D$. The angle of the asymptotes is $180/(5-2) = 60°$.

The addition of the derivative (lead-lag) unit makes the system more stable. It pulls the root locus away from the imaginary axis. The ZN value of τ_D and gain are shown in Fig. 10.8d. They give a damping coefficient of only 0.175. Table 10.2 summarizes the effects of increasing τ_D. There is an optimum value of τ_D around 0.9 that gives the highest gain with small second- and first-order time constants when designing for a damping coefficient of 0.316.

Figure 10.9 shows the time-domain performance of these PI and PID controllers. The disturbance is a step change in C_{AD}. Note the improved dynamic performance of the PID controllers.

PROBLEMS

10.1. Consider the nonisothermal CSTR system of Example 9.6.
   (a) Use the Routh stability criterion to find the openloop stability requirements for the linearized system.
   (b) Use the numerical values of the parameters given in Table 5.5 to see if the system is openloop stable.
   (c) Add to the system a proportional controller that changes cooling-jacket temperature by the relationship:

$$T_J = \bar{T}_J + K_c(T^{set} - T)$$

   What are the closedloop stability requirements?

10.2. Find the ultimate gain and period of a fourth-order system given below. The controller is proportional and the system openloop transfer function is

$$G_{M(s)} = \frac{(0.047)(112)(2)(0.12)}{(0.083s + 1)(0.017s + 1)(0.432s + 1)(0.024s + 1)}$$

10.3. Use the Routh stability criterion to find the ultimate gain of the closedloop three-CSTR system with a PI controller:
   (a) For $\tau_1 = 3.03$
   (b) For $\tau_1 = 4.5$

10.4. Find the ultimate gain and period of a closedloop system with a proportional controller and an openloop transfer function:

$$G_{M(s)} = \frac{1}{(s + 1)(5s + 1)(\frac{1}{2}s + 1)}$$

10.5. Find the value of feedback controller gain that gives a closedloop damping coefficient of 0.8 for the system with a proportional controller and an openloop transfer function:

$$G_{M(s)} = \frac{s + 4}{s(s + 2)}$$

10.6. The liquid level $h_{(t)}$ in a tank is held by a PI controller that changes the flow rate $F_{(t)}$ out of the tank. The flow rate into the tank $F_{0(t)}$ and the level setpoint $h_{(t)}^{set}$ are disturbances. The vertical cylindrical tank is 10 ft^2 in cross-sectional area. The transfer function of the feedback controller plus the control valve is

$$B_{(s)} = \frac{F_{(s)}}{E_{(s)}} = -K_c\left(1 + \frac{1}{\tau_I s}\right) \quad \text{with units of } \frac{\text{ft}^3/\text{min}}{\text{ft}}$$

(a) Write the equations describing the openloop system.
(b) Write the equations describing the closedloop system.
(c) Derive the openloop transfer functions of the system:

$$G_{M(s)} = \frac{H_{(s)}}{F_{(s)}} \quad \text{and} \quad G_{L(s)} = \frac{H_{(s)}}{F_{0(s)}}$$

(d) Derive the two closedloop transfer functions of the system:

$$\frac{H_{(s)}}{H_{(s)}^{\text{set}}} \quad \text{and} \quad \frac{H_{(s)}}{F_{0(s)}}$$

(e) Make a root locus plot of the closedloop system with a value of integral time $\tau_I = 10$ minutes.
(f) What value of gain K_c gives a closedloop system with a damping coefficient of 0.707? What is the closedloop time constant at this gain?
(g) What gain gives critical damping? What is the time constant with this gain?

10.7. Make a root locus plot of a system with an openloop transfer function:

$$B_{(s)} G_{M(s)} = \frac{K_c}{(s+1)(5s+1)(\frac{1}{2}s+1)} \frac{\tau_D s + 1}{(\frac{1}{20}\tau_D s + 1)}$$

(a) For $\tau_D = 2.5$
(b) For $\tau_D = 5$
(c) For $\tau_D = 7.5$

10.8. Find the ultimate gain and period of the closedloop three CSTR system with a PID controller tuned at $\tau_I = \tau_D = 1$. Make a root locus plot of the system.

10.9. Make a root locus plot of a system with an openloop transfer function:

$$B_{(s)} G_{M(s)} = \frac{K_c e^{-2s}}{s+1}$$

Use the first-order Pade approximation of deadtime. Find the ultimate gain.

10.10. When a system has poles that are widely different in value, it is difficult to plot them all on a root locus plot using conventional rectangular coordinates in the s plane. It is sometimes more convenient to make the root locus plots in the log s plane. Instead of using the conventional axis Re s and Im s, an ordinate of the arg s and an abscissa of the log $|s|$ are used, since the natural logarithm of a complex number is defined:

$$\ln s = \ln |s| + i \arg s$$

(a) Show that the region of stability becomes a horizontal band in the log s plane.
(b) Show that lines of constant time constant in the s plane transform into vertical lines in the log s plane.
(c) Show that lines of constant damping coefficient in the s plane transform into horizontal lines in the log s plane.

(d) Make log s plane root locus plots of systems with the openloop transfer functions $B_{(s)} G_{M(s)}$:

(i) $$\frac{\frac{1}{8}K_c}{(s+1)^3}$$

(ii) $$\frac{K_c}{(s+1)(5s+1)(\frac{1}{2}s+1)}$$

(iii) $$\frac{K_c(-3s+1)}{(s+1)(5s+1)}$$

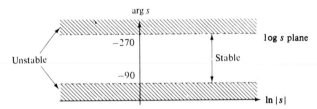

FIGURE P10.10

10.11. A two-tank system with recycle is sketched below. Liquid levels are held by proportional controllers $F_1 = K_1 h_1$ and $F_2 = K_2 h_2$. Flow into the system F_0 and recycle flow F_R can be varied by the operator.

(a) Derive the four closedloop transfer functions relating the two levels and the two load disturbances:

$$\left(\frac{H_1}{F_0}\right)_{(s)} \quad \left(\frac{H_1}{F_R}\right)_{(s)} \quad \left(\frac{H_2}{F_0}\right)_{(s)} \quad \left(\frac{H_2}{F_R}\right)_{(s)}$$

(b) Does the steadystate level in the second tank vary with the recycle flow rate F_R? Use the final-value theorem of Laplace transforms.

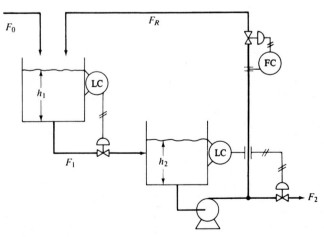

FIGURE P10.11

10.12. The system of Prob. 10.4 is modified by using the cascade control system sketched below.

(a) Find the value of the gain K_s in the proportional controller that gives a 0.707 damping coefficient for the closedloop slave loop.

(b) Using this value of K_s in the slave loop, find the maximum closedloop stable value of the master controller gain K_m. Compare this with the ultimate gain found without cascade control in Prob. 10.4. Also compare ultimate periods.

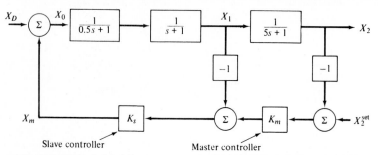

FIGURE P10.12

10.13. Repeat Prob. 10.6 using a proportional feedback controller [parts (b) and (d)]. Will there be a steadystate error in the closedloop system for (a) a step change in set-point h^{set} or (b) a step change in feed rate F_0?

10.14. We would like to compare the closedloop dynamic performance of two types of reboilers.

(a) In the first type, there is a control valve on the steam line to the reboiler and a steam trap on the condensate line leaving the reboiler. The flow transmitter acts like a first-order lag with a 6-second time constant. The control valve also acts like a 6-second lag. These are the only dynamic elements in the steam flow control loop. If a PI flow control is used with $\tau_I = 0.1$ minutes, calculate the closedloop time constant of the steam flow loop when a closedloop damping coefficient of 0.3 is used.

(b) In the second type of reboiler, the control valve is on the condensate line leaving the reboiler. There is no valve in the steam line so the tubes in the reboiler see full steam-header pressure.

Changes in steam flow are achieved by increasing or decreasing the area used for condensing steam in the reboiler. This variable-area flooded reboiler is used in some processes because it permits the use of lower-pressure steam. However, as you will show in your calculations (I hope), the dynamic performance of this configuration is distinctly poorer than direct manipulation of steam flow.

The steam flow meter still acts like a first-order lag with a 6-second time constant, but the smaller control valve on the liquid condensate can be assumed to be instantaneous.

The condensing temperature of the steam is 300°F. The process into which the heat is transferred is at a constant temperature of 200°F. The overall heat transfer coefficient is 300 Btu/h °F ft². The reboiler has 509 tubes that are 10 feet long and 1 inch inside diameter. The steam and condensate are inside the tubes. The density of the condensate is 62.4 lb$_m$/ft³ and the latent heat of condensation of the steam is 900 Btu/lb$_m$. Neglect any sensible heat transfer.

Derive a dynamic mathematical model of the flooded-condenser system. Calculate the transfer function relating steam flow rate to condensate flow rate. Using
a PI controller with $\tau_I = 0.1$ minute, calculate the closedloop time constant of the
steam flow control loop when a closedloop damping coefficient of 0.3 is used.
Compare this with the result found in (a).

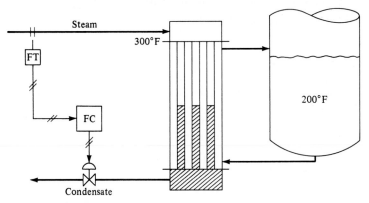

FIGURE P10.14

10.15. A chemical reactor is cooled by both jacket cooling water and condenser cooling
water. A mathematical model of the system has yielded the following openloop
transfer functions (time is in minutes):

$$\frac{T}{F_c} = \frac{-1}{s+1} \qquad (°F/gpm)$$

$$\frac{T}{F_J} = \frac{-5}{10s+1} \qquad (°F/gpm)$$

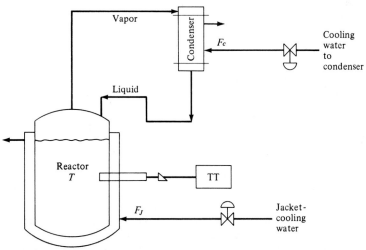

FIGURE P10.15

The span of the temperature transmitter is 100–200°F. Control valves have linear trim and constant pressure drop, and are half open under normal conditions. Normal condenser flow is 30 gpm. Normal jacket flow is 20 gpm. A temperature measurement lag of 12 seconds is introduced into the system by the thermowell.

If a proportional feedback temperature controller is used, calculate the controller gain K_c that yields a closedloop damping coefficient of 0.707 and calculate the closedloop time constant of the system when

(a) Jacket water only is used.

(b) Condenser water only is used.

Derive the closedloop characteristic equation for the system when both jacket and condenser water are used.

10.16 Oil and water are mixed together and then decanted. Oil flow rate is ratioed to water flow rate F_W. Interface is controlled by oil flow F_L from the decanter with a proportional level controller. Water flow (F_H) from the decanter, which is liquid full, is on pressure control (PI). Steadystate flow rates are:

$$\text{Oil} = 177.8 \text{ gpm}$$

$$\text{Water} = 448.2 \text{ gpm}$$

Water and oil holdups in the decanter are each 1130 gallons at steadystate.

(a) Derive the closedloop transfer function between F_H and F_W in terms of level controller gain K_c.

(b) Determine the transient response of F_H for a step change in F_W of 1 gpm when level controller gain is 6.36 gpm/gal.

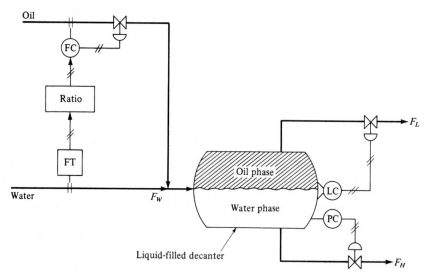

FIGURE P10.16

10.17. The openloop transfer function relating steam flow rate to temperature in a feed preheater has been found to consist of a steadystate gain K_p and a first-order lag with the time constant τ_p. The lag associated with temperature measurement is τ_m. A proportional-only temperature controller is used.

(a) Derive an expression for the roots of the closedloop characteristic equation in terms of the parameters τ_p, K_p, τ_m, and K_c.

(b) Solve for the value of controller gain that will give a critically damped closedloop system when $K_p = 1$, $\tau_p = 10$ and

    (i) $\tau_m = 1$

    (ii) $\tau_m = 5$

10.18. The liquid flow rate from a vertical cylindrical tank, 10 feet in diameter, is flow-controlled. The liquid flow into the tank is manipulated to control liquid level in the tank. The control valve on the inflow stream has linear installed characteristics and can pass 1000 gpm when wide open. The level transmitter has a span of 6 feet of liquid. A proportional controller is used with a gain of 2. Liquid density is constant.

(a) Should the control valve be A-O or A-C?

(b) Should the controller be reverse or direct acting?

(c) What is the dimensionless openloop system transfer function relating liquid height and inflow rate?

(d) Solve for the time response of the inflow rate to a step change in the outflow rate from 500 to 750 gpm with the tank initially half full.

10.19. A process has a positive pole located at $(+1,0)$ in the s plane (with time in minutes). The process steadystate gain is 2. An addition lag of 20 seconds exists in the control loop. Sketch root locus plots and calculate controller gains which give a closedloop damping coefficient of 0.707 when

(a) A proportional feedback controller is used.

(b) A proportional-derivative feedback controller is used with the derivative time set equal to the lag in the control loop. The ratio of numerator to denominator time constants in the derivative unit is 6.

10.20. A process has an openloop transfer function that is a first-order lag with a time constant τ_p and a steadystate gain K_p. If a PI feedback controller is used with a reset time τ_I, sketch root locus plots for the cases where:

(a) $\tau_p < \tau_I$

(b) $\tau_p = \tau_I$

(c) $\tau_p > \tau_I$

What value of the τ_I/τ_p ratio gives a closedloop system that has a damping coefficient of $\frac{1}{2}\sqrt{2}$ for only one value of controller gain?

10.21. The openloop process transfer functions relating the manipulated and load variables (M and L) to the controlled variable (X) are first-order lags with identical time constants (τ) but with different gains (K_M and K_L). Derive equations for the closedloop steadystate error and the closedloop time constant for step disturbances in load if a proportional feedback controller is used.

10.22. The liquid level in a tank is controlled by manipulating the flow out of the tank, using a PI controller. The outflow rate is a function of only the valve position. The valve has linear installed characteristics and passes 20 ft^3/min wide open.

    The tank is vertical and cylindrical with a cross-sectional area of 25 ft^2 and a 2 ft level transmitter span.

(a) Derive the relationship between the feedback controller gain K_c and the reset time τ_I which gives a critically damped closedloop system.

(b) For a critically damped system with $\tau_I = 5$ minutes, calculate the closedloop time constant.

10.23. A process has an openloop transfer function G_M relating controlled and manipulated variables that is a first-order lag τ_p and steadystate gain K_p. There is an additional first-order lag τ_m in the measurement of the controlled variable. A proportional-only feedback controller is used.

Derive an expression relating the controller gain K_c to the parameters τ_p, τ_m, and K_p such that the closedloop system damping coefficient is 0.707. What happens to K_c as τ_m gets very small or very large? What is the value of τ_m that provides the smallest value of K_c?

10.24. A fixed-gain relay is used as a low base-level override controller on a distillation column. The column is 7 ft in diameter and has a base level transmitter span of 3 ft. The density of the liquid in the base is 50 lb_m/ft^3. Its heat of vaporization is 200 Btu/lb_m.

The reboiler steam valve has linear installed characteristics and passes 30,000 lb_m/h when wide open. Steam latent heat is 1000 Btu/lb_m.

There is a first-order dynamic lag of τ minutes between a change in the signal to the steam valve and vapor boilup. The low base-level override controller pinches the reboiler steam valve over the lower 25 percent of the level transmitter span.

Solve for the value of gain that should be used in the relay as a function of τ to give a closedloop damping coefficient of 0.5 for the override level loop.

10.25. A process has G_M and G_L openloop transfer functions that are:

$$G_M = \frac{K_M}{(\tau s + 1)^2} \qquad G_L = \frac{K_L}{(\tau s + 1)^2}$$

If a PI controller is used with τ_I set equal to τ, calculate:
(a) The value of controller gain that gives a closedloop damping coefficient of 0.707
(b) The closedloop time constant, using this value of gain
(c) The closedloop transfer function between the load variable and the output variable
(d) The steadystate error for a step change in the load variable

10.26. Two tanks are connected by a pipe through which liquid can flow in each direction, depending on the difference in liquid levels.

$$F_c = K_B(h_1 - h_2)$$

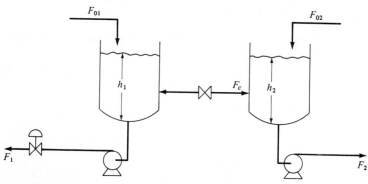

FIGURE P10.26

where K_B is a constant with units of ft^2/min. Disturbances are the flow rates F_{01}, F_{02}, and F_2. The manipulated variable is the flow rate F_1. Cross-sectional areas of the vertical cylindrical tanks are A_1 and A_2.

(a) Derive the openloop transfer function between the height of liquid in tank 1 (h_1) and F_1.

(b) What are the poles and zeros of the openloop transfer function? What is the openloop characteristic equation?

(c) If a proportional-only level controller is used, derive the closedloop characteristic equation and sketch a root locus plot for the case where $A_1 = A_2$.

10.27. Our old friend the three-isothermal CSTR process has the following openloop transfer function relating controlled and manipulated variables:

$$G_{M(s)} = \frac{\frac{1}{8}}{(s + 1)^3}$$

When a proportional feedback controller is used, this process has an ultimate gain of 64 and an ultimate frequency of $\sqrt{3}$.

(a) Find the ultimate gain and ultimate frequency when a proportional-derivative (PD) controller is used with derivative time set equal to 1.

$$B_{(s)} = K_c \frac{\tau_D s + 1}{0.1\tau_D s + 1}$$

(b) Sketch root locus plots using:
  (i) A proportional controller.
  (ii) A PD controller with $\tau_D = 1$.
  (iii) A PD controller with τ_D much less than 1.
  (iv) A PD controller with τ_D much greater than 1.

CHAPTER

11

LAPLACE-DOMAIN ANALYSIS OF ADVANCED CONTROL SYSTEMS

In the last chapter we used Laplace-domain techniques to study the dynamics and stability of simple closedloop control systems. In this chapter we want to apply these same methods to more complex systems: cascade control, feedforward control, openloop unstable processes, and processes with inverse response. Finally we will discuss an alternative way to look at controller design that is called "model-based" control.

The tools are those developed in Chaps. 9 and 10. We use transfer functions to design feedforward controllers or to develop the characteristic equation of the system and to find the location of its roots in the s plane.

11.1 CASCADE CONTROL

Cascade control was discussed qualitatively in Sec. 8.2. There is a secondary (or "slave") loop and a primary (or "master") loop. Both load rejection and performance can sometimes be improved by using cascade control.

376

There are two types of process structures for which cascade control can be applied. If the manipulated variable affects one variable and then this variable affects a second controlled variable, the structure leads to *series* cascade control. If the manipulated variable affects both variables directly, the structure leads to *parallel* cascade.

11.1.1 Series Cascade

Figure 11.1a shows an openloop process in which two transfer functions G_1 and G_2 are connected in series. The manipulated variable M enters G_1 and produces a change in X_1. The X_1 variable then enters G_2 and changes X_2.

Figure 11.1b shows the conventional feedback control system where a single controller senses the controlled variable X_2 and changes the manipulated variable M. The closedloop characteristic equation for this system was developed in Chap. 10.

$$1 + G_{1(s)} G_{2(s)} B_{(s)} = 0 \tag{11.1}$$

Figure 11.1c shows a series cascade system. There are now two controllers. The secondary controller B_1 adjusts M to control the secondary variable X_1. The setpoint signal X_1^{set} to the B_1 controller comes from the primary controller, i.e., the output of the primary controller B_2 is the setpoint for the B_1 controller. The B_2 controller setpoint is X_2^{set}.

The closedloop characteristic equation for this system is *not* the same as that given in Eq. (11.1). To derive what it is, let us first look at the secondary loop by itself. From the analysis presented in Chap. 10, the equation that describes this closedloop system is

$$X_1 = \frac{G_1 B_1}{1 + G_1 B_1} X_1^{\text{set}} \tag{11.2}$$

So to design the secondary controller B_1 we use the closedloop characteristic equation

$$1 + G_1 B_1 = 0 \tag{11.3}$$

(a) Openloop process

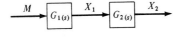

(b) Conventional feedback control

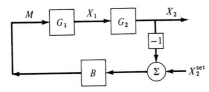

FIGURE 11.1
Series cascade.

Next we look at the controlled output variable X_2. Figure 11.1d shows the reduced block diagram of the system in the conventional form. We can deduce the closedloop characteristic equation of this system by inspection.

$$1 + G_2 B_2\left(\frac{G_1 B_1}{1 + G_1 B_1}\right) = 0 \tag{11.4}$$

However, let us derive it rigorously.

$$X_2 = G_2 X_1 \tag{11.5}$$

Substituting for X_1 from Eq. (11.2) gives

$$X_2 = G_2 \frac{G_1 B_1}{1 + G_1 B_1} X_1^{set} \tag{11.6}$$

(c) Series cascade

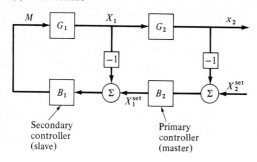

Secondary controller (slave)

Primary controller (master)

(d) Reduced block diagram

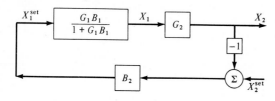

(e) Example 11.1

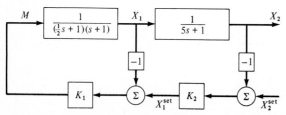

FIGURE 11.1
Series cascade.

But X_1^{set} is the output from the B_2 controller.

$$X_1^{\text{set}} = B_2(X_2^{\text{set}} - X_2) \tag{11.7}$$

Combining Eqs. (11.6) and (11.7) gives

$$X_2 = G_2\left(\frac{G_1 B_1}{1 + G_1 B_1}\right) B_2(X_2^{\text{set}} - X_2)$$

$$X_2\left[1 + G_2 B_2\left(\frac{G_1 B_1}{1 + G_1 B_1}\right)\right] = G_2 B_2\left(\frac{G_1 B_1}{1 + G_1 B_1}\right) X_2^{\text{set}}$$

Rearranging gives

$$X_2 = \frac{G_2 B_2[G_1 B_1/(1 + G_1 B_1)]}{1 + G_2 B_2[G_1 B_1/(1 + G_1 B_1)]} X_2^{\text{set}} \tag{11.8}$$

So Eq. (11.4) gives the closedloop characteristic equation of this series cascade system. A little additional rearrangement leads to a completely equivalent form of Eq. (11.8).

$$X_2 = \frac{G_1 G_2 B_1 B_2}{1 + G_1 B_1(1 + G_2 B_2)} X_2^{\text{set}} \tag{11.9}$$

An alternative and equivalent closedloop characteristic equation is

$$1 + G_1 B_1(1 + G_2 B_2) = 0 \tag{11.10}$$

The roots of this equation will dictate the dynamics of the series cascade system. Note that both of the openloop transfer functions are involved as well as both of the controllers. Equation (11.4) is a little more convenient to use than Eq. (11.10) because we can make conventional root locus plots, varying the gain of the B_2 controller, after the parameters of the B_1 controller have been specified.

Example 11.1. Consider the process with a series cascade control system sketched in Fig. 11.1e. The secondary controller B_1 and the primary controller B_2 are both proportional only.

$$B_1 = K_1 \qquad B_2 = K_2$$

In this example

$$G_1 = \frac{1}{(\tfrac{1}{2}s + 1)(s + 1)} \qquad G_2 = \frac{1}{5s + 1}$$

(a) *Conventional control.* First we will look at a conventional single proportional controller (K_c) that manipulates M to control X_2^{set}. The closedloop characteristic equation is

$$1 + \frac{1}{(\tfrac{1}{2}s + 1)(s + 1)(5s + 1)} K_c = 0 \tag{11.11}$$

$$\tfrac{5}{2}s^3 + 8s^2 + \tfrac{13}{2}s + 1 + K_c = 0 \tag{11.12}$$

To solve for the ultimate gain and ultimate frequency, we substitute $i\omega$ for s.

$$-i\tfrac{5}{2}\omega^3 - 8\omega^2 + i\tfrac{13}{2}\omega + 1 + K_c = 0$$

$$[-8\omega^2 + 1 + K_c] + i[\tfrac{13}{2}\omega - \tfrac{5}{2}\omega^3] = 0 + i0 \tag{11.13}$$

Solving the two equations simultaneously for the two unknowns gives

$$K_u = \frac{99}{5} \quad \text{and} \quad \omega_u = \sqrt{\frac{13}{5}}$$

(b) *Designing the secondary (slave) loop.* We will pick a closedloop damping coefficient specification for the secondary loop of 0.707 and calculate the required value of K_1. The closedloop characteristic equation for the slave loop is

$$1 + K_1 \frac{1}{(\tfrac{1}{2}s + 1)(s + 1)} = 0 = \tfrac{1}{2}s^2 + \tfrac{3}{2}s + 1 + K_1 \tag{11.14}$$

Solving for the closedloop roots gives

$$s = -\tfrac{3}{2} \pm i\tfrac{1}{2}\sqrt{8K_1 - 1} \tag{11.15}$$

To have a damping coefficient of 0.707 the roots must lie on a radial line whose angle with the real axis is arccos $0.707 = 45°$. On this line the real and imaginary parts of the roots are equal. So for a closedloop damping coefficient of 0.707

$$\tfrac{3}{2} = \tfrac{1}{2}\sqrt{8K_1 - 1} \quad \Rightarrow \quad K_1 = \tfrac{5}{4} \tag{11.16}$$

Now the closedloop relationship between X_1 and X_1^{set} is

$$X_1 = \frac{G_1 B_1}{1 + G_1 B_1} X_1^{\text{set}} = \frac{[1/(\tfrac{1}{2}s + 1)(s + 1)]\tfrac{5}{4}}{1 + [1/(\tfrac{1}{2}s + 1)(s + 1)]\tfrac{5}{4}} X_1^{\text{set}} \tag{11.17}$$

$$X_1 = \frac{\tfrac{5}{2}}{s^2 + 3s + \tfrac{9}{2}} X_1^{\text{set}} \tag{11.18}$$

(c) *Designing the primary (master) loop.* The closedloop characteristic equation for the master loop is

$$1 + G_2 B_2 \left(\frac{G_1 B_1}{1 + G_1 B_1} \right) = 1 + \frac{K_2}{5s + 1} \frac{\tfrac{5}{2}}{s^2 + 3s + \tfrac{9}{2}} = 0 \tag{11.19}$$

$$5s^3 + 16s^2 + \tfrac{51}{2}s + \tfrac{9}{2} + \tfrac{5}{2}K_2 = 0 \tag{11.20}$$

Solving for the ultimate gain K_u and ultimate frequency ω_u by substituting $i\omega$ for s gives

$$K_u = 30.8 \quad \omega_u = \sqrt{5.1} = 2.26$$

It is useful to compare these values with those found for conventional control: $K_u = 19.8$ and $\omega_u = 1.61$. We can see that cascade control results in higher controller gain and smaller closedloop time constant (the reciprocal of the frequency). Figure 11.2b gives a root locus plot for the primary controller with the secondary controller gain set at $\tfrac{5}{4}$. Two of the loci start at the complex poles $s = -\tfrac{3}{2} \pm i\tfrac{3}{2}$ which come from the closedloop secondary loop. The other curve starts at the pole $s = -\tfrac{1}{5}$.

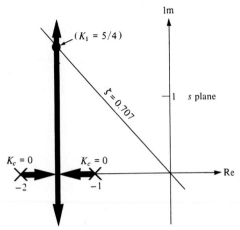

(a) Root locus for secondary loop

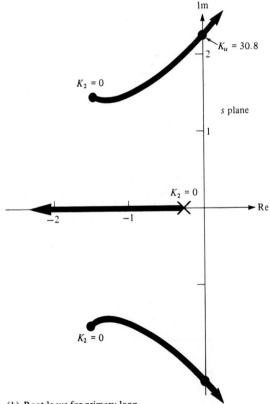

(b) Root locus for primary loop

FIGURE 11.2

11.1.2   Parallel Cascade

Figure 11.3a shows a process where the manipulated variable affects the two controlled variables X_1 and X_2 in parallel. An important example is in distillation column control where reflux flow affects both distillate composition and a tray temperature. The process has a parallel structure and this leads to a parallel cascade control system.

If only a single controller B_2 is used to control X_2 by manipulating M, the closedloop characteristic equation is the conventional

$$1 + G_{2(s)} B_{2(s)} = 0 \tag{11.21}$$

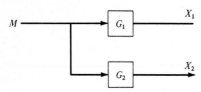

(a) Openloop process

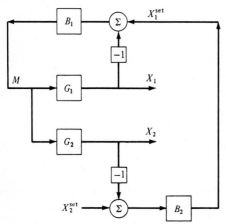

(b) Parallel cascade control

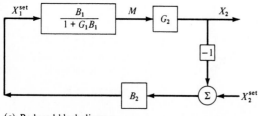

(c) Reduced block diagram

FIGURE 11.3
Parallel cascade.

If, however, a cascade control system is used, as sketched in Fig. 11.3b, the closedloop characteristic equation is not that given in Eq. (11.21). To derive it, let us start with the secondary loop.

$$X_1 = G_1 M = G_1 B_1 (X_1^{set} - X_1) \tag{11.22}$$

$$X_1 = \frac{G_1 B_1}{1 + G_1 B_1} X_1^{set} \tag{11.23}$$

Combining Eqs. (11.22) and (11.23) gives the closedloop relationship between M and X_1^{set}.

$$M = \frac{1}{G_1} X_1 = \frac{1}{G_1} \frac{G_1 B_1}{1 + G_1 B_1} X_1^{set} = \frac{B_1}{1 + G_1 B_1} X_1^{set} \tag{11.24}$$

Now we solve for the closedloop transfer function for the primary loop with the secondary loop on automatic. Figure 11.3c shows the simplified block diagram. By inspection we can see that the closedloop characteristic equation is

$$1 + G_2 B_2 \left(\frac{B_1}{1 + G_1 B_1} \right) = 0 \tag{11.25}$$

Note the difference between the series cascade [Eq. (11.4)] and the parallel cascade [Eq. (11.25)] characteristic equations.

11.2 FEEDFORWARD CONTROL

11.2.1 Fundamentals

Most of the control systems we have discussed, simulated, and designed thus far in this book have been feedback control devices. A deviation of an output variable from a setpoint is detected. This error signal is fed into a feedback controller that changes the manipulated variable. The controller makes no use of any information about the source, magnitude, or direction of the disturbance that has caused the output variable to change.

The basic notion of feedforward control is to detect disturbances as they enter the process and make adjustments in manipulated variables so that output variables are held constant. We do not wait until the disturbance has worked its way through the process and has disturbed everything to produce an error signal. If a disturbance can be detected as it enters the process, it makes sense to take immediate action in order to compensate for its effect on the process.

Feedforward control systems have gained wide acceptance in chemical engineering in the past two decades. They have demonstrated their ability to improve control, sometimes quite spectacularly. We will illustrate this improvement in this section by comparing the responses of systems with feedforward control and with conventional feedback control when load disturbances occur.

Feedforward control is probably used more in chemical engineering systems than in any other field of engineering. Our systems are often slow-moving, nonlinear, and multivariable, and contain appreciable deadtime. All these characteristics make life miserable for feedback controllers. Feedforward controllers can handle all these with relative ease.

11.2.2 Linear Feedforward Control

A block diagram of a simple openloop process is sketched in Fig. 11.4a. The load disturbance $L_{(s)}$ and the manipulated variable $M_{(s)}$ affect the controlled variable

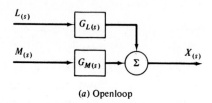

(a) Openloop

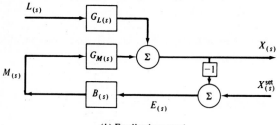

(b) Feedback control

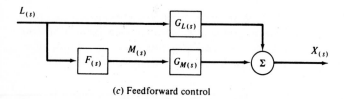

(c) Feedforward control

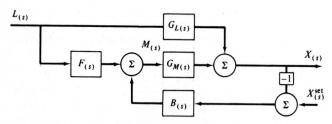

(d) Combined feedforward/feedback control

FIGURE 11.4
Block diagrams.

$X_{(s)}$. A conventional feedback control system is shown in Fig. 11.4b. The error signal $E_{(s)}$ is fed into a feedback controller $B_{(s)}$ that changes the manipulated variable $M_{(s)}$.

Figure 11.4c shows the feedforward control system. The load disturbance $L_{(s)}$ still enters the process through the $G_{L(s)}$ process transfer function. The load disturbance is also fed into a feedforward control device that has a transfer function $F_{(s)}$. The feedforward controller detects changes in the load $L_{(s)}$ and makes changes in the manipulated variable $M_{(s)}$.

Thus the transfer function of a feedforward controller is a relationship between a manipulated variable and a disturbance variable (usually a load change).

$$F_{(s)} = \left(\frac{M}{L}\right)_{(s)} = \left(\frac{\text{manipulated variable}}{\text{disturbance}}\right)_{X \text{ is constant}} \tag{11.26}$$

To design a feedforward controller, that is, to find $F_{(s)}$, we must know both $G_{L(s)}$ and $G_{M(s)}$. The objective of most feedforward controllers is to hold the controlled variable constant at its steadystate value. Therefore the change or perturbation in $X_{(s)}$ should be zero. The output $X_{(s)}$ is given by the equation

$$X_{(s)} = G_{L(s)} L_{(s)} + G_{M(s)} M_{(s)} \tag{11.27}$$

Setting $X_{(s)}$ equal to zero and solving for the relationship between $M_{(s)}$ and $L_{(s)}$ give the feedforward controller transfer function

$$\left(\frac{M_{(s)}}{L_{(s)}}\right)_{(X=0)} \equiv F_{(s)} = \left(\frac{-G_L}{G_M}\right)_{(s)} \tag{11.28}$$

Example 11.2. Suppose we have a distillation column with the process transfer functions $G_{M(s)}$ and $G_{L(s)}$ relating bottoms composition x_B to steam flow rate F_s and to feed flow rate F_L.

$$\left(\frac{x_B}{F_s}\right)_{(s)} = G_{M(s)} = \frac{K_M}{\tau_M s + 1}$$

$$\left(\frac{x_B}{F_L}\right)_{(s)} = G_{L(s)} = \frac{K_L}{\tau_L s + 1} \tag{11.29}$$

All these variables are perturbations from steadystate. These transfer functions could have been derived from a mathematical model of the column or found experimentally.

We want to use a feedforward controller $F_{(s)}$ to make adjustments in steam flow to the reboiler, whenever the feed rate to the column changes, so that bottoms composition is held constant. The feedforward-controller design equation [Eq. (11.28)] gives

$$F_{(s)} = \left(\frac{-G_L}{G_M}\right)_{(s)} = \frac{-K_L/(\tau_L s + 1)}{K_M/(\tau_M s + 1)} = \frac{-K_L}{K_M} \frac{\tau_M s + 1}{\tau_L s + 1} \tag{11.30}$$

The feedforward controller contains a steadystate gain and dynamic terms. For this system the dynamic element is a first-order lead-lag. The unit step response of this lead-lag is an initial change to a value that is $(-K_L/K_M)(\tau_M/\tau_L)$, followed by an exponential rise or decay to the final steadystate value $-K_L/K_M$.

The advantage of feedforward control over feedback control is that perfect control can, in theory, be achieved. A disturbance will produce no error in the controlled output variable if the feedforward controller is perfect. The disadvantages of feedforward control are:

1. The disturbance must be detected. If we cannot measure it, we cannot use feedforward control. This is one reason why feedforward control for throughput changes is commonly used, whereas feedforward control for feed composition disturbances is only occasionally used. The former requires a flow-measurement device, which is usually available. The latter requires a composition analyzer, which may or may not be available.
2. We must know how the disturbance and manipulated variables affect the process. The transfer functions $G_{L(s)}$ and $G_{M(s)}$ must be known, at least approximately. One of the nice features of feedforward control is that even crude, inexact feedforward controllers can be quite effective in reducing the upset caused by a disturbance.

In practice, many feedforward control systems are implemented by using ratio control systems, as discussed in Chap. 8. Most feedforward control systems are installed as combined feedforward-feedback systems. The feedforward controller takes care of the large and frequent measurable disturbances. The feedback controller takes care of any errors that come through the process because of inaccuracies in the feedforward controller or other unmeasured disturbances. Figure 11.4d shows the block diagram of a simple linear combined feedforward-feedback system. The manipulated variable is changed by both the feedforward controller and the feedback controller.

The addition of the feedforward controller has *no* effect on the closedloop stability of the system for linear systems. The denominators of the closedloop transfer functions are unchanged.

With feedback control

$$X_{(s)} = \frac{G_{L(s)}}{1 + G_{M(s)} B_{(s)}} L_{(s)} + \frac{G_{M(s)} B_{(s)}}{1 + G_{M(s)} B_{(s)}} X_{(s)}^{\text{set}} \qquad (11.31)$$

With feedforward-feedback control

$$X_{(s)} = \frac{G_{L(s)} + F_{(s)} G_{M(s)}}{1 + G_{M(s)} B_{(s)}} L_{(s)} + \frac{G_{M(s)} B_{(s)}}{1 + G_{M(s)} B_{(s)}} X_{(s)}^{\text{set}} \qquad (11.32)$$

In a nonlinear system the addition of a feedforward controller often permits tighter tuning of the feedback controller because the magnitude of the disturbances that the feedback controller must cope with is reduced.

Figure 11.5a shows a typical implementation of feedforward controller. A distillation column provides the specific example. Steam flow to the reboiler is ratioed to the feed flow rate. The feedforward controller gain is set in the ratio device. The dynamic elements of the feedforward controller are provided by the lead-lag unit.

Figure 11.5b shows a combined feedforward-feedback system where the feedback signal is added to the feedforward signal in a summing device. Figure 11.5c shows another combined system where the feedback signal is used to change the feedforward controller gain in the ratio device. Figure 11.6 shows a combined feedforward-feedback control system for a distillation column where feed-rate disturbances are detected and both steam flow and reflux flow are changed to hold both overhead and bottoms compositions constant. Two feedforward controllers are required.

Figure 11.7 shows some typical results of using feedforward control. A first-order lag is used in the feedforward controller so that the change in the manipulated variable is not instantaneous. The feedforward action is not perfect because the dynamics are not perfect, but there is a significant improvement over just feedback control.

It is not always possible to achieve perfect feedforward control. If the $G_{M(s)}$ transfer function has a deadtime that is larger than the deadtime in the $G_{L(s)}$

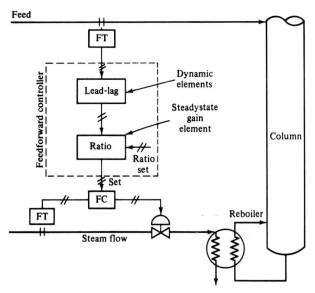

(a) Feedforward control

FIGURE 11.5
Feedforward systems.

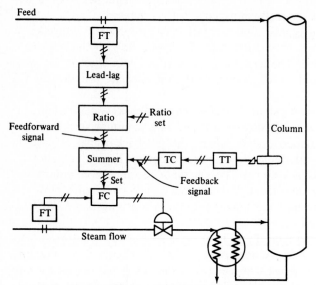

(b) Feedforward-feedback control with additive signals

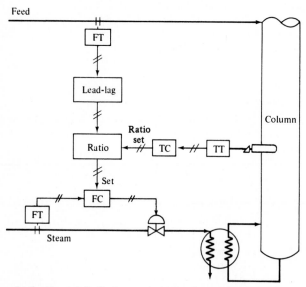

(c) Feedforward-feedback control with feedforward gain modified

FIGURE 11.5
Feedforward systems.

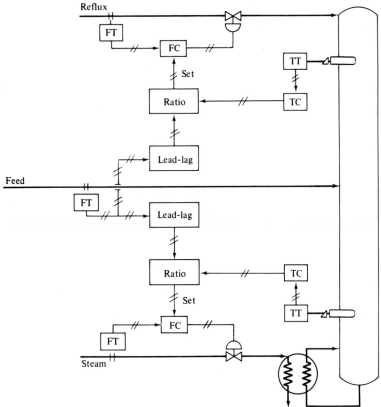

FIGURE 11.6
Combined feedforward-feedback system with two controlled variables.

transfer function, the feedforward controller will be physically unrealizable because it requires predictive action. Also, if the $G_{M(s)}$ transfer function is of higher order than the $G_{L(s)}$ transfer function, the feedforward controller will be physically unrealizable [see Eq. (11.28)].

11.2.3 Nonlinear Feedforward Control

There are no inherent linear limitations in feedforward control. Nonlinear feedforward controllers can be designed for nonlinear systems. The concepts are illustrated in Example 11.3.

Example 11.3. The nonlinear ODEs describing the constant holdup, nonisothermal CSTR system are

$$\frac{dC_A}{dt} = \frac{F}{V}(C_{A0} - C_A) - C_A \alpha e^{-E/RT} \tag{11.33}$$

$$\frac{dT}{dt} = \frac{F}{V}(T_0 - T) - \left(\frac{\lambda}{\rho C_p}\right)C_A \alpha e^{-E/RT} - \left(\frac{UA}{C_p V \rho}\right)(T - T_J) \tag{11.34}$$

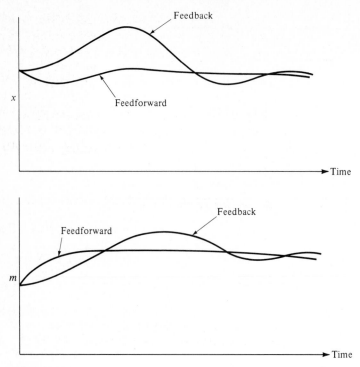

FIGURE 11.7
Feedforward control performance for load disturbance.

Let us choose a feedforward control system that holds both reactor temperature T and reactor concentration C_A constant at their steadystate values, $\bar{T}$ and $\bar{C}_A$. The feed flow rate F and the jacket temperature T_J are the manipulated variables. Disturbances are feed concentration C_{A0} and feed temperature T_0.

Remembering that we are dealing with total variables now and not perturbations, the feedforward control objectives are

$$C_{A(t)} = \bar{C}_A \qquad \text{and} \qquad T_{(t)} = \bar{T} \tag{11.35}$$

Substituting these into Eqs. (11.33) and (11.34) gives

$$\frac{d\bar{C}_A}{dt} = 0 = \frac{F_{(t)}}{V}(C_{A0(t)} - \bar{C}_A) - \bar{C}_A \bar{k} \tag{11.36}$$

$$\frac{d\bar{T}}{dt} = 0 = \frac{F_{(t)}}{V}(T_{0(t)} - \bar{T}) - \left(\frac{\lambda}{\rho C_p}\right)\bar{C}_A \bar{k} - \left(\frac{UA}{C_p V \rho}\right)(\bar{T} - T_{J(t)}) \tag{11.37}$$

Rearranging Eq. (11.36) to find $F_{(t)}$, the manipulated variable, in terms of the disturbance $C_{A0(t)}$, gives the nonlinear feedforward controller relating the load variable C_{A0} to the manipulated variable F.

$$F_{(t)} = \frac{\bar{C}_A \bar{k} V}{C_{A0(t)} - \bar{C}_A} \tag{11.38}$$

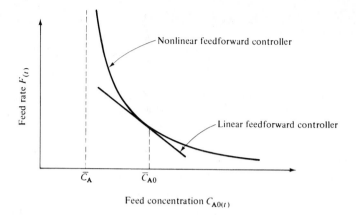

FIGURE 11.8
Nonlinear relationship between feed rate and feed concentration.

The relationship is hyperbolic, as shown in Fig. 11.8. Feed rate must be decreased as feed concentration increases. This increases the holdup time, with constant volume, so that the additional reactant is consumed. Equation (11.38) tells us that feed flow rate does *not* have to be changed when feed temperature T_0 changes.

Substituting Eq. (11.38) into Eq. (11.37) and solving for the other manipulated variable T_J give

$$T_{J(t)} = \bar{T} + \frac{\bar{C}_A \bar{k} V}{UA}\left[\lambda + \frac{C_p(\bar{T} - T_{0(t)})}{C_{A0(t)} - \bar{C}_A}\right] \tag{11.39}$$

This is a second nonlinear feedforward relationship that shows how cooling-jacket temperature $T_{J(t)}$ must be changed as both feed concentration $C_{A0(t)}$ and feed temperature $T_{0(t)}$ change. Notice that the relationship between T_J and C_{A0} is nonlinear, but the relationship between T_J and T_0 is linear.

The above nonlinear feedforward controller equations were found analytically. In more complex systems, analytical methods become too complex, and numerical techniques must be used to find the required nonlinear changes in manipulated variables. The nonlinear steadystate changes can be found by using the nonlinear algebraic equations describing the process. The dynamic portion can often be approximated by linearizing around various steadystates.

11.3 OPENLOOP UNSTABLE PROCESSES

We remarked earlier in this book that one of the most interesting processes that chemical engineers have to control is the exothermic chemical reactor. This process can be openloop unstable.

Openloop instability means that reactor temperature will take off when there is no feedback control of cooling rate. It is easy to visualize qualitatively how this can occur. The reaction rate increases as the temperature climbs and more heat is given off. This heats the reactor to an even higher temperature, at which the reaction rate is still faster and even more heat is generated.

There is also an openloop unstable mechanical system: the *inverted pendulum*. This is the problem of balancing a broom on the palm of your hand. You must keep moving your hand to keep the broom balanced. If you put your brain on manual and hold your hand still, the broom will topple over. So the process is openloop unstable.

We will explore this phenomenon quantitatively in the s plane. We will discuss linear systems in which instability means that the reactor temperature would theoretically go off to infinity. Actually, in any real system, reactor temperature will not go to infinity because the real system is nonlinear. The nonlinearity makes the reactor temperature climb to some high temperature at which it levels out. The concentration of reactant becomes so low that the reaction rate is limited.

Nevertheless linear techniques are very useful in looking at stability near some operating level. Mathematically, if the system is openloop unstable, it must have an openloop transfer function $G_{M(s)}$ that has at least one pole in the RHP.

11.3.1   Simple Systems

As a simple example let us look at just the energy equation of the nonisothermal CSTR process of Example 9.6. We will neglect any changes in C_A for the moment.

$$\frac{dT}{dt} = a_{22} T + a_{26} T_J + \cdots \tag{11.40}$$

Laplace transforming gives

$$(s - a_{22})T_{(s)} = a_{26} T_{J(s)} + \cdots$$

$$T_{(s)} = \frac{a_{26}}{s - a_{22}} T_{J(s)} + \cdots \tag{11.41}$$

Thus the stability of the system depends on the location of the pole a_{22}. If this pole is positive, the system is openloop unstable. The value of a_{22} is given in Eqs. (9.82).

$$a_{22} = \frac{-\lambda \bar{k} E \bar{C}_A}{\rho C_p R \bar{T}^2} - \frac{\bar{F}}{V} - \frac{UA}{V \rho C_p} \tag{11.42}$$

For the system to be openloop stable $a_{22} < 0$

$$\frac{-\lambda \bar{k} E \bar{C}_A}{\rho C_p R \bar{T}^2} - \frac{\bar{F}}{V} - \frac{UA}{V \rho C_p} < 0$$

$$\frac{-\lambda \bar{k} E \bar{C}_A}{\rho C_p R \bar{T}^2} < \frac{\bar{F}}{V} + \frac{UA}{V \rho C_p} \tag{11.43}$$

The left side of Eq. (11.43) represents the heat generation. The right side represents heat removal. Thus our simple linear analysis tells us that the heat-removal capacity must be greater than the heat generation if the system is to be stable.

The actual stability requirement for the nonisothermal CSTR system is a little more complex than Eq. (11.43) because the concentration C_A does change.

A. FIRST-ORDER OPENLOOP UNSTABLE PROCESS. Suppose we have a first-order process with the openloop transfer function

$$G_{M(s)} = \frac{K_p}{\tau_p s - 1} \qquad (11.44)$$

Note that this is *not* a first-order lag because of the negative sign in the denominator. The system has an openloop pole in the RHP at $s = +1/\tau_p$. The unit step response of this system is an exponential that goes off to infinity as time increases.

Can we make the system stable by using feedback control? That is, can an openloop unstable process be made closedloop stable by appropriate design of the feedback controller? Let us try a proportional controller: $B_{(s)} = K_c$. The closedloop characteristic equation is

$$1 + G_{M(s)} B_{(s)} = 1 + \frac{K_p}{\tau_p s - 1} K_c = 0$$

$$s = \frac{1 - K_c K_p}{\tau_p} \qquad (11.45)$$

There is a single closedloop root. The root locus plot is given in Fig. 11.9a. It starts at the openloop pole in the RHP. The system is closedloop unstable for small values of controller gain. When the controller gain equals $1/K_p$, the closedloop root is located right at the origin. For gains greater than this, the root is in the LHP so the system is closedloop stable.

Thus in this system there is a *minimum* stable gain. Some of the systems studied up to now have had maximum values of gain K_{max} (or ultimate gain K_u) *beyond* which the system is closedloop unstable. Now we have a case that has a minimum gain K_{min} *below* which the system is closedloop unstable.

B. SECOND-ORDER OPENLOOP UNSTABLE PROCESS. Consider the process given in Eq. (11.44) with a first-order lag added.

$$G_{M(s)} = \frac{K_p}{(\tau_{p1} s + 1)(\tau_{p2} s - 1)} \qquad (11.46)$$

One of the roots of the openloop characteristic equation lies in the RHP at $s = +1/\tau_{p2}$.

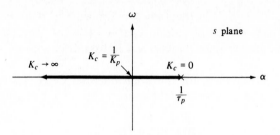

$$(a) \quad \text{First-order:} \quad B_{(s)} G_{M(s)} = \frac{K_c K_p}{\tau_p s - 1}$$

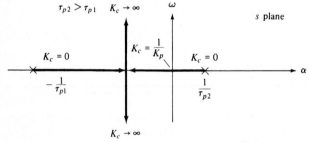

$$(b) \quad \text{Second-order:} \quad B_{(s)} G_{M(s)} = \frac{K_c K_p}{(\tau_{p1} s + 1)(\tau_{p2} - 1)}$$

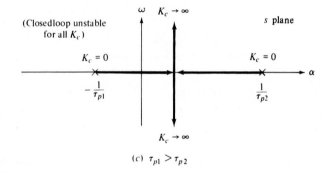

$$(c) \quad \tau_{p1} > \tau_{p2}$$

FIGURE 11.9
Root locus curves for openloop unstable processes (positive poles).

Can we make this system closedloop stable? A proportional feedback controller gives a closedloop characteristic equation:

$$1 + G_{M(s)} B_{(s)} = 1 + \frac{K_p}{(\tau_{p1} s + 1)(\tau_{p2} s - 1)} K_c = 0$$

$$\tau_{p1} \tau_{p2} s^2 + (\tau_{p2} - \tau_{p1})s + K_c K_p - 1 = 0$$

(11.47)

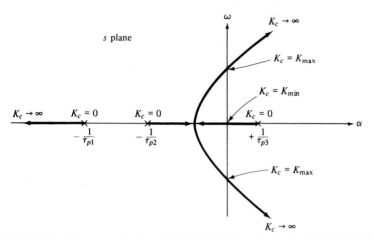

(d) Third-order: $B_{(s)}G_{M(s)} = \dfrac{K_c K_p}{(\tau_{p1}s + 1)(\tau_{p2}s + 1)(\tau_{p3}s - 1)}$

There are two conditions that must be satisfied if there are to be no sign changes in the first column of the Routh array.

$$\tau_{p2} > \tau_{p1} \quad \text{and} \quad K_c > \frac{1}{K_p} \tag{11.48}$$

Therefore, if $\tau_{p2} < \tau_{p1}$ a proportional controller cannot make the system closed-loop stable. A controller with derivative action might be able to stabilize the system. Figure 11.9b,c gives the root locus plots for the two cases: $\tau_{p2} > \tau_{p1}$ and $\tau_{p2} < \tau_{p1}$. In the latter case there is always at least one closedloop root in the RHP, so the system is always unstable.

C. THIRD-ORDER OPENLOOP UNSTABLE PROCESS. If an additional lag is added to the system and a proportional controller is used, the closedloop characteristic equation becomes

$$1 + G_{M(s)} B_{(s)} = 1 + \frac{K_p}{(\tau_{p1}s + 1)(\tau_{p2}s + 1)(\tau_{p3}s - 1)} K_c = 0 \tag{11.49}$$

Figure 11.9d gives a sketch of a typical root locus plot for this type of system. We now have a case of *conditional stability*. Below K_{min} the system is closedloop unstable. Above K_{max} the system is again closedloop unstable. A range of stable values of controller gain exists between these limits.

$$K_{min} < K_c < K_{max} \tag{11.50}$$

Clearly, the closer the values of K_{max} and K_{min} are to each other, the less controllable the system will be.

Example 11.4. The transfer function relating process temperature T to cooling-water flow rate F_w in an openloop unstable chemical reactor is

$$G_{M(s)} = \frac{-0.7(^\circ\text{F/gpm})}{(s + 1)(\tau s - 1)} \tag{11.51}$$

where the time constants 1 and τ are in minutes. The temperature measurement has a dynamic first-order lag of 30 seconds. The range of the analog electronic (4 to 20 mA) temperature transmitter is 200 to 400°F. The control valve on the cooling water has linear installed characteristics and passes 500 gpm when wide open. The temperature controller is proportional.

(a) What is the closedloop characteristic equation of the system?

We must include the 0.5 minute lag of the temperature transmitter and the gains for both the transmitter and the valve.

$$1 + G_{M(s)} G_{T(s)} G_{V(s)} B_{(s)} = 0$$

$$1 + \left[\frac{-0.7(^\circ\text{F/gpm})}{(s + 1)(\tau s - 1)}\right]\left[\frac{16/200(\text{mA}/^\circ\text{F})}{0.5s + 1}\right][-500/16(\text{gpm/mA})][K_c] = 0$$

Note that the gain of the controller is chosen to be positive (reverse-acting) so the controller output decreases as temperature increases, which increases cooling-water flow through the AC valve (this makes the gain of the control valve negative).

$$(0.5\tau)s^3 + (1.5\tau - 0.5)s^2 + (\tau - 1.5)s + (1.75K_c - 1) = 0 \tag{11.52}$$

(b) What is the minimum value of controller gain, K_{min}, that gives a closedloop stable system?

Letting $s = i\omega$ in Eq. (11.52) gives two equations in two unknowns: K_c and ω.

From the real part:

$$0.5\omega^2 - 1.5\omega^2\tau + 1.75K_c - 1 = 0 \tag{11.53}$$

From the imaginary part:

$$\omega\tau - 1.5\omega - 0.5\tau\omega^3 = 0 \tag{11.54}$$

There are two solutions for Eq. (11.54):

$$\omega = 0 \quad \text{and} \quad \omega = \sqrt{\frac{\tau - 1.5}{0.5\tau}} \tag{11.55}$$

Using $\omega = 0$ gives the minimum value of gain.

$$K_{min} = 1/1.75$$

(c) Derive a relationship between τ (the positve pole) and the maximum closedloop stable gain, K_{max}.

Using the second value of ω in Eq. (11.55) gives K_{max}.

$$K_{max} = \frac{1.5 - 4.5\tau + 3\tau^2}{1.75\tau} \tag{11.56}$$

(d) Calculate K_{max} when $\tau = 5$ minutes and 10 minutes.

For $\tau = 5$, $K_{max} = 6.17$
For $\tau = 10$, $K_{max} = 14.7$

Note that this result shows that the smaller the value of τ (i.e., the closer the positive pole is to the value of the negative poles: $s = -1$ and -2), the more difficult it is to stabilize the system.

(e) At what value of τ will a proportional-only controller be unable to stabilize the system?

When $K_{max} = K_{min}$ the system will always be unstable.

$$\frac{1.5 - 4.5\tau + 3\tau^2}{1.75\tau} = \frac{1}{1.75} \qquad \Rightarrow \qquad \tau = 1.5 \text{ min}$$

Note that there are actually two values of τ that satisfy the equation above, but the limiting one is the larger of the two.

11.3.2 Effects of Lags

The systems explored above illustrate a very important point about the control of openloop unstable systems: the control of these systems becomes more difficult as the order of the system is increased and as the magnitudes of the first-order lags increase.

Our examples above demonstrated this quantitatively. For this reason, it is vital to design a reactor control system with very fast measurement dynamics and very fast heat-removal dynamics. If the thermal lags in the temperature sensor and in the cooling jacket are not small, it may not be possible to stabilize the reactor with feedback control.

Bare-bulb thermocouples and oversized cooling-water valves are often used to improve controllability.

11.3.3 PD Control

Up to this point we have looked at using proportional controllers on openloop unstable systems. The controllability can often be improved by using derivative action in the controller. An example will illustrate the point.

Example 11.5. Let us take the same third-order process analyzed in Example 11.4. For a $\tau = 5$ minutes and a proportional controller, the ultimate gain was 6.17 and the ultimate frequency was 1.18 radians per minute.

Now we will use a PD controller with τ_D set equal to 0.5 minutes (just to make the algebra work out nicely; this is not necessarily the optimum value of τ_D). The closedloop characteristic equation becomes

$$1 + \left[\frac{1.75}{(s+1)(5s-1)(0.5s+1)} \right] K_c \frac{\tau_D s + 1}{0.1\tau_D s + 1} = 0$$

$$0.25s^3 + 5.2s^2 + 3.95s + 1.75K_c - 1 = 0 \qquad (11.57)$$

Solving for the ultimate gain and frequency gives $K_u = 47.5$ and $\omega_u = 3.97$. Comparing these with the results for P control shows a significant increase in gain and reduction in closedloop time constant.

11.3.4 Effects of Reactor Scale-up On Controllability

One of the classical problems in scaling-up a jacketed reactor is the decrease in the ratio of heat-transfer area to reactor volume as size is increased. This has a profound effect on the controllability of the system. Table 11.1 gives some results that quantify the effects for reactors varying from 5 gallons (typical pilot-plant size) to 5000 gallons. Table 11.2 gives parameter values that are held constant as the reactor is scaled up.

Notice that the temperature difference between the cooling jacket and the reactor must be increased as the size of the reactor increases. The flow rate of cooling water also increases rapidly as reactor size increases.

The ratio of K_{max} to K_{min}, which is a measure of the controllability of the system, decreases from 124 for a 5-gallon reactor to 33 for a 5000-gallon reactor.

11.4 PROCESSES WITH INVERSE RESPONSE

Another interesting type of process is one that exhibits *inverse response*. This phenomenon, which occurs in a number of real systems, is sketched in Fig.

TABLE 11.1
Effect of scale-up on controllability

Reactor volume (gal)	5	500	5000
Feed rate (lb$_m$/h)	27.8	2780	27,800
Heat transfer (10^6 Btu/h)	0.0028	0.28	2.8
Reactor height (ft)	1.504	6.98	15.04
Reactor diameter (ft)	0.752	3.49	7.52
Heat transfer area (ft^2)	3.99	86.15	400
Cooling water flow (gpm)	0.086	11.58	240
Jacket temperature (°F)	135.3	118.3	93.3
Controller gains			
Max	169	100	144
Min	1.37	1.95	4.41
Ratio	124	51	33

TABLE 11.2
Reactor parameters

Reactor holdup time	1.2 h
Jacket holdup time	0.077 h
Overall heat transfer coefficient	150 Btu/h ft^2 °F
Heat capacity products and feeds	0.75 Btu/lb$_m$ °F
Heat capacity cooling water	1.0 Btu/lb$_m$ °F
Density of products and feeds	50 lb$_m$/ft^3
Density of cooling water	62.3 lb$_m$/ft^3
Inlet cooling-water temperature	70°F
Temperature measurement lag	30 s
Feed concentration	0.50 lb · mol A/ft^3
Feed temperature	70°F
Reactor temperature	140°F
Preexponential factor	7.08 × 10^{10} h^{-1}
Activation energy	30,000 Btu/lb · mol
Heat of reaction	−30,000 Btu/lb · mol
Steadystate concentration	0.245 lb · mol A/ft^3
Specific reaction rate	0.8672 h^{-1}
Temperature transmitter span	100°F
Cooling-water valve size	Twice normal design flow rate

11.10b. The initial response of the output variable $x_{(t)}$ is in the opposite direction to where it eventually ends up. Thus the process starts out in the wrong direction. You can imagine what this sort of behavior would do to a poor feedback controller in such a loop. We will show quantitatively how inverse response degrades control-loop performance.

An important example of a physical process that shows inverse response is the base of a distillation column. The responses of bottoms composition and base level to a change in vapor boilup can show inverse behavior. In a binary distillation column, we know that an increase in vapor boilup V must drive more low-boiling material up the column and therefore decrease the mole fraction of light component in the bottoms x_B. However, the tray hydraulics can produce some unexpected results. When the vapor rate through a tray is increased, it tends to (1) back up more liquid in the downcomer to overcome the increase in pressure drop through the tray, and (2) reduce the density of the liquid and vapor froth on the active part of the tray. The first effect momentarily reduces the liquid flow rates through the column while the liquid holdup in the downcomer is building up. The second effect tends to momentarily increase the liquid rates since there is more height over the weir.

Which of these two opposing effects dominates depends on the tray design and operating level. The pressure drops through valve trays change little with vapor rates unless the valves are completely lifted. Therefore the second effect is sometimes larger than the first. If this occurs, an increase in vapor boilup produces a transient increase in liquid rates down the column. This increase in liquid rate carries material that is richer in light component into the reboiler and momentarily increases x_B.

Eventually, of course, the liquid rates will return to normal when the liquid inventory on the trays has dropped to the new steadystate levels. Then the effect of the increase in vapor boilup will drive x_B down.

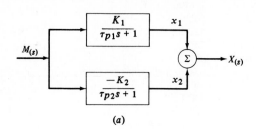

(a)

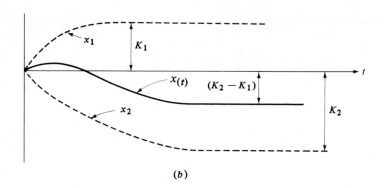

(b)

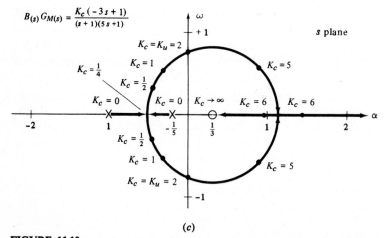

(c)

FIGURE 11.10

Process with inverse response. (a) Block diagram; (b) step response; (c) root locus plot.

Thus the vapor-liquid hydraulics can produce inverse response in the effect of V on x_B (and also on the liquid holdup in the base).

Mathematically, inverse response can be represented by a system that has a transfer function with a *positive zero*, a zero in the RHP. Consider the system sketched in Fig. 11.10a. There are two parallel first-order lags with gains of opposite sign. The transfer function for the overall system is

$$
\frac{X_{(s)}}{M_{(s)}} = \frac{K_1}{\tau_{p1}s + 1} - \frac{K_2}{\tau_{p2}s + 1}
$$

$$
= \frac{(K_1\tau_{p2} - K_2\tau_{p1})s - (K_2 - K_1)}{(\tau_{p1}s + 1)(\tau_{p2}s + 1)} \tag{11.58}
$$

If the K's and τ_p's are such that

$$
\frac{\tau_{p2}}{\tau_{p1}} > \frac{K_2}{K_1} > 1
$$

the system will show inverse response as sketched in Fig. 11.10b. Equation (11.58) can be rearranged

$$
\frac{X_{(s)}}{M_{(s)}} = -(K_2 - K_1) \frac{\left[-\left(\dfrac{K_1\tau_{p2} - K_2\tau_{p1}}{K_2 - K_1} \right)s + 1 \right]}{(\tau_{p1}s + 1)(\tau_{p2}s + 1)} \tag{11.59}
$$

Thus the system has a positive zero at

$$
s = \frac{K_2 - K_1}{K_1\tau_{p2} - K_2\tau_{p1}}
$$

Keep in mind that the positive zero does not make the system openloop unstable. Stability depends on the poles of the transfer function, not on the zeros. Positive zeros in a system do, however, affect closedloop stability as the example below illustrates.

Example 11.6. Let us take the same system used in Example 10.7 and add a positive zero at $s = +\frac{1}{3}$.

$$
G_{M(s)} = \frac{-3s + 1}{(s + 1)(5s + 1)} \tag{11.60}
$$

With a proportional feedback controller the closedloop characteristic equation is

$$
1 + G_{M(s)}B_{(s)} = 1 + \frac{-3s + 1}{(s + 1)(5s + 1)}K_c
$$

$$
5s^2 + (6 - 3K_c)s + 1 + K_c = 0 \tag{11.61}
$$

The root locus curves are shown in Fig. 11.10c. The loci start at the poles of the openloop transfer function: $s = -1$ and $s = -\frac{1}{5}$. Since the loci must end at the zeros of the openloop transfer function ($s = +\frac{1}{3}$) the curves swing over into the RHP. Therefore the system is closedloop unstable for gains greater than 2.

Remember that in Example 10.8 adding a lead or a negative zero made the closedloop system more stable. In this example we have shown that adding a positive zero has just the reverse effect.

11.5 MODEL-BASED CONTROL

Up to this point we have usually chosen a type of controller (P, PI, or PID) and determined the tuning constants that gave some desired performance (closedloop damping coefficient). We have used a model of the process to calculate the controller settings, but the structure of the model has not been explicitly involved in the controller design.

There are several alternative controller design methods that make more explicit use of a process model. We will discuss two of these below.

11.5.1 Minimal Prototype Design

This method originated in sampled-data control and we will discuss it in that context in Chap. 20. However, the same thinking can be applied in continuous systems.

In this approach, the desired closedloop response for a given input is specified. Then, knowing the model of the process, the required form and tuning of the feedback controller is back-calculated. These steps can be clarified by a simple example.

Example 11.7. Suppose we have a process with the openloop transfer function

$$G_{M(s)} = \frac{K_0}{\tau_0 s + 1} \tag{11.62}$$

where K_0 and τ_0 are the openloop gain and time constant. Let us assume that we want to specify the closedloop servo transfer function to be

$$\frac{X_{(s)}}{X_{(s)}^{set}} = \frac{1}{\tau_c s + 1} \tag{11.63}$$

We are saying that we want the process to respond to a step change in setpoint as a first-order process with a closedloop time constant τ_c. The steadystate gain between the controlled variable and the setpoint is specified as unity, so there will be no offset.

Now, knowing the process model and having specified the desired closedloop servo transfer function, we can solve for the feedback controller transfer function $B_{(s)}$. We define the closedloop servo transfer function as $S_{(s)}$.

$$S_{(s)} \equiv \frac{X_{(s)}}{X_{(s)}^{set}} = \frac{G_{M(s)} B_{(s)}}{1 + G_{M(s)} B_{(s)}} \tag{11.64}$$

Equation (11.64) contains only one unknown (i.e., the feedback controller transfer function $B_{(s)}$). Solving for $B_{(s)}$ in terms of the known values of $G_{M(s)}$ and $S_{(s)}$ gives

$$B_{(s)} = \frac{S_{(s)}}{(1 - S_{(s)})G_{M(s)}} \tag{11.65}$$

Equation (11.65) is a general solution for any process and for any desired closedloop servo transfer function. Plugging in the values for $G_{M(s)}$ and $S_{(s)}$ for the specific example gives

$$B_{(s)} = \frac{1/(\tau_c s + 1)}{[1 - 1/(\tau_c s + 1)]K_0/(\tau_0 s + 1)} = \frac{\tau_0 s + 1}{K_0 \tau_c s} \tag{11.66}$$

Equation (11.66) can be rearranged to look just like a PI controller if K_c is set equal to $\tau_0/\tau_c K_0$ and the reset time τ_I is set equal to τ_0.

$$B_{(s)} = K_c \frac{\tau_I s + 1}{\tau_I s} = \left(\frac{\tau_0}{\tau_c K_0}\right)\frac{\tau_0 s + 1}{\tau_0 s} \tag{11.67}$$

Thus we have found that the appropriate structure for the controller is PI, and we have solved analytically for the gain and reset time in terms of the parameters of the process model and the desired closedloop response.

Before we leave this example, it is important to make sure that you understand the limitations of the method. Suppose the process openloop transfer function also contained a deadtime.

$$G_{M(s)} = \frac{K_0 e^{-Ds}}{\tau_0 s + 1} \tag{11.68}$$

Using this $G_{M(s)}$ in Eq. (11.65) gives a new feedback controller:

$$B_{(s)} = \frac{(\tau_0 s + 1)e^{+Ds}}{K_0 \tau_c s} \tag{11.69}$$

This controller is *not* physically realizable. The negative deadtime implies that we can change the output of the device D minutes before the input changes, which is impossible.

This last case illustrates that the desired closedloop relationship *cannot* be chosen arbitrarily. You cannot make a jumbo jet behave like a jet fighter! We must select the desired response such that the controller is physically realizable. In this case all we need to do is modify the specified closedloop servo transfer function $S_{(s)}$ to include the deadtime.

$$S_{(s)} = \frac{e^{-Ds}}{\tau_c s + 1} \tag{11.70}$$

Using this $S_{(s)}$ in Eq. (11.65) gives exactly the same $B_{(s)}$ as found in Eq. (11.66), which is physically realizable.

As an additional case, suppose we had a second-order process transfer function.

$$G_{M(s)} = \frac{K_0}{(\tau_0 s + 1)^2}$$

Specifying the original closedloop servo transfer function [Eq. (11.63)] and solving for the feedback controller using Eq. (11.65) gives

$$B_{(s)} = \frac{(\tau_0 s + 1)^2}{K_0 \tau_c s}$$

(11.71)

Again, this controller is physically unrealizable because the order of the numerator is greater than the order of the denominator. We would have to modify our specified $S_{(s)}$ so as to make this controller realizable.

This type of controller design has been around for many years. The "pole-placement" methods that are used in aerospace systems use the same basic idea: the controller is designed so as to position the poles of the closedloop transfer function at the desired location in the s plane. This is exactly what we do when we specify the closedloop time constant in Eq. (11.63).

11.5.2 Internal Model Control

Morari and coworkers (Garcia and Morari, *Ind. Eng. Chem. Process Des. Dev.* Vol. 21, p. 308, 1982) have used a similar approach in developing "Internal Model Control" (IMC). The method is useful in that it gives the control engineer a different perspective on the controller design problem.

The basic idea of IMC is to use a model of the process openloop $G_{M(s)}$ transfer function in such a way that the selection of the specified closedloop response yields a physically realizable feedback controller.

Figure 11.11 gives the IMC structure. The model of the process $\tilde{G}_{M(s)}$ is run in parallel with the actual process. The output of the model $\tilde{X}$ is subtracted from the actual output of the process X and this signal is fed back into the controller $C_{(s)}$. This signal is, in a sense, the effect of load disturbances on the output (since we have subtracted the effect of the manipulated variable M). Thus we are "inferring" the load disturbance without having to measure it. This signal is X_L, the output of the process load transfer function, and is equal to $G_{L(s)} L_{(s)}$. We know from our studies of feedforward control [Eq. (11.28)] that if we change the manipulated variable $M_{(s)}$ by the relationship

$$M_{(s)} = \left(\frac{-G_L}{G_M} \right)_{(s)} L_{(s)}$$

(11.72)

we would get perfect control of the output $X_{(s)}$. This tells us that if we could set the controller

$$C_{(s)} = \frac{1}{G_{M(s)}}$$

(11.73)

we would get perfect control for load disturbances.

In addition, this choice of $C_{(s)}$ will also give perfect control for setpoint disturbances: the total transfer function between the setpoint X^{set} and X is simply unity.

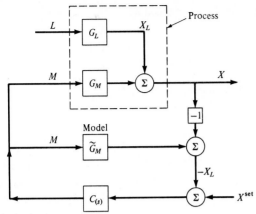

(*a*) Basic structure

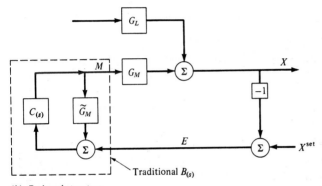

(*b*) Reduced structure

FIGURE 11.11
IMC.

However, there are two practical problems with this ideal choice of the feedback controller $C_{(s)}$. First, it assumes that the model is perfect. More importantly it assumes that the inverse of the plant model $G_{M(s)}$ is physically realizable. This is almost *never* true since most plants have deadtime and/or numerator polynomials that are of lower order than denominator polynomials.

So if we cannot attain perfect control, what do we do? From the IMC perspective we simply break up the controller transfer function $C_{(s)}$ into two parts. The first part is the inverse of $G_{M(s)}$. The second part, which Morari calls a "filter," is chosen to make the total $C_{(s)}$ physically realizable. As we will show below, this second part turns out to be the closedloop servo transfer function that we defined as $S_{(s)}$ in Eq. (11.64).

Referring to Fig. 11.11 and assuming that $G_M = \tilde{G}_M$, we see that

$$X = G_L L + G_M M = X_L + G_M C_{(s)}(X^{\text{set}} - X_L) \tag{11.74}$$

$$X = [G_M C_{(s)}]X^{\text{set}} + [1 - G_M C_{(s)}]X_L \tag{11.75}$$

Now if the controller transfer function $C_{(s)}$ is selected as

$$C_{(s)} = \frac{1}{G_{M(s)}} S_{(s)} \tag{11.76}$$

Eq. (11.75) becomes

$$X_{(s)} = [S_{(s)}]X_{(s)}^{set} + [1 - S_{(s)}]G_{L(s)} L_{(s)} \tag{11.77}$$

So the closedloop servo transfer function $S_{(s)}$ must be chosen such that $C_{(s)}$ is physically realizable.

Example 11.8. Let's take the same process as studied in Example 11.7. The process openloop transfer function is

$$G_{M(s)} = \frac{K_0}{\tau_0 s + 1}$$

We want to design $C_{(s)}$ using Eq. (11.76).

$$C_{(s)} = \frac{1}{G_{M(s)}} S_{(s)} = \frac{1}{K_0/(\tau_0 s + 1)} S_{(s)} = \frac{\tau_0 s + 1}{K_0} S_{(s)} \tag{11.78}$$

Now the logical choice of $S_{(s)}$ that will make $C_{(s)}$ physically realizable is the same as that chosen in Eq. (11.63).

$$S_{(s)} = \frac{X_{(s)}}{X_{(s)}^{set}} = \frac{1}{\tau_c s + 1}$$

So the IMC controller becomes a PD controller.

$$C_{(s)} = \frac{\tau_0 s + 1}{K_0(\tau_c s + 1)} \tag{11.79}$$

The IMC structure is an alternative way of looking at controller design. The model of the process is clearly indicated in the block diagram. The tuning of the $C_{(s)}$ controller reduces to selecting a reasonable closedloop servo transfer function.

The reduced block diagram for the IMC structure in Fig. 11.11 shows that there is a precise relationship between the traditional feedback controller $B_{(s)}$ and the $C_{(s)}$ controller used in IMC.

$$\frac{M_{(s)}}{E_{(s)}} = B_{(s)} = \frac{C_{(s)}}{1 - C_{(s)} \tilde{G}_{M(s)}} \tag{11.80}$$

The negative sign in the denominator of Eq. (11.80) comes from the positive feedback in the internal loop in the controller. Applying this equation to Example 11.8 gives

$$B_{(s)} = \frac{C_{(s)}}{1 - C_{(s)} \tilde{G}_{M(s)}} = \frac{(\tau_0 s + 1)/[K_0(\tau_c s + 1)]}{1 - \{(\tau_0 s + 1)/[K_0(\tau_c s + 1)]\}\{K_0/(\tau_0 s + 1)\}}$$

$$= \frac{\tau_0 s + 1}{K_0 \tau_c s} \tag{11.81}$$

This is exactly the same result (a PI controller) that we found in Eq. (11.66).

Example 11.9. Apply the IMC design to the process with the openloop transfer function

$$G_{M(s)} = \frac{K_0 e^{-Ds}}{\tau_0 s + 1} \qquad (11.82)$$

Using Eq. (11.76) and substituting Eq. (11.82) give

$$C_{(s)} = \frac{1}{G_{M(s)}} S_{(s)} = \frac{\tau_0 s + 1}{K_0 e^{-Ds}} S_{(s)}$$

Clearly the best way to select the closedloop servo transfer function $S_{(s)}$ to make $C_{(s)}$ physically realizable is

$$S_{(s)} = \frac{e^{-Ds}}{\tau_c s + 1} \qquad (11.83)$$

The response of X to a step change in setpoint will be a deadtime of D minutes followed by an exponential rise. The IMC controller becomes a PD controller

$$C_{(s)} = \frac{\tau_0 s + 1}{K_0 e^{-Ds}} \frac{e^{-Ds}}{\tau_c s + 1} = \frac{\tau_0 s + 1}{K_0 (\tau_c s + 1)} \qquad (11.84)$$

It should be noted that the equivalent conventional controller $B_{(s)}$ does *not* have the standard P, PI, or PID form

$$B_{(s)} = \frac{C_{(s)}}{1 - C_{(s)} \tilde{G}_{M(s)}} = \frac{(\tau_0 s + 1)/[K_0(\tau_c s + 1)]}{1 - \{(\tau_0 s + 1)/[K_0(\tau_c s + 1)]\}[K_0 e^{-Ds}/(\tau_0 s + 1)]}$$

$$= \frac{\tau_0 s + 1}{K_0(\tau_c s + 1 - e^{-Ds})} \qquad (11.85)$$

This controller has a uniquely new transfer function.

Maurath, Mellichamp, and Seborg (*I&EC Research* 1988, Vol. 27, p. 956) give guidelines for selecting parameter values in IMC designs.

One final comment should be made about model-based control before we leave the subject. These model-based controllers depend quite strongly on the validity of the model. If we have a poor model or if the plant parameters change, the performance of a model-based controller is usually seriously affected. Model-based controllers are less "robust" than the more conventional PI controllers. This lack of robustness can be a problem in the single-input–single-output (SISO) loops that we have been examining. It is an even more serious problem in multivariable systems, as we will find out in Chaps. 16 and 17.

PROBLEMS

11.1. The load and manipulated-variable transfer functions of a process are

$$\frac{X_{(s)}}{M_{(s)}} = G_{M(s)} = \frac{1}{(s+1)(5s+1)}$$

$$\frac{X_{(s)}}{L_{(s)}} = G_{L(s)} = \frac{2}{(s+1)(5s+1)(\frac{1}{2}s+1)}$$

Derive the feedforward-controller transfer function that will keep the process output $X_{(s)}$ constant with load changes $L_{(s)}$.

11.2. Repeat Prob. 11.1 with

$$G_{M(s)} = \frac{\frac{1}{2}s+1}{(s+1)(5s+1)}$$

11.3. The transfer functions of a binary distillation column between distillate composition x_D and feed rate F, reflux rate R, and feed composition z are

$$\frac{x_D}{F} = \frac{K_F e^{-D_{FS}}}{(\tau_F s + 1)^2} \qquad \frac{x_D}{z} = \frac{K_z e^{-D_z s}}{(\tau_z s + 1)^2} \qquad \frac{x_D}{R} = \frac{K_R e^{-D_{RS}}}{\tau_R s + 1}$$

Find the feedforward-controller transfer functions that will keep x_D constant, by manipulating R, despite changes in z and F. For what values of parameters are these feedforward controllers physically realizable?

11.4. Greg Shinskey has suggested that the steadystate distillate and bottoms compositions in a binary distillation column can be approximately related by

$$\frac{x_D/(1-x_D)}{x_B/(1-x_B)} = S$$

where S is a separation factor. At total reflux it is equal to α^{N_T+1} where α is the relative volatility and N_T is the number of theoretical trays. Assuming S is a constant, derive the nonlinear steadystate relationship showing how distillate drawoff rate D must be manipulated, as feed rate F and feed composition z vary, in order to hold distillate composition x_D constant. Sketch this relationship for several values of S and x_D.

11.5. Make root locus plots of first- and second-order openloop unstable processes with PI feedback controllers.

11.6. Find the closedloop stability requirements for a third-order openloop unstable process with a proportional controller:

$$B_{(s)} G_{M(s)} = \frac{K_c K_p}{(\tau_{p1}s + 1)(\tau_{p2}s + 1)(\tau_{p3}s - 1)}$$

11.7. Find the value of feedback controller gain K_c that gives a closedloop system with a damping coefficient of 0.707 for a second-order openloop unstable process with $\tau_{p2} > \tau_{p1}$:

$$B_{(s)} G_{M(s)} = \frac{K_p K_c}{(\tau_{p1}s + 1)(\tau_{p2}s - 1)}$$

11.8. What is the ultimate gain and period of the system with a positive zero:

$$G_{M(s)} = \frac{-3s + 1}{(s + 1)(5s + 1)}$$

(a) With a proportional controller?
(b) With a PI controller for $\tau_I = 2$?

11.9. (a) Sketch the root locus plot of a system with an openloop transfer function

$$B_{(s)} G_{M(s)} = \frac{K_c}{(s + 1)(s + 5)(s - 0.5)}$$

(b) For what values of gain K_c is the system closedloop stable?
(c) Make a root locus plot of this system in the log s plane (see Prob. 10.10).

11.10. Design a feedforward controller for the two-heated-tank process considered in Example 10.1. The load disturbance is inlet feed temperature T_0.

11.11. Modify your feedforward controller design of Prob. 11.10 so that it can handle both feed temperature and feed flow rate changes and uses a feedback temperature controller to trim up the steam flow.

11.12. A "valve position controller" is used to minimize operating pressure in a distillation column. Assume that the openloop process transfer function between column pressure and cooling-water flow $G_M = P/F_w$ is known.
(a) Sketch a block diagram of the closedloop system.
(b) What is the closedloop characteristic equation of the system?

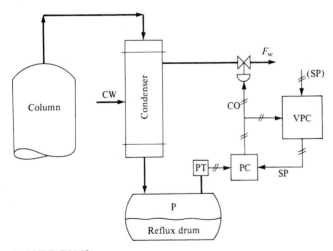

FIGURE P11.12

11.13. A proportional-only controller is used to control the liquid level in a tank by manipulating the outflow. It has been proposed that the steadystate offset of the proportional-only controller could be eliminated by using the combined feedforward-feedback system sketched below.

The flow rate into the tank is measured. The flow signal is sent through a first-order lag with time constant τ_F. The output of the lag is added to the output of the level controller. The sum of these two signals sets the outflow rate. Assume that the flow rate follows the setpoint signal to the flow controller exactly.

(a) Derive the closedloop transfer function between liquid level h and inflow rate F_0.

(b) Show that there is no steadystate offset of level from the setpoint for step changes in inflow rate.

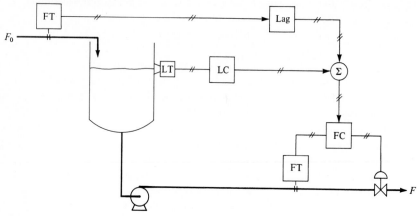

FIGURE P11.13

11.14. A process has a positive zero

$$G_M = \frac{-3s + 1}{(s + 1)(5s + 1)}$$

When a proportional-only feedback controller is used, the ultimate gain is 2. Outline your procedure for finding the optimum value of τ_D if a proportional-derivative controller is used. The optimum τ_D will give the maximum value for the ultimate gain.

11.15. Draw a block diagram of a process that has two manipulated variable inputs (M_1 and M_2) that each affect the output (X). A feedback controller B_1 is used to control X by manipulating M_1 since the transfer function between M_1 and X (G_{M1}) has a small time constant and small deadtime.

However, since M_1 is more expensive than M_2, we wish to minimize the long-term steadystate use of M_1. Therefore, a "valve position controller" B_2 is used to control M_1 at M_1^{set}. What is the closedloop characteristic equation of the system?

11.16. Make root locus plots for the two processes given below and calculate the ultimate gains and the gains which give closedloop damping coefficients of 0.707 for both processes.

(a) $BG_M = \dfrac{K_c}{(10s + 1)(50s + 1)}$

(b) $BG_M = \dfrac{K_c(-\tau s + 1)}{(10s + 1)(50s + 1)}$ with $\tau = 6$

For part b derive your expression for the ultimate gain as a general function of τ.

11.17. An openloop unstable process has a transfer function containing a positive pole at $+1/\tau$ and a negative pole at $-1/a\tau$. Its steadystate gain is unity. If a proportional-only controller is used, what is the value of a that gives a closedloop damping coefficient of 0.5 when the controller gain is ten times the minimum gain?

11.18. The "Smith Predictor" for deadtime compensation is a feedback controller that has been modified by feeding back the controller output into the controller input through a transfer function $F(s) = G_{MWD} - G_M$. The transfer function G_{MWD} is the portion of the process openloop transfer function G_M which does not contain the deadtime D. What is the closedloop characteristic equation of the system?

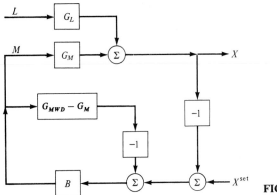

FIGURE P11.18

11.19. We want to analyze the floating pressure "valve position control" system proposed by Shinskey. Let the transfer function between cooling water flow and column pressure be

$$G_{M(s)} = \frac{0.1 \text{ psi/gpm}}{s(s + 1)}$$

The span of the pressure transmitter is 50 psi. The control valve has linear installed characteristics and passes 600 gpm when wide open. A proportional-only pressure controller is used.

$$B_1 = K_1$$

An integral-only valve position controller is used.

$$B_2 = \frac{K_2}{s}$$

(a) Draw a block diagram of this VPC closedloop system.

(b) What is the closedloop characteristic equation of this system?

(c) Considering only the pressure control loop, determine the value of the pressure controller gain K_1 that gives a closedloop damping coefficient of 0.707 for the pressure control loop. Sketch a root locus plot for the pressure control loop.

(d) Using the value of K_1 found above and with the pressure controller on automatic, determine the VPC tuning constant (K_2) that makes the entire VPC/pressure controller system critically damped, i.e., closedloop damping coefficient of unity. Sketch a root locus plot for the VPC controller with the pressure controller on automatic.

11.20. Two openloop transfer functions $G_{M1(s)}$ and $G_{M2(s)}$ are connected in parallel. They both have the same input $M_{(s)}$ but each has its own output: $X_{1(s)}$ and $X_{2(s)}$, respectively. In the closedloop system, a proportional controller K_1 is installed to control X_1 by changing M.

However, a cascade system is used where another proportional controller K_2 is used to control X_2 by changing the setpoint of the K_1 controller. Thus we have a parallel cascade system.

(a) Draw a block diagram of the system.

(b) Derive the closedloop transfer function between X_2 and X_2^{set}.

(c) What is the closedloop characteristic equation?

Suppose the openloop process transfer functions are

$$G_{M1(s)} = \frac{1}{(s + 1)^2} \qquad G_{M2(s)} = \frac{1}{5s + 1}$$

(d) Determine the value of K_1 that gives a closedloop damping coefficient of 0.5 in the secondary loop.

(e) Determine the ultimate gain in the primary K_2 loop when the secondary loop gain is $K_1 = 3$.

PART
V

FREQUENCY-DOMAIN DYNAMICS AND CONTROL

Our language lessons are coming along nicely. You should be fairly fluent in both English (differential equations) and Russian (Laplace transforms) by this time. We have found that only a small vocabulary is needed to handle our controller design problems. We must know the meaning and the pronunciation of only nine "words" in the two languages: first-order lag, first-order lead, dead-time, gain, second-order underdamped lag, integrator, derivative, positive zero, and positive pole.

We have found that dynamics can be more conveniently handled in the Russian transfer-function language than in the English ODE language. However, the manipulation of the algebraic equations becomes more and more difficult as the system becomes more complex and higher in order. If the system is Nth-order, an Nth-order polynomial in s must be factored into its N roots. For N greater than 2, we usually abandon analytical methods and turn to numerical root-solving techniques. Also, deadtime in the transfer function cannot be handled easily by the Russian Laplace-domain methods.

To overcome these problems, we must learn another language: Chinese. This is what we will call the frequency-domain methods. These methods are a little more removed from our mother tongue of English and a little more abstract. But they are extremely powerful and very useful in dealing with realistically complex processes. Basically this is because the manipulation of transfer functions becomes a problem of combining complex numbers numerically (addition, multiplication, etc.). This is easily done on a digital computer.

In Chap. 12 we will learn this new "Chinese" language (including several dialects: Bode, Nyquist, and Nichols plots). In Chap. 13 we will use frequency-domain methods to design closedloop feedback control systems. Finally, in Chap. 14, we will briefly discuss some frequency-domain and other methods for experimentally identifying a process.

As with Russian, you will have to learn only a limited Chinese vocabularly. The learning of 2000 to 3000 Chinese characters is *not* required (thank goodness)! We have to learn nine words in each of the three dialects. It takes a little practice to get the hang of handling the complex numbers in various coordinate systems, but let me assure you that the effort is well worth it.

CHAPTER
12

FREQUENCY-DOMAIN DYNAMICS

12.1 DEFINITION

The *frequency response* of most processes is defined as the steadystate behavior of the system when forced by a sinusoidal input. Suppose the input $m_{(t)}$ to the process is a sine wave $m_{s(t)}$ of amplitude $\bar{m}$ and frequency ω as shown in Fig. 12.1.

$$m_{s(t)} = \bar{m} \sin (\omega t) \qquad (12.1)$$

The period of one complete cycle is T units of time. Frequency is expressed in a variety of units. The electrical engineers usually use frequency in "hertz" (cycles per second):

$$\omega \text{ (hertz)} = \frac{1}{T} \qquad (12.2)$$

We chemical engineers find it more convenient to use frequency units of radians per time:

$$\omega \text{ (radians/time)} = \frac{2\pi}{T} \qquad (12.3)$$

The reason for this preference will become clear later in this chapter.

In a linear system, if the input is a sine wave with frequency ω, the output will also be a sine wave with the same frequency. The output will, however, have

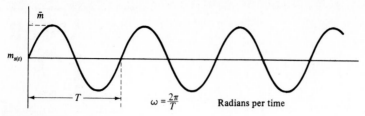

FIGURE 12.1
Sine-wave input.

a different amplitude and will lag (fall behind) or lead (rise ahead of) the input. Figure 12.2a shows the output $x_{s(t)}$ lagging the input $m_{s(t)}$ by T_x units of time. Figure 12.2b shows the output leading the input by T_x. The *phase angle* θ is defined as the angular difference between the input and the output. In equation form,

$$x_{s(t)} = \bar{x} \sin (\omega t + \theta) \tag{12.4}$$

where $x_{s(t)}$ = output resulting from the sine-wave input of frequency ω
   $\bar{x}$ = maximum amplitude of the output x_s
   θ = phase angle in radians

If the output lags the input, θ is negative. If the output leads the input, θ is positive.

$$\theta = \frac{T_x}{T} 2\pi \quad \text{(in radians)} \qquad \theta = \frac{T_x}{T} 360 \quad \text{(in degrees)} \tag{12.5}$$

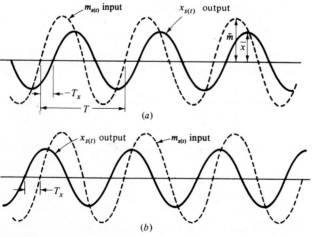

FIGURE 12.2
Sinusoidal input-output. (a) Output lags; (b) output leads.

The magnitude ratio MR is defined as the ratio of the maximum amplitude of the output over the maximum amplitude of the input:

$$\text{MR} \equiv \frac{\bar{x}}{\bar{m}} \qquad (12.6)$$

For a given process, both the phase angle θ and the magnitude ratio MR will change if frequency ω is changed. We must find out how θ and MR vary as ω goes from zero to infinity. Then we know the system's *frequency response* (its Chinese translation).

Different processes have different MR and θ dependence on ω. Since each process is unique, the frequency-response curves are like fingerprints. By merely looking at curves of MR and θ we can tell the kind of system (order and damping) and the values of parameters (time constants, steadystate gain, and damping coefficient).

There are a number of ways to obtain the frequency response of a process. Experimental methods, discussed in Chap. 14, are used when a mathematical model of the system is not available. If equations can be developed that adequately describe the system, the frequency response can be obtained directly from the system transfer function.

12.2 BASIC THEOREM

As we will show below, the frequency response of a system can be found by simply substituting $i\omega$ for s in the system transfer function $G_{(s)}$. Making the substitution $s = i\omega$ gives a complex number $G_{(i\omega)}$ that has the following:

1. A magnitude $|G_{(i\omega)}|$ that is the same as the magnitude ratio MR that would be obtained by forcing the system with a sine wave input of frequency ω
2. A phase angle or *argument*, arg $G_{(i\omega)}$, that is equal to the phase angle θ that would be obtained when forcing the system with a sine wave of frequency ω

$$|G_{(i\omega)}| = \text{MR}_{(\omega)} \qquad (12.7)$$

$$\arg G_{(i\omega)} = \theta_{(\omega)} \qquad (12.8)$$

$G_{(i\omega)}$ is a complex number, so it can be represented in terms of a real part and an imaginary part:

$$G_{(i\omega)} = \text{Re}\,[G_{(i\omega)}] + i\,\text{Im}\,[G_{(i\omega)}] \qquad (12.9)$$

In polar form, the complex number $G_{(i\omega)}$ is represented

$$G_{(i\omega)} = |\,G_{(i\omega)}\,|\,e^{i\,\arg\,G_{(i\omega)}} \qquad (12.10)$$

where

$$|G_{(i\omega)}| = \text{absolute value of } G_{(i\omega)} = \sqrt{(\text{Re}\,[G_{(i\omega)}])^2 + (\text{Im}\,[G_{(i\omega)}])^2} \quad (12.11)$$

$$\arg G_{(i\omega)} = \text{argument of } G_{(i\omega)} = \arctan\left(\frac{\text{Im}\,[G_{(i\omega)}]}{\text{Re}\,[G_{(i\omega)}]}\right) \qquad (12.12)$$

This very remarkable result [Eqs. (12.7) and (12.8)] permits us to go from the Laplace domain to the frequency domain with ease.

$$\text{Russian Laplace domain } G_{(s)} \xrightarrow{\; s=i\omega \;} \text{Chinese frequency domain } G_{(i\omega)}$$

Before we prove that this simple substitution is valid, let's illustrate its application in a specific example.

Example 12.1. Suppose we want to find the frequency response of a first-order process with the transfer function

$$G_{(s)} = \frac{K_p}{\tau_p s + 1} \tag{12.13}$$

Substituting $s = i\omega$ gives

$$G_{(i\omega)} = \frac{K_p}{1 + i\omega\tau_p} \tag{12.14}$$

Multiplying numerator and denominator by the complex conjugate of the denominator gives

$$G_{(i\omega)} = \frac{K_p}{1 + i\omega\tau_p} \frac{1 - i\omega\tau_p}{1 - i\omega\tau_p} = \frac{K_p(1 - i\omega\tau_p)}{1 + \omega^2\tau_p^2}$$

$$= \left[\frac{K_p}{1 + \omega^2\tau_p^2}\right] + i\left[\frac{-K_p\omega\tau_p}{1 + \omega^2\tau_p^2}\right] \tag{12.15}$$

Therefore

$$\text{Re}\,[G_{(i\omega)}] = \frac{K_p}{1 + \omega^2\tau_p^2} \tag{12.16}$$

$$\text{Im}\,[G_{(i\omega)}] = \frac{-K_p\omega\tau_p}{1 + \omega^2\tau_p^2} \tag{12.17}$$

Therefore

$$\text{MR} = |G_{(i\omega)}| = \sqrt{\left(\frac{K_p}{1 + \omega^2\tau_p^2}\right)^2 + \left(\frac{-K_p\omega\tau_p}{1 + \omega^2\tau_p^2}\right)^2}$$

$$= \frac{K_p}{\sqrt{1 + \omega^2\tau_p^2}} \tag{12.18}$$

$$\theta = \arg G_{(i\omega)} = \arctan\left[\frac{-K_p\omega\tau_p/(1 + \omega^2\tau_p^2)}{K_p/(1 + \omega^2\tau_p^2)}\right] = \arctan\,(-\omega\tau_p) \tag{12.19}$$

Notice that both MR and θ vary with frequency ω.

Now let us prove that this simple substitution $s = i\omega$ really works. Let $G_{(s)}$ be the transfer function of any arbitrary Nth-order system. The only restriction

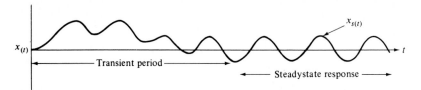

FIGURE 12.3
Response of a system to a sine-wave input.

we will place on the system is that it is stable. If it were unstable and we forced it with a sine wave input, the output would go off to infinity. So we cannot experimentally get the frequency response of an unstable system. This does *not* mean that we cannot use frequency-domain methods for openloop unstable systems. We will return to this subject in Chap. 13.

If the system is initially at rest (all derivatives equal zero) and we start to force it with a sine wave $m_{s(t)}$, the output $x_{(t)}$ will go through some transient period as shown in Fig. 12.3 and then settle down to a steady sinusoidal oscillation. In the Laplace domain, the output is by definition

$$X_{(s)} = G_{(s)} M_{(s)} \qquad (12.20)$$

For the sine-wave input $m_{(t)} = \bar{m} \sin(\omega t)$. Laplace-transforming,

$$M_{(s)} = \bar{m} \frac{\omega}{s^2 + \omega^2} \qquad (12.21)$$

Therefore the output with this sine-wave input is

$$X_{(s)} = G_{(s)} M_{(s)} = G_{(s)} \frac{\bar{m}\omega}{s^2 + \omega^2}$$

$G_{(s)}$ is a ratio of polynomials in s that can be factored into poles and zeros.

$$G_{(s)} = \frac{(s - z_1)(s - z_2)\cdots(s - z_M)}{(s - p_1)(s - p_2)\cdots(s - p_N)}$$

$$X_{(s)} = \frac{(s - z_1)(s - z_2)\cdots(s - z_M)}{(s - p_1)(s - p_2)\cdots(s - p_N)} \frac{\bar{m}\omega}{s^2 + \omega^2} \qquad (12.22)$$

$$= \frac{(s - z_1)(s - z_2)\cdots(s - z_M)}{(s - p_1)(s - p_2)\cdots(s - p_N)} \frac{\bar{m}\omega}{(s + i\omega)(s - i\omega)}$$

Expanding in partial fractions expansion gives

$$X_{(s)} = \frac{A}{s + i\omega} + \frac{B}{s - i\omega} + \frac{C}{s - p_1} + \cdots + \frac{W}{s - p_N} \qquad (12.23)$$

where

$$A = \lim_{s \to -i\omega} [(s + i\omega)X_{(s)}] = \lim_{s \to -i\omega} \left[\frac{\omega \bar{m} G_{(s)}}{s - i\omega} \right] = -\frac{\bar{m}}{2i} G_{(-i\omega)}$$

$$B = \lim_{s \to i\omega} [(s - i\omega)X_{(s)}] = \lim_{s \to i\omega} \left[\frac{\omega \bar{m} G_{(s)}}{s + i\omega} \right] = \frac{\bar{m}}{2i} G_{(i\omega)}$$

$$C = \lim_{s \to p_1} [(s - p_1)X_{(s)}]$$

Substituting into Eq. (12.23) and inverting to the time domain,

$$x_{(t)} = \left(\frac{-\bar{m}}{2i} G_{(-i\omega)} \right) e^{-i\omega t} + \left(\frac{\bar{m}}{2i} G_{(i\omega)} \right) e^{i\omega t} + \sum_{j=1}^{N} c_j e^{p_j t} \tag{12.24}$$

Now we are interested only in the steadystate response after the initial transients have died out and the system has settled into a sustained oscillation. As time goes to infinity, all the exponential terms in the summation shown in Eq. (12.24) decay to zero. The system is stable so all the poles p_j must be negative. The steadystate output with a sine-wave input, which we called $x_{s(t)}$, is

$$x_{s(t)} = \frac{\bar{m}}{2i} [G_{(i\omega)} e^{i\omega t} - G_{(-i\omega)} e^{-i\omega t}] \tag{12.25}$$

The $G_{(i\omega)}$ and $G_{(-i\omega)}$ terms are complex numbers and can be put into polar form:

$$G_{(i\omega)} = |G_{(i\omega)}| e^{i \text{ arg } G_{(i\omega)}}$$

$$G_{(-i\omega)} = |G_{(-i\omega)}| e^{i \text{ arg } G_{(-i\omega)}} = |G_{(i\omega)}| e^{-i \text{ arg } G_{(i\omega)}} \tag{12.26}$$

Equation (12.25) becomes

$$x_{s(t)} = \bar{m} |G_{(i\omega)}| \left[\frac{e^{i(\omega t + \text{arg } G_{(i\omega)})} - e^{-i(\omega t + \text{arg } G_{(i\omega)})}}{2i} \right]$$

Therefore

$$\frac{x_{s(t)}}{\bar{m}} = |G_{(i\omega)}| \sin(\omega t + \text{arg } G_{(i\omega)}) \tag{12.27}$$

Therefore we have proved what we set out to prove: (1) the magnitude ratio **MR** is the absolute value of $G_{(s)}$ with s set equal to $i\omega$ and (2) the phase angle is the argument of $G_{(s)}$ with s set equal to $i\omega$.

12.3   REPRESENTATION

There are three different kinds of plots that are commonly used to show how magnitude ratio (absolute magnitude) and phase angle (argument) vary with frequency ω. They are called *Nyquist*, *Bode* (pronounced "Bow-dee"), and *Nichols* plots. After defining what each of them is, we will show what some common transfer functions look like in the three different plots.

12.3.1 Nyquist Plots

A Nyquist plot (also called a *polar plot* or a *G-plane plot*) is generated by plotting the complex number $G_{(i\omega)}$ in a two-dimensional diagram whose ordinate is the imaginary part of $G_{(i\omega)}$ and whose abscissa is the real part of $G_{(i\omega)}$. The real and imaginary parts of $G_{(i\omega)}$ at a specific value of frequency ω_1 define a point in this coordinate system. As shown in Fig. 12.4a, either rectangular (real versus imaginary) or polar (absolute magnitude versus phase angle) can be used to locate the point.

As frequency is varied continuously from zero to infinity a curve is formed in the G plane, as shown in Fig. 12.4b. Frequency is thus a parameter along this curve. The shape and location of the curve are unique characteristics of the system.

Let us show what the Nyquist plots of some simple transfer functions look like.

A. FIRST-ORDER LAG

$$G_{(s)} = \frac{K_p}{\tau_p s + 1}$$

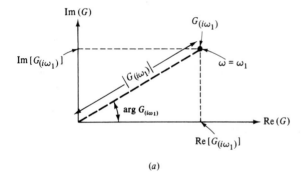

(a)

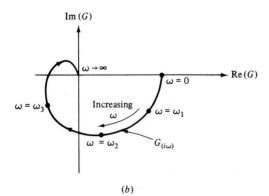

(b)

FIGURE 12.4
Nyquist plots in the G plane.
(a) Single point $G_{(i\omega_1)}$;
(b) complete curve $G_{(i\omega)}$.

We developed $G_{(i\omega)}$ for this transfer function in Example 12.1 [Eqs. (12.18) and (12.19)].

$$|G_{(i\omega)}| = \frac{K_p}{\sqrt{1 + \omega^2\tau_p^2}} \qquad \arg G_{(i\omega)} = \arctan(-\omega\tau_p) \qquad (12.28)$$

When frequency is zero, $|G|$ is equal to K_p and arg G is equal to zero. So the Nyquist plot starts ($\omega = 0$) on the positive real axis at Re $[G] = K_p$.

When frequency is equal to the reciprocal of the time constant ($\omega = 1/\tau_p$),

$$|G_{(i\omega)}| = \frac{K_p}{\sqrt{1 + (1/\tau_p)^2\tau_p^2}} = \frac{K_p}{\sqrt{2}}$$

$$\arg G_{(i\omega)} = \arctan[-(1/\tau_p)\tau_p] = -45° = -\frac{\pi}{4} \text{ radians}$$

The above illustrates why we use frequency in radians per time: there is a convenient relationship between the time constant of the process and frequency if these units are used.

As frequency goes to infinity, $|G_{(i\omega)}|$ goes to zero and arg $G_{(i\omega)}$ goes to $-90°$ or $-\pi/2$ radians. All of these points are shown in Fig. 12.5. The complete Nyquist plot is a semicircle. This is a unique curve. Anytime you see it, you know you are dealing with a first-order lag.

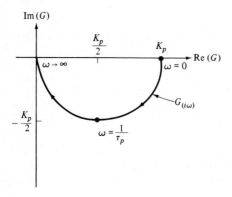

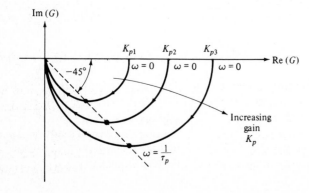

FIGURE 12.5
Nyquist plot of first-order lag.

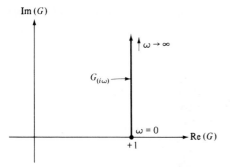

FIGURE 12.6
Nyquist plot for first-order lead.

The effect of changing the gain K_p is also shown in Fig. 12.5. The magnitude of each point is changed but the phase angle is not affected.

B. FIRST-ORDER LEAD

$$G_{(s)} = \tau_z s + 1 \qquad \Rightarrow \qquad G_{(i\omega)} = 1 + i\omega\tau_z$$

The real part is constant at $+1$. The imaginary part increases directly with frequency.

$$|G_{(i\omega)}| = \sqrt{1 + \omega^2\tau_z^2} \qquad \arg G_{(i\omega)} = \arctan(\omega\tau_z) \qquad (12.29)$$

When $\omega = 0$, $\arg G = 0$ and $|G| = 1$. As ω goes to infinity, $|G|$ becomes infinite and $\arg G$ goes to $+90°$ or $+\pi/2$ radians. The Nyquist plot is shown in Fig. 12.6.

C. DEADTIME

$$G_{(s)} = e^{-Ds} \qquad \Rightarrow \qquad G_{(i\omega)} = e^{-i\omega D}$$

This is a complex number with magnitude of one and argument equal to $-\omega D$.

$$|G_{(i\omega)}| = 1 \qquad \arg G_{(i\omega)} = -\omega D \qquad (12.30)$$

Thus deadtime changes the phase angle but has no effect on the magnitude. The magnitude is unity at all frequencies. The Nyquist plot is shown in Fig. 12.7. The curve moves around the unit circle as ω increases.

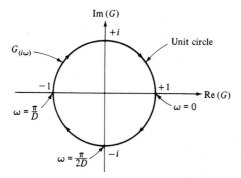

FIGURE 12.7
Nyquist plot for deadtime.

D. DEADTIME AND FIRST-ORDER LAG. Combining these two transfer functions gives

$$G_{(s)} = \frac{K_p e^{-Ds}}{\tau_p s + 1}$$

$$G_{(i\omega)} = \frac{K_p e^{-Di\omega}}{1 + i\omega\tau_p} = \left(\frac{K_p}{\sqrt{1 + \omega^2\tau_p^2}} \, e^{i \arctan (-\omega\tau_p)} \right) e^{-i\omega D}$$

$$= \frac{K_p}{\sqrt{1 + \omega^2\tau_p^2}} \, e^{i[\arctan (-\omega\tau_p) - D\omega]}$$

Therefore

$$|G_{(i\omega)}| = \frac{K_p}{\sqrt{1 + \omega^2\tau_p^2}} \qquad \arg G_{(i\omega)} = \arctan (-\omega\tau_p) - D\omega \qquad (12.31)$$

Note that the magnitude is exactly the same as the first-order lag alone. Phase angle is decreased by the deadtime contribution. Figure 12.8 shows that the Nyquist plot is a spiral that wraps around the origin as it shrinks in magnitude.

This example illustrates a very important property of complex numbers. The magnitude of the product of two complex numbers is the *product* of the magnitudes of each. The argument of the product of two complex numbers is the *sum* of arguments of each.

E. INTEGRATOR. The transfer function for a pure integrator is $G_{(s)} = 1/s$. Going into the frequency domain by substituting $s = i\omega$ gives

$$G_{(i\omega)} = \frac{1}{i\omega} = -\frac{1}{\omega} i$$

$G_{(i\omega)}$ is a pure imaginary number (its real part is zero), lying on the imaginary axis. It starts at minus infinity when ω is zero and goes to the origin as $\omega \to \infty$.

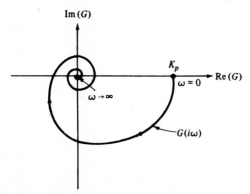

FIGURE 12.8
Nyquist plot for deadtime with first-order lag.

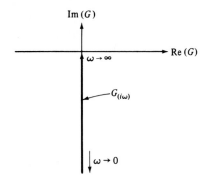

FIGURE 12.9
Nyquist plot for an integrator.

The Nyquist plot is sketched in Fig. 12.9.

$$|G_{(i\omega)}| = \frac{1}{\omega}$$

$$\arg G_{(i\omega)} = \arctan\left(\frac{-1/\omega}{0}\right) = -90° = -\frac{\pi}{2} \text{ radians} \qquad (12.32)$$

F. INTEGRATOR AND FIRST-ORDER LAG. Combining these two transfer functions gives

$$G_{(s)} = \frac{K_p}{s(\tau_p s + 1)}$$

$$G_{(i\omega)} = \frac{K_p}{-\omega^2 \tau_p + i\omega} = \frac{-K_p \tau_p \omega - K_p i}{\omega(\omega^2 \tau_p^2 + 1)}$$

$$|G_{(i\omega)}| = \frac{K_p}{\omega\sqrt{1 + \omega^2 \tau_p^2}} \qquad \arg G_{(i\omega)} = \arctan\left(\frac{-1}{-\omega\tau_p}\right) \qquad (12.33)$$

The Nyquist curve is shown in Fig. 12.10. Note that the results given in Eq. (12.33) could have been derived by combining the magnitudes and arguments of an integrator [Eq. (12.32)] and a first-order lag [Eq. (12.28)].

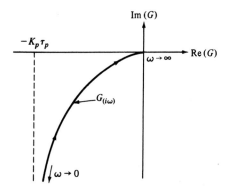

FIGURE 12.10
Nyquist plot of an integrator and first-order lag.

G. SECOND-ORDER UNDERDAMPED SYSTEM. This is probably the most important transfer function that we need to translate into the frequency domain. Since we often design for a desired closedloop damping coefficient, we need to know what the Nyquist plot of such a system looks like.

$$G_{(s)} = \frac{K_p}{\tau_p^2 s^2 + 2\zeta\tau_p s + 1}$$

$$G_{(i\omega)} = \frac{K_p}{(1 - \tau_p^2\omega^2) + i(2\zeta\tau_p\omega)}$$

$$= \frac{K_p(1 - \tau_p^2\omega^2) - iK_p(2\zeta\tau_p\omega)}{(1 - \tau_p^2\omega^2)^2 + (2\zeta\tau_p\omega)^2}$$

(12.34)

$$|G_{(i\omega)}| = \frac{K_p}{\sqrt{(1 - \tau_p^2\omega^2)^2 + (2\zeta\tau_p\omega)^2}}$$

$$\arg G_{(i\omega)} = \arctan\left(\frac{-2\zeta\tau_p\omega}{1 - \tau_p^2\omega^2}\right)$$

(12.35)

Figure 12.11 shows the Nyquist plot. It starts at K_p on the positive real axis. It intersects the imaginary axis (arg $G = -\pi/2$) when $\omega = 1/\tau_p$. At this point, $|G| = K_p/2\zeta$. Therefore the smaller the damping coefficient, the farther out on the negative imaginary axis the curve will cross. This shape is unique to an underdamped system. Anytime you see it you know the damping coefficient must be less than unity. As ω goes to infinity, the magnitude goes to zero and the phase angle goes to $-\pi$ radians ($-180°$).

The results of the examples above show that adding lags (poles) to the transfer function moves the Nyquist plot clockwise around the origin in the G plane. Adding leads (zeros) moves it counterclockwise. We will return to this generalization in the next chapter when we start designing controllers that shift these curves in the desired way.

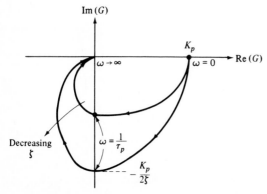

FIGURE 12.11
Nyquist plot of a second-order system.

12.3.2 Bode Plots

The Nyquist plot discussed in the previous section presents all the frequency information in a compact, one-curve form. Bode plots require that two curves be plotted instead of one. This increase is well worth the trouble because complex transfer functions can be handled much more easily using Bode plots. The two curves show how magnitude ratio and phase angle (argument) vary with frequency.

Phase angle is usually plotted against the log of frequency, using semilog graph paper as illustrated in Fig. 12.12. The magnitude ratio is sometimes plotted against the log of frequency on a log-log plot. However, usually it is more convenient to convert magnitude to *log modulus* defined by the equation

$$L \equiv \text{log modulus} \equiv 20 \log_{10} |G_{(i\omega)}| \qquad (12.36)$$

Then semilog graph paper can be used to plot both phase angle and log modulus versus the log of frequency, as shown in Fig. 12.12. There are very practical reasons for using these kinds of graphs, as we will find out shortly.

The units of log modulus are decibels (dB), a term originally used in communications engineering to indicate the ratio of two values of power. Figure 12.13 is convenient to use to convert back and forth from magnitude to decibels.

Now let us look at the Bode plots of some common transfer functions. We have already calculated the magnitudes and phase angles for most of them in the previous section. The job now is to plot them in this new coordinate system.

A. GAIN. If $G_{(s)}$ is just a constant K_p, $|G_{(i\omega)}| = K_p$ and phase angle = arg $G_{(i\omega)} = 0$. Neither magnitude nor phase angle vary with frequency. The log modulus is

$$L = 20 \log_{10} |K_p| \qquad (12.37)$$

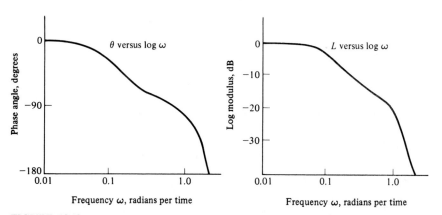

FIGURE 12.12
Bode plots of phase angle and log modulus versus the logarithm of frequency.

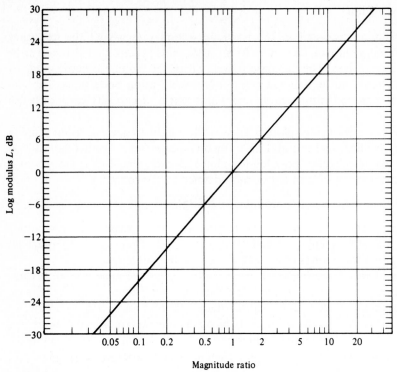

FIGURE 12.13
Conversion between magnitude ratio and log modulus.

Both the phase angle and log modulus curves are horizontal lines on a Bode plot, as shown in Fig. 12.14.

If K_p is less than 1, L is negative. If K_p is unity, L is zero. If K_p is greater than 1, L is positive. Increasing K_p by a factor of 10 (a "decade") increases L by a factor of 20 dB. Increasing K_p moves the L curve up in the Bode plot.

B. FIRST-ORDER LAG

$$G_{(s)} = \frac{1}{\tau_p s + 1}$$

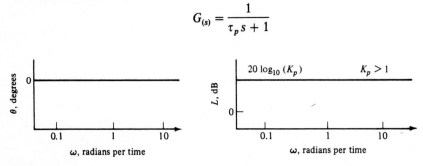

FIGURE 12.14
Bode plots of gain.

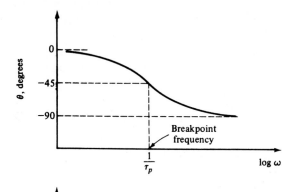

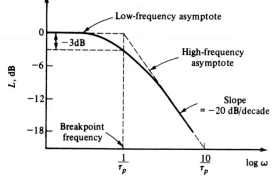

FIGURE 12.15
Bode plots for first-order lag.

From Eq. (12.28) (with $K_p = 1$)

$$|G_{(i\omega)}| = \frac{1}{\sqrt{1 + \omega^2 \tau_p^2}} \qquad \arg G_{(i\omega)} = \arctan\,(-\omega\tau_p) \qquad (12.38)$$

$$L = 20 \log_{10} \frac{1}{\sqrt{1 + \omega^2 \tau_p^2}} = -10 \log_{10} [1 + \omega^2 \tau_p^2] \qquad (12.39)$$

The Bode plots are shown in Fig. 12.15. One of the most convenient features of Bode plots is that the L curves can be easily sketched by considering the low- and high-frequency asymptotes. As ω goes to zero, L goes to zero. As ω becomes very large, Eq. (12.39) reduces to

$$\lim_{\omega \to \infty} L = -10 \log_{10} [\omega^2 \tau_p^2] = -20 \log_{10} (\omega) - 20 \log_{10} (\tau_p) \qquad (12.40)$$

This is the equation of a straight line of L versus log ω. It has a slope of -20. L will decrease 20 dB as log ω increases by 1 (or as ω increases by 10, a decade). Therefore the slope of the high-frequency asymptote is -20 dB/decade.

The high-frequency asymptote intersects the $L = 0$ line at $\omega = 1/\tau_p$. This is called the *breakpoint frequency*. The log modulus is "flat" (horizontal) out to this point and then begins to drop off.

Thus the L curve can be easily sketched by drawing a line with slope of -20 dB/decade from the breakpoint frequency on the $L = 0$ line. Notice also that the phase angle is $-45°$ at the breakpoint frequency, which is the reciprocal of the time constant. The L curve has a value of -3 dB at the breakpoint frequency, as shown in Fig. 12.15.

C. FIRST-ORDER LEAD

$$G_{(s)} = \tau_z s + 1 \qquad G_{(i\omega)} = 1 + i\omega\tau_z$$

From Eq. (12.29)

$$|G_{(i\omega)}| = \sqrt{1 + \omega^2\tau_z^2} \qquad \arg G_{(i\omega)} = \arctan(\omega\tau_z)$$
$$L = 20 \log_{10} \sqrt{1 + \omega^2\tau_z^2} \tag{12.41}$$

These curves are shown in Fig. 12.16. The high-frequency asymptote has a slope of $+20$ dB/decade. The breakpoint frequency is $1/\tau_z$. The phase angle goes from zero to $+90°$ and is $+45°$ at $\omega = 1/\tau_z$. Thus a lead contributes positive phase angle. A lag contributes negative phase angle. A gain doesn't change the phase angle.

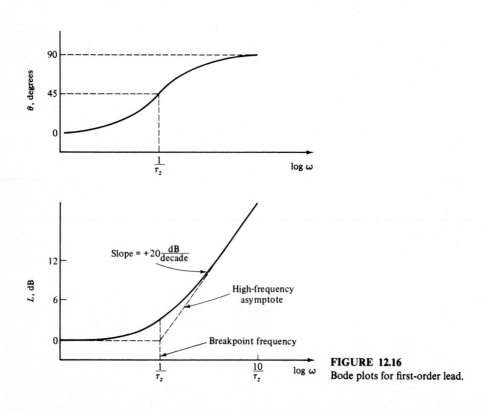

FIGURE 12.16
Bode plots for first-order lead.

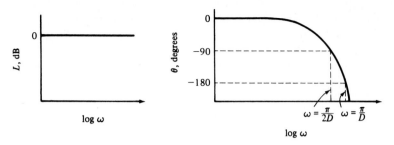

FIGURE 12.17
Bode plots for deadtime.

D. DEADTIME

$$G_{(i\omega)} = e^{-iD\omega}$$

$$L = 20 \log_{10} |G_{(i\omega)}| = 20 \log_{10} (1) = 0 \qquad (12.42)$$

$$\arg G_{(i\omega)} = -\omega D \qquad (12.43)$$

As shown in Fig. 12.17, the deadtime transfer function has a flat L at 0 dB curve for all frequencies, but the phase angle drops off to minus infinity. The phase angle is down to $-180°$ when the frequency is π/D. So the bigger the deadtime, the lower the frequency at which the phase angle drops off rapidly.

E. nth POWER OF s. Under this general category are included a differentiator ($n = +1$) and an integrator ($n = -1$).

$$G_{(s)} = s^n \qquad n = \pm 1, \pm 2, \dots \qquad G_{(i\omega)} = \omega^n i^n$$

$$L = 20 \log_{10} (\omega^n) = 20n \log_{10} \omega \qquad (12.44)$$

The L curve is a straight line in the L-log ω plane (Bode plot) with a slope of $20n$. See Fig. 12.18.

$$\arg G_{(i\omega)} = \arctan \left(\frac{\text{Re } (G)}{\text{Im } (G)} \right)$$

If n is odd, $G_{(i\omega)}$ is a pure imaginary number and the phase angle is the arctan of infinity ($\theta = \pm 90°, \pm 270°, \dots$). If n is even, $G_{(i\omega)}$ is a real number and the phase angle is the arctan of zero ($\theta = \pm 0°, \pm 180°, \dots$). Therefore phase angle changes by $90°$ or $\pi/2$ radians for each successive integer value of n.

$$\arg G_{(i\omega)} = n \left(\frac{\pi}{2} \right) \text{ radians} \qquad (12.45)$$

The important specific values of n are:

1. $n = 1$. This is the transfer function of an ideal derivative.

$$G_{(i\omega)} = i\omega \qquad \Rightarrow \qquad L = 20 \log_{10} \omega \qquad (12.46)$$

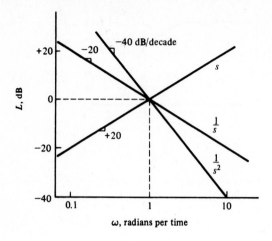

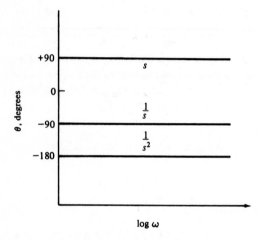

FIGURE 12.18
Bode plots for s^n.

This is a straight line with a slope of $+20$ dB/decade.

$$\theta = \arctan\left(\frac{\omega}{0}\right) = +90° \qquad (12.47)$$

So an ideal derivative *increases* phase angle by 90°.

2. $n = -1$. This is the transfer function of an integrator.

$$G_{(i\omega)} = \frac{1}{i\omega} = \left(-\frac{1}{\omega}\right)i \qquad \Rightarrow \qquad L = -20\log_{10}\omega \qquad (12.48)$$

This is a straight line with a slope of -20 dB/decade.

$$\arg G_{(i\omega)} = \arctan\left(\frac{-1/\omega}{0}\right) = -90° \qquad (12.49)$$

Thus an integrator *decreases* phase angle by 90°.

3. $n = -2$. Two integrators in series would produce a straight-line L curve with a slope of -40 dB/decade and a straight-line θ curve at $-180°$.

F. SECOND-ORDER UNDERDAMPED LAG. Equations (12.34) and (12.35) give (with $K_p = 1$)

$$|G_{(i\omega)}| = \frac{1}{\sqrt{(1 - \tau_p^2\omega^2)^2 + (2\zeta\tau_p\omega)^2}} \qquad \arg G_{(i\omega)} = \arctan\left(\frac{-2\zeta\tau_p\omega}{1 - \tau_p^2\omega^2}\right)$$

$$L = 20 \log_{10}\left[\frac{1}{\sqrt{(1 - \tau_p^2\omega^2)^2 + (2\zeta\tau_p\omega)^2}}\right] \qquad (12.50)$$

Figure 12.19 shows the Bode plots for several values of damping coefficient ζ. The breakpoint frequency is the reciprocal of the time constant. The high-frequency

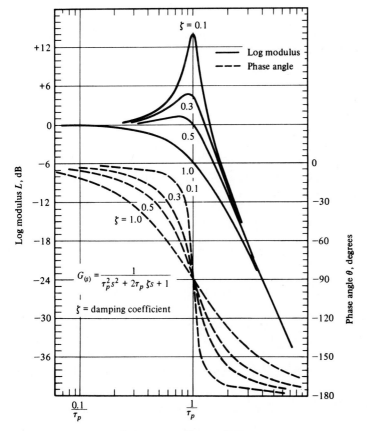

FIGURE 12.19
Second-order system Bode plots.

asymptote has a slope of -40 dB/decade.

$$\lim_{\omega \to \infty} L = 20 \log_{10} \left(\frac{1}{\tau_p^2 \omega^2} \right) = -40 \log_{10} (\omega \tau_p)$$

Note the very unique shape of the log modulus curves in Fig. 12.19. The lower the damping coefficient, the higher the peak in the L curve. A damping coefficient of about 0.4 gives a peak of about $+2$ dB. We will use this property extensively in our tuning of feedback controllers. We will adjust the controller gain to give a maximum peak of $+2$ dB in the log modulus curve for the closedloop servo transfer function X/X^{set}.

G. GENERAL TRANSFER FUNCTIONS IN SERIES.
The historical reason for the widespread use of Bode plots is that, before the use of computers, they made it possible to handle complex processes fairly easily. A complex transfer function can be broken down into its simple elements: leads, lags, gains, deadtimes, etc. Then each of these is plotted on the same Bode plots. Finally the total complex transfer function is obtained by adding the individual log modulus curves and the individual phase curves at each value of frequency.

Consider a general transfer function $G_{(s)}$ that can be broken up into two simple transfer functions $G_{1(s)}$ and $G_{2(s)}$:

$$G_{(s)} = G_{1(s)} G_{2(s)}$$

In the frequency domain

$$G_{(i\omega)} = G_{1(i\omega)} G_{2(i\omega)} \tag{12.51}$$

Each of the G's is a complex number and can be expressed in polar form:

$$G_{1(i\omega)} = |G_{1(i\omega)}| e^{i \text{ arg } G_{1(i\omega)}}$$

$$G_{2(i\omega)} = |G_{2(i\omega)}| e^{i \text{ arg } G_{2(i\omega)}}$$

$$G_{(i\omega)} = |G_{(i\omega)}| e^{i \text{ arg } G_{(i\omega)}}$$

Combining

$$G_{(i\omega)} = |G_{1(i\omega)}| e^{i \text{ arg } G_{1(i\omega)}} |G_{2(i\omega)}| e^{i \text{ arg } G_{2(i\omega)}}$$

$$|G_{(i\omega)}| e^{i \text{ arg } G_{(i\omega)}} = |G_{1(i\omega)}| \, \| G_{2(i\omega)}| e^{i [\text{arg } G_{1(i\omega)} + \text{arg } G_{2(i\omega)}]} \tag{12.52}$$

Taking the logarithm of both sides gives

$$\ln |G| + i \text{ arg } G = \ln |G_1| + \ln|G_2| + i (\text{arg } G_1 + \text{arg } G_2)$$

Therefore the log modulus curves and phase-angle curves of the individual components are simply added at each value of frequency to get the total L and θ curves for the complex transfer function.

$$20 \log_{10} |G| = 20 \log_{10} |G_1| + 20 \log_{10} |G_2| \tag{12.53}$$

$$\text{arg } G = \text{arg } G_1 + \text{arg } G_2 \tag{12.54}$$

Example 12.2. Consider the transfer function $G_{(s)}$:

$$G_{(s)} = \frac{1}{(\tau_{p1}s + 1)(\tau_{p2}s + 1)}$$

Bode plots of the individual transfer functions G_1 and G_2 are sketched in Fig. 12.20 and added to give $G_{(i\omega)}$

$$G_{1(s)} = \frac{1}{\tau_{p1}s + 1} \qquad G_{2(s)} = \frac{1}{\tau_{p2}s + 1}$$

Note that the total phase angle drops down to $-180°$ and the slope of the high-frequency asymptote of the log modulus line is -40 dB/decade since the process is net second-order.

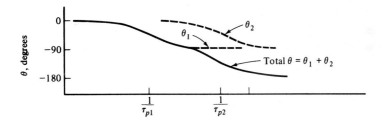

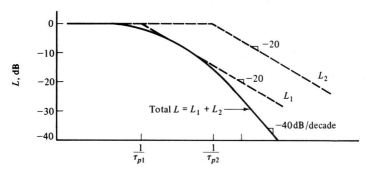

FIGURE 12.20
Bode plots for two lags.

Example 12.3. Bode plots for the transfer function

$$G_{(s)} = \frac{\tau_z s + 1}{(\tau_{p1}s + 1)(\tau_{p2}s + 1)}$$

are sketched in Fig. 12.21. Note that the phase angle goes to $-90°$ and the slope of the log modulus line is -20 dB/decade at high frequencies because the system is net first-order ($M = 1$ and $N = 2$).

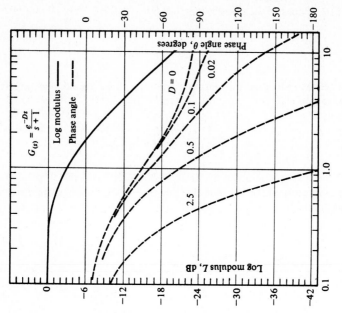

$$G_{(s)} = \frac{e^{-Ds}}{s+1}$$

Log modulus

Phase angle

$D = 0$

0.02

0.1

0.5

2.5

Frequency ω, radians per time

FIGURE 12.22
Bode plots for lag and deadtime.

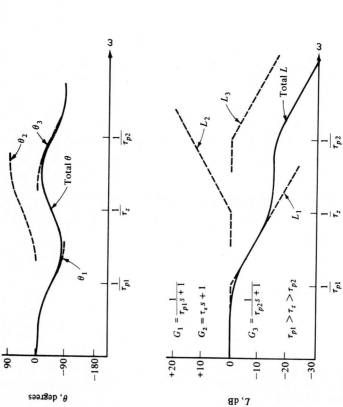

θ_2

θ_3

Total θ

θ_1

L_2

L_3

Total L

L_1

$$G_1 = \frac{1}{\tau_{p1}s+1}$$

$$G_2 = \tau_z s + 1$$

$$G_3 = \frac{1}{\tau_{p2}s+1}$$

$$\tau_{p1} > \tau_z > \tau_{p2}$$

FIGURE 12.21
Bode plots for two lags and one lead.

Example 12.4. Figure 12.22 gives Bode plots for a transfer containing a first-order lag and deadtime.

$$G_{(s)} = \frac{e^{-Ds}}{s+1}$$

Several different values of deadtime D are shown. Only the phase-angle curve is changed as D changes. The larger the value of D, the lower the frequency at which the phase angle drops to $-180°$. As we will learn in Chap. 13, the lower the phase angle curve, the poorer the control. We will show quantitatively in Chap. 13 how increasing deadtime degrades feedback control performance. Note that the L curve is the same for all values of D.

Up to this point we have stressed graphical methods for quickly sketching Bode plots. Since the combining of simple transfer functions is just numerical manipulation of complex numbers, a digital computer can be easily programmed to generate any of the desired forms of the complex number: real and imaginary parts, magnitude, log modulus, and phase angle.

Table 12.1 gives a FORTRAN program that calculates the frequency response of several simple systems. Figure 12.23 gives Bode plots for four different transfer functions.

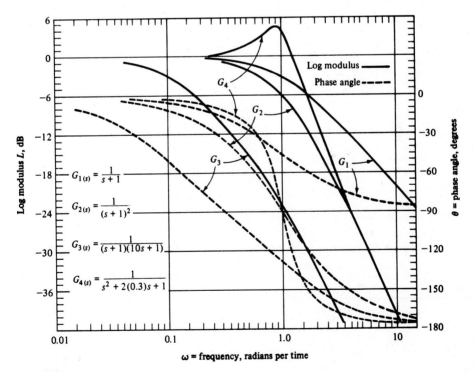

FIGURE 12.23
Bode plots of several systems.

TABLE 12.1

Digital program to calculate frequency response

```
      DIMENSION G(4),DB(4),DEG(4)
      COMPLEX G
      WRITE(6,1)
      WRITE(6,2)
    1 FORMAT(5X,'FREQ REG1 IMG1  REG2 IMG2  REG3 IMG3 REG4 IMG4')
    2 FORMAT(5X,'   DBG1 DEGG1 DGG2 DEGG2 DBG3 DEGG3 DBG4 DEG4')
C   W IS FREQUENCY IN RADIANS PER MINUTE
      W=0.01
      DW=10.**(0.1)
      DO 100 I=1,31
C   G1 IS FIRST ORDER LAG WITH TAU = 1
      G(1)=1./CMPLX(1.,1.0*W)
C   G2 IS TWO FIRST-ORDER LAGS WITH TAU'S = 1
      G(2)=G(1)*G(1)
C   G3 IS TWO FIRST-ORDER LAGS WITH TAU'S = 1 AND 10
      G(3)=G(1)/CMPLX(1.,10.*W)
C   G4 IS SECOND-ORDER UNDERDAMPED LAG WITH TAU = 1 AND ZETA = 0.3
      G(4)=1./CMPLX(1.-W**2,2.*W*0.3)
      DO 10 J=1,4
   10 DB(J)=20.*ALOG10(CABS(G(J)))
      DEG(1)=ATAN(-W)*180./3.1416
      DEG(2)=DEG(1)+DEG(1)
      DEG(3)=DEG(1)+ATAN(-10.*W)*180./3.1416
      DEG(4)=ATAN2(-2.*W*0.3,1.-W**2)*180./3.1416
      WRITE(6,4)W,G
      WRITE(6,3)(DB(J),DEG(J),J=1,4)
    4 FORMAT(1X,9F7.3)
    3 FORMAT(10X,8F7.1)
  100 W=W*DW
      STOP
      END
```

Results

```
 FREQ  REG1  IMG1   REG2  IMG2  REG3  IMG3  REG4 IMG4
       DBG1 DEGG1 DGG2 DEGG2 DBG3 DEGG3 DBG4 DEG4

0.010 1.000 -0.010 1.000 -0.020 0.989 -0.109 1.000 -0.006
        0.0  -0.6   0.0  -1.1   0.0  -6.3   0.0  -0.3
0.013 1.000 -0.013 1.000 -0.025 0.983 -0.136 1.000 -0.008
        0.0  -0.7   0.0  -1.4  -0.1  -7.9   0.0  -0.4
0.016 1.000 -0.016 0.999 -0.032 0.973 -0.170 1.000 -0.010
        0.0  -0.9   0.0  -1.8  -0.1  -9.9   0.0  -0.5
0.020 1.000 -0.020 0.999 -0.040 0.958 -0.211 1.000 -0.012
        0.0  -1.1   0.0  -2.3  -0.2 -12.4   0.0  -0.7
0.025 0.999 -0.025 0.998 -0.050 0.934 -0.260 1.000 -0.015
        0.0  -1.4   0.0  -2.9  -0.3 -15.5   0.0  -0.9
0.032 0.999 -0.032 0.997 -0.063 0.899 -0.316 1.001 -0.019
        0.0  -1.8   0.0  -3.6  -0.4 -19.4   0.0  -1.1
0.040 0.998 -0.040 0.995 -0.079 0.848 -0.377 1.001 -0.024
        0.0  -2.3   0.0  -4.6  -0.6 -24.0   0.0  -1.4
```

TABLE 12.1 (*continued*)

```
0.050  0.997 -0.050  0.992 -0.100  0.777 -0.440  1.002 -0.030
       0.0   -2.9    0.0   -5.7   -1.0  -29.5    0.0   -1.7
0.063  0.996 -0.063  0.988 -0.125  0.684 -0.494  1.003 -0.038
       0.0   -3.6    0.0   -7.2   -1.5  -35.9    0.0   -2.2
0.079  0.994 -0.079  0.981 -0.157  0.571 -0.532  1.004 -0.048
       0.0   -4.5   -0.1   -9.1   -2.2  -43.0    0.0   -2.7
0.100  0.990 -0.099  0.970 -0.196  0.446 -0.545  1.006 -0.061
       0.0   -5.7   -0.1  -11.4   -3.1  -50.7    0.1   -3.5
0.126  0.984 -0.124  0.954 -0.244  0.320 -0.527  1.010 -0.078
      -0.1   -7.2   -0.1  -14.4   -4.2  -58.7    0.1   -4.4
0.158  0.975 -0.155  0.928 -0.302  0.208 -0.484  1.016 -0.099
      -0.1   -9.0   -0.2  -18.0   -5.6  -66.8    0.2   -5.6
0.200  0.962 -0.192  0.888 -0.369  0.116 -0.424  1.026 -0.128
      -0.2  -11.3   -0.3  -22.6   -7.1  -74.7    0.3   -7.1
0.251  0.941 -0.236  0.829 -0.445  0.047 -0.356  1.040 -0.167
      -0.3  -14.1   -0.5  -28.2   -8.9  -82.4    0.5   -9.1
0.316  0.909 -0.287  0.744 -0.523  0.000 -0.287  1.064 -0.224
      -0.4  -17.5   -0.8  -35.1  -10.8  -90.0    0.7  -11.9
0.398  0.863 -0.344  0.627 -0.593 -0.030 -0.224  1.100 -0.312
      -0.6  -21.7   -1.3  -43.4  -12.9  -97.6    1.2  -15.8
0.501  0.799 -0.401  0.478 -0.640 -0.046 -0.169  1.150 -0.462
      -1.0  -26.6   -1.9  -53.2  -15.1 -105.3    1.9  -21.9
0.631  0.715 -0.451  0.308 -0.646 -0.052 -0.122  1.190 -0.749
      -1.5  -32.3   -2.9  -64.5  -17.6 -113.2    3.0  -32.2
0.794  0.613 -0.487  0.139 -0.597 -0.051 -0.084  1.016 -1.312
      -2.1  -38.5   -4.2  -76.9  -20.2 -121.3    4.4  -52.2
1.000  0.500 -0.500  0.000 -0.500 -0.045 -0.054  0.000 -1.667
      -3.0  -45.0   -6.0  -90.0  -23.1 -129.3    4.4  -90.0
1.259  0.387 -0.487 -0.088 -0.377 -0.036 -0.034 -0.641 -0.828
      -4.1  -51.5   -8.2 -103.1  -26.2 -137.0    0.4 -127.8
1.585  0.285 -0.451 -0.123 -0.257 -0.027 -0.020 -0.474 -0.298
      -5.5  -57.7  -10.9 -115.5  -29.5 -144.1   -5.0 -147.8
1.995  0.201 -0.401 -0.120 -0.161 -0.020 -0.011 -0.289 -0.116
      -7.0  -63.4  -13.9 -126.8  -33.0 -150.5  -10.1 -158.1
2.512  0.137 -0.344 -0.099 -0.094 -0.013 -0.006 -0.174 -0.049
      -8.6  -68.3  -17.3 -136.6  -36.6 -156.0  -14.8 -164.2
3.162  0.091 -0.287 -0.074 -0.052 -0.009 -0.003 -0.106 -0.022
     -10.4  -72.5  -20.8 -144.9  -40.4 -160.6  -19.3 -168.1
3.981  0.059 -0.236 -0.052 -0.028 -0.006 -0.002 -0.066 -0.011
     -12.3  -75.9  -24.5 -151.8  -44.3 -164.5  -23.5 -170.9
5.012  0.038 -0.192 -0.035 -0.015 -0.004 -0.001 -0.041 -0.005
     -14.2  -78.7  -28.3 -157.4  -48.2 -167.6  -27.7 -172.9
6.310  0.025 -0.155 -0.023 -0.008 -0.002  0.000 -0.026 -0.002
     -16.1  -81.0  -32.2 -162.0  -52.1 -170.1  -31.8 -174.4
7.943  0.016 -0.124 -0.015 -0.004 -0.002  0.000 -0.016 -0.001
     -18.1  -82.8  -36.1 -165.6  -56.1 -172.1  -35.9 -175.6
10.000 0.010 -0.099 -0.010 -0.002 -0.001  0.000 -0.010 -0.001
     -20.0  -84.3  -40.1 -168.6  -60.0 -173.7  -39.9 -176.5
```

The variables must be declared COMPLEX at the beginning of the program. Note in the program how the phase angles are calculated by summing up the arguments of each of the individual components. This is particularly useful when there is a deadtime element in the transfer function. If you try to calculate the phase angle from the final total complex number, the FORTRAN subroutines ATAN and ATAN2 cannot determine what quadrant you are in. ATAN only has one argument and therefore can only track the complex number in the first or fourth quadrants. So phase angles between $-90°$ and $-180°$ will be reported as $+90°$ to $0°$. The subroutine ATAN2, since it has two arguments (the imaginary and the real part of the number), can accurately track the phase angle between $+180°$ and $-180°$, but not beyond. Getting the phase angle by summing the angles of the components eliminates all these problems.

A complex number G always has two parts: a real part and an imaginary part. These parts can be specified by using the statement

$$G = CMPLX(X,Y)$$

where G = complex number
   X = real part of G
   Y = imaginary part of G

Complex numbers can be added $(G = G1 + G2)$, multiplied $(G = G1*G2)$ and divided $(G = G1/G2)$. The magnitude of a complex number can be found by using the statement

$$XX = CABS(G)$$

where XX = the real number that is the magnitude of G

Knowing the complex number G, its real and imaginary parts can be found by using the statements

$$X = REAL(G) \qquad Y = AIMAG(G)$$

A deadtime element $G_{(s)} = e^{-Ds}$

$$G_{(i\omega)} = e^{-iD\omega} = \cos (D\omega) - i \sin (D\omega) \qquad (12.55)$$

can be calculated

$$G = CMPLX(COS(D*W), -SIN(D*W))$$

where D = deadtime
   W = frequency in radians per minute in the FORTRAN program

The program in Table 12.1 illustrates the use of some of these complex FORTRAN statements.

Commercial software can also be used to generate these frequency response curves (CC, CONSYD, MATRIX-X).

12.3.3 Nichols Plots

The final plot that we need to learn how to make is called a Nichols plot. It is a single curve in a coordinate system with phase angle as the abscissa and log modulus as the ordinate. Frequency is a parameter along the curve. Figure 12.24 gives Nichols plots of some simple transfer functions.

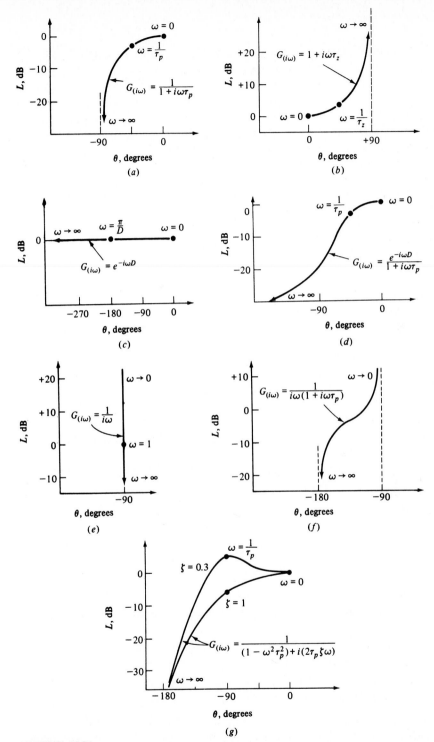

FIGURE 12.24
Nichols plots. (a) First-order lag; (b) first-order lead; (c) deadtime; (d) deadtime and lag; (e) integrator; (f) integrator and lag; (g) second-order underdamped lag.

441

At this point you may be asking why we need another type of plot. After all, both the Nyquist and the Bode plots are simply different ways to plot complex numbers. Well, as we will see in Chap. 13, all of these plots (Nichols, Bode, and Nyquist) are very useful for designing control systems. Each has its own individual application, so we have to learn all three of them. Keep in mind, however, that the main workhorse of our Chinese language is the Bode plot. We usually make it first since it is easy to construct from its individual simple elements. Then we use the Bode plot to sketch the Nyquist and Nichols plots.

12.4 FREQUENCY-DOMAIN SOLUTION TECHNIQUES

In the preceding sections of this chapter we assumed that the system transfer function $G_{(s)}$ was known. Then the simple substitution $s = i\omega$ into $G_{(s)}$ gave the frequency response of the system.

However, it is sometimes difficult to find $G_{(s)}$ analytically by solving all the algebraic equations in s that result when the linearized ODEs describing the system are Laplace-transformed. For example, the equations of the simple second-order nonisothermal CSTR of Example 9.6 got a little complicated when we tried to solve analytically for the eight transfer functions as explicit functions of s. As the system gets more and more complex and higher-order, analytical solution for the system transfer functions becomes practically impossible. For example, a typical distillation column may have three components and 50 trays. This gives a model with 150 ODEs and an analytical transfer function that would have a denominator polynomial that was 150th-order in s.

These large systems of equations can be solved fairly easily by going into the frequency domain. The procedure is:

1. The linear ODEs are Laplace-transformed.

2. Then $i\omega$ is immediately substituted for s.

3. A specific numerical value of frequency ω is chosen and substituted into the equations.

4. The algebraic equations, which are now in terms of complex variables, are solved numerically to obtain the desired transfer-function relationships. The $G_{(i\omega)}$'s will be complex numbers that are points on a Nyquist plot that correspond to the specific frequency chosen.

5. Another numerical value of ω is specified and step 4 is repeated. Picking a number of frequencies over the range of interest for the process gives the complete frequency-response curves.

Let us illustrate the procedure with two examples.

Example 12.5. Consider the nonisothermal CSTR described by the linearized ODEs

$$\frac{dC_A}{dt} = a_{11}C_A + a_{12}T + a_{13}C_{A0} + a_{15}F \tag{12.56}$$

$$\frac{dT}{dt} = a_{21}C_A + a_{22}T + a_{24}T_0 + a_{25}F + a_{26}T_J \tag{12.57}$$

where the a_{ij}'s are all constants. Laplace-transforming and substituting $s = i\omega$ give

$$(i\omega - a_{11})C_{A(i\omega)} = a_{12}T_{(i\omega)} + a_{13}C_{A0(i\omega)} + a_{15}F_{(i\omega)} \tag{12.58}$$

$$(i\omega - a_{22})T_{(i\omega)} = a_{21}C_{A(i\omega)} + a_{24}T_{0(i\omega)} + a_{25}F_{(i\omega)} + a_{26}T_{J(i\omega)} \tag{12.59}$$

All the variables are now complex numbers with real and imaginary parts.

$$C_{A(i\omega)} = \text{Re } [C_A] + i \text{ Im } [C_A] \equiv C_A^R + iC_A^I$$

$$T_{(i\omega)} = \text{Re } [T] + i \text{ Im } [T] \equiv T^R + iT^I$$

$$C_{A0(i\omega)} = \text{Re } [C_{A0}] + i \text{ Im } [C_{A0}] \equiv C_{A0}^R + iC_{A0}^I$$

$$F_{(i\omega)} = \text{Re } [F] + i \text{ Im } [F] \equiv F^R + iF^I$$

$$T_{0(i\omega)} = \text{Re } [T_0] + i \text{ Im } [T_0] \equiv T_0^R + iT_0^I$$

$$T_{J(i\omega)} = \text{Re } [T_J] + i \text{ Im } [T_J] \equiv T_J^R + iT_J^I$$

Picking a specific numerical value of frequency ω gives two algebraic complex-variable equations that must be rearranged to get the output variables $C_{A(i\omega)}$ and $T_{(i\omega)}$ in terms of the input variables $C_{A0(i\omega)}$, $T_{0(i\omega)}$, $F_{(i\omega)}$, and $T_{J(i\omega)}$.

$$C_{A(i\omega)} = G_{11(i\omega)}C_{A0(i\omega)} + G_{12(i\omega)}T_{0(i\omega)} + G_{13(i\omega)}F_{(i\omega)} + G_{14(i\omega)}T_{J(i\omega)} \tag{12.60}$$

$$T_{(i\omega)} = G_{21(i\omega)}C_{A0(i\omega)} + G_{22(i\omega)}T_{0(i\omega)} + G_{23(i\omega)}F_{(i\omega)} + G_{24(i\omega)}T_{J(i\omega)} \tag{12.61}$$

The solution of Eqs. (12.58) and (12.59) is made easier if we take one input variable at a time. First we set C_{A0} equal to 1 and all other input variables equal to zero.

$$C_{A0} = C_{A0}^R + iC_{A0}^I = 1 + i0 \tag{12.62}$$

$$T_0 = F = T_J = 0 + i0 \tag{12.63}$$

Now Eqs. (12.58) and (12.59) become

$$(i\omega - a_{11})C_{A(i\omega)} = a_{12}T_{(i\omega)} + a_{13} \tag{12.64}$$

$$(i\omega - a_{22})T_{(i\omega)} = a_{21}C_{A(i\omega)} \tag{12.65}$$

These two equations can then be solved for C_A and T. Since there are only two equations in this simple example, they can be solved analytically to give C_A and T in terms of any value of frequency.

$$C_{A(i\omega)} = \frac{a_{13}(i\omega - a_{22})}{(i\omega - a_{22})(i\omega - a_{11}) - a_{21}a_{12}} \tag{12.66}$$

$$T_{(i\omega)} = \frac{a_{21}a_{13}}{(i\omega - a_{22})(i\omega - a_{11}) - a_{21}a_{12}} \tag{12.67}$$

However, in a more complex case with many equations where analytical solution would be difficult, the equations can be solved numerically. All the a_{ij}'s and ω are just numbers. Standard simultaneous equation-solving packages available in most digital-computer libraries can be used. Keep in mind, however, that these are complex-variable simultaneous equations. A special-purpose method for solving the equations for a distillation column will be presented in the next example.

The numerical values of the complex variables $C_{A(i\omega)}$ and $T_{(i\omega)}$ are equal to the transfer functions $G_{11(i\omega)}$ and $G_{21(i\omega)}$ in Eqs. (12.60) and (12.61) because we set C_{A0} equal to 1 originally.

$$(C_{A(i\omega)})_{C_{A0}=1} = (C_A^R + iC_A^I)_{C_{A0}=1} = G_{11(i\omega)} \tag{12.68}$$

$$(T_{(i\omega)})_{C_{A0}=1} = (T^R + iT^I)_{C_{A0}=1} = G_{21(i\omega)} \tag{12.69}$$

Now T_0 is set equal to 1, with all other input variables zero, and the equations are solved again for new values of $C_{A(i\omega)}$ and $T_{(i\omega)}$. These are the transfer functions G_{12} and G_{22}.

$$(C_{A(i\omega)})_{T_0=1} = G_{12(i\omega)} \tag{12.70}$$

$$(T_{(i\omega)})_{T_0=1} = G_{22(i\omega)} \tag{12.71}$$

Repeating the same process with $F = 1$ gives G_{13} and G_{23}, and with $T_J = 1$ gives G_{14} and G_{24}. Then a new value of frequency is selected and the complete process is repeated.

Example 12.6. Let us consider a much more complex system where the advantages of frequency-domain solution will be apparent. Rippin and Lamb showed how a frequency-domain "stepping" technique could be used to find the frequency response of a binary, equimolal-overflow distillation column. The column has many trays and therefore the system is of very high order.

The linearized equations for the column are:

Reboiler (assuming holdup M_B is constant):

$$\frac{dx_B}{dt} = a_{B1}L_1 - a_{B2}V + a_{B3}x_1 - a_{B4}x_B \tag{12.72}$$

$$0 = L_1 - V - B$$

Tray 1:

$$\frac{dx_1}{dt} = a_{11}L_2 - a_{12}V + a_{13}x_2 - a_{14}x_1 + a_{15}x_B \tag{12.73}$$

$$\frac{dL_1}{dt} = \frac{1}{\beta}(L_2 - L_1) \tag{12.74}$$

Tray n:

$$\frac{dx_n}{dt} = a_{n1}L_{n+1} - a_{n2}V + a_{n3}x_{n+1} - a_{n4}x_n + a_{n5}x_{n-1} + a_{n6}z + a_{n7}F \tag{12.75}$$

$$\frac{dL_n}{dt} = \frac{1}{\beta}(L_{n+1} - L_n + F) \tag{12.76}$$

Top tray ($n = NT$):

$$\frac{dx_{NT}}{dt} = a_{NT,1}R - a_{NT,2}V + a_{NT,3}x_D - a_{NT,4}x_{NT} + a_{NT,5}x_{NT-1} \quad (12.77)$$

$$\frac{dL_{NT}}{dt} = \frac{1}{\beta}(R - L_{NT}) \quad (12.78)$$

Reflux drum (holdup M_D is assumed constant; total condenser):

$$\frac{dx_D}{dt} = a_{D1}x_{NT} - a_{D2}x_D$$

$$0 = V - R - D \quad (12.79)$$

The a_{ij}'s are all constants made up of the steadystate holdups, flow rates and compositions. Table 12.2 gives their values. The variables in Eqs. (12.72) to (12.79) are all perturbation variables. β is the hydraulic constant, the linearized relationship between a perturbation in liquid holdup on a tray, M_n, and the perturbation in the liquid flow rate L_n leaving the tray.

$$M_n = \beta L_n \quad (12.80)$$

K_n is the linearized relationship between the perturbations in vapor composition y_n and liquid composition x_n. Note that this is *not* the same "K value" used in VLE calculations which relates total x and y variables. The K's in Table 12.2 are the slopes of the equilibrium line and relate perturbation variables.

$$y_n = K_n x_n \quad (12.81)$$

As the name implies, the "stepping technique" involves stepwise calculation from the base of the column to the top. All the equations are Laplace-transformed, s is set equal to $i\omega$, and a value of frequency is specified. One of the variables at the base or at the feed tray (z, F, V, B, or x_B) is set equal to $1 + i0$ and the remaining set equal to zero. For example, let us set $x_B = 1 + i0$.

Starting with the reboiler equations, the composition x_1 and the liquid flow rate L_1 can be calculated.

$$L_1 = V_B + B = 0$$

$$x_1 = [(i\omega + a_{B4})x_B - a_{B1}L_1 + a_{B2}V]\frac{1}{a_{B3}} = \frac{i\omega + a_{B4}}{a_{B3}} \quad (12.82)$$

These variables x_1 and L_1 are complex numbers.

Then the equations for tray 1 can be used to calculate x_2 and L_2. This procedure is continued up the column, tray by tray. The top-tray equations give complex values for x_D and R, which are stored as the complex numbers g_{11} and g_{21}.

$$x_D \equiv g_{11}^R + ig_{11}^I \equiv g_{11}$$

$$R \equiv g_{21}^R + ig_{21}^I \equiv g_{21} \quad (12.83)$$

The first condenser equation [Eq. (12.79)] gives a second value of x_D, which is stored as g_{31}. These g's are just intermediate numbers that will be used shortly to calculate the desired transfer functions.

TABLE 12.2

Coefficients of linearized ODEs for binary distillation column

$$a_{B1} = \frac{\bar{x}_1 - \bar{x}_B}{\bar{M}_B} \qquad a_{B2} = \frac{\bar{y}_B - \bar{x}_B}{\bar{M}_B} \qquad\qquad a_{n5} = \frac{\bar{V}K_{n-1}}{\bar{M}_n} \qquad a_{n6} = \frac{\bar{F}}{\bar{M}_n}$$

$$a_{B3} = \frac{\bar{L}_1}{\bar{M}_B} \qquad a_{B4} = \frac{\bar{B} + \bar{V}K_B}{\bar{M}_B} \qquad\qquad a_{n7} = \frac{\bar{z} - \bar{x}_n}{\bar{M}_n}$$

$$a_{11} = \frac{\bar{x}_2 - \bar{x}_1}{\bar{M}_1} \qquad a_{12} = \frac{\bar{y}_1 - \bar{y}_B}{\bar{M}_1} \qquad a_{NT,1} = \frac{\bar{x}_D - \bar{x}_{NT}}{\bar{M}_{NT}} \qquad a_{NT,2} = \frac{\bar{y}_{NT} - \bar{y}_{NT-1}}{\bar{M}_{NT}}$$

$$a_{13} = \frac{\bar{L}_2}{\bar{M}_1} \qquad a_{14} = \frac{\bar{L}_1 + \bar{V}K_1}{\bar{M}_1} \qquad a_{NT,3} = \frac{\bar{R}}{\bar{M}_{NT}} \qquad a_{NT,4} = \frac{\bar{L}_{NT} + \bar{V}K_{NT}}{\bar{M}_{NT}}$$

$$a_{15} = \frac{\bar{V}K_B}{\bar{M}_1}$$

$$a_{n1} = \frac{\bar{x}_{n+1} - \bar{x}_n}{\bar{M}_n} \qquad a_{n2} = \frac{\bar{y}_n - \bar{y}_{n-1}}{\bar{M}_n} \qquad a_{NT,5} = \frac{\bar{V}K_{NT-1}}{\bar{M}_{NT}}$$

$$a_{n3} = \frac{\bar{L}_{n+1}}{\bar{M}_n} \qquad a_{n4} = \frac{\bar{L}_n + \bar{V}K_n}{\bar{M}_n} \qquad a_{D1} = \frac{\bar{V}K_{NT}}{\bar{M}_D} \qquad a_{D2} = \frac{\bar{V}}{\bar{M}_D}$$

Then V is set equal to $1 + i0$, and x_B, B, z, and F are set equal to zero. Stepping up the column again gives three new g's: g_{12}, g_{22}, and g_{32}. Then the procedure is repeated with $B = 1$, then with $z = 1$, and finally with $F = 1$.

These calculations would be very tedious to do by hand, but they can be done easily on a digital computer because they are all numerical calculations.

The resulting g's from the five cycles up the column form three equations:

$$x_D = g_{11}x_B + g_{12}V + g_{13}B + g_{14}z + g_{15}F$$

$$R = g_{21}x_B + g_{22}V + g_{23}B + g_{24}z + g_{25}F \qquad (12.84)$$

$$x_D = g_{31}x_B + g_{32}V + g_{33}B + g_{34}z + g_{35}F$$

These three equations can then be rearranged to get the output variables x_D and x_B in terms of the input variables z, F, R, and V.

$$x_D = P_{11(i\omega)}z + P_{12(i\omega)}F + P_{13(i\omega)}R + P_{14(i\omega)}V$$

$$x_B = P_{21(i\omega)}z + P_{22(i\omega)}F + P_{23(i\omega)}R + P_{24(i\omega)}V \qquad (12.85)$$

The P_{ij}'s are points on Bode, Nyquist, or Nichols plots of the distillation-column transfer functions.

A new value of frequency is specified and the calculations repeated. Table 12.3 gives a FORTRAN program that performs all these calculations. The initial part of the program solves for all the steadystate compositions and flow rates, given feed composition and feed flow rate and the desired bottoms and distillate compositions, by converging on the correct value of vapor boilup V_S. Next the coefficients for the linearized equations are calculated. Then the stepping technique is used to calculate the intermediate g's and the final P_{ij} transfer functions in the frequency domain.

TABLE 12.3

Frequency-domain solution for binary distillation

```
      DIMENSION SSKP(2,4),DBP(2,4),DEGP(2,4),G(3,5),A(20,7),P(2,4),
    + X(20),K(20),Y(20)
      COMPLEX P,XP1,XP2,Z1,ZL,G,XP,XLP
      REAL LR,LS,MB,MS,MR,MD,KB,K
      COMMON ALPHA
      DATA WMIN,WMAX,WN/.1,11.,5/
      DATA NT,NF,XB,XD,Z,F,Q/20,10,.02,.98,.5,100.,1./
      DATA VS,BETA,MB,MS,MR,MD/178.,.1,100.,10.,10.,100./
      WRITE(6,12) NT,NF,F,Z,Q
   12 FORMAT(' NT = ',I3,' NF = ',I3,' F = ',F8.2,' Z = ',F8.6,
    + ' Q = ',F8.6)
      ALPHA=2.
      WRITE(6,13) XB,XD,ALPHA,BETA
   13 FORMAT(' XB = ',F8.6,' XD = ',F8.6,' ALPHA = ',F6.2,
    + ' BETA = ',F6.2)
      WRITE(6,14)MB,MS,MR,MD
   14 FORMAT(' MB = ',F8.4,' MS = ',F8.4,' MR = ',F8.4,' MD = ',F8.4)
C................
C STEADYSTATE RATING PROGRAM
C................
      DVS=VS/100.
      YB=EQUIL(XB)
      LOOP=0
      FLAGP=1.
      FLAGM=1.
      B=((XD-Z)*F)/(XD-XB)
      D=F-B
  100 LS=VS+B
      LR=LS-Q*F
      VR=VS+(1.-Q)*F
      R=LR
      LOOP=LOOP+1
      IF(LOOP.GT.50 ) GO TO 41
C   STRIPPING TRAYS
      X(1)=(VS*YB+B*XB)/LS
      Y(1)=EQUIL(X(1))
      DO 30 N=2,NF
      X(N)=(VS*Y(N-1)+B*XB)/LS
      Y(N)=EQUIL(X(N))
   30 CONTINUE
C   FEED TRAY
      X(NF+1)=(VR*Y(NF)+B*XB-F*Z)/LR
      Y(NF+1)=EQUIL(X(NF+1))
      IF(X(NF+1).GT.XD) GO TO 35
      IF(X(NF+1).LE.X(NF)) GO TO 31
C   RECTIFYING TRAYS
      DO 40 N=NF+2,NT
      X(N)=(VR*Y(N-1)+B*XB-F*Z)/LR
      Y(N)=EQUIL(X(N))
   40 IF(X(N).GT.XD) GO TO 35
      IF(ABS(XD-Y(NT)).LT..0000001) GO TO 50
      IF(XD-Y(NT))35,31,31
   41 WRITE(6,42)
   42 FORMAT(1X,4HLOOP)
      STOP
   31 IF(FLAGP)32,32,33
```

TABLE 12.3 (*continued*)

```
   32 DVS=DVS/2.
   33 VS=VS+DVS
      FLAGM=-1.
      GO TO 100
   35 IF(FLAGM)36,36,37
   36 DVS=DVS/2.
   37 VS=VS-DVS
      FLAGP=-1.
      GO TO 100
   50 WRITE(6,45)VS,R
   45 FORMAT(' VS = ',F10.4,' R = ',F10.4)
      WRITE(6,51)
   51 FORMAT('   N       X       Y        SLOPE')
      KB=SLOPE(XB)
      N=0
      WRITE(6,55)N,XB,YB,KB
   55 FORMAT(3X,I3,5X,7F10.7 )
      DO 56 N=1,NT
      K(N)= SLOPE(X(N))
   56 WRITE(6,55)N,X(N),Y(N),K(N)
C ................
C CALCULATE COEFFICIENTS
C ................
      N=0
      AB1=(X(1)-XB)/MB
      AB2=(YB-XB)/MB
      AB3=LS/MB
      AB4=(B+VS*KB)/MB
      WRITE(6,60)
   60 FORMAT(' N      A1       A2      A3      A4      A5
     +    A6      A7')
      WRITE(6,57)N,AB1,AB2,AB3,AB4
   57 FORMAT(1X,I3,3X,2F10.6,4F10.3,F10.6)
      A(1,1)=(X(2)-X(1))/MS
      A(1,2)=(Y(1)-YB)/MS
      A(1,3)=LS/MS
      A(1,4)=(LS+VS*K(1))/MS
      A(1,5)=VS*KB/MS
      A(1,6)=0.
      A(1,7)=0.
      N=1
      WRITE(6,57)N,A(1,1),A(1,2),A(1,3),A(1,4),A(1,5),A(1,6),A(1,7)
      DO 70 N=2,NF-1
      A(N,1)=(X(N+1)-X(N))/MS
      A(N,2)=(Y(N)-Y(N-1))/MS
      A(N,3)=LS/MS
      A(N,4)=(LS+VS*K(N))/MS
      A(N,5)=VS*K(N-1)/MS
      A(N,6)=0.
      A(N,7)=0.
   70 WRITE(6,57)N,A(N,1),A(N,2),A(N,3),A(N,4),A(N,5),A(N,6),A(N,7)
      A(NF,1)=(X(NF+1)-X(NF))/MS
      A(NF,2)=(Y(NF)-Y(NF-1))/MS
      A(NF,3)=LR/MS
      A(NF,4)=(LS+VR*K(NF))/MS
      A(NF,5)=VS*K(NF-1)/MS
      A(NF,7)=(Z-X(NF))/MS
      A(NF,6)=F/MS
```

TABLE 12.3 (*continued*)

```
      N=NF
      WRITE(6,57)NF,A(N,1),A(N,2),A(N,3),A(N,4),A(N,5),A(N,6),A(N,7)
      DO 80 N=NF+1,NT
      IF(N.NE.NT) A(N,1)=(X(N+1)-X(N))/MR
      A(N,2)=(Y(N)-Y(N-1))/MR
      A(N,3)=LR/MR
      A(N,4)=(LR+VR*K(N))/MR
      A(N,5)=VR*K(N-1)/MR
      A(N,6)=0.
      A(N,7)=0.
      IF(N.EQ.NT) A(NT,1)=(XD-X(NT))/MR
   80 WRITE(6,57)N,A(N,1),A(N,2),A(N,3),A(N,4),A(N,5),A(N,6),A(N,7)
      AD1=VR*K(NT)/MD
      AD2=VR/MD
      WRITE(6,81)AD1,AD2
   81 FORMAT(7X,6F10.6)
C*****************************************************************
C  STEPPING UP COLUMN
C*****************************************************************
      DW=10.**(1./WN)
      W=0.
  200 IF(W.GT.WMAX) STOP
      DO 300 KK=1,5
      XBP=0.
      VP=0.
      BP=0.
      ZP=0.
      FP=0.0
      GO TO  (201,202,203,204,205),KK
  201 XBP=1.
      GO TO 210
  202 VP=1.
      GO TO 210
  203 BP=1.
      GO TO 210
  204 ZP=1.
      GO TO 210
  205 FP=1.
  210 XLP=CMPLX((BP+VP),0.)
      Z1=CMPLX(AB4,W)
      XP1=(Z1*XBP-AB1*XLP+AB2*VP)/AB3
      XP2=CMPLX(XBP,0.)
      ZL=CMPLX(1.,W*BETA)
C  STRIPPING TRAYS
      DO 220 N=1,NF-1
      XLP=XLP*ZL
      Z1=CMPLX(A(N,4),W)
      XP=(Z1*XP1-A(N,1)*XLP-A(N,5)*XP2+A(N,2)*VP)/A(N,3)
      XP2=XP1
      XP1=XP
  220 CONTINUE
C  FEED TRAY
      XLP=ZL*XLP-FP
      Z1=CMPLX(A(NF,4),W)
      XP=(Z1*XP1-A(NF,1)*XLP-A(NF,5)*XP2-A(NF,6)*ZP-A(NF,7)*FP
     1  +A(NF,2)*VP)/A(NF,3)
      XP2=XP1
      XP1=XP
```

TABLE 12.3 (*continued*)

```
C   RECTIFYING TRAYS
      DO 240 N=NF+1,NT
      XLP=XLP*ZL
      Z1=CMPLX(A(N,4),W)
      XP=(Z1*XP1-A(N,1)*XLP-A(N,5)*XP2+A(N,2)*VP)/A(N,3)
      XP2=XP1
  240 XP1=XP
      Z1=CMPLX(AD2,W)
      XP=XP2*AD1/Z1
      G(1,KK)=XP1
      G(2,KK)=XLP
      G(3,KK)=XP
  300 CONTINUE
      P(2,1)=(G(1,4)-G(3,4))/(G(3,1)-G(1,1))
      P(2,2)=(G(1,5)-G(3,5)+(G(3,3)-G(1,3))*G(2,5)/G(2,3))/(G(3,1)-G(1,1
     1  ))
      P(2,3)=(G(1,3)-G(3,3))/((G(3,1)-G(1,1))*G(2,3))
      P(2,4)=(G(1,2)-G(3,2)+G(2,2)*(G(3,3)-G(1,3))/G(2,3))/(G(3,1)-
     1 G(1,1))
      P(1,1)=G(1,1)*P(2,1)+G(1,4)
      P(1,3)=G(1,3)/G(2,3) + G(1,1)*P(2,3)
      P(1,2)=G(1,1)*P(2,2)+G(1,5)-G(1,3)*G(2,5)/G(2,3)
      P(1,4)=G(1,1)*P(2,4)+G(1,2)-G(1,3)*G(2,2)/G(2,3)
      IF(W.GT.0.) GO TO 305
      DO 306 L=1,2
      DO 307 M=1,4
  307 SSKP(L,M)=REAL(P(L,M))
  306 CONTINUE
      WRITE(6,320)
  320 FORMAT(' W  XD/Z  XD/F  XD/R  XD/V  XB/Z  XB/F
     +XB/R  XB/V')
      WRITE(6,310) W,(SSKP(1,M),M=1,4),(SSKP(2,M),M=1,4)
  310 FORMAT(1X,F6.3,8F8.4)
      W=WMIN
      GO TO 200
  305 DO 315 L=1,2
      DO 316 M=1,4
      DBP(L,M)=20.*ALOG10(CABS(P(L,M)/SSKP(L,M)))
      DEGP(L,M)=ATAN(AIMAG(P(L,M))/REAL(P(L,M)))*180./3.1416
      IF((REAL( P(L,M))/SSKP(L,M)).LT.0.)DEGP(L,M)=DEGP(L,M)-180.
  316 CONTINUE
  315 CONTINUE
      WRITE(6,504)W,((DBP (L,M),M=1,4),L=1,2)
  504 FORMAT(1X,F6.3,8F8.3)
      WRITE(6,503)((DEGP(L,M),M=1,4),L=1,2)
  503 FORMAT(7X,8F8.3)
  317 W=W*DW
      GO TO 200
      END
C...............
      FUNCTION EQUIL(T)
      COMMON ALPHA
      EQUIL=ALPHA*T/(1.+(ALPHA-1.)*T)
      RETURN
      END
C...............
      FUNCTION SLOPE(T)
      COMMON ALPHA
```

TABLE 12.3 (*continued*)

```
SLOPE=ALPHA/((1.+(ALPHA-1.)*T)**2)
RETURN
END
```

Results

NT = 20 NF = 10 F = 100.00 Z = 0.500000 Q = 1.000000
XB = 0.020000 XD = 0.980000 ALPHA = 2.00 BETA = 0.10
MB = 100.0000 MS = 10.0000 MR = 10.0000 MD = 100.0000
VS = 178.0089 R = 128.0089

N	X	Y	SLOPE
0	0.0200000	0.0392157	1.9223380
1	0.0350019	0.0676364	1.8670150
2	0.0571902	0.1081928	1.7894670
3	0.0888531	0.1632049	1.6869080
4	0.1318015	0.2329057	1.5613110
5	0.1862177	0.3139688	1.4213510
6	0.2495045	0.3993655	1.2810150
7	0.3161745	0.4804447	1.1545240
8	0.3794739	0.5501720	1.0510010
9	0.4339107	0.6052130	0.9727154
10	0.4768819	0.6457955	0.9169348
11	0.5152560	0.6800910	0.8710798
12	0.5629472	0.7203662	0.8187313
13	0.6189537	0.7646343	0.7630643
14	0.6805128	0.8098871	0.7081843
15	0.7434413	0.8528435	0.6579840
16	0.8031763	0.8908461	0.6151111
17	0.8560228	0.9224271	0.5805817
18	0.8999392	0.9473347	0.5540520
19	0.9345757	0.9661816	0.5343903
20	0.9607842	0.9800000	0.5202001

N	A1	A2	A3	A4	A5	A6	A7
0	0.000150	0.000192	2.280	3.922			
1	0.002219	0.002842	22.801	56.035	34.219	0.000	0.000000
2	0.003166	0.004056	22.801	54.655	33.235	0.000	0.000000
3	0.004295	0.005501	22.801	52.829	31.854	0.000	0.000000
4	0.005442	0.006970	22.801	50.594	30.028	0.000	0.000000
5	0.006329	0.008106	22.801	48.102	27.793	0.000	0.000000
6	0.006667	0.008540	22.801	45.604	25.301	0.000	0.000000
7	0.006330	0.008108	22.801	43.352	22.803	0.000	0.000000
8	0.005444	0.006973	22.801	41.510	20.552	0.000	0.000000
9	0.004297	0.005504	22.801	40.116	18.709	0.000	0.000000
10	0.003837	0.004058	12.801	39.123	17.315	10.000	0.002312
11	0.004769	0.003430	12.801	28.307	16.322	0.000	0.000000
12	0.005601	0.004028	12.801	27.375	15.506	0.000	0.000000
13	0.006156	0.004427	12.801	26.384	14.574	0.000	0.000000
14	0.006293	0.004525	12.801	25.407	13.583	0.000	0.000000
15	0.005974	0.004296	12.801	24.514	12.606	0.000	0.000000
16	0.005285	0.003800	12.801	23.750	11.713	0.000	0.000000
17	0.004392	0.003158	12.801	23.136	10.950	0.000	0.000000
18	0.003464	0.002491	12.801	22.664	10.335	0.000	0.000000
19	0.002621	0.001885	12.801	22.314	9.863	0.000	0.000000
20	0.001922	0.001382	12.801	22.061	9.513	0.000	0.000000
	0.926002	1.780089					

W	XD/Z	XD/F	XD/R	XD/V	XB/Z	XB/F	XB/R	XB/V
0.000	0.9226	0.0038	0.0092	-0.0088	1.0774	0.0058	0.0100	-0.0104
0.100	-4.311	-4.409	-4.383	-4.339	-4.127	-3.814	-4.096	-4.082
	-65.225	-74.126	-62.644	-61.048	-55.368	-47.684	-59.485	-50.487

TABLE 12.3 (*continued*)

```
0.158 -7.362 -7.604 -7.537 -7.429 -6.922 -6.220 -6.847 -6.817
      -84.591 -98.552 -80.320 -77.894 -69.445 -58.418 -76.045 -61.831
0.251 -11.235 -11.819 -11.639 -11.387 -10.245 -8.845 -10.076 -10.014
      -104.980-126.569 -97.549 -94.091 -82.556 -68.247 -93.257 -70.864
0.398 -15.864 -17.210 -16.701 -16.173 -13.871 -11.489 -13.518 -13.417
      -127.465-159.838-113.506-109.318 -96.245 -80.038-113.871 -78.632
0.631 -21.332 -24.174 -22.642 -21.802 -17.853 -14.348 -17.178 -17.040
      -153.535-199.499-125.931-122.512-112.989 -96.898-142.497 -86.785
1.000 -27.823 -33.155 -28.861 -28.101 -22.470 -17.717 -21.314 -21.073
      -184.771-245.401-133.921-131.809-134.574-120.766-184.731 -95.778
1.585 -35.637 -44.614 -35.309 -34.595 -28.048 -21.841 -26.415 -25.653
      -222.548  63.212-142.063-136.309-161.844-154.121-248.038-104.813
2.512 -45.079 -59.293 -42.689 -40.784 -34.898 -27.174 -33.555 -30.780
      -267.810  10.147-144.890-138.632-195.103-201.163  17.059-112.566
3.981 -56.507 -76.608 -48.964 -46.923 -43.251 -34.789 -45.148 -36.300
       37.979 -31.439-142.534-142.595-234.497-266.886-121.490-117.937
6.310 -70.497 -83.038 -55.447 -53.264 -53.295 -46.783 -65.833 -41.997
      -26.299-153.435-143.808-148.299  79.004   7.796  51.605-120.703
10.000 -84.490 -90.028 -61.777 -53.086 -65.352 -65.646-102.055 -47.713
      -56.310-180.000-142.875-168.690  23.608 -82.682-160.298-121.265
```

The numerical case given is for a 20-tray column with 10 trays in the stripping section. A constant relative volatility of 2 is used. The column steadystate profile is given in Table 12.3, together with the values of coefficients and the transfer functions in terms of log modulus (decibels) and phase angle (degrees) at frequencies from 0 to 10 radians per minute. The values at zero frequency are the steadystate gains of the transfer functions.

PROBLEMS

12.1. Sketch Nyquist, Bode, and Nichols plots for the following transfer functions:

(a) $G_{(s)} = \dfrac{1}{(s + 1)^3}$

(b) $G_{(s)} = \dfrac{1}{(s + 1)(10s + 1)(100s + 1)}$

(c) $G_{(s)} = \dfrac{1}{s^2(s + 1)}$

(d) $G_{(s)} = \dfrac{\tau s + 1}{(\tau/6)s + 1}$

(e) $G_{(s)} = \dfrac{s}{2s + 1}$

(f) $G_{(s)} = \dfrac{1}{(10s + 1)(s^2 + s + 1)}$

12.2. Draw the Bode plots for the transfer functions:
   (a) $G_{(s)} = 0.5$
   (b) $G_{(s)} = 5.0$

12.3. Sketch Nyquist, Bode, and Nichols plots for the proportional-integral feedback controller $B_{(s)}$:

$$B_{(s)} = K_c\left(1 + \frac{1}{\tau_I s}\right)$$

12.4. Sketch Nyquist, Bode, and Nichols plots for a system with the transfer function

$$G_{(s)} = \frac{-3s + 1}{(s + 1)(5s + 1)}$$

12.5. Draw the Bode plots of the transfer function

$$G_{(s)} = \frac{7.5(s + 0.2)}{s(s + 1)^3}$$

12.6. Write a digital computer program that gives the real and imaginary parts, log modulus, and phase angle for the transfer functions:

(a) $G_{(s)} = B_{(s)} \dfrac{P_{11}P_{22} - P_{12}P_{21}}{P_{22}}$

where

$$B_{(s)} = 8\left(1 + \frac{1}{400s}\right)$$

$$P_{11(s)} = \frac{1}{(1 + 167s)(1 + s)(1 + 0.1s)^4}$$

$$P_{12(s)} = \frac{0.85}{(1 + 83s)(1 + s)^2}$$

$$P_{21(s)} = \frac{0.85}{(1 + 167s)(1 + 0.5s)^3(1 + s)}$$

$$P_{22(s)} = \frac{1}{(1 + 167s)(1 + s)^2}$$

(b) $G_{(s)} = \dfrac{e^{-0.1s}}{s + 1 + e^{-0.1s}}$

12.7. Draw Bode, Nyquist, and Nichols plots for the transfer functions:

(a) $A_{L(s)} = \dfrac{G_{(s)}}{1 + B_{(s)}G_{(s)}}$

(b) $A_{s(s)} = \dfrac{B_{(s)}G_{(s)}}{1 + B_{(s)}G_{(s)}}$

where

$$B_{(s)} = K_c\left(1 + \frac{1}{\tau_I s}\right) \qquad K_c = 6 \qquad \tau_I = 6$$

$$G_{(s)} = \frac{1}{\tau s + 1} \qquad\qquad \tau = 10$$

12.8. Draw the Bode plot of

$$G_{(s)} = \frac{1 - e^{-Ds}}{s}$$

12.9. A process is forced by sinusoidal input $m_{s(t)}$. The output is a sine wave $x_{s(t)}$. If these two signals are connected to an $x - y$ recorder, we get a Lissajous plot. Time is the parameter along the curve, which repeats itself with each cycle. The shape of the curve will change if the frequency is changed and will be different for different kinds of processes.

(a) How can the magnitude ratio MR and phase angle θ be found from this curve?

(b) Sketch Lissajous curves for the following systems:

   (i) $G_{(s)} = K_p$

   (ii) $G_{(s)} = \dfrac{1}{s}$ at $\omega = 1$ radian per time

   (iii) $G_{(s)} = \dfrac{1}{\tau_p s + 1}$ at $\omega = \dfrac{1}{\tau_p}$ radians per time

CHAPTER
13

FREQUENCY-DOMAIN ANALYSIS OF CLOSEDLOOP SYSTEMS

The design of feedback controllers in the frequency domain is the subject of this chapter. The Chinese language that we learned in Chap. 12 is now put to use to tune controllers. Frequency-domain methods are widely used because they have the significant advantage of being easier to use for high-order systems than the time- and Laplace-domain methods.

We will show in Sec. 13.1 that *closedloop stability* can be determined from the frequency-response plot of the total *openloop* transfer function of the system (process openloop transfer function *and* feedback controller $G_{M(i\omega)} B_{(i\omega)}$). This means that a Bode plot of $G_{M(i\omega)} B_{(i\omega)}$ is all we need. As you remember from Chap. 12, the total frequency-response curve of a complex system is easily obtained on a Bode plot by splitting the system into its simple elements, plotting each of these, and merely adding log moduli and phase angles together. Therefore the graphical generation of the required $G_{M(i\omega)} B_{(i\omega)}$ curve is relatively easy. Of course all this algebraic manipulation of complex numbers can be even more easily performed on a digital computer.

13.1 NYQUIST STABILITY CRITERION

The Nyquist stability criterion is a method for determining the stability of systems in the frequency domain. It is almost always applied to closedloop systems. A working, but not completely general, statement of the Nyquist stability criterion is:

If a polar plot of the total openloop transfer function of the system $G_{M(i\omega)} B_{(i\omega)}$ wraps around the $(-1, 0)$ point in the $G_M B$ plane as frequency ω goes from zero to infinity, the system is closedloop unstable.

The two polar plots sketched in Fig. 13.1a show that system A is closedloop unstable whereas system B is closedloop stable.

The Nyquist stability criterion is, on the surface, quite remarkable. We are able to deduce something about the stability of the *closedloop* system by making a frequency response plot of the *openloop* system! And the encirclement of the mystical, magical $(-1, 0)$ point somehow tells us that the system is closedloop unstable. This all looks like blue smoke and mirrors! However, as we will prove below, it all goes back to finding out if there are any roots of the closedloop characteristic equation in the RHP.

13.1.1 Proof

A. COMPLEX VARIABLE THEOREM. The Nyquist stability criterion is derived from a theorem of complex variables that says:

If a complex function $F_{(s)}$ has Z zeros and P poles inside a certain area of the s plane, the number N of encirclements of the origin that a mapping of a closed contour around the area makes in the F plane is equal to $Z - P$.

$$Z - P = N \tag{13.1}$$

Consider the hypothetical function $F_{(s)}$ of Eq. (13.2) with two zeros at $s = z_1$ and $s = z_2$ and one pole at $s = p_1$.

$$F_{(s)} = \frac{(s - z_1)(s - z_2)}{s - p_1} \tag{13.2}$$

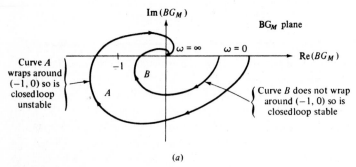

(a)

FIGURE 13.1a
Polar plots showing closedloop stability or instability.

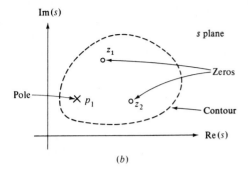

FIGURE 13.1b
s-plane location of zeros and poles.

(b)

The locations of the zeros and the pole are sketched in the s plane in Fig. 13.1b.
The argument of $F_{(s)}$ is

$$\arg F_{(s)} = \arg \left[\frac{(s - z_1)(s - z_2)}{s - p_1} \right]$$

(13.3)

$$\arg F_{(s)} = \arg (s - z_1) + \arg (s - z_2) - \arg (s - p_1)$$

Remember, the argument of the product of two complex numbers z_1 and z_2 is the sum of the arguments.

$$z_1 z_2 = (r_1 e^{i\theta_1})(r_2 e^{i\theta_2}) = r_1 r_2 e^{i(\theta_1 + \theta_2)}$$

$$\arg (z_1 z_2) = \theta_1 + \theta_2$$

And the argument of the quotient of two complex numbers is the difference between the arguments.

$$\frac{z_1}{z_2} = \frac{r_1 e^{i\theta_1}}{r_2 e^{i\theta_2}} = \frac{r_1}{r_2} e^{i(\theta_1 - \theta_2)}$$

$$\arg \left(\frac{z_1}{z_2} \right) = \theta_1 - \theta_2$$

Let us pick an arbitrary point s on the contour and draw a line from the zero z_1 to this point (see Fig. 13.1c). The angle between this line and the horizontal, θ_{z_1}, is equal to the argument of $(s - z_1)$. Now let the point s move completely around the contour. The angle θ_{z_1} or $\arg (s - z_1)$ will increase by 2π radians. Therefore the arg $F_{(s)}$ will *increase* by 2π radians for each zero inside the contour.

A similar development shows that the arg $F_{(s)}$ *decreases* by 2π for each pole inside the contour because of the negative sign in Eq. (13.3). Two zeros and one pole will mean that the arg $F_{(s)}$ must show a net increase of $+2\pi$. Thus a plot of $F_{(s)}$ in the complex F plane (real part of $F_{(s)}$ versus imaginary part of $F_{(s)}$) must encircle the origin once as s goes completely around the contour.

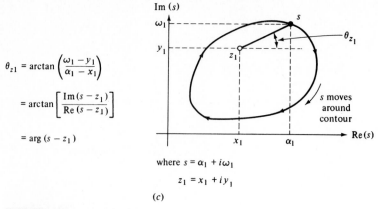

$$\theta_{z1} = \arctan\left(\frac{\omega_1 - y_1}{\alpha_1 - x_1}\right)$$

$$= \arctan\left[\frac{\operatorname{Im}(s - z_1)}{\operatorname{Re}(s - z_1)}\right]$$

$$= \arg(s - z_1)$$

where $s = \alpha_1 + i\omega_1$

$z_1 = x_1 + iy_1$

(c)

FIGURE 13.1c
Argument of $(s - z_1)$.

In this system $Z = 2$ and $P = 1$, and we have found that $N = Z - P = 2 - 1 = 1$. Generalizing to a system with Z zeros and P poles gives the desired theorem [Eq. (13.1)].

If any of the zeros or poles are repeated, of order M, they contribute $2\pi M$ radians. Thus Z is the number of zeros inside the contour with Mth-order zeros counted M times. And P is the number of poles inside the contour with Nth-order poles counted N times.

B. APPLICATION OF THEOREM TO CLOSEDLOOP STABILITY. To check the stability of a system, we are interested in the roots or zeros of the characteristic equation. If any of them lie in the right half of the s plane, the system is unstable. For a closedloop system, the characteristic equation is

$$1 + G_{M(s)}B_{(s)} = 0 \tag{13.4}$$

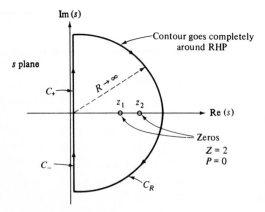

FIGURE 13.2a
s-plane area of interest.

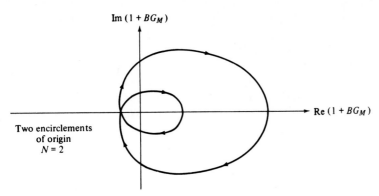

FIGURE 13.2b
$(1 + B_{(s)} G_{M(s)})$ plane.

So for a closedloop system, the function we are interested in is

$$F_{(s)} = 1 + G_{M(s)} B_{(s)} \tag{13.5}$$

If this function has any zeros in the RHP, the closedloop system is unstable.

If we pick a contour that goes completely around the entire right half of the s plane and plot $1 + G_{M(s)} B_{(s)}$, Eq. (13.1) tells us that the number of encirclements of the origin in this $(1 + G_M B)$ plane will be equal to the difference between the zeros and poles of $1 + G_M B$ that lie in the RHP. Figure 13.2 shows a case where there are two zeros in the RHP and no poles. There are two encirclements of the origin in the $(1 + G_M B)$ plane.

We are familiar with making plots of complex functions like $G_{M(i\omega)} B_{(i\omega)}$ in the $G_M B$ plane. It is therefore easier (but more confusing unless you are careful to keep track of what you are doing) to use the $G_M B$ plane instead of the $(1 + G_M B)$

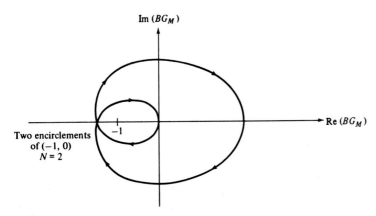

FIGURE 13.2c
$B_{(s)} G_{M(s)}$ plane.

plane. The origin in the $(1 + G_M B)$ plane maps into the $(-1, 0)$ point in the $G_M B$ plane since the real part of every point is moved to the left one unit.

We therefore look at the encirclements of the $(-1, 0)$ point in the $G_M B$ plane, instead of encirclements of the origin in the $(1 + G_M B)$ plane.

After we map the contour into the $G_M B$ plane and count the number N of encirclements of the $(-1, 0)$ point, we know the difference between the number of zeros Z and the number of poles P that lie in the RHP. We want to find out if there are any zeros of the function $F_{(s)} = 1 + G_{M(s)} B_{(s)}$ in the RHP. Therefore, we must find the number of poles of $F_{(s)}$ in the RHP before we can determine the number of zeros.

$$Z = N + P \tag{13.6}$$

The poles of the function $F_{(s)} = 1 + G_{M(s)} B_{(s)}$ are the same as the poles of $G_{M(s)} B_{(s)}$. If the process is openloop stable, there are no poles of $G_{M(s)} B_{(s)}$ in the RHP. Thus an openloop stable process means that $P = 0$. Therefore the number N of encirclements of the $(-1, 0)$ point is equal to the number of zeros of $1 + G_{M(s)} B_{(s)}$ in the RHP for an openloop stable process. Any encirclement means the closedloop system is unstable.

If the process is openloop *unstable*, $G_{M(s)}$ will have one or more poles in the RHP, so $F_{(s)} = 1 + G_{M(s)} B_{(s)}$ will also have one or more poles in the RHP. We can find out how many poles there are by solving for the roots of the *openloop characteristic equation* or by using the Routh stability criterion on the openloop characteristic equation (the denominator of $G_{M(s)}$). Once the number of poles P is known, the number of zeros can be found from Eq. (13.6).

13.1.2 Examples

Let us illustrate the mapping of the contour that goes around the entire RHP of the s plane using some examples.

Example 13.1. Consider our old friend the 3-CSTR process:

$$G_{M(s)} = \frac{\frac{1}{8}}{(s + 1)^3}$$

With a proportional feedback controller, the total openloop transfer function (process and controller) is

$$G_{M(s)} B_{(s)} = \frac{\frac{1}{8} K_c}{(s + 1)^3} \tag{13.7}$$

This system is openloop stable (with three poles that are all in the left half of the s plane and none in the RHP), so $P = 0$.

The contour around the entire RHP is shown in Fig. 13.2a. Let us split it up into three parts: C_+, the path up the positive imaginary axis from the origin to $+\infty$; C_R, the path around the infinitely large semicircle; and C_-, the path back up the negative imaginary axis from $-\infty$ to the origin.

(a) C_+ *contour.* On the C_+ contour the variable s is a pure imaginary number. Thus $s = i\omega$ as ω goes from 0 to $+\infty$. Substituting $i\omega$ for s in the total openloop system transfer function gives

$$G_{M(i\omega)} B_{(i\omega)} = \frac{\frac{1}{8}K_c}{(i\omega + 1)^3} \tag{13.8}$$

We now let ω take on values from 0 to $+\infty$ and plot the real and imaginary parts of $G_{M(i\omega)} B_{(i\omega)}$. This, of course, is just a polar plot of $G_{M(s)} B_{(s)}$ as sketched in Fig. 13.3a. The plot starts ($\omega = 0$) at $\frac{1}{8}K_c$ on the positive real axis. It ends at the origin, as ω goes to infinity, with a phase angle of $-270°$.

(b) C_R *contour.* On the C_R contour,

$$s = Re^{i\theta} \tag{13.9}$$

R will go to infinity and θ will take on values from $+\pi/2$ through 0 to $-\pi/2$ radians. Substituting Eq. (13.9) into $G_{M(s)} B_{(s)}$ gives

$$G_{M(s)} B_{(s)} = \frac{\frac{1}{8}K_c}{(Re^{i\theta} + 1)^3} \tag{13.10}$$

As R becomes large, the $+1$ term in the denominator can be neglected.

$$\lim_{R \to \infty} G_M B = \lim_{R \to \infty} \left(\frac{K_c}{8R^3} e^{-3\theta i} \right) \tag{13.11}$$

The magnitude of $G_M B$ goes to zero as R goes to infinity. Thus the infinitely large semicircle in the s plane maps into a point (the origin) in the $G_M B$ plane (Fig. 13.3b). The argument of $G_M B$ goes from $-3\pi/2$ through 0 to $+3\pi/2$ radians.

(c) C_- *contour.* On this contour s is again equal to $i\omega$, but now ω takes on values from $-\infty$ to 0. The $G_M B$ on this path is just the complex conjugate of the path with positive values of ω. See Fig. 13.3c.

The complete contour is shown in Fig. 13.3d. The bigger the value of K_c, the farther out on the positive real axis the $G_M B$ plot starts, and the farther out on the negative real axis is the intersection of the $G_M B$ plot with the axis.

If the $G_M B$ plot crosses the negative real axis beyond (to the left of) the critical $(-1, 0)$ point, the system is closedloop unstable. There would then be two encirclements of the $(-1, 0)$ point and therefore $N = 2$. Since we know P is zero, there must be two zeros in the RHP.

If the $G_M B$ plot crosses the negative real axis between the origin and the $(-1, 0)$ point, the system is closedloop stable. Now $N = 0$ and therefore $Z = N = 0$. There are no zeros of the closedloop characteristic equation in the RHP.

There is some critical value of gain K_c at which the $G_M B$ plot goes right through the $(-1, 0)$ point. This is the limit of closedloop stability. See Fig. 13.3e. The value of K_c at this limit should be the ultimate gain K_u that we have dealt with before in making root locus plots of this system. We found in Chap. 10 that $K_u = 64$ and $\omega_u = \sqrt{3}$. Let us see if the frequency-domain Nyquist stability criterion studied in this chapter gives the same results.

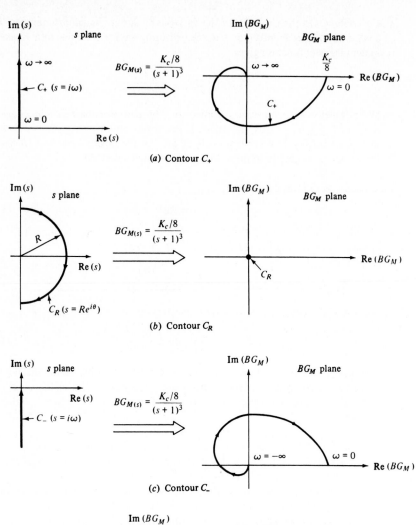

(a) Contour C_+

(b) Contour C_R

(c) Contour C_-

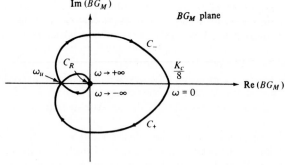

(d) Complete contour

FIGURE 13.3
Nyquist plots of three-CSTR system with proportional controller.

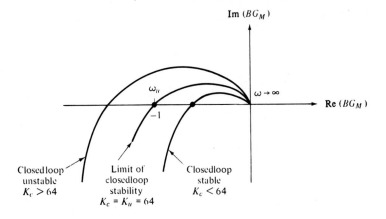

(e) Intersections on negative real axis

At the limit of closedloop stability

$$G_{M(i\omega)} B_{(i\omega)} = -1 + i0 \tag{13.12}$$

$$\left[\frac{\frac{1}{8}K_c}{s^3 + 3s^2 + 3s + 1}\right]_{s=i\omega} = \frac{\frac{1}{8}K_c}{(1 - 3\omega^2) + i(3\omega - \omega^3)}$$

$$= \frac{(\frac{1}{8}K_c)(1 - 3\omega^2)}{(1 - 3\omega^2)^2 + (3\omega - \omega^3)^2} + i\frac{(\frac{1}{8}K_c)(\omega^3 - 3\omega)}{(1 - 3\omega^2)^2 + (3\omega - \omega^3)^2} \tag{13.13}$$

Equating the imaginary part of the above to zero gives

$$\frac{(\frac{1}{8}K_c)(\omega^3 - 3\omega)}{(1 - 3\omega^2)^2 + (3\omega - \omega^3)^2} = 0$$

$$\omega = \sqrt{3} = \omega_u$$

This is exactly what we found from our root locus plot. This is the value of frequency at the intersection of the $G_M B$ plot with the negative real axis.

Equating the real part of Eq. (13.13) to -1 gives

$$\frac{(\frac{1}{8}K_c)(1 - 3\omega^2)}{(1 - 3\omega^2)^2 + (3\omega - \omega^3)^2} = -1$$

$$\frac{(\frac{1}{8}K_c)[1 - 3(3)]}{[1 - 3(3)]^2 + [(3\sqrt{3} - 3\sqrt{3})]^2} = -1$$

$$\frac{-K_c}{64} = -1 \quad \Rightarrow \quad K_c = 64 = K_u$$

This is the same ultimate gain that we found from the root locus plot.

Remember also that for gains greater than the ultimate gain, the root locus plot showed two roots of the closedloop characteristic equation in the

RHP. This is exactly the result we get from the Nyquist stability criterion $(N = 2 = Z)$.

Example 13.2. The system of Example 10.7 is second-order.

$$G_{M(s)} B_{(s)} = \frac{K_c}{(s + 1)(5s + 1)} \tag{13.14}$$

It has two poles, both in the LHP: $s = -1$ and $s = -\frac{1}{5}$. Thus the number of poles of $G_M B$ in the RHP is zero: $P = 0$. Let us break up the contour around the entire RHP into the same three parts used in the previous example.

(a) *On C_+ contour.* $s = i\omega$ goes from 0 to $+\infty$. This is just the polar plot of $G_{M(i\omega)} B_{(i\omega)}$. See Fig. 13.4a.

(b) *On C_R contour.* $s = Re^{i\theta}$ as $R \to \infty$ and θ goes from $\pi/2$ to $-\pi/2$.

$$G_{M(s)} B_{(s)} = \frac{K_c}{(Re^{i\theta} + 1)(5Re^{i\theta} + 1)}$$

$$\lim_{R \to \infty} G_{M(s)} B_{(s)} = \lim_{R \to \infty} \left[\frac{K_c}{5R^2} e^{-2\theta i} \right] = 0 \tag{13.15}$$

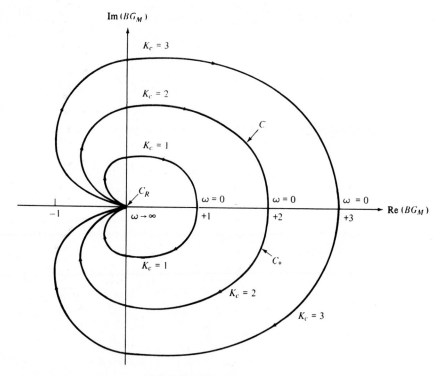

(a) Second-order system

FIGURE 13.4
(a) Nyquist plot of the second-order system; (b) s-plane contour to avoid pole at origin; (c) Nyquist plot of system with integrator.

Thus the infinite semicircle in the s plane again maps into the origin in the $G_M B$ plane. This happens for all transfer functions where the order of the numerator is greater than the order of the denominator.

(c) *On C_- contour.* $s = i\omega$ as ω goes from $-\infty$ to 0. The $G_{M(i\omega)} B_{(i\omega)}$ curve for negative values of ω is the reflection over the real axis of the curve for positive values of ω. So we really don't need to plot the C_- contour. The C_+ contour gives us all the information we need.

The complete Nyquist plot is shown in Fig. 13.4a for several values of gain K_c. Notice that the curves will *never* encircle the $(-1, 0)$ point, even as the gain is made infinitely large. This says that this second-order system can never be closedloop unstable. This is exactly what our root locus curves showed in Chap. 10.

As the gain is increased, the $G_M B$ curve gets closer and closer to the $(-1, 0)$ point. Later in this chapter we will use the closeness to the $(-1, 0)$ point as a specification for designing controllers.

Example 13.3. If the openloop transfer function of the system has poles that lie on the imaginary axis, the s-plane contour must be modified slightly to exclude these poles. A system with an integrator is a common example.

$$G_{M(s)} B_{(s)} = \frac{K_c}{s(\tau_{p1} s + 1)(\tau_{p2} s + 1)} \tag{13.16}$$

This system has a pole at the origin. We pick a contour in the s plane that goes counterclockwise around the origin, excluding the pole from the area enclosed by the contour. As shown in Fig. 13.4b, the contour C_0 is a semicircle of radius r_0. And r_0 is made to approach zero.

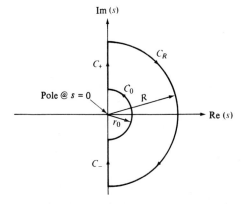

FIGURE 13.4 (*b*) *s*-plane contour with integrator in system

(a) *On C_+ contour.* $s = i\omega$ *as ω goes from r_0 to R, with r_0 going to 0 and R going to $+\infty$.*

$$G_{M(i\omega)} B_{(i\omega)} = \frac{K_c}{i\omega(1 + i\omega\tau_{p1})(1 + i\omega\tau_{p2})}$$

$$= \frac{-K_c \omega(\tau_{p1} + \tau_{p2}) - iK_c(1 - \tau_{p1}\tau_{p2}\omega^2)}{\omega^3(\tau_{p1} + \tau_{p2})^2 + \omega(1 - \tau_{p1}\tau_{p2}\omega^2)^2} \qquad (13.17)$$

The polar plot is shown in Fig. 13.4c.

(b) *On C_R contour.* $s = Re^{i\theta}$

$$G_{M(s)} B_{(s)} = \frac{K_c}{Re^{i\theta}(\tau_{p1} Re^{i\theta} + 1)(\tau_{p2} Re^{i\theta} + 1)} \qquad (13.18)$$

$$\lim_{R \to \infty} [G_{M(s)} B_{(s)}] = \lim_{R \to \infty} \left(\frac{K_c}{R^3 \tau_{p1} \tau_{p2}} e^{-3\theta i} \right) = 0$$

The C_R contour maps into the origin in the $G_M B$ plane.

(c) *On C_- contour. The $G_M B$ curve is the reflection over the real axis of the $G_M B$ curve for the C_+ contour.*

(d) *On C_0 contour. On this small semicircular contour*

$$s = r_0 e^{i\theta} \qquad (13.19)$$

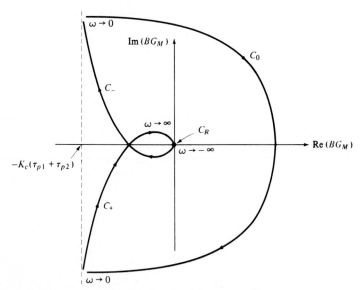

FIGURE 13.4   (c) Nyquist plot: $B_{(s)} G_{M(s)} = \dfrac{K_c}{s(\tau_{p1}s + 1)(\tau_{p2}s + 1)}$

The radius r_0 goes to zero and θ goes from $-\pi/2$ through 0 to $+\pi/2$ radians. The system transfer function becomes

$$G_{M(s)} B_{(s)} = \frac{K_c}{r_0 e^{i\theta}(\tau_{p1} r_0 e^{i\theta} + 1)(\tau_{p2} r_0 e^{i\theta} + 1)} \tag{13.20}$$

As r_0 gets very small, the $\tau_{p1} r_0 e^{i\theta}$ and $\tau_{p2} r_0 e^{i\theta}$ terms become negligible compared with unity.

$$\lim_{r_0 \to 0} (G_{M(s)} B_{(s)}) = \lim_{r_0 \to 0} \left(\frac{K_c}{r_0 e^{i\theta}}\right) = \lim_{r_0 \to 0} \left(\frac{K_c}{r_0} e^{-i\theta}\right) \tag{13.21}$$

Thus the C_0 contour maps into a semicircle in the $G_M B$ plane that has a radius that goes to infinity and a phase angle that goes from $+\pi/2$ through 0 to $-\pi/2$. See Fig. 13.4c.

The Nyquist plot does not encircle the $(-1, 0)$ point if the polar plot of $G_{M(i\omega)} B_{(i\omega)}$ crosses the negative real axis inside the unit circle. The system would then be closedloop *stable*.

The maximum value of K_c for which the system is still closedloop stable can be found by setting the real part of $G_{M(i\omega)} B_{(i\omega)}$ equal to -1 and the imaginary part equal to 0. The results are

$$K_u = \frac{\tau_{p1} + \tau_{p2}}{\tau_{p1} \tau_{p2}} \qquad \omega_u = \frac{1}{\sqrt{\tau_{p1} \tau_{p2}}} \tag{13.22}$$

As we have seen in the three examples above, the C_+ contour usually is the only one that we need to map into the $G_M B$ plane. Therefore from now on we will make only polar (or Bode or Nichols) plots of $G_{M(i\omega)} B_{(i\omega)}$.

Example 13.4. Figure 13.5a shows the polar plot of an interesting system that has conditional stability. The system openloop transfer function has the form

$$G_{M(s)} B_{(s)} = \frac{K_c(\tau_{z1} s + 1)}{(\tau_{p1} s + 1)(\tau_{p2} s + 1)(\tau_{p3} s + 1)(\tau_{p4} s + 1)} \tag{13.23}$$

If the controller gain K_c is such that the $(-1, 0)$ point is in the stable region indicated in Fig. 13.5a, the system is closedloop stable. Let us define three values of controller gain:

$$K_1 = \text{value of } K_c \text{ when } |G_{M(i\omega_1)} B_{(i\omega_1)}| = 1$$

$$K_2 = \text{value of } K_c \text{ when } |G_{M(i\omega_2)} B_{(i\omega_2)}| = 1$$

$$K_3 = \text{value of } K_c \text{ when } |G_{M(\omega_3)} B_{(i\omega_3)}| = 1$$

The system is closedloop stable for two ranges of feedback controller gain:

$$K_c < K_1 \quad \text{and} \quad K_2 < K_c < K_3 \tag{13.24}$$

This conditional stability is shown on a root locus plot for this system sketched in Fig. 13.5b.

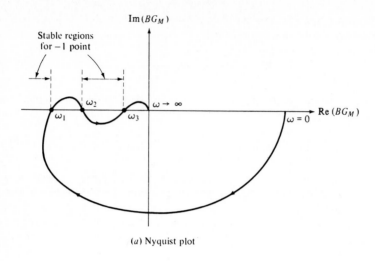

(a) Nyquist plot

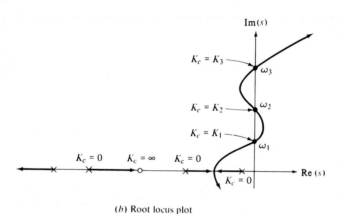

(b) Root locus plot

FIGURE 13.5
System with conditional stability.

13.1.3 Representation

In Chap. 12 we presented three different kinds of graphs that were used to represent the frequency response of a system: Nyquist, Bode, and Nichols plots. The Nyquist stability criterion was developed in the previous section for Nyquist or polar plots. The critical point for closedloop stability was shown to be the $(-1, 0)$ point on the Nyquist plot.

Naturally we also can show closedloop stability or instability on Bode and Nichols plots. The $(-1, 0)$ point has a phase angle of $-180°$ and a magnitude of unity or a log modulus of 0 decibels. The stability limit on Bode and Nichols plots is, therefore, the $(0 \text{ dB}, -180°)$ point. At the limit of closedloop stability

$$L = 0 \text{ dB} \quad \text{and} \quad \theta = -180° \tag{13.25}$$

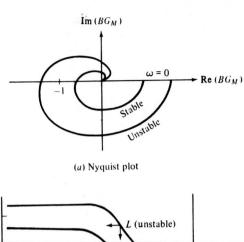

(*a*) Nyquist plot

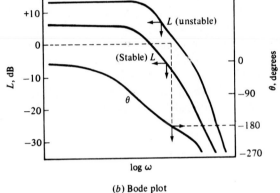

(*b*) Bode plot

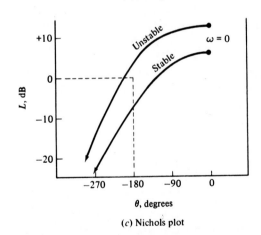

(*c*) Nichols plot

FIGURE 13.6
Stable and unstable closedloop systems in Nyquist, Bode, and Nichols plots.

The system is closedloop stable if

$$L < 0 \text{ dB} \qquad \text{at} \qquad \theta = -180°$$

$$\theta > -180° \qquad \text{at} \qquad L = 0 \text{ dB}$$

Figure 13.6 illustrates stable and unstable closedloop systems on the three types of plots.

Keep in mind that we are talking about *closedloop* stability and that we are studying it by making frequency-response plots of the total *openloop* system transfer function. We are also considering openloop stable systems most of the time. We will show how to deal with openloop unstable processes in Sec. 13.4.

13.2 CLOSEDLOOP SPECIFICATIONS IN THE FREQUENCY DOMAIN

There are two basic types of specifications that are commonly used in the frequency domain. The first type (*phase margin* and *gain margin*) specifies how near the *openloop* $G_{M(i\omega)} B_{(i\omega)}$ polar plot is to the critical $(-1, 0)$ point. The second type (*maximum closedloop log modulus*) specifies the height of the resonant peak on the log modulus Bode plot of the *closedloop* servo transfer function. So keep the apples and the oranges straight. We make *openloop* transfer function plots and look at the $(-1, 0)$ point. We make *closedloop* servo transfer function plots and look at the peak in the log modulus curve (indicating an underdamped system). But in both cases we are concerned with *closedloop* stability.

These specifications are easy to use, as we will show with some examples in Sec. 13.4. They can be related qualitatively to time-domain specifications such as damping coefficient.

13.2.1   Phase Margin

Phase margin (PM) is defined as the angle between the negative real axis and a radial line drawn from the origin to the point where the $G_M B$ curve intersects the unit circle. See Fig. 13.7. The definition is more compact in equation form.

$$\text{PM} = 180° + (\arg G_M B)_{|G_M B| = 1} \tag{13.26}$$

If the $G_M B$ polar plot goes through the $(-1, 0)$ point, the phase margin is zero. If the $G_M B$ polar plot crosses the negative real axis to the right of the $(-1, 0)$ point, the phase margin will be some positive angle. The bigger the phase margin, the more stable the closedloop system will be. A negative phase margin means an unstable closedloop system.

Phase margins of around 45° are often used. Figure 13.7 shows how phase margin is found on Bode and Nichols plots.

13.2.2   Gain Margin

Gain margin (GM) is defined as the reciprocal of the intersection of the $G_M B$ polar plot on the negative real axis.

$$\text{GM} = \frac{1}{|G_M B|_{\arg G_M B = -180°}} \tag{13.27}$$

Figure 13.8 shows gain margins on Nyquist, Bode, and Nichols plots. Gain margins are sometimes reported in decibels.

If the $G_M B$ curve goes through the critical $(-1, 0)$ point, the gain margin is unity (0 dB). If the $G_M B$ curve crosses the negative real axis between the origin and -1, the gain margin will be greater than 1. Therefore, the bigger the gain margin, the more stable the system, i.e., the farther away from -1 the curve crosses the real axis. Gain margins of around 2 are often used.

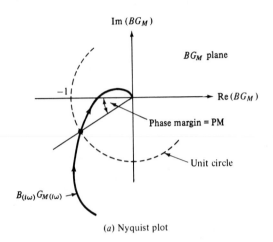

(a) Nyquist plot

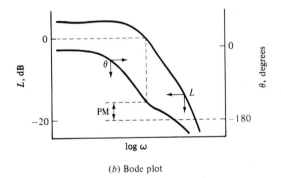

(b) Bode plot

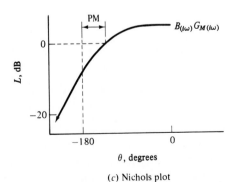

(c) Nichols plot

FIGURE 13.7
Phase margin.

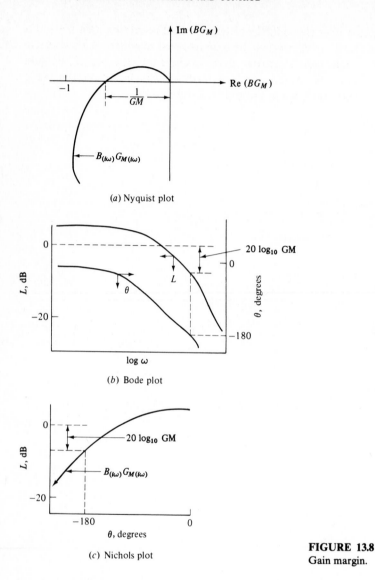

(a) Nyquist plot

(b) Bode plot

(c) Nichols plot

FIGURE 13.8
Gain margin.

A system must be third- or higher-order (or have deadtime) to have a meaningful gain margin. Polar plots of first- and second-order systems do not intersect the negative real axis.

13.2.3 Maximum Closedloop Log Modulus ($L_c^{\max}$)

The most useful frequency-domain specification is the maximum closedloop log modulus. The phase margin and gain margin specifications can sometimes give poor results when the shape of the frequency-response curve is unusual.

For example, consider the Nyquist plot of a process sketched in Fig. 13.9a where the shape of the $G_M B$ curve gives a good phase margin but the curve still passes very close to the $(-1, 0)$ point. The damping coefficient of this system would be quite low. This type of $G_M B$ curve is commonly encountered when the process has a large deadtime. Figure 13.9b gives a $G_M B$ curve that shows a good gain margin but passes too close to the $(-1, 0)$ point. These two cases illustrate that using phase or gain margins does not necessarily give the desired degree of damping. This is because these criteria measure the closeness of the $G_M B$ curve to the $(-1, 0)$ point at only one particular spot.

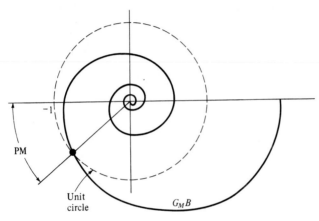

(a) Phase margin doesn't work

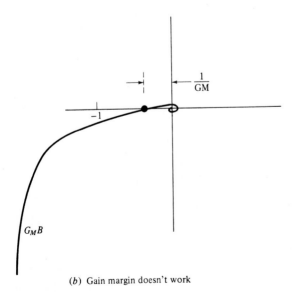

(b) Gain margin doesn't work

FIGURE 13.9

The maximum closedloop log modulus does not have these problems since it measures directly the closeness of the $G_M B$ curve to the $(-1, 0)$ point at all frequencies. The closedloop log modulus refers to the closedloop servo transfer function:

$$\frac{X_{(s)}}{X_{(s)}^{\text{set}}} = \frac{G_{M(s)} B_{(s)}}{1 + G_{M(s)} B_{(s)}} \tag{13.28}$$

The feedback controller is designed to give a maximum resonant peak or hump in the closedloop log modulus plot.

All the Nyquist, Bode, and Nichols plots discussed in previous sections have been for *openloop* system transfer functions $G_{M(i\omega)} B_{(i\omega)}$. Frequency-response plots can be made for any type of system, openloop or closedloop. The two closedloop transfer functions that we derived in Chap. 10 show how the output $X_{(s)}$ is affected in a closedloop system by a setpoint input $X_{(s)}^{\text{set}}$ and by a load $L_{(s)}$. Equation (13.28) gives the closedloop servo transfer function. Equation (13.29) gives the closedloop load transfer function.

$$\frac{X_{(s)}}{L_{(s)}} = \frac{G_{L(s)}}{1 + G_{M(s)} B_{(s)}} \tag{13.29}$$

Typical log modulus Bode plots of these two closedloop transfer functions are shown in Fig. 13.10a. If it were possible to achieve perfect or ideal control, the two ideal closedloop transfer functions would be

$$\frac{X_{(s)}}{L_{(s)}} = 0 \quad \text{and} \quad \frac{X_{(s)}}{X_{(s)}^{\text{set}}} = 1 \tag{13.30}$$

Equation (13.30) says that we want the output to track the setpoint perfectly for all frequencies, and we want the output to be unaffected by the load disturbance for all frequencies. Log modulus curves for these ideal (but unattainable) closedloop systems are shown in Fig. 13.10b.

In most systems, the closedloop servo log modulus curves move out to higher frequencies as the gain of the feedback controller is increased. This is desirable since it means a faster closedloop system. Remember, the breakpoint frequency is the reciprocal of the closedloop time constant.

But the height of the resonant peak also increases as the controller gain is increased. This means that the closedloop system becomes more underdamped. The effects of increasing controller gain are sketched in Fig. 13.10c.

A commonly used maximum closedloop log modulus specification is $+2$ dB. The controller parameters are adjusted to give a maximum peak in the closedloop servo log modulus curve of $+2$ dB. This corresponds to a magnitude ratio of 1.3 and is approximately equivalent to an underdamped system with a damping coefficient of 0.4.

Both the openloop and the closedloop frequency-response curves can be easily generated on a digital computer by using the complex variables and functions discussed in Chap. 12. The frequency-response curves for the closedloop

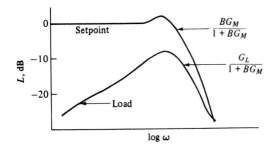

(a) Load and setpoint closedloop transfer functions

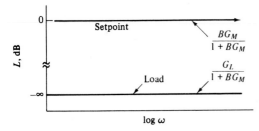

(b) Ideal load and setpoint closedloop transfer functions

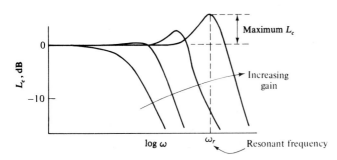

(c) Typical setpoint closedloop transfer functions

FIGURE 13.10
Closedloop log modulus curves.

servo transfer function can also be found fairly easily graphically by using a Nichols chart. This chart was developed many years ago before computers were available, and was widely used because it greatly facilitated the conversion of openloop frequency response into closedloop frequency response.

A Nichols chart is a graph that shows what the closedloop log modulus L_c and closedloop phase angle θ_c are for any given openloop log modulus L_0 and openloop phase angle θ_0. See Fig. 13.11a. The graph is a completely general one

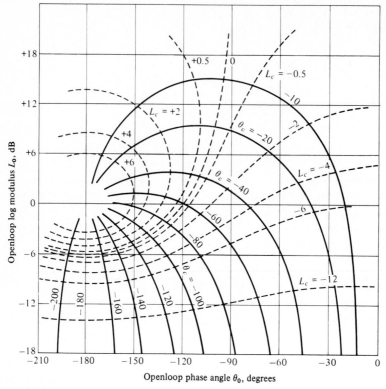

FIGURE 13.11a
Nichols chart.

and can be used for any system. To prove this, let us choose any arbitrary open-loop $G_{M(i\omega)} B_{(i\omega)}$. In polar form the openloop complex function is

$$G_{M(i\omega)} B_{(i\omega)} = r_0 e^{i\theta_0} \tag{13.31}$$

where r_0 = magnitude of the openloop complex function at frequency ω
$\quad\quad \theta_0$ = argument of the openloop complex function at frequency ω

The closedloop servo transfer function is

$$\frac{G_{M(s)} B_{(s)}}{1 + G_{M(s)} B_{(s)}} = \frac{r_0 e^{i\theta_0}}{1 + r_0 e^{i\theta_0}} \tag{13.32}$$

Putting this complex function into polar form gives

$$r_c e^{i\theta_c} = \frac{r_0 e^{i\theta_0}}{1 + r_0 e^{i\theta_0}} \tag{13.33}$$

where r_c = magnitude of the closedloop complex function at frequency ω
$\quad\quad \theta_c$ = argument of the closedloop complex function at frequency ω

Equation (13.33) can be rearranged to get r_c and θ_c as explicit functions of r_0 and θ_0.

$$r_c = \frac{r_0}{\sqrt{1 + 2r_0 \cos \theta_0 + r_0^2}} \tag{13.34}$$

$$\theta_c = \arctan \left(\frac{\sin \theta_0}{r_0 + \cos \theta_0} \right) \tag{13.35}$$

Thus for any arbitrary system with the given openloop parameters θ_0 and L_0, Eqs. (13.34) and (13.35) give the closedloop parameters θ_c and L_c. The Nichols chart is a plot of these relationships.

To use a Nichols chart, we first construct the openloop $G_M B$ Bode plots. Then we drawn an openloop Nichols plot of $G_{M(i\omega)} B_{(i\omega)}$. Finally we sketch this openloop curve of L_0 versus θ_0 onto a Nichols chart. At each point on this curve (which corresponds to a certain value of frequency), the values of the closedloop log modulus L_c can be read off.

Figure 13.11b is a Nichols chart with two $G_M B$ curves plotted on it. They are from the three-CSTR system with a proportional controller.

$$G_{M(s)} B_{(s)} = \frac{\frac{1}{8} K_c}{(s + 1)^3} \tag{13.36}$$

The two curves have two different values of controller gain: $K_c = 8$ and $K_c = 20$. The openloop Bode plots of $G_M B$ and the closedloop Bode plots of $G_{M(s)} B_{(s)} / (1 + G_{M(s)} B_{(s)})$, with $K_c = 20$, are given in Fig. 13.12.

The lines of constant closedloop log modulus L_c are part of the Nichols chart. If we are designing a closedloop system for an $L_c^{\max}$ specification, we merely have to adjust the controller type and settings so that the openloop $G_M B$ curve is tangent to the desired L_c line on the Nichols chart. For example, the $G_M B$ curve in Fig. 13.11b with $K_c = 20$ is just tangent to the $+2$ dB L_c line of the Nichols chart. The value of frequency at the point of tangency, 1.1 radians per minute, is the closedloop resonant frequency ω_r. The peak in the log modulus plot is clearly seen in the closedloop curves given in Fig. 13.12.

There are two aspects about using the maximum closedloop log modulus specification that you should be made aware of. First, the L_c curves can display multiple peaks. We always are looking for the highest peak, so make sure you cover the entire frequency range of interest when you plot the L_c curve. This multiple peak phenomenon can be a particularly confusing problem when you are using a computer program to determine the controller gain that gives a desired $L_c^{\max}$. Keep in mind that multiple peaks can occur in some systems.

The second item that you should be alert to is that a plot of $G_{M(s)} B_{(s)} / (1 + G_{M(s)} B_{(s)})$ only tells you how close you are to the $(-1, 0)$ point. As you make the gain bigger and bigger, approaching the ultimate gain, the peak in the curve increases. At the ultimate gain the peak height is infinite. However, if you continue to make the same plot for gains greater than the

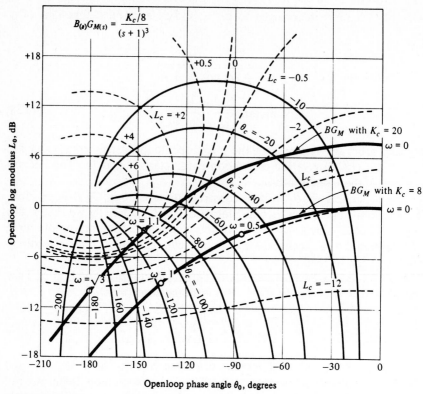

FIGURE 13.11b
Nichols chart with a three-CSTR system openloop $B_{(i\omega)} G_{M(i\omega)}$ plotted.

ultimate gain, the peak height will *decrease*. This is because the function we are plotting only measures the closeness of the $G_M B$ curve to the $(-1, 0)$ point. After the $(-1, 0)$ point has been encircled, increasing the controller gain moves the $G_M B$ curve farther away from the $(-1, 0)$ point on the other side. So be sure to check that the $G_M B$ curve does not encircle the $(-1, 0)$ point.

13.3 FREQUENCY RESPONSE OF FEEDBACK CONTROLLERS

Before we give some examples of the design of feedback controllers in the frequency domain, it would be wise to show what the common P, PI, and PID controllers look like in the frequency domain. These will be the $B_{(i\omega)}$'s that we will add to the process $G_{M(i\omega)}$ to get the total openloop Bode plots of $G_{M(i\omega)} B_{(i\omega)}$.

13.3.1 Proportional Controller (P)

The transfer function of a P controller is $B_{(s)} = K_c$. Substituting $s = i\omega$ gives

$$B_{(i\omega)} = K_c \tag{13.37}$$

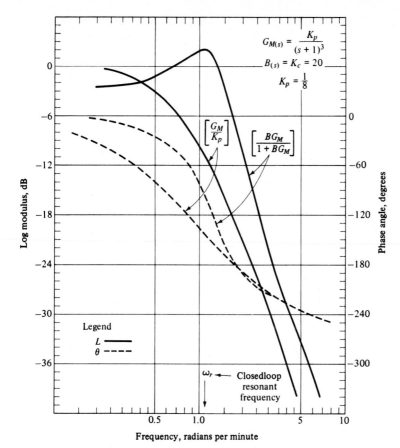

FIGURE 13.12
Openloop and closedloop Bode plots for a three-CSTR system.

A proportional controller merely multiplies the magnitude of $G_{M(i\omega)}$ at every frequency by a constant K_c. On a Bode plot, this means a proportional controller raises the log modulus curve by $20 \log_{10} K_c$ decibels but has no effect on the phase-angle curve. See Fig. 13.13a.

13.3.2 Proportional-Integral Controller (PI)

The transfer function of a PI controller is

$$B_{(s)} = K_c\left(1 + \frac{1}{\tau_I s}\right) = K_c\left(\frac{\tau_I s + 1}{\tau_I s}\right)$$

$$B_{(i\omega)} = K_c\left(\frac{\tau_I i\omega + 1}{\tau_I i\omega}\right)$$

(13.38)

The Bode plot of this combination of an integrator and a first-order lead is shown in Fig. 13.13b. At low frequencies, a PI controller amplifies magnitudes and contributes $-90°$ of phase-angle lag. This loss of phase angle is undesirable from a dynamic standpoint since it moves the $G_M B$ polar plot closer to the $(-1, 0)$ point.

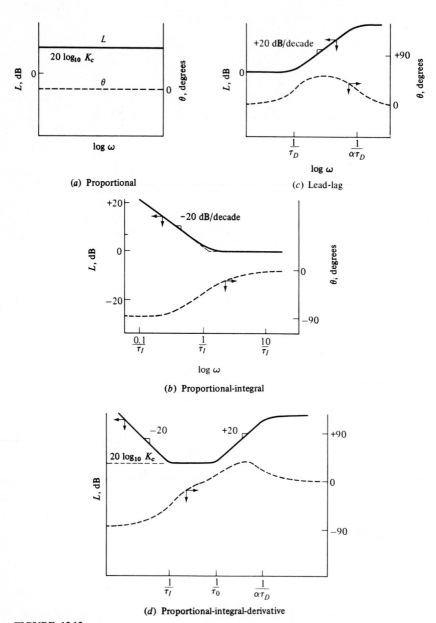

(a) Proportional

(c) Lead-lag

(b) Proportional-integral

(d) Proportional-integral-derivative

FIGURE 13.13
Bode plots of controllers.

13.3.3 Proportional-Integral-Derivative Controller (PID)

$$B_{(s)} = K_c\left(\frac{\tau_I s + 1}{\tau_I s}\right)\left(\frac{\tau_D s + 1}{\alpha \tau_D s + 1}\right) \tag{13.39}$$

The Bode plot for the lead-lag element is sketched in Fig. 13.13c. It contributes positive phase-angle advance over a range of frequencies between $1/\tau_D$ and $1/\alpha\tau_D$.

The lead-lag element can move the $G_M B$ curve away from the $(-1, 0)$ point and improve stability. When the derivative setting on a PID controller is tuned, the location of the phase-angle advance is shifted so that it occurs near the critical $(-1, 0)$ point.

13.4 EXAMPLES

13.4.1 Three-CSTR System

The process openloop transfer function is

$$G_{M(s)} = \frac{\frac{1}{8}}{(s + 1)^3}$$

Before we design controllers in the frequency domain, it might be interesting to see what the frequency-domain indicators of closedloop performance turn out to be when the Ziegler-Nichols settings are used on this system. Table 13.1 shows the phase and gain margins and the maximum closedloop log moduli that the Ziegler-Nichols settings give. Also shown in Table 13.1 are the results when the settings for a damping coefficient of 0.316 are used.

The Ziegler-Nichols settings give quite small phase and gain margins and large maximum closedloop log moduli. The $\zeta = 0.316$ settings are more conservative. Figure 13.14 shows the closedloop and openloop Bode plots for the PI controllers with the two different settings.

TABLE 13.1
Frequency-domain indicators that result from Ziegler-Nichols settings and 0.316 damping coefficient settings

	Ziegler-Nichols			0.316 Damping coefficient			
	P	**PI**	**PID**	**P**	**PI**	**PID**	**PID**
K_c	32	29.1	37.6	17	13	30	17
τ_I	...	3.03	1.82	...	3.03	1.82	1.82
τ_D	...	...	0.45	...	...	0.9	0.45
Phase margin, degrees	28	13	22	64	52	38	41
Gain margin	2	1.6	7	3.8	3.5	10	15
L_c^{max}, dB	6.9	13	8.3	0.5	1.9	3.8	3.2
Resonant frequency ω_r, radians per minute	1.3	1.3	1.6	1.0	0.8	1.6	1.0

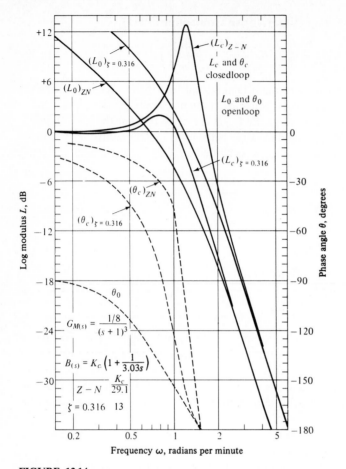

FIGURE 13.14
Bode plots of openloop BG_M and closedloop $BG_M/(1 + BG_M)$ for a three-CSTR system and PI controllers.

Now we are ready to find the controller settings required to give various frequency-domain specifications with P, PI, and PID controllers.

A. PROPORTIONAL CONTROLLER

1. Gain margin. Suppose we want to find the value of feedback controller gain K_c that gives a gain margin GM = 2. We must find the value of K_c that makes the Nyquist plot of

$$G_{M(s)} B_{(s)} = \frac{\frac{1}{8} K_c}{(s + 1)^3}$$

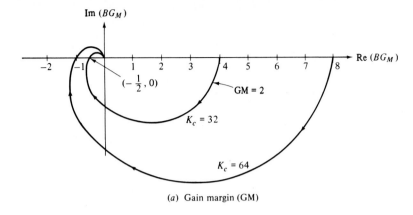

(a) Gain margin (GM)

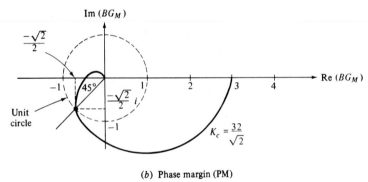

(b) Phase margin (PM)

FIGURE 13.15
Nyquist plots for a three-CSTR system with proportional controllers.

cross the negative real axis at $(-0.5, 0)$. As shown in Fig. 13.15a, the ultimate gain is 64. Thus a gain of 32 will reduce the magnitude of each point by one-half and make the $G_M B$ polar curve pass through the $(-0.5, 0)$ point.

Figure 13.16 shows the same result in Bode-plot form. When the phase angle is $-180°$ (at frequency $\omega_u = \sqrt{3}$), the magnitude must be 0.5 or the log modulus must be -6 dB. Thus the log modulus curve must be raised $+12$ dB (gain 4) above its position when the controller gain is 8. Therefore the total gain must be 32 for a GM = 2. Notice that this is the Ziegler-Nichols setting.

2. Phase margin. To get a 45° phase margin we must find the value of K_c that makes the Nyquist plot pass through the unit circle when the phase angle is $-135°$, as shown in Fig. 13.15b. The real and imaginary parts of $G_M B$ must both be equal to $-\frac{1}{2}\sqrt{2}$ at this point on the unit circle. Solving the two simultaneous equations gives

$$K_c = \frac{32}{\sqrt{2}} = 22.6 \quad \text{and} \quad \omega = 1 \text{ radian per minute}$$

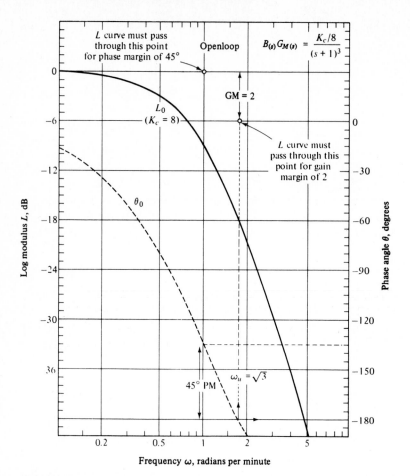

FIGURE 13.16
Bode plots of a three-CSTR system with proportional controller.

On a Bode plot (Fig. 13.16), the log modulus curve of $G_M B$ must pass through the 0-dB point when the phase-angle curve is at $-135°$. This occurs at $\omega = 1$ radian per minute. The log modulus curve for $K_c = 8$ must be raised $+9$ dB (gain 2.82). Therefore the controller gain must be $(8)(2.82) = 22.6$.

Notice that this gain is lower than that needed to give a gain margin of 2. The gain margin with a $K_c = 22.6$ can be easily found from the Bode plot. When the phase angle is $-180°$, the log modulus is -18 dB (for $K_c = 8$). If the gain of 22.6 is used, the log modulus is raised $+9$ dB. The log modulus is now -9 dB at the $-180°$ frequency, giving a gain margin of 2.82.

3. Maximum closedloop log modulus. We have already designed in Sec. 13.2.3 a proportional controller that gave an L_c^{max} of $+2$ dB. Figure 13.11b gives a

Nichols chart with the $G_M B$ curve for this system. A gain of 20 makes the open-loop $G_M B$ curve tangent to the $+2$-dB L_c curve on the Nichols chart.

From the three cases above we can conclude that, for this third-order system with three equal first-order lags, the $+2$-dB L_c^{max} specification is the most conservative, the $45°$ PM is next, and the 2 GM gives the controller gain that is closest to instability.

B. PROPORTIONAL-INTEGRAL CONTROLLERS. A PI controller has two adjustable parameters, and therefore we should, theoretically, be able to set two frequency-domain specifications and find the values of τ_I and K_c that satisfy them. We cannot make this choice of specifications completely arbitrary. For example, we cannot achieve a $45°$ phase margin and a gain margin of 2 with a PI controller in this three-CSTR system. A PI controller cannot reshape the Nyquist plot to make it pass through both the $(-\frac{1}{2}\sqrt{2}, -\frac{1}{2}\sqrt{2})$ point and the $(-0.5, 0)$ point because of the loss of phase angle at low frequencies.

Let us design a PI controller for a $+2$-dB L_c^{max} specification. For proportional controllers, all we have to do is find the value of K_c that makes the $G_M B$ curve on a Nichols chart tangent to the $+2$-dB L_c line. For a PI controller there are two parameters to find. Design procedures and guides have been developed over the years for finding the values of τ_I. The procedure has the following steps:

1. Plot the openloop $G_{M(i\omega)}$ on a Bode plot and then on a Nichols plot (see Figs. 13.11b and 13.12).
2. Move the $G_{M(i\omega)}$ curve vertically on the Nichols chart until it is tangent to the $+2$-dB L_c line. Read off the resonant frequency ω_r. (Figure 13.11b shows $\omega_r = 1.1$ radians per minute.)
3. Set the integral time constant at

$$\frac{1}{\tau_I} = 0.2\omega_r \qquad (13.40)$$

The idea is to make τ_I big enough to remove most of the integrator phase-angle lag from the total phase-angle curve at the frequency where the resonant peak occurs. (For our example $(0.2)(1.1) = 0.22 = 1/\tau_I \quad \Rightarrow \quad \tau_I = 4.5$ minutes).

4. Plot $B_{(i\omega)} = (1 + i\omega\tau_I)/i\omega\tau_I$ on the Bode plot and add it to $G_{M(i\omega)}$ to get the total $G_M B$ curves (Fig. 13.17).
5. Plot the openloop $G_{M(i\omega)} B_{(i\omega)}$ curve on a Nichols chart (Fig. 13.18 shows the new $G_M B$ curves for $K_c = 8$ and $K_c = 16$).
6. Move the $G_M B$ curve vertically until it is tangent to the $+2$-dB L_c line on the Nichols chart. The decibels that the plot must be moved give the required change in the controller gain (a controller gain of 16 in Fig. 13.18 gives the desired tangency to the $+2$-dB L_c line).

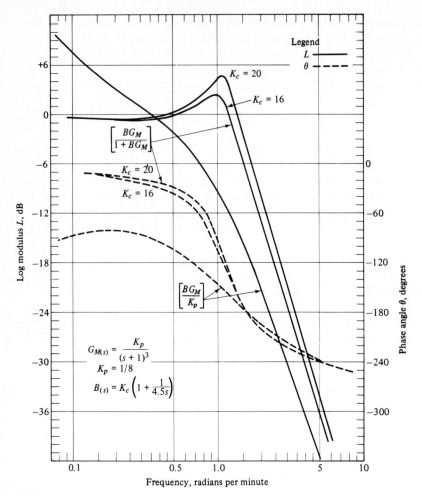

FIGURE 13.17
Openloop and closedloop Bode plots with PI controllers.

7. Find the new resonant frequency (the frequency at the point of tangency). If it has changed appreciably, repeat steps 3 to 6.

The time-domain performance of this PI controller ($K_c = 16$ and $\tau_I = 4.5$ min) is the $+2$-dB curve shown back in Chap. 7 (Fig. 7.16).

C. PROPORTIONAL-INTEGRAL-DERIVATIVE CONTROLLERS. PID controllers provide three adjustable parameters. We should theoretically be able to satisfy three specifications. A practical design procedure that I have used with good success for many years is outlined below:

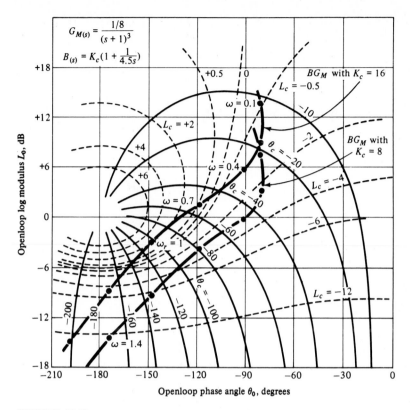

FIGURE 13.18
Nichols chart with PI controllers.

1. Determine the ultimate gain and ultimate frequency of the process from either the transfer function or experimentally. For our three-CSTR example, these values are $K_u = 64$ and $\omega_u = \sqrt{3}$.

2. Calculate the Ziegler-Nichols value for τ_I for a PI controller. Hold this value constant for the rest of the design. We only add integral action to eliminate steadystate offset, so it is not too critical what value is used, as long as it is reasonable, i.e., about the same magnitude as the process time constant. For our example, the ZN value for τ_I with a PI controller is 3.03 minutes.

3. Pick a value of τ_D and find the value of K_c that gives $+2$-dB L_c^{max}.

4. Repeat step 3 for a whole range of τ_D values, with a new value of K_c calculated at each new value of τ_D such that an $L_c^{max} = +2$ dB is achieved.

5. Select the value of τ_D that gives the maximum value of K_c. This τ_D^{opt} gives the largest gain and therefore the smallest closedloop time constant for a specified closedloop damping coefficient (as inferred from the L_c^{max} specification). Figure 13.19 gives the plot of K_c versus τ_D for the three-CSTR process. The optimum value for the derivative time is 1.0 min, giving a controller gain of 25.

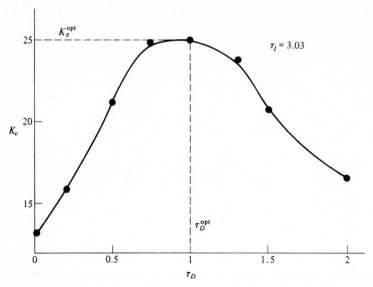

FIGURE 13.19
Optimum τ_D.

The above procedure leads to the controller settings: $K_c = 25$, $\tau_I = 3.03$ min, and $\tau_D = 1$ min. The gain and phase margins with these settings are 6.3 and 48°, respectively. The Ziegler-Nichols are $K_c = 37.6$, $\tau_I = 1.82$ min, and $\tau_D = 0.45$ min. The maximum closedloop log modulus for the ZN settings is $+8.3$ dB (Table 13.1), which is too underdamped for most chemical engineering systems.

13.4.2 First-Order Lag With Deadtime

Many chemical engineering systems can be modeled by this type of transfer function. Let us consider a typical transfer function

$$G_{M(s)} = \frac{K_p e^{-Ds}}{\tau_p s + 1} \tag{13.41}$$

We will look at several values of deadtime D. For all cases the values of K_p and τ_p will be set equal to unity. Other values of τ_p simply modify the frequency and time scales. Other values of K_p modify the controller gain.

The Bode plot of $G_{M(i\omega)}$ is given in Fig. 13.20 for $D = 0.5$. The ultimate gain is 3.9 (11.6 dB), and the ultimate frequency is 3.7 radians per minute. The ZN controller settings for P and PI controllers and the corresponding phase and gain margins and log moduli are shown in Table 13.2 for several values of deadtime D. Also shown are the K_c values for a proportional controller that give $+2$-dB maximum closedloop log modulus.

Notice that the ZN settings give very large phase margins for large dead-times. This illustrates that the phase-margin criterion would result in poor control for large deadtime processes. The ZN settings also give L_c^{max} that are too large when the deadtime is small, but too small when the deadtime is large. The $L_c^{max} = +2$ dB specification gives reliable controller settings for all values of deadtime.

Figure 13.20 shows the closedloop servo transfer function Bode plots for P and PI controllers with the ZN settings for a deadtime of 0.5 min. The effect of

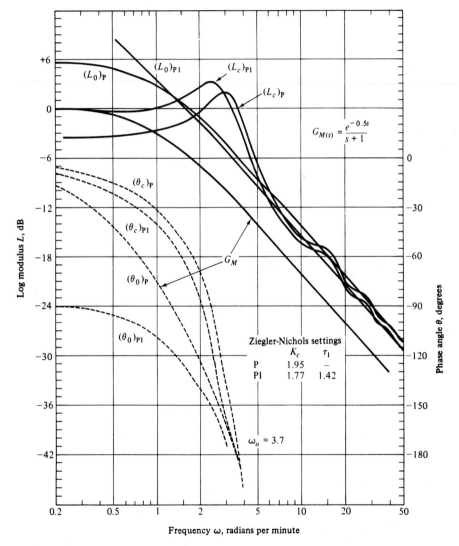

FIGURE 13.20
Openloop and closedloop plots for deadtime with lag process.

TABLE 13.2
Settings for first-order lag process
$(K_p = \tau_p = 1)$

Deadtime D	0.1	0.5	2
ZN (P)			
K_c	8.18	1.90	0.760
GM	2.0	2.0	2.0
PM, degrees	51	71	180
L_c^{max}, dB	2.8	1.6	0.34
ZN (PI)			
K_c	7.43	1.73	0.690
τ_I	0.321	1.42	4.58
GM	1.9	2.0	2.1
PM, degrees	30	53	98
L_c^{max}, dB	6.0	2.9	0.43
+2 dB tuning (P)			
K_c	7.70	1.95	0.833
GM	2.1	2.0	1.8
PM, degrees	54	73	180
L_c^{max}, dB	2.0	2.0	2.0

the deadtime on the first-order lag is to drop the phase angle below $-180°$. The system can be made closedloop unstable if the gain is high enough. Since there is always some deadtime in any real system, all real processes can be made closedloop unstable by making the feedback controller gain high enough.

Notice in Fig. 13.20 that the L_c curve for the P controller does not approach 0 dB at low frequencies. This shows that there is a steadystate offset with a proportional controller. The L_c curve for the PI controller does go to 0 dB at low frequencies because the integrator drives the closedloop servo transfer function to unity (i.e., no offset).

13.4.3 Openloop Unstable Processes

The Nyquist stability criterion can be used for openloop unstable processes, but we have to use the complete, rigorous version with P (the number of poles of the closedloop characteristic equation in the RHP) no longer equal to zero.

Consider the simple openloop unstable process

$$G_{M(s)} = \frac{K_p}{\tau_p s - 1} \tag{13.42}$$

We found in Chap. 11 that we could make this system closedloop stable by using a proportional controller with a gain K_c that was greater than $1/K_p$. Let us see if

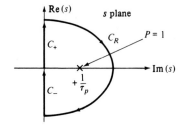

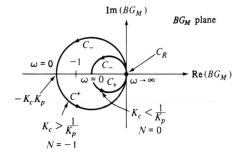

FIGURE 13.21
Nyquist stability criterion applied to an open-loop unstable process.

the Nyquist stability criterion leads us to the same conclusion. It certainly should if it is any good because a table in Chinese must be a table in Russian!

First of all, we know immediately that the openloop system transfer function $G_{M(s)} B_{(s)}$ has one pole (at $s = +1/\tau_p$) in the RHP. Therefore the closedloop characteristic equation

$$1 + G_{M(s)} B_{(s)} = 0$$

must also have one pole in the RHP; so $P = 1$.

On the C_+ contour up the imaginary axis, $s = i\omega$. We must make a polar plot of $G_{M(i\omega)} B_{(i\omega)}$.

$$G_{M(i\omega)} B_{(i\omega)} = \frac{K_c K_p}{\tau_p i\omega - 1} = \frac{K_c K_p(-1 - i\omega\tau_p)}{1 + \omega^2\tau_p^2} \tag{13.43}$$

Figure 13.21 shows that the curve starts ($\omega = 0$) at $-K_c K_p$ on the negative real axis where the phase angle is $-180°$. It ends at the origin, coming in with angle of $-90°$.

The C_R contour maps into the origin. The C_- contour is the reflection of the C_+ contour over the real axis.

If $K_c > 1/K_p$, the $(-1, 0)$ point is encircled. *But* the encirclement is in a counterclockwise direction! You will recall that all the curves considered up to now have encircled the $(-1, 0)$ point in a clockwise direction. A clockwise encirclement is a positive N. A counterclockwise encirclement is a negative N.

Therefore $N = -1$ for this example if $K_c > 1/K_p$. The number of zeros of the closedloop characteristic equation in the RHP is then

$$Z = P + N = 1 + (-1) = 0$$

Thus the system is closedloop stable if $K_c > 1/K_p$. This is exactly the conclusion we reached using root locus methods. So the Chinese frequency-domain conclusions are the same as the Russian Laplace-domain conclusions.

If $K_c < 1/K_p$, the $(-1, 0)$ point is not encircled and $N = 0$. The number of zeros of the closedloop characteristic equation is

$$Z = P + N = 1 + 0 = 1$$

The closedloop system has one zero in the RHP and is unstable if $K_c < 1/K_p$.

Figure 13.22 gives Nyquist plots for higher-order systems. For the third-order system, conditional stability can occur: the closedloop system is stable for controller gains between K_{min} and K_{max}. For gains greater than K_{max} there is one positive encirclement of the $(-1, 0)$ point, so $N = 1$ and $Z = P + N = 1 + 1 = 2$. The system is closedloop unstable.

For gains less than K_{min} there is *no* encirclement, and $N = 0$. This makes $Z = 1$ and the system is again closedloop unstable.

But for gains between K_{min} and K_{max} there is one negative encirclement of the $(-1, 0)$ point, so $N = -1$ and $Z = P + N = 1 - 1 = 0$. The system is closed-

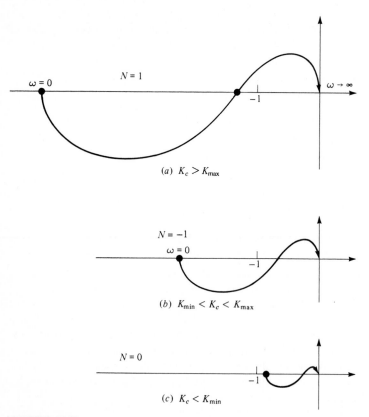

(a) $K_c > K_{max}$

(b) $K_{min} < K_c < K_{max}$

(c) $K_c < K_{min}$

FIGURE 13.22

loop stable since its closedloop characteristic equation has no zeros in the right half of the s plane.

PROBLEMS

13.1. (a) Make Bode, Nyquist, and Nichols plots of the system with $K_c = 1$:

$$B_{(s)} G_{M(s)} = \frac{K_c}{(s + 1)(5s + 1)(\frac{1}{2}s + 1)}$$

(b) Find the value of gain K_c that gives a phase margin of 45°. What is the gain margin?

(c) Find the value of gain K_c that gives a gain margin of 2. What is the phase margin?

(d) Find the value of gain K_c that gives a maximum closedloop log modulus of $+2$ dB. What are the gain and phase margins with this value of gain?

(e) Find the Ziegler-Nichols settings for this process and calculate the gain and phase margins and maximum closedloop log moduli that they give for P, PI, and PID controllers.

13.2. (a) Make Bode, Nyquist, and Nichols plots of the system with $K_c = 1$:

$$B_{(s)} G_{M(s)} = \frac{K_c(\frac{1}{2}s + 1)}{(s + 1)(5s + 1)}$$

(b) Find the value of gain K_c that gives a phase margin of 45°. What is the maximum closedloop log modulus with this value of gain?

(c) Find the value of gain K_c that gives a maximum closedloop log modulus of $+2$ dB. What is the phase margin with this value of gain?

13.3. (a) Make Bode, Nyquist, and Nichols plots of the system with $K_c = 1$:

$$B_{(s)} G_{M(s)} = \frac{K_c(-3s + 1)}{(s + 1)(5s + 1)}$$

(b) Find the ultimate gain and frequency.

(c) Find the value of K_c that gives a phase margin of 45°.

(d) Find the value of K_c that gives a gain margin of 2.

(e) Find the value of K_c that gives a maximum closedloop log modulus of $+2$ dB.

13.4. Repeat Prob. 13.3 for the system

$$B_{(s)} G_{M(s)} = K_c \left(1 + \frac{1}{2s}\right) \frac{-3s + 1}{(s + 1)(5s + 1)}$$

13.5. How would you use the "$Z - P = N$" theorem to develop a test for openloop stability?

13.6. A process has G_M and G_L openloop transfer functions that are first-order lags and gains: τ_M, τ_L, K_M, and K_L. Assume τ_M is twice τ_L. Sketch the log modulus Bode plot for the closedloop load transfer function when:

(a) A proportional-only feedback controller is used with $K_c K_M = 8$.

(b) A PI controller is used, with $\tau_I = \tau_M$ and the same gain as above.

13.7. (*a*) Sketch Bode, Nichols, and Nyquist plots of the closedloop servo and closedloop load transfer functions of the process

$$G_{L(s)} = G_{M(s)} = \frac{1}{10s + 1} \qquad B_{(s)} = 6\left(1 + \frac{1}{6s}\right)$$

(*b*) Calculate the phase margin and maximum closedloop log modulus for the system.

13.8. Using a first-order Pade approximation of deadtime, find the ultimate gain and frequency of the system:

$$B_{(s)} G_{M(s)} = \frac{K_c e^{-0.5s}}{s + 1}$$

Compare your answer with Sec. 13.4.2.

13.9. (*a*) Draw Bode, Nyquist, and Nichols plots of the system

$$B_{(s)} G_{M(s)} = \frac{K_c}{(s + 1)(s + 5)(s - 0.5)}$$

(*b*) Use the Nyquist stability criterion to find the values of K_c for which the system is closedloop stable.

13.10. (*a*) Make Nyquist and Bode plots of the openloop transfer function

$$B_{(s)} G_{M(s)} = \frac{K_c}{s^2(s + 1)}$$

(*b*) Is this system closedloop stable? Will using a PI controller stabilize it?

(*c*) Will a lead-lag element used as a feedback controller provide enough phase-angle advance to meet a 45° phase-margin specification?

(*d*) Find the values of τ_D and K_c that give a 45° phase margin. What is the gain margin?

$$B_{(s)} = K_c\left[\frac{\tau_D s + 1}{(\tau_D/20)s + 1}\right]$$

13.11. Find the largest value of deadtime D that can be tolerated in a process

$$G_{M(s)} = e^{-Ds}/s$$

and still achieve a 45° phase margin with a feedback controller having a reset time constant $\tau_I = 1$ minute. Find the value of gain K_c that gives the 45° of phase margin with the value of deadtime found above.

13.12. A process consists of two transfer functions in series. The first, G_{M1}, relates the manipulated variable M to the variable x_1 and is a steadystate gain of 1 and two first-order lags in series with equal time constants of 1 minute.

$$G_{M1(s)} = \frac{1}{(s + 1)^2}$$

The second, G_{M2}, relates x_1 to the controlled variable x_2 and is a steadystate gain of 1 and a first-order lag with a time constant of 5 minutes.

$$G_{M2(s)} = \frac{1}{5s + 1}$$

If a single proportional controller is used to control x_2 by manipulating M, determine the gain that gives a phase margin of 45 degrees. What is the maximum closedloop log modulus when this gain is used?

13.13. Suppose we want to use a cascade control system in the process considered above. The secondary or slave loop will control x_1 by manipulating M. The primary or master loop will control x_2 by changing the setpoint x_1^{set} of the secondary controller.

(a) Design a proportional secondary controller (K_1) that gives a phase margin of 45 degrees for the secondary loop.

(b) Using a value of gain for the secondary loop of $K_1 = 6.82$, design the master proportional controller (K_2) that gives a phase margin of 45 degrees for the primary loop.

(c) What is the maximum closedloop log modulus for the primary loop when this value of gain is used?

13.14. A process has the following transfer function:

$$G_M = \frac{1}{(\tau s + 1)^2}$$

It is controlled using a PI controller with τ_I set equal to τ.

(a) Sketch a root locus plot and calculate the controller gain that gives a closed-loop damping coefficient of 0.4.

(b) Sketch Bode, Nyquist, and Nichols plots of $G_M B$.

(c) Calculate analytically the gain that gives a phase margin of 45° and check your answer graphically.

(d) Determine the values of the maximum closedloop log modulus for the two values of gains from (a) and (c).

13.15. An openloop unstable, second-order process has one positive pole at $+1/\tau_1$ and one negative pole at $-1/\tau_2$. If a proportional controller is used and if $\tau_1 < \tau_2$, show by using a root locus plot and then by using the Nyquist stability criterion that the system is always unstable.

13.16. A process has the openloop transfer function $G_M = K_p/[s(\tau s + 1)]$. A proportional-only controller is used. Calculate analytically the closedloop damping coefficient that is equivalent to a phase margin of 45 degrees. What is the maximum closedloop log modulus when the controller gain that gives a 45-degree phase margin is used?

13.17. A process has the openloop transfer function

$$G_{M(s)} = \frac{K_p e^{-Ds}}{\tau_p s + 1}$$

where $K_p = 1, \tau_p = 1, D = 0.3$.

(a) Draw a Bode plot for the openloop system.

(b) What is the ultimate gain and ultimate frequency of this system?

(c) Using Ziegler-Nichols settings, draw a Bode plot for $G_M B$ when a PI controller is used on this process.

13.18. 250 gpm of cold 70°F liquid is fed into a 500-gallon perfectly mixed tank. The tank is heated by steam which condenses in a jacket surrounding the vessel. The heat of condensation of the steam is 950 Btu/lb$_m$. The liquid in the tank is heated to 180°F under steadystate conditions and continuously withdrawn from the tank to maintain a constant level. Heat capacity of the liquid is 0.9 Btu/lb_m °F; density is 8.33 lb_m/gal.

The control valve on the steam has linear installed characteristics and passes 500 lb_m/min when wide open. An electronic temperature transmitter (range: 50–250°F) is used. A temperature measurement lag of 10 seconds and a heat transfer lag of 20 seconds can be assumed. A proportional-only temperature controller is used.

(a) Derive a mathematical model of the system.

(b) Derive the openloop transfer functions between the output variable temperature (T) and the two input variables steam flow rate (F_s) and liquid inlet temperature (T_0).

(c) Sketch a root locus plot for the closedloop system.

(d) What are the ultimate gain and ultimate frequency ω_u?

(e) If a PI controller were used with $\tau_I = 5/\omega_u$, what value of controller gain would give a maximum closedloop modulus of +2 decibels?

(f) What are the gain margin and phase margin with the controller settings of part (e)?

13.19. Prepare a plot of closedloop damping coefficient versus phase margin for a process with an openloop transfer function

$$G_M B = \frac{K_c}{(\tau s + 1)^2}$$

13.20. A process has the following transfer function relating controlled and manipulated variables:

$$G_M = \frac{1}{(s + 1)^2}$$

If a proportional-only controller is used, what value of controller gain will give a maximum closedloop log modulus of +2 dB?

(a) Solve this problem analytically.

(b) Check your answer graphically using Bode and Nichols plots.

(c) Draw a root locus plot and determine the closedloop damping coefficient for the value of gain found in (a) and (b) above.

13.21. A process has an openloop process transfer function:

$$G_M = \frac{1}{(s - 1)(\frac{1}{10}s + 1)^2}$$

(a) Plot Bode, Nyquist, and Nichols plots for this system.

(b) If a proportional-only controller is used, over what range of controller gains K_c will the system be closedloop stable? Use frequency domain methods to determine your answer, but confirm them with a root locus plot.

13.22. Derive an analytical relationship between openloop maximum log modulus and damping coefficient for a second-order underdamped openloop system with a gain of unity. Show that a damping coefficient of 0.4 corresponds to a maximum log modulus of +2.7 decibels.

13.23. A first-order lag process with a time constant of 1 minute and a steadystate gain of 5°F/10^3 lb_m/h is controlled with a PI feedback controller.

(a) Sketch Bode plots of $G_M B$ when the reset time τ_I is very much smaller than 1 minute and when it is much larger than 1 minute.

(b) Find the biggest value of reset time τ_I for which a phase margin of 45 degrees is feasible.

13.24. A process has an openloop transfer function that is approximately a pure deadtime of D minutes. A proportional-derivative controller is to be used with a value of α equal to 0.1. What is the optimum value of the derivative time constant τ_D? Note that part of this problem involves defining what you mean by optimum.

13.25. A process has the following openloop transfer function:

$$G_M = \frac{0.5}{(10s + 1)(50s + 1)}$$

A proportional-only feedback controller is used.

(a) Make a root locus plot for the closedloop system.

(b) Calculate the value of controller gain that will give a closedloop damping coefficient of 0.707.

(c) Using this value of gain, what is the phase margin?

(d) What is the maximum closedloop modulus?

13.26. The frequency response data given below were obtained by pulse-testing a closedloop system that contained a proportional-only controller with a proportional band of 25. Controller setpoint was pulsed and the process measurement signal was recorded as the output signal.

(a) What is the openloop frequency response of the process?

(b) What is the openloop process transfer function?

ω, radians per minute	Log modulus, dB	Phase angle, degrees
0.1	2.50	-2
0.4	2.54	-6
0.8	2.65	-12
1.0	2.74	-16
2.0	3.39	-34
4.0	3.90	-90
6.3	-2.31	-151
8	-7.1	-172
10	-12.1	-187
16	-22.3	-211
20	-27.6	-221
40	-44.5	-244
80	-64.6	-256

13.27. 1000 kg/h of gas at 100°C flow through two pressurized cylindrical vessels in series. The first tank is 2 meters in diameter and 5 meters high. The second is 3 meters in diameter and 8 meters high. The molecular weight of the gas is 30.

The first tank operates at 2000 kPa at the initial steadystate. There is a pressure drop between the vessels which varies linearly with gas flow rate F_1. This pressure drop is 100 kPa when the flow rate is 1000 kg/h.

$$P_1 - P_2 = K_1 F_1$$

Assume the perfect gas law can be used ($R = 8.314$ kPa m^3/kg · mol K). Assume the pressure transmitter range is 1800–2000 kPa and that the valve has linear installed characteristics with a maximum flow rate of 2000 kg/h.

(a) Derive a mathematical model for the system.

(b) Determine the openloop transfer function between P_1 and the two inputs F_0 and F_2.

(c) Assuming a proportional-only controller is used to manipulate F_2 to control P_1, make a root locus plot and calculate the controller gain that gives a closed-loop damping coefficient of 0.3.

(d) Draw Bode, Nyquist, and Nichols plots for the openloop process transfer function P_1/F_2.

(e) Calculate the controller gain that gives 45 degrees of phase margin.

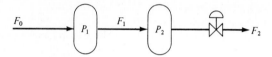

FIGURE P13.27

13.28. A process has the following openloop transfer function G_M relating controlled to manipulated variables:

$$G_{M(s)} = \frac{e^{-Ds}}{\tau s - 1}$$

(a) Sketch Bode, Nyquist, and Nichols plots for the system.

(b) If a proportional feedback controller is used, what is the largest ratio of D/τ for which a phase margin of 45° can be obtained?

13.29. A process has a transfer function $G_M = 1/[s(\tau_p s + 1)]$. It is controlled by a PI controller with reset time τ_I. Sketch the $G_M B$ plot of phase angle versus frequency, the Nyquist plot of $G_M B$, and a root locus plot for the case where

(a) $\tau_p > \tau_I$

(b) $\tau_p < \tau_I$

13.30. A process has an openloop transfer function that contains a positive pole at $+1/\tau$, a negative pole at $-10/\tau$ and a gain of unity. If a proportional-only controller is used, find the two values of controller gain that give a maximum closedloop log modulus of $+2$ decibels.

(a) Do this problem graphically.

(b) Solve it analytically.

13.31. A two-pressurized-tank process has an openloop transfer function

$$G_{M(s)} = \frac{K_p}{(\tau_p s + 1)s}$$

If a PI controller is used, find the smallest value of the ratio of the reset time τ_I to, the process time constant τ_p for which a maximum closedloop log modulus of $+2$ decibels is attainable.

13.32. A process with an openloop transfer function consisting of a steadystate gain, deadtime, and first-order lag is to be controlled by a PI controller. The deadtime (D) is one-fifth the magnitude of the time constant (τ).

Sketch Bode, Nyquist, and Nichols plots of the total openloop transfer function $(G_M B)$ when
(a) $\tau_I > 5\tau$
(b) $\tau_I = \tau$
(c) $\tau_I = 2D/\tau$

13.33. Write a short computer program that will calculate the feedback controller gain K_c that gives a maximum closedloop log modulus of $+2$ decibels for a process with the openloop transfer function

$$G_M = \frac{K_p e^{-Ds}}{\tau_p s + 1}$$

and a PI controller with the reset time set equal to $2D/\pi$.

13.34. A process has an openloop transfer function

$$G_M = \frac{1}{(s-1)(0.1s+1)}$$

Sketch phase angle plots for $G_M B$ and root locus plots when:
(a) A proportional controller is used.
(b) A PI controller is used with
  (i) $\tau_I \gg 0.1$
  (ii) $\tau_I \ll 0.1$
  (iii) $\tau_I = 0.1$
How would you find the maximum value of τ_I for which a $45°$ phase margin is attainable?

13.35. The frequency response Bode plot of the output of a closedloop system for setpoint changes, using a proportional controller with a gain of 10, shows the following features:
(a) The low-frequency asymptote on the log modulus plot is -0.828 dB.
(b) The breakpoint frequency is 11 radians per minute.
(c) The slope of the high-frequency asymptote is -20 dB per decade.
(d) The phase angle goes to $-90°$ as frequency becomes large.
Calculate the form and the constants in the openloop transfer function relating the controlled and the manipulated variables.

13.36. A process has the following transfer function:

$$G_{M(s)} = \frac{24.85(5s+1)e^{-5s}}{(2.93s+1)(16.95s+1)}$$

(a) Make a Bode plot of the openloop system.
(b) Determine the ultimate gain and frequency if a proportional controller is used.
(c) What value of gain for a proportional controller gives a maximum closedloop log modulus of $+2$ dB?
(d) If a PI controller is used with Ziegler-Nichols settings, what are the phase margin, gain margin, and maximum closedloop log modulus?

13.37. The block diagram of a dc motor is given in Fig. P13.37.
(a) Show that the openloop transfer function between the manipulated variable M (voltage to the armature) and the controlled variable X (angular position of the

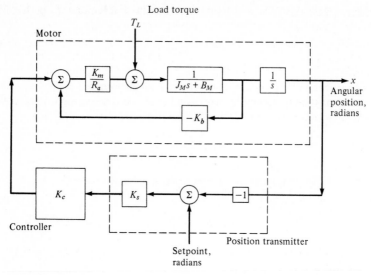

FIGURE P13.37

motor) is

$$G_M = \frac{16.13}{s(0.129s + 1)}$$

(b) Draw a root locus plot for this system if a proportional feedback controller is used.

(c) What value of gain gives a closedloop damping coefficient of 0.707?

(d) What value of gain gives a phase margin of 45°?

(e) What is the maximum closedloop log modulus using this gain?

R_a = armature resistance = 1 ohm	B_M = motor viscous friction
K_m = motor torque constant = 10 oz · in/A	= 0.1 oz · in/rad/s
K_b = back emf constant = 0.052 V/rad/s	K_s = transmitter gain = 1 V/rad
J_M = motor inertia = 0.08 oz · in/rad/s^2	K_c = feedback controller gain (V/V)

13.38. An ethanol-water distillation column is controlled by manipulating heat input to the reboiler to control the temperature on tray 6. The openloop process transfer function relating the tray 6 temperature transmitter output signal (PM) to the temperature controller output signal going to the steam valve on the reboiler is given below.

$$G_{M(s)} = \frac{PM}{CO} = \frac{360e^{-0.2s}}{(1.2s + 1)(590s + 1)}$$

(a) Make a Bode plot of $G_{M(i\omega)}$.

(b) If a proportional temperature controller is used, what are its ultimate gain K_u and ultimate frequency ω_u?

(c) Using a proportional controller, calculate the value of controller gain that gives
  (i) A phase margin of 45 degrees
  (ii) A gain margin of 2

(d) Using a PI controller with reset time set equal to 12.5 min, calculate the gain that gives a maximum closedloop log modulus of +2 dB.

13.39. The Superconducting Super Collider includes an enormous helium refrigeration system to cool the magnets to 4.3 K. Boiling helium at about 1 atmosphere and 4.1 K provides cooling. Vapor from the coolers is compressed, cooled by liquid nitrogen, and expanded to achieve the low temperatures required.

Suppose the openloop process transfer function relating controlled variable "magnet temperature" and manipulated variable "compressor speed" is

$$G_{M(s)} = \frac{1}{s(s + 1)}$$

(a) Make a Bode plot of $G_{M(i\omega)}$.
(b) Using a proportional controller, show that a controller gain of $\sqrt{2}$ gives a phase margin of 45 degrees.
(c) Using this controller gain, sketch a plot of the log modulus of the closedloop servo transfer function of the system.
(d) Will there be any steadystate error for setpoint changes in this system, despite the fact that only a proportional controller has been used?

13.40. If a proportional controller is used in the three-isothermal CSTR process, a controller gain of 22.6 gives a phase margin of 45°. A gain of 20 gives a maximum closedloop log modulus of +2 dB with a closedloop resonant frequency of 1.1 radian per minute.

(a) Calculate the controller gain that gives a phase margin of 45° when a PD controller is used with a derivative time set equal to 1.
(b) Calculate the PD controller gain (with $\tau_D = 1$) that gives a maximum closedloop log modulus of +2 dB. What is the closedloop resonant frequency?
(c) *Qualitatively* show why there is an optimum value of derivative time τ_D by *sketching* openloop phase angle Bode plots of this process with PD controllers for a range of values of derivative times from very small to very large.

CHAPTER
14

PROCESS IDENTIFICATION

14.1 PURPOSE

The dynamic relationships discussed thus far in this book were determined from mathematical models of the process. Mathematical equations, based on fundamental physical and chemical laws, were developed to describe the time-dependent behavior of the system. We assumed that the values of all parameters, such as holdups, reaction rates, heat transfer coefficients, etc., were known. Thus the dynamic behavior was predicted on essentially a theoretical basis.

For a process that is already in operation, there is an alternative approach that is based on experimental dynamic data obtained from plant tests. The experimental approach is sometimes used when the process is thought to be too complex to model from first principles. More often, it is used to find the values of some parameters in the model that are unknown. Many of the parameters can be calculated from steadystate plant data, but some parameters must be found from dynamic tests (e.g., holdups in nonreactive systems).

A third and very important use of dynamic experiments is to confirm the predictions of a theoretical mathematical model. As we indicated in Part I, the verification of the model is a very desirable step in its development and application.

Experimental identification of process dynamics has been an active area of research for many years by workers in several areas of engineering. The literature

502

is extensive, and entire books have been devoted to the subject. For a recent summary of the state of the art in this field see *System Identification*, L. Ljung, Prentice-Hall, Inc. 1987.

A number of techniques have been proposed. We will discuss only the more conventional methods that are widely used in the chemical and petroleum industries. Only the identification of linear transfer-function models will be discussed. Nonlinear identification is beyond the scope of this book.

There are several computer software packages that are quite helpful in applying some of the computationally intensive methods. The PC-MATLAB: System Identification Toolbox (The Math Works, Inc., Sherborn, Mass.) is an easy-to-use, powerful software package that provides an array of alternative tools.

14.2 DIRECT METHODS

14.2.1 Time-Domain "Eyeball" Fitting of Step Test Data

The most direct way of obtaining an empirical linear dynamic model of a process is to find the parameters (deadtime, time constant, and damping coefficient) that fit the experimentally obtained step response data. The process being identified is usually openloop, but experimental testing of closedloop systems is also possible.

We put in a step disturbance $m_{(t)}$ and record the output variable $x_{(t)}$ as a function of time, as illustrated in Fig. 14.1. The quick-and-dirty engineering approach is to simply look at the shape of the $x_{(t)}$ curve and find some approximate transfer function $G_{(s)}$ that would give the same type of step response.

Probably 80 percent of all chemical engineering openloop processes can be modeled by a gain, deadtime, and one lag.

$$G_{(s)} = K_p \frac{e^{-Ds}}{\tau_p s + 1} \tag{14.1}$$

The steadystate gain K_p is easily obtained from the ratio of the final steadystate change in the output $\overline{\Delta x}$ over the size of the step input $\overline{\Delta m}$. The deadtime can be easily read from the $x_{(t)}$ curve. The time constant can be estimated from the time it takes the output $x_{(t)}$ to reach 62.3 percent of the final steadystate change.

Closedloop processes are usually tuned to be somewhat underdamped, so a second-order underdamped model must be used.

$$G_{(s)} = K_p \frac{e^{-Ds}}{\tau^2 s^2 + 2\tau \zeta s + 1} \tag{14.2}$$

As shown in Fig. 14.1, the steadystate gain and deadtime are obtained in the same way as with a first-order model. The damping coefficient ζ can be calculated from the "peak overshoot ratio," POR (see Prob. 6.11), using Eq. (14.3).

$$POR = e^{-\pi \cot \phi} \tag{14.3}$$

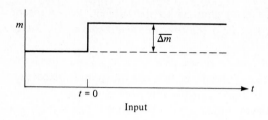

Input

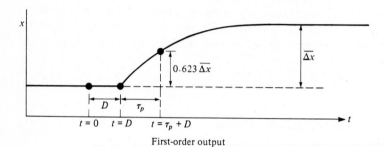

First-order output

Second-order output

FIGURE 14.1
Step response.

where

$$POR = \frac{\Delta x_{(t_p)} - \overline{\Delta x}}{\overline{\Delta x}}$$

(14.4)

$$\phi = \arccos \zeta$$

(14.5)

$\Delta x_{(t_p)}$ = change in $x_{(t)}$ at the peak overshoot
t_p = time to reach the peak overshoot (excluding the deadtime)

Then the time constant τ can be calculated from Eq. (14.6).

$$\frac{t_R}{\tau} = \frac{\pi - \phi}{\sin \phi} \qquad (14.6)$$

where t_R = time it takes the output to first reach the final steadystate value (see Fig. 14.1).

These "eyeball" estimation methods are simple and easy to use. They can provide a rough model that is adequate for many engineering purposes. For example, an approximate model can be used to get preliminary values for controller settings.

However, these crude methods cannot provide a precise, higher-order model and they are quite sensitive to nonlinearity. Most chemical engineering processes are fairly nonlinear. A step test drives the process away from the initial steadystate, and the values of the parameters of a linear transfer-function model may be significantly in error. If the magnitude of the step change could be made very small (sometimes as small as 10^{-4} to 10^{-6} percent of the normal value of the input), nonlinearity would not be a problem. But in most plant situations, such small changes would give output responses that could not be seen because of the normal noise on the signals. Thus step testing has definite limitations for plant testing.

14.2.2   Direct Sine-Wave Testing

The next level of dynamic testing is direct sine-wave testing. The input of the plant, which is usually a control-valve position or a flow-controller setpoint, is varied sinusoidally at a fixed frequency ω. After waiting for all transients to die out and for a steady oscillation in the output to be established, the amplitude ratio and phase angle are found by recording input and output data. See Fig. 12.2. The data point at this frequency is plotted on a Nyquist, Bode, or Nichols plot. Then the frequency is changed to another value and a new amplitude ratio and phase angle are determined. Thus the complete frequency-response curves are found experimentally by varying frequency over the range of interest.

Once the $G_{(i\omega)}$ curves have been found, they can be used directly to examine the dynamics and stability of the system or to design controllers in the frequency domain (see Chap. 13).

If a transfer-function model is desired, approximate transfer functions can be fitted to the experimental $G_{(i\omega)}$ curves. First the log modulus Bode plot is used. The low-frequency asymptote gives the steadystate gain. The time constants can be found from the breakpoint frequency and the slope of the high-frequency asymptote. The damping coefficient can be found from the resonant peak.

Once the log modulus curve has been adequately fitted by an approximate transfer function $G_{(i\omega)}^A$, the phase angle of $G_{(i\omega)}^A$ is compared with the experimental phase-angle curve. The difference is usually the contribution of deadtime. The procedure is illustrated in Fig. 14.2.

It is usually important to get an accurate fit of the frequency response of the model to the experimental frequency *only* near the critical region where the phase

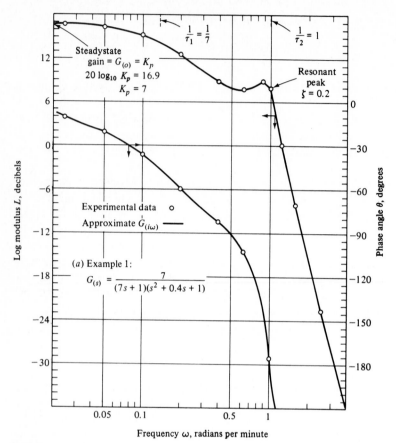

FIGURE 14.2a

Fitting approximate transfer function to experimental frequency-response data.

angle is between $-135°$ and $-180°$. It doesn't matter how well or how poorly the approximate transfer function fits the data once the phase angle has dropped below $-180°$. So the fitting of the approximate transfer function should weigh heavily the differences between the model and the data over this frequency range.

Direct sine-wave testing is an extremely useful way to obtain precise dynamic data. Damping coefficients, time constants, and system order can all be quite accurately found. Direct sine-wave testing is particularly useful for processes with signals that are noisy. Since you are putting in a sine wave signal with a known frequency and the output signal has this same frequency, you can easily filter out all of the noise signals at other frequencies and obtain an output signal with a much higher signal-to-noise ratio.

The main disadvantage of direct sine-wave testing is that it can be very time-consuming when applied to typical large time-constant chemical process equipment. The steadystate oscillation must be established at each value of frequency. It can take days to generate the complete frequency-response curves of a slow process.

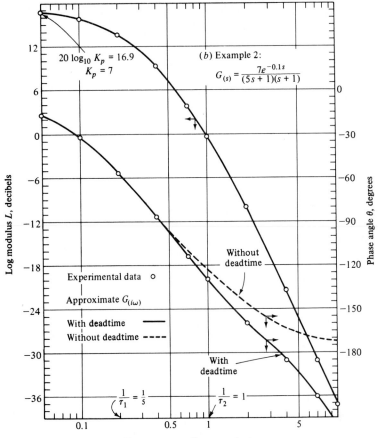

FIGURE 14.2b

When the test is being conducted over this long period of time, other disturbances and changes in operating conditions can occur that can affect the results of the test. Therefore direct sine-wave testing is only rarely used to get the complete frequency response.

However, it can be very useful for getting accurate data at one or two important frequencies. For example, it can be used to get amplitude and phase angle data near the critical $-180°$ point.

14.3 PULSE TESTING

One of the most useful and practical methods for obtaining experimental dynamic data from many chemical engineering processes is pulse testing. It yields reasonably accurate frequency-response curves and requires only a fraction of the time that direct sine-wave testing takes.

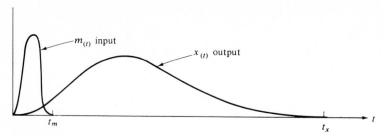

FIGURE 14.3
Pulse test input and output curves.

An input pulse $m_{(t)}$ of fairly arbitrary shape is put into the process. This pulse starts and ends at the same value and is often just a square pulse (i.e., a step up at time zero and a step back to the original value at a later time t_m). See Fig. 14.3. The response of the output is recorded. It typically returns eventually to its original steadystate value. If $x_{(t)}$ and $m_{(t)}$ are perturbations from steadystate, they start and end at zero. The situation where the output does not return to zero will be discussed in Sec. 14.3.4.

The input and output functions are then Fourier-transformed and divided to give the system transfer function in the frequency domain $G_{(i\omega)}$. The details of one procedure for accomplishing this Fourier transformation are discussed in the following sections, and a little digital computer program that does this job is given in Table 14.1. Alternative methods include the use of "Fast Fourier Transforms," which are available in most computing centers.

In theory only one pulse input is required to generate the entire frequency-response curve. In practice several pulses are usually needed to establish the required size and duration of the input pulse. Some tips on the practical aspects of pulse testing are discussed in Sec. 14.3.3.

14.3.1   Calculation of $G_{(i\omega)}$ From Pulse Test Data

Consider a process with an input $m_{(t)}$ and an output $x_{(t)}$. By definition, the transfer function of the process is

$$G_{(s)} = \frac{X_{(s)}}{M_{(s)}} = \frac{\displaystyle\int_0^\infty x_{(t)} e^{-st}\, dt}{\displaystyle\int_0^\infty m_{(t)} e^{-st}\, dt} \tag{14.7}$$

We now go into the frequency domain by substituting $s = i\omega$.

$$G_{(i\omega)} = \frac{\displaystyle\int_0^\infty x_{(t)} e^{-i\omega t}\, dt}{\displaystyle\int_0^\infty m_{(t)} e^{-i\omega t}\, dt} \tag{14.8}$$

TABLE 14.1

Pulse test program

```
      DIMENSION XIN(200),TIN(200),XOUT(200),TOUT(200)
      COMPLEX GNUM,GDENOM,G1,G2,G3,G4,G5,G
      OPEN(6,FILE='XXX')
      OPEN(7,FILE='PDATA')
C   READ INPUT AND OUTPUT DATA
      READ(7,1)NIN,NOUT,WO,WMAX,WNUM
    1 FORMAT(2I5,3F10.5)
      DO 5 I=1,NIN
    2 FORMAT (2F10.5)
    5 READ(7,2)TIN(I),XIN(I)
      WRITE(6,3)
    3 FORMAT('    TIN    XIN')
      DO 10 I=1,NIN
   10 WRITE(6,4)TIN(I),XIN(I)
    4 FORMAT(1X,2F10.5)
      DO 15 I=1,NOUT
   15 READ(7,2)TOUT(I),XOUT(I)
      WRITE(6,6)
    6 FORMAT('    TOUT    XOUT')
      DO 20 I=1,NOUT
   20 WRITE(6,4) TOUT(I),XOUT(I)
      DW=10.**(1./WNUM)
      W=0.
  100 IF(W.GT.WMAX) STOP
      IF(NIN.GT.1) GO TO 30
C   CALCULATE FIT FOR RECTANGULAR PULSE INPUT
      IF(W.EQ.0.) GO TO 25
      G1=CMPLX(0.,W)
      G2=CMPLX(0.,-W*TIN(1))
      GDENOM=XIN(1)*(1.-CEXP(G2))/G1
      GO TO 50
C   FOR ZERO FREQUENCY
   25 GDENOM= CMPLX(XIN(1)*TIN(1),0.)
      GO TO 50
C   CALCULATE FIT FOR ARBITRARY INPUT
   30 IF(W.EQ.0.) GO TO 40
      G1=CMPLX(0.,W)
      G2=CMPLX(0.,-W*TIN(1))
      GDENOM=XIN(1)*((CEXP(G2)-1.)/(TIN(1)*W**2)-CEXP(G2)/G1)
      DO 35 N=2,NIN
      DELTA=TIN(N)-TIN(N-1)
      G2=CMPLX(0.,-W*DELTA)
      G3=CMPLX(0.,-W*TIN(N-1))
      G4=CEXP(G2)
      G5=(G4-1.)/(DELTA*W**2)
      GDENOM=GDENOM+CEXP(G3)*(XIN(N)*(G5-G4/G1)-XIN(N-1)*(G5-1./G1))
   35 CONTINUE
      GO TO 50
   40 AREA=XIN(1)*TIN(1)/2.
      DO 41 N=2,NIN
      DELTA=TIN(N)-TIN(N-1)
   41 AREA=AREA+(XIN(N)+XIN(N-1))*DELTA/2.
      GDENOM=CMPLX(AREA,0.)
C   CALCULATE FIT FOR ARBITRARY OUTPUT
   50 IF(W.EQ.0.) GO TO 60
      G2=CMPLX(0.,-W*TOUT(1))
      GNUM=XOUT(1)*((CEXP(G2)-1.)/(TOUT(1)*W**2)-CEXP(G2)/G1)
```

TABLE 14.1 (*continued*)

```
   DO 55 N=2,NOUT
   DELTA=TOUT(N)-TOUT(N-1)
   G2=CMPLX(0.,-W*DELTA)
   G3=CMPLX(0.,-W*TOUT(N-1))
   G4=CEXP(G2)
   G5=(G4-1.)/(DELTA*W**2)
   GNUM=GNUM+CEXP(G3)*(XOUT(N)*(G5-G4/G1)-XOUT(N-1)*(G5-1./G1))
55 CONTINUE
   GO TO 70
60 AREA =XOUT(1)*TOUT(1)/2.
   DO 61 N=2,NOUT
   DELTA=TOUT(N)-TOUT(N-1)
61 AREA=AREA+(XOUT(N)+XOUT(N-1))*DELTA/2.
   GNUM=CMPLX(AREA,0.)
C  CALCULATE TRANSFER FUNCTION
70 G=GNUM/GDENOM
   IF(W.EQ.0.) GO TO 90
   DB=20.*ALOG10(CABS(G)/ABS(GAIN))
   DEG=ATAN(AIMAG(G)/REAL(G))*180./3.1416
   IF((REAL(G)/GAIN).LT.0.) DEG=DEG-180.
   WRITE(6,75)W,G,DB,DEG
75 FORMAT(1X,F10.3,2F10.5,2F10.2)
   W=W*DW
   GO TO 100
90 GAIN=REAL(G)
   WRITE(6,91)GAIN
91 FORMAT(' STEADYSTATE GAIN = ',F10.3)
   WRITE(6,92)
92 FORMAT(' FREQUENCY   REAL   IMAGINARY   LOG MODULUS  ANGLE')
   WRITE(6,93)
93 FORMAT(' (RADIANS/TIME)            (DB)    (DEGREES)')
   W=WO
   GO TO 100
   END
```

Results

```
    TIN      XIN
  2.00000   2.00000
   TOUT     XOUT
  0.50000   8.00000
  1.00000  14.00000
  1.50000  18.00000
  2.00000  20.00000
  2.50000  15.00000
  3.00000  10.00000
  4.00000   8.00000
  6.00000   4.00000
  8.00000   1.00000
 10.00000   0.00000
STEADYSTATE GAIN =    16.750
FREQUENCY    REAL    IMAGINARY   LOG MODULUS  ANGLE
(RADIANS/TIME)              (DB)    (DEGREES)
   0.100  16.15245  -3.25677   -0.14   -11.40
   0.112  16.00172  -3.62761   -0.18   -12.77
   0.126  15.81426  -4.03306   -0.23   -14.31
   0.141  15.58188  -4.47321   -0.28   -16.02
   0.158  15.29486  -4.94672   -0.36   -17.92
   0.178  14.94228  -5.44998   -0.45   -20.04
```

TABLE 14.1 (*continued*)

0.200	14.51186	-5.97646	-0.57	-22.38
0.224	13.99071	-6.51557	-0.71	-24.97
0.251	13.36624	-7.05156	-0.89	-27.81
0.282	12.62782	-7.56243	-1.12	-30.92
0.316	11.76938	-8.01916	-1.41	-34.27
0.355	10.79302	-8.38571	-1.77	-37.85
0.398	9.71363	-8.62092	-2.21	-41.59
0.447	8.56384	-8.68293	-2.76	-45.40
0.501	7.39782	-8.53761	-3.42	-49.09
0.562	6.29103	-8.17123	-4.21	-52.41
0.631	5.33182	-7.60556	-5.12	-54.97
0.708	4.60094	-6.91001	-6.10	-56.34
0.794	4.13917	-6.19975	-7.03	-56.27
0.891	3.91400	-5.60731	-7.78	-55.08
1.000	3.81150	-5.22483	-8.27	-53.89
1.122	3.68337	-5.04455	-8.57	-53.86
1.259	3.44283	-4.95996	-8.86	-55.23
1.413	3.12179	-4.86728	-9.24	-57.32
1.585	2.78309	-4.76782	-9.64	-59.73
1.778	2.39771	-4.68237	-10.06	-62.88
1.995	1.97872	-4.53856	-10.59	-66.44
2.239	1.61549	-4.45273	-10.97	-70.06
2.512	0.88746	-4.48344	-11.28	-78.80
2.818	-0.09755	-3.62007	-13.30	-91.54
3.162	0.37503	-2.53385	-16.31	-81.58
3.548	0.51496	-3.06162	-14.64	-80.45
3.981	-0.31710	-2.58472	-16.17	-96.99
4.467	-0.37518	-1.89573	-18.76	-101.19
5.012	-0.19475	-1.25643	-22.39	-98.81
5.623	0.50922	-0.86185	-24.47	-59.42
6.310	-23.95585	-0.00365	3.11	-179.99
7.079	-0.26390	-0.69573	-27.05	-110.77
7.943	0.20991	-0.98951	-24.38	-78.02
8.913	-0.10367	-1.27008	-22.37	-94.67
10.000	-0.17422	-1.13571	-23.27	-98.72

The numerator is the Fourier transformation of the time function $x_{(t)}$. The denominator is the Fourier transformation of the time function $m_{(t)}$. Therefore the frequency response of the system $G_{(i\omega)}$ can be calculated from the experimental pulse test data $x_{(t)}$ and $m_{(t)}$ as shown in Fig. 14.3.

$$G_{(i\omega)} = \frac{\int_0^\infty x_{(t)} \cos(\omega t)\, dt - i \int_0^\infty x_{(t)} \sin(\omega t)\, dt}{\int_0^\infty m_{(t)} \cos(\omega t)\, dt - i \int_0^\infty m_{(t)} \sin(\omega t)\, dt} \tag{14.9}$$

$$= \frac{A - iB}{C - iD} = \frac{(AC + BD) + i(AD - BC)}{C^2 + D^2} \tag{14.10}$$

$$= \text{Re}\,[G_{(i\omega)}] + i\,\text{Im}\,[G_{(i\omega)}] \tag{14.11}$$

where

$$A = \int_0^{t_x} x_{(t)} \cos (\omega t) \, dt \qquad B = \int_0^{t_x} x_{(t)} \sin (\omega t) \, dt \qquad (14.12)$$

$$C = \int_0^{t_m} m_{(t)} \cos (\omega t) \, dt \qquad D = \int_0^{t_m} m_{(t)} \sin (\omega t) \, dt \qquad (14.13)$$

The problem reduces to being able to evaluate the integrals A, B, C, and D given in Eqs. (14.12) and (14.13) for known functions $x_{(t)}$ and $m_{(t)}$. The integrations are with respect to time between the definite limits of zero and the times that the experimental time functions go to zero: t_x for the output $x_{(t)}$ and t_m for the input $m_{(t)}$.

A specific numerical value of frequency ω is picked. The integrations are performed numerically (see Sec. 14.3.2 below), giving one point on the frequency-response curves. Then frequency is changed and the integrations repeated, using the same experimental time functions $x_{(t)}$ and $m_{(t)}$ but a new value of frequency ω. Repeating for frequencies over the range of interest gives the complete $G_{(i\omega)}$. The $x_{(t)}$ and $m_{(t)}$ data are used over and over again.

The integrations shown in Eqs. (14.12) and (14.13) are performed on a digital computer, and the problem of numerical integration again rears its ugly head. The problem is made particularly difficult by the oscillatory behavior of the sine and cosine terms at high values of frequency.

14.3.2 Digital Evaluation of Fourier Transformations

The experimental data from a pulse test are usually two continuous curves of x and m recorded as functions of time. A reasonable number of points are selected from these curves and fed into the digital computer. We will discuss later what a reasonable number is.

Let the value of $x_{(t)}$ at the kth increment in time be x_k as shown in Fig. 14.4. We feed into the computer values of x at specified points in time:

$$(t_1, x_1), (t_2, x_2), \ldots, (t_k, x_k), \ldots, (t_N, x_N)$$

Notice that at $t_N = t_x$, the value of x (x_N) is zero. These data points need *not* be selected at equally spaced intervals Δt_k. The total number of intervals is N.

From Eq. (14.8), we wish to evaluate the Fourier integral transform (FIT) of $x_{(t)}$.

$$\text{FIT} \equiv \int_0^\infty x_{(t)} e^{-i\omega t} \, dt \qquad (14.14)$$

We can break up the total interval (0 to t_x) into a number of unequal subintervals of length Δt_k. Then the FIT can be written, with no loss of rigor, as a sum of integrals:

$$\text{FIT} = \sum_{k=1}^N \left(\int_{t_{k-1}}^{t_k} x_{(t)} e^{-i\omega t} \, dt \right) \qquad (14.15)$$

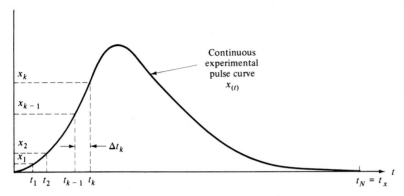

FIGURE 14.4
Discrete data points picked from an experimental pulse curve.

Over each interval t_{k-1} to t_k the true time function $x_{(t)}$ is now approximated by some polynomial approximating functions $\phi_{k(t)}$. A number of types can be used, but the simplest is a first-order approximation. This corresponds to using a straight line between the data points.

$$x_{(t)} \simeq \phi_{k(t)} \qquad \text{for} \qquad t_{k-1} < t < t_k$$

$$\phi_{k(t)} = \alpha_{0k} + \alpha_{1k}(t - t_{k-1}) \tag{14.16}$$

Equation (14.16) is the equation of a straight line. The constant α_{1k} is the slope of the line over the kth interval.

$$\alpha_{1k} = \frac{x_k - x_{k-1}}{\Delta t_k} \tag{14.17}$$

where $\Delta t_k = t_k - t_{k-1}$. The constant α_{0k} is the value of ϕ_k at the beginning of the interval.

$$\alpha_{0k} = x_{k-1} \tag{14.18}$$

The constants α_{0k} and α_{1k} change with each interval. Equation (14.15) can thus be approximated by

$$\text{FIT} \simeq \sum_{k=1}^{N} \left(\int_{t_{k-1}}^{t_k} [\alpha_{0k} + \alpha_{1k}(t - t_{k-1})] e^{-i\omega t}\, dt \right) \tag{14.19}$$

$$\text{FIT} = \sum_{k=1}^{N} I_k \tag{14.20}$$

Each of the I_k integrals above can be evaluated analytically.

$$I_k = \int_{t_{k-1}}^{t_k} [\alpha_{0k} + \alpha_{1k}(t - t_{k-1})] e^{-i\omega t}\, dt$$

$$= \left[\frac{-\alpha_{0k}}{i\omega} e^{-i\omega t} \right]_{t_{k-1}}^{t_k} + \alpha_{1k} \int_{t_{k-1}}^{t_k} (t - t_{k-1}) e^{-i\omega t}\, dt$$

Integrating the integral by parts gives

$$u = t - t_{k-1} \qquad dv = e^{-i\omega t}\, dt$$

$$du = dt \qquad v = -\frac{e^{-i\omega t}}{i\omega}$$

$$I_k = \frac{\alpha_{0k}}{i\omega}\left(e^{-i\omega t_{k-1}} - e^{-i\omega t_k}\right) - \left[\alpha_{1k}(t - t_{k-1})\frac{e^{-i\omega t}}{i\omega}\right]_{t_{k-1}}^{t_k} + \frac{\alpha_{1k}}{i\omega}\int_{t_{k-1}}^{t_k} e^{-i\omega t}\, dt$$

$$I_k = \frac{\alpha_{0k}}{i\omega}\left(e^{-i\omega t_{k-1}} - e^{-i\omega t_k}\right) - \frac{\alpha_{1k}}{i\omega}(t_k - t_{k-1})e^{-i\omega t_k} + \frac{\alpha_{1k}}{\omega^2}\left(e^{-i\omega t_k} - e^{-i\omega t_{k-1}}\right)$$

(14.21)

Substituting for α_{0k} and α_{1k} from Eqs. (14.17) and (14.18) gives

$$I_k = \frac{x_{k-1}}{i\omega}\left(e^{-i\omega t_{k-1}} - e^{-i\omega t_k}\right) - \frac{x_k - x_{k-1}}{\Delta t_k}\frac{\Delta t_k}{i\omega}e^{-i\omega t_k}$$

$$+ \frac{x_k - x_{k-1}}{\Delta t_k \omega^2}\left(e^{-i\omega t_k} - e^{-i\omega t_{k-1}}\right)$$

$$= \left[x_{k-1}e^{-i\omega t_{k-1}} - x_{k-1}e^{-i\omega t_k} - x_k e^{-i\omega t_k} + x_{k-1}e^{-i\omega t_k}\right]\frac{1}{i\omega}$$

$$+ \frac{x_k - x_{k-1}}{\Delta t_k \omega^2}\left(e^{-i\omega t_k} - e^{-i\omega t_{k-1}}\right)$$

$$= x_k\left(\frac{-e^{-i\omega t_k}}{i\omega} + \frac{e^{-i\omega t_k} - e^{-i\omega t_{k-1}}}{\omega^2\, \Delta t_k}\right) + x_{k-1}\left(\frac{e^{-i\omega t_{k-1}}}{i\omega} - \frac{e^{-i\omega t_k} - e^{-i\omega t_{k-1}}}{\omega^2\, \Delta t_k}\right)$$

$$I_k = e^{-i\omega t_{k-1}}\left\{x_k\left(\frac{e^{-i\omega\, \Delta t_k} - 1}{\omega^2\, \Delta t_k} - \frac{e^{-i\omega\, \Delta t_k}}{i\omega}\right) - x_{k-1}\left(\frac{e^{-i\omega\, \Delta t_k} - 1}{\omega^2\, \Delta t_k} - \frac{1}{i\omega}\right)\right\} \quad (14.22)$$

Finally, the Fourier transformation of $x_{(t)}$ becomes

$$\int_0^\infty x_{(t)}\, e^{-i\omega t}\, dt \simeq \sum_{k=1}^N e^{-i\omega t_{k-1}}\left\{x_k\left(\frac{e^{-i\omega\, \Delta t_k} - 1}{\omega^2\, \Delta t_k} - \frac{e^{-i\omega\, \Delta t_k}}{i\omega}\right)\right.$$

$$\left. - x_{k-1}\left(\frac{e^{-i\omega\, \Delta t_k} - 1}{\omega^2\, \Delta t_k} - \frac{1}{i\omega}\right)\right\} \quad (14.23)$$

Equation (14.23) looks a little complicated but it is easily programmed on a digital computer. Table 14.1 gives a FORTRAN program that reads input and output time functions from a file, calculates the Fourier transformations of the input and of the output, divides the two to get the transfer function $G_{(i\omega)}$, and prints out log modulus and phase angle at different values of frequency.

Notice that the computed values of log modulus and phase angle begin to oscillate at the higher values of frequency. This is due to numerical-integration problems and to the limited "frequency content" of the input-forcing function (i.e., the Fourier transformation of the input becomes smaller and smaller as frequency is increased). We will discuss this problem further in the next section. The computed results are meaningless at these high frequencies and should be ignored.

The steadystate gain of the transfer function is $G_{(0)}$ or just the ratio of the areas under the input and output curves.

$$K_p = G_{(0)} = \frac{\int_0^{t_x} x_{(t)}\, dt}{\int_0^{t_m} m_{(t)}\, dt} \tag{14.24}$$

If the input pulse is a rectangular pulse of height h and duration D, its Fourier transformation is simply

$$\int_0^\infty m_{(t)} e^{-i\omega t}\, dt = h \int_0^D e^{-i\omega t}\, dt = -\frac{h}{i\omega}\, [e^{-i\omega t}]_{t=0}^{t=D}$$

$$= \frac{h}{i\omega}\, (1 - e^{-i\omega D}) \tag{14.25}$$

This, of course, involves no approximation whatsoever. The special case is included in the program in Table 14.1. The general formula given in Eq. (14.23) gives exact Fourier transformations of any functions that have straight-line segments. Therefore triangles, trapezoids, etc., are handled with no approximation at all.

When reading the literature on pulse testing you should exercise some caution when using the equations presented since some assume equally spaced data points. There is also considerable confusion about the number of data points required to give a meaningful $G_{(i\omega)}$ out to a desired frequency. The criterion for picking the number of data points should be that enough points are used to make the approximating function $\phi_{(t)}$ match the real function $x_{(t)}$ over the intervals. For example, Eq. (14.25) gives the Fourier transformation of a rectangular pulse out to *any* frequency, and only one data point is required.

14.3.3 Practical Tips on Pulse Testing

Theoretically, the best possible input pulse would be an impulse or a Dirac function $\delta_{(t)}$. The Fourier transformation of $\delta_{(t)}$ is equal to unity at all frequencies.

$$\int_0^\infty \delta_{(t)} e^{-i\omega t}\, dt = [e^{-i\omega t}]_{t=0} = 1 \tag{14.26}$$

Therefore $G_{(i\omega)}$ would be simply the Fourier transformation of the output function. No division by a small number would be required.

Practically, however, we can never have an infinitely high pulse with zero width. In general, we need to keep the width of the pulse fairly small to keep its "frequency content" from becoming too small at higher frequencies. For example, Eq. (14.25) gives the Fourier transform of a rectangular pulse of width D.

$$\text{FIT} = \frac{h}{i\omega}(1 - e^{-i\omega D}) = \frac{h}{i\omega}[1 - \cos(\omega D) + i\sin(\omega D)]$$

$$= h\frac{\sin(\omega D)}{\omega} - i\frac{h}{\omega}[1 - \cos(\omega D)] \tag{14.27}$$

When frequency ω is equal to $2\pi/D$, the FIT of the input pulse goes to zero. Since we are dividing by this term, the calculation of the transfer function becomes meaningless at frequencies near $2\pi/D$. Therefore the smaller D can be made, the higher is the frequency to which $G_{(i\omega)}$ can be accurately found.

A good rule of thumb is to keep the width of the pulse less than about half the smallest time constant of interest. If the dynamics of the process are completely unknown, it takes a few trials to establish a reasonable pulse width. If the width of the pulse is too small, for a given pulse height, the system is disturbed very little, and it becomes difficult to separate the real output signal from noise and experimental error. The height of the pulse can be increased to "kick" the process more, but there is a limit here also.

We want to obtain an experimental linear dynamic model of the system in the form of $G_{(i\omega)}$. It must be a linear model since the notion of a transfer function applies only to a linear system. The process is usually nonlinear, and we are obtaining a model that is linearized around the steadystate operation level. If the height of the pulse is too high we may drive the process out of the linear range. Therefore pulses of various heights should be tried. It is also a good idea to make both positive and negative pulses in the input (increase and decrease). The computed $G_{(i\omega)}$'s should be identical if the region of linearity is not exceeded. For highly nonlinear processes, this is difficult to do. Therefore pulse testing does not work very well with highly nonlinear processes.

14.3.4 Processes With Integration

The input disturbance for a pulse test begins and ends at the same value. In terms of perturbation variables, the input $m_{(t)}$ is initially zero and is returned to zero after some time t_m.

The perturbation in the output variable $x_{(t)}$ will usually also return to zero. Once in a while, however, a process will give an output curve that will not return to zero. This occurs if the process contains a pure integration element. For example, suppose the process is a tank, with the output variable the liquid level in the tank and the input variable the flow rate into the tank. Assume the flow rate out of the tank is constant. If a positive pulse in feed flow rate is made, the liquid level will rise by the total incremental amount of material added during the pulse. The liquid level will stay at this new higher level, as sketched in Fig. 14.5.

If the output signal does not return to zero, the integral in the numerator of Eq. (14.8) becomes infinitely large. This situation can be handled by separating

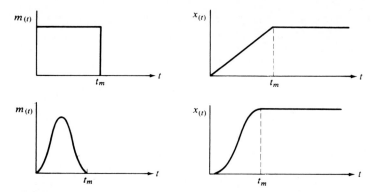

FIGURE 14.5
Pulse test of an integrator.

out the effect of the integrator. We assume the total output signal is the sum of the output of an integrator plus another transfer function $G^*_{(s)}$.

First the gain of the integration element is found from the final steadystate offset of $x_{(t)}$ from zero, $\overline{\Delta x}$, and the integral of the input pulse.

$$K_I = \frac{\overline{\Delta x}}{\displaystyle\int_0^{t_m} m_{(t)} \, dt} \tag{14.28}$$

The integrator contributes to the total output signal only when the input signal is not equal to zero. Therefore, integrating the input signal times the integrator gain K_I and subtracting this function from the original output function at each instant in time gives a function $x^*_{(t)}$ that does return to zero at some time t_x, as shown in Fig. 14.6.

$$x^*_{(t)} = x_{(t)} - K_I \int_0^t m_{(t)} \, dt \tag{14.29}$$

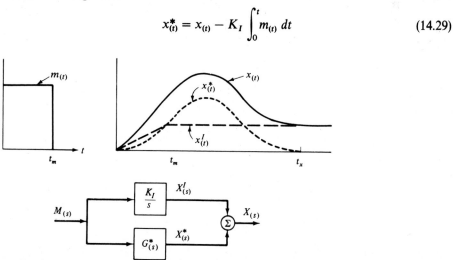

FIGURE 14.6
Separation of the output signal into contributions from the integrator and G^* transfer function.

The $G^*_{(s)}$ transfer function is found by using $x^*_{(t)}$ and the input $m_{(t)}$ in the integrals of Eq. (14.8). Then the total system transfer function is

$$G_{(s)} = \frac{X_{(s)}}{M_{(s)}} = \frac{K_I}{s} + G^*_{(s)} \tag{14.30}$$

$$G_{(i\omega)} = \frac{K_I}{i\omega} + G^*_{(i\omega)} = \text{Re} \left[G^*_{(i\omega)} \right] + i \left\{ \text{Im} \left[G^*_{(i\omega)} \right] - \frac{K_I}{\omega} \right\} \tag{14.31}$$

14.4 STEP TESTING

A plant operator makes changes from time to time in various input variables such as feed flow rate, steam flow rate, and reflux flow rate from one operating level to a new level. These changes can be either instantaneous or gradual, as sketched in Fig. 14.7. These step-response data are often easily obtained by merely recording the variables of interest for a few hours or days of plant operation.

These data can be converted into frequency-response curves, basically by differentiating both input and output curves in the frequency domain. The process transfer function $G_{(s)}$ is

$$G_{(s)} = \frac{X_{(s)}}{M_{(s)}}$$

We multiply both the numerator and denominator by s, which is equivalent to differentiation in the time domain.

$$G_{(s)} = \frac{sX_{(s)}}{sM_{(s)}} = \frac{\int_0^\infty \left(\frac{dx}{dt} \right) e^{-st} \, dt}{\int_0^\infty \left(\frac{dm}{dt} \right) e^{-st} \, dt}$$

Going into the frequency domain,

$$G_{(i\omega)} = \frac{\int_0^\infty \left(\frac{dx}{dt} \right) e^{-i\omega t} \, dt}{\int_0^\infty \left(\frac{dm}{dt} \right) e^{-i\omega t} \, dt} \tag{14.32}$$

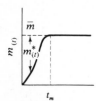 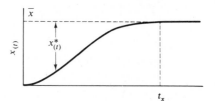

FIGURE 14.7
Step test data.

Two new variables are defined as the departure of the input and output variables from their *final* steadystate values.

$$m^*_{(t)} = \bar{m} - m_{(t)} \qquad x^*_{(t)} = \bar{x} - x_{(t)} \tag{14.33}$$

Both $m^*_{(t)}$ and $x^*_{(t)}$ go to zero after some finite time. Their derivatives are

$$\frac{dm^*}{dt} = -\frac{dm}{dt} \qquad \frac{dx^*}{dt} = -\frac{dx}{dt} \tag{14.34}$$

Then Eq. (14.32) becomes

$$G_{(i\omega)} = \frac{-\displaystyle\int_0^\infty \left(\frac{dx^*}{dt}\right) e^{-i\omega t}\, dt}{-\displaystyle\int_0^\infty \left(\frac{dm^*}{dt}\right) e^{-i\omega t}\, dt} \tag{14.35}$$

Integrating the numerator by parts gives

$$u = e^{-i\omega t} \qquad\qquad dv = \frac{dx^*}{dt}\, dt$$

$$du = -i\omega e^{-i\omega t}\, dt \qquad v = x^*$$

$$\int_0^\infty \left(\frac{dx^*}{dt}\right) e^{-i\omega t}\, dt = [x^* e^{-i\omega t}]_{t=0}^{t=\infty} + i\omega \int_0^\infty x^*_{(t)} e^{-i\omega t}\, dt$$

$$= -x^*_{(0)} + i\omega \int_0^\infty x^*_{(t)} e^{-i\omega t}\, dt$$

$$= -\bar{x} + i\omega \int_0^\infty x^*_{(t)} e^{-i\omega t}\, dt \tag{14.36}$$

The denominator can be handled in exactly the same way. Equation (14.35) becomes

$$G_{(i\omega)} = \frac{\bar{x} - i\omega \displaystyle\int_0^\infty x^*_{(t)} e^{-i\omega t}\, dt}{\bar{m} - i\omega \displaystyle\int_0^\infty m^*_{(t)} e^{-i\omega t}\, dt} \tag{14.37}$$

The integrals can be evaluated in the same way as for pulse tests.

Since the operations to get frequency response from step-test data involve numerical differentiation of the data, the results are less reliable than pulse test data as frequency is increased.

14.5 ATV IDENTIFICATION

Pulse testing does not work well on processes that are highly nonlinear because the pulse tends to drive the process away from the steadystate into a nonlinear region unless the pulse height is made very small.

Pulse testing also has problems in situations where load disturbances occur at the same time as the pulse is being performed. These other disturbances can effect the shape of the output response and produce poor results. The output of the process may not return to its original value because of load disturbances.

We are trying to extract a lot of information from one pulse test, i.e., the whole frequency response curve. This is asking a lot from one experiment.

Once we get the frequency response results, we often really only use the information in the frequency range where the phase angle gets near $-180°$. For example, we might use the ultimate frequency ω_u and the magnitude of $G_{(i\omega)}$ at $\omega = \omega_u$ to calculate the ultimate gain. Then we calculate the Ziegler-Nichols settings. So in many situations what we really need is not an accurate frequency-response curve over the entire frequency range but only the ultimate gain and ultimate frequency.

14.5.1 Autotuning

Åström and Hagglund (Proceedings of the 1983 IFAC Conference, San Francisco) suggested an "autotune" procedure that is a very attractive technique for determining the ultimate frequency and ultimate gain. We call this method "ATV" (*autotune variation*). The acronym also stands for "all-terrain vehicle" which makes it easy to remember and is not completely inappropriate since ATV does provide a useful tool for the rough and rocky road of process identification.

ATV is illustrated in Fig. 14.8. A relay of height h is inserted as a feedback controller. The manipulated variable m is increased by h above the steadystate value. When the controlled variable x crosses the setpoint, the relay reduces m to a value h below the steadystate value. The system will respond to this "bang-bang" control by producing a limit cycle, provided the system phase angle drops below $-180°$, which is true for all real processes.

The period of the limit cycle is the ultimate period (P_u) for the transfer function relating the controlled variable x and the manipulated variable m. So the ultimate frequency is

$$\omega_u = \frac{2\pi}{P_u} \tag{14.38}$$

The ultimate gain of the same transfer function is given by

$$K_u = \frac{4h}{a\pi} \tag{14.39}$$

where h = height of the relay
a = amplitude of the primary harmonic of the output x.

It should be noted that Eqs. (14.38) and (14.39) give approximate values for ω_u and K_u because the relay feedback introduces a nonlinearity into the system. However, for most systems, the approximation is close enough for engineering purposes.

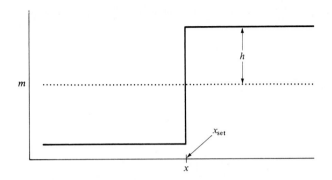

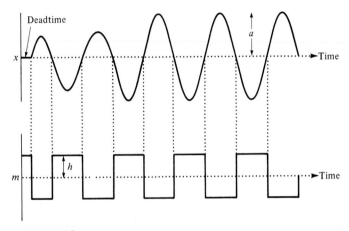

FIGURE 14.8

Åström's autotune method has several distinct advantages over openloop pulse testing:

1. No a priori knowledge of the system time constants is needed. The method automatically results in a sustained oscillation at the critical frequency of the process. The only parameter that has to be specified is the height of the relay step. This would typically be set at 2 to 10 percent of the manipulated variable range.

2. ATV is a closedloop test, so the process will not drift away from the setpoint. This keeps the process in the linear region where we are trying to get transfer functions. This is precisely why the method works well on highly nonlinear processes. The process is never pushed very far away from the steadystate conditions.

3. Accurate information is obtained around the important frequency, i.e., near phase angles of $-180°$. In contrast, pulse testing tries to extract information for a range of frequencies. It therefore is inherently less accurate than a

method which concentrates on a specific frequency. Remember, however, that we do not have to specify the frequency. The relay feedback automatically finds it.

14.5.2   Approximate Transfer Functions

Once the test has been performed and the ultimate gain and ultimate frequency have been determined, we may simply use it to calculate Ziegler-Nichols settings. Alternatively, it is possible to use this information, along with other easily determined data, to calculate approximate transfer functions. The idea is to pick some simple forms of transfer functions (gains, deadtime, first- or second-order lags) and find the parameter values that fit the ATV results.

The other data needed are the steadystate gain and the deadtime.

A. STEADYSTATE GAIN. The steadystate gain can be calculated from the steadystate equations describing the process. A small change is made in the manipulated variable $\overline{\Delta m}$ from its normal steadystate value. The algebraic steadystate equations are solved for the new value of the controlled variable. The change in the controlled variable $\overline{\Delta x}$ divided by $\overline{\Delta m}$ gives the steadystate gain K_p.

$$K_p = \overline{\Delta x} \big/ \overline{\Delta m} \tag{14.40}$$

This kind of calculation is called a "rating" calculation (equipment size is fixed), as opposed to a "design" calculation in which equipment is sized.

Caution must be taken to make sure that the sizes of the changes in m are made small enough so that the gains truly represent the linear gain. In a highly nonlinear process, these changes are typically 10^{-3} to 10^{-5} percent of the range of the manipulated variable! Such small changes would only be feasible using a mathematical model. Trying to obtain reliable steadystate gains from plant data step tests is usually impractical.

B. DEADTIME. The deadtime D in the transfer function can be easily read off the initial part of the ATV test. It is simply the time it takes x to start responding to the initial change in m.

Now we are ready to find an approximate model. We assume one of the forms given in Eqs. (14.41) and (14.42) given below.

$$G_{(s)} = \frac{K_p e^{-Ds}}{(\tau_p s + 1)^n} \qquad n = 1, 2, \text{ or } 3 \tag{14.41}$$

$$G_{(s)} = \frac{K_p e^{-Ds}}{(\tau_{p1} s + 1)(\tau_{p2} s + 1)^n} \qquad n = 1 \text{ or } 2 \tag{14.42}$$

These transfer functions have either one or two unknown parameters since we know the gain K_p and the deadtime D.

From the autotune test, the frequency (ω_u) and the argument of $G_{(i\omega)}$ (A) and the magnitude of $G_{(i\omega)}$ (M) are known at this frequency.

$$A = -\pi \quad \text{radians} \qquad M = \frac{1}{K_u} \tag{14.43}$$

The best model and the power n is determined by selecting the form that best satisfies the equations given below. These equations come from two equations that give the argument and the magnitude of the complex number.

Model 1 (first-order lag):

$$\tau_p = \frac{1}{\omega} \sqrt{\left(\frac{K_p}{M}\right)^2 - 1} \tag{14.44}$$

$$\tau_p = \frac{1}{\omega} \tan (A - \omega D) \tag{14.45}$$

Model 2 (two equal first-order lags):

$$\tau_p = \frac{1}{\omega} \sqrt{\left(\frac{K_p}{M}\right) - 1} \tag{14.46}$$

$$\tau_p = \frac{1}{\omega} \tan \left(\frac{A - \omega D}{2}\right) \tag{14.47}$$

Model 3 (three equal first-order lags):

$$\tau_p = \frac{1}{\omega} \sqrt{\left(\frac{K_p}{M}\right)^{1.5} - 1} \tag{14.48}$$

$$\tau_p = \frac{1}{\omega} \tan \left(\frac{A - \omega D}{3}\right) \tag{14.49}$$

Model 4 (two unequal first-order lags):

$$M = \frac{K_p}{\sqrt{1 + (\omega\tau_{p1})^2} \sqrt{1 + (\omega\tau_{p2})^2}} \tag{14.50}$$

$$A = -\omega D + \arctan (-\omega\tau_{p1}) + \arctan (-\omega\tau_{p2}) \tag{14.51}$$

Model 5 (one first-order lag and two equal first-order lags):

$$M = \frac{K_p}{\sqrt{1 + (\omega\tau_{p1})^2} \, [1 + (\omega\tau_{p2})^2]} \tag{14.52}$$

$$A = -\omega D + \arctan (-\omega\tau_{p1}) + 2 \arctan (-\omega\tau_{p2}) \tag{14.53}$$

where M = magnitude of $G_{(i\omega)}$
$\quad\quad\ A$ = argument of $G_{(i\omega)}$

To see if model 1 fits the data, Eqs. (14.44) and (14.45) are solved for their τ_p values. If the two values are approximately the same, it means that model 1 fits the data well. Similarly model 2 is tested by using Eqs. (14.46) and (14.47) and model 3 with Eqs. (14.48) and (14.49). The lowest-order model that fits the data should be used. Note that if the deadtime in the model is zero, a third-order model must be used. The phase angle must drop below $-180°$ as frequency increases.

The evaluations of models 4 and 5 are a little more involved. The two equations for each model must be solved simultaneously for the two unknowns τ_{p1} and τ_{p2}. Sometimes there is no physical solution. Model 5, quite interestingly, can sometimes give two solutions (see Example 14.2).

One solution technique is to guess a value for τ_{p1}. Then Eq. (14.50) for model 4 [(14.52) for model 5] is solved for τ_{p2}. Finally the right-hand side of Eq. (14.51) [(14.53)] is calculated; let us call this A_{calc}. If the actual argument of $G_{(i\omega)}$ (A) is equal to A_{calc}, the correct values of τ_{p1} and τ_{p2} have been found. Interval-halving can be used to reguess τ_{p1} if A and A_{calc} are not sufficiently close.

Example 14.1. Suppose $K_p = 34.16$, $D = 0.3$, $M = 0.198$, $A = -\pi$, and $\omega = 0.542$.

Model 1 results: $\quad \tau_{p1} = 318.3$ and $\tau_{p2} = -0.303$
So model 1 does not fit the data.
Model 2 results: $\quad \tau_{p1} = 24.16$ and $\tau_{p2} = 22.64$
Model 2 fits the data fairly well.
Model 3 results: $\quad \tau_{p1} = 10.10$ and $\tau_{p2} = 2.83$
So model 3 does not fit the data.
Model 4 results: $\quad \tau_{p1} = 16.98$ and $\tau_{p2} = 34.33$
So model 4 can be used.
Model 5 results: $\quad \tau_{p1} = 186.2$ and $\tau_{p2} = 1.554$
So model 5 could also be used.

Model 4 could be used to describe this process.

$$G_{(s)} = \frac{34.16e^{-0.3s}}{(16.98s + 1)(34.33s + 1)}$$

Example 14.2. Suppose $K_p = 45.51$, $D = 0$, $M = 0.281$, $A = -\pi$, and $\omega = 1$. Note that the deadtime is zero, so a third-order model must be used (the phase angle must drop below $-180°$). Therefore only model 3 (with $n = 3$) or model 5 can be used.

Model 3 results: $\quad \tau_{p1} = 5.358$ and $\tau_{p2} = 1.732$
So model 3 does not fit the data.
Model 5 results: There are two different transfer functions that fit the data:
(a) $\tau_{p1} = 0.162$ and $\tau_{p2} = 12.6$
(b) $\tau_{p1} = 81.28$ and $\tau_{p2} = 0.996$
Both of these transfer functions pass through exactly the same zero frequency and ultimate frequency points. So we could use either

$$G_{(s)} = \frac{45.51}{(0.162s + 1)(12.6s + 1)^2} \quad \text{or} \quad G_{(s)} = \frac{45.51}{(81.28s + 1)(0.996s + 1)^2}$$

The important feature of the ATV method is that it gives transfer function models that fit the frequency-response data very well near the important frequencies of zero (steadystate gains) and the ultimate frequency (which determines closedloop stability).

This method can be extended to evaluate models with more parameters and with different kinds of transfer functions (e.g., underdamped second-order lag) by using hysteresis in the relay feedback or by inserting an additional deadtime in the loop to produce a limit cycle with a different frequency. The two autotune tests give four equations so four parameters can be evaluated.

14.6 LEAST-SQUARES METHOD

Instead of converting the step or pulse responses of a system into frequency response curves, it is fairly easy to use classical least-squares methods to solve for the "best" values of parameters of a model that fit the time-domain data.

Any type of input-forcing function can be used: steps, pulses, or a sequence of positive and negative pulses. Figure 14.9a shows some typical input/output data from a process. The specific example is a heat exchanger in which the manipulated variable is steam flow rate and the output variable is the temperature of the process steam leaving the exchanger.

A very popular sequence of inputs is the "pseudorandom binary sequence" (PRBS). It is easy to generate and has some attractive statistical properties. See *System Identification For Self-Adaptive Control*, W. D. T. Davies, London, Wiley-Interscience, 1970.

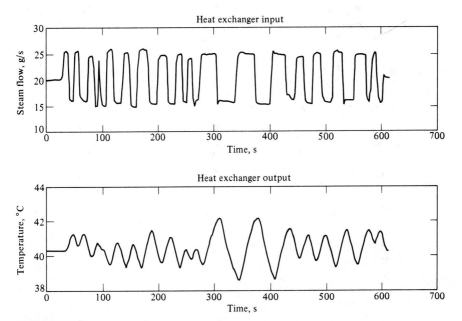

FIGURE 14.9a

Whatever the form of the input, the basic idea is to use a difference-equation model for the process in which the current output x_n is related to previous values of the output $(x_{n-1}, x_{n-2}, \ldots)$ and present and past values of the input $(m_n, m_{n-1}, \ldots)$. The relationship is a linear one, so classical least squares can be used to solve for the best values of the unknown coefficients. These difference-equation models occur naturally in sampled-data system (see Chap. 18) and can be easily converted into Laplace-domain transfer function models.

Let us assume a model of the form

$$\bar{x}_n = a_0 m_n + a_1 m_{n-1} + a_2 m_{n-2} + \cdots + a_M m_{n-M}$$
$$- b_1 x_{n-1} - b_2 x_{n-2} - \cdots - b_N x_{n-N} \quad (14.54)$$

where $\bar{x}_n$ = the predicted value of the current output of the process

The unknown parameters are $a_0, a_1, \ldots, a_M, b_1, b_2, \ldots, b_N$. In a typical model, there may be two a_i's and three b_i's to be determined. It is convenient to define a vector $\underline{\theta}$ that contains these unknown parameters.

$$\underline{\theta} \equiv \begin{bmatrix} a_0 \\ a_1 \\ \vdots \\ a_M \\ b_1 \\ \vdots \\ b_N \end{bmatrix} \quad (14.55)$$

There are $N + M + 1$ unknown parameters. Let us define NC as this total. So the vector has NC rows.

Now suppose we have NP data points that give values of the output x_n for known values of $x_{n-1}, x_{n-2}, \ldots, x_{n-N}, m_n, m_{n-1}, \ldots, m_{n-M}$. The data would look like:

	x_n values	x_{n-1} values	$\cdots$	x_{n-N} values	m_n values	m_{n-1} values	$\cdots$	m_{n-M} values
1	$x_{1,n}$	$x_{1,n-1}$		$x_{1,n-N}$	$m_{1,n}$	$m_{1,n-1}$		$m_{1,n-M}$
2	$x_{2,n}$	$x_{2,n-1}$		$x_{2,n-N}$	$m_{2,n}$	$m_{2,n-1}$		$m_{2,n-M}$
$\vdots$								
NP	$x_{NP,n}$	$x_{NP,n-1}$		$x_{NP,n-N}$	$m_{NP,n}$	$m_{NP,n-1}$		$m_{NP,n-M}$

Our objective is to minimize the sum of the squares of the differences between the actual measured data points $(x_{i,n})$ and those predicted by our model equation $(\bar{x}_{i,n})$.

$$J = \sum_{i=1}^{NP} (x_{i,n} - \bar{x}_{i,n})^2 \quad (14.56)$$

This problem should look very familiar to you. It is exactly the problem we discussed in Sec. 8.9.1. And the solution is exactly the same. We take partial derivatives of J with respect to each one of the unknown parameters (the NC elements of the $\underline{\theta}$ vector) and set these partial derivatives equal to zero. This gives NC equations in NC unknowns. The solution compactly written in matrix form:

$$\underline{\theta} = [\underline{A}^T\underline{A}]^{-1}\underline{A}^T\underline{x} \tag{14.57}$$

where the $\underline{A}$ matrix (with NP rows and NC columns) and the $\underline{x}$ vector (with NP rows) are defined to represent the data points:

$$\underline{A} = \begin{bmatrix} x_{1,n-1} & x_{1,n-2} & \cdots & x_{1,n-N} & m_{1,n} & \cdots & m_{1,n-M} \\ x_{2,n-1} & x_{2,n-2} & & x_{2,n-N} & m_{2,n} & & m_{2,n-M} \\ \vdots & & & & & & \\ x_{NP,n-1} & & & \cdots & & & m_{NP,n-M} \end{bmatrix} \tag{14.58}$$

$$\underline{x} \equiv \begin{bmatrix} x_{1,n} \\ x_{2,n} \\ x_{3,n} \\ x_{4,n} \\ \vdots \\ x_{NP,n} \end{bmatrix} \tag{14.59}$$

Several commercial software packages are available that make it easy to perform this type of identification. The PC-MATLAB: System Identification Toolbox (The Math Works, Inc., Sherborn, Mass.) was used in the example given below.

One very important point should be remembered when using this discrete data, i.e., data that is collected only at discrete points in time. As we will develop in Chap. 18, there is an important basic sampling theorem that says that you must keep the sampling period T_s small enough so that the sampling frequency ($\omega_s \equiv 2\pi/T_s$) is greater than twice the frequency of any significant component of the signal being sampled. For example, if a signal contains components out to 10 radians per minute, the sampling period should be no larger than $T_s = 2\pi/\omega_s = 2\pi/20 = 0.314$ min.

Failure to follow this rule can lead to meaningless discrete models. The problem is due to "aliasing" which will be discussed in Chap. 18. Therefore it is strongly recommended that an analog filter be used to attenuate any high-frequency components in the signal before sampling. Digital filtering after sampling will do *nothing* to solve the problem. So be alert to this problem. Remember, what is important is the highest frequency in the signal you are sampling, *not* the highest frequency of interest to you from a control standpoint.

Example 14.3. PRBS testing of a simulated distillation column was performed. Data were sampled every 1.8 seconds and stored in two files. These data were then fed into the commercial software package "MATLAB." You must specify the order of the model and the deadtime. Then the parameters of the model are calculated using least squares.

The results show that the accuracy of the model is very sensitive to the deadtime assumed but less sensitive to the order of the model. The coefficients in a discrete model like that given in Eq. (14.54) are listed below.

Order of Model	Second				Third	
Deadtime $T_s = k$ Coefficients	2	3	4	5	3	4
a_{1+k}	−0.0022	−0.0065	−0.132	−0.028	−0.0068	−0.132
a_{2+k}	0.0019	−0.129	−0.037	−0.010	−0.132	−0.046
a_{3+k}	....	....	....	....	−0.041	−0.014
b_1	−1.67	−1.66	−1.58	−1.59	−1.55	−1.51
b_2	0.735	0.712	0.628	0.655	0.575	0.553
b_3	....	....	....	....	0.027	0.018

The best model in terms of fitting the frequency response results (obtained in MATLAB by fast-Fourier-transforming the input and output signals) is the third-order with a deadtime equal to four sampling periods. Fig. 14.9b compares the model frequency response with the data.

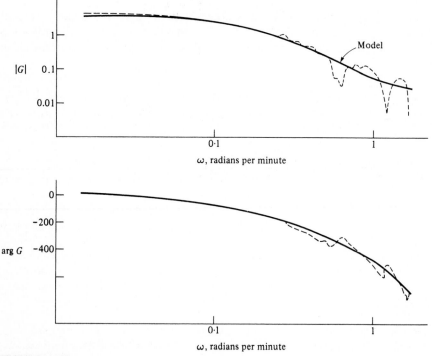

FIGURE 14.9b

The distillation column used in this example separated a binary mixture of propylene and propane. Because of the low relative volatility and large number of trays, the dominant time constant is very large (500 minutes). Despite this large time constant, a sampling period of 9.6 minutes gave poor results. The period had to be reduced to 1.8 minutes to get good identification, both dynamic and steadystate gain.

14.7 STATE ESTIMATORS

The last decade has seen the development of considerable interest in "state estimators." The focus of these methods is somewhat different than conventional identification, but the two areas have many similarities. That is the justification for a brief discussion of state estimators in this chapter.

A rigorous discussion of this extensive subject is beyond the scope of this book, and an extensive treatment is also probably not justified at this point in time. There has been a lot of interest by academia and by industry in these methods, but the jury is still out as to their real practical utility in chemical engineering processes with their very high order, nonlinearity, unknown noise properties, and frequent and unmeasurable load disturbances. Several industrial researchers have reported successful applications. But other industrial reports indicate a significant number of failures, or at least a lack of significant improvement in control over more simple conventional methods.

The books by T. Kailath (*Linear Systems*, 1980, Prentice-Hall) and H. Kwakernaak and R. Sivan (*Linear Optimal Control Systems*, 1972, Wiley) can be consulted for detailed information. A recent paper by MacGregor, et al. ("State Estimation For Polymerization Reactors," 1986 IFAC Symposium, Bournemouth, U.K.) discusses some of the problems of applying state estimators to chemical engineering systems.

State estimators are used to provide on-line predictions of those variables that describe the system dynamic behavior but cannot be directly measured. For example, suppose we have a chemical reactor and can measure the temperature in the reactor but not the compositions of reactants or products. A state estimator could be used to predict these compositions.

State estimators are basically just mathematical models of the system that are solved on-line. These models usually assume linear ODEs, but nonlinear equations can be incorporated. The actual measured inputs to the process (manipulated variables) are fed into the model equations, and the model equations are integrated. Then the available measured output variables are compared with the predictions of the model. The differences between the actual measured output variables and the predictions of the model for these same variables are used to change the model estimates through some sort of feedback. As these differences between the predicted and measured variables are driven to zero, the model predictions of all the state variables are changed.

The most popular of these estimators is the "Kalman Filter." Use of this estimator requires a level of knowledge in statistics and stochastic control that is beyond the scope of this text. The books by A. H. Jazwinski (*Stochastic Processes and Filtering Theory*, 1970, Academic Press) and R. F. Stengel (*Stochastic Optimal Control*, 1986, Wiley) cover this subject.

14.8 RELATIONSHIPS AMONG TIME, LAPLACE, AND FREQUENCY DOMAINS

At this point it might be useful to pull together some of the concepts that you have waded through in the last several chapters. We now know how to look at and think about dynamics in three languages: time (English), Laplace (Russian) and frequency (Chinese). For example, a third-order, underdamped system would have the time-domain step responses sketched in Fig. 14.10 for two different values of the real root. In the Laplace domain, the system is represented by a transfer function $G_{(s)}$ or by plotting the poles of the transfer function (the roots of the system's characteristic equation) in the s plane, as shown in Fig. 14.10. In the frequency domain, the system could be represented by a Bode plot of $G_{(i\omega)}$.

Do we know how to convert from one domain to another? These conversions are summarized below.

14.8.1 Laplace to Frequency Domain

This is the easiest of all the conversions. We simply substitute $i\omega$ for s in the system transfer function.

14.8.2 Frequency to Laplace Domain

To go the other direction, the Bode plots can be approximated, as discussed in Sec. 12.3.

14.8.3 Time to Laplace Domain

Laplace transformation converts time functions into Laplace transforms and converts ordinary differential equations into transfer functions $G_{(s)}$.

14.8.4 Laplace to Time Domain

Inversion of the Laplace transformation gives the time function. If we have a transfer function $G_{(s)}$, the unit step function is $\mathcal{L}^{-1}[G_{(s)}/s]$ and the impulse response is $\mathcal{L}^{-1}[G_{(s)}]$.

To convert the transfer function into differential equations, we simply cross-multiply the input/output relationship and substitute d/dt everywhere there is an s. For example, suppose we have the transfer function given below:

$$G_{(s)} = \frac{X_{(s)}}{M_{(s)}} = \frac{K_p(\tau_z s + 1)}{(\tau_{p1} s + 1)(\tau_{p2} s + 1)} \tag{14.60}$$

Cross-multiplying gives

$$X_{(s)}\left[(\tau_{p1}\tau_{p2})s^2 + (\tau_{p1} + \tau_{p2})s + 1\right] = M_{(s)}K_p(\tau_z s + 1) \qquad (14.61)$$

Converting into a differential equation yields

$$\tau_{p1}\tau_{p2}\frac{d^2x}{dt^2} + (\tau_{p1} + \tau_{p2})\frac{dx}{dt} + x = K_p\left(\tau_z\frac{dm}{dt} + m\right) \qquad (14.62)$$

Time domain: $(\tau_1\tau_2^2)\dfrac{d^3x}{dt^3} + (\tau_2^2 + 2\tau_1\tau_2\zeta)\dfrac{d^2x}{dt^2} + (\tau_1 + 2\tau_2\zeta)\dfrac{dx}{dt} + x = K_p m_{(t)}$

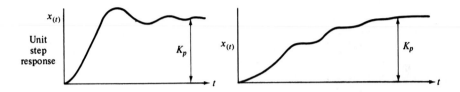

Laplace domain: $G_{(s)} = \dfrac{K_p}{(\tau_1 s + 1)(\tau_2^2 s^2 + 2\tau_2\zeta s + 1)}$

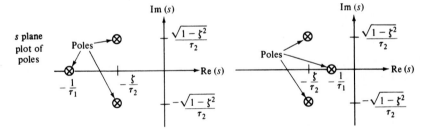

Frequency domain: $G(i\omega)$

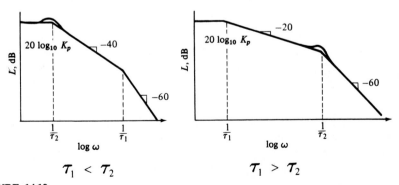

FIGURE 14.10
Third-order process in the time, Laplace, and frequency domains.

14.8.5   Time to Frequency Domain

The ordinary differential equations could be solved in the time domain with a sinusoidal forcing function, but it is easier to go into the Laplace domain first and then substitute $s = i\omega$.

14.8.6   Frequency to Time Domain

We could go through the Laplace domain by approximating $G_{(i\omega)}$ and then inverting. However, there is a direct conversion (V. V. Solodovnikov, *Introduction to Statistical Dynamics of Automatic Control*, Dover, 1960). Suppose we want to find the impulse response of a stable system (defined as $g_{(t)}$), given the system's frequency response $G_{(i\omega)}$. Since the Laplace transformation of the impulse input is unity,

$$g_{(t)} = \mathcal{L}^{-1}[G_{(s)}] \tag{14.63}$$

Using the inversion formula gives

$$g_{(t)} = \frac{1}{2\pi i} \int_{\alpha - i\infty}^{\alpha + i\infty} e^{st} G_{(s)} \, ds \tag{14.64}$$

If we pick the path of this contour integration along the imaginary axis in the s plane, $s = i\omega$.

$$g_{(t)} = \frac{1}{2\pi i} \int_{-i\infty}^{i\infty} e^{i\omega t} G_{(i\omega)} \, d(i\omega) \tag{14.65}$$

$$g_{(t)} = \frac{1}{2\pi} \int_{-\infty}^{\infty} [\mathrm{Re}\,(G_{(i\omega)}) + i\,\mathrm{Im}\,(G_{(i\omega)})][\cos\,(\omega t) + i\sin\,(\omega t)]\, d\omega$$

$$= \frac{1}{2\pi} \int_{-\infty}^{\infty} [\mathrm{Re}\,(G_{(i\omega)})\cos\,(\omega t) - \mathrm{Im}\,(G_{(i\omega)})\sin\,(\omega t)]\, d\omega$$

$$+ i\frac{1}{2\pi} \int_{-\infty}^{\infty} [\mathrm{Re}\,(G_{(i\omega)})\sin\,(\omega t) + \mathrm{Im}\,(G_{(i\omega)})\cos\,(\omega t)]\, d\omega \tag{14.66}$$

Now

$$\mathrm{Re}\,(G_{(i\omega)}) = |G_{(i\omega)}|\cos\,(\arg G_{(i\omega)})$$

is an "even" function; i.e., it satisfies $f_{(x)} = f_{(-x)}$.

$$\mathrm{Im}\,(G_{(i\omega)}) = |G_{(i\omega)}|\sin\,(\arg G_{(i\omega)})$$

is an "odd" function: i.e., it satisfies $f_{(x)} = -f_{(-x)}$.

Therefore

$$\text{Re } (G_{(i\omega)}) \cos (\omega t) \quad \text{and} \quad \text{Im } (G_{(i\omega)}) \sin (\omega t) \quad \text{are even functions}$$
$$\text{Re } (G_{(i\omega)}) \sin (\omega t) \quad \text{and} \quad \text{Im } (G_{(i\omega)}) \cos (\omega t) \quad \text{are odd functions}$$

Integrating an even function from $-\infty$ to $+\infty$ is the same as integrating it from 0 to $+\infty$ and multiplying the result by 2.

$$\int_{-\infty}^{\infty} (\text{even function}) \, dt = 2 \int_{0}^{\infty} (\text{even function}) \, dt \tag{14.67}$$

Integrating an odd function from $-\infty$ to $+\infty$ gives zero.

$$\int_{-\infty}^{\infty} (\text{odd function}) \, dt = 0 \tag{14.68}$$

Therefore Eq. (14.66) reduces to

$$g_{(t)} = \frac{1}{\pi} \int_{-\infty}^{\infty} [\text{Re } (G_{(i\omega)}) \cos (\omega t) - \text{Im } (G_{(i\omega)}) \sin (\omega t)] \, d\omega \tag{14.69}$$

Now $g_{(t)}$ must, by definition, be equal to zero for time less than zero. For this to be true, the integrand must be equal zero for negative values of t.

$$\text{Re } (G_{(i\omega)}) \cos (-\omega t) - \text{Im } (G_{(i\omega)}) \sin (-\omega t) = 0$$
$$\text{Re } (G_{(i\omega)}) \cos (\omega t) + \text{Im } (G_{(i\omega)}) \sin (\omega t) = 0$$
$$\text{Re } (G_{i\omega)}) \cos (\omega t) = -\text{Im } (G_{(i\omega)}) \sin (\omega t)$$

Equation (14.69) can be written in two ways:

$$g_{(t)} = \frac{2}{\pi} \int_{-\infty}^{\infty} \text{Re } (G_{(i\omega)}) \cos (\omega t) \, d\omega \tag{14.70}$$

$$g_{(t)} = -\frac{2}{\pi} \int_{-\infty}^{\infty} \text{Im } (G_{(i\omega)}) \sin (\omega t) \, d\omega \tag{14.71}$$

Therefore the time-domain impulse response can be calculated from the frequency domain $G_{(i\omega)}$ by evaluating either of these integrals.

PROBLEMS

14.1. Write a digital computer program that will calculate $G_{(i\omega)}$ from step test data.

14.2. Write a digital computer program that will calculate $G_{(i\omega)}$ from pulse test data when the process contains an integrator.

14.3. The frequency-response data given below were obtained from direct sine-wave tests of a chemical plant. Fit an approximate transfer function $G_{(s)}$ to these data.

Frequency, radians per minute	Real part	Imaginary part	Log modulus, dB	Phase angle, degrees
0.01	6.964	−0.522	16.88	−4.2
0.02	6.859	−1.028	16.82	−8.5
0.04	6.467	−1.942	16.59	−16.7
0.08	5.254	−3.202	15.78	−31.3
0.10	4.568	−3.554	15.25	−37.9
0.20	2.096	−3.673	12.52	−60.2
0.40	0.324	−2.741	8.82	−83.2
0.63	−0.557	−2.234	7.49	−103
0.80	−1.462	−2.083	8.11	−125
1.00	−2.472	−0.104	7.87	−177
1.41	−0.282	0.547	−4.21	−243
2.00	−0.021	0.160	−15.81	−262
4.00	0.004	0.016	−35.5	−285
8.00	0.001	0.001	−53.9	−312

14.4. Show how the ATV method can be used to determine the time constants and damping coefficient of a third-order model: one first-order lag and a second-order underdamped lag. Note that there are three unknown parameters and only two equations, but the third relationship is that a model with the largest possible damping coefficient is desired.

14.5. Simulate several first-order lag and second-order lag plus deadtime processes on a digital computer with a relay feedback. Compare the ultimate gains and frequencies obtained by the auto-tune method with the real values of ω_u and K_u obtained from the transfer function.

VI

MULTIVARIABLE
PROCESSES

Perhaps the area of process control that has changed the most drastically in the last decade is multivariable control. This change has been driven by the increasingly frequent occurrence of highly complex and interacting processes that have resulted from the design of more energy-efficient plants. The tenfold increase in energy prices in the 1970s spurred activity to make chemical and petroleum processes more efficient. The result has been more and more plants with complex interconnections of both material flows and energy exchange. Thus the control engineer must be equipped to design control systems that give effective control in this multivariable environment.

The next three chapters are devoted to this subject. Chapter 15 reviews and summarizes some basic mathematics and useful notation. We need to learn yet another language! In previous chapters we have found the perspectives of time (English), Laplace (Russian), and frequency (Chinese) to be useful. Now we must learn some matrix methods and a little about their use in the "state space" approach to control systems design. Let's call this state-space methodology the "Greek" language in honor of the many workers in the multivariable control area of Greek extraction. We will use both Greek (state space) and Russian (Laplace transfer function) methods in these studies of multivariable systems.

Chapter 16 covers the analysis of multivariable processes: stability, robustness, performance. Chapter 17 presents a practical procedure for designing conventional multiloop SISO controllers (the diagonal control structure) and briefly discusses some of the full-blown multivariable controller structures that have been developed in recent years.

It should be emphasized that the area of multivariable control is still in an early stage of development. Many active research programs are underway around the world studying this problem and every year brings many new developments. Therefore the methods and procedures presented in this book should be viewed as a summary of some of the practical tools developed so far. The weaknesses and limitations of the existing methods will be pointed out in the discussion. Improved methods will undoubtedly grow from current and future research.

CHAPTER
15

MATRIX
PROPERTIES
AND
STATE
VARIABLES

15.1 MATRIX MATHEMATICS

Many books have been written on matrix notation and mathematics. Its elegance has great appeal to many mathematically inclined individuals. Many hard-nosed engineers, however, are not interested in elegance but in useful tools. I have attempted in this chapter to weed out most of the chaff, blue smoke, and mirrors. Only those aspects that I have found to have useful engineering applications are summarized in this chapter. For a more extensive treatment, the readable book *Control System Design: An Introduction To State-Space Methods*, 1986, McGraw-Hill, by Bernard Friedland is recommended. Chemical engineering applications can be found in the pioneering book *Mathematical Methods In Chemical Engineering: Matrices and Their Application*, by N. R. Amundson, Prentice-Hall, 1966.

We will use the symbolism of placing either brackets ($[A]$) around or a double underline ($\underline{\underline{A}}$) under terms that are matrices. A single underline ($\underline{x}$) will be used to indicate a vector, i.e., a matrix with only one column. This should help us keep track of which quantities are matrices, which are vectors, and which are scalar terms.

15.1.1   Matrix Addition

The sum of two matrices is defined as the matrix whose elements are the sum of the elements of the two original matrices. Thus if we add matrix A with elements a_{ij} to matrix B with elements b_{ij}, we get a matrix C whose elements c_{ij} are given by Eq. (15.2).

$$A + B = C \tag{15.1}$$

$$c_{ij} = a_{ij} + b_{ij} \tag{15.2}$$

Of course, the matrices must have the same order, i.e., they must have the same number of rows and columns. Most of the matrices used in this book will be square (same number of rows and columns), and they will usually be the same order as system (the number of inputs and outputs). If there are N manipulated variables and N controlled variables, the process is an Nth-order system and the process transfer function matrix will be an $N \times N$ square matrix.

Example 15.1. Add the 2×2 A and B matrices given below.

$$A = \begin{bmatrix} a_{11} & a_{12} \\ a_{21} & a_{22} \end{bmatrix} \qquad B = \begin{bmatrix} b_{11} & b_{12} \\ b_{21} & b_{22} \end{bmatrix}$$

Solution

$$
\begin{aligned}
C = A + B &= \begin{bmatrix} a_{11} & a_{12} \\ a_{21} & a_{22} \end{bmatrix} + \begin{bmatrix} b_{11} & b_{12} \\ b_{21} & b_{22} \end{bmatrix} \\
&= \begin{bmatrix} (a_{11} + b_{11}) & (a_{12} + b_{12}) \\ (a_{21} + b_{21}) & (a_{22} + b_{22}) \end{bmatrix}
\end{aligned} \tag{15.3}
$$

15.1.2   Matrix Multiplication

The product of two matrices is defined by Eq. (15.5) below. If A and B matrices are multiplied to give a matrix C,

$$C = AB \tag{15.4}$$

each element of the C is given by

$$c_{ij} = \sum_{k=1}^{N} a_{ik} b_{kj} \tag{15.5}$$

The number of columns N in the first matrix must be equal to the number of rows in the second matrix. Nonsquare matrices can be multiplied. The order of multiplication is important, that is, AB is not equal to BA. Matrix multiplication does *not* "commute" as the mathematicians say. There is a difference between "premultiplying" and "postmultiplying" a matrix by another matrix.

Example 15.2. Find the product of the A and B matrices given in Example 15.1 (a) when A is multiplied by B and (b) when B is multiplied by A.

Solution

(a) $AB = \begin{bmatrix} a_{11} & a_{12} \\ a_{21} & a_{22} \end{bmatrix} \begin{bmatrix} b_{11} & b_{12} \\ b_{21} & b_{22} \end{bmatrix}$

$$= \begin{bmatrix} (a_{11}b_{11} + a_{12}b_{21}) & (a_{11}b_{12} + a_{12}b_{22}) \\ (a_{21}b_{11} + a_{22}b_{21}) & (a_{21}b_{12} + a_{22}b_{22}) \end{bmatrix} \tag{15.6}$$

(b) $BA = \begin{bmatrix} b_{11} & b_{12} \\ b_{21} & b_{22} \end{bmatrix} \begin{bmatrix} a_{11} & a_{12} \\ a_{21} & a_{22} \end{bmatrix}$

$$= \begin{bmatrix} (a_{11}b_{11} + a_{21}b_{12}) & (a_{12}b_{11} + a_{22}b_{12}) \\ (a_{11}b_{21} + a_{21}b_{22}) & (a_{12}b_{21} + a_{22}b_{22}) \end{bmatrix} \tag{15.7}$$

This example illustrates that matrix multiplication does *not* commute.

15.1.3   Transpose of a Matrix

The transpose of a matrix is another matrix which has columns that are the same as the rows of the original matrix.

If the matrix A^T is defined as the transpose of the matrix A, each element in A^T is given by

$$(a^T)_{ij} = a_{ji} \tag{15.8}$$

The rows and the columns are simply interchanged.

Example 15.3. What is the transpose of the A matrix given in Example 15.1?

Solution

$$A = \begin{bmatrix} a_{11} & a_{12} \\ a_{21} & a_{22} \end{bmatrix} \qquad A^T = \begin{bmatrix} a_{11} & a_{21} \\ a_{12} & a_{22} \end{bmatrix}$$

When we are dealing with complex matrices, each of the elements have both real and imaginary parts. The "conjugate transpose" of a complex matrix is a new matrix which has (a) columns that were the rows in the original matrix and (b) imaginary parts that are the negative of the imaginary parts of the original matrix. The notion is similar to the complex conjugate of a scalar imaginary number.

If A^{CT} is defined as the conjugate transpose of A, each element in A^{CT} is given by

$$(a^{CT})_{kj} = \text{Re} \, [a_{jk}] - i \, \text{Im} \, [a_{jk}] \tag{15.9}$$

where $i = \sqrt{-1}$

Example 15.4. Determine the conjugate transpose of the complex matrix $\underset{\sim}{C}$:

$$\underset{\sim}{C} = \begin{bmatrix} (a_{11} + ib_{11}) & (a_{12} + ib_{12}) \\ (a_{21} + ib_{21}) & (a_{22} + ib_{22}) \end{bmatrix}$$

$$\underset{\sim}{C}^{CT} = \begin{bmatrix} (a_{11} - ib_{11}) & (a_{21} - ib_{21}) \\ (a_{12} - ib_{12}) & (a_{22} - ib_{22}) \end{bmatrix} \tag{15.10}$$

15.1.4 Matrix Inversion

The inverse of a matrix is defined as a matrix which, when multiplied by the original matrix, gives the "identity matrix." This is a diagonal matrix (a square matrix with terms on the diagonal but zeros on all the off-diagonal positions) with "1" terms on the diagonal. The 2×2 identity matrix is $\underset{\sim}{I}$:

$$\underset{\sim}{I} = \begin{bmatrix} 1 & 0 \\ 0 & 1 \end{bmatrix} \tag{15.11}$$

The 3×3 identity matrix is

$$\underset{\sim}{I} = \begin{bmatrix} 1 & 0 & 0 \\ 0 & 1 & 0 \\ 0 & 0 & 1 \end{bmatrix} \tag{15.12}$$

The symbolism $\underset{\sim}{A}^{-1}$ is used to indicate the inverse of the matrix $\underset{\sim}{A}$.

$$\underset{\sim}{A}\underset{\sim}{A}^{-1} = \underset{\sim}{A}^{-1}\underset{\sim}{A} = \underset{\sim}{I} \tag{15.13}$$

The inverse of a matrix is usually calculated numerically, particularly in realistically complex engineering applications. Standard computer library subroutines are readily available. We use the IMSL subroutine "LEQ2C" in this book to calculate the inverse of a complex matrix.

Inverses of simple matrices can be calculated analytically from the equation

$$\underset{\sim}{A}^{-1} = \frac{\text{Adj } \underset{\sim}{A}}{\text{Det } \underset{\sim}{A}} = \frac{[\text{Cofactor } \underset{\sim}{A}]^T}{\text{Det } \underset{\sim}{A}} \tag{15.14}$$

where "Adj" stands for *adjoint* and "Det" for *determinant*. The determinant of a matrix is a single scalar quantity; it is *not* a matrix. This is an important fact, and we will use it often in our analysis of multivariable processes.

Example 15.5. The determinant of the 2×2 $\underset{\sim}{A}$ matrix from Example 15.1 is

$$\text{Det } \underset{\sim}{A} = \text{Det} \begin{bmatrix} a_{11} & a_{12} \\ a_{21} & a_{22} \end{bmatrix} = a_{11}a_{22} - a_{12}a_{21} \tag{15.15}$$

The adjoint of a matrix is the transpose of the matrix which is formed by replacing each element with its "cofactor." A cofactor is the determinant formed by eliminating the row and column in which each element lies and using the remaining $N - 1$ rows and columns and the appropriate sign. All of these definitions are easier to understand by looking at an example.

Example 15.6. The inverse of the A matrix of Example 15.1 is found from the following steps:

$$A = \begin{bmatrix} a_{11} & a_{12} \\ a_{21} & a_{22} \end{bmatrix} \quad [\text{Cofactor } A] = \begin{bmatrix} a_{22} & -a_{21} \\ -a_{12} & a_{11} \end{bmatrix} \tag{15.16}$$

$$[\text{Cofactor } A]^T = \begin{bmatrix} a_{22} & -a_{12} \\ -a_{21} & a_{11} \end{bmatrix} \tag{15.17}$$

$$A^{-1} = \frac{\text{Adj } A}{\text{Det } A} = \frac{[\text{Cofactor } A]^T}{\text{Det } A}$$

$$= \frac{\begin{bmatrix} a_{22} & -a_{12} \\ -a_{21} & a_{11} \end{bmatrix}}{a_{11}a_{22} - a_{12}a_{21}}$$

$$= \begin{bmatrix} \dfrac{a_{22}}{a_{11}a_{22} - a_{12}a_{21}} & \dfrac{-a_{12}}{a_{11}a_{22} - a_{12}a_{21}} \\ \dfrac{-a_{21}}{a_{11}a_{22} - a_{12}a_{21}} & \dfrac{a_{11}}{a_{11}a_{22} - a_{12}a_{21}} \end{bmatrix}$$

To verify that Eq. (15.17) does indeed give the inverse of A, multiple A^{-1} and A together (in either order) and we should get the identity matrix.

$$AA^{-1} = \begin{bmatrix} a_{11} & a_{12} \\ a_{21} & a_{22} \end{bmatrix} \frac{\begin{bmatrix} a_{22} & -a_{12} \\ -a_{21} & a_{11} \end{bmatrix}}{a_{11}a_{22} - a_{12}a_{21}}$$

$$= \frac{\begin{bmatrix} (a_{11}a_{22} - a_{12}a_{21}) & (-a_{11}a_{12} + a_{11}a_{12}) \\ (a_{21}a_{22} - a_{21}a_{22}) & (-a_{12}a_{21} + a_{11}a_{22}) \end{bmatrix}}{a_{11}a_{22} - a_{12}a_{21}}$$

$$= \begin{bmatrix} 1 & 0 \\ 0 & 1 \end{bmatrix}$$

Now, for a little practice, verify for yourself that $A^{-1}A$ also equals I.

15.2 MATRIX PROPERTIES

There are a host of properties that have been studied by the mathematicians. We will discuss only the notions of "eigenvalues," "singular values," and "canonical form." These will prove valuable in our design methods for multivariable systems. Eigenvalues, as we will show, are simply another name for the roots of the characteristic equation of the system. Singular values give us a measure of the size of the matrix and an indication of how close it is to being "singular." A matrix is singular if its determinant is zero. Since the determinant appears in the denominator when taking the inverse of a matrix [see Eq. (15.14)], the inverse will not exist if the matrix is singular.

15.2.1   Eigenvalues

The eigenvalues of a square $N \times N$ matrix are the N roots of the scalar equation

$$\text{Det } [\lambda I - A] = 0 \qquad (15.18)$$

where λ is a scalar quantity. Since there are N eigenvalues, it is convenient to define the vector λ, of length N, that consists of the eigenvalues: $\lambda_1, \lambda_2, \lambda_3, \ldots, \lambda_N$.

$$\lambda = \begin{bmatrix} \lambda_1 \\ \lambda_2 \\ \lambda_3 \\ \vdots \\ \lambda_N \end{bmatrix} \qquad (15.19)$$

We will use the following notation. The expression $\lambda_{[A]}$ means the vector of eigenvalues of the matrix A. Thus, the eigenvalues of a matrix $[I + GB]$ will be written $\lambda_{[I + GB]}$.

Example 15.7. Calculate the eigenvalues of the following A matrix.

$$A = \begin{bmatrix} -2 & 0 \\ 2 & -4 \end{bmatrix} \qquad (15.20)$$

We use Eq. (15.18)

Det $[\lambda I - A] = 0$

$$\lambda I - A = \lambda \begin{bmatrix} 1 & 0 \\ 0 & 1 \end{bmatrix} - \begin{bmatrix} -2 & 0 \\ 2 & -4 \end{bmatrix} = \begin{bmatrix} (\lambda + 2) & 0 \\ -2 & (\lambda + 4) \end{bmatrix}$$

$$\text{Det } [\lambda I - A] = \text{Det} \begin{bmatrix} (\lambda + 2) & 0 \\ -2 & (\lambda + 4) \end{bmatrix} = (\lambda + 2)(\lambda + 4) - (0)(-2)$$

$$\lambda^2 + 6\lambda + 8 = 0 = (\lambda + 2)(\lambda + 4)$$

The roots of this equation are $\lambda = -2$ and $\lambda = -4$. Therefore the eigenvalues of the matrix given in Eq. (15.20) are $\lambda_1 = -2$ and $\lambda_2 = -4$.

$$\lambda_{[A]} = \begin{bmatrix} -2 \\ -4 \end{bmatrix} \qquad (15.21)$$

It might be useful at this point to give the motivation for defining such seemingly abstract quantities as eigenvalues. Consider a system N linear ordinary differential equations

$$\frac{dx}{dt} = \underline{A}\underline{x} + \underline{C}\underline{m}$$
(15.22)

where $\underline{x}$ = vector of the N controlled output variables of the system
$\underline{A}$ = $N \times N$ matrix of constants
$\underline{C}$ = a different $N \times N$ matrix of constants
$\underline{m}$ = vector of the N manipulated variables of the system

We will show in Sec. 15.3 that the eigenvalues of the $\underline{A}$ matrix are the roots of the characteristic equation of the system. Thus the eigenvalues tell us whether the system is stable or unstable, fast or slow, overdamped or underdamped. They are essential for the analysis of dynamic systems.

15.2.2   Canonical Transformation

There are several types of transformations that can be applied to matrices. One of the most useful is the canonical transformation. To transform a matrix, you pre-multiply by a matrix of constants and postmultiply by another matrix of constants. The canonical transformation is one which converts a matrix into another matrix that is diagonal and has the eigenvalues of the original matrix as its diagonal elements.

Consider the system described by the linear, homogeneous ordinary differential equations

$$\frac{dx}{dt} = \underline{A}\underline{x}$$
(15.23)

where

$$\underline{x} = \begin{bmatrix} x_1 \\ x_2 \\ x_3 \\ \vdots \\ x_N \end{bmatrix} \qquad \underline{A} = \begin{bmatrix} a_{11} & a_{12} & a_{13} & \cdots & a_{1N} \\ a_{21} & a_{22} & a_{23} & & \\ a_{31} & & & & \\ \cdots\cdots\cdots\cdots\cdots\cdots\cdots\cdots\cdots \\ a_{N1} & & \cdots & & a_{NN} \end{bmatrix}$$

The eigenvalues of the $\underline{A}$ matrix will be the roots of the characteristic equation of the system, and they can be calculated from Eq. (15.18).

Now suppose we change variables by defining a new vector of variables $\underline{z}$ according to the following equation.

$$\underline{x} = \underline{W}\underline{z}$$
(15.24)

where $\underline{W}$ is a matrix of constants which we will define later. This matrix is chosen so that a diagonal matrix $\underline{\Delta}$ is produced when we postmultiply $\underline{A}$ by $\underline{W}$ and premultiply $\underline{A}$ by $\underline{W}^{-1}$. In addition to being diagonal, the $\underline{\Delta}$ matrix has diagonal elements that are the eigenvalues of the original $\underline{A}$ matrix, as we will illustrate shortly.

$$\underline{W}^{-1}\underline{A}\underline{W} = \underline{\Delta} \tag{15.25}$$

where

$$\underline{W} = \begin{bmatrix} W_{11} & W_{12} & W_{13} & \cdots & W_{1N} \\ W_{21} & W_{22} & W_{23} & & \\ W_{31} & & & & \\ \cdots\cdots\cdots\cdots\cdots\cdots\cdots\cdots\cdots \\ W_{N1} & & \cdots & & W_{NN} \end{bmatrix} \tag{15.26}$$

$$\underline{\Delta} = \begin{bmatrix} \lambda_1 & 0 & 0 & \cdots & 0 \\ 0 & \lambda_2 & 0 & & \\ 0 & 0 & \lambda_3 & 0 & \\ \cdots\cdots\cdots\cdots\cdots\cdots \\ 0 & & \cdots & 0 & \lambda_N \end{bmatrix} \tag{15.27}$$

Actually, the $\underline{W}$ matrix consists of the "eigen vectors" of the $\underline{A}$ matrix. The interested reader is referred to the texts mentioned in Sec. 15.1 for details. For our purposes, we need only to accept the notion that such a matrix of constants can be found that will convert the nondiagonal $\underline{A}$ into the diagonal $\underline{\Delta}$.

Example 15.8. Show that the $\underline{W}$ matrix given below canonically transforms the $\underline{A}$ matrix given in Example 15.7 into a diagonal matrix with the eigenvalues of -2 and -4 on the diagonal.

$$\underline{W} = \begin{bmatrix} 1 & 0 \\ 1 & 1 \end{bmatrix} \quad \text{for the matrix} \quad \underline{A} = \begin{bmatrix} -2 & 0 \\ 2 & -4 \end{bmatrix}$$

$$\underline{\Delta} = \underline{W}^{-1}\underline{A}\underline{W} = \begin{bmatrix} 1 & 0 \\ 1 & 1 \end{bmatrix}^{-1} \begin{bmatrix} -2 & 0 \\ 2 & -4 \end{bmatrix} \begin{bmatrix} 1 & 0 \\ 1 & 1 \end{bmatrix}$$

$$= \begin{bmatrix} 1 & 0 \\ -1 & 1 \end{bmatrix} \begin{bmatrix} -2 & 0 \\ -2 & -4 \end{bmatrix} = \begin{bmatrix} -2 & 0 \\ 0 & -4 \end{bmatrix} \tag{15.28}$$

The eigenvalues of the new $\underline{\Delta}$ matrix are exactly the same as the eigenvalues of the original $\underline{A}$ matrix. To demonstrate this for any $\underline{A}$ and $\underline{\Delta}$ matrices we use Eq. (15.18). The eigenvalues of the $\underline{\Delta}$ matrix are $\lambda_{[\Delta]}$ and are the roots of the equation:

$$\text{Det } [\lambda\underline{I} - \underline{\Delta}] = 0 \tag{15.29}$$

$$\text{Det } [\lambda\underline{W}^{-1}\underline{I}\underline{W} - \underline{W}^{-1}\underline{A}\underline{W}] = 0 \tag{15.30}$$

$$\text{Det } [\underline{W}^{-1}\{\lambda\underline{I} - \underline{A}\}\underline{W}] = 0 \tag{15.31}$$

$$\text{Det } [\underline{W}^{-1}] \text{ Det } [\lambda\underline{I} - \underline{A}] \text{ Det } [\underline{W}] = 0 \tag{15.32}$$

Since the determinants of the $\underline{W}^{-1}$ and $\underline{W}$ matrices are not zero,

$$\text{Det } [\lambda\underline{I} - \underline{A}] = 0 \tag{15.33}$$

The roots of this equation are the eigenvalues of $\underline{A}$. Thus the eigenvalues of $\underline{A}$ and of $\underline{\Delta}$ are the same.

Example 15.9. Show that the eigenvalues of the $\underline{\Delta}$ matrix in Example 15.8 are the eigenvalues of the $\underline{A}$ matrix.

$$\text{Det } [\lambda\underline{I} - \underline{\Delta}] = 0$$

$$\text{Det } \left[\lambda \begin{bmatrix} 1 & 0 \\ 0 & 1 \end{bmatrix} - \begin{bmatrix} -2 & 0 \\ 0 & -4 \end{bmatrix} \right] = 0 \tag{15.34}$$

$$\text{Det } \begin{bmatrix} (\lambda + 2) & 0 \\ 0 & (\lambda + 4) \end{bmatrix} = (\lambda + 2)(\lambda + 4) = 0$$

$\lambda = -2, -4$ which are the eigenvalues of the original $\underline{A}$ matrix.

The canonical transformation is useful in a number of ways. For example, it can be used to solve the system of ordinary differential equations

$$\frac{dx}{dt} = \underline{A}x$$

by finding the suitable canonical transformation $\underline{x} = \underline{W}\underline{z}$ and substituting for $\underline{x}$.

$$\underline{W} \frac{d\underline{z}}{dt} = \underline{A}\underline{W}\underline{z} \tag{15.35}$$

Premultiplying by $\underline{W}^{-1}$ gives

$$\underline{W}^{-1}\underline{W} \frac{d\underline{z}}{dt} = \underline{W}^{-1}\underline{A}\underline{W}\underline{z}$$

$$\tag{15.36}$$

$$\frac{d\underline{z}}{dt} = \underline{\Delta}\underline{z}$$

Since $\underline{\Delta}$ is diagonal, the N first-order differential equations of the system are

$$\frac{dz_1}{dt} = \lambda_1 z_1$$

$$\frac{dz_2}{dt} = \lambda_2 z_2$$

$$\frac{dz_3}{dt} = \lambda_3 z_3 \tag{15.37}$$

$$\vdots$$

$$\frac{dz_N}{dt} = \lambda_N z_N$$

And the solutions of these completely independent differential equations can be written down by inspection.

$$z_1 = c_1 e^{\lambda_1 t}$$

$$z_2 = c_2 e^{\lambda_2 t}$$

$$z_3 = c_3 e^{\lambda_3 t} \qquad (15.38)$$

$$\vdots$$

$$z_N = c_N e^{\lambda_N t}$$

where the c_i's are constants of integration

Finally, the solution in terms of the original x variables can be found by multiplying the vector of z's by the $\underline{W}$ matrix.

$$\underline{x} = \underline{W} \underline{z}$$

$$
\begin{bmatrix} x_1 \\ x_2 \\ \vdots \\ x_N \end{bmatrix}
=
\begin{bmatrix}
W_{11} & W_{12} & \cdots & W_{1N} \\
W_{21} & W_{22} & & \\
W_{31} & & & \\
\hdotsfor{4} \\
W_{N1} & & \cdots & W_{NN}
\end{bmatrix}
\begin{bmatrix} z_1 \\ z_2 \\ \vdots \\ z_N \end{bmatrix}
\qquad (15.39)
$$

$$x_1 = W_{11} z_1 + W_{12} z_2 + W_{13} z_3 + \cdots + W_{1N} z_N$$

$$x_2 = W_{21} z_1 + W_{22} z_2 + W_{23} z_3 + \cdots + W_{2N} z_N$$

$$\vdots \qquad (15.40)$$

$$x_N = W_{N1} z_1 + W_{N2} z_2 + W_{N3} z_3 + \cdots + W_{NN} z_N$$

Thus the canonical transformation can be used to solve differential equations. It is also useful for other things, as we will see later.

15.2.3 Singular Values

The singular values of a matrix are a measure of how close the matrix is to being "singular," i.e., having a determinant that is zero. It has been used for a number of years in numerical analysis. It is used in control for several purposes, one of which is to give a measure of the size of the matrix.

A matrix that is $N \times N$ has N singular values. We use the symbol σ_i for singular values. The σ_i that is the biggest in magnitude is called the maximum singular value and the notation σ^{max} is used. The σ_i that is the smallest in magnitude is called the minimum singular value and the notation σ^{min} is used.

The N singular values of a *real* $N \times N$ matrix are defined as the square root of the eigenvalues of the matrix formed by multiplying the transpose of the original matrix by itself.

$$\sigma_{i[\underline{A}]} = \sqrt{\lambda_{i[\underline{A}^T \underline{A}]}} \qquad i = 1, 2, \ldots, N \qquad (15.41)$$

Example 15.10. Find the singular values of the A matrix from Example 15.8.

$$A = \begin{bmatrix} -2 & 0 \\ 2 & -4 \end{bmatrix} \quad A^T = \begin{bmatrix} -2 & 2 \\ 0 & -4 \end{bmatrix}$$

$$A^T A = \begin{bmatrix} -2 & 2 \\ 0 & -4 \end{bmatrix} \begin{bmatrix} -2 & 2 \\ 0 & -4 \end{bmatrix} = \begin{bmatrix} 8 & -8 \\ -8 & 16 \end{bmatrix}$$

(15.42)

Now to get the eigenvalues of this matrix we use Eq. (15.18).

$$\text{Det } [\lambda I - A^T A] = 0$$

$$\text{Det } \begin{bmatrix} (\lambda - 8) & 8 \\ 8 & (\lambda - 16) \end{bmatrix} = 0 = (\lambda - 8)(\lambda - 16) - 64$$

$$\lambda^2 - 24\lambda + 128 - 64 = 0 = \lambda^2 - 24\lambda + 64 \qquad (15.43)$$

$$\lambda_1 = 20.94 \qquad \lambda_2 = 3.06$$

$$\sigma_1 = \sqrt{20.94} = 4.58 \qquad \sigma_2 = \sqrt{3.06} = 1.75$$

Neither of these singular values is small, so the matrix is not close to being singular. The determinant of A is 8, so it is indeed not singular.

Example 15.11. Calculate the singular values of the matrix

$$A = \begin{bmatrix} -1 & 1 \\ 1 & -1 \end{bmatrix}$$

Note that the determinant of this matrix is zero, so it is singular.

$$A^T = \begin{bmatrix} -1 & 1 \\ 1 & -1 \end{bmatrix}$$

$$A^T A = \begin{bmatrix} -1 & 1 \\ 1 & -1 \end{bmatrix} \begin{bmatrix} -1 & 1 \\ 1 & -1 \end{bmatrix} = \begin{bmatrix} 2 & -2 \\ -2 & 2 \end{bmatrix}$$

$$\text{Det } [\lambda I - A^T A] = 0$$

$$\text{Det } \begin{bmatrix} (\lambda - 2) & 2 \\ 2 & (\lambda - 2) \end{bmatrix} = 0 = (\lambda - 2)(\lambda - 2) - 4$$

$$\lambda^2 - 4\lambda + 4 - 4 = \lambda(\lambda - 4) = 0$$

$$\lambda_1 = 0 \qquad \lambda_2 = 4 \qquad \sigma_1 = 0 \qquad \sigma_2 = 2$$

The singular value of zero tells us that the matrix is singular.

The singular values of a *complex* matrix are similar to those of a *real* matrix. The only difference is that we use the conjugate transpose.

$$\sigma_{i[A]} = \sqrt{\lambda_{i[A^{CT}A]}} \qquad i = 1, 2, \ldots, N \qquad (15.44)$$

First we calculate the conjugate transpose [see Eq. (15.9)]. Then we multiply A by it. Then we calculate the eigenvalues. These can be found using Eq. (15.18) for simple systems. In more realistic problems the IMSL subroutine "EIGCC" can be used to calculate the eigenvalues of a complex matrix. Note that the product of a complex matrix with its conjugate transpose gives a complex matrix (called a *hermitian* matrix) that has real elements on the diagonal and has real eigenvalues. Thus all the singular values of a complex matrix are real numbers.

Example 15.12. Calculate the singular values of the complex matrix

$$\underline{A} = \begin{bmatrix} (1+i) & (1+i) \\ (2+i) & (1+i) \end{bmatrix} \tag{15.45}$$

$$\underline{A}^{CT} = \begin{bmatrix} (1-i) & (2-i) \\ (1-i) & (1-i) \end{bmatrix}$$

$$\underline{A}^{CT}\underline{A} = \begin{bmatrix} (1-i) & (2-i) \\ (1-i) & (1-i) \end{bmatrix} \begin{bmatrix} (1+i) & (1+i) \\ (2+i) & (1+i) \end{bmatrix}$$

$$= \begin{bmatrix} 7 & (5+i) \\ (5-i) & 4 \end{bmatrix}$$

$$\text{Det}\,[\lambda\underline{I} - \underline{A}^{CT}\underline{A}] = 0$$

$$\text{Det}\left[\lambda\begin{bmatrix} 1 & 0 \\ 0 & 1 \end{bmatrix} - \begin{bmatrix} 7 & (5+i) \\ (5-i) & 4 \end{bmatrix}\right] = 0$$

$$\text{Det}\begin{bmatrix} (\lambda-7) & (-5-i) \\ (-5+i) & (\lambda-4) \end{bmatrix} = 0$$

$$\lambda^2 - 11\lambda + 28 - 25 - 1 = \lambda^2 - 11\lambda + 2 = 0$$

$$\lambda_1 = 10.63 \qquad \lambda_2 = 0.185 \qquad \sigma_1 = 3.29 \qquad \sigma_2 = 0.430$$

Note that the singular values are real.

15.3 REPRESENTATION OF MULTIVARIABLE PROCESSES

15.3.1 Transfer Function Matrix

A. OPENLOOP SYSTEM. Let us first consider an openloop process with N controlled variables, N manipulated variables, and one load disturbance. The system can be described in the Laplace domain by N equations that show how all of the manipulated variables and the load disturbance affect each of the controlled variables through their appropriate transfer functions.

$$x_1 = G_{M11}m_1 + G_{M12}m_2 + \cdots + G_{M1N}m_N + G_{L1}L$$

$$x_2 = G_{M21}m_1 + G_{M22}m_2 + \cdots + G_{M2N}m_N + G_{L2}L$$

$$\vdots \tag{15.46}$$

$$x_N = G_{MN1}m_1 + G_{MN2}m_2 + \cdots + G_{MNN}m_N + G_{LN}L$$

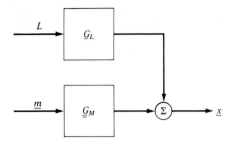

FIGURE 15.1

All the variables are in the Laplace domain, as are all of the transfer functions. This set of N equations is very conveniently represented by one matrix equation.

$$\underline{x} = \underline{G}_{M(s)}\underline{m}_{(s)} + \underline{G}_{L(s)}L_{(s)} \tag{15.47}$$

where $\underline{x}$ = vector of N controlled variables

$\underline{G}_M$ = $N \times N$ matrix of process openloop transfer functions relating the controlled variables and the manipulated variables

$\underline{m}$ = vector of N manipulated variables

$\underline{G}_L$ = vector of process openloop transfer functions relating the controlled variables and the load disturbance

$L_{(s)}$ = load disturbance

These relationships are shown pictorially in Fig. 15.1. We have chosen to use only one load variable in this development in order to keep things as simple as possible. Clearly, there could be several load disturbances. We would just add additional terms to Eqs. (15.46). Then the $L_{(s)}$ in Eq. (15.47) becomes a vector and G_L becomes a matrix with N rows and as many columns as there are load disturbances. Since the effects of each of the load disturbances can be considered one at a time, we will do it that way to simplify the mathematics. Note that the effects of each of the manipulated variables can also be considered one at a time if we were looking only at the *openloop* system or if we were considering controlling only one variable. However, when we go to a multivariable *closedloop* system, the effects of all manipulated variables must be considered simultaneously.

B. CLOSEDLOOP SYSTEM. Figure 15.2 gives the matrix block diagram description of the openloop system with a feedback control system added. The $\underline{I}$ matrix is the identity matrix. The $\underline{B}_{(s)}$ matrix contains the feedback controllers. Most industrial processes use conventional single-input–single-output (SISO) feedback controllers. One controller is used in each loop to control one controlled variable by changing one manipulated variable. In this case the $\underline{B}_{(s)}$ matrix has only diagonal elements. All the off-diagonal elements are zero.

$$\underline{B}_{(s)} = \begin{bmatrix} B_1 & 0 & \cdots & 0 \\ 0 & B_2 & 0 & \\ \multicolumn{4}{c}{\dotfill} \\ 0 & \cdots & 0 & B_N \end{bmatrix} \tag{15.48}$$

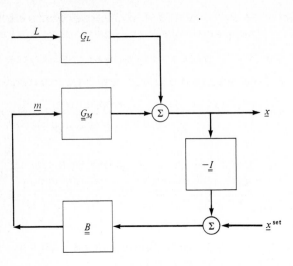

FIGURE 15.2

where B_1, B_2, ..., B_N are the individual controllers in each of the N loops. We call this multiloop SISO system a *diagonal controller* structure. It is important to recognize right from the beginning that having multiple SISO controllers does *not* mean that we can tune each controller independently. As we will soon see, the dynamics and stability of this multivariable closedloop process depend on the settings of all controllers.

Controller structures that are not diagonal but have elements in all positions in the $\underline{B}_{(s)}$ matrix are called multivariable controllers.

$$\underline{B}_{(s)} = \begin{bmatrix} B_{11} & B_{12} & \cdots & B_{1N} \\ B_{21} & B_{22} & & \\ B_{31} & & & \\ \cdots\cdots\cdots\cdots\cdots\cdots\cdots\cdots \\ B_{N1} & & \cdots & B_{NN} \end{bmatrix} \qquad (15.49)$$

The feedback controller matrix gives the transfer functions between the manipulated variables and the errors.

$$\begin{bmatrix} m_1 \\ m_2 \\ \vdots \\ m_N \end{bmatrix} = \begin{bmatrix} B_{11} & B_{12} & \cdots & B_{1N} \\ B_{21} & B_{22} & & \\ B_{31} & & & \\ \cdots\cdots\cdots\cdots\cdots\cdots\cdots\cdots \\ B_{N1} & & \cdots & B_{NN} \end{bmatrix} \begin{bmatrix} E_1 \\ E_2 \\ \vdots \\ E_N \end{bmatrix} \qquad (15.50)$$

In compact form this equation becomes

$$\underline{m} = \underline{B}_{(s)}\,\underline{E} \tag{15.51}$$

Since the errors are the differences between setpoints and controlled variables

$$\begin{bmatrix} E_1 \\ E_2 \\ \vdots \\ E_N \end{bmatrix} = \begin{bmatrix} (x_1^{\text{set}} - x_1) \\ (x_2^{\text{set}} - x_2) \\ \vdots \\ (x_N^{\text{set}} - x_N) \end{bmatrix} \tag{15.52}$$

Substituting into Eq. (15.51) gives

$$\underline{m} = \underline{B}_{(s)}[\underline{x}^{\text{set}} - \underline{x}] \tag{15.53}$$

We will use this matrix of transfer function representation extensively in the rest of our work with multivariable processes.

15.3.2 State Variables

The time-domain differential equation description of systems can be used instead of the Laplace-domain transfer function description. Naturally the two are related, and we will derive these relationships later in this chapter. State variables are very popular in electrical and mechanical engineering control problems which tend to be of lower order (fewer differential equations) than chemical engineering control problems. Transfer function representations are more useful in practical process control problems because the matrices are of lower order than would be required by a state variable representation.

The "states" of a dynamic system are simply the variables that appear in the time differential. For example, if we have a chemical reactor in which the concentration of reactant C_A and the temperature T change with time, the material balance for component A and the energy balance would give two differential equations:

$$\frac{dC_A}{dt} = f_1 \qquad \frac{dT}{dt} = f_2$$

The *state variables* of the system are C_A and T.

State variables appear very naturally in the differential equations describing chemical engineering systems because our mathematical models are based on a number of first-order differential equations: component balances, energy equations, etc. If there are N such equations, they can be linearized (if necessary) and written in matrix form

$$\frac{d\underline{x}}{dt} = \underline{A}\underline{x} + \underline{C}\underline{m} + \underline{D}L \tag{15.54}$$

where $\underline{x}$ = vector of the N state variables of the system

$$\underline{x} = \begin{bmatrix} x_1 \\ x_2 \\ x_3 \\ \vdots \\ x_N \end{bmatrix}$$

$$\underline{A} = N \times N \text{ matrix of constants} = \begin{bmatrix} a_{11} & a_{12} & a_{13} & \cdots & a_{1N} \\ a_{21} & a_{22} & a_{23} & & \\ a_{31} & & & & \\ & & \cdots & & \\ a_{N1} & & \cdots & & a_{NN} \end{bmatrix} \quad (15.55)$$

$\underline{C}$ = a different $N \times R$ matrix of constants
$\underline{m}$ = vector of the R manipulated variables of the system
$\underline{D}$ = vector of constants
L = load disturbance

There are usually many more state variables that manipulated variables. In a distillation column there are typically over 100 state variables ($N = 100$), while there are only 5 manipulated variables ($R = 5$). There is only one load disturbance shown in Eq. (15.54) for simplicity. If there were more than one, the effects of each could be determined individually.

Example 15.13. The irreversible chemical reaction A → B takes place in two perfectly mixed reactors connected in series as shown in Fig. 15.3. The reaction rate is proportional to the concentration of reactant. Let x_1 be the concentration of reactant A in the first tank and x_2 the concentration in the second tank. The concentration of reactant in the feed is x_0. The feed flow rate is F. Both x_0 and F can be manipulated. Assume the specific reaction rates k_1 and k_2 in each tank are constant (isothermal operation). Assume constant volumes V_1 and V_2.

The component balances for the system are

$$V_1 \frac{dx_1}{dt} = F(x_0 - x_1) - k_1 V_1 x_1$$

$$V_2 \frac{dx_2}{dt} = F(x_1 - x_2) - k_2 V_2 x_2 \quad (15.56)$$

FIGURE 15.3

Linearizing around the initial steadystate gives two linear ODEs.

$$V_1 \frac{dx_1}{dt} = -(\bar{F} + k_1 V_1)x_1 + (\bar{F})x_0 + (\bar{x}_0 - \bar{x}_1)F$$

$$V_2 \frac{dx_2}{dt} = (\bar{F})x_1 - (\bar{F} + k_2 V_2)x_2 + (\bar{x}_1 - \bar{x}_2)F$$

(15.57)

These two equations in matrix form are

$$\frac{d}{dt}\begin{bmatrix} x_1 \\ x_2 \end{bmatrix} = \begin{bmatrix} \left(-k_1 - \dfrac{F}{V_1}\right) & 0 \\ \left(\dfrac{F}{V_2}\right) & \left(-k_2 - \dfrac{F}{V_2}\right) \end{bmatrix}\begin{bmatrix} x_1 \\ x_2 \end{bmatrix} + \begin{bmatrix} \left(\dfrac{F}{V_1}\right) & \dfrac{(\bar{x}_0 - \bar{x}_1)}{V_1} \\ 0 & \dfrac{(\bar{x}_1 - \bar{x}_2)}{V_2} \end{bmatrix}\begin{bmatrix} x_0 \\ F \end{bmatrix}$$

(15.58)

The state variables are the two concentrations. The feed concentration x_0 and the feed flow rate F are the manipulated variables. To take a specific numerical case, let $k_1 = 1$ min^{-1}, $k_2 = 2$ min^{-1}, $V_1 = 100$ ft^3, and $V_2 = 50$ ft^3. The initial steadystate conditions are $\bar{F} = 100$ ft^3/min, $\bar{x}_0 = 0.5$ mol A/ft^3, $\bar{x}_1 = 0.25$ mol A/ft^3, and $\bar{x}_2 = 0.125$ mol A/ft^3. This gives the $\underline{A}$ matrix that we used in Example 15.7.

$$\underline{A} = \begin{bmatrix} -2 & 0 \\ 2 & -4 \end{bmatrix}$$

The $\underline{C}$ matrix is

$$\underline{C} = \begin{bmatrix} 1 & 0.0025 \\ 0 & 0.0025 \end{bmatrix}$$

State variable representation can be transformed into transfer function representation by simply Laplace-transforming the set of N linear ordinary differential equations [Eq. (15.54)]

$$\frac{dx}{dt} = \underline{A}x + \underline{C}m + \underline{D}L$$

$$s\underline{x}_{(s)} = \underline{A}\underline{x}_{(s)} + \underline{C}\underline{m}_{(s)} + \underline{D}L_{(s)}$$

(15.59)

$$[s\underline{I} - \underline{A}]\underline{x}_{(s)} = \underline{C}\underline{m}_{(s)} + \underline{D}L_{(s)}$$

$$\underline{x}_{(s)} = [[s\underline{I} - \underline{A}]^{-1}\underline{C}]\underline{m}_{(s)} + [[s\underline{I} - \underline{A}]^{-1}\underline{D}]L_{(s)}$$

Comparing this with Eq. (15.47), we can see how the transfer function matrix $\underline{G}_{M_{(s)}}$ and transfer function vector $\underline{G}_{L_{(s)}}$ are related to the $\underline{A}$ and $\underline{C}$ matrices and to the $\underline{D}$ vector.

Example 15.14. Determine the transfer function matrix $\underline{G}_{M_{(s)}}$ for system described in Example 15.13.

$$\frac{d}{dt}\begin{bmatrix} x_1 \\ x_2 \end{bmatrix} = \begin{bmatrix} -2 & 0 \\ 2 & -4 \end{bmatrix}\begin{bmatrix} x_1 \\ x_2 \end{bmatrix} + \begin{bmatrix} 1 & 0.0025 \\ 0 & 0.0025 \end{bmatrix}\begin{bmatrix} x_0 \\ F \end{bmatrix}$$

$$\underline{G}_{M_{(s)}} = [s\underline{I} - \underline{A}]^{-1}\underline{C} = \left[s\begin{bmatrix} 1 & 0 \\ 0 & 1 \end{bmatrix} - \begin{bmatrix} -2 & 0 \\ 2 & -4 \end{bmatrix}\right]^{-1}\begin{bmatrix} 1 & 0.0025 \\ 0 & 0.0025 \end{bmatrix}$$

$$= \begin{bmatrix} (s+2) & 0 \\ -2 & (s+4) \end{bmatrix}^{-1}\begin{bmatrix} 1 & 0.0025 \\ 0 & 0.0025 \end{bmatrix}$$

$$= \frac{\begin{bmatrix} (s+4) & 0 \\ 2 & (s+2) \end{bmatrix}}{(s+2)(s+4)}\begin{bmatrix} 1 & 0.0025 \\ 0 & 0.0025 \end{bmatrix}$$

$$\begin{bmatrix} x_1 \\ x_2 \end{bmatrix} = \underline{G}_{M_{(s)}}\begin{bmatrix} x_0 \\ F \end{bmatrix} = \begin{bmatrix} \dfrac{1}{s+2} & \dfrac{0.0025}{s+2} \\ \dfrac{2}{(s+2)(s+4)} & \dfrac{0.0025}{s+2} \end{bmatrix}\begin{bmatrix} x_0 \\ F \end{bmatrix} \tag{15.60}$$

The system considered in the example above has a characteristic equation that is the denominator of the transfer function set equal to zero. This is true, of course, for any system. Since the system is uncontrolled, the openloop characteristic equation is [using Eq. (15.60)]

$$(s+2)(s+4) = 0$$

The roots of the openloop characteristic equation are $s = -2$ and $s = -4$. These are exactly the values we calculated for the eigenvalues of the $\underline{A}$ matrix of this system!

> The eigenvalues of the $\underline{A}$ matrix are equal to the roots of the characteristic equation of the system.

The $\underline{A}$ matrix is the matrix that multiplies the $\underline{x}$ vector when the differential equations are in the standard form $d\underline{x}/dt = \underline{A}\underline{x}$.

We have considered openloop systems up to this point, but the mathematics applies to *any* system, openloop or closedloop, as we will see in the next section.

15.4 OPENLOOP AND CLOSEDLOOP SYSTEMS

15.4.1 Transfer Function Representation

Figure 15.2 shows a closedloop multivariable system. The process is always described by the equation

$$\underline{x}_{(s)} = \underline{G}_{M_{(s)}}\underline{m}_{(s)} + \underline{G}_{L_{(s)}}L_{(s)}$$

If the system is openloop, $\underline{m}$ can be changed independently. If the system is closedloop, $\underline{m}$ is determined by the feedback controller. Substituting for $\underline{m}$ from Eq. (15.53) gives

$$\underline{x} = \underline{G}_{M_{(s)}} \underline{B}_{(s)} \underline{E}_{(s)} + \underline{G}_{L_{(s)}} L_{(s)}$$

$$\underline{x} = \underline{G}_{M_{(s)}} \underline{B}_{(s)} [\underline{x}^{set} - \underline{x}] + \underline{G}_{L_{(s)}} L_{(s)}$$

(15.61)

Bringing all the terms with $\underline{x}$ to the left side gives

$$\underline{I}\underline{x} + \underline{G}_{M_{(s)}} \underline{B}_{(s)} \underline{x} = \underline{G}_{M_{(s)}} \underline{B}_{(s)} \underline{x}^{set} + \underline{G}_{L_{(s)}} L_{(s)} \qquad (15.62)$$

$$[\underline{I} + \underline{G}_{M_{(s)}} \underline{B}_{(s)}]\underline{x} = \underline{G}_{M_{(s)}} \underline{B}_{(s)} \underline{x}^{set} + \underline{G}_{L_{(s)}} L_{(s)} \qquad (15.63)$$

$$\underline{x} = [[\underline{I} + \underline{G}_{M_{(s)}} \underline{B}_{(s)}]^{-1} \underline{G}_{M_{(s)}} \underline{B}_{(s)}]\underline{x}^{set} + [[\underline{I} + \underline{G}_{M_{(s)}} \underline{B}_{(s)}]^{-1} \underline{G}_{L_{(s)}}]L_{(s)} \qquad (15.64)$$

Equation (15.64) gives the effects of setpoint and load changes on the controlled variables in the closedloop multivariable environment. The matrix (of order $N \times N$) multiplying the vector of setpoints is the closedloop servo transfer function matrix. The matrix ($N \times 1$) multiplying the load disturbance is the closedloop regulator transfer function vector.

It is clear that the matrix equation [Eq. (15.64)] is very similar to the scalar equation describing a closedloop system derived back in Chap. 10 for SISO systems.

$$x = \left[\frac{G_M B}{1 + G_M B} \right] x^{set} + \left[\frac{G_L}{1 + G_M B} \right] L \qquad (15.65)$$

Now we have matrix inverses to worry about, but the structure is essentially the same.

Remember that the inverse of a matrix has the determinant of the matrix in the denominator of each element. Therefore the denominators of all of the transfer functions in Eq. (15.64) will contain Det $[\underline{I} + \underline{G}_{M_{(s)}} \underline{B}_{(s)}]$. Now we know that the characteristic equation of any system is the denominator set equal to zero. Therefore the closedloop characteristic equation of the multivariable system with feedback controllers is the simple *scalar* equation

$$\boxed{\text{Det } [\underline{I} + \underline{G}_{M_{(s)}} \underline{B}_{(s)}] = 0} \qquad (15.66)$$

We will use this extensively in Chap. 16 to study the stability of closedloop multivariable processes.

Example 15.15. Determine the closedloop characteristic equation for the system whose openloop transfer function matrix was derived in Example 15.14. Use a diagonal controller structure (two SISO controllers) that are proportional only.

$$B_{(s)} = \begin{bmatrix} K_1 & 0 \\ 0 & K_2 \end{bmatrix} \qquad G_{M_{(s)}} = \begin{bmatrix} \dfrac{1}{s+2} & \dfrac{0.0025}{s+2} \\ \dfrac{2}{(s+2)(s+4)} & \dfrac{0.0025}{s+2} \end{bmatrix}$$

$$G_{M_{(s)}} B_{(s)} = \begin{bmatrix} \dfrac{K_1}{s+2} & \dfrac{0.0025K_2}{s+2} \\ \dfrac{2K_1}{(s+2)(s+4)} & \dfrac{0.0025K_2}{s+2} \end{bmatrix}$$

$$I + G_{M_{(s)}} B_{(s)} = \begin{bmatrix} \left(1 + \dfrac{K_1}{s+2}\right) & \dfrac{0.0025K_2}{s+2} \\ \dfrac{2K_1}{(s+2)(s+4)} & \left(1 + \dfrac{0.0025K_2}{s+2}\right) \end{bmatrix}$$

$$\text{Det } [I + G_{M_{(s)}} B_{(s)}] = 0$$

$$\left(1 + \frac{K_1}{s+2}\right)\left(1 + \frac{0.0025K_2}{s+2}\right) - \left(\frac{0.0025K_2}{s+2}\right)\left(\frac{2K_1}{(s+2)(s+4)}\right) = 0 \quad (15.67)$$

$$1 + \frac{K_1}{s+2} + \frac{0.0025K_2}{s+2} + \frac{0.0025K_1K_2}{(s+2)^2} - \frac{0.005K_1K_2}{(s+2)^2(s+4)} = 0$$

$$1 + \frac{K_1}{s+2} + \frac{0.0025K_2}{s+2} + \frac{0.0025K_1K_2}{(s+2)^2}\left(1 - \frac{2}{s+4}\right) = 0$$

$$1 + \frac{K_1}{s+2} + \frac{0.0025K_2}{s+2} + \frac{0.0025K_1K_2}{(s+2)^2}\left(\frac{s+4-2}{s+4}\right) = 0$$

$$1 + \frac{K_1}{s+2} + \frac{0.0025K_2}{s+2} + \frac{0.0025K_1K_2}{(s+2)(s+4)} = 0$$

$$s^2 + 6s + 8 + K_1(s+4) + 0.0025K_2(s+4) + 0.0025K_1K_2 = 0$$

$$s^2 + s(6 + K_1 + 0.0025K_2) + (8 + 4K_1 + 0.01K_2 + 0.0025K_1K_2) = 0 \quad (15.68)$$

Remember that these values of s are the roots of the closedloop characteristic equation.

15.4.2   State Variable Representation

Suppose an openloop system is described by

$$\frac{dx}{dt} = Ax + Cm + DL \tag{15.69}$$

The eigenvalues of the A matrix, $\lambda_{[A]}$, will be the *openloop eigenvalues* and will be equal to the roots of the openloop characteristic equation. In order to help us keep ourselves straight on what are "apples" and what are "oranges," we will call the openloop eigenvalues λ_{OL}.

Now suppose a feedback controller is added to the system. The manipulated variables m will now be set by the feedback controller. To keep things as simple as possible, let us make two assumptions that are not very good ones, but permit us to illustrate an important point. We assume that the feedback controller matrix $B_{(s)}$ consists of just constants (gains) K, and we assume that there are as many manipulated variables m as state variables x.

$$m = K[x^{\text{set}} - x] \tag{15.70}$$

Substituting into Eq. (15.69) gives

$$\frac{dx}{dt} = Ax + CK[x^{\text{set}} - x] + DL$$

Rearranging to put the differential equations in the standard form gives

$$\frac{dx}{dt} = [A - CK]x + CKx^{\text{set}} + DL \tag{15.71}$$

This equation describes the closedloop system. Let us define the matrix that multiplies x as the "closedloop A" matrix and use the symbol A_{CL}.

$$\frac{dx}{dt} = A_{CL}x + CKx^{\text{set}} + DL \tag{15.72}$$

Thus the characteristic matrix for this closedloop system is the A_{CL} matrix. Its eigenvalues will be the *closedloop eigenvalues*, and they will be the roots of the closedloop characteristic equation.

The purpose of the discussion above is to contrast openloop eigenvalues and closedloop eigenvalues. We must use the appropriate eigenvalue for the system we are studying. The "Greek" state-space language uses the term *eigenvalue* instead of the Russian language Laplace-transfer-function term *root of the characteristic equation*. But in any language they are exactly the same thing. So we have openloop and closedloop eigenvalues, or we have roots of the openloop and closedloop characteristic equations.

Example 15.16. The openloop eigenvalues for the two-reactor system studied in Example 15.13 were $\lambda_{OL} = -2, -4$. Calculate the closedloop eigenvalues if two proportional controllers are used. B_1 manipulates x_0 to control x_1, and B_2 manipulates F to control x_2.

$$\begin{aligned}
A_{CL} = A - CK &= \begin{bmatrix} -2 & 0 \\ 2 & -4 \end{bmatrix} - \begin{bmatrix} 1 & 0.0025 \\ 0 & 0.0025 \end{bmatrix}\begin{bmatrix} K_1 & 0 \\ 0 & K_2 \end{bmatrix} \\
&= \begin{bmatrix} (-2 - K_1) & -0.0025K_2 \\ 2 & (-4 - 0.0025K_2) \end{bmatrix}
\end{aligned} \tag{15.73}$$

Using Eq. (15.18) to solve for the eigenvalues of this matrix gives the closedloop eigenvalues λ_{CL}.

$$\text{Det}\,[\lambda_{CL}\boldsymbol{I} - \boldsymbol{A}_{CL}] = 0 \quad (15.74)$$

$$\text{Det}\left[\lambda_{CL}\begin{bmatrix}1 & 0\\ 0 & 1\end{bmatrix} - \begin{bmatrix}(-2 - K_1) & -0.0025K_2\\ 2 & (-4 - 0.0025K_2)\end{bmatrix}\right] = 0$$

$$\text{Det}\begin{bmatrix}(\lambda_{CL} + 2 + K_1) & 0.0025K_2\\ -2 & (\lambda_{CL} + 4 + 0.0025K_2)\end{bmatrix} = 0$$

$$\lambda_{CL}^2 + \lambda_{CL}(6 + K_1 + 0.0025K_2) + (8 + 4K_1 + 0.01K_2 + 0.0025K_1K_2) = 0 \quad (15.75)$$

Note that this is exactly the same characteristic equation that we found using the transfer function notation [see Eq. (15.68)].

For $K_1 = 1$ and $K_2 = 100$, the closedloop eigenvalues are $\lambda_{CL} = -3.62 \pm i0.331$. For $K_1 = 5$ and $K_2 = 500$, $\lambda_{CL} = -6.12 \pm i1.32$. Figure 15.4 is a plot of the closedloop eigenvalues as a function of the two controller gains. Note that this is *not* a traditional SISO root locus plot, so some of the traditional rules do not apply. Both gains are changing along the curves. The shapes of the curves are quite unusual. For example, the two loci both run out the negative real axis as the gains become large.

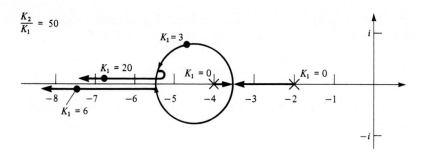

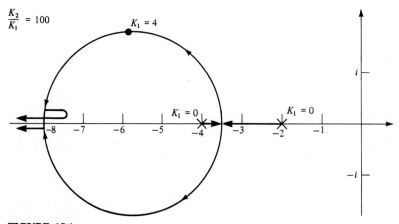

FIGURE 15.4
Closedloop eigenvalues.

15.5 COMPUTER PROGRAMS
FOR MATRIX CALCULATIONS

Table 15.1 gives some simple FORTRAN subroutines for handling matrices of complex numbers. These will be used in later chapters in programs to examine the stability, performance, and robustness of multivariable processes.

MADD   adds complex matrices A and B to give C
MMULT   multiplies complex matrices A and B to give C
IDENT   forms an $N \times N$ identity matrix I
CONJT   takes the conjugate transpose of matrix A
MTRAN   takes the transpose of matrix A
DET   calculates the scalar determinant of a complex matrix A

TABLE 15.1
FORTRAN programs for basic matrix operations

A. Matrix Addition: $C = A + B$

```
SUBROUTINE MADD(A,B,C,N)
COMPLEX A(4,4),B(4,4),C(4,4)
DO 10 I=1,N
DO 10 J=1,N
10 C(I,J)=A(I,J)+B(I,J)
RETURN
END
```

B. Form the Identity Matrix I

```
SUBROUTINE IDENT(B,N)
COMPLEX B(4,4)
DO 20 I=1,N
DO 10 J=1,N
B(I,J)=CMPLX(0.0,0.0)
IF(I.EQ.J) B(I,J)=CMPLX(1.,0.0)
10 CONTINUE
20 CONTINUE
RETURN
END
```

C. Matrix Multiplication: $C = A B$

```
SUBROUTINE MMULT(A,B,C,N)
COMPLEX A(4,4),B(4,4),C(4,4)
DO 30 I=1,N
DO 20 J=1,N
C(I,J)=CMPLX(0.,0.)
DO 10 K=1,N
10 C(I,J)=C(I,J)+A(I,K)*B(K,J)
20 CONTINUE
30 CONTINUE
RETURN
END
```

TABLE 15.1 (*continued*)

D. Form Conjugate Transpose $\underline{B} = \underline{A}^{CT}$

```
SUBROUTINE CONJT(A,B,N)
COMPLEX A(4,4),B(4,4)
DO 10 I=1,N
DO 10 J=1,N
B(I,J)=A(J,I)
Z=AIMAG(B(I,J))
10 B(I,J)=CMPLX(REAL(B(I,J)),-Z)
RETURN
END
```

E. Transpose of $\underline{A} = \underline{A}^{T}$

```
SUBROUTINE MTRAN(A,AT,N)
COMPLEX A,AT
DIMENSION A(4,4),AT(4,4)
DO 10 I=1,N
DO 10 J=1,N
10 AT(I,J)=A(J,I)
RETURN
END
```

F. Determinant: XDETER = Det $\underline{A}$

```
SUBROUTINE DETER(N,G,A,C,XDETER)
EXTERNAL D2,D3,D4
COMPLEX D2,D3,D4
COMPLEX C(4,4),G(4,4),XDETER,A(4,4)
IF(N.EQ.2)THEN
XDETER=D2(N,G)
RETURN
ELSE IF(N.EQ.3)THEN
XDETER=D3(N,G,A)
RETURN
ELSE IF(N.EQ.4)THEN
XDETER=D4(N,G,A,C)
RETURN
ENDIF
WRITE(6,*)' DETERMINANT ABORT'
STOP777
END
C
COMPLEX FUNCTION D2(N,A)
COMPLEX A(4,4)
D2=A(1,1)*A(2,2)-A(1,2)*A(2,1)
RETURN
END
C
COMPLEX FUNCTION D3(N,G,A)
COMPLEX D2,A(4,4),G(4,4)
EXTERNAL D2
D3=CMPLX(0.,0.)
DO 100 L=1,3
DO 10 I=1,2
DO 10 J=1,2
M=I+1
K=L+J
```

TABLE 15.1 (*continued*)

```
       IF(K.EQ.4)K=1
       IF(K.EQ.5)K=2
   10 A(I,J)=G(M,K)
       D3=D3+G(1,L)*D2(N,A)
  100 CONTINUE
       RETURN
       END
C
       COMPLEX FUNCTION D4(N,G,A,C)
       COMPLEX D3,C(4,4),A(4,4),G(4,4)
       EXTERNAL D3
       D4=CMPLX(0.,0.)
       DO 100 L=1,4
       DO 10 I=1,3
       DO 10 J=1,3
       M=I+1
       K=L+J
       IF(K.EQ.5)K=1
       IF(K.EQ.6)K=2
       IF(K.EQ.7)K=3
   10 A(I,J)=G(M,K)
       IF(L.EQ.1 .OR. L.EQ.3) D4=D4+G(1,L)*D3(N,A,C)
       IF(L.EQ.2 .OR. L.EQ.4) D4=D4-G(1,L)*D3(N,A,C)
  100 CONTINUE
       RETURN
       END
```

PROBLEMS

15.1. Find the determinant of the matrix.

(a) $\quad A = \begin{bmatrix} 2 & 1.5 \\ 1.5 & 2 \end{bmatrix}$

(b) $\quad A = \begin{bmatrix} 1 & -2.4 \\ 0.5 & 1 \end{bmatrix}$

(c) $\quad A = \begin{bmatrix} -3 & -5 & 2 \\ 1 & -5 & 1 \\ 1 & 2 & 3 \end{bmatrix}$

15.2. Calculate the inverse of each the matrices given in Prob. 15.1.

15.3. Calculate the eigenvalues of each of the matrices given in Prob. 15.1.

15.4. Calculate the singular values of each of the matrices given in Prob. 15.1.

CHAPTER

16

ANALYSIS OF MULTIVARIABLE SYSTEMS

In Chap. 15 we reviewed a little matrix mathematics and notation. Now that the tools are available, we will apply them in this chapter to the analysis of multivariable processes. Our primary concern is with closedloop systems. Given a process with its matrix of openloop transfer functions, we want to be able to see the effects of using various feedback controllers. Therefore we must be able to find out if the entire closedloop multivariable system is stable. And if it is stable, we want to know how stable it is. The last question considers the "robustness" of the controller, i.e., the tolerance of the controller to changes in parameters. If the system becomes unstable for small changes in process gains, time constants, or deadtimes, the controller is not robust.

16.1 STABILITY

16.1.1 Openloop and Closedloop Characteristic Equations

The characteristic equation of any system, closedloop or openloop, is the equation that you get when you take the denominator of the transfer function describing the system and set it equal to zero. The resulting Nth-order polynomial equation in s (or in λ) has N roots, and these dictate the stability, damping, and speed of response of the system.

For openloop systems, the denominator of the transfer functions in the $\mathbf{G}_{M(s)}$ matrix gives the openloop characteristic equation. In Example 15.14 the denominator of the elements in $\mathbf{G}_{M(s)}$ was $(s + 2)(s + 4)$. Therefore the openloop characteristic equation was

$$s^2 + 6s + 8 = 0$$

The system was openloop stable since both roots were in the left half of the s plane. Putting it another way, both openloop eigenvalues were in the left half of the s plane.

For closedloop systems, the denominator of the transfer functions in the closedloop servo and load transfer function matrices gives the closed-loop characteristic equation. This denominator was shown in Chap. 15 to be Det $[\mathbf{I} + \mathbf{G}_{M(s)} \mathbf{B}_{(s)}]$, which is a scalar Nth-order polynomial in s. Therefore, the closedloop characteristic equation for a multivariable process is

$$\boxed{\text{Det } [\mathbf{I} + \mathbf{G}_{M(s)} \mathbf{B}_{(s)}] = 0} \tag{16.1}$$

This is the most important equation in multivariable control. It applies for any type of controller, diagonal (multiloop SISO) or full multivariable controller. If any of the roots of this equation are in the right half of the s plane, the system is closedloop unstable.

Example 16.1. Determine the closedloop characteristic equation for a 2×2 process with a diagonal feedback controller.

$$\mathbf{G}_{M(s)} = \begin{bmatrix} G_{M_{11}} & G_{M_{12}} \\ G_{M_{21}} & G_{M_{22}} \end{bmatrix} \qquad \mathbf{B}_{(s)} = \begin{bmatrix} B_1 & 0 \\ 0 & B_2 \end{bmatrix} \tag{16.2}$$

$$\text{Det } [\mathbf{I} + \mathbf{G}_{M(s)} \mathbf{B}_{(s)}] = \text{Det } \left[\begin{bmatrix} 1 & 0 \\ 0 & 1 \end{bmatrix} + \begin{bmatrix} G_{M_{11}} & G_{M_{12}} \\ G_{M_{21}} & G_{M_{22}} \end{bmatrix} \begin{bmatrix} B_1 & 0 \\ 0 & B_2 \end{bmatrix} \right] = 0$$

$$\text{Det } \left[\begin{bmatrix} 1 & 0 \\ 0 & 1 \end{bmatrix} + \begin{bmatrix} B_1 G_{M_{11}} & B_2 G_{M_{12}} \\ B_1 G_{M_{21}} & B_2 G_{M_{22}} \end{bmatrix} \right] = 0 \tag{16.3}$$

$$\text{Det } \begin{bmatrix} (1 + B_1 G_{M_{11}}) & (B_2 G_{M_{12}}) \\ (B_1 G_{M_{21}}) & (1 + B_2 G_{M_{22}}) \end{bmatrix} = 0$$

$$(1 + B_1 G_{M_{11}})(1 + B_2 G_{M_{22}}) - B_2 G_{M_{12}} B_1 G_{M_{21}} = 0$$

$$1 + B_1 G_{M_{11}} + B_2 G_{M_{22}} + B_1 B_2 (G_{M_{11}} G_{M_{22}} - G_{M_{12}} G_{M_{21}}) = 0$$

Notice that the closedloop characteristic equation depends on the tuning of *both* feedback controllers.

The following sections discuss several methods for determining the location of the roots of Eq. (16.1).

16.1.2 Multivariable Nyquist Plot

The Nyquist stability criterion that we developed in Chap. 13 can be directly applied to multivariable processes. As you should recall, the procedure is based on a complex variable theorem which says that the difference between the number of zeros and poles that a function has inside a closed contour can be found by plotting the function and looking at the number of times it encircles the origin.

We can use this theorem to find out if the closedloop characteristic equation has any roots or zeros in the right half of the s plane. The s variable follows a closed contour that completely surrounds the entire right half of the s plane. Since the closedloop characteristic equation is given in Eq. (16.1), the function that we are interested in is

$$F_{(s)} = \text{Det} \left[\underline{I} + \underline{G}_{M(s)} \underline{B}_{(s)} \right] \tag{16.4}$$

The contour of $F_{(s)}$ is plotted in the F plane. The number of encirclements of the origin made by this plot is equal to the difference between the number of zeros and the number of poles of $F_{(s)}$ in the right half of the s plane.

If the process is openloop stable, none of the transfer functions in $\underline{G}_{M(s)}$ will have any poles in the right half of the s plane. And the feedback controllers in $\underline{B}_{(s)}$ are always chosen to be openloop stable (P, PI, or PID action), so $\underline{B}_{(s)}$ has no poles in the right half of the s plane. Clearly, the poles of $F_{(s)}$ are the poles of $\underline{G}_{M(s)} \underline{B}_{(s)}$. Thus if the process is openloop stable, the $F_{(s)}$ function has *no* poles in the right half of the s plane. So the number of encirclements of the origin made by the $F_{(s)}$ function is equal to the number of zeros in the right half of the s plane.

Thus the Nyquist stability criterion for a multivariable openloop-stable process is:

> If a plot of Det $\left[\underline{I} + \underline{G}_{M(s)} \underline{B}_{(s)} \right]$ encircles the origin, the system is closedloop unstable!

The usual way to use the Nyquist stability criterion in SISO systems is to *not* plot $1 + G_{M(i\omega)} B_{(i\omega)}$ and look at encirclements of the origin. Instead we simply plot just $G_{M(i\omega)} B_{(i\omega)}$ and look at encirclements of the $(-1, 0)$ point. Therefore to use a similar plot in multivariable systems we define a function $W_{(i\omega)}$ as follows:

$$W_{(i\omega)} = -1 + \text{Det} \left[\underline{I} + \underline{G}_{M(i\omega)} \underline{B}_{(i\omega)} \right] \tag{16.5}$$

Then the number of encirclements of the $(-1, 0)$ point made by $W_{(i\omega)}$ as ω varies from 0 to ∞ gives the number of zeros of the closedloop characteristic equation in the right half of the s plane.

Example 16.2. The Wood and Berry (*Chem. Eng. Science*, 1973, Vol. 28, p. 1707) distillation column is a 2×2 system with the following openloop process transfer functions:

$$\mathcal{G}_{M(s)} = \begin{bmatrix} G_{M11} & G_{M12} \\ G_{M21} & G_{M22} \end{bmatrix} = \begin{bmatrix} \dfrac{12.8e^{-s}}{16.7s+1} & \dfrac{-18.9e^{-3s}}{21s+1} \\ \dfrac{6.6e^{-7s}}{10.9s+1} & \dfrac{-19.4e^{-3s}}{14.4s+1} \end{bmatrix} \tag{16.6}$$

The process is openloop stable with no poles in the right half of the s plane. The authors used a diagonal controller structure with PI controllers and found, by empirical tuning, the following settings: $K_{c1} = 0.20$, $K_{c2} = -0.04$, $\tau_{I1} = 4.44$, and $\tau_{I2} = 2.67$. The feedback controller matrix was

$$\mathcal{B}_{(s)} = \begin{bmatrix} \dfrac{K_{c1}(\tau_{I1}s+1)}{\tau_{I1}s} & 0 \\ 0 & \dfrac{K_{c2}(\tau_{I2}s+1)}{\tau_{I2}s} \end{bmatrix} \tag{16.7}$$

Table 16.1 gives a FORTRAN program which generates the multivariable Nyquist plot for the Wood and Berry column. The subroutines given in Chap. 15 are used for manipulating the matrices with complex elements. The subroutines PROCTF and FEEDBC calculate the $\mathcal{G}_{M(i\omega)}$ and $\mathcal{B}_{(i\omega)}$ complex numbers at each value of frequency. A general process transfer function is used for each of the elements in the $\mathcal{G}_M$ matrix that has the form

$$G_{M(s)} = \frac{K(\tau_1 s+1)e^{-Ds}}{(\tau_2 s+1)(\tau_3 s+1)(\tau_4 s+1)} \tag{16.8}$$

The diagonal controller structure is assumed in $\mathcal{B}$ with P or PI controllers.

Figure 16.1 gives the W plane plots when the empirical settings are used and when the Ziegler-Nichols (ZN) settings for each individual controller are used ($K_{c1} = 0.960$, $K_{c2} = -0.19$, $\tau_{I1} = 3.25$ and $\tau_{I2} = 9.2$). The curve with the empirical settings does not encircle the $(-1, 0)$ point, and therefore the system is closedloop stable. Figure 16.2 gives the response of the system to a unit step change in x_1^{set}, verifying that the multivariable system is indeed closedloop stable.

However, the W plane curve using the ZN settings does encircle the $(-1, 0)$ point, showing that the system is closedloop unstable with these settings. This example illustrates that tuning up each loop independently with the other loops on manual does not necessarily give a stable system when all loops are on automatic.

Also shown in Fig. 16.1 is the W plot when only proportional controllers are used. Note that the curves with P controllers start on the positive real axis. However, with PI controllers the curves start on the negative real axis. This is due to the two integrators, one in each controller, which give 180 degrees of phase angle lag at low frequencies. As shown in Eq. (16.3), the product of the B_1 and B_2 controllers appears in the closedloop characteristic equation.

TABLE 16.1

Multivariable Nyquist plot for Wood and Berry column

```
      REAL KP(4,4),KC(4)
      DIMENSION RESET(4),TAU(4,4,4),D(4,4)
      COMPLEX GM(4,4),B(4,4),Q(4,4),YID(4,4),CHAR(4,4),A(4,4),C(4,4)
      COMPLEX XDETER,WFUNC
      EXTERNAL D2,D3,D4
C PROCESS TRANSFER FUNCTION DATA
      DATA KP(1,1),KP(1,2),KP(2,1),KP(2,2)/12.8,-18.9,6.6,-19.4/
      DATA TAU(1,1,1),TAU(1,1,2),TAU(1,2,1),TAU(1,2,2)/4*0./
      DATA TAU(2,1,1),TAU(2,1,2),TAU(2,2,1),TAU(2,2,2)
     +  /16.7,21.,10.9,14.4/
      DO 1 I=1,2
      DO 1 J=1,2
      TAU(3,I,J)=0.
    1 TAU(4,I,J)=0.
      DATA D(1,1),D(1,2),D(2,1),D(2,2)/1.,3.,7.,3./
C EMPIRICAL CONTROLLER SETTINGS
      DATA KC(1),KC(2)/0.2,-0.04/
      DATA RESET(1),RESET(2)/4.44,2.67/
C SET INITIAL FREQUENCY AND NUMBER OF POINTS PER DECADE
      W=0.01
      DW=(10.)**(1./10.)
      N=2
C MAIN LOOP FOR EACH FREQUENCY
      WRITE(6,10)
   10 FORMAT(' W      REAL     IMAG      DB      DEG')
  100 CALL PROCTF(GM,W,N,KP,TAU,D)
      CALL FEEDBK(B,W,N,KC,RESET)
      CALL MMULT(GM,B,Q,N)
      CALL IDENT(YID,N)
      CALL MADD(YID,Q,CHAR,N)
      CALL DETER(N,CHAR,A,C,XDETER)
      DB=20.*ALOG10(CABS(XDETER))
      WFUNC=XDETER-1.
      DEG=ATAN(AIMAG(WFUNC)/REAL(WFUNC))*180./3.1416
      IF(REAL(WFUNC).LT.0.)DEG=DEG-180.
      WRITE(6,2)W,WFUNC,DB,DEG
    2 FORMAT(1X,5F10.5)
      W=W*DW
C STOP IF FREQUENCY IS GREATER THAN 10
      IF(W.LT.10.)GO TO 100
      STOP
      END
C
      SUBROUTINE FEEDBK(B,W,N,KC,RESET)
      DIMENSION KC(4),RESET(4)
      COMPLEX B(4,4),ZINT,ZL
      REAL KC
      ZINT(X)=CMPLX(0.,W*X)
      ZL(X)=CMPLX(1.,X*W)
      DO 10 I=1,N
      DO 10 J=1,N
      B(I,J)=CMPLX(0.,0.)
      IF(I.EQ.J) THEN
      B(I,J)=CMPLX(KC(I),0.)
      IF(RESET(I).GT.0.) B(I,J)=B(I,J)*ZL(RESET(I))/ZINT(RESET(I))
      ENDIF
   10 CONTINUE
```

TABLE 16.1 (*continued*)

```
     RETURN
     END
C
     SUBROUTINE PROCTF(GM,W,N,KP,TAU,D)
     DIMENSION KP(4,4),TAU(4,4,4),D(4,4)
     COMPLEX GM(4,4),ZL,ZD
     REAL KP
     ZL(X)=CMPLX(1.,X*W)
     ZD(X)=CMPLX(COS(W*X),-SIN(W*X))
     DO 10 I=1,N
     DO 10 J=1,N
  10 GM(I,J)=KP(I,J)*ZD(D(I,J))*ZL(TAU(1,I,J))/ZL(TAU(2,I,J))
    + /ZL(TAU(3,I,J))/ZL(TAU(4,I,J))
     RETURN
     END
```

Wood and Berry

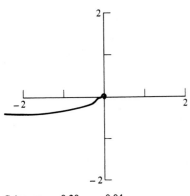

Gains =   0.20   −0.04
Resets =   4.44   2.67

Wood and Berry

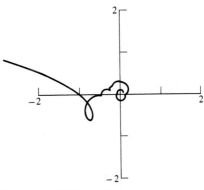

Gains =   0.96   −0.19
Resets =   3.25   9.20

Wood and Berry

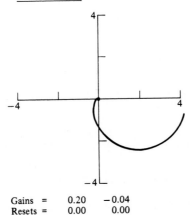

Gains =   0.20   −0.04
Resets =   0.00   0.00

FIGURE 16.1

Wood and Berry

| Gains = | 0.20 | −0.04 |
| Resets = | 4.44 | 2.67 |

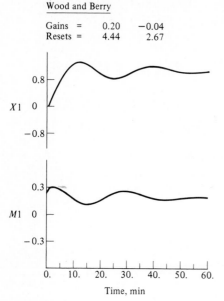

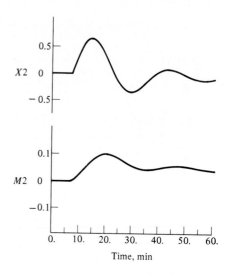

Time, min

FIGURE 16.2

16.1.3   Characteristic Loci Plots

The multivariable Nyquist plots discussed above give one curve. These curves can be quite complex, particularly with high-order systems and with multiple deadtimes. Loops can appear in the curves, making it difficult sometimes to see if the $(−1, 0)$ point is being encircled.

A different kind of plot, called a "characteristic loci plot," is sometimes easier to understand. The method is as follows:

(a) Specify the controller $B_{(s)}$ (this means both its structure and tuning)

(b) Pick a specific numerical value of frequency ω.

(c) Calculate the complex matrix $Q_{(i\omega)}$ which is the product of the $G_{M(i\omega)}$ and $B_{(i\omega)}$ matrices.

$$Q_{(i\omega)} = G_{M(i\omega)} B_{(i\omega)} \qquad (16.9)$$

(d) Calculate the eigenvalues of $Q_{(i\omega)}$. If the system is $N \times N$ (N controlled variables and N manipulated variables) there will be N complex eigenvalues. The IMSL subroutine "EIGCC" is used in the program given in Table 16.2.

(e) Plot these N eigenvalues as frequency is varied from 0 to ∞.

(f) If *any* of these curves encircle the $(−1, 0)$ point, the system is closedloop unstable (for an openloop stable process).

The eigenvalues of the $Q_{(i\omega)}$ matrix are called $q_1, q_2, q_3, \ldots, q_N$.

$$q_k = \lambda_{k[Q]} = \lambda_{k[G_M B]} \qquad (16.10)$$

TABLE 16.2
Characteristic loci plots for Wood and Berry column

```
      REAL KP(4,4),KC(4)
      DIMENSION RESET(4),TAU(4,4,4),D(4,4)
      COMPLEX GM(4,4),B(4,4),Q(4,4),A(4,4),C(4,4)
      COMPLEX CEIGEN(4),WK(12)
C PROCESS TRANSFER FUNCTION DATA
      DATA KP(1,1),KP(1,2),KP(2,1),KP(2,2)/12.8,-18.9,6.6,-19.4/
      DATA TAU(1,1,1),TAU(1,1,2),TAU(1,2,1),TAU(1,2,2)/4*0./
      DATA TAU(2,1,1),TAU(2,1,2),TAU(2,2,1),TAU(2,2,2)
     +  /16.7,21.,10.9,14.4/
      DO 1 I=1,2
      DO 1 J=1,2
      TAU(3,I,J)=0.
    1 TAU(4,I,J)=0.
      DATA D(1,1),D(1,2),D(2,1),D(2,2)/1.,3.,7.,3./
C EMPIRICAL CONTROLLER SETTINGS
      DATA KC(1),KC(2)/0.2,-0.04/
      DATA RESET(1),RESET(2)/4.44,2.67/
C SET INITIAL FREQUENCY AND NUMBER OF POINTS PER DECADE
      W=0.01
      DW=(10.)**(1./10.)
C SET PARAMETERS FOR IMSL SUBROUTINE "EIGCC"
      N=2
      IZ=4
      IA=4
      IJOB=0
      WRITE(6,10)
   10 FORMAT('     W     RE(Q1)   IM(Q1)   RE(Q2)   IM(Q2)')
C MAIN LOOP FOR EACH FREQUENCY
  100 CALL PROCTF(GM,W,N,KP,TAU,D)
      CALL FEEDBK(B,W,N,KC,RESET)
      CALL MMULT(GM,B,Q,N)
      CALL EIGCC(Q,N,IA,IJOB,CEIGEN,A,IZ,WK,IER)
      WRITE(6,2)W,(CEIGEN(J),J=1,N)
    2 FORMAT(1X,5F10.5)
      W=W*DW
C STOP IF FREQUENCY IS GREATER THAN 10
      IF(W.LT.10.)GO TO 100
      STOP
      END
C
      SUBROUTINE FEEDBK(B,W,N,KC,RESET)
      DIMENSION KC(4),RESET(4)
      COMPLEX B(4,4),ZINT,ZL
      REAL KC
      ZINT(X)=CMPLX(0.,W*X)
      ZL(X)=CMPLX(1.,X*W)
      DO 10 I=1,N
      DO 10 J=1,N
      B(I,J)=CMPLX(0.,0.)
      IF(I.EQ.J) THEN
      B(I,J)=CMPLX(KC(I),0.)
      IF(RESET(I).GT.0.) B(I,J)=B(I,J)*ZL(RESET(I))/ZINT(RESET(I))
      ENDIF
   10 CONTINUE
      RETURN
      END
```

TABLE 16.2 (*continued*)

```
C
      SUBROUTINE PROCTF(GM,W,N,KP,TAU,D)
      DIMENSION KP(4,4),TAU(4,4,4),D(4,4)
      COMPLEX GM(4,4),ZL,ZD
      REAL KP
      ZL(X)=CMPLX(1.,X*W)
      ZD(X)=CMPLX(COS(W*X),-SIN(W*X))
      DO 10 I=1,N
      DO 10 J=1,N
   10 GM(I,J)=KP(I,J)*ZD(D(I,J))*ZL(TAU(1,I,J))/ZL(TAU(2,I,J))
    +  /ZL(TAU(3,I,J))/ZL(TAU(4,I,J))
      RETURN
      END
```

Results

W	RE(Q1)	IM(Q1)	RE(Q2)	IM(Q2)
.01000	-10.93861	-73.93397	-.69893	-10.97331
.01259	-10.77362	-57.94015	-.70154	-8.69403
.01585	-10.52296	-45.07131	-.70426	-6.87520
.01995	-10.15061	-34.67506	-.70560	-5.41860
.02512	-9.61544	-26.25261	-.70209	-4.24563
.03162	-8.88105	-19.43858	-.68745	-3.29489
.03981	-7.93375	-13.98019	-.65334	-2.52193
.05012	-6.80296	-9.70591	-.59328	-1.89904
.06310	-5.56933	-6.48106	-.50896	-1.41125
.07943	-4.34690	-4.16391	-.41292	-1.04679
.10000	-3.24441	-2.58527	-.32303	-.78823
.12589	-2.33036	-1.56020	-.25321	-.61188
.15849	-1.62209	-.91593	-.20913	-.49410
.19953	-1.09771	-.51329	-.19052	-.41728
.25119	-.71786	-.25031	-.19364	-.37372
.31623	-.19787	-.35982	-.45773	-.06077
.39811	-.19106	-.33411	-.29072	.05551
.50119	-.19649	-.28201	-.16390	.10739
.63096	-.20761	-.20640	-.06256	.10977
.79433	-.19927	-.11268	.00200	.07250
1.00000	-.14416	-.03580	.00868	.03019
1.25893	-.10863	-.03576	.02118	.04213
1.58489	-.10893	.01474	.04596	-.01127
1.99526	-.05624	.03026	-.00732	-.01991
2.51189	-.02969	.06183	-.02432	-.01433

Note that these eigenvalues are neither the openloop eigenvalues nor the closed-loop eigenvalues of the system! They are eigenvalues of a completely different matrix, not the $\underline{A}$ or the $\underline{A}_{CL}$ matrices.

Characteristic loci plots for the Wood and Berry column are shown in Fig. 16.3. They show that the empirical controllers settings give a stable closedloop system, but the ZN settings do not since the q_1 eigenvalue goes through the $(-1, 0)$ point.

A brief justification for the characteristic loci method (thanks to C. C. Yu) is sketched below. For a more rigorous treatment see McFarland and Belletrutti (*Automatica* 1973, Vol. 8, p. 455). We assume an openloop stable system so the closedloop characteristic equation has no poles in the right half of the s plane.

Wood and Berry

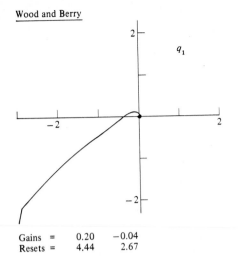

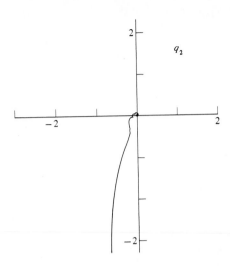

Gains = 0.20 −0.04
Resets = 4.44 2.67

Wood and Berry

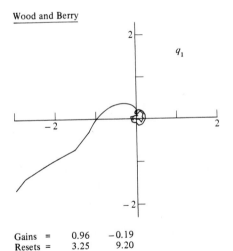

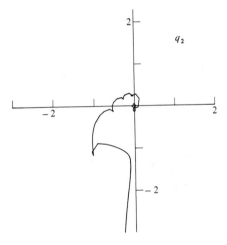

Gains = 0.96 −0.19
Resets = 3.25 9.20

FIGURE 16.3

The closedloop characteristic equation for a multivariable closedloop process is

$$\text{Det}\ [\underline{I} + \underline{G}_{M_{(s)}}\ \underline{B}_{(s)}] = \text{Det}\ [\underline{I} + \underline{Q}] = 0$$

Let us canonically transform the $\underline{Q}$ matrix into a diagonal matrix Δ_Q which has as its diagonal elements the eigenvalues of $\underline{Q}$.

$$\underline{Q} = \underline{W}\,\Delta_Q\,\underline{W}^{-1} \tag{16.11}$$

where

$$\Delta_Q = \begin{bmatrix} q_1 & 0 & \\ 0 & q_2 & 0 \\ \cdots\cdots\cdots\cdots \\ 0 & & 0 & q_N \end{bmatrix} \tag{16.2}$$

where $q_i = i$th eigenvalue Δ_Q

$$\mathrm{Det}\,[\underline{I} + \underline{Q}] = 0$$

$$\mathrm{Det}\,[\underline{I} + \underline{W}\Delta_Q\,\underline{W}^{-1}] = 0$$

$$\mathrm{Det}\,[\underline{W}\underline{I}\underline{W}^{-1} + \underline{W}\Delta_Q\,\underline{W}^{-1}] = 0$$

$$(\mathrm{Det}\,\underline{W})\,(\mathrm{Det}\,[\underline{I} + \Delta_Q])\,(\mathrm{Det}\,\underline{W}^{-1}) = 0 \tag{16.13}$$

Since the $\underline{W}$ matrix is not singular, its determinant is not zero. And the determinant of its inverse is also not zero. Thus the closedloop characteristic equation becomes

$$\mathrm{Det}\,[\underline{I} + \Delta_Q] = 0 = \mathrm{Det} \begin{bmatrix} (1 + q_1) & 0 & \\ 0 & (1 + q_2) & 0 \\ \cdots\cdots\cdots\cdots\cdots \\ 0 & \cdots & 0 & 1 + q_N \end{bmatrix} \tag{16.14}$$

$$(1 + q_1)(1 + q_2) \cdots (1 + q_N) = 0 \tag{16.15}$$

Now, the argument (phase angle) of the product of a series of complex numbers is the sum of the arguments of the individual numbers.

$$\mathrm{arg}\,\{\mathrm{Det}\,[\underline{I} + \underline{Q}]\} = \mathrm{arg}\,(1 + q_1) + \mathrm{arg}\,(1 + q_2) + \cdots + \mathrm{arg}\,(1 + q_N) \tag{16.16}$$

If the argument of any of the $(1 + q_i)$ functions on the right-hand side of Eq. (16.16) increases by 2π as ω goes from 0 to ∞ (i.e., encircles the origin), the closedloop characteristic equation has a zero in the right half of the s plane. Therefore the system is closedloop unstable. Instead of plotting $(1 + q_i)$ and looking at encirclements of the origin, we plot just the q_i's and look at encirclements of the $(-1, 0)$ point.

16.1.4 Niederlinski Index

A fairly useful stability analysis method is the Niederlinski index. It can be used to eliminate unworkable pairings of variables at an early stage in the design. The settings of the controllers do not have to be known, but it applies only when integral action is used in all loops. It uses only the steadystate gains of the process transfer function matrix.

The method is a "necessary but not sufficient condition" for stability of a closedloop system with integral action. If the index is negative, the system *will* be unstable for any controller settings (this is called "integral instability"). If the index is positive, the system may or may not be stable. Further analysis is necessary.

$$\text{Niederlinski index} = \text{NI} = \frac{\text{Det } [\underline{K}_P]}{\prod\limits_{j=1}^{N} K_{P_{jj}}} \tag{16.17}$$

where $\underline{K}_P = \underline{G}_{M(0)}$ = matrix of steadystate gains from the process openloop $\underline{G}_M$ transfer function

$K_{P_{jj}}$ = diagonal elements in steadystate gain matrix

Example 16.3. Calculate the Niederlinski index for the Wood and Berry column.

$$\underline{K}_P = \underline{G}_{M(0)} = \begin{bmatrix} 12.8 & -18.9 \\ 6.6 & -19.4 \end{bmatrix} \tag{16.18}$$

$$\text{NI} = \frac{\text{Det } [\underline{K}_P]}{\prod\limits_{j=1}^{N} K_{P_{jj}}} = \frac{(12.8)(-19.4) - (-18.9)(6.6)}{(12.8)(-19.4)} = 0.498 \tag{16.19}$$

Since the NI is positive, the closedloop system with the specified pairing *may* be stable.

Notice that the pairing assumes distillate composition x_D is controlled by reflux R and bottoms composition x_B is controlled by vapor boilup V.

$$\begin{bmatrix} x_D \\ x_B \end{bmatrix} = \begin{bmatrix} 12.8 & -18.9 \\ 6.6 & -19.4 \end{bmatrix} \begin{bmatrix} R \\ V \end{bmatrix} \tag{16.20}$$

If the pairing had been reversed, the steadystate gain matrix would be

$$\begin{bmatrix} x_D \\ x_B \end{bmatrix} = \begin{bmatrix} -18.9 & 12.8 \\ -19.4 & 6.6 \end{bmatrix} \begin{bmatrix} V \\ R \end{bmatrix} \tag{16.21}$$

and the NI for this pairing would be

$$\text{NI} = \frac{\text{Det } [\underline{K}_P]}{\prod\limits_{j=1}^{N} K_{P_{jj}}} = \frac{(-18.9)(6.6) - (12.8)(-19.4)}{(-18.9)(6.6)} = -0.991$$

Therefore, the pairing of x_D with V and x_B with R gives a closedloop system that is "integrally unstable" for any controller tuning.

16.2 RESILIENCY

Some processes are easier to control than others. Some choices of manipulated and controlled variables produce systems that are easier to control than others. This inherent property of ease of controllability is called *resiliency*.

Morari (*Chemical Eng. Science*, 1983, Vol. 38, p. 1881) developed a very useful measure of this property. The Morari resiliency index (MRI) gives an indication of the inherent controllability of a process. It depends on the controlled and manipulated variables, but it does not depend on the pairing of these variables or on the tuning of the controllers. Thus it is a useful tool for comparing alternative processes and alternative choices of manipulated variables.

The MRI is the minimum singular value of the process openloop transfer function matrix $G_{M(i\omega)}$. It can be evaluated over a range of frequencies ω or just at zero frequency. If the latter case, only the steadystate gain matrix K_P is needed.

$$\text{MRI} = \sigma^{\min}_{[G_{M(i\omega)}]} \tag{16.22}$$

The larger the value of the MRI, the more controllable or resilient the process. Without going into an elaborate mathematical proof, we can intuitively understand why a big MRI is good. The larger the minimum singular value of a matrix, the farther it is from being singular and the easier it is to find its inverse. Now a feedback controller is an approximation of the inverse of the process. Remember back in Chap. 11 where we discussed Internal Model Control and showed that perfect control could be achieved if we could make the controller the inverse of the plant. Thus the more invertible the plant, the easier it is to build a controller that does a good control job.

Example 16.4. Calculate the MRI for the Wood and Berry column at zero frequency.

We use Eq. (15.44) for the singular values and select the smallest for the MRI.

$$\text{MRI} = \sigma^{\min}_{[G_{M(i\omega)}]} = \sigma^{\min}_{[K_{P(0)}]} = \min \sqrt{\lambda_{i[K_P^T K_P]}} \tag{16.23}$$

$$K_P^T K_P = \begin{bmatrix} 12.8 & 6.6 \\ -18.9 & -19.4 \end{bmatrix}\begin{bmatrix} 12.8 & -18.9 \\ 6.6 & -19.4 \end{bmatrix} = \begin{bmatrix} 207.4 & -369.96 \\ -369.96 & 733.57 \end{bmatrix}$$

To get the eigenvalues of this matrix we solve the equation Det $[\lambda I - K_P^T K_P] = 0$

$$\text{Det}\begin{bmatrix} (\lambda - 207.4) & 369.96 \\ 369.96 & (\lambda - 733.57) \end{bmatrix} = \lambda^2 - 940.97\lambda + 15,272 = 0$$

$$\lambda = 916.19, \ 16.52$$

Therefore the minimum singular value of K_P is $\sqrt{16.52} = 4.06 = \text{MRI}$.

Example 16.5. Yu and Luyben (*Ind. Eng. Chem. Process Des. Dev.*, 1986, Vol. 25, p. 498) give the following steadystate gain matrices for three alternative choices of manipulated variables: reflux and vapor boilup $(R - V)$, distillate and vapor boilup $(D - V)$, and reflux ratio and vapor boilup $(RR - V)$.

	R − V	D − V	RRV
$G_{M_{11}}$	16.3	−2.3	1.5
$G_{M_{12}}$	−18.0	1.04	−1.12
$G_{M_{21}}$	−26.2	3.1	−2.06
$G_{M_{22}}$	28.63	1.8	4.73

Calculate the MRIs for each choice of manipulated variables.

Using Eq. (16.23), we calculate MRIs for each system: $R - V = 0.114$, $D - V = 1.86$, and $RR - V = 0.89$. Based on these results, the $D - V$ system should be easier to control (more resilient).

One important aspect of the MRI calculations should be emphasized at this point. The singular values depend on the scaling use in the steadystate gains of the transfer functions. If different engineering units are used for the gains, different singular values will be calculated. For example, as we change from time units of minutes to hours or change from temperature units in Fahrenheit to Celsius, the values of the singular values will change. Thus the comparison of alternative control structures and processes can be obscured by this effect.

The practical solution to the problem is to always use dimensionless gains in the transfer functions. The gains with engineering units should be divided by the appropriate transmitter spans and multiplied by the appropriate valve gains. Typical transmitter spans are 100°F, 50 gpm, 250 psig, 2 mol % benzene, and the like. Typical valve gains are perhaps twice the normal steadystate flow rate of the manipulated variable. The slope of the installed characteristics of the control valve at the steadystate operating conditions should be used. Thus the gains that should be used are those that the control system will see. It gets inputs in mA or psig from transmitters and puts out mA or psig to valves or to flow controller setpoints.

16.3 INTERACTION

Interaction among control loops in a multivariable system has been the subject of much research over the last 20 years. Various types of decouplers were explored to separate the loops. Rosenbrock presented the inverse Nyquist array (INA) to quantify the amount of interaction. Bristol, Shinskey, and McAvoy developed the relative gain array (RGA) as an index of loop interaction.

All of this work was based on the premise that interaction was undesirable. This is true for setpoint disturbances. One would like to be able to change a setpoint in one loop without affecting the other loops. And if the loops do not interact, each individual loop can be tuned by itself and the whole system should be stable if each individual loop is stable.

Unfortunately much of this interaction analysis work has clouded the issue of how to design an effective control system for a multivariable process. In most process control applications the problem is not setpoint responses but load responses. We want a system that holds the process at the desired values in the face of load disturbances. Interaction is therefore not necessarily bad, and in fact in some systems it helps in rejecting the effects of load disturbances. Niederlinski (*AIChE J* 1971, Vol. 17, p. 1261) showed in an early paper that the use of decouplers made the load rejection worse.

Therefore the discussions of the RGA, INA, and decoupling techniques will be quite brief. I include them, not because they are all that useful, but because they are part of the history of multivariable control. You should be aware of what they are and of their limitations.

16.3.1 Relative Gain Array (Bristol Array)

Undoubtedly the most discussed method for studying interaction is the RGA. It was proposed by Bristol (*IEEE Trans. Autom. Control AC-II*, 1966, p. 133) and has been extensively applied (and in my opinion, often misapplied) by many workers. Detailed discussions are presented by Shinskey (*Process Control Systems*, McGraw-Hill, New York, 1967) and McAvoy (*Interaction Analysis*, Instr. Soc. America, Research Triangle Park, N.C., 1983).

The RGA has the advantage of being easy to calculate and only requires steadystate gain information. Let's define it, give some examples, show how it is used, and point out its limitations.

A. DEFINITION. The RGA is a matrix of numbers. The ijth element in the array is called β_{ij}. It is the ratio of the steadystate gain between the ith controlled variable and the jth manipulated variable when all other manipulated variables are constant, divided by the steadystate gain between the same two variables when all other controlled variables are constant.

$$\beta_{ij} = \frac{[x_i/m_j]_{\bar{m}_k}}{[x_i/m_j]_{\bar{x}_k}} \tag{16.24}$$

For example, suppose we have a 2×2 system with the steadystate gains $K_{p_{ij}}$.

$$x_1 = K_{p_{11}} m_1 + K_{p_{12}} m_2$$
$$x_2 = K_{p_{21}} m_1 + K_{p_{22}} m_2 \tag{16.25}$$

For this system, the gain between x_1 and m_1 when m_2 is constant is

$$\left[\frac{x_1}{m_1}\right]_{\bar{m}_2} = K_{p_{11}}$$

The gain between x_1 and m_1 when x_2 is constant ($x_2 = 0$) is found from solving the equations

$$x_1 = K_{p_{11}} m_1 + K_{p_{12}} m_2$$
$$0 = K_{p_{21}} m_1 + K_{p_{22}} m_2 \tag{16.26}$$

$$x_1 = K_{p_{11}} m_1 + K_{p_{12}} \left[\frac{-K_{p_{21}} m_1}{K_{p_{22}}}\right]$$

$$x_1 = \left[\frac{K_{p_{11}} K_{p_{22}} - K_{p_{12}} K_{p_{21}}}{K_{p_{22}}}\right] m_1 \tag{16.27}$$

$$\left[\frac{x_1}{m_1}\right]_{\bar{x}_2} = \left[\frac{K_{p_{11}} K_{p_{22}} - K_{p_{12}} K_{p_{21}}}{K_{p_{22}}}\right] \tag{16.28}$$

Therefore the β_{11} term in the RGA is

$$\beta_{11} = \frac{K_{p11} K_{p22}}{K_{p11} K_{p22} - K_{p12} K_{p21}} \qquad (16.29)$$

$$\beta_{11} = \frac{1}{1 - \dfrac{K_{p12} K_{p21}}{K_{p11} K_{p22}}} \qquad (16.30)$$

Example 16.6. Calculate the β_{11} element of the RGA for the Wood and Berry column.

$$\underline{K}_P = \underline{G}_{M(0)} = \begin{bmatrix} 12.8 & -18.9 \\ 6.6 & -19.4 \end{bmatrix}$$

$$\beta_{11} = \frac{1}{1 - K_{p12} K_{p21}/(K_{p11} K_{p22})} = \frac{1}{1 - (-18.9)(6.6)/[(12.8)(-19.4)]} = 2.01$$

Equation (16.30) applies to only a 2×2 system. The elements of the RGA can be calculated for any size system by using the following equation.

$$\beta_{ij} = (ij\text{th element of } \underline{K}_P)(ij\text{th element of } [\underline{K}_P^{-1}]^T) \qquad (16.31)$$

Note that Eq. (16.31) does *not* say that we take the ijth element of the product of the $\underline{K}_P$ and $[\underline{K}_P^{-1}]^T$ matrices.

Example 16.7. Use Eq. (16.31) to calculate all of the elements of the Wood and Berry column.

$$\underline{K}_P = \begin{bmatrix} 12.8 & -18.9 \\ 6.6 & -19.4 \end{bmatrix}$$

$$\underline{K}_P^{-1} = \frac{\begin{bmatrix} -19.4 & 18.9 \\ -6.6 & 12.8 \end{bmatrix}}{-123.58} \qquad [\underline{K}_P^{-1}]^T = \frac{\begin{bmatrix} -19.4 & -6.6 \\ 18.9 & 12.8 \end{bmatrix}}{-123.58}$$

$$\beta_{11} = (12.8)\left(\frac{-19.4}{-123.58}\right) = 2.01$$

This is the same result we obtained using Eq. (16.30).

$$\beta_{12} = (-18.9)\left(\frac{-6.6}{-123.58}\right) = -1.01$$

$$\beta_{21} = (6.6)\left(\frac{18.9}{-123.58}\right) = -1.01$$

$$\beta_{22} = (-19.4)\left(\frac{12.8}{-123.58}\right) = 2.01$$

$$\underline{\underline{RGA}} = \begin{bmatrix} \beta_{11} & \beta_{12} \\ \beta_{21} & \beta_{22} \end{bmatrix} = \begin{bmatrix} 2.01 & -1.01 \\ -1.01 & 2.01 \end{bmatrix} \qquad (16.32)$$

Note that the sum of the elements in each row is 1. The sum of the elements in each column is also 1. This property holds for any RGA, so in the 2 × 2 case we only have to calculate one element.

Example 16.8. Calculate the RGA for the 3 × 3 system studied by Ogunnaike and Ray (*AIChE J*, 1979, Vol. 25, p. 1043).

$$G_{M(s)} = \begin{vmatrix} \dfrac{0.66e^{-2.6s}}{6.7s + 1} & \dfrac{-0.61e^{-3.5s}}{8.64s + 1} & \dfrac{-0.0049e^{-s}}{9.06s + 1} \\[2mm] \dfrac{1.11e^{-6.5s}}{3.25s + 1} & \dfrac{-2.36e^{-3s}}{5s + 1} & \dfrac{-0.012e^{-1.2s}}{7.09s + 1} \\[2mm] \dfrac{-34.68e^{-9.2s}}{8.15s + 1} & \dfrac{46.2e^{-9.4s}}{10.9s + 1} & \dfrac{0.87(11.61s + 1)e^{-s}}{(3.89s + 1)(18.8s + 1)} \end{vmatrix} \tag{16.33}$$

$$K_P = \begin{bmatrix} 0.66 & -0.61 & -0.0049 \\ 1.11 & -2.36 & -0.012 \\ -34.68 & 46.2 & 0.87 \end{bmatrix}$$

TABLE 16.3
RGA for the Ogunnaike and Ray column

```
      REAL KP(4,4)
      COMPLEX G(4,4),GIN(4,4),GINT(4,4),WA(24)
      DIMENSION WK(4),BETA(4,4)
C PROCESS STEADYSTATE GAIN DATA
      DATA KP(1,1),KP(1,2),KP(1,3)/0.66,-0.61,-0.0049/
      DATA KP(2,1),KP(2,2),KP(2,3)/1.11,-2.36,-0.012/
      DATA KP(3,1),KP(3,2),KP(3,3)/-34.68,46.2,0.87/
C SET PARAMETERS FOR IMSL SUBROUTINE "LEQ2C"
      N=3
      IB=4
      M=N
      IA=4
      IJOB=0
      WRITE(6,10)
 10 FORMAT('   RGA FOR OR 3X3 COLUMN')
      DO 30 I=1,N
      DO 30 J=1,N
 30 G(I,J)= CMPLX(KP(I,J),0.)
      CALL IDENT(GIN,N)
      CALL LEQ2C(G,N,IA,GIN,M,IB,IJOB,WA,WK,IER)
      CALL MTRAN(GIN,GINT,N)
C CALCULATE EACH BETA ELEMENT OF THE RGA
      DO 40 I=1,N
      DO 40 J=1,N
 40 BETA(I,J)=G(I,J)*GINT(I,J)
      DO 50 I=1,N
 50 WRITE(6,2)(BETA(I,J),J=1,N)
  2 FORMAT(1X,4F10.3)
      STOP
      END
```

Results of RGA calculations

```
RGA FOR OR 3X3 COLUMN
  1.962    -.665    -.297
  -.670   1.892    -.222
  -.292    -.227   1.519
```

The program given in Table 16.3 uses Equation 16.31 to calculate the elements of the 3×3 RGA matrix. The IMSL subroutine LEQ2C is used to get the inverse of the matrix. The other subroutines are from Chapter 15. The results are

$$\underline{\underline{RGA}} = \begin{bmatrix} 1.96 & -0.66 & -0.30 \\ -0.67 & 1.89 & -0.22 \\ -0.29 & -0.23 & 1.52 \end{bmatrix} \tag{16.34}$$

Note that the sums of the elements in all rows and all columns are 1.

B. USES AND LIMITATIONS. The elements in the RGA can be numbers that vary from very large negative values to very large positive values. The closer the number is to 1, the less difference closing the other loop makes on the loop being considered. Therefore there should be less interaction, so the proponents of the RGA claim that variables should be paired so that they have RGA elements near 1. Numbers around 0.5 indicate interaction. Numbers that are very large indicate interaction. Numbers that are negative indicate that the sign of the controller may have to be different when other loops are on automatic.

As pointed out earlier, the problem with pairing on the basis of avoiding interaction is that interaction is not necessarily a bad thing. Therefore, the use of the RGA to decide how to pair variables is not an effective tool for process control applications. Likewise the use of the RGA to decide what control structure (choice of manipulated and controlled variables) is best is not effective. What is important is the ability of the control system to keep the process at setpoint in the face of load disturbances. Thus, load rejection is the most important criterion on which to make the decision of what variables to pair, and what controller structure is best.

The RGA is useful for avoiding poor pairings. If the diagonal element in the RGA is negative, the system may show integral instability: the same situation that we discussed in the use of the Niederlinski index. Very large values of the RGA indicate that the system can be quite sensitive to changes in the parameter values.

16.3.2 Inverse Nyquist Array (INA)

Rosenbrock (*Computer-Aided Control System Design*, Academic Press, 1974) was one of the early workers in the area of multivariable control. He proposed the use of INA plots to indicate the amount of interaction among the loops.

In a SISO system we normally make a Nyquist plot of the total openloop transfer function $G_M B$. If the system is closedloop stable, the $(-1, 0)$ point will not be encircled positively (clockwise). Alternatively, we could plot $(1/G_M B)$. This inverse plot should encircle the $(-1, 0)$ point negatively (counterclockwise) if the system is closedloop stable. See Fig. 16.4.

An INA plot for an Nth-order multivariable system consists of N curves, one for each of the diagonal elements of the matrix that is the inverse of the

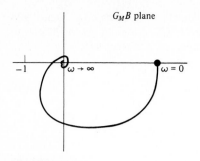
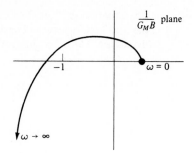

FIGURE 16.4
SISO Nyquist and inverse Nyquist plots.

product of the $\underline{G}_{M(s)}$ and $\underline{B}_{(s)}$ matrices. If this product is defined as $\underline{Q}_{(i\omega)} = \underline{G}_{M(i\omega)} \underline{B}_{(i\omega)}$, then the INA plots show the diagonal elements of $\underline{Q}_{(i\omega)}^{-1}$. It is convenient to use the nomenclature

$$\underline{\hat{Q}}_{(i\omega)} = \underline{Q}_{(i\omega)}^{-1} \tag{16.35}$$

and let the ijth element of the $\underline{\hat{Q}}$ matrix be $\hat{q}_{ij}$.

$$\underline{\hat{Q}} = \begin{bmatrix} \hat{q}_{11} & \hat{q}_{12}\cdots\cdots \\ \hat{q}_{21} & \cdots\cdots\cdots \\ \cdots\cdots\cdots\cdots \\ \hat{q}_{N1} & \cdots & \hat{q}_{NN} \end{bmatrix} \tag{16.36}$$

The three plots for a 3×3 system of $\hat{q}_{11}$, $\hat{q}_{22}$, and $\hat{q}_{33}$ are sketched in Fig. 16.5.

Then the sum of the magnitudes of the off-diagonal elements in a given row of the $\underline{\hat{Q}}$ matrix is calculated at one value of frequency and a circle is drawn with this radius. This is done for several frequency values and for each diagonal element.

$$r_1 = |\hat{q}_{12}| + |\hat{q}_{13}| + \cdots + |\hat{q}_{1N}|$$

$$r_2 = |\hat{q}_{21}| + |\hat{q}_{23}| + \cdots + |\hat{q}_{2N}| \tag{16.37}$$

$$\vdots$$

$$r_N = |\hat{q}_{N1}| + |\hat{q}_{N2}| + \cdots + |\hat{q}_{N, N-1}|$$

The resulting circles are sketched in Fig. 16.5 for the $\hat{q}_{11}$ plot. The circles are called *Gershgorin rings*. The bands that the circles sweep out are Gershgorin bands. If all the off-diagonal elements were zero, the circles would have zero radius and no interaction would be present. Therefore the bigger the circles, the more interaction is present in the system.

If all of the Gershgorin bands encircle the $(-1, 0)$ point, the system is closedloop stable as shown in Fig. 16.6a. If some of the bands do not encircle the

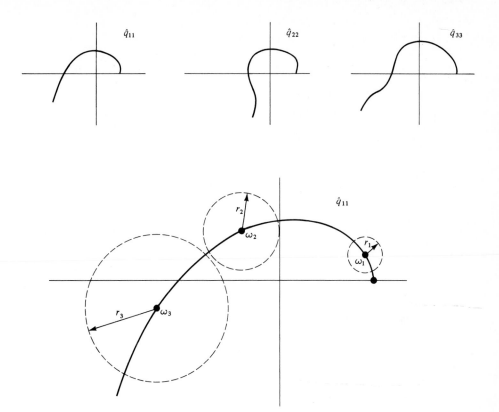

FIGURE 16.5
Multivariable INA

$(-1, 0)$ point, the system *may be* closedloop unstable (Fig. 16.6*b*). The INA plots
for the Wood and Berry column are shown in Fig. 16.6*c*.

Usually the INA is a very conservative measure of stability. Compensators
are found by trial and error to reshape the INA plots so that the circles are small
and the system is "diagonally dominant." The INA method strives for the elimi-
nation of interaction among the loops and therefore has limited usefulness in
process control where load rejection is the most important question.

16.3.3 Decoupling

Some of the earliest work in multivariable control involved the use of decouplers
to remove the interaction between the loops. Figure 16.7 gives the basic structure
of the system. The decoupling matrix $\underline{D}_{(s)}$ is chosen such that each loop does not
affect the others. Figure 16.8 shows the details of a 2×2 system. The decoupling
element D_{ij} can be selected in a number of ways. One of the most straightforward
is to set $D_{11} = D_{22} = 1$ and design the D_{12} and D_{21} elements so that they cancel

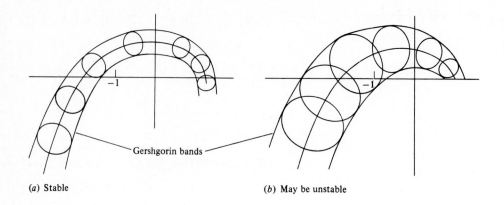

(a) Stable   (b) May be unstable

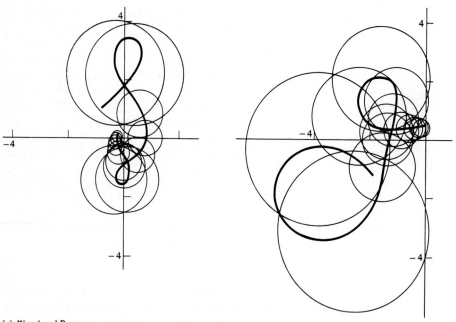

(c) Wood and Berry

FIGURE 16.6
INA plots.

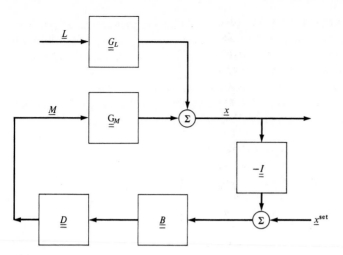

FIGURE 16.7
Decoupling.

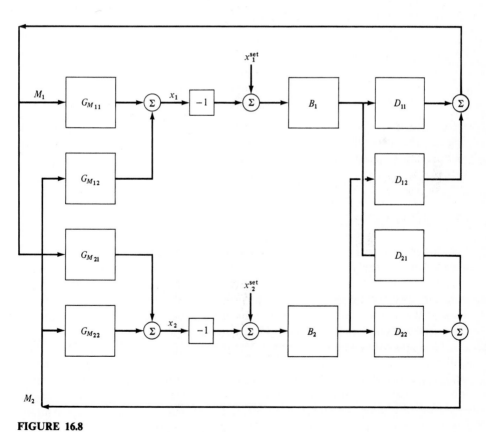

FIGURE 16.8

(in a feedforward way) the effect of each manipulated variable in the other loop. For example, suppose x_1 is not at its setpoint but x_2 is. The B_1 controller will change m_1 in order to drive x_1 back to x_1^{set}. But the change in m_1 will disturb x_2 through the $G_{M_{21}}$ transfer function.

If, however, the D_{21} decoupler element was set equal to $(-G_{M_{21}}/G_{M_{22}})$, there would be a change in m_2 that would come through the $G_{M_{22}}$ transfer function and would cancel out the effect of the change in m_1 on x_2.

$$D_{21} = \frac{-G_{M_{21}}}{G_{M_{22}}} \tag{16.38}$$

Using the same arguments for the other loop, the D_{12} decoupler could be set equal to

$$D_{21} = \frac{-G_{M_{12}}}{G_{M_{11}}} \tag{16.39}$$

This so-called "simplified decoupling" splits the two loops so that they can be independently tuned. Note, however, that the closedloop characteristic equations for the two loops are *not* $1 + G_{M_{11}} B_1 = 0$ and $1 + G_{M_{22}} B_2 = 0$. The presence of the decouplers changes the closedloop characteristic equations to

$$1 + B_1 \frac{G_{M_{11}} G_{M_{22}} - G_{M_{12}} G_{M_{21}}}{G_{M_{22}}} = 0 \tag{16.40}$$

$$1 + B_2 \frac{G_{M_{11}} G_{M_{22}} - G_{M_{12}} G_{M_{21}}}{G_{M_{11}}} = 0 \tag{16.41}$$

Other choices of decouplers are also possible. However, since decoupling may degrade the load rejection capability of the system, the use of decouplers is not recommended except in those cases where setpoint changes are the major disturbances.

16.4 ROBUSTNESS

The design of a controller depends on a model of the process. If the controller is installed on a process whose parameters are not exactly the same as the model used to design the controller, what happens to its performance? Is the system unstable? Is it too underdamped or too slow? These are very practical questions because the parameters of any industrial process always change somewhat due to nonlinearities, changes in operating conditions, etc.

If a control system is tolerant to changes in process parameters, it is called "robust." In previous chapters for SISO systems, we designed for reasonable damping coefficients or reasonable gain or phase margins or maximum closedloop log moduli so that the system would not be right at the limit of stability and could tolerate some changes in parameters. In particular, $+2$ dB for a maximum closedloop log modulus was often used as a design criterion.

In multivariable systems the question of robustness is very important. One method developed by Doyle and Stein (*IEEE Trans.*, 1981, Vol. AC-26, p. 4) is quite useful and easy to use. It has the added advantage that it is quite similar to the maximum closedloop log modulus criterion used in SISO systems.

Another procedure, recommended by Skogestad and Morari (*Chemical Engineering Science*, Vol. 24, 1987, p. 1765), involves designing the controller so that the closedloop resonant peak occurs at a frequency region where the uncertainty is not too large.

16.4.1 Trade-off Between Performance and Robustness

Before we discuss these two methods of robustness analysis, let us consider the general relationship between performance and robustness.

In SISO systems, performance is improved (closedloop time constants are decreased giving tighter control) as controller gain is increased. However, increasing the gain also typically decreases the closedloop damping coefficient, which makes the system more sensitive to parameter changes. This is an example of the trade-off that always exists in any system between performance and robustness. In multivariable systems the same trade-offs occur: the tighter the control, the less robust the system.

A good example of this is in jet fighter control systems. During attack periods when performance is very important, the pilot switches to the "attack mode" where the controller type and settings are such that time constants are small and performance is high. However, the plane is operating close to the stability region. Underdamped behavior can result in large overshoots.

When the pilot is trying to land the aircraft, this type of performance is undesirable. An overshoot of position may mean the plane misses the deck of the aircraft carrier. Also during landing it is not important to have lightning-fast response to "joystick" commands. Therefore the pilot switches the control system into the "landing mode" which has lower performance but is more robust.

16.4.2 Doyle-Stein Criterion

A. SCALAR SISO SYSTEMS. Remember in the scalar SISO case we looked at the closedloop servo transfer function $G_M B/(1 + G_M B)$. The peak in this curve, the maximum closedloop log modulus L_c^{max} (as shown in Fig. 16.9a), is a measure of the damping coefficient of the system. The higher the peak, the more underdamped the system and the less margin for changes in parameter values. Thus, in SISO systems the peak in the closedloop log modulus curve is a measure of robustness.

Now let us rearrange the closedloop servo transfer function.

$$\frac{G_M B}{1 + G_M B} = \frac{1}{1 + 1/G_M B}$$

(16.42)

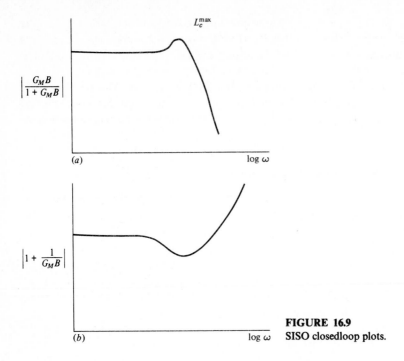

FIGURE 16.9

SISO closedloop plots.

If we plotted just the denominator of the right-hand side of Eq. 16.42, the curve would be the reciprocal of the closedloop servo transfer function and would look something like that shown in Fig. 16.9b. The lower the dip in the curve, the lower the damping coefficient and the less robust the system.

B. MATRIX MULTIVARIABLE SYSTEMS. For multivariable systems, the Doyle-Stein criterion for robustness is very similar to the reciprocal plot discussed above. The minimum singular value of the matrix given in Eq. (16.43) is plotted as a function of frequency ω. This gives a measure of the robustness of a closedloop multivariable system.

$$\text{Doyle-Stein criterion: minimum singular value of } [\underline{I} + \underline{Q}_{(i\omega)}^{-1}] \qquad (16.43)$$

The $\underline{Q}$ matrix is defined as before $\underline{Q}_{(i\omega)} = \underline{G}_{M(i\omega)}\,\underline{B}_{(i\omega)}$.

The tuning and/or structure of the feedback controller matrix $\underline{B}_{(i\omega)}$ is changed until the minimum dip in the curve is something reasonable. Doyle and Stein gave no definite recommendations, but a value of about -12 dB seems to give good results.

The procedure is summarized in the program given in Table 16.4 which calculates the minimum singular value of the $[\underline{I} + \underline{Q}_{(i\omega)}^{-1}]$ matrix for the Wood and Berry column with the empirical controller settings. Figure 16.10 gives plots of the singular values as a function of frequency. The lowest dip occurs at 0.23 radians per minute and is about -10.4 dB.

TABLE 16.4
Doyle-Stein criterion for Wood and Berry column

```
      REAL KP(4,4),KC(4)
      DIMENSION RESET(4),TAU(4,4,4),D(4,4),WK(4),WKEIG(12),SVAL(4)
      COMPLEX GM(4,4),B(4,4),Q(4,4),QIN(4,4),YID(4,4)
      COMPLEX A(4,4),ASTAR(4,4),H(4,4),CEIGEN(4),WA(24),VECT(4,4)
C PROCESS TRANSFER FUNCTION DATA
      DATA KP(1,1),KP(1,2),KP(2,1),KP(2,2)/12.8,-18.9,6.6,-19.4/
      DATA TAU(1,1,1),TAU(1,1,2),TAU(1,2,1),TAU(1,2,2)/4*0./
      DATA TAU(2,1,1),TAU(2,1,2),TAU(2,2,1),TAU(2,2,2)
     +  /16.7,21.,10.9,14.4/
      DO 1 I=1,2
      DO 1 J=1,2
      TAU(3,I,J)=0.
    1 TAU(4,I,J)=0.
      DATA D(1,1),D(1,2),D(2,1),D(2,2)/1.,3.,7.,3./
C EMPIRICAL CONTROLLER SETTINGS
      DATA KC(1),KC(2)/0.2,-0.04/
      DATA RESET(1),RESET(2)/4.44,2.67/
C SET INITIAL FREQUENCY AND NUMBER OF POINTS PER DECADE
      W=0.01
      DW=(10.)**(1./20.)
      N=2
      IA=4
      M=N
      IB=4
      IJOB=0
C MAIN LOOP FOR EACH FREQUENCY
      WRITE(6,10)
   10 FORMAT(' FREQUENCY        SINGULAR VALUES')
  100 CALL PROCTF(GM,W,N,KP,TAU,D)
      CALL FEEDBK(B,W,N,KC,RESET)
      CALL MMULT(GM,B,Q,N)
      CALL IDENT(QIN,N)
      CALL LEQ2C(Q,N,IA,QIN,M,IB,IJOB,WA,WK,IER)
      CALL IDENT(YID,N)
      CALL MADD(YID,QIN,A,N)
      CALL CONJT(A,ASTAR,N)
      CALL MMULT(ASTAR,A,H,N)
      IZ=4
      JOBN=0
      CALL EIGCC(H,N,IA,JOBN,CEIGEN,VECT,IZ,WKEIG,IER)
      DO 6 I=1,N
    6 SVAL(I)=SQRT(REAL(CEIGEN(I)))
      WRITE(6,7)W, (20.*ALOG10(SVAL(I)),I=1,N)
    7 FORMAT(1X,4F10.5)
      W=W*DW
      IF(W.LT.1.) GO TO 100
      STOP
      END
```

Results

FREQUENCY	SINGULAR VALUES	
.01000	-.00281	-.02777
.01122	-.00410	-.03454
.01259	-.00573	-.04311
.01413	-.00775	-.05402
.01585	-.01024	-.06790
.01778	-.01323	-.08560

TABLE 16.4 (*continued*)

.01995	-.01674	-.10818
.02239	-.02068	-.13696
.02512	-.02476	-.17357
.02818	-.02836	-.21995
.03162	-.03018	-.27835
.03548	-.02781	-.35125
.03981	-.01686	-.44131
.04467	.01029	-.55131
.05012	.06633	-.68430
.05623	.17099	-.84406
.06310	.35212	-1.03602
.07079	.64339	-1.26838
.07943	1.07633	-1.55325
.08913	1.66805	-1.90751
.10000	2.41108	-2.35387
.11220	3.27344	-2.92264
.12589	4.20928	-3.65467
.14125	5.17272	-4.60511
.15849	6.12728	-5.84469
.17783	7.04909	-7.43694
.19953	7.92596	-9.27595
.22387	8.75518	-10.43061
.25119	9.54200	-9.13860
.28184	10.29958	-6.10422
.31623	11.05093	-2.98828
.35481	11.83274	-.26051
.39811	12.69999	2.05383
.44668	13.72702	3.98402
.50119	5.55631	14.99744
.56234	6.81744	16.58355
.63096	7.85957	18.54112
.70795	8.81277	20.94697
.79433	9.81271	23.97799
.89125	10.95702	28.01324

16.4.3 Skogestad-Morari Method

The Doyle-Stein criterion discussed in the previous section gives us a simple way to evaluate quantitatively the robustness of a multivariable system. However, it assumes random variability in all the parameters of the system.

Skogestad and Morari recommend the use of "uncertainty" models for the design of robust controllers. The idea is easy to visualize for an SISO system. Suppose we have a process with the following openloop transfer function:

$$G_{M(s)} = \frac{K_p e^{-Ds}}{\tau_p s + 1} \tag{16.44}$$

Let us assume that there is some uncertainty in the value of the deadtime D. It can vary $\pm \delta$ fraction of D.

$$G_{M(s)} = \frac{K_p e^{-D(1+\delta)s}}{\tau_p s + 1} = \frac{K_p e^{-Ds}}{\tau_p s + 1} e^{-D\delta s}$$

$$= \frac{K_p e^{-Ds}}{\tau_p s + 1} (e^{-D\delta s} + 1 - 1) = \frac{K_p e^{-Ds}}{\tau_p s + 1} (1 + \Delta) \tag{16.45}$$

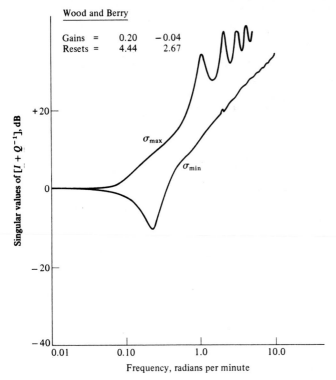

Wood and Berry

Gains =	0.20	−0.04
Resets =	4.44	2.67

FIGURE 16.10

where

$$\Delta \equiv e^{-D\delta s} - 1 \qquad (16.46)$$

The Δ function is a measure of the model uncertainty. If we plot Δ as a function of frequency (see Fig. 16.11) we can see that its magnitude is quite small at low frequencies, but increases as frequency is increased. The larger the uncertainty (the bigger δ), the lower the frequency at which the magnitude of the uncertainty becomes large.

The basic idea proposed by Skogestad and Morari is to design the controller so that the closedloop resonant frequency of the system occurs at a frequency that is lower than the region where the uncertainty becomes significant. For example, suppose $K_p = \tau_p = 1$ in Eq. (16.44). If a proportional controller is used and a tuning criterion of $+2$ dB maximum closedloop log modulus is assumed, the value of the controller gains and closedloop resonant frequencies for different deadtimes are listed below.

D	K_c	ω_r
0.1	8.03	9.55
0.2	4.12	6.31
0.5	1.95	3.02
1.0	1.21	1.82

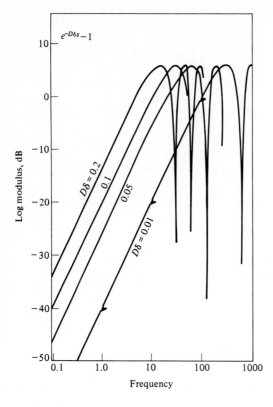

FIGURE 16.11
Uncertainty plots.

If the uncertainty in the deadtime is 0.01 min, Fig. 16.11 shows that its magnitude becomes significant (magnitude greater than -20 dB, that is, 10 percent) at frequencies above 10 radians per minute. Therefore this much uncertainty would not require controller detuning for deadtimes greater than 0.2, but the 0.1 deadtime may have to be detuned somewhat to maintain stability despite the uncertainty.

However, if the uncertainty is 0.1 min, its magnitude becomes significant at frequencies greater than 1 radian per second. Therefore lower gains would have to be used in all of the cases.

For the multivariable case, an uncertainty matrix $\underline{\Delta}$ must be defined which contains the uncertainty in the various elements of the plant transfer function (and also could include uncertainties in setting control valves or in measurements).

$$\underline{G}_M = \tilde{\underline{G}}_M [\underline{I} + \underline{\Delta}] \qquad (16.47)$$

where $\tilde{\underline{G}}_M$ is the nominal model of the system.

In the SISO case we look at magnitudes. In the multivariable case we look at singular values. Thus plots of the maximum singular value of the $\underline{\Delta}$ matrix will show the frequency region where the uncertainties become significant. Then the controllers should be designed so that the minimum dip in the Doyle-Stein criterion curves (the minimum singular value of the matrix $[I + Q^{-1}]$) occurs in a frequency range where the uncertainty is not significant.

If the uncertainty has a known structure, "structured singular values" are used. In the book by Morari and Zafiriou (*Robust Process Control*, 1989, Prentice-Hall) this topic is discussed in detail.

PROBLEMS

16.1. Wardle and Wood (*I. Chem. E. Symp. Series*, 1969, No. 32, p. 1) give the following transfer function matrix for an industrial distillation column:

$$G_M = \begin{bmatrix} \dfrac{0.126e^{-6s}}{60s + 1} & \dfrac{-0.101e^{-12s}}{(48s + 1)(45s + 1)} \\ \dfrac{0.094e^{-8s}}{38s + 1} & \dfrac{-0.12e^{-8s}}{35s + 1} \end{bmatrix}$$

The empirical PI controller settings reported were:

$$K_c = 18/-24 \qquad \tau_I = 19/24$$

(*a*) Use a multivariable Nyquist plot and characteristic loci plots to see if the system is closedloop stable.

(*b*) Calculate the values of the RGA, the Niederlinski index, and the Morari resiliency index.

16.2. A distillation column has the following transfer function matrix:

$$G_M = \begin{bmatrix} \dfrac{34}{(54s + 1)(0.5s + 1)^2} & \dfrac{-44.7}{(114s + 1)(0.5s + 1)^2} \\ \dfrac{31.6}{(78s + 1)(0.5s + 1)^2} & \dfrac{-45.2}{(42s + 1)(0.5s + 1)^2} \end{bmatrix}$$

Empirical PI diagonal controller settings are:

$$K_c = 1.6/-1.6 \qquad \tau_I = 20/9 \text{ min}$$

(*a*) Check the closedloop stability of the system using a multivariable Nyquist plot and characteristic loci plots.

(*b*) Calculate values of the RGA, the Niederlinski index, and the Morari resiliency index.

16.3. A distillation column is described by the following linear ODEs:

$$\frac{dx_D}{dt} = -4.74x_D + 5.99x_B + 0.708R - 0.472V$$

$$\frac{dx_B}{dt} = 10.84x_D - 18.24x_B + 1.28R - 1.92V + 4z$$

(*a*) Use state-variable matrix methods to derive the openloop transfer function matrix.

(*b*) What are the openloop eigenvalues of the system?

(c) If the openloop steadystate gain matrix is

$$K_p = \begin{bmatrix} 0.958 & -0.936 \\ 0.639 & -0.661 \end{bmatrix}$$

calculate the RGA, the Niederlinski index, and the Morari resiliency index.

16.4. A 2×2 process has the openloop transfer function matrix

$$G_M = \frac{1}{(s + 1)} \begin{bmatrix} 2 & -1 \\ 2 & 2 \end{bmatrix}$$

A diagonal proportional feedback controller is used with both gains set equal to K_c. Time is in minutes.

(a) What is the openloop time constant of the system?
(b) Calculate the closedloop eigenvalues as functions of K_c.
(c) What value of K_c will give a closedloop time constant of 0.1 min?

16.5. Air and water streams are fed into a pressurized tank through two control valves. Air flows out of the top of the tank through a restriction, and water flows out the bottom through another restriction. The linearized equations describing the system are

$$\frac{dP}{dt} = -0.8P + 0.5F_a + 0.1F_w$$

$$\frac{dh}{dt} = -0.4P - 0.1h + 0.5F_w$$

where P = pressure
   h = liquid height
   F_a = air flow rate into tank
   F_w = water flow rate into tank

Use state variable methods to calculate:

(a) The openloop eigenvalues and the openloop transfer function matrix.
(b) The closedloop eigenvalues if two proportional SISO controllers are used with gains of 5 (for the pressure controller manipulating air flow) and 2 (for the level controller manipulating water flow).
(c) Calculate the RGA and the Niederlinski index for this system.

16.6. A 2×2 multivariable process has the following openloop plant transfer function matrix:

$$G_M = \frac{1}{s + 1} \begin{bmatrix} 2 & 1 \\ 1 & 2 \end{bmatrix}$$

A diagonal feedback controller is used with proportional controllers.

$$B = \begin{bmatrix} K_1 & 0 \\ 0 & K_2 \end{bmatrix}$$

(a) What is the closedloop characteristic equation of this system?
(b) What are the RGA, Morari resiliency index, and Niederlinski index?

16.7. The state-space description of a process is given below:

$$\dot{x} = Ax + Cm$$

where $x = \begin{bmatrix} x_1 \\ x_2 \end{bmatrix}$ $\quad m = \begin{bmatrix} m_1 \\ m_2 \end{bmatrix}$

$$A = \begin{bmatrix} -1 & 1 \\ 1 & -2 \end{bmatrix} \quad C = \begin{bmatrix} 2 & -1 \\ 1 & 2 \end{bmatrix}$$

Derive the openloop plant transfer function matrix relating controlled variables x_i and manipulated variables m_j.

CHAPTER
17

DESIGN OF CONTROLLERS FOR MULTIVARIABLE PROCESSES

In the last two chapters we have developed some mathematical tools and some methods of analyzing multivariable closedloop systems. This chapter studies the development of control structures for these processes.

The first part of this chapter deals with the conventional diagonal structure: multiloop SISO controllers. Full-blown multivariable controllers are briefly discussed at the end of the chapter.

17.1 PROBLEM DEFINITION

Most industrial control systems use the multiloop SISO diagonal control structure. It is the most simple and understandable structure. Operators and plant engineers can use it and modify it when necessary. It does not require an expert in applied mathematics to design and maintain. In addition, the performance of these diagonal controller structures is usually quite adequate for process control applications. In fact there has been little quantitative, unbiased data showing that the performances of the more sophisticated controller structures are really that much better! Some comparisons of performance are given in Sec. 17.7, and you can be the judge as to whether the slight (if any) improvement is worth the price of the additional complexity and engineering cost of implementation and maintenance.

So the multiloop SISO diagonal controller remains an important structure. It is the base case against which the other structures should be compared. The procedure discussed in this chapter was developed to provide a workable, stable, simple SISO system with only a modest amount of engineering effort. The resulting diagonal controller can then serve as a realistic benchmark, against which the more complex multivariable controller structures can be compared.

The limitations of the procedure should be pointed out. It does not apply to openloop unstable systems. It also does not work well when the time constants of the different transfer functions are quite different; i.e., some parts of the process much faster than others. The fast and slow sections should be designed separately in this case.

The procedure has been tested primarily on realistic distillation column models. This choice was deliberate because most industrial processes have similar gain, deadtime, and lag transfer functions. Undoubtedly some pathological transfer functions can be found that the procedure cannot handle. But we are interested in a practical engineering tool, not elegant, rigorous all-inclusive mathematical theorems.

The procedure is called the "LACEY" procedure (not because it is full of holes, as claimed by some!) from its developers: Luyben, Alatiqi, Chiang, Elaahi, and Yu. The steps are summarized below. Each step is discussed in more detail in later sections of this chapter.

1. *Select controlled variables.* Use primarily engineering judgment based on process understanding.
2. *Select manipulated variables.* Find the set of manipulated variables that gives the largest Morari resiliency index (MRI).
3. *Eliminate unworkable variable pairings.* Eliminate pairings with negative Niederlinski indexes.
4. *Find the best pairing from the remaining sets.*
   (a) Tune all combinations using BLT tuning and check for stability (characteristic loci plots) and robustness (Doyle-Stein and Skogestad-Morari criteria).
   (b) Select the pairing that gives the lowest magnitude closedloop load transfer function: Tyreus load-rejection criterion (TLC).

You can see from the acronyms used above that I like to use ones that are easy to remember: BLT (bacon, lettuce, and tomato) and TLC (tender loving care). Remember we also discussed the "ATV" (all-terrain vehicle) identification method back in Chap. 14.

17.2 SELECTION OF CONTROLLED VARIABLES

17.2.1 Engineering Judgment

Engineering judgment is the principal tool for deciding what variables to control. A good understanding of the process leads in most cases to a logical choice of what needs to be controlled. Considerations of economics, safety, constraints, availability and reliability of sensors, etc. must be factored into this decision.

For example, in a distillation column we are usually interested in controlling the purities of the distillate and bottoms product streams. In chemical reactors, heat exchangers, and furnaces the usual controlled variable is temperature. Levels and pressures must be controlled. In most cases these choices are fairly obvious.

It should be remembered that controlled variables need not be simple directly measured variables. They can also be computed from a number of sensor inputs. Common examples are heat removal rates, mass flow rates, ratios of flow rates, etc.

However, sometimes the selection of the appropriate controlled variable is not so easy. For example, in a distillation column it is frequently difficult and expensive to measure product compositions directly with sensors such as gas chromatographs. Instead, temperatures on various trays are controlled. The selection of the best control tray to use requires a considerable amount of knowledge about the column, its operation, and its performance. Varying amounts of non-key components in the feed can affect very importantly the best choice of control trays.

17.2.2 Singular Value Decomposition

The use of singular value decomposition (SVD), introduced into chemical engineering by Moore and Downs (*Proc. JACC*, paper WP-7C, 1981) can give some guidance in the question of what variables to control. They used SVD to select the best tray temperatures. SVD involves expressing the matrix of plant transfer function steadystate gains $\underline{K}_p$ as the product of three matrices: a $\underline{U}$ matrix, a diagonal $\underline{\Sigma}$ matrix, and a $\underline{V}^T$ matrix.

$$\underline{K}_p = \underline{U}\underline{\Sigma}\underline{V}^T \tag{17.1}$$

It is somewhat similar to canonical transformation. But it is different in that the diagonal $\underline{\Sigma}$ matrix contains as its diagonal elements, not the eigenvalues of the $\underline{K}_p$ matrix, but its singular values.

The biggest elements in each column of the $\underline{U}$ matrix indicate which outputs of the process are the most sensitive. Thus SVD can be used to help select which tray temperatures in a distillation column should be controlled. Example 17.1 from the Moore and Downs paper illustrates the procedure.

Example 17.1. A nine-tray distillation column separating isopropanol and water has the following steadystate gains between tray temperatures and the manipulated variables reflux R and heat input Q.

Tray number	$\dfrac{\Delta T_n}{\Delta R}$	$\dfrac{\Delta T_n}{\Delta Q}$
9	−0.00773271	0.0134723
8	−0.2399404	0.2378752
7	−2.5041590	2.4223120
6	−5.9972530	5.7837800
5	−1.6773120	1.6581630
4	0.0217166	0.0259478
3	0.1976678	−0.1586702
2	0.1289912	−0.1068900
1	0.0646059	−0.0538632

The elements in this table are the elements in the steadystate gain matrix of the column $\underline{K}_p$, which has 9 rows and 2 columns.

$$\begin{bmatrix} \Delta T_9 \\ \Delta T_8 \\ \vdots \\ \Delta T_1 \end{bmatrix} = \underline{K}_p \begin{bmatrix} R \\ Q \end{bmatrix} \tag{17.2}$$

Now $\underline{K}_p$ is decomposed into the product of three matrices.

$$\underline{K}_p = \underline{U}\underline{\Sigma}\underline{V}^T \tag{17.3}$$

$$\underline{U} = \begin{bmatrix} -0.0015968 & -0.0828981 \\ -0.0361514 & -0.0835548 \\ -0.3728142 & -0.0391486 \\ -0.8915611 & 0.1473784 \\ -0.2523673 & -0.6482796 \\ -0.0002581 & -0.6482796 \\ 0.0270092 & -0.4463671 \\ 0.0178741 & -0.2450451 \\ 0.0089766 & -0.1182182 \end{bmatrix} \tag{17.4}$$

$$\underline{\Sigma} = \begin{bmatrix} 9.3452 & 0 \\ 0 & 0.052061 \end{bmatrix} \tag{17.5}$$

$$\underline{V}^T = \begin{bmatrix} 0.7191619 & -0.6948426 \\ -0.6948426 & -0.7191619 \end{bmatrix} \tag{17.6}$$

The largest element in the first column in $\underline{U}$ is -0.8915611 which corresponds to tray 6. Therefore SVD would suggest the control of tray 6 temperature.

The software to do the SVD calculations is readily available (*Computer Methods For Mathematical Computations*, Forsythe, Malcolm, and Moler, Prentice-Hall, 1977).

17.3 SELECTION OF MANIPULATED VARIABLES

Once the controlled variables have been specified, the control structure depends only on the choice of manipulated variables. For a given process, selecting different manipulated variables will produce different control structure alternatives. These *control structures* are independent of the *controller structure*, i.e., pairing of variables in a diagonal multiloop SISO structure or one full-blown multivariable controller.

For example, in a distillation column the manipulated variables could be the flow rates of reflux and vapor boilup $(R - V)$ to control distillate and bottoms compositions. This choice gives one possible control structure. Alternatively we could have chosen to manipulate the flow rates of distillate and vapor boilup $(D - V)$. This yields another control structure for the same basic distillation process.

The Morari resiliency index (MRI) discussed in Chap. 16 can be used to choose the best set of manipulated variables. The set of manipulated variables that gives the *largest* minimum singular value over the frequency range of interest is the best. The differences in MRI values should be fairly large to be meaningful. If one set of manipulated variables gives an MRI of 10 and another set gives an MRI of 1, you can conclude that the first set is better. However, if the two sets have MRIs that are 10 and 8, there is probably little difference from a resiliency standpoint and either set of manipulated variables could be equally effective.

This selection of control structure is independent of variable pairing and controller tuning. The MRI is a measure of the inherent ability of the process (with the specified choice of manipulated variables) to handle disturbances, changes in operating conditions, etc.

The problem of the effect of scaling on singular values is handled by expressing the gains of all the plant transfer functions in dimensionless form. The gains with engineering units are divided by transmitter spans and multiplied by valve gains. This yields the dimensionless gain that the controller sees and has to cope with.

17.4 ELIMINATION OF POOR PAIRINGS

The Niederlinski index (Sec. 16.1.4) can be used to eliminate some of the pairings. Negative values of this index mean unstable pairings, independent of controller tuning. As illustrated in Example 16.3, pairing x_D with V and x_B with R gives a negative Niederlinski index, so this pairing should not be used.

For a 2×2 system, a negative Niederlinski index is equivalent to pairing on a negative RGA element. For example, in Example 16.7 the β_{11} RGA element is positive (2.01). This says that the x_D to R pairing is okay. However, the β_{12} element is negative (-1.01), telling us that the x_D to V pairing is not okay.

17.5 BLT TUNING

One of the major questions in multivariable control is how to tune controllers in a diagonal multiloop SISO system. If PI controllers are used, there are $2N$ tuning parameters to be selected. The gains and reset times must be specified so that the overall system is stable and gives acceptable load responses. Once a consistent and rational tuning procedure is available, the pairing problem can be attacked.

The tuning procedure discussed below (called *BLT*, biggest log-modulus tuning) provides such a standard tuning methodology. It satisfies the objective of arriving at reasonable controller settings with only a small amount of engineering and computational effort. It is not claimed that the method will produce the best results or that some other tuning or controller structure will not give superior performance. What is claimed is that the method is easy to use, is easily understandable by control engineers, and leads to settings that compare very favorably with the empirical settings found by exhaustive and expensive trial-and-error tuning methods used in many studies.

The method should be viewed in the same light as the classical SISO Ziegler-Nichols method. It gives reasonable settings which provide a starting point for further tuning and a benchmark for comparative studies.

BLT tuning involves the following four steps:

1. Calculate the Ziegler-Nichols settings for each individual loop. The ultimate gain and ultimate frequency ω_u of each diagonal transfer function $G_{jj(s)}$ are calculated in the classical SISO way. To do this numerically, a value of frequency ω is guessed. The phase angle is calculated, and the frequency is varied to find the point where the Nyquist plot of $G_{jj(i\omega)}$ crosses the negative real axis (phase angle is -180 degrees). The frequency where this occurs is ω_u. The reciprocal of the real part of $G_{jj(i\omega)}$ is the ultimate gain. Table 17.1 gives a FORTRAN program with a subroutine ZNT which does this iterative calculation using simple interval halving.

2. A detuning factor F is assumed. F should always be greater than 1. Typical values are between 1.5 and 4. The gains of *all* feedback controllers K_{ci} are calculated by *dividing* the Ziegler-Nichols gains K_{ZNi} by the factor F.

$$K_{ci} = \frac{K_{ZNi}}{F} \tag{17.7}$$

where $K_{ZNi} = K_{ui}/2.2$.

Then all feedback controller reset times τ_{Ii} are calculated by multiplying the Ziegler-Nichols reset times τ_{ZNi} by the same factor F.

$$\tau_{Ii} = \tau_{ZNi} F \tag{17.8}$$

where $\tau_{ZNi} = 2\pi/1.2\omega_{ui}$. $\tag{17.9}$

The F factor can be considered as a detuning factor which is applied to all loops. The larger the value of F, the more stable the system will be but the

TABLE 17.1
BLT tuning for Wood and Berry column

```
      REAL KP(4,4),KC(4)
      DIMENSION RESET(4),TAU(4,4,4),D(4,4)
      COMPLEX GM(4,4),B(4,4),Q(4,4),YID(4,4),CHAR(4,4),A(4,4),C(4,4)
      COMPLEX XDETER,WFUNC
      EXTERNAL D2,D3,D4
C PROCESS TRANSFER FUNCTION DATA
      DATA KP(1,1),KP(1,2),KP(2,1),KP(2,2)/12.8,-18.9,6.6,-19.4/
      DATA TAU(1,1,1),TAU(1,1,2),TAU(1,2,1),TAU(1,2,2)/4*0./
      DATA TAU(2,1,1),TAU(2,1,2),TAU(2,2,1),TAU(2,2,2)
     +   /16.7,21.,10.9,14.4/
      DO 1 I=1,2
      DO 1 J=1,2
      TAU(3,I,J)=0.
    1 TAU(4,I,J)=0.
      DATA D(1,1),D(1,2),D(2,1),D(2,2)/1.,3.,7.,3./
      N=2
      CALL BLT(N,KP,TAU,D,KC,RESET,FBLT)
      STOP
      END
C
      SUBROUTINE BLT(N,KP,TAU,D,KC,RESET,FBLT)
      COMPLEX G(4,4),B(4,4),Q(4,4),YID(4,4),CHAR(4,4),A(4,4),C(4,4)
      COMPLEX XDETER,CLTF
      REAL KC,KP,KU
      DIMENSION KP(4,4),TAU(4,4,4),D(4,4),KC(4),RESET(4),KU(4),WU(4)
      FLAGP=1.
      FLAGM=1.
      LOOP=0
      FBLT=10.
      DF=2.
C FIND ZIEGLER-NICHOLS SETTINGS
      DW=10.**(1./80.)
      CALL ZNT(N,KP,TAU,D,KU,WU)
      WRITE(6,77)
   77 FORMAT(' F      DBMAX      WMAX    ')
 1000 DO 10 K=1,N
      KC(K)=KU(K)/2.2/FBLT
   10 RESET(K)=FBLT *2.*3.1416/WU(K)/1.2
      DBMAX=-200.
      W=.01
      LOOP=LOOP+1
      IF(LOOP.GT.30)THEN
      WRITE(6,*)' BLT LOOP'
      STOP
      ENDIF
C EVALUATE PROCESS TRANSFER FUNCTION MATRIX
C    GAIN KP(I), DEADTIME D(I), LEAD TAU(1,I,J)
C      THREE LAGS TAU(2,I,L), TAU(3,I,J), TAU(4,I,J)
  100 CALL PROCTF(G,W,N,KP,TAU,D)
C EVALUATE DIAGONAL PI CONTROLLER MATRIX
      CALL FEEDBK(B,W,N,KC,RESET)
      CALL IDENT(YID,N)
      CALL MMULT(G,B,Q,N)
      CALL MADD(YID,Q,CHAR,N)
      CALL DETER(N,CHAR,A,C,XDETER)
C FORM W FUNCTION = DET(I+GB) - 1.
      XDETER=XDETER-1.
```

TABLE 17.1 (*continued*)

```
      CLTF=XDETER/(1.+XDETER)
      DBCL=20.*ALOG10(CABS(CLTF))
      IF(DBCL.GT.DBMAX)THEN
      DBMAX=DBCL
      WMAX=W
      ENDIF
      W=W*DW
      IF(W.LT.10.)GO TO 100
      WRITE(6,2)FBLT,DBMAX,WMAX
    2 FORMAT(1X,3F10.5)
C DBMAX SPECIFICATION IS 2*N
      IF(ABS(DBMAX-2.*N).LT. .01)GO TO 50
C USE INTERVAL HALVING TO CHANGE FBLT
      IF(DBMAX-2.*N)20,20,30
   20 IF(FLAGM.LT.0.)DF=DF/2.
      FBLT=FBLT-DF
      IF(FBLT.LT..5) GO TO 30
      FLAGP=-1.
      GO TO 1000
   30 IF(FLAGP.LT.0.)DF=DF/2.
      FBLT=FBLT+DF
      FLAGM=-1.
      GO TO 1000
   50 WRITE(6,9)(KC(I),I=1,N)
    9 FORMAT(' BLT GAINS  = ',4F8.4)
      WRITE(6,8)(RESET(I),I=1,N)
    8 FORMAT('  BLT RESET TIMES  = ',4F8.2)
      RETURN
      END
C
      SUBROUTINE ZNT(N,KP,TAU,D,KU,WU)
      COMPLEX G(4,4),B(4,4),Q(4,4),YID(4,4),CHAR(4,4),A(4,4),C(4,4)
      COMPLEX XDETER,CLTF
      REAL KC,KP,KU,KZN
      DIMENSION KP(4),TAU(4,4,4),D(4,4),KC(4),RESET(4),KU(4),WU(4)
      DO 1000 K=1,N
      FLAGP=1.
      FLAGM=1.
      LOOP=0
C GUESS FREQUENCY
      W2=1./TAU(2,K,K)
      IF(TAU(3,K,K).LE.0.) THEN
      W=W2
      GO TO 5
      ELSE
      W1=1./TAU(3,K,K)
      W=W2
      IF(W2.LT.W1)W=W1
      ENDIF
    5 DW=W
C CALCULATE PHASE ANGLE
  100 DEG=-W*D(K,K)+ATAN(W*TAU(1,K,K))-ATAN(W*TAU(2,K,K))
     +-ATAN(W*TAU(3,K,K))-ATAN(W*TAU(4,K,K))
      DEG=DEG*180./3.1416
      LOOP=LOOP+1
      IF(LOOP.GT.200)THEN
      WRITE(6,*)'Z-N LOOP'
      STOP
```

TABLE 17.1 (*continued*)

```
      ENDIF
C CHECK IF PHASE ANGLE IS -180
      IF(ABS(DEG+180.).LT..1) GO TO 50
C USE INTERVAL HALVING TO REGUESS FREQUENCY
      IF(DEG+180.)20,20,30
  20 IF(FLAGM)21,21,22
  21 DW=DW*.5
  22 W=W-DW
      FLAGP=-1.
      GO TO 100
  30 IF(FLAGP)31,31,32
  31 DW=DW*.5
  32 W=W+DW
      FLAGM=-1.
      GO TO 100
  50 WU(K)=W
C CALCULATE ULTIMATE GAIN
      CALL PROCTF(G,W,N,KP,TAU,D)
      KU(K)=-1./REAL(G(K,K))
1000 CONTINUE
      WRITE(6,*)' ZIEGLER-NICHOLS TUNING'
      WRITE(6,*)' N      KU    WU    KZN    TZN'
      DO 200 K=1,N
      KZN=KU(K)/2.2
      TZN=2.*3.1416/WU(K)/1.2
 200 WRITE(6,99)K,KU(K),WU(K),KZN,TZN
  99 FORMAT(1X,I2,3X,4F8.2)
      RETURN
      END
```

Results

ZIEGLER-NICHOLS TUNING

N	KU	WU	KZN	TZN
1	2.10	1.61	.96	3.25
2	-.42	.56	-.19	9.28

F	DBMAX	WMAX
10.00000	-.19344	.01000
8.00000	.07800	.01000
6.00000	.18621	.01000
4.00000	.79220	.17783
2.00000	6.47123	.37584
3.00000	2.62291	.28184
2.50000	4.16342	.32546
2.75000	3.32153	.30726
2.62500	3.72360	.31623
2.56250	3.93576	.32546
2.53125	4.04989	.32546
2.54688	3.99289	.32546

BLT GAINS =   .3750 -.0753
BLT RESET TIMES =   8.29 23.63

more sluggish will be the setpoint and load responses. The method yields settings that give a reasonable compromise between stability and performance in multivariable systems.

3. Using the guessed value of F and the resulting controller settings, a multivariable Nyquist plot of the scalar function $W_{(i\omega)} = -1 + \text{Det}\,[\underline{I} + \underline{G}_{M(i\omega)}\,\underline{B}_{(i\omega)}]$ is made. See Sec. 16.1.2, Eq. (16.5). The closer this contour is to the $(-1, 0)$ point, the closer the system is to instability. Therefore the quantity $W/(1 + W)$ will be similar to the closedloop servo transfer function for a SISO loop $G_M B/(1 + G_M B)$. Therefore, based on intuition and empirical grounds, we define a multivariable closedloop log modulus L_{cm}.

$$L_{cm} = 20 \log_{10} \left| \frac{W}{1 + W} \right| \qquad (17.10)$$

The peak in the plot of L_{cm} over the entire frequency range is the biggest log modulus L_{cm}^{max}.

4. The F factor is varied until L_{cm}^{max} is equal to $2N$, where N is the order of the system. For $N = 1$, the SISO case, we get the familiar $+2$ dB maximum closedloop log modulus criterion. For a 2×2 system, a $+4$ dB value of L_{cm}^{max} is used; for a 3×3, $+6$ dB; and so forth. This empirically determined criterion has been tested on a large number of cases and gives reasonable performance, which is a little on the conservative side.

      This tuning method should be viewed as giving preliminary controller settings which can be used as a benchmark for comparative studies. Note that the procedure guarantees that the system is stable with all controllers on automatic and also that each individual loop is stable if all others are on manual (the F factor is limited to values greater than one so the settings are always more conservative than the Ziegler-Nichols values). Thus a portion of the integrity question is automatically answered. However, further checks of stability would have to be made for other combinations of manual/automatic operation.

      The method weighs each loop equally, i.e., each loop is equally detuned. If it is important to keep tighter control of some variables than others, the method can be easily modified by using different weighting factors for different controlled variables. The less-important loop could be detuned more than the more-important loop.

      Table 17.1 gives a FORTRAN program that calculates the BLT tuning for the Wood and Berry column. The subroutine BLT is called from the main program where the process parameters are given. The resulting controller settings are compared with the empirical setting in Table 17.2. The BLT settings usually have larger gains and larger reset times than the empirical. Time responses using the two sets of controller tuning parameters are compared in Fig. 17.1.

      Results for the 3×3 Ogunnaike and Ray column are given in Table 17.2 and in Fig. 17.2. The "$+4$" and "$+6$" refer to the value of L_{cm}^{max} used. Both of these cases illustrate that the BLT procedure gives reasonable controller settings.

TABLE 17.2
BLT, Ziegler-Nichols, and empirical controller tuning

	Wood and Berry	Ogunnaike and Ray
Empirical		
K_c	0.2/−0.04	1.2/−0.15/0.6
τ_I	4.44/2.67	5/10/4
Z-N		
K_c	0.96/−0.19	3.24/−0.63/5.66
τ_I	3.25/9.2	7.62/8.36/3.08
BLT		
L_{cm}^{max} (dB)	+4	+6
F factor	2.55	2.15
K_c	0.375/−0.075	1.51/−0.295/2.63
τ_I	8.29/23.6	16.4/18/6.61

    If the process transfer functions have greatly differing time constants, the BLT procedure does not work well. It tends to give a response that is too oscillatory. The problem can be handled by breaking up the system into fast and slow sections and applying BLT to each smaller subsection.

    The BLT procedure discussed above was applied with PI controllers. The method can be extended to include derivative action (PID controllers) by using two detuning factors: F detunes the ZN reset and gain values and F_D detunes the ZN derivative value. The optimum value of F_D is that which gives the minimum

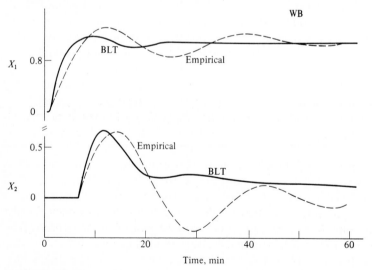

FIGURE 17.1
Wood and Berry column: BLT and empirical tuning.

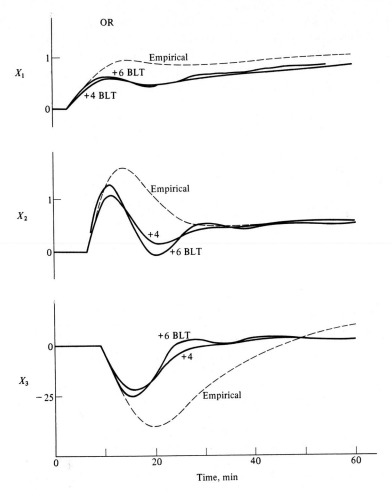

FIGURE 17.2
Ogunnaike and Ray column: BLT and empirical tuning.

value of F and still satisfies the $+2N$ maximum closedloop log modulus criterion (see the paper by Monica, Yu, and Luyben in *IEC Research*, Vol. 27, 1988, p. 969).

17.6 LOAD REJECTION PERFORMANCE

In most chemical processes the principal control problem is load rejection. We want a control system that can keep the controlled variables at or near their setpoints in the face of load disturbances. Thus the closedloop regulator transfer function is the most important.

The ideal closedloop relationship between the controlled variable and the load is zero. Of course this can never be achieved, but the smaller the magnitude of the closedloop regulator transfer function, the better the control. Thus a rational criterion for selecting the best pairing of variables is to choose the pairing that gives the smallest peaks in a plot of the elements of the closedloop regulator transfer function matrix. This criterion was suggested by Tyreus (Paper presented at the Lehigh University Short Course on Distillation Control, Bethlehem, Pa., 1984), so we call it the *Tyreus load-rejection criterion* (TLC).

The closedloop relationships for a multivariable process were derived in Chap. 15 [Eq. (15.64)]

$$\underline{x} = [[\underline{I} + \underline{G}_{M_{(s)}} \underline{B}_{(s)}]^{-1} \underline{G}_{M_{(s)}} \underline{B}_{(s)}] \underline{x}^{set} + [[\underline{I} + \underline{G}_{M_{(s)}} \underline{B}_{(s)}]^{-1} \underline{G}_{L_{(s)}}] L_{(s)} \quad (17.11)$$

For a 3×3 system there are three elements in the $\underline{x}$ vector, so three curves are plotted of each of the three elements in the vector $[[\underline{I} + \underline{G}_{M_{(i\omega)}} \underline{B}_{(i\omega)}]^{-1} \underline{G}_{L_{(i\omega)}}]$ for the specified load variable L. Table 17.3 gives a program that calculates the TLC curves for the 3×3 Ogunnaike and Ray column.

17.7 MULTIVARIABLE CONTROLLERS

Several multivariable controllers have been proposed during the last few decades. The optimal control research of the 1960s used variational methods to produce multivariable controllers that minimized some quadratic performance index. The method is called *linear quadratic* (LQ). The mathematics are elegant but very few chemical engineering industrial applications grew out of this work. Our systems are too high-order and nonlinear for successful application of LQ methods.

In the last decade several other multivariable controllers have been proposed. We will briefly discuss two of the most popular in the sections below. Other multivariable controllers that will not be discussed but are worthy of mention are "minimum variance controllers" (see Bergh and MacGregor, *IEC Research*, Vol. 26, 1987, p. 1558) and "extended horizon controllers" (see Ydstie, Kershenbaum, and Sargent, *AIChE J.*, Vol. 31, 1985, p. 1771).

17.7.1 Multivariable DMC

Undoubtedly the most popular multivariable controller is the multivariable extension of dynamic matrix control. We developed DMC for a SISO loop in Chap. 8. The procedure was a fairly direct least-squares computational one that solved for the future values of the manipulated variable such that some performance index was minimized.

The procedure can be extended to the multivariable case with little conceptual effort. You simply expand the $\underline{A}$ matrix discussed in Chap. 8 to include all the individual $\underline{A}$ matrices relating all of the outputs to all of the inputs. The performance index is expanded to include the errors in all the controlled variables plus weighting the moves in all the manipulated variables. The bookkeeping and the number-crunching become more extensive but the concept is the same.

TABLE 17.3

TLC for Ogunnaike and Ray column

```
      REAL KP(4,4),KC(4),KL(4)
      DIMENSION RESET(4),TAU(4,4,4),D(4,4),TAUL(4,4),DL(4)
      COMPLEX GM(4,4),B(4,4),Q(4,4),YID(4,4),CHAR(4,4),A(4,4),C(4,4)
      COMPLEX GL(4)
C GM PROCESS TRANSFER FUNCTION DATA
      DATA KP(1,1),KP(1,2),KP(1,3)/0.66,-0.61,-0.0049/
      DATA KP(2,1),KP(2,2),KP(2,3)/1.11,-2.36,-0.012/
      DATA KP(3,1),KP(3,2),KP(3,3)/-34.68,46.2,0.87/
      DATA TAU(1,1,1),TAU(1,1,2),TAU(1,1,3)/3*0./
      DATA TAU(1,2,1),TAU(1,2,2),TAU(1,2,3)/3*0./
      DATA TAU(1,3,1),TAU(1,3,2),TAU(1,3,3)/2*0., 11.61/
      DATA TAU(2,1,1),TAU(2,1,2),TAU(2,1,3)/6.7,8.64,9.06/
      DATA TAU(2,2,1),TAU(2,2,2),TAU(2,2,3)/3.25,5.0,7.09/
      DATA TAU(2,3,1),TAU(2,3,2),TAU(2,3,3)/8.15,10.9,3.89/
      DATA TAU(3,1,1),TAU(3,1,2),TAU(3,1,3)/3*0./
      DATA TAU(3,2,1),TAU(3,2,2),TAU(3,2,3)/3*0./
      DATA TAU(3,3,1),TAU(3,3,2),TAU(3,3,3)/2*0.,18.8/
      DO 1 I=1,3
      DO 1 J=1,3
    1 TAU(4,I,J)=0.
      DATA D(1,1),D(1,2),D(1,3)/2.6,3.5,1./
      DATA D(2,1),D(2,2),D(2,3)/6.5,3.,1.2/
      DATA D(3,1),D(3,2),D(3,3)/9.2,9.4,1./
C GL PROCESS TRANSFER FUNCTION DATA
      DATA KL(1),KL(2),KL(3)/0.14,0.53,-11.54/
      DATA TAUL(1,1),TAUL(1,2),TAUL(1,3)/3*0./
      DATA TAUL(2,1),TAUL(2,2),TAUL(2,3)/6.2,6.9,7.01/
      DO 2 I=1,3
      TAUL(3,I)=0.
    2 TAUL(4,I)=0.
      DATA DL(1),DL(2),DL(3)/12.,10.5,0.6/
C CONTROLLER DATA - EMPIRICAL SETTINGS
      DATA KC(1),KC(2),KC(3)/1.2,-0.15,0.6/
      DATA RESET(1),RESET(2),RESET(3)/5.,10.,4./
      N=3
      CALL TLC(N,KP,TAU,D,KL,TAUL,DL,KC,RESET)
      STOP
      END
C
      SUBROUTINE TLC(N,KP,TAU,D,KL,TAUL,DL,KC,RESET)
      REAL KP(4,4),KC(4),KL(4)
      DIMENSION RESET(4),TAU(4,4,4),D(4,4),TAUL(4,4),DL(4)
      COMPLEX GM(4,4),B(4,4),Q(4,4),HIN(4,4),H(4,4)
      COMPLEX GL(4),WA(24),XTLC(4)
      DIMENSION DB(4),WK(4)
      W=.01
      DW=(10.)**(1./10.)
      NN=N
      IA=4
      M=N
      IB=4
      IJOB=0
      WRITE(6,77)
   77 FORMAT(' W    TLC1      TLC2      TCL3')
  100 CALL PROCTF(GM,W,N,KP,TAU,D)
      CALL FEEDBK(B,W,N,KC,RESET)
      CALL MMULT(GM,B,Q,N)
```

TABLE 17.3 *(continued)*

```
    CALL IDENT(HIN,N)
    CALL MADD(HIN,Q,H,N)
    CALL LEQ2C(H,NN,IA,HIN,M,IB,IJOB,WA,WK,IER)
    CALL LOADTF(GL,W,N,KL,TAUL,DL)
    CALL MVMULT(HIN,GL,XTLC,W,N)
    DO 6 I=1,N
  6 DB(I)=20.*ALOG10(CABS(XTLC(I)))
    WRITE(6,7)W,(DB(I),I=1,N)
  7 FORMAT(1X,F7.4,4F9.3)
    W=W*DW
    IF(W.LT.10.)GO TO 100
    RETURN
    END
C
    SUBROUTINE LOADTF(GL,W,N,KL,TAUL,DL)
    COMPLEX GL(4),ZL,ZD
    DIMENSION DL(4),TAUL(4,4)
    REAL KL(4)
    ZL(X)=CMPLX(1.,X*W)
    ZD(X)=CMPLX(COS(W*X),-SIN(W*X))
    DO 10 I=1,N
    GL(I)=KL(I)*ZD(DL(I))*ZL(TAUL(1,I))/ZL(TAUL(2,I))/
   + ZL(TAUL(3,I))/ZL(TAUL(4,I))
 10 CONTINUE
    RETURN
    END
C
C MATRIX-VECTOR MULTIPLICATION
    SUBROUTINE MVMULT(A,B,C,W,N)
    COMPLEX A(4,4),B(4),C(4)
    DO 30 I=1,N
    C(I)=CMPLX(0.,0.)
    DO 20 J=1,N
 20 C(I)=A(I,J)*B(J)+C(I)
 30 CONTINUE
    RETURN
    END
```

Results

W	TLC1	TLC2	TCL3	W	TLC1	TLC2	TCL3
.0100	-51.187	-18.386	-7.112	.3981	-21.309	-11.531	15.413
.0126	-47.967	-17.091	-3.642	.5012	-24.020	-16.312	8.943
.0158	-45.022	-16.026	-.240	.6310	-27.079	-19.446	9.338
.0200	-42.393	-15.209	3.093	.7943	-29.796	-19.764	6.413
.0251	-40.119	-14.645	6.361	1.0000	-32.102	-21.345	5.627
.0316	-38.231	-14.331	9.540	1.2589	-35.142	-24.907	3.107
.0398	-36.722	-14.266	12.547	1.5849	-37.406	-26.733	.995
.0501	-35.442	-14.440	15.252	1.9953	-39.624	-28.397	-1.342
.0631	-34.002	-14.781	17.553	2.5119	-41.039	-29.725	-3.389
.0794	-32.012	-15.072	19.407	3.1623	-42.692	-32.305	-5.623
.1000	-29.484	-14.857	20.770	3.9811	-45.035	-34.280	-7.692
.1259	-26.696	-13.620	21.450	5.0119	-46.915	-36.055	-9.727
.1585	-23.911	-11.456	20.922	6.3096	-49.044	-38.224	-11.729
.1995	-21.406	-9.153	17.734	7.9433	-50.828	-40.280	-13.540
.2512	-19.702	-7.744	7.871	10.0000	-52.814	-42.443	-15.796
.3162	-19.614	-8.386	15.111				

Figure 17.3 gives some comparisons of the performance of the multivariable DMC structure with the diagonal structure. Three linear transfer-function models are presented, varying from the 2×2 Wood and Berry column to the 4×4 sidestream column/stripper complex configuration. The DMC tuning constants used for these three examples are $NP = 40$ and $NC = 15$. See Chap. 8, Sec. 8.9.

The weighting factors (f) for the changes in the manipulated variables were 10, 0.1, and 0.1 for the three cases. These were chosen so that the maximum magnitudes of the changes in the manipulated variables did not exceed those experienced when the diagonal structure was used. If these weighting factors were made smaller, tighter control could be achieved, but the changes in the manipulated variables would be larger. In most chemical engineering systems, we cannot permit rapid and large swings in manipulated variables. For example, such large swings in steam to the reboiler of a distillation column could flood the column or even blow out trays in the lower section of the tower.

Using the multivariable DMC produces some improvement for setpoint disturbances. However, as Fig. 17.3 shows, there are only slight differences between DMC and the diagonal structure (with BLT tuning) for load disturbances. Thus the real economic incentive for the more complex multivariable configuration has yet to be demonstrated by impartial evaluators.

17.7.2 Multivariable IMC

In theory, the internal model control methods discussed for SISO systems in Chap. 11 can be extended to multivariable systems (see the paper by Garcia and Morari in *IEC Process Design and Development*, Vol. 24, 1985, p. 472).

In practice, however, this extension is not as straightforward as in DMC. In multivariable DMC, there is a definite design procedure to follow. In multivariable IMC, there are steps in the design procedure that are not quantitative but involve some "art." The problem is in the selection of the invertible part of the process transfer function matrix. Since there are many possible choices, the design procedure becomes cloudy.

The basic idea in multivariable IMC is the same as in single-loop IMC. The ideal controller would be the inverse of the plant transfer function matrix. This would give perfect control. However, the inverse of the plant transfer function matrix is not physically realizable because of deadtimes, higher-order denominators than numerators, and RHP zeros (which would give an openloop unstable controller).

So the openloop plant transfer function matrix $\underline{G}_M$ is split up into two parts: an invertible part and a noninvertible part.

$$\underline{G}_M = \underline{G}_+ \underline{G}_- \tag{17.12}$$

Then the multivariable IMC controller is set equal to the invertible part times a "filter" matrix which slows up the closedloop response to give the system more robustness. The filter acts as a tuning parameter (like setting the closedloop time constant in the SISO case in Chap. 8).

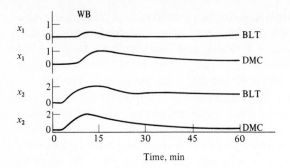

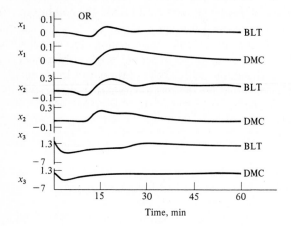

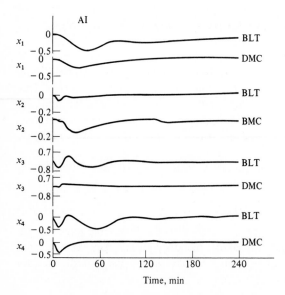

FIGURE 17.3

However, the procedure for specifying the invertible part G_- is not a definite, step-by-step recipe. There are several invertible parts that could be chosen. These problems are discussed in the book by Morari and Zafiriou (*Robust Process Control*, Prentice-Hall, 1989).

PROBLEMS

17.1. Calculate the BLT settings for the Wardle and Wood column (see Prob. 16.1). Using these controller settings, calculate the minimum singular value from the Doyle-Stein method.

17.2. Determine the BLT settings for the 2×2 multivariable system given in Prob. 16.2.

17.3. Alatiqi presented (*I&EC Process Design Dev.* 1986, Vol. 25, p. 762) the transfer functions for a 4×4 multivariable complex distillation column with sidestream stripper for separating a ternary mixture into three products. There are four controlled variables: purities of the three product streams (x_{D1}, x_{S2}, and x_{B3}) and a temperature difference ΔT to minimize energy consumption. There are four manipulated variables: reflux R, heat input to the reboiler Q_R, heat input to the stripper reboiler Q_S, and flow rate of feed to the stripper L_S. The 4×4 matrix of openloop transfer functions relating controlled and manipulated variables is:

$$\begin{vmatrix} \dfrac{4.09e^{-1.3s}}{(33s+1)(8.3s+1)} & \dfrac{-6.36e^{-1.2s}}{(31.6s+1)(20s+1)} & \dfrac{-0.25e^{-1.4s}}{21s+1} & \dfrac{-0.49e^{-6s}}{(22s+1)^2} \\[2mm] \dfrac{-4.17e^{-5s}}{45s+1} & \dfrac{6.93e^{-1.02s}}{44.6s+1} & \dfrac{-0.05e^{-6s}}{(34.5s+1)^2} & \dfrac{1.53e^{-3.8s}}{48s+1} \\[2mm] \dfrac{1.73e^{-18s}}{(13s+1)^2} & \dfrac{5.11e^{-12s}}{(13.3s+1)^2} & \dfrac{4.61e^{-1.01s}}{18.5s+1} & \dfrac{-5.49e^{-1.5s}}{15s+1} \\[2mm] \dfrac{-11.2e^{-2.6s}}{(43s+1)(6.5s+1)} & \dfrac{14(10s+1)e^{-0.02s}}{(45s+1)(17.4s^2+3s+1)} & \dfrac{0.1e^{-0.05s}}{(31.6s+1)(5s+1)} & \dfrac{4.49e^{-0.6s}}{(48s+1)(6.3s+1)} \end{vmatrix}$$

The vector of openloop transfer functions relating the controlled variables to the load disturbance (feed composition: mole fraction of intermediate component) is:

$$G_L = \begin{vmatrix} \dfrac{-0.86e^{-6s}}{(19.2s+1)^2} \\[2mm] \dfrac{-1.06e^{-5s}}{35s+1} \\[2mm] \dfrac{1.2e^{-9s}}{24s+1} \\[2mm] \dfrac{-0.86e^{-0.2s}}{16s^2+4s+1} \end{vmatrix}$$

Calculate the BLT settings and the TLC curves using these settings.

17.4 A distillation column has the following openloop transfer function matrix relating controlled variables (x_D and x_B) to manipulated variables (reflux ratio RR and reboiler heat input Q_R):

$$
G_M = \begin{bmatrix}
\dfrac{3.7e^{-1.2s}}{(6.9s + 1)(162s + 1)} & \dfrac{-5.8e^{-1.3s}}{(1.8s + 1)(1190s + 1)} \\[3mm]
\dfrac{-170e^{-0.2s}}{(0.6s + 1)(127s + 1)^2} & \dfrac{323e^{-0.2s}}{(1.3s + 1)(400s + 1)}
\end{bmatrix}
$$

(a) Calculate the RGA and Niederlinski index for this pairing.

(b) Calculate the Morari resiliency index at zero frequency.

(c) Calculate the Ziegler-Nichols settings for the individual diagonal loops.

(d) Check the closedloop stability of the system when these ZN settings are used by making a multivariable Nyquist plot.

(e) Use the BLT procedure to determine a detuning F factor that gives a maximum closedloop log modulus of $+4$ dB.

PART
VII

SAMPLED-DATA CONTROL SYSTEMS

All the control systems that we have studied in previous parts of this book used continuous analog devices. All control signals were continuously generated by transmitters, fed continuously to analog controllers, and sent continuously to control valves.

The development of digital control computers and of chromatographic composition analyzers has resulted in a large number of control systems that have discontinuous, intermittent components. The nature of operation of both of these devices is such that their input and output signals are discrete.

A chromatograph injects a sample into a chromatographic column every few minutes (2 to 10 minutes or more). The sample works its way through the column and is detected as it emerges a few minutes later. Thus a composition signal is produced only once every few minutes. The time between composition signals is called the sampling period, T_s.

Digital computers are used in process control systems on a "time-shared" basis. A single digital computer (or microprocessor) services a number of control loops. At a given instant in time, the computer looks at one loop, checking the value of the controlled variable and computing a new signal to send to the control valve (in "direct digital control," DDC) or to the setpoint of a continuous analog controller (in "supervisory control"). The controller output signal for this loop is then held constant as the computer moves on to look at all the other loops. The control signal is changed only at discrete moments in time.

To analyze systems with discontinuous control elements we will need to learn another new "language." The mathematical tool of z *transformation* is used to design control systems for discrete systems. z transforms are to sampled-data systems what Laplace transforms are to continuous systems. The mathematics in the z domain and in the Laplace domain are very similar. We have to learn how to translate our small list of words from English and Russian into the language of z transforms, which we will call German.

In Chap. 18 we will define mathematically the sampling process, derive the z transforms of common functions (learn our German vocabulary) and develop transfer functions in the z domain. These fundamentals are then applied to basic controller design in Chap. 19 and to advanced controllers in Chap. 20. We will find that practically all the stability-analysis and controller-design techniques that we used in the Laplace and frequency domains can be directly applied in the z domain for sampled-data systems.

CHAPTER
18

SAMPLING AND
z TRANSFORMS

18.1 INTRODUCTION

18.1.1 Definition

Sampled-data systems are systems in which signals are discontinuous or discrete. Figure 18.1 shows a continuous analog signal or function $f_{(t)}$ being fed into a sampler. Every T_s minutes the sampler closes for a brief instant. The output of the sampler $f_{s(t)}$ is, therefore, an intermittent series of pulses. Between sampling times, the sampler output is zero. At the instant of sampling the output of the sampler is equal to the input function.

$$f_{s(t)} = f_{(nT_s)} \qquad \text{for } t = nT_s$$
$$f_{s(t)} = 0 \qquad \text{for } t \neq nT_s$$
$$(18.1)$$

18.1.2 Occurrence of Sampled-Data Systems in Chemical Engineering

As mentioned in the introduction to Part VII, chromatographs and digital control computers are the principal gadgets that produce sampled-data systems in chemical engineering processes.

Figure 18.2 sketches a typical chromatograph system. The process output variable $x_{(t)}$ is sampled every T_s minutes. The sample is injected into a chromatographic column that has a retention time of D_c minutes, which is essentially a

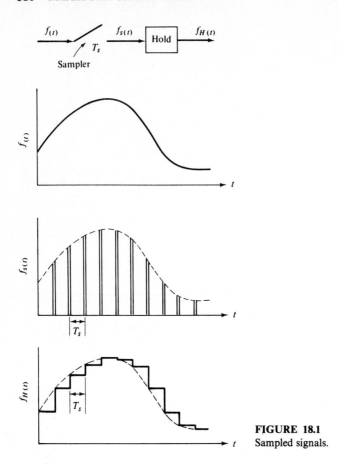

FIGURE 18.1
Sampled signals.

pure time delay or deadtime. The sampling period T_s is usually set equal to the chromatograph cycle time D_c. The detector on the output of the column produces a signal that can be related to composition. The "peak picker" (or an area integrator) converts the detector signal into a composition signal. The maximum value or peak on the chromatograph curve can sometimes be used directly, but usually the areas under the curves are integrated and converted into a composition signal.

This signal is generated only every T_s minutes. It is fed into a device called a *hold* that clamps the signal until the next sample comes along; i.e., the output of the hold is maintained at a constant value over the sampling period. The hold converts the sampled-data, discrete $x_{s(t)}$ signal, which is a series of pulses, into a continuous signal $x_{H(t)}$ that is a stair-step function. In Fig. 18.1 this continuous signal is the input to a conventional continuous analog feedback controller. In most modern systems a microprocessor feedback controller would be used. The equivalent block diagram of this system is shown in Fig. 18.2 at the bottom. The transfer function of the hold is $H_{(s)}$.

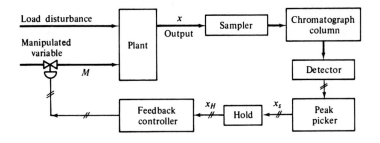

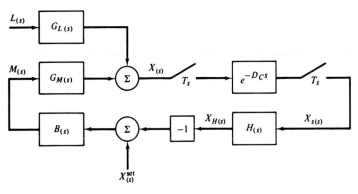

FIGURE 18.2
Chromatograph loop.

Figure 18.3*a* shows a "supervisory" digital control computer. Output variables $x_1, x_2, \ldots, x_N$ are sensed and converted into signals by transmitters T_1, $T_2, \ldots, T_N$. Continuous analog controllers $B_1, B_2, \ldots, B_N$ send signals to control valves on the manipulated variables $m_1, m_2, \ldots, m_N$. The setpoints of the analog controllers come from the digital computer and are clamped between sampling times by holds $H_1, H_2, \ldots, H_N$. Data enter the digital computer through a multi-plexed analog-to-digital (A/D) converter. Setpoint signals are sent to the holds through a multiplexed digital-to-analog (D/A) converter. A block diagram of one loop is shown in the bottoms of Fig. 18.3*a*. The digital computer is designated as $D_{(s)}^*$.

Figure 18.3*b* shows a DDC (direct digital control) system. All the control calculations are done in the digital computer. The computer outputs go, through holds, directly to the control valves.

The sampling rate of these digital control computers can vary from several times a second to only several times an hour. The dynamics of the process dictate the sampling time required. The faster the process, the smaller the sampling period T_s must be for good control. One of the important questions that we will explore in these three chapters is what should the sampling rate be for a given

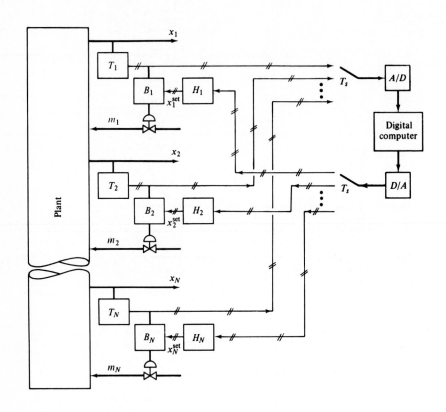

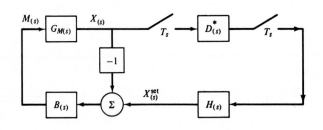

(a) Supervisory

FIGURE 18.3a
Digital control computers.

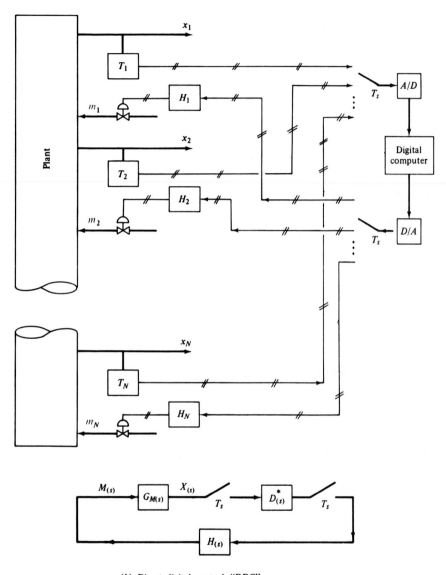

(*b*) Direct digital control "DDC"

FIGURE 18.3*b*

process. For a given number of loops, the smaller the value of T_s specified the faster the computer and the input-output equipment must be. This increases the cost of the digital hardware.

18.2   IMPULSE SAMPLER

A real sampler, as shown in Fig. 18.1, is closed for a finite period of time. This time of closure is usually small compared with the sampling period T_s. Therefore the real sampler can be closely approximated by an *impulse sampler*. An impulse sampler is a device that converts a continuous input signal into a sequence of impulses or delta functions. Remember, these are impulses, not pulses. The height of each of these impulses is infinite. The width of each is zero. The area of the impulse or the "strength" of the impulse is equal to the magnitude of the input function at the sampling instant.

$$\int_{-nT_s}^{+nT_s} f^*_{(t)} \, dt = f_{(nT_s)} \tag{18.2}$$

If the units of $f_{(t)}$ are, for example, kilograms, the units of $f^*_{(t)}$ are kilograms per minute.

The impulse sampler is, of course, a mathematical fiction; an impulse sampler is not physically realizable. But the behavior of a real sampler and hold circuit is practically identical to that of the idealized impulse sampler and hold circuit. The impulse sampler is used in the analysis of sampled-data systems and in the design of sampled-data controllers because it greatly simplifies these calculations.

Let us now define an infinite sequence of unit impulses $\delta_{(t)}$ or Dirac delta functions whose strengths are all equal to unity. One unit impulse occurs at every sampling time. We will call this series of unit impulses, shown in Fig. 18.4, the function $I_{(t)}$.

$$I_{(t)} = \delta_{(t)} + \delta_{(t-T_s)} + \delta_{(t-2T_s)} + \delta_{(t-3T_s)} + \cdots$$

$$I_{(t)} \equiv \sum_{n=0}^{\infty} \delta_{(t-nT_s)} \tag{18.3}$$

Thus the sequence of impulses $f^*_{(t)}$ that comes out of an impulse sampler can be expressed:

$$f^*_{(t)} = f_{(t)} I_{(t)} = f_{(0)} \delta_{(t)} + f_{(T_s)} \delta_{(t-T_s)} + f_{(2T_s)} \delta_{(t-2T_s)} + \cdots$$

$$f^*_{(t)} = \sum_{n=0}^{\infty} f_{(nT_s)} \delta_{(t-nT_s)} \tag{18.4}$$

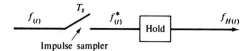

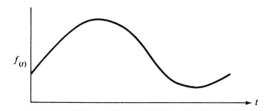

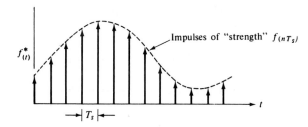

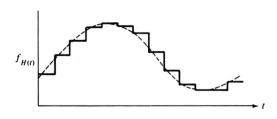

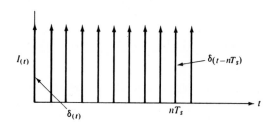

FIGURE 18.4
Impulse sampler.

Laplace-transforming Eq. (18.4) gives

$$\mathcal{L}[f^*_{(t)}] = \mathcal{L}\left[\sum_{n=0}^{\infty} f_{(nT_s)}\, \delta_{(t-nT_s)}\right] = \sum_{n=0}^{\infty} f_{(nT_s)}\,\mathcal{L}[\delta_{(t-nT_s)}]$$

$$= \sum_{n=0}^{\infty} f_{(nT_s)}\, e^{-nT_s s}\,\mathcal{L}[\delta_{(t)}]$$

$$F^*_{(s)} \equiv \sum_{n=0}^{\infty} f_{(nT_s)}\, e^{-nT_s s} \tag{18.5}$$

Equation (18.4) expresses the sequence of impulses that comes out of an impulse sampler in the time domain. Equation (18.5) gives the sequence in the Laplace domain. Substituting $i\omega$ for s gives the impulse sequence in the frequency domain.

$$F^*_{(i\omega)} \equiv \sum_{n=0}^{\infty} f_{(nT_s)}\, e^{-in\omega T_s} \tag{18.6}$$

The sequence of impulses $f^*_{(t)}$ can also be represented in an alternative manner. The $I_{(t)}$ function is a periodic function (see Fig. 18.4) with period T_s and a frequency ω_s in radians per minute.

$$\omega_s = \frac{2\pi}{T_s} \tag{18.7}$$

Since $I_{(t)}$ is periodic, it can be represented as a complex Fourier series:

$$I_{(t)} = \sum_{n=-\infty}^{n=+\infty} C_n\, e^{in\omega_s t} \tag{18.8}$$

where

$$C_n = \frac{1}{T_s} \int_{-T_s/2}^{+T_s/2} I_{(t)}\, e^{-in\omega_s t}\, dt \tag{18.9}$$

Over the interval from $-T_s/2$ to $+T_s/2$ the function $I_{(t)}$ is just $\delta_{(t)}$. Therefore Eq. (18.9) becomes

$$C_n = \frac{1}{T_s} \int_{-T_s/2}^{+T_s/2} \delta_{(t)}\, e^{-in\omega_s t}\, dt = \frac{1}{T_s}\,[e^{-in\omega_s t}]_{t=0} = \frac{1}{T_s} \tag{18.10}$$

Remember, multiplying a function $f_{(t)}$ by the Dirac delta function and integrating give $f_{(0)}$. Therefore, $I_{(t)}$ becomes

$$I_{(t)} = \frac{1}{T_s} \sum_{n=-\infty}^{n=+\infty} e^{in\omega_s t}$$

The sequence of impulses $f^*_{(t)}$ can be expressed as a doubly infinite series:

$$f^*_{(t)} = f_{(t)}\, I_{(t)} = \frac{1}{T_s} \sum_{n=-\infty}^{n=+\infty} f_{(t)}\, e^{in\omega_s t} \tag{18.11}$$

Remember that the Laplace transformation of a function multiplied by an exponential e^{at} is simply the Laplace transform of the function with $s - a$ substituted for s.

$$\mathcal{L}[f_{(t)}\, e^{at}] = \int_0^\infty f_{(t)}\, e^{at} e^{-st}\, dt = \int_0^\infty f_{(t)}\, e^{-(s-a)t}\, dt \equiv F_{(s-a)}$$

So Laplace-transforming Eq. (18.11) gives

$$F^*_{(s)} = \frac{1}{T_s} \sum_{n=-\infty}^{n=+\infty} F_{(s-in\omega_s)} = \frac{1}{T_s} \sum_{n=-\infty}^{n=+\infty} F_{(s+in\omega_s)} \tag{18.12}$$

Substituting $i\omega$ for s gives

$$F^*_{(i\omega)} = \frac{1}{T_s} \sum_{n=-\infty}^{n=+\infty} F_{[i(\omega + n\omega_s)]} \tag{18.13}$$

Equation (18.4) is completely equivalent to Eq. (18.11) in the time domain. Equation 18.5 is equivalent to Eq. (18.12) in the Laplace domain. Equation (18.6) is equivalent to Eq. (18.13) in the frequency domain. We will use these alternative forms of representation in several ways later.

18.3 BASIC SAMPLING THEOREM

A very important theorem of sampled-data systems is:

> To obtain dynamic information about a plant from a signal that contains components out to a frequency ω_{max}, the sampling frequency ω_s must be set at a rate greater than twice ω_{max}.
>
> $$\omega_s > 2\omega_{max} \tag{18.14}$$

Example 18.1. Suppose we have a signal that has components out to 100 radians per minute. We must set the sampling frequency at a rate greater than 200 radians per minute.

$$\omega_s > 200 \text{ rad/min}$$

$$T_s = \frac{2\pi}{\omega_s} = \frac{2\pi}{200} = 0.0314 \text{ min}$$

This basic sampling theorem has profound implications. It says that any high-frequency components in the signal (for example, 60-cycle-per-second electrical noise) can necessitate very fast sampling, even if the basic process is quite slow. It is, therefore, always recommended that signals be analog-filtered *before* they are sampled. This eliminates the unimportant high-frequency components.

To prove the sampling theorem let us consider a continuous $f_{(t)}$ that is a sine wave with a frequency ω_0 and an amplitude A_0.

$$f_{(t)} = A_0 \sin(\omega_0 t) \tag{18.15}$$

$$f_{(t)} = A_0 \frac{e^{i\omega_0 t} - e^{-i\omega_0 t}}{2i} \tag{18.16}$$

Suppose we sample this $f_{(t)}$ with an impulse sampler. The sequence of impulses $f^*_{(t)}$ coming out of the impulse sampler will be, according to Eq. (18.11),

$$f^*_{(t)} = \frac{1}{T_s} \sum_{n=-\infty}^{n=+\infty} f_{(t)} e^{in\omega_s t} = \frac{1}{T_s} \sum_{n=-\infty}^{n=+\infty} A_0 \left(\frac{e^{i\omega_0 t} - e^{-i\omega_0 t}}{2i} \right) e^{in\omega_s t}$$

$$= \frac{A_0}{2iT_s} \sum_{n=-\infty}^{n=+\infty} (e^{i(\omega_0 + n\omega_s)t} - e^{-i(\omega_0 - n\omega_s)t})$$

Now we write out a few of the terms, grouping together some of the positive and negative n terms.

$$f^*_{(t)} = \frac{A_0}{T_s} \left(\frac{e^{i\omega_0 t} - e^{-i\omega_0 t}}{2i} + \frac{e^{i(\omega_0 + \omega_s)t} - e^{-i(\omega_0 + \omega_s)t}}{2i} \right.$$

$$+ \frac{e^{i(\omega_0 + 2\omega_s)t} - e^{-i(\omega_0 + 2\omega_s)t}}{2i} + \frac{e^{i(\omega_0 - \omega_s)t} - e^{-i(\omega_0 - \omega_s)t}}{2i}$$

$$\left. + \frac{e^{i(\omega_0 - 2\omega_s)t} - e^{-i(\omega_0 - 2\omega_s)t}}{2i} + \cdots \right)$$

$$f^*_{(t)} = \frac{A_0}{T_s} \{ \sin(\omega_0 t) + \sin[(\omega_0 + \omega_s)t] + \sin[(\omega_0 + 2\omega_s)t]$$

$$+ \sin[(\omega_0 - \omega_s)t] + \sin[(\omega_0 - 2\omega_s)t] + \cdots \} \quad (18.17)$$

Thus the sampled function $f^*_{(t)}$ contains a primary component at frequency ω_0 plus an infinite number of complementary components at frequencies $\omega_0 + \omega_s$, $\omega_0 + 2\omega_s$, ..., $\omega_0 - \omega_s$, $\omega_0 - 2\omega_s$, ... The amplitude of each component is the amplitude of the original sine wave $f_{(t)}$ attenuated by $1/T_s$. The sampling process produces a signal that has components at frequencies that are multiples of the sampling frequency plus the original frequency of the continuous signal before sampling. Figure 18.5a illustrates this in terms of the frequency spectrum of the signal. This is referred to by electrical engineers as "aliasing."

Now suppose we have a continuous function $f_{(t)}$ that contains components over a range of frequencies. Figure 18.5b shows its frequency spectrum $f_{(\omega)}$. If this signal is sent through an impulse sampler, the output $f^*_{(t)}$ will have a frequency spectrum $f^*_{(\omega)}$, as shown in Fig. 18.5b. If the sampling rate or sampling frequency ω_s is high, there will be no overlap between the primary and complementary components. Therefore $f^*_{(t)}$ can be filtered to remove all the high-frequency complementary components, leaving just the primary component. This can then be related to the original continuous function. Therefore, if the sampling frequency is greater than twice the highest frequency in the original signal, the original signal can be determined from the sampled signal.

If, however, the sampling frequency is less than twice the highest frequency in the original signal, the primary and complementary components will overlap. Then the sampled signal cannot be filtered to recover the original signal and the sampled signal will incorrectly predict the steadystate gain and the dynamic components of the original signal.

Figure 18.5*b* shows that $f^*_{(\omega)}$ is a periodic function of frequency ω. Its period is ω_s.

$$f^*_{(\omega)} = f^*_{(\omega + \omega_s)} = f^*_{(\omega + 2\omega_s)} = \cdots \tag{18.18}$$

This equation can also be written

$$f^*_{(i\omega)} = f^*_{(i\omega + i\omega_s)} = f^*_{(i\omega + i2\omega_s)} = \cdots \tag{18.19}$$

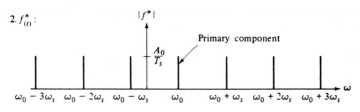

(a) Single frequency $f_{(t)} = A_0 \sin \omega_0 t$

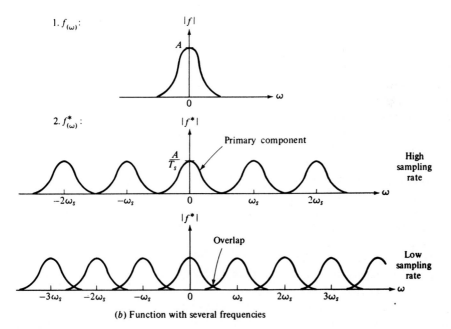

(b) Function with several frequencies

FIGURE 18.5
Frequency spectrum of continuous and sampled signals.

Going into the Laplace domain by substituting s for $i\omega$ gives

$$F^*_{(s)} = F^*_{(s+i\omega_s)} = F^*_{(s+i2\omega_s)} = \cdots \quad (18.20)$$

Thus $F^*_{(s)}$ is a periodic function of s with period $i\omega_s$. We will use this periodicity property to develop pulse transfer functions in Sec. 18.7.

18.4 z TRANSFORMATION

18.4.1 Definition

Sequences of impulses, such as the output of an impulse sampler, can be z-transformed. For a specified sampling period T_s, the z transformation of an impulse-sampled signal $f^*_{(t)}$ is defined by the equation

$$\mathscr{L}[f^*_{(t)}] \equiv f_{(0)} + f_{(T_s)}z^{-1} + f_{(2T_s)}z^{-2} + f_{(3T_s)}z^{-3} + \cdots + f_{(nT_s)}z^{-n} + \cdots \quad (18.21)$$

The notation $\mathscr{L}[\ ]$ means the z-transformation operation. The $f_{(nT_s)}$ values are the magnitudes of the continuous function $f_{(t)}$ (before impulse-sampling) at the sampling periods. We will use the notation that the z transform of $f^*_{(t)}$ is $F_{(z)}$.

$$\boxed{\mathscr{L}[f^*_{(t)}] \equiv F_{(z)} = \sum_{n=0}^{\infty} f_{(nT_s)}z^{-n}} \quad (18.22)$$

The z variable can be considered an "ordering" variable whose exponent represents the position of the impulse in the infinite sequence $f^*_{(t)}$.

Comparing Eqs. (18.5) and (18.22), we can see that the s and z variables are related by

$$\boxed{z = e^{T_s s}} \quad (18.23)$$

We will make frequent use of this very important relationship between these two complex variables.

Keep in mind the concept that we always take z transforms of impulse-sampled signals, *not* continuous functions. We will also use the notation

$$\mathscr{L}[F^*_{(s)}] = \mathscr{L}[f^*_{(t)}] \equiv F_{(z)} \quad (18.24)$$

This means exactly the same thing that is shown in Eq. (18.22). We can go directly from the time domain $f^*_{(t)}$ to the z domain. Or we can go from the time domain $f^*_{(t)}$ to the Laplace domain $F^*_{(s)}$ and on to the z domain $F_{(z)}$.

18.4.2 Derivation of z Transforms of Common Functions

Just as we did in learning Russian (Laplace transforms), we need to develop a small German vocabulary of z transforms.

A. STEP FUNCTION

$$f_{(t)} = Ku_{(t)}$$

Passing the step function through an impulse sampler gives $f^*_{(t)} = Ku_{(t)}I_{(t)}$ where $I_{(t)}$ is the sequence of unit impulses defined in Eq. (18.4). Using the definition of z transformation [Equation (18.22)] gives

$$\mathscr{L}[f^*_{(t)}] = \sum_{n=0}^{\infty} f_{(nT_s)}z^{-n} = f_{(0)} + f_{(T_s)}z^{-1} + f_{(2T_s)}z^{-2} + \cdots$$

$$= K + Kz^{-1} + Kz^{-2} + Kz^{-3} + \cdots$$

$$= K(1 + z^{-1} + z^{-2} + z^{-3} + \cdots) = K\frac{1}{1 - z^{-1}}$$

provided $|z^{-1}| < 1$. This requirement is analogous to the requirement in Laplace transformation that s be large enough so that the integral converges. Since $z^{-1} = e^{-T_s s}$, s must be large enough to keep the exponential less than 1.

The z transform of the impulse-sampled step function is

$$\mathscr{L}[Ku_{(t)}I_{(t)}] = K\frac{z}{z-1} \tag{18.25}$$

B. RAMP FUNCTIONS

$$f_{(t)} = Kt \qquad \Rightarrow \qquad f^*_{(t)} = KtI_{(t)}$$

$$\mathscr{L}[f^*_{(t)}] = \sum_{n=0}^{\infty} f_{(nT_s)}z^{-n} = f_{(0)} + f_{(T_s)}z^{-1} + f_{(2T_s)}z^{-2} + \cdots$$

$$= 0 + KT_s z^{-1} + 2KT_s z^{-2} + 3KT_s z^{-3} + \cdots$$

$$= KT_s z^{-1}(1 + 2z^{-1} + 3z^{-2} + \cdots) = \frac{KT_s z^{-1}}{(1 - z^{-1})^2}$$

for $|z^{-1}| < 1$. The z transform of the impulse sampled ramp function is

$$\mathscr{L}[KtI_{(t)}] = \frac{KT_s z}{(z-1)^2} \tag{18.26}$$

Notice the similarity between the Laplace domain and the z domain. The Laplace transformation of a constant (K) is K/s and of a ramp (Kt) is K/s^2. The z transformation of a constant is $Kz/(z-1)$ and of a ramp is $KT_s z/(z-1)^2$. Thus the s in the denominator of a Laplace transformation and the $(z-1)$ in the denominator of a z transformation behave somewhat similarly.

You should now be able to guess what the z transformation of t^2 would be. We know there would be an s^3 term in the denominator of the Laplace transformation of this function. So we can extrapolate our results to predict that there would be a $(z-1)^3$ in the denominator of the z transformation.

We will find later in this chapter that a $(z-1)$ in the denominator of a transfer function in the z domain means that there is an integrator in the system, just like the presence of an s in the denominator in the Laplace domain told us there was an integrator.

C. EXPONENTIAL

$$f_{(t)} = Ke^{-at}$$

$$F_{(z)} = \sum_{n=0}^{\infty} (Ke^{-anT_s})z^{-n}$$

$$= K[1 + (e^{-aT_s}z^{-1}) + (e^{-aT_s}z^{-1})^2 + (e^{-aT_s}z^{-1})^3 + \cdots]$$

$$= K \frac{1}{1 - e^{-aT_s}z^{-1}} \quad \text{for} \quad |e^{-aT_s}z^{-1}| < 1$$

The z transform of the impulse-sampled exponential function is

$$\mathscr{Z}[Ke^{-at}I_{(t)}] = \frac{Kz}{z - e^{-aT_s}} \tag{18.27}$$

Remember that the Laplace transformation of the exponential was $K/(s + a)$. So the $(s + a)$ term in the denominator of a Laplace transformation is similar to the $(z - e^{-aT_s})$ term in a z transformation. Both indicate an exponential function. In the s plane, we have a pole at $s = -a$. In the z plane we will find later in this chapter that we have a pole at $z = e^{-aT_s}$. So we can immediately conclude that poles on the negative real axis in the s plane "map" (to use the complex variable term) onto the positive real axis between 0 and $+1$. More about this later.

D. EXPONENTIAL MULTIPLIED BY TIME. In the Laplace domain we found that repeated roots $1/(s + a)^2$ occur when we have the exponential multiplied by time. We can guess that similar repeated roots should occur in the z domain. Let us consider a very general function:

$$f_{(t)} = \frac{K}{p!} t^p e^{-at} \tag{18.28}$$

This function can be expressed in the alternative form:

$$f_{(t)} = (-1)^p \frac{K}{p!} \frac{\partial^p(e^{-at})}{\partial a^p} \tag{18.29}$$

The z transformation of this function after impulse sampling is

$$F_{(z)} = \sum_{n=0}^{\infty} (-1)^p \frac{K}{p!} \frac{\partial^p(e^{-anT_s})}{\partial a^p} z^{-n}$$

$$= (-1)^p \frac{K}{p!} \frac{\partial^p}{\partial a^p} \left[\sum_{n=0}^{\infty} (z^{-1}e^{-aT_s})^n \right]$$

$$= (-1)^p \frac{K}{p!} \frac{\partial^p}{\partial a^p} \left(\frac{z}{z - e^{-aT_s}} \right) \tag{18.30}$$

Example 18.2. Take the case where $p = 1$.

$$\mathscr{L}[Kte^{-at}I_{(t)}] = -K\frac{\partial}{\partial a}\left(\frac{z}{z - e^{-aT_s}}\right) = \frac{KT_s e^{-aT_s}z}{(z - e^{-aT_s})^2} \tag{18.31}$$

So we get a repeated root in the z plane, just as we did in the s plane.

E. SINE

$$f_{(t)} = \sin(\omega t)$$

$$F_{(z)} = \sum_{n=0}^{\infty}\left(\frac{e^{in\omega T_s} - e^{-in\omega T_s}}{2i}\right)z^{-n}$$

$$= \frac{1}{2i}\left(\frac{1}{1 - e^{i\omega T_s}z^{-1}} - \frac{1}{1 - e^{-i\omega T_s}z^{-1}}\right)$$

$$= \frac{1}{2i}\left(\frac{e^{i\omega T_s}z^{-1} - e^{-i\omega T_s}z^{-1}}{1 + z^{-2} - e^{i\omega T_s}z^{-1} - e^{-i\omega T_s}z^{-1}}\right)$$

$$= \frac{1}{2i}\left(\frac{z^{-1}(2i)\sin(\omega T_s)}{1 + z^{-2} - 2z^{-1}\cos(\omega T_s)}\right) = \frac{z\sin(\omega T_s)}{z^2 + 1 - 2z\cos(\omega T_s)} \tag{18.32}$$

F. UNIT IMPULSE FUNCTION. By definition, the z transformation of an impulse-sampled function is

$$F_{(z)} = f_{(0)} + f_{(T_s)}z^{-1} + f_{(2T_s)}z^{-2} + \cdots$$

If $f_{(t)}$ is a unit impulse, putting it through an impulse sampler should give an $f_{(t)}^*$ that is still just a unit impulse $\delta_{(t)}$. But Eq. (18.4) says that

$$f_{(t)}^* = f_{(0)}\delta_{(t)} + f_{(T_s)}\delta_{(t-T_s)} + f_{(2T_s)}\delta_{(t-2T_s)} + \cdots$$

But if $f_{(t)}^*$ must be equal to just $\delta_{(t)}$, the term $f_{(0)}$ in the equation above must be equal to 1 and all the other terms $f_{(T_s)}, f_{(2T_s)}, \ldots$ must be equal to zero. Therefore the z transformation of the unit impulse is unity.

$$\mathscr{L}[\delta_{(t)}] = 1 \tag{18.33}$$

18.4.3 Effect of Deadtime

Deadtime in a sampled-data system is very easily handled, particularly if the deadtime D is an integer multiple of the sampling period T_s. Let us assume that

$$D = kT_s \tag{18.34}$$

where k is an integer. We will handle the case where k is not an integer later in this chapter.

Consider an arbitrary function $f_{(t-D)}$. The original function $f_{(t)}$ before the time delay is assumed to be zero for time less than zero. Running the delayed function through an impulse sampler and z-transforming give

$$\mathcal{Z}[f^*_{(t-D)}] = \sum_{n=0}^{\infty} f_{(nT_s - kT_s)}z^{-n}$$

Now we let $x \equiv n - k$

$$\mathcal{Z}[f^*_{(t-D)}] = \sum_{x=-k}^{\infty} f_{(xT_s)}z^{-x-k} = \left[\sum_{x=0}^{\infty} f_{(xT_s)}z^{-x}\right]z^{-k}$$

since $f_{(xT_s)} = 0$ for $x < 0$. The term in the brackets is just the z transform of $f^*_{(t)}$ since x is a dummy variable of summation.

$$\boxed{\mathcal{Z}[f^*_{(t-D)}] = F_{(z)}z^{-k}} \tag{18.35}$$

Therefore the deadtime transfer function in the z domain is z^{-k}.

18.4.4 z-Transform Theorems

Just as in Laplace transforms, there are several useful theorems in z transforms that are given below.

A. LINEARITY. The linearity property is easily proved from the definition of z transformation.

$$\mathcal{Z}[f^*_{1(t)} + f^*_{2(t)}] = \mathcal{Z}[f^*_{1(t)}] + \mathcal{Z}[f^*_{2(t)}] \tag{18.36}$$

B. SCALE CHANGE

$$\mathcal{Z}[e^{-at}f^*_{(t)}] = F_{(z\,e^{aT_s})} = F_{(z_1)} = \mathcal{Z}_1[f^*_{(t)}] \tag{18.37}$$

The notation $\mathcal{Z}_1[\ ]$ means z-transforming using the z_1 variable where $z_1 \equiv ze^{aT_s}$. This theorem is proved by going back to the definition of z transformation.

$$\mathcal{Z}[e^{-at}f^*_{(t)}] = \sum_{n=0}^{\infty} e^{-anT_s}f_{(nT_s)}z^{-n} = \sum_{n=0}^{\infty} f_{(nT_s)}(ze^{aT_s})^{-n}$$

Now substitute $z_1 = ze^{aT_s}$ into the equation above.

$$\mathcal{Z}[e^{-at}f^*_{(t)}] = \sum_{n=0}^{\infty} f_{(nT_s)}(z_1)^{-n} = F_{(z_1)}$$

Example 18.3. Suppose we want to take the z transformation of the function

$$f^*_{(t)} = Kte^{-at}I_{(t)}.$$

Using Eqs. (18.26) and (18.37) gives

$$\mathcal{Z}[Kte^{-at}I_{(t)}] = \mathcal{Z}_1[KtI_{(t)}] = \frac{KT_s z_1}{(z_1 - 1)^2}$$

Substituting $z_1 = ze^{aT_s}$ gives

$$\mathscr{L}[Kte^{-at}I_{(t)}] = \frac{KT_s ze^{aT_s}}{(ze^{aT_s} - 1)^2} = \frac{KT_s ze^{-aT_s}}{(z - e^{-aT_s})^2} \tag{18.38}$$

This is exactly what we found in Example 18.2.

C. FINAL-VALUE THEOREM

$$\lim_{t \to \infty} f_{(t)} = \lim_{z \to 1} \left(\frac{z - 1}{z} F_{(z)} \right) \tag{18.39}$$

To prove this theorem, let $f_{(t)}$ be the step response of an arbitrary, openloop-stable Nth-order system:

$$f_{(t)} = Ku_{(t)} + \sum_{i=1}^{N} K_i e^{-a_i t}$$

The steadystate value of $f_{(t)}$ or the limit as $f_{(t)}$ as times goes to infinity is K. Running $f_{(t)}$ through an impulse sampler and z-transforming give

$$\mathscr{L}[f^*_{(t)}] = \mathscr{L}[Ku_{(t)}I_{(t)}] + \mathscr{L}\left[\sum_{i=1}^{N} K_i e^{-a_i t}I_{(t)} \right]$$

$$F_{(z)} = \frac{Kz}{z - 1} + \sum_{i=1}^{N} K_i \frac{z}{z - e^{-a_i T_s}}$$

Multiplying both sides by $(z - 1)/z$ and letting $z \to 1$ give

$$\lim_{z \to 1} \left(\frac{z - 1}{z} F_{(z)} \right) = K = \lim_{t \to \infty} f_{(t)}$$

D. INITIAL-VALUE THEOREM

$$\lim_{t \to 0} f_{(t)} = \lim_{z \to \infty} F_{(z)} \tag{18.40}$$

The definition of the z transform of $f^*_{(t)}$ is

$$F_{(z)} = f_{(0)} + f_{(T_s)}z^{-1} + f_{(2T_s)}z^{-2} + \cdots$$

Letting z go to infinity (for $|z^{-1}| < 1$) in this equation gives $f_{(0)}$, which is the limit of $f_{(t)}$ as $t \to 0$.

18.4.5 Inversion

We sometimes want to invert from the z domain back into the time domain. The inversion will give the values of the function $f_{(t)}$ only at the sampling instants.

$$\mathscr{L}^{-1}[F_{(z)}] = f_{(nT_s)} \quad \text{for} \quad n = 0, 1, 2, 3, \ldots \tag{18.41}$$

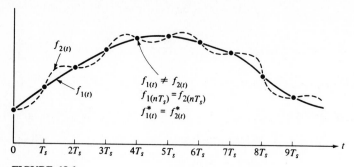

FIGURE 18.6
Continuous functions with identical values at sampling time.

The z transformation of an impulse-sampled function is unique; i.e., there is only one $F_{(z)}$ that is the z transform of a given $f^*_{(t)}$. The inverse z transform of any $F_{(z)}$ is also unique; i.e., there is only one $f_{(nT_s)}$ that corresponds to a given $F_{(z)}$.

However, keep in mind the fact that there are more than one continuous function $f_{(t)}$ that will give the same impulse-sampled function $f^*_{(t)}$. The sampled function $f^*_{(t)}$ contains information about the original continuous function $f_{(t)}$ only at the sampling times. This nonuniqueness between $f^*_{(t)}$ (and $F_{(z)}$) and $f_{(t)}$ is illustrated in Fig. 18.6. Both continuous functions $f_{1(t)}$ and $f_{2(t)}$ pass through the same points at the sampling times but are different in between the sampling instants. They would have exactly the same z transformation.

There are several ways to invert z transforms.

A. COMPLEX INTEGRATION. The inversion formula of this seldom-used method is

$$f_{(nT_s)} = \frac{1}{2\pi i} \int F_{(z)} z^{n-1} \, dz \qquad (18.42)$$

B. PARTIAL-FRACTIONS EXPANSION. The linearity theorem [Eq. (18.36)] permits us to expand the function $F_{(z)}$ into a sum of simple terms and invert each individually. This is completely analogous to Laplace-transformation inversion. Let $F_{(z)}$ be a ratio of polynomials in z, Mth-order in the numerator and Nth-order in the denominator. We factor the denominator into its N roots: $p_1, p_2, p_3, \cdots, p_N$.

$$F_{(z)} = \frac{Z_{(z)}}{(z - p_1)(z - p_2)(z - p_3) \cdots (z - p_N)} \qquad (18.43)$$

where $Z_{(z)} = M$th-order numerator polynomial. Each root p_i can be expressed in terms of the sampling period:

$$p_i = e^{-a_i T_s} \qquad (18.44)$$

Using partial-fractions expansion, Eq. (18.43) becomes

$$F_{(z)} = \frac{Az}{z - p_1} + \frac{Bz}{z - p_2} + \frac{Cz}{z - p_3} + \cdots + \frac{Wz}{z - p_N}$$

$$= \frac{Az}{z - e^{-a_1 T_s}} + \frac{Bz}{z - e^{-a_2 T_s}} + \frac{Cz}{z - e^{-a_3 T_s}} + \cdots + \frac{Wz}{z - e^{-a_N T_s}} \quad (18.45)$$

The coefficients $A, B, C, \ldots, W$ are found and $F_{(z)}$ is inverted term by term to give

$$\mathscr{L}^{-1}[F_{(z)}] = f_{(nT_s)} = Ae^{-a_1 nT_s} + Be^{-a_2 nT_s} + \cdots + We^{-a_N nT_s} \quad (18.46)$$

Example 18.4. We will show later that the closedloop response to a unit step change in load with a sampled-data proportional controller and a first-order process is

$$X_{(z)} = \frac{K_p(1 - b)z}{z^2 + z[K_c K_p(1 - b) - (1 + b)] + [b - K_c K_p(1 - b)]} \quad (18.47)$$

where $b \equiv e^{T_s/\tau_p}$
   K_c = feedback controller gain
   K_p = process steadystate gain
   τ_p = process time constant

For the numerical values of $K_p = \tau_p = K_c = 1$ and $T_s = 0.2$, $X_{(z)}$ becomes

$$X_{(z)} = \frac{0.181z}{z^2 - 1.638z + 0.638} = \frac{0.181z}{(z - 1)(z - 0.638)}$$

$$= \frac{0.5z}{z - 1} - \frac{0.5z}{z - 0.638} \quad (18.48)$$

The pole at 0.638 can be expressed:

$$0.638 = e^{-0.451} = e^{-aT_s}$$

$$X_{(z)} = \frac{0.5z}{z - 1} - \frac{0.5z}{z - e^{-0.451}}$$

Inverting each of the terms above by inspection gives

$$x_{(nT_s)} = x_{(0.2n)} = 0.5 - 0.5e^{-naT_s} = 0.5(1 - e^{-0.451n}) \quad (18.49)$$

Table 18.1 and Fig. 18.7 give the calculated results of $x_{(nT_s)}$ as a function of time.

TABLE 18.1
Results for Example 18.4

t	n	$x_{(nT_s)}$
0	0	0
0.2	1	0.181
0.4	2	0.297
0.6	3	0.372
0.8	4	0.421
1.0	5	0.451

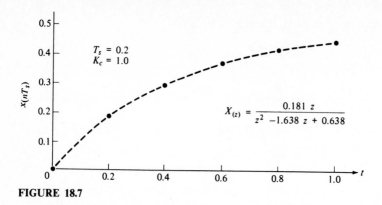

FIGURE 18.7

C. LONG DIVISION. The most interesting and most useful z-transform inversion technique is simple long division of the numerator by the denominator of $F_{(z)}$. The ease with which z transforms can be inverted by this technique is one of the reasons why z transforms are often used.

By definition,

$$F_{(z)} = f_{(0)} + f_{(T_s)} z^{-1} + f_{(2T_s)} z^{-2} + f_{(3T_s)} z^{-3} + \cdots$$

If we can get $F_{(z)}$ in terms of an infinite series of powers of z^{-1}, the coefficients in front of all the terms give the values of $f_{(nT_s)}$. The infinite series is obtained by merely dividing the numerator of $F_{(z)}$ by the denominator of $F_{(z)}$.

$$F_{(z)} = \frac{Z_{(z)}}{P_{(z)}} = f_{(0)} + f_{(T_s)} z^{-1} + f_{(2T_s)} z^{-2} + f_{(3T_s)} z^{-3} + \cdots \qquad (18.50)$$

where $Z_{(z)}$ and $P_{(z)}$ are polynomials in z. The method is easily understood by looking at a specific example.

Example 18.5. The function considered in Example 18.4 was

$$X_{(z)} = \frac{0.181z}{z^2 - 1.638z + 0.638}$$

Long division gives

$$
\begin{array}{r}
0.181z^{-1} + 0.297z^2 + 0.372z^{-3} + \\
z^2 - 1.638z + 0.638 \overline{)\,0.181z} \\
\underline{0.181z - 0.297 + 0.115z^{-1}} \\
0.297 - 0.115z^{-1} \\
\underline{0.297 - 0.487z^{-1} + 0.19z^{-2}} \\
0.372z^{-1} - 0.19z^{-2} \\
\cdots\cdots\cdots\cdots\cdots\cdots
\end{array}
$$

Therefore
$$f_{(0)} = 0$$

$$f_{(T_s)} = f_{(0.2)} = 0.181$$

$$f_{(2T_s)} = f_{(0.4)} = 0.297$$

$$f_{(3T_s)} = f_{(0.6)} = 0.372$$

$$\cdots\cdots\cdots\cdots\cdots$$

These are, of course, exactly the same results we found by partial fractions expansion in Example 18.4.

Example 18.6. If the value of K_c in Example 18.4 is changed to 15, we will show later in this chapter that $X_{(z)}$ becomes

$$X_{(z)} = \frac{0.181z}{z^2 + 0.90z - 1.9}$$

Inverting by long division gives

$$
\begin{array}{r}
0.181z^{-1} - 0.163z^{-2} + 0.491z^{-3} - 0.751z^{-4} + 1.61z^{-5} - \cdots \\
\hline
z^2 + 0.90z - 1.9 \overline{)0.181z} \\
0.181z + 0.163 - 0.344z^{-1} \\
\hline
-0.163 + 0.344z^{-1} \\
-0.163 - 0.147z^{-1} + 0.31z^{-2} \\
\hline
0.491z^{-1} - 0.31z^{-2} \\
0.491z^{-1} + 0.441z^{-2} - 0.932z^{-3} \\
\hline
-0.751z^{-2} + 0.932z^{-3} \\
-0.751z^{-2} - 0.676z^{-3} + 1.42z^{-4} \\
\hline
1.61z^{-3} - 1.42z^{-4}
\end{array}
$$

$$\cdots\cdots\cdots\cdots\cdots\cdots\cdots\cdots$$

These results are plotted in Fig. 18.8. The system is closedloop unstable for this value of gain ($K_c = 15$), as we will see in Chap. 19. Notice that this example shows that a first-order process which is controlled by a sampled-data proportional controller can be made unstable if the gain is high enough. If an analog controller had been used, the first-order process could *never* be made closedloop unstable. Thus

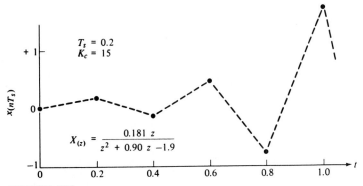

FIGURE 18.8

$f_{(nT_s)}$ from inversion of $F_{(z)}$.

there is a very important difference between continuous and discrete closedloop systems. We will expand on this in Chap. 19.

Inversion of z transforms by long division is very easily accomplished numerically by a digital computer. The FORTRAN subroutine LONGD given in Table 18.2 performs this long division. The output variable X is calculated for NT sampling times, given the coefficients $A0$, $A(1)$, $A(2)$, ..., $A(M)$ of the numerator and the coefficients $B(1)$, $B(2)$, ..., $B(N)$ of the denominator.

$$X_{(z)} = X0 + X(1)z^{-1} + X(2)z^{-2} + X(3)z^{-3} + \cdots$$
$$= \frac{A0 + A(1)z^{-1} + A(2)z^{-2} + \cdots + A(M)z^{-M}}{1 + B(1)z^{-1} + B(2)z^{-2} + \cdots + B(N)z^{-N}} \tag{18.51}$$

18.5   PULSE TRANSFER FUNCTIONS

We know how to find the z transformations of functions. Let us now turn to the problem of expressing input-output transfer-function relationships in the z domain. Figure 18.9a shows a system with samplers on the input and on the output of the process. Time, Laplace, and z-domain representations are shown. $G_{(z)}$ is called a *pulse transfer function*. It will be defined below.

A sequence of impulses $m^*_{(t)}$ comes out of the impulse sampler on the input of the process. Each of these impulses will produce a response from the process.

TABLE 18.2
Long-division subroutine

```
SUBROUTINE LONGD (A0,A,B,X0,X,N,M,NT)
DIMENSION A(10),B(10),X(100),D(10)
NMAX=N
IF(M.GT.N) NMAX=M
DO 10 I=1,NMAX
D(I)=A(I)
IF(I.GT.N) B(I)=0.
10 IF(I.GT.M) D(I)=0.
D(NMAX+1)=0.
IF(A0 .EQ.0.) GO TO 30
X0=A0
DO 20 K=1,NMAX
20 D(K)=D(K )-X0 *B(K)
X(1)=D(1)
GO TO 40
30 X0=0.
X(1)=A(1)
40 DO 100 J=2,NT
DO 50 K=1,NMAX
50 D(K)=D(K+1)-X(J-1)*B(K)
100 X(J)=D(1)
RETURN
END
```

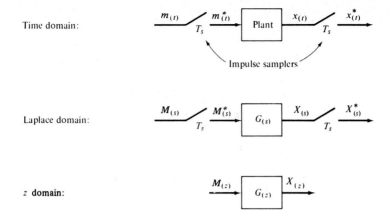

Time domain:

Laplace domain:

z domain:

(a) Representation of process

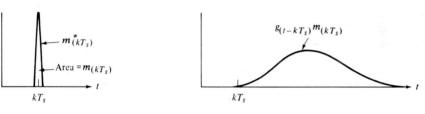

(b) Effect of kth impulse

FIGURE 18.9
Pulse transfer functions.

Consider the kth impulse $m^*_{(kT_s)}$. Its area or strength is equal to $m_{(nT_s)}$. Its effect on the continuous output of the plant $x_{(t)}$ will be

$$x_{k(t)} = g_{(t-kT_s)} m_{(kT_s)} \tag{18.52}$$

where $x_{k(t)}$ = response of the process to the kth impulse
  $g_{(t)}$ = unit impulse response of the process = $\mathcal{L}^{-1}[G_{(s)}]$

Figure 18.9b shows these functions.
  The system is linear, so the total output $x_{(t)}$ is the sum of all the x_k's.

$$x_{(t)} = \sum_{k=0}^{\infty} x_{k(t)} = \sum_{k=0}^{\infty} g_{(t-kT_s)} m_{(kT_s)} \tag{18.53}$$

At the sampling times, the value of $x_{(t)}$ is $x_{(nT_s)}$.

$$x_{(nT_s)} = \sum_{k=0}^{\infty} g_{(nT_s-kT_s)} m_{(kT_s)} \tag{18.54}$$

The continuous function $x_{(t)}$ coming out of the process is then impulse-sampled, producing a sequence of impulses $x_{(t)}^*$. If we z-transform $x_{(t)}^*$ we get

$$\mathscr{Z}[x_{(t)}^*] = \sum_{n=0}^{\infty} x_{(nT_s)} z^{-n} = X_{(z)}$$

$$X_{(z)} = \sum_{n=0}^{\infty} \left(\sum_{k=0}^{\infty} g_{(nT_s - kT_s)} m_{(kT_s)} \right) z^{-n} \tag{18.55}$$

Letting $p = n - k$ and remembering that $g_{(t)} = 0$ for $t < 0$ give

$$X_{(z)} = \sum_{p=0}^{\infty} \sum_{k=0}^{\infty} g_{(pT_s)} m_{(kT_s)} z^{-(p+k)}$$

$$= \left(\sum_{p=0}^{\infty} g_{(pT_s)} z^{-p} \right) \left(\sum_{k=0}^{\infty} m_{(kT_s)} z^{-k} \right) \tag{18.56}$$

$$X_{(z)} = G_{(z)} M_{(z)} \tag{18.57}$$

The pulse transfer function $G_{(z)}$ is defined as the first term in Eq. (18.56).

$$G_{(z)} \equiv \sum_{p=0}^{\infty} g_{(pT_s)} z^{-p} \tag{18.58}$$

Defining $G_{(z)}$ in this way permits us to use transfer functions in the z domain [Eq. (18.57)] just as we use transfer functions in the Laplace domain. $G_{(z)}$ is the z transform of the impulse-sampled response $g_{(t)}^*$ of the process to a unit impulse function $\delta_{(t)}$. In z-transforming functions, we used the notation $\mathscr{Z}[f_{(t)}^*] = \mathscr{Z}[F_{(s)}^*] = F_{(z)}$. In handling pulse transfer functions, we will use similar notation.

$$\mathscr{Z}[g_{(t)}^*] = \mathscr{Z}[G_{(s)}^*] = G_{(z)} \tag{18.59}$$

where $G_{(s)}^*$ is the Laplace transform of the impulse-sampled response $g_{(t)}^*$ of the process to a unit impulse input.

$$G_{(s)}^* = \mathscr{L}[g_{(t)}^*] \tag{18.60}$$

$G_{(s)}^*$ can also be expressed, using Eq. (18.12), as

$$G_{(s)}^* = \frac{1}{T_s} \sum_{n=-\infty}^{n=+\infty} G_{(s + in\omega_s)} \tag{18.61}$$

We will show how these pulse transfer functions are applied to openloop and closedloop systems in Sec. 18.7.

18.6 HOLD DEVICES

A hold device is always needed in a sampled-data process control system. The hold converts the sequence of impulses of an impulse-sampled function $f_{(t)}^*$ into a continuous (usually staircase) function $f_{H(t)}$. There are several types of mathematical holds, but the only one that is of any practical interest is called a *zero-order hold*. This type of hold generates the stair-step function described above.

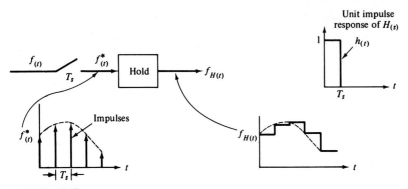

FIGURE 18.10
Zero-order hold.

The hold must convert an impulse $f^*_{(t)}$ of area or strength $f_{(nT_s)}$ at time $t = nT_s$ into a square *pulse* (not an impulse) of height $f_{(nT_s)}$ and width T_s. See Fig. 18.10. Let the unit *impulse* response of the hold be defined as $h_{(t)}$. If the hold is to do what we want it to do (i.e., convert an impulse into a step up and then a step down after T_s minutes), its unit impulse response must be

$$h_{(t)} = u_{(t)} - u_{(t-T_s)} \tag{18.62}$$

where $u_{(t)}$ is the unit step function. Therefore the Laplace-domain transfer function $H_{(s)}$ of a zero-order hold is

$$H_{(s)} = \mathcal{L}[h_{(t)}] = \mathcal{L}[u_{(t)} - u_{(t-T_s)}] = \frac{1}{s} - \frac{e^{-T_s s}}{s}$$

$$\boxed{H_{(s)} = \frac{1 - e^{-T_s s}}{s}} \tag{18.63}$$

18.7 OPENLOOP AND CLOSEDLOOP SYSTEMS

We are now ready to use the concepts of impulse-sampled functions, pulse transfer functions, and holds to study the dynamics of sampled-data systems.

18.7.1 Openloop Systems

Consider the sampled-data system shown in Fig. 18.11*a* in the Laplace domain. The input enters through an impulse sampler. The continuous output of the process $X_{(s)}$ is

$$X_{(s)} = G_{(s)} M^*_{(s)} \tag{18.64}$$

$X_{(s)}$ is then impulse-sampled to give $X^*_{(s)}$. Equation (18.12) says that $X^*_{(s)}$ is

$$X^*_{(s)} = \frac{1}{T_s} \sum_{n=-\infty}^{n=+\infty} X_{(s+in\omega_s)}$$

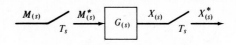

(a) Single element

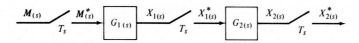

(b) Series elements with intermediate sampler

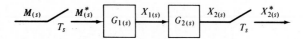

(c) Series elements that are continuous

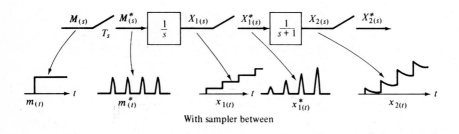

With sampler between

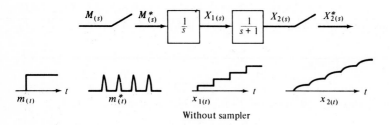

Without sampler

(d) System of Example 18.7

FIGURE 18.11
Openloop sampled-data systems.

Substituting for $X_{(s+in\omega_s)}$, using Eq. (18.64) gives

$$X^*_{(s)} = \frac{1}{T_s} \sum_{n=-\infty}^{n=+\infty} G_{(s+in\omega_s)} M^*_{(s+in\omega_s)} \tag{18.65}$$

We showed [Eq. (18.20)] that the Laplace transform of an impulse-sampled function is periodic.

$$M^*_{(s)} = M^*_{(s+i\omega_s)} = M^*_{(s-i\omega_s)} = M^*_{(s+i2\omega_s)} = \cdots \tag{18.66}$$

Therefore the $M^*_{(s+in\omega_s)}$ terms can be factored out of the summation in Eq. (18.65) to give

$$X^*_{(s)} = \left(\frac{1}{T_s} \sum_{n=-\infty}^{n=+\infty} G_{(s+in\omega_s)} \right) M^*_{(s)}$$

The term in the parentheses is $G^*_{(s)}$ according to Eq. (18.61), and therefore the output of the process in the Laplace domain is

$$X^*_{(s)} = G^*_{(s)} M^*_{(s)} \tag{18.67}$$

By z-transforming this equation, using Eq. (18.59), the output in the z domain is

$$X_{(z)} = G_{(z)} M_{(z)} \tag{18.68}$$

Now consider the system shown in Fig. 18.11b where there are two elements separated by a sampler. The continuous output $X_{1(s)}$ is

$$X_{1(s)} = G_{1(s)} M^*_{(s)}$$

When $X_{1(s)}$ goes through the impulse sampler it becomes $X^*_{1(s)}$, which can be expressed [see Eq. (18.67)]:

$$X^*_{1(s)} = \frac{1}{T_s} \sum_{n=-\infty}^{n=+\infty} G_{1(s+in\omega_s)} M^*_{(s+in\omega_s)} = G^*_{1(s)} M^*_{(s)} \tag{18.69}$$

The continuous function $X_{2(s)}$ is

$$X_{2(s)} = G_{2(s)} X^*_{1(s)}$$

The impulse-sampled $X^*_{2(s)}$ is

$$X^*_{2(s)} = G^*_{2(s)} X^*_{1(s)} = G^*_{2(s)} (G^*_{1(s)} M^*_{(s)})$$
$$X^*_{2(s)} = G^*_{1(s)} G^*_{2(s)} M^*_{(s)} \tag{18.70}$$

In the z domain, the equation above becomes

$$X_{2(z)} = G_{1(z)} G_{2(z)} M_{(z)} \tag{18.71}$$

Thus the overall transfer function of the process can be expressed as a product of the two individual pulse transfer functions if there is an impulse sampler between the elements.

Consider now the system shown in Fig. 18.11c where the two continuous elements $G_{1(s)}$ and $G_{2(s)}$ do *not* have a sampler between them. The continuous output $X_{2(s)}$ is

$$X_{2(s)} = G_{2(s)} X_{1(s)} = G_{2(s)} G_{1(s)} M^*_{(s)} \tag{18.72}$$

Sampling the output gives

$$X^*_{2(s)} = \frac{1}{T_s} \sum_{n=-\infty}^{n=+\infty} G_{1(s+in\omega_s)} G_{2(s+in\omega_s)} M^*_{(s+in\omega_s)}$$

$$= \left(\frac{1}{T_s} \sum_{n=-\infty}^{n=+\infty} G_{1(s+in\omega_s)} G_{2(s+in\omega_s)} \right) M^*_{(s)} \tag{18.73}$$

The term in parentheses is the Laplace transformation of the impulse-sampled response of the *total combined* process to a unit impulse input. We will call this $(G_1 G_2)^*_{(s)}$ in the Laplace domain and $(G_1 G_2)_{(z)}$ in the z domain.

$$X^*_{2(s)} = (G_1 G_2)^*_{(s)} M^*_{(s)} \tag{18.74}$$

$$X_{2(z)} = (G_1 G_2)_{(z)} M_{(z)} \tag{18.75}$$

Equations (18.70) and (18.71) look somewhat like Eqs. (18.74) and (18.75), but they are not at all the same. The two processes are physically different: one has a sampler between the G_1 and G_2 elements; the other does not have a sampler between them.

$$(G_1 G_2)^*_{(s)} \neq G^*_{1(s)} G^*_{2(s)}$$

$$(G_1 G_2)_{(z)} \neq G_{1(z)} G_{2(z)} \tag{18.76}$$

Let us take a specific example to illustrate the difference between these two systems.

Example 18.7. Suppose the system has two elements shown in Fig. 18.11d.

$$G_{1(s)} = \frac{1}{s} \qquad G_{2(s)} = \frac{1}{s+1} \tag{18.77}$$

With an impulse sampler between the elements, the overall system transfer function is, from Eq. (18.71),

$$G_{1(z)} G_{2(z)} = \mathscr{Z}[g^*_{1(t)}] \mathscr{Z}[g^*_{2(t)}] = \mathscr{Z}\left[I_{(t)} \mathscr{L}^{-1}\left(\frac{1}{s}\right) \right] \mathscr{Z}\left[I_{(t)} \mathscr{L}^{-1}\left(\frac{1}{s+1}\right) \right]$$

$$= \mathscr{Z}[I_{(t)} u_{(t)}] \mathscr{Z}[I_{(t)} e^{-t}] = \left(\frac{z}{z-1}\right)\left(\frac{z}{z-e^{-T_s}}\right) \tag{18.78}$$

In the calculation above, we went through the time domain, getting $g_{(t)}$ by inverting $G_{(s)}$ and then z-transforming $g^*_{(t)}$. The operation can be represented more concisely by going directly from the Laplace domain to the z domain.

$$G_{1(z)} G_{2(z)} = \mathscr{Z}[G^*_{1(s)}] \mathscr{Z}[G^*_{2(s)}] = \mathscr{Z}\left[\frac{1}{s}\right] \mathscr{Z}\left[\frac{1}{s+1}\right] \tag{18.79}$$

This equation is a shorthand expression for Eq. (18.78). The inversion to the impulse response $g_{(t)}$ and the impulse sampling to get $g_{(t)}^*$ is implied in the notation $\mathscr{Z}[1/s]$ and $\mathscr{Z}[1/(s+1)]$.

$$G_{1(z)} G_{2(z)} = \mathscr{Z}\left[\frac{1}{s}\right] \mathscr{Z}\left[\frac{1}{s+1}\right] = \left(\frac{z}{z-1}\right)\left(\frac{z}{z-e^{-T_s}}\right) \tag{18.80}$$

The responses of $x_{(t)}^*$, $x_{1(t)}$, and $x_{2(t)}$ to a unit step change in $m_{(t)}$ are sketched in Fig. 18.11d.

Without a sampler between the elements, the overall system transfer function is [Eq. (18.75)]

$$(G_1 G_2)_{(z)} = \mathscr{Z}\left[I_{(t)} \mathcal{L}^{-1}\left(\frac{1}{s(s+1)}\right)\right] = \mathscr{Z}\left[I_{(t)} \mathcal{L}^{-1}\left(\frac{1}{s} - \frac{1}{s+1}\right)\right]$$

$$= \mathscr{Z}[I_{(t)} u_{(t)} - I_{(t)} e^{-t}] = \frac{z}{z-1} - \frac{z}{z-e^{-T_s}} = \frac{z(z-e^{-T_s})}{(z-1)(z-e^{-T_s})} \tag{18.81}$$

Using the shorthand notation,

$$(G_1 G_2)_{(z)} = \mathscr{Z}\left[\frac{1}{s(s+1)}\right] = \mathscr{Z}\left[\frac{1}{s} - \frac{1}{s+1}\right] = \frac{z}{z-1} - \frac{z}{z-e^{-T_s}}$$

$$= \frac{z(1-e^{-T_s})}{(z-1)(z-e^{-T_s})}$$

From now on we will use the shorthand, Laplace-domain notation, but keep in mind what is implied in its use.

Notice that Eq. (18.78) is not equal to Eq. (18.81). The responses of the two systems $x_{2(t)}$'s are not the same, as shown in Fig. 18.11d, because the systems are physically different.

18.7.2 Closedloop Systems

The methods used in the previous section can be easily extended to closedloop sampled-data systems. Let us first consider a chromatograph control loop, as shown in Fig. 18.12a. The output of the process $X_{(s)}$ is

$$X_{(s)} = G_{L(s)} L_{(s)} + G_{M(s)} B_{(s)}(X_{(s)}^{\text{set}} - X_{H(s)}) \tag{18.82}$$

The output of the hold $X_{H(s)}$ is

$$X_{H(s)} = H_{(s)} X_{c(s)}^* \tag{18.83}$$

The output of the chromatograph column (the deadtime element) is

$$X_{c(s)} = e^{-D_c s} X_{(s)}^* \tag{18.84}$$

Substituting Eq. (18.83) into Eq. (18.82) and sampling $X_{(s)}$ give

$$X_{(s)}^* = [G_L L]_{(s)}^* + [G_M B X^{\text{set}}]_{(s)}^* - [G_M BH]_{(s)}^* X_{c(s)}^* \tag{18.85}$$

z-transforming gives

$$X_{(z)} = [G_L L]_{(z)} + [G_M B X^{\text{set}}]_{(z)} - [G_M BH]_{(z)} X_{c(z)} \tag{18.86}$$

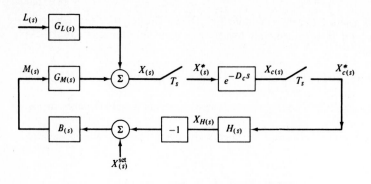

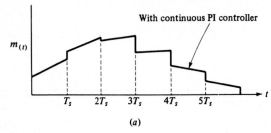

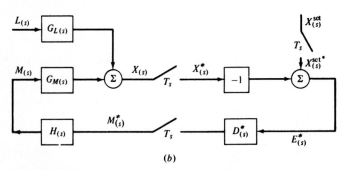

FIGURE 18.12

Closedloop systems. (a) Chromatograph loop; (b) DDC loop.

Let us assume that the chromatograph deadtime is an integer multiple of the sampling period.

$$D_c = kT_s$$

$$X_{c(s)} = e^{-kT_s s} X^*_{(s)}$$

z-transforming $X^*_{c(s)}$ gives [using Eq. (18.35)]

$$X_{c(z)} = \mathscr{Z}[e^{-kT_s s} X^*_{(s)}] = z^{-k} \mathscr{Z}[X^*_{(s)}] = z^{-k} X_{(z)} \qquad (18.87)$$

Combining this with Eq. (18.86) gives

$$X_{(z)} = [G_L L]_{(z)} + [G_M B X^{\text{set}}]_{(z)} - [G_M BH]_{(z)} z^{-k} X_{(z)}$$

$$X_{(z)} = \frac{[G_L L]_{(z)} + [G_M B X^{\text{set}}]_{(z)}}{1 + z^{-k}[G_M BH]_{(z)}} \qquad (18.88)$$

In this system explicit input-output transfer-function relationships are not obtained. The $(G_L L)_{(z)}$ and $(G_M BX^{\text{set}})_{(z)}$ terms in the numerator of Eq. (18.88) are z transforms of functions. The inputs L and X^{set} cannot be separated out because they are directly connected to the output $X_{(s)}$ by continuous elements.

A specific example will illustrate how the output of the closedloop system can be obtained in the z domain.

Example 18.8. Consider a first-order process with the transfer functions of Fig. 18.12a:

$$G_{M(s)} = G_{L(s)} = \frac{K_p}{\tau_p s + 1} \tag{18.89}$$

A zero-order hold and a proportional controller are used.

$$B_{(s)} = K_c \tag{18.90}$$

$$H_{(s)} = \frac{1 - e^{-T_s s}}{s} \tag{18.91}$$

We want to find the response of the closedloop system to a unit step change in load.

$$L_{(s)} = \frac{1}{s} \tag{18.92}$$

There is no change in setpoint, so $X^{\text{set}}_{(s)}$ is zero. We need to find $(G_L L)_{(z)}$ and $(G_M BX^{\text{set}})_{(z)}$ to plug into Eq. (18.88)

$$\mathscr{Z}[G_L L] = \mathscr{Z}\left[\frac{K_p}{\tau_p s + 1} \frac{1}{s}\right] = \mathscr{Z}\left[\frac{K_p}{s} - \frac{K_p}{s + 1/\tau_p}\right]$$

$$= K_p\left(\frac{z}{z - 1} - \frac{z}{z - b}\right) = \frac{K_p(1 - b)z}{(z - 1)(z - b)} \tag{18.93}$$

where b is defined

$$b \equiv e^{-T_s/\tau_p} \tag{18.94}$$

$$\mathscr{Z}[G_M BH] = \mathscr{Z}\left[K_c \frac{K_p}{\tau_p s + 1} \frac{1 - e^{-T_s s}}{s}\right] = K_c K_p \mathscr{Z}\left[\frac{1 - e^{-T_s s}}{s(\tau_p s + 1)}\right]$$

$$= K_c K_p(1 - z^{-1})\mathscr{Z}\left[\frac{1}{s(\tau_p s + 1)}\right]$$

since

$$\mathscr{Z}[(1 - e^{-T_s s})F^*_{(s)}] = \mathscr{Z}[F^*_{(s)}] - \mathscr{Z}[e^{-T_s s}F^*_{(s)}]$$

$$= F_{(z)} - z^{-1}F_{(z)} = (1 - z^{-1})F_{(z)}$$

Continuing with our evaluation of $(G_M BH)_{(z)}$,

$$(G_M BH)_{(z)} = \mathscr{Z}[G_M BH] = K_c K_p\left(1 - \frac{1}{z}\right)\frac{(1 - b)z}{(z - 1)(z - b)}$$

$$= \frac{K_c K_p(1 - b)}{z - b} \tag{18.95}$$

Therefore $X_{(z)}$ becomes

$$X_{(z)} = \frac{(G_L L)_{(z)}}{1 + z^{-k}(G_M BH)_{(z)}} = \frac{\dfrac{K_p(1-b)z}{(z-1)(z-b)}}{1 + z^{-k}\dfrac{K_c K_p(1-b)}{z-b}}$$

$$= \frac{K_p(1-b)z}{(z-b)(z-1) + z^{-k}K_c K_p(1-b)(z-1)}$$

$$= \frac{K_p(1-b)z}{z^2 + z[K_c K_p(1-b)z^{-k} - (1+b)] + [b - z^{-k}K_c K_p(1-b)]} \qquad (18.96)$$

When there is no deadtime ($k = 0$), this equation reduces to Eq. (18.47) used in Example 18.4. When the deadtime is equal to the sampling period ($k = 1$), Eq. (18.96) becomes

$$X_{(z)} = \frac{K_p(1-b)z^2}{z^3 - (1+b)z^2 + [K_c K_p(1-b) + b]z - K_c K_p(1-b)} \qquad (18.97)$$

When the deadtime is equal to two sampling periods ($k = 2$), $X_{(z)}$ becomes

$$X_{(z)} = \frac{K_p(1-b)z^3}{z^4 - (1+b)z^3 + [K_c K_p(1-b) + b]z^2 - K_c K_p(1-b)} \qquad (18.98)$$

The order of the system in the z domain (the highest power of z in the denominator) increases with increasing deadtime or k.

Example 18.9. Let us take the same system as above but use a PI analog controller.

$$B_{(s)} = K_c\left(1 + \frac{1}{\tau_I s}\right) \qquad (18.99)$$

The chromatograph deadtime is assumed equal to the sampling period, so $D_c = T_s$ and $k = 1$. The disturbance is again a unit step change in load so $(G_L L)_{(z)}$ is the same as used in Example 18.8. However, we must find the new $(G_M BH)_{(z)}$.

$$\mathscr{Z}[G_M BH] = \mathscr{Z}\left[K_c \frac{s + 1/\tau_I}{s} \frac{K_p}{\tau_p s + 1} \frac{1 - e^{-T_s s}}{s}\right]$$

$$= K_c K_p(1 - z^{-1})\mathscr{Z}\left[\frac{(s + 1/\tau_I)(1/\tau_p)}{s^2(s + 1/\tau_p)}\right]$$

$$= K_c K_p \frac{z-1}{z} \mathscr{Z}\left[\frac{1/\tau_I}{s^2} + \frac{1 - \tau_p/\tau_I}{s} - \frac{1 - \tau_p/\tau_I}{s + 1/\tau_p}\right]$$

$$= K_c K_p \frac{z-1}{z}\left[\frac{(T_s/\tau_I)z}{(z-1)^2} - \frac{(\tau_p/\tau_I - 1)z}{z-1} + \frac{(\tau_p/\tau_I - 1)z}{z - e^{-T_s/\tau_p}}\right]$$

$$= K_c K_p \frac{(T_s/\tau_I)(z - b) + (\tau_p/\tau_I - 1)(b - 1)(z - 1)}{(z-1)(z-b)} \qquad (18.100)$$

$$X_{(z)} = \frac{(G_L L)_{(z)}}{1 + z^{-1}(G_M BH)_{(z)}} = \frac{\dfrac{K_p(1 - b)z}{(z - 1)(z - b)}}{1 + z^{-1}K_c K_p \dfrac{(T_s/\tau_I)(z - b) + (\tau_p/\tau_I - 1)(b - 1)(z - 1)}{(z - 1)(z - b)}}$$

$$(18.101)$$

$$X_{(z)} = \frac{K_p(1 - b)z^2}{z^3 - (1 + b)z^2 + z(b_1) + b_2} \qquad (18.102)$$

where
$$b_1 \equiv b - K_c K_p(\tau_p/\tau_I - 1)(1 - b) + K_c K_p(T_s/\tau_I)$$
$$b_2 \equiv K_c K_p[(\tau_p/\tau_I - 1)(1 - b) - bT_s/\tau_I]$$

The order of the system is increased by 1 when a PI controller is used instead of a proportional controller.

The feedback controller is continuous, and it sees a constant error between samples. Therefore the manipulated variable $m_{(t)}$ is ramped up or down during the sampling period by the integral action, as sketched in Fig. 18.12a.

For our second sampled-data closedloop system, let us consider the direct digital control (DDC) loop sketched in Fig. 18.12b. The equations describing the system are

$$X_{(s)} = G_{L(s)} L_{(s)} + H_{(s)} G_{M(s)} M^*_{(s)} \qquad (18.103)$$

$$M_{(s)} = D^*_{(s)}(X^{\text{set}*}_{(s)} - X^*_{(s)}) \qquad (18.104)$$

Sampling $X_{(s)}$ and $M_{(s)}$ gives

$$X^*_{(s)} = [G_L L]^*_{(s)} + [HG_M]^*_{(s)} M^*_{(s)}$$

$$M^*_{(s)} = D^*_{(s)} X^{\text{set}*}_{(s)} - D^*_{(s)} X^*_{(s)}$$

z-transforming and combining give

$$X_{(z)} = (G_L L)_{(z)} + (HG_M)_{(z)}(D_{(z)} X^{\text{set}}_{(z)} - D_{(z)} X_{(z)})$$

$$X_{(z)} = \frac{(G_L L)_{(z)} + (HG_M)_{(z)} D_{(z)} X^{\text{set}}_{(z)}}{1 + (HG_M)_{(z)} D_{(z)}} \qquad (18.105)$$

In this system we obtain an explicit input-output transfer-function relationship between $X_{(z)}$ and $X^{\text{set}}_{(z)}$:

$$\frac{X_{(z)}}{X^{\text{set}}_{(z)}} = \frac{(HG_M)_{(z)} D_{(z)}}{1 + (HG_M)_{(z)} D_{(z)}} \qquad (18.106)$$

For proportional controllers, the continuous $B_{(s)}$ and the sampled-data $D_{(z)}$ give the same results since

$$(G_M BH)_{(z)} = K_c(G_M H)_{(z)}$$

$$(HG_M)_{(z)} D_{(z)} = K_c(HG_M)_{(z)} \qquad (18.107)$$

Example 18.10. One form of $D_{(z)}$ that gives approximately the same response as a continuous PI controller is (as we will show in Chap. 19)

$$D_{(z)} = \frac{K_c}{\alpha} \frac{z - \alpha}{z - 1} \tag{18.108}$$

where

$$\alpha \equiv \frac{\tau_I}{T_s + \tau_I} \tag{18.109}$$

For the same process and disturbance of Example 18.8 with no deadtime, the output $X_{(z)}$ becomes

$$X_{(z)} = \frac{(G_L L)_{(z)}}{1 + (HG_M)_{(z)} D_{(z)}} = \frac{\dfrac{K_p(1 - b)z}{(z - 1)(z - b)}}{1 + \dfrac{K_p(1 - b)}{z - b} \dfrac{K_c}{\alpha} \dfrac{z - \alpha}{z - 1}}$$

$$= \frac{K_p(1 - b)z}{z^2 + z\left[\dfrac{K_c K_p(1 - b)}{\alpha} - (1 + b)\right] + [b - K_c K_p(1 - b)]} \tag{18.110}$$

18.8 DISCRETE APPROXIMATION OF CONTINUOUS TRANSFER FUNCTIONS

Sometimes it is desired to have a discrete device (a digital computer) respond in a manner that is similar to a continuous (analog) device. For example, we may want to do some filtering of signals as they enter the computer, so developing a digital filter that approximates what an analog filter would do might be convenient. The objective is to convert a transfer function $G_{(s)}$ in the Laplace domain into an equivalent transfer function $G_{(z)}$ in the z domain.

The basic strategy to accomplish this goal is to develop discrete approximations for integration (like rectangular or trapezoidal) in terms of functions of z. These are called *z forms*, and they are different than z transforms. Then we substitute the z form for $1/s$ in $G_{(s)}$. So the first thing we must do is develop discrete approximations for integration ($1/s$).

Let the input to a pure integrator be $m_{(t)}$ and the output be $x_{(t)}$. In the Laplace domain, x and m are related by the transfer function $G_{(s)}$.

$$G_{(s)} = \frac{X_{(s)}}{M_{(s)}} = \frac{1}{s} \tag{18.111}$$

In the time domain

$$x_{(nT_s)} = \int_0^{nT_s} m_{(t)} \, dt = x_{[(n-1)T_s]} + \int_{(n-1)T_s}^{nT_s} m_{(t)} \, dt \tag{18.112}$$

A. RECTANGULAR. Rectangular integration uses the present value of the input times the sampling period to approximate the integral of the input.

$$x_{(nT_s)} \simeq x_{[(n-1)T_s]} + m_{(nT_s)} T_s$$

In the z domain

$$X_{(z)} = z^{-1} X_{(z)} + T_s M_{(z)}$$

Rearranging to get the input-output transfer function gives

$$\frac{X_{(z)}}{M_{(z)}} = \frac{T_s}{1 - z^{-1}} \qquad (18.113)$$

Example 18.11. Suppose we want to find a discrete approximation for a first-order lag. This would be called a first-order *digital filter*. Let $x_{(t)}$ be the output of the filter and $m_{(t)}$ be the input.

$$G_{(s)} = \frac{1}{\tau s + 1} = \frac{X_{(s)}}{M_{(s)}} \qquad (18.114)$$

Cross-multiplying gives

$$M_{(s)} = \tau s X_{(s)} + X_{(s)}$$

Dividing by s gives an equation that contains $1/s$ terms.

$$\frac{1}{s} M_{(s)} = \tau X_{(s)} + \frac{1}{s} X_{(s)}$$

Then we go into the z domain by substituting for $1/s$ using Eq. (18.113).

$$\left(\frac{T_s}{1 - z^{-1}}\right) M_{(z)} = \tau X_{(z)} + \left(\frac{T_s}{1 - z^{-1}}\right) X_{(z)}$$

Rearranging to get the transfer gives

$$\frac{X_{(z)}}{M_{(z)}} = \frac{(T_s/\tau)z}{(T_s/\tau + 1)z - 1} \qquad (18.115)$$

If we want to convert this into difference-equation form so that it can be programmed into a control computer, we cross-multiply Eq. (18.115).

$$X_{(z)}\left(\frac{T_s}{\tau} + 1\right)z - X_{(z)} = \left(\frac{T_s}{\tau}\right)z M_{(z)}$$

Dividing by z gives

$$X_{(z)}\left(\frac{T_s}{\tau} + 1\right) - z^{-1}X_{(z)} = \left(\frac{T_s}{\tau}\right)M_{(z)} \qquad (18.116)$$

Converting to difference equation form yields

$$x_n\left(\frac{T_s}{\tau} + 1\right) - x_{n-1} = \left(\frac{T_s}{\tau}\right)m_n \qquad (18.117)$$

where x_n = calculated value of the output of the filter at the present sampling time
m_n = the present value of the input of the filter
x_{n-1} = value of the output of the filter at the previous sampling time since $z^{-1}X_{(z)}$ represents the signal delayed by one sampling period

Rearranging gives the computer algorithm for a first-order filter using the rectangular integration approximation.

$$x_n = \left(\frac{\tau}{T_s + \tau}\right)x_{n-1} + \left(\frac{T_s}{T_s + \tau}\right)m_n \qquad (18.118)$$

Note that if the filter time constant τ is made very large, the present value of the input (m_n) has little effect on the output. But as τ is made very small, the output changes directly with the input.

B. TRAPEZOIDAL INTEGRATION. Trapezoidal integration draws a straight line between the values of the input.

$$x_{(nT_s)} \simeq x_{[(n-1)T_s]} + [m_{(nT_s)} + m_{[(n-1)T_s]}]\frac{T_s}{2}$$

In the z domain

$$X_{(z)} = z^{-1}X_{(z)} + [M_{(z)} + z^{-1}M_{(z)}]\frac{T_s}{2}$$

Rearranging to get the input-output transfer function gives

$$\frac{X_{(z)}}{M_{(z)}} = \frac{T_s}{2}\frac{1 + z^{-1}}{1 - z^{-1}} \qquad (18.119)$$

Example 18.12. Repeating Example 18.11 for a digital filter, using trapezoidal approximation gives

$$G_{(s)} = \frac{1}{\tau s + 1} = \frac{X_{(s)}}{M_{(s)}}$$

$$\frac{1}{s}M_{(s)} = \tau X_{(s)} + \frac{1}{s}X_{(s)}$$

Substituting for the $1/s$ terms, using Eq. (18.119), gives

$$\frac{T_s}{2}\frac{1 + z^{-1}}{1 - z^{-1}}M_{(z)} = \tau X_{(z)} + \frac{T_s}{2}\frac{1 + z^{-1}}{1 - z^{-1}}X_{(z)}$$

Rearranging gives the transfer function for the digital filter using trapezoidal approximation

$$\frac{X_{(z)}}{M_{(z)}} = \frac{a_0 + a_1 z^{-1}}{1 + b_1 z^{-1}} \qquad (18.120)$$

where $a_0 = a_1 = \dfrac{T_s}{2\tau + T_s}$

$$b_1 = \frac{T_s - 2\tau}{2\tau + T_s} \qquad (18.121)$$

This same approach will be used in Chap. 19 to develop approximations of analog PI controllers.

18.9 MODIFIED z TRANSFORMS

When the deadtime in a process is an integer multiple of the sampling period, the function e^{-Ds} in the Laplace domain converts easily into z^{-k} in the z domain, where deadtime $D = kT_s$. When the deadtime is *not* an integer multiple of the sampling period, we can use modified z transforms to handle the situation.

First we must define modified z transformation. This is a dialect of our German language, like Pennsylvania Dutch! So we will learn a few words of "Dutch." Then we will give an example which illustrates how modified z transforms can be used to handle noninteger values of deadtime.

18.9.1 Definition

Figure 18.13 sketches a sampled-data system of input $m_{(t)}$ and output $x_{(t)}$. Knowing $X_{(z)}$ gives us knowledge about $x_{(t)}$ only at the sampling instants. Therefore we only know $x_{(nT_s)}$.

Suppose we add the fictitious deadtime element and impulse sampler sketched with dashed lines in Fig. 18.13. Then by letting the deadtime vary continuously between 0 and 1, we could obtain a description of $x_{(t)}$ at any time in between the sampling periods.

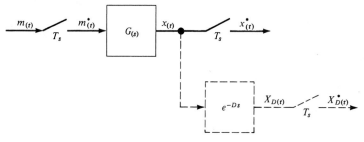

FIGURE 18.13
Modified z transform.

The output of the deadtime is defined as $x_{D(t)}$.

$$x_{D(t)} = x_{(t-D)} \tag{18.122}$$

Impulse sampling this signal gives $x_{D(t)}^*$.

$$x_{D(t)}^* = x_{(-D)}\delta_{(t)} + x_{(T_s - D)}\delta_{(t - T_s)} + x_{(2T_s - D)}\delta_{(t - 2T_s)} + \cdots \tag{18.123}$$

Now we let the deadtime vary continuously between 0 and T_s by defining a new variable m that varies between 1 and 0.

$$D \equiv (1 - m)T_s \tag{18.124}$$

This makes x_D a continuous function of both t and m. Equation (18.123) can be written

$$x_{D(t, m)}^* = \sum_{n=0}^{\infty} x_{[nT_s - (1-m)T_s]}\delta_{(t - nT_s)} \tag{18.125}$$

z-transforming gives

$$X_{D(z, m)} = z^{-1} \sum_{n=0}^{\infty} x_{[nT_s + mT_s]} z^{-n} \tag{18.126}$$

We define the right-hand side of Eq. (18.126) as the "modified z transformation" of the original function $x_{(t)}$: $X_{(z, m)}$. The defining equation for any function $f_{(t)}$ is

$$F_{(z, m)} \equiv z^{-1} \sum_{n=0}^{\infty} f_{[nT_s + mT_s]} z^{-n} \equiv \mathscr{L}_m[F_{(s)}^*] \tag{18.127}$$

where the symbol "$\mathscr{L}_m$" means *modified z transformation*.

The most important use of modified z transforms is to handle deadtimes that are not integer multiples of the sampling period. From the development above, it should be clear that the regular z transformation of a function with a deadtime that is a fraction of the sampling period is just the modified z-transformation of the function with no deadtime.

$$\boxed{\mathscr{L}[e^{-(1-m)T_s}F_{(s)}^*] = \mathscr{L}_m[F_{(s)}^*]} \tag{18.128}$$

This is the relationship that we will use later in an example.

18.9.2 Modified z Transforms of Common Functions

We need to derive the modified z transformation of a few simple functions.

A. STEP

$$f_{(t)} = Ku_{(t)} \quad \Rightarrow \quad f_{(nT_s + mT_s)} = K$$

Using the definition of the modified z transformation given in Eq. (18.127) gives

$$\mathscr{L}_m[Ku_{(t)}I_{(t)}] \equiv z^{-1} \sum_{n=0}^{\infty} f_{[nT_s+mT_s]} z^{-n}$$

$$= z^{-1} \sum_{n=0}^{\infty} Kz^{-n} = Kz^{-1}[1 + z^{-1} + z^{-2} + z^{-3} + \cdots] \quad (18.129)$$

$$F_{(z,\,m)} = Kz^{-1}\left(\frac{1}{1 - z^{-1}}\right) = \frac{K}{z - 1}$$

Note that this modified z transform does not depend on the variable m since the function is constant in between the sample periods.

B. RAMP

$$f_{(t)} = Kt \quad \Rightarrow \quad f_{(nT_s+mT_s)} = K(nT_s + mT_s)$$

$$\mathscr{L}_m[KtI_{(t)}] \equiv z^{-1} \sum_{n=0}^{\infty} K(n + m)T_s z^{-n}$$

$$= z^{-1}KT_s[m + (m + 1)z^{-1} + (m + 2)z^{-2} + (m + 3)z^{-3} + \cdots]$$

$$= z^{-1}KT_s[m(1 + z^{-1} + z^{-2} + z^{-3} + \cdots)$$

$$+ z^{-1}(1 + 2z^{-1} + 3z^{-2} + 4z^{-3} + \cdots)]$$

$$= z^{-1}KT_s\left[\frac{m}{1 - z^{-1}} + \frac{z^{-1}}{(1 - z^{-1})^2}\right]$$

$$F_{(z,\,m)} = \frac{KmT_s}{z - 1} + \frac{KT_s}{(z - 1)^2} \qquad (18.130)$$

Notice that this modified z transformation is a function of both m and z.

C. EXPONENTIAL.
As with all the transforms, this is the most important function.

$$f_{(t)} = e^{-at} \quad \Rightarrow \quad f_{(nT_s+mT_s)} = e^{-(n+m)aT_s}$$

$$\mathscr{L}_m[e^{-at}I_{(t)}] = z^{-1} \sum_{n=0}^{\infty} e^{-(n+m)aT_s} z^{-n}$$

$$= z^{-1}[e^{-amT_s} + e^{-a(1+m)T_s}z^{-1} + e^{-a(2+m)T_s}z^{-2} + \cdots]$$

$$= z^{-1}e^{-amT_s}[1 + e^{-aT_s}z^{-1} + e^{-2aT_s}z^{-2} + e^{-3T_s}z^{-3} + \cdots]$$

$$F_{(z,\,m)} = \frac{z^{-1}e^{-amT_s}}{1 - z^{-1}e^{-aT_s}} = \frac{e^{-amT_s}}{z - e^{-aT_s}} \qquad (18.131)$$

This transformation also depends on both m and z.

Pulse transfer functions for modified z transforms are defined in the same way as for regular z transforms. For a system with input $m_{(t)}$ and output $x_{(t)}$, the pulse transfer function is

$$X_{(z, m)} = G_{(z, m)} M_{(z)} \qquad (18.132)$$

where
$$G_{(z, m)} \equiv z^{-1} \sum_{n=0}^{\infty} g_{(nT_s + mT_s)} z^{-n} \qquad (18.133)$$

$g_{(t)} = $ impulse response of the system

Example 18.13. Consider the system with an openloop process transfer function

$$G_{(s)} = \frac{0.027 e^{-1.4s}}{12.5s + 1}$$

Suppose the sampling period T_s is 1 minute. We want to find the $HG_{M(z)}$ for this system, using a zero-order hold. Let $D = kT_s + (1 - m_1)T_s$, where m_1 is a specific numerical value for m. For this example, $T_s = 1$. Therefore $k = 1$ and $m_1 = 0.6$.

$$HG_{M(z)} = \mathscr{Z}\left[\frac{1 - e^{-T_s s}}{s} \frac{0.027 e^{-0.4s} e^{-s}}{12.5s + 1} \right]$$

$$HG_{M(z)} = (1 - z^{-1})(0.027)z^{-1} \mathscr{Z}\left[\frac{e^{-(1-0.6)s}}{s(12.5s + 1)} \right] \qquad (18.134)$$

Now we use Eq. (18.128).

$$\mathscr{Z}\left[\frac{e^{-(1-0.6)s}}{s(12.5s + 1)} \right] = \mathscr{Z}_m\left[\frac{1}{s(12.5s + 1)} \right]_{m=0.6}$$

$$= \mathscr{Z}_m\left[\frac{1}{s} - \frac{1}{s + 1/12.5} \right]_{m=0.6}$$

$$= \left[\frac{1}{z - 1} - \frac{e^{-mT_s/12.5}}{z - e^{-T_s/12.5}} \right]_{m=0.6}$$

$$= \left[\frac{1}{z - 1} - \frac{e^{-0.6(1)/12.5}}{z - e^{-1/12.5}} \right]$$

$$= \left[\frac{1}{z - 1} - \frac{0.9531}{z - 0.9231} \right]$$

Finally Eq. (18.134) becomes

$$HG_{M(z)} = \frac{0.027(z - 1)}{z^2} \left[\frac{z - 0.9231 - (z - 1)(0.9531)}{(z - 1)(z - 0.9231)} \right]$$

$$= \frac{0.00127(z + 0.638)}{z^2(z - 0.9231)} \qquad (18.135)$$

PROBLEMS

18.1. Derive the *z* transforms of the functions:

(a) $f^*_{(t)} = I_{(t)} t^2$

(b) $f^*_{(t)} = I_{(t)} t^2 e^{-at}$

(c) $f^*_{(t)} = I_{(t)} \cos(\omega t)$

(d) $f^*_{(t)} = I_{(t)} e^{-\zeta t/\tau_p} \left[\cos\left(\frac{\sqrt{1-\zeta^2}}{\tau_p} t\right) + \frac{\zeta}{\sqrt{1-\zeta^2}} \sin\left(\frac{\sqrt{1-\zeta^2}}{\tau_p} t\right) \right]$

(e) $f^*_{(t)} = I_{(t)} K_p \tau_p \left(\frac{t}{\tau_p} - 1 + e^{-t/\tau_p}\right)$

(f) $f^*_{(t)} = I_{(t)} e^{-a(t - kT_s)}$ where *k* is an integer

(g) $f^*_{(t)} = I_{(t)} \dfrac{K}{\tau_{p1} - \tau_{p2}} [e^{-t/\tau_{p1}} - e^{-t/\tau_{p2}}]$

18.2. Find the pulse transfer functions in the *z* domain $(HBG_M)_{(z)}$ for the systems ($H_{(s)}$ is zero-order hold):

(a) $G_{M(s)} = \dfrac{K_p}{(\tau_{p1} s + 1)(\tau_{p2} s + 1)}$ $B_{(s)} = K_c$

(b) $G_{M(s)} = \dfrac{K_p e^{-kT_s s}}{(\tau_{p1} s + 1)(\tau_{p2} s + 1)}$ $B_{(s)} = K_c$

(c) $G_{M(s)} = \dfrac{K_p e^{-kT_s s}}{(\tau_{p1} s + 1)(\tau_{p2} s + 1)}$ $B_{(s)} = K_c\left(1 + \dfrac{1}{\tau_I s}\right)$

18.3. Find $x_{(nT_s)}$ for a unit step input in $m_{(t)}$ for the system in part (a) of Prob. 18.2 by partial fractions expansion and by long division. Use the numerical values of parameters given below:

$$K_p = \tau_{p2} = K_c = 1 \qquad \tau_{p1} = 5 \qquad T_s = 0.5$$

18.4. Repeat Prob. 18.3 for part (c) of Prob. 18.2. Use $\tau_I = 2$ and $k = 3$.

18.5. Use the subroutine LONGD given in Table 18.2 to find the response of the closedloop system of Example 18.4 to a unit step load disturbance. Use values of $\tau_p = K_p = 1$.

(a) With $T_s = 0.2$ and $K_c = 2, 4, 6, 8, 10, 12$

(b) With $T_s = 0.4$ and $K_c = 2, 4, 6, 8, 10, 12$

(c) With $T_s = 0.6$ and $K_c = 2, 4, 6, 8, 10, 12$

What do you conclude about the effect of sampling time on stability from these results?

18.6. Find the outputs $x_{2(nT_s)}$ of the two systems of Example 18.7 for a unit step input in $m_{(t)}$. Use partial fractions expansion and long division.

18.7. Repeat Prob. 18.6 for a ramp input in $m_{(t)}$.

18.8. Find the output $x_{(nT_s)}$ of the system of Example 18.9 with a continuous PI controller for values of integral time given below. Use $\tau_p = K_p = 1$ and $K_c = 4$ and $T_s = 0.2$.

(a) $\tau_I = 2$

(b) $\tau_I = 1$
(c) $\tau_I = 0.5$
(d) $\tau_I = 0.25$

18.9. Repeat Prob. 18.8 for the system of Example 18.10 with a sampled-data PI controller.

18.10. A distillation column has an approximate transfer function between overhead composition x_D and reflux flow rate R of

$$G_{M(s)} = \frac{x_{D(s)}}{R_{(s)}} = \frac{0.0092}{(5s + 1)^2} \frac{\text{mol fraction}}{\text{mol/min}}$$

A chromatograph must be used to detect x_D. A continuous PI controller is used with a gain of 1000 and an integral time of 5 min. Calculate the response of x_D to a unit step change in setpoint for different chromatograph cycle times D_c (5, 10, and 20 minutes). The sampling period T_s is set equal to the chromatograph cycle time.

18.11. A tubular chemical reactor's response to a change in feed concentration is found to be essentially a pure deadtime D with attenuation K_p. A DDC computer monitors the outlet concentration $C_{AL(t)}$ and changes the feed concentration $C_{A0(t)}$, through a zero-order hold, using proportional action. The sampling period T_s can be adjusted to an integer multiple of D. Calculate the response of C_{AL} for a unit step change in setpoint C_{AL}^{set} for $D/T_s = 1$ and $D/T_s = 2$
(a) With $K_c = 1/K_p$
(b) With $K_c = 1/2K_p$

CHAPTER
19

STABILITY
ANALYSIS
OF
SAMPLED-
DATA
SYSTEMS

We developed the mathematical tool of z transformation in the last chapter. Now we are ready to apply it to analyze the dynamics of sampled-data systems. Our primary task is to design sampled-data feedback controllers for these systems. We will explore the very important impact of sampling period T_s on these designs.

First we will look at the question of stability in the z plane. Then root locus and frequency response methods will be used to analyze sampled-data systems. Various types of processes and controllers will be studied.

19.1 STABILITY IN THE z PLANE

The stability of any system is determined by the location of the roots of its characteristic equation (or the poles of its transfer function). The characteristic equation of a continuous system is a polynomial in the complex variable s. If all the roots of this polynomial are in the left half of the s plane, the system is stable. For a continuous closedloop system, all the roots of $1 + G_{M(s)}B_{(s)}$ must lie in the left half of the s plane. Thus the region of stability in continuous systems is the left half of the s plane.

The stability of a sampled-data system is determined by the location of the roots of a characteristic equation that is a polynomial in the complex variable z. This characteristic equation is the denominator of the system transfer function set equal to zero. The roots of this polynomial (the poles of the system transfer function) are plotted in the z plane. The ordinate is the imaginary part of z, and the abscissa is the real part of z.

The region of stability in the z plane can be found directly from the region of stability in the s plane by using the basic relationship between the complex variables s and z. [Equation (18.23).]

$$z = e^{T_s s} \tag{19.1}$$

Figure 19.1 shows the s plane. Let the real part of s be α and the imaginary part of s be ω.

$$s = \alpha + i\omega \tag{19.2}$$

The stability region in the s plane is where α, the real part of s, is negative. Substituting Eq. (19.2) into Eq. (19.1) gives

$$z = e^{T_s(\alpha + i\omega)} = (e^{\alpha T_s})e^{i\omega T_s} \tag{19.3}$$

The absolute value of z, $|z|$, is $e^{\alpha T_s}$. When α is negative, $|z|$ is less than 1. When α is positive, $|z|$ is greater than 1. Therefore the left half of the s plane maps into the inside of the unit circle in the z plane, as shown in Fig. 19.1.

A sampled-data system is stable if all the roots of its characteristic equation (the poles of its transfer function) lie inside the unit circle in the z plane.

First let's consider an openloop system with the openloop transfer function

$$HG_{M(z)} = \frac{(z - z_1)(z - z_2)\cdots(z - z_M)}{(z - p_1)(z - p_2)\cdots(z - p_N)} \tag{19.4}$$

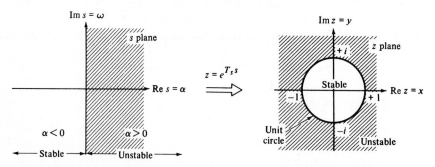

FIGURE 19.1
Stability regions in the s plane and in the z plane.

The stability of this openloop system will depend on the values of the poles of the openloop transfer function. If all the p_i lie inside the unit circle, the system is openloop stable.

The more important problem is closedloop stability. The equation describing the closedloop chromatograph system of Sec. 18.7 was

$$X_{(z)} = \frac{(G_L L)_{(z)} + (G_M B X^{\text{set}})_{(z)}}{1 + z^{-k}(G_M BH)_{(z)}} \tag{19.5}$$

The closedloop stability of this system depends on the location of the roots of the characteristic equation:

$$1 + z^{-k}(G_M BH)_{(z)} = 0 \tag{19.6}$$

If all the roots lie inside the unit circle, the system is closedloop stable.

The equation describing the closedloop digital-computer control system of Sec. 18.7 was

$$X_{(z)} = \frac{(G_L L)_{(z)} + (HG_M)_{(z)} D_{(z)} X^{\text{set}}_{(z)}}{1 + (HG_M)_{(z)} D_{(z)}} \tag{19.7}$$

The closedloop stability of this system depends on the location of the roots of the characteristic equation:

$$1 + (HG_M)_{(z)} D_{(z)} = 0 \tag{19.8}$$

If all the roots lie inside the unit circle, the system is closedloop stable.

It is sometimes convenient to write the closedloop characteristic equation of a general sampled-data system as

$$1 + A_{(z)} = 0 \tag{19.9}$$

where $A_{(z)}$ is either $(HG_M)_{(z)} D_{(z)}$ or $z^{-k}(G_M BH)_{(z)}$.

Example 19.1. Consider a first-order process with a zero-order hold and proportional sampled-data controller.

$$G_{M(s)} = \frac{K_p}{\tau_p s + 1}$$

As we developed in Chap. 18, the openloop pulse transfer function for this process is

$$(HG_M)_{(z)} = \mathcal{Z}\left[\frac{1 - e^{-T_s s}}{s} \frac{K_p}{\tau_p s + 1}\right] = \frac{K_p(1 - b)}{z - b} \tag{19.10}$$

where $b \equiv e^{-T_s/\tau_p}$ \hfill (19.11)

The *openloop* characteristic equation is

$$z - b = 0$$

So the root of the openloop characteristic equation is b. Since b is less than 1, this root lies inside the unit circle and the system is *openloop* stable.

The *closedloop* characteristic equation for this system is

$$1 + A_{(z)} = 1 + (HG_M)_{(z)} D_{(z)} = 1 + \frac{K_c K_p(1 - b)}{z - b} = 0 \tag{19.12}$$

since $D_{(z)} = K_c$. Solving for the closedloop root gives

$$z = b - K_c K_p(1 - b) \tag{19.13}$$

There is a single root. It lies on the real axis in the z plane and its location depends on the value of the feedback controller gain K_c. When the feedback controller gain is zero (the openloop system), the root lies at $z = b$. As K_c is increased, the closedloop root moves to the left along the real axis in the z plane. We will return to this example in the next section.

19.2   ROOT LOCUS DESIGN METHODS

With continuous systems we made root locus plots in the s plane. Controller gain was varied from zero to infinity, and the roots of the closedloop characteristic equation were plotted. Time constants, damping coefficients, and stability could be easily determined from the positions of the roots in the s plane. The limit of stability was the imaginary axis. Lines of constant closedloop damping coefficient were radial straight lines from the origin. The closedloop time constant was the reciprocal of the distance from the origin.

With sampled-data systems, root locus plots can be made in the z plane in almost exactly the same way. Controller gain is varied from zero to infinity, and the roots of the closedloop characteristic equation $1 + A_{(z)} = 0$ are plotted. When the roots lie inside the unit circle, the system is closedloop stable. When the roots lie outside the unit circle, the system is closedloop unstable.

Sometimes other planes beside the z plane are used. The "log z" and "$\mathcal{W}$" planes offer some advantages to the z plane for some systems. We will discuss these later in this chapter.

19.2.1   *z*-Plane Root Locus Plots

Lines of constant damping coefficient ζ in the s plane are radial lines from the origin as sketched in Fig. 19.2.

$$\zeta = \cos \theta \tag{19.14}$$

where θ is the angle between the radial line and the negative real axis. Along a line of constant ζ in the s plane, the tangent of θ is

$$\tan \theta = \frac{\omega}{\alpha} = \frac{(\sqrt{1 - \zeta^2})/\tau}{-\zeta/\tau} = \frac{\sqrt{1 - \zeta^2}}{-\zeta} \tag{19.15}$$

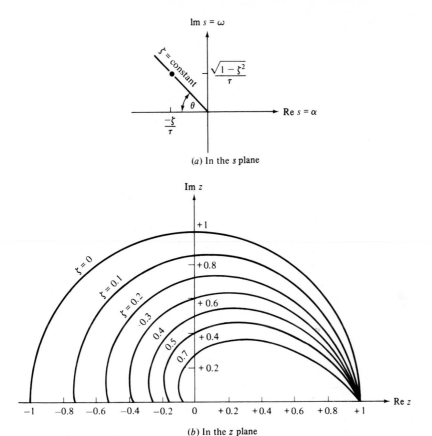

FIGURE 19.2
Lines of constant damping coefficient.

Using Eq. (19.15), the real part of s, α, can be expressed in terms of the imaginary part of s, ω, and the damping coefficient ζ.

$$\alpha = \frac{-\zeta\omega}{\sqrt{1-\zeta^2}} \tag{19.16}$$

These lines of constant damping coefficient can be mapped into the z plane. The z variable along a line of constant damping coefficient is

$$z = e^{T_s s} = e^{T_s(\alpha + i\omega)} = e^{\alpha T_s} e^{i\omega T_s}$$

$$z = \exp\left(-\frac{\zeta\omega T_s}{\sqrt{1-\zeta^2}}\right) e^{i\omega T_s} \tag{19.17}$$

Lines of constant damping coefficient in the z plane can be generated by picking a value of ζ and varying ω in Eq. (19.17) from 0 to $\omega_s/2$. See Sec. 19.4 for a discussion of the required range of ω.

Figure 19.2*b* shows these curves in the *z* plane. On the unit circle, the damping coefficient is zero. On the positive real axis, the damping coefficient is greater than one. At the origin, the damping coefficient is exactly unity.

Notice the very significant result that the damping coefficient is less than one on the *negative* real axis. This means that in sampled-data systems a real root can give underdamped response. This can never happen in a continuous system; the roots must be complex to give underdamped response.

So we can design sampled-data controllers for a desired closedloop damping coefficient by adjusting the controller gain to position the roots on the desired damping line. Some examples will illustrate the method and will point out the differences and the similarities between continuous systems and sampled-data systems.

Example 19.2. Let us make a root locus plot for the first-order system considered in Example 19.1.

$$G_{M(s)} = \frac{K_p}{\tau_p s + 1} \quad \Rightarrow \quad (HG_M)_{(z)} = \frac{K_p(1 - b)}{z - b}$$

Using a proportional sampled-data controller, the closedloop characteristic equation for this system was

$$1 + A_{(z)} = 1 + (HG_M)_{(z)} D_{(z)} = 1 + \frac{K_c K_p(1 - b)}{z - b} = 0$$

The closedloop root was

$$z = b - K_c K_p(1 - b) \tag{19.18}$$

Figure 19.3 shows the root locus plot. It starts ($K_c = 0$) at $z = b$ on the positive real axis inside the unit circle, so it is openloop stable. As K_c is increased, the closedloop root moves to the left.

At the origin (where $z = 0$) the damping coefficient is unity. Solving Eq. (19.18) for the value of controller gain that give this "critically damped" system

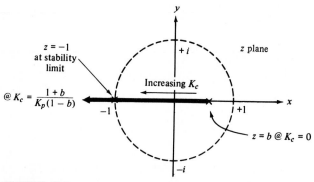

FIGURE 19.3
z-plane location of roots (Example 19.1).

yields

$$0 = b - K_c K_p(1 - b) \qquad \Rightarrow \qquad (K_c)_{\zeta=1} = \frac{b}{K_p(1 - b)} \qquad (19.19)$$

For gains less than this, the system is overdamped. For gains greater than this, the system will be underdamped! This is a distinct difference between sampled-data and continuous systems. A first-order continuous system can never by underdamped. But this is not true for a sampled-data system.

If we wanted to design for a closedloop damping of 0.3, we would use Eq. (19.17) with $\zeta = 0.3$ and with $\omega T_s = \pi$ because we are on the negative real axis where the argument of z is π. Equation (19.17) shows that the argument of z on a line of constant damping coefficient is ωT_s.

$$\arg z = \omega T_s = \pi \qquad (19.20)$$

$$z = \exp\left(-\frac{0.3\pi}{\sqrt{1 - (0.3)^2}}\right) e^{i\pi} = (0.372)(\cos \pi + i \sin \pi) = -0.372 \quad (19.21)$$

So positioning the closedloop root on the negative real axis at -0.372 will give a closedloop system with a damping coefficient of 0.3. Solving for the required gain gives

$$-0.372 = b - K_c K_p(1 - b) \qquad \Rightarrow \qquad (K_c)_{\zeta=0.3} = \frac{b + 0.372}{K_p(1 - b)} \qquad (19.22)$$

The system reaches the limit of closedloop stability when the root crosses the unit circle at $z = -1$. The value of controller gain at this limit is the ultimate gain K_u.

$$-1 = b - K_u K_p(1 - b) \qquad \Rightarrow \qquad K_u = \frac{1 + b}{K_p(1 - b)} \qquad (19.23)$$

Let us take some numerical values to show the effect of changing the sampling period. Let $K_p = \tau_p = 1$. Table 19.1 gives values for the critical gain ($\zeta = 1$),

TABLE 19.1
Effect of sampling period

		First-order lag		
T_s	b	$(K_c)_{\zeta=1}$	$(K_c)_{\zeta=0.3}$	K_u
0.1	0.905	9.51	13.4	20.0
0.2	0.819	4.52	6.57	10.0
0.5	0.606	1.54	2.49	4.08
1.0	0.368	0.582	1.17	2.16

First-order lag with deadtime $D = T_s$		
T_s	$(K_c)_{\zeta=1}$	K_u
0.1	2.15	10.5
0.2	0.925	5.52
0.5	0.234	2.54
1.0	0.054	1.58

the gain that gives $\zeta = 0.3$ and the ultimate gain for different sampling periods. Note that the gains decrease as T_s increases, showing that control gets worse as the sampling period gets bigger. Remember, we found in Example 18.6 with $T_s = 0.2$ that a K_c of 15 gave a closedloop unstable response. Table 19.1 shows that $K_u = 10$ for this T_s.

This little example has demonstrated several extremely important facts about sampled-data control. This simple first-order system, which could never be made closedloop unstable in a *continuous* control system, *can* become closedloop unstable in a *sampled-data* system. This is an extremely important difference between continuous control and sampled-data control. It points out the fact that continuous control is almost always *better* than sampled-data control!

It is logical to ask at this point why we use sampled-data, digital-computer control if it is inherently worse than continuous analog. There are a number of other advantages that computer control offers that outweigh its theoretical dynamic disadvantages: cost, ease of maintenance, reliability, data-acquisition capability, ability to handle nonstandard, complex, and nonlinear control algorithms, etc. But probably the main reason computer control is often chosen over analog control is management politics. The flashy CRT displays and high-tech hardware impress management and make it easy to sell a project.

But keep in mind that from a purely technical control performance standpoint, an analog controller can do a better job than a sampled-data controller in most applications. The process of sampling results in some loss of information: we don't know what is going on in between the sampling periods. Thus there is an inherent degradation of dynamic performance that no amount of high-tech sales hype and blue smoke can change.

Example 19.3. Consider the first-order lag process with a deadtime of one sampling period.

$$G_{M(s)} = \frac{K_p e^{-T_s s}}{\tau_p s + 1} \quad \Rightarrow \quad (HG_M)_{(z)} = \frac{K_p(1 - b)}{z(z - b)} \tag{19.24}$$

The addition of the one sampling-period deadtime increases the order of the denominator polynomial to two. Using a proportional sampled-data controller gives the closedloop characteristic equation

$$1 + A_{(z)} = 1 + (HG_M)_{(z)} D_{(z)} = 1 + \frac{K_c K_p(1 - b)}{z(z - b)} = 0$$

$$z^2 - bz + K_c K_p(1 - b) = 0 \tag{19.25}$$

There are two root loci, as shown in Fig. 19.4a for the numerical values $K_p = \tau_p = 1$ and $T_s = 0.2$ ($b = 0.819$).

$$z = \frac{b \pm \sqrt{b^2 - 4K_c K_p(1 - b)}}{2} \tag{19.26}$$

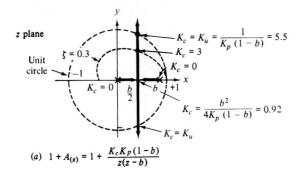

$$(a) \quad 1 + A_{(z)} = 1 + \frac{K_c K_p (1 - b)}{z(z - b)}$$

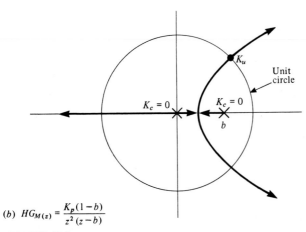

$$(b) \quad HG_{M(z)} = \frac{K_p (1 - b)}{z^2 (z - b)}$$

FIGURE 19.4

The paths start ($K_c = 0$) at $z = 0$ and $z = b$. They come together on the positive real axis at $z = b/2$ when

$$K_c = \frac{b^2}{4K_p(1 - b)} = 0.925$$

This value of controller gain gives a critically damped system. For larger controller gains, the system is underdamped. The complex roots are

$$z = \frac{b}{2} \pm i\tfrac{1}{2}\sqrt{4K_c K_p(1 - b) - b^2} \tag{19.27}$$

The value of gain that gives a closedloop damping coefficient of 0.3 can be found by using Eqs. (19.17) and (19.27) with $\zeta = 0.3$.

$$\exp\left(-\frac{0.3\omega T_s}{\sqrt{1 - (0.3)^2}}\right)e^{i\omega T_s} = \frac{b}{2} \pm i\tfrac{1}{2}\sqrt{4K_c K_p(1 - b) - b^2}$$

$$e^{-0.3145\omega T_s}[\cos(\omega T_s) + i\sin(\omega T_s)] = \frac{b}{2} \pm i\tfrac{1}{2}\sqrt{4K_c K_p(1 - b) - b^2}$$

Equating the real and imaginary parts of the left- and right-hand sides of the above equation gives two equations in the two unknowns, ω and K_c, for a given sampling period T_s.

The maximum K_c for which the closedloop system is stable occurs when the paths cross the unit circle. At that point the magnitude of z, $|z|$, is unity,

$$|z| = \sqrt{\left(\frac{b}{2}\right)^2 + \left[\frac{\sqrt{4K_u K_p(1-b) - b^2}}{2}\right]^2}$$

$$1 = \tfrac{1}{2}\sqrt{4K_u K_p(1-b)}$$

$$K_u = \frac{1}{K_p(1-b)} \tag{19.28}$$

Table 19.1 shows numerical values for different sampling periods. The addition of the deadtime to the lag reduces the gains.

Example 19.4. Adding a deadtime to the first-order lag process that is equal to *two* sampling periods gives a third-order system in the z plane.

$$1 + A_{(z)} = 1 + (HG_M)_{(z)} D_{(z)} = 1 + \frac{K_c K_p(1-b)}{z^2(z-b)} = 0$$

$$z^3 - bz^2 + K_c K_p(1-b) = 0 \tag{19.29}$$

Figure 19.4b shows the root locus plot in the z plane. There are now three loci. The ultimate gain can occur on either the real-root path (at $z = -1$) or on the complex-conjugate roots path. We can solve for these two values of controller gain and see which is smaller.

At $z = -1$: Using Eq. (19.29)

$$-1 - b + K_u K_p(1-b) = 0 \quad \Rightarrow \quad K_u = \frac{1+b}{K_p(1-b)} \tag{19.30}$$

For the numerical case $K_p = \tau_p = 1$ and $T_s = 0.5$, the result is 4.09.

On the complex-conjugate path:
At the unit circle $z = e^{i\theta}$. Substituting into Eq. (19.29) gives

$$e^{i3\theta} - be^{i2\theta} + K_u K_p(1-b) = 0$$

$$\cos(3\theta) + i\sin(3\theta) - b[\cos(2\theta) + i\sin(2\theta)] + K_u K_p(1-b) = 0$$

$$\{\cos(3\theta) - b\cos(2\theta) + K_u K_p(1-b)\} + i\{\sin(3\theta) - b\sin(2\theta)\} = 0 + i0$$

Equating real and imaginary parts of both sides of the equation above gives two equations and two unknowns: θ and K_u.

$$\cos(3\theta) - b\cos(2\theta) + K_u K_p(1-b) = 0 \tag{19.31}$$

$$\sin(3\theta) - b\sin(2\theta) = 0 \tag{19.32}$$

Equation (19.32) can be solved for θ. For the numerical case $K_p = \tau_p = 1$ and $T_s = 0.5$, the result is $\theta = 0.83$ radians. Then Eq. (19.31) can be solved for K_u ($=1.88$ for the numerical case). Since this is smaller than that calculated from Eq. (19.30), the ultimate gain is 1.88. Note that this is lower than the ultimate gain for the process with the smaller deadtime.

Example 19.5. Now let's look at a second-order system.

$$G_{M(s)} = \frac{K_p}{(\tau_1 s + 1)(\tau_2 s + 1)} \tag{19.33}$$

Using a zero-order hold gives an openloop transfer function

$$HG_{M(z)} = \frac{K_p a_0(z - z_1)}{(z - p_1)(z - p_2)} \tag{19.34}$$

where

$$p_1 = e^{-T_s/\tau_1} \quad \text{and} \quad p_2 = e^{-T_s/\tau_2} \tag{19.35}$$

$$a_0 = 1 + \frac{p_2 \tau_2 - p_1 \tau_1}{\tau_1 - \tau_2} \tag{19.36}$$

$$z_1 = \frac{p_1 p_2(\tau_2 - \tau_1) + p_2 \tau_1 - p_1 \tau_2}{\tau_1 - \tau_2 + p_2 \tau_2 - p_1 \tau_1} \tag{19.37}$$

As a specific numerical example, consider the two-heated tank process from Example 10.1 with the openloop transfer function

$$G_{M(s)} = \frac{2.315}{(s + 1)(5s + 1)} \tag{19.38}$$

Using a zero-order hold and a sampling period $T_s = 0.5$ min gives

$$HG_{M(z)} = \frac{2.315(0.0209)(z + 0.8133)}{(z - 0.607)(z - 0.905)} \tag{19.39}$$

Several important features should be noted. The first-order process considered in Example 19.1 gave a pulse transfer function that was also first-order, i.e., the denominator of the transfer function was first-order in z. The second-order process considered in this example gave a sampled-data pulse transfer function that had a second-order denominator polynomial. These results can be generalized to an Nth-order system. The order of s in the continuous transfer function is the same as the order of z in the corresponding sampled-data transfer function.

However, note that in the case of the second-order system, the sampled-data transfer function has a zero, whereas the continuous transfer function does not. So in this respect the analogy between continuous and sampled-data transfer functions does not hold.

The closedloop characteristic equation for the second-order system with a proportional sampled-data controller is

$$1 + A_{(z)} = 1 + HG_{M(z)} D_{(z)} = 1 + \frac{K_c K_p a_0(z - z_1)}{(z - p_1)(z - p_2)} = 0$$

For the numerical example

$$1 + \frac{0.0479K_c(z + 0.8133)}{(z - 0.607)(z - 0.905)} = 0 \tag{19.40}$$

$$z^2 + z(0.0479K_c - 1.512) + 0.549 + 0.039K_c = 0 \tag{19.41}$$

To find the ultimate gain for the second-order system, we set $|z| = 1$ and solve for K_u.

$$z = \frac{-(0.0479K_u - 1.512)}{2} \pm i \frac{\sqrt{4(0.549 + 0.039K_u) - (0.0479K_u - 1.512)^2}}{2}$$

$$1 = \sqrt{\left(\frac{0.0479K_u - 1.512}{2}\right)^2 + \left(\frac{\sqrt{4(0.549 + 0.039K_u) - (0.0479K_u - 1.512)^2}}{2}\right)^2}$$

$$K_u = 11.6$$

This is the ultimate gain if the sampling period is 0.5 min. If a $T_s = 2$ min is used, the openloop pulse transfer function becomes

$$HG_{M(z)} = \frac{0.4535(z + 0.455)}{(z - 0.135)(z - 0.67)} \tag{19.42}$$

The location of the zero is closer to the origin. This reduces the radius of the circular part of the root locus plot (see Fig. 19.5) and reduces the ultimate gain to 4.45.

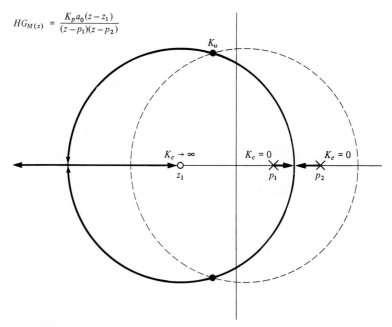

FIGURE 19.5
Second-order process.

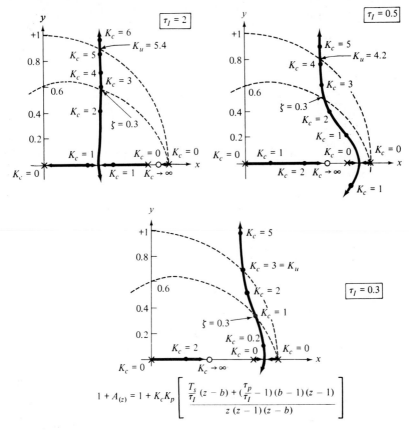

$$1 + A_{(z)} = 1 + K_c K_p \left[\frac{\frac{T_s}{\tau_I}(z - b) + (\frac{\tau_p}{\tau_I} - 1)(b - 1)(z - 1)}{z(z - 1)(z - b)} \right]$$

FIGURE 19.6
Root locus plots.

Example 19.6. The chromatographic system studied in Example 18.9 had a first-order lag openloop process transfer function and a deadtime of one sampling period. The closedloop characteristic equation was [see Eq. (18.100)]

$$1 + A_{(z)} = 1 + K_c K_p \left[\frac{(T_s/\tau_I)(z - b) + (\tau_p/\tau_I - 1)(b - 1)(z - 1)}{z(z - 1)(z - b)} \right] \quad (19.43)$$

There are three loci (when $\tau_I \neq \tau_p$). They begin at $z = 0$, $z = +1$ and $z = b$. One path ends at the zero of $A_{(z)}$. The root loci for several values of τ_I are shown in Fig. 19.6. Decreasing the reset time τ_I decreases stability and ultimate gain K_u.

19.2.2 Log-z Plane Root Locus Plots

Instead of making root locus plots in the z plane, it is sometimes convenient to make them in the log-z plane. In the z plane, the ordinate is the imaginary part of z and the abscissa is the real part of z. In the log-z plane, the ordinate is the argument of z and the abscissa is the natural logarithm of the absolute magnitude of z.

$$\ln z \equiv \ln |z| + i \text{ arg } z \tag{19.44}$$

There are two effects of using this new coordinate system:

1. The limit of stability becomes the imaginary axis in the log-z plane and the region of stability is the left half of the log-z plane. This is analogous to the situation in the continuous s plane.
2. Lines of constant damping coefficient become radial straight lines from the origin in the log-z plane. This is analogous to what occurs in the s plane.

The first effect is obvious from the definition of the logarithm of z as given in Eq. (19.44). Inside the unit circle, the magnitude of z is less than 1. Therefore the $\ln |z|$ is negative. On the unit circle, $\ln |z| = 0$. So the unit circle in the z plane maps into the left half of the log-z plane.

The second effect is easily shown by recalling that on a line of constant damping coefficient in the z plane [Eq. (19.17)]

$$z = \exp\left(-\frac{\zeta \omega T_s}{\sqrt{1 - \zeta^2}}\right) e^{i\omega T_s} = |z| e^{i \text{ arg } z}$$

Taking the natural logarithm of the equation above gives

$$\ln z = \ln |z| + i \text{ arg } z = \left[-\frac{\zeta \omega T_s}{\sqrt{1 - \zeta^2}}\right] + i[\omega T_s] \tag{19.45}$$

Thus the real part of $\ln z$ is the first term in brackets and the imaginary part of $\ln z$ is ωT_s.

Consider a point in the log-z plane that has these real and imaginary parts. If we draw a straight line from the origin through this point, this radial line will make an angle θ with the horizontal axis whose tangent is the imaginary part divided by the real part (see Fig. 19.7a).

$$\tan \theta = \frac{\text{Im } [\ln z]}{\text{Re } [\ln z]} = \frac{\omega T_s}{-\zeta \omega T_s / \sqrt{1 - \zeta^2}} = \frac{-\sqrt{1 - \zeta^2}}{\zeta} \tag{19.46}$$

For a constant value of damping coefficient ζ, $\tan \theta$ is constant. This means that a line of constant damping coefficient in the log-z plane is a radial straight line. The distance from the origin out to the point (the hypotenuse of the triangle) is

$$\text{Hypotenuse} = \sqrt{[\text{Re } (\ln z)]^2 + [\text{Im } (\ln z)]^2} = \sqrt{[\omega T_s]^2 + \left[-\frac{\zeta \omega T_s}{\sqrt{1 - \zeta^2}}\right]^2}$$

$$= \frac{\omega T_s}{\sqrt{1 - \zeta^2}}$$

(a) Log-z plane:

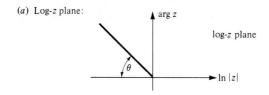

(b) First-order lag: $HG_{M(z)}D_{(z)} = \dfrac{K_C K_p(1-b)}{z-b}$

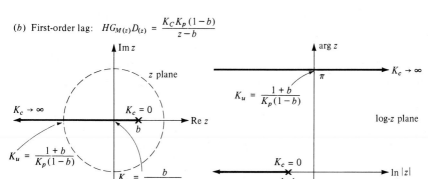

(c) Second-order lag: $HG_{M(z)}D_{(z)} = \dfrac{K_c a_0(z-z_1)}{(z-p_1)(z-p_2)}$

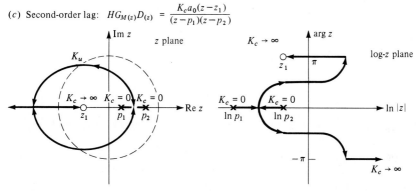

FIGURE 19.7
log z plane root locus.

So the cosine of the angle θ is

$$\cos\theta = \frac{-\text{Re}\,(\ln z)}{\text{hypotenuse}} = \frac{\zeta\omega T_s/\sqrt{1-\zeta^2}}{\omega T_s/\sqrt{1-\zeta^2}} = \zeta \qquad (19.47)$$

This is exactly the same relationship we found in the s plane. These radial lines in the log-z plane are much easier to draw than the ellipsoidal lines given by Eq. (19.17) for the z plane.

Figure 19.7b,c compare z-plane and log-z-plane root locus plots for first-order and second-order systems. For the first-order system, the single root moves to minus infinity in the log-z plane as z goes to zero. Then the root comes back along a horizontal line at $+\pi$, and crosses the imaginary axis at the ultimate gain K_u. It is easy to design for a desired damping coefficient by simply finding the gain that positions the root on the desired straight damping coefficient line.

The second-order system has two loci. As K_c is increased the roots move into the right half of the log-z plane at K_u.

19.3 BILINEAR TRANSFORMATION

The bilinear transformation is another change of variables. We convert from the z variable into the $\mathcal{W}$ variable. The transformation maps the unit circle in the z plane into the left half of the $\mathcal{W}$ plane. This mapping converts the stability region back to the familiar LHP region. The Routh criterion can then be used. Root locus plots can be made in the $\mathcal{W}$ plane with the system going closedloop unstable when the loci cross over into the RHP.

The transformation $z = e^{T_s s}$ maps the left half of the s plane into the unit circle. Actually, a number of horizontal strips in the s plane are each mapped into the unit circle, one on top of the other. The mapping is, therefore, not unique. The "primary strip" from $\omega = -\omega_s/2$ to $\omega = \omega_s/2$, sketched in Fig. 19.8a, is the only one that we are interested in since $A^*_{(s)}$ is periodic.

The bilinear transformation is defined in two ways:

$$\mathcal{W} = \frac{z-1}{z+1} \tag{19.48}$$

$$\mathcal{W} = \frac{z+1}{z-1} \tag{19.49}$$

These two equations differ only in the sign of z. Since we are interested in the interior and exterior of the unit circle, switching signs or quadrants in the z plane does not matter.

To prove that this transformation maps the unit circle in the z plane into the left half of the $\mathcal{W}$ plane, let us express the complex variables $\mathcal{W}$ and z in the following rectangular and polar forms:

$$\mathcal{W} \equiv u + iv \tag{19.50}$$

$$z = re^{i\theta} \tag{19.51}$$

Where u is the real part of $\mathcal{W}$, v is the imaginary part of $\mathcal{W}$, r is the magnitude of z, and θ is the argument or phase angle of z.

Substituting these into Eq. (19.48) gives

$$\mathcal{W} = \frac{z-1}{z+1} = \frac{-1 + r\cos\theta + ir\sin\theta}{1 + r\cos\theta + ir\sin\theta}$$

$$u + iv = \frac{r^2 - 1}{1 + r^2 + 2r\cos\theta} + i\frac{2r\sin\theta}{1 + r^2 + 2r\cos\theta} \tag{19.52}$$

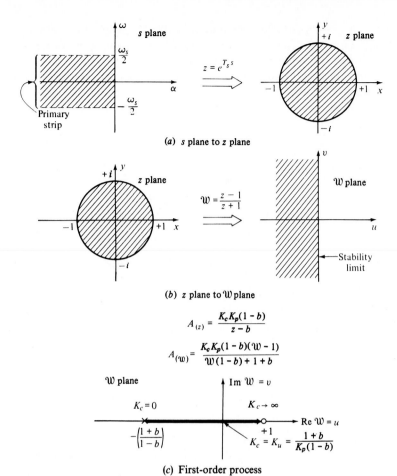

FIGURE 19.8
Bilinear transformation.

When r is less than 1, u is negative. Thus a point inside the unit circle in the z plane maps into a point in the left half of the $\mathcal{W}$ plane.

On the stability boundary in the z plane, r is equal to 1 and θ goes from 0 to π and then from $-\pi$ to 0 radians. When $r = 1$, u is zero. This is the imaginary axis in the $\mathcal{W}$ plane. Also when $r = 1$,

$$v = \frac{\sin \theta}{1 + \cos \theta}$$

At $\theta = 0$, $v = 0$. At $\theta = \pi/2$, $v = 1$. As θ goes to π, we must take the limit of $\sin \theta / (1 + \cos \theta)$ using L'Hôpital's rule.

$$\lim_{\theta \to \pi} v = \lim_{\theta \to \pi} \frac{\sin \theta}{1 + \cos \theta} = \lim_{\theta \to \pi} \frac{\cos \theta}{-\sin \theta} = +\infty$$

Thus the path along the upper half of the unit circle in the z plane maps into the positive imaginary axis in the $\mathcal{W}$ plane. As θ goes from $-\pi$ to 0, v goes from $-\infty$ to 0, which is the negative imaginary axis in the $\mathcal{W}$ plane.

The examples below illustrate the use of the bilinear transformation to analyze the stability of sampled-data systems. We can use all the classical methods that we are used to employing in the s plane. The price that we pay is the additional algebra to convert to $\mathcal{W}$ from z.

It should be noted that the stability limits in the s plane, the $\mathcal{W}$ plane, and the log-z plane are all the same: the imaginary axis. However, lines of constant damping coefficients in the $\mathcal{W}$ plane are *not* radial straight lines as they are in the s and log-z planes.

Example 19.7. The first-order lag process, zero-order hold, and proportional sampled-data controller from Example 19.1 gave an openloop system transfer function

$$A_{(z)} = (HG_M)_{(z)} D_{(z)} = \frac{K_c K_p(1 - b)}{z - b} \tag{19.53}$$

Equation (19.49) can be solved for z to give

$$z = \frac{\mathcal{W} + 1}{\mathcal{W} - 1} \tag{19.54}$$

Substituting for z in Eq. (19.53) gives

$$A_{(\mathcal{W})} = \frac{K_c K_p(1 - b)}{[(\mathcal{W} + 1)/(\mathcal{W} - 1)] - b} = \frac{K_c K_p(1 - b)(\mathcal{W} - 1)}{\mathcal{W}(1 - b) + 1 + b} \tag{19.55}$$

The closedloop characteristic equation for this system is

$$1 + A_{(\mathcal{W})} = 1 + \frac{K_c K_p(1 - b)(\mathcal{W} - 1)}{\mathcal{W}(1 - b) + 1 + b} = 0$$

$$\mathcal{W} = \frac{[(1 + b)/(1 - b)] - K_c K_p}{1 + K_c K_p} \tag{19.56}$$

Varying K_c from 0 to ∞ and plotting the location of this single root in the $\mathcal{W}$ plane give a root locus plot of the closedloop system. Figure 19.8c shows that the path starts on the negative real axis at $u = -(1 + b)/(1 - b)$ when $K_c = 0$. The path moves along the real axis toward the origin. As long as it is in the LHP, the system is closedloop stable. The limit of stability occurs when the root passes through the origin ($\mathcal{W} = 0$). The value of controller gain at this limit is

$$K_u = \frac{1 + b}{K_p(1 - b)} \tag{19.57}$$

This is the same value we found in Example 19.2. As K_c goes to infinity, the path goes to the zero of the openloop transfer function at $\mathcal{W} = +1$.

Example 19.8. Let us consider the system of Example 19.6. We will pick a τ_I that is equal to τ_p. This simplifies the closedloop characteristic equation to

$$1 + A_{(z)} = 1 + \frac{K_c K_p T_s/\tau_p}{z(z-1)} = 0 \tag{19.58}$$

Substituting for z gives

$$1 + \frac{K_c K_p T_s/\tau_p}{[(\mathcal{W}+1)/(\mathcal{W}-1)]\{[(\mathcal{W}+1)/(\mathcal{W}-1)]-1\}} = 0$$

$$\mathcal{W}^2 + \mathcal{W}\left(\frac{2\tau_p}{K_c K_p T_s} - 2\right) + \left(1 + \frac{2\tau_p}{K_c K_p T_s}\right) = 0 \tag{19.59}$$

We could make a root locus plot in the $\mathcal{W}$ plane. Or we could use the direct-substitution method (let $\mathcal{W} = iv$) to find the maximum stable value of K_c. Let us use the Routh stability criterion. This criterion cannot be applied in the z plane because it gives the number of positive roots, not the number of roots outside the unit circle. The Routh array is

$$\begin{bmatrix} 1 & 1 + \dfrac{2\tau_p}{K_c K_p T_s} \\[3ex] \dfrac{2\tau_p}{K_c K_p T_s} - 2 & \\[3ex] 1 + \dfrac{2\tau_p}{K_c K_p T_s} & \end{bmatrix}$$

All the terms in the first column must have the same sign for the system to have no zeros in the RHP of the $\mathcal{W}$ plane. Therefore, the two requirements are

$$\frac{\tau_p}{K_c K_p T_s} - 1 > 0 \quad \text{and} \quad 1 + \frac{2\tau_p}{K_c K_p T_s} > 0$$

The first establishes the upper limit on K_c.

$$K_u = \frac{\tau_p}{K_p T_s} \tag{19.60}$$

19.4 FREQUENCY-DOMAIN DESIGN TECHNIQUES

Sampled-data control systems can be designed in the frequency domain by using the same techniques that we employed for continuous systems. The Nyquist stability criterion is applied to the appropriate closedloop characteristic equation to find the number of zeros outside the unit circle.

19.4.1 Nyquist Stability Criterion

The closedloop characteristic equation of a sampled-data system is

$$1 + A_{(z)} = 0$$

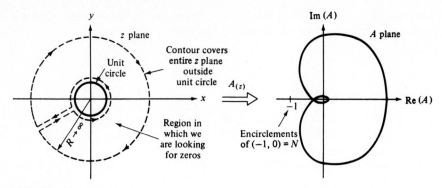

FIGURE 19.9
Nyquist stability criterion.

We want to find out if the function $F_{(z)} \equiv 1 + A_{(z)}$ has any zeros or roots outside the unit circle in the z plane. If it does, the system is closedloop unstable.

We can apply the $Z - P = N$ theorem of Chap. 13 to this new problem. We pick a contour that goes completely around the area in the z plane that is outside the unit circle, as shown in Fig. 19.9. We then plot $A_{(z)}$ in the A plane and look at the encirclements of the $(-1, 0)$ point to find N.

If the number of poles of $A_{(z)}$ outside the unit circle is known, the number of zeros outside the unit circle can be calculated from

$$Z = N + P \qquad (19.61)$$

If the system is openloop stable there will be no poles of $A_{(z)}$ outside the unit circle and $P = 0$.

Thus the Nyquist stability criterion can be applied directly to sampled-data systems.

19.4.2 Rigorous Method

The relationship between z and s is

$$z = e^{T_s s}$$

Going from the Laplace domain to the frequency domain by substituting $s = i\omega$ gives

$$z = e^{i\omega T_s} \qquad (19.62)$$

To make a Nyquist plot, we merely substitute $e^{i\omega T_s}$ for z in the $A_{(z)}$ function and make a polar plot of $A^*_{(i\omega)}$ as frequency ω goes from 0 to $\omega_s/2$, where

$$\omega_s = \frac{2\pi}{T_s} \qquad (19.63)$$

The reason why we only have to cover this frequency range will be demonstrated in the example below.

Example 19.9. Let's consider the first-order lag process with a proportional sampled-data controller.

$$A_{(z)} = \frac{K_c K_p(1 - b)}{z - b} \tag{19.64}$$

Note that this function has *no* poles outside the unit circle, so $P = 0$.

We must let z move around a closed contour in the z plane that completely encloses the area outside the unit circle. Figure 19.10a sketches such a contour. The C_+ contour starts at $z = +1$ and moves along the top of the unit circle to $z = -1$. The C_{out} contour goes from $z = -1$ to $-\infty$. The C_R contour is a circle of infinite radius going all the way around the z plane. Finally the C_{in} and C_- contours get us back to our starting point at $z = +1$.

On the C_+ contour $z = e^{i\theta}$ since the magnitude of z is unity on the unit circle. The angle θ goes from 0 to $+\pi$. Now from Eq. (19.62)

$$z = e^{i\omega T_s}$$

Therefore on the C_+ contour

$$\omega T_s = \theta$$

As θ goes from 0 to $+\pi$, ω must go from 0 to π/T_s. Using Eq. (19.63) gives

$$\frac{\pi}{T_s} = \frac{\pi}{(2\pi/\omega_s)} = \frac{\omega_s}{2}$$

Therefore, moving around the top of the unit circle is equivalent to letting frequency ω vary from 0 to $\omega_s/2$.

Substituting $z = e^{i\omega T_s}$ into Eq. (19.64) gives

$$A^*_{(i\omega)} = \frac{K_c K_p(1 - b)}{e^{i\omega T_s} - b} \tag{19.65}$$

Figure 19.10b shows the Nyquist plot. At $\omega = 0$ (where $z = +1$), the plot starts on the positive real axis at $K_c K_p$. When $\omega = \omega_s/2$ (where $z = -1$)

$$A^*_{(i\omega_s/2)} = \frac{K_c K_p(1 - b)}{-1 - b} = -\frac{K_c K_p(1 - b)}{1 + b} \tag{19.66}$$

Thus the Nyquist plot of the sampled-data system does not end at the origin as the Nyquist plot of the continuous system does. It ends on the negative real axis at the value given in Eq. (19.66).

As we will show in a minute when we have completed the rest of the contours, this means that if the controller gain is made big enough, the Nyquist plot *will* encircle the $(-1, 0)$ point. If $N = 1$, $Z = 1$ for this system since $P = 0$. Thus there will be one zero or root of the closedloop characteristic equation outside the unit circle.

The limiting gain is when the $A^*_{(i\omega)}$ curve ends right at -1. From Eq. (19.66)

$$-1 = -\frac{K_u K_p(1 - b)}{1 + b} \qquad \Rightarrow \qquad K_u = \frac{1 + b}{K_p(1 - b)}$$

This is exactly what our root locus analysis showed.

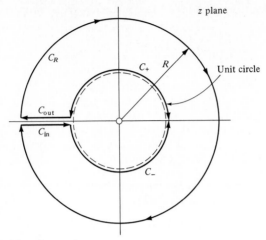

(a) z-plane contours

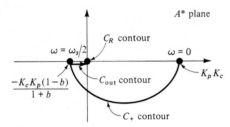

(b) First-order lag

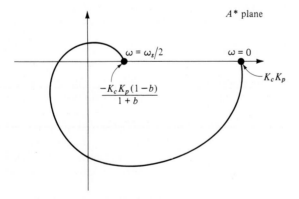

(c) First-order lag with deadtime ($D = T_s$)

FIGURE 19.10
Frequency-domain methods.

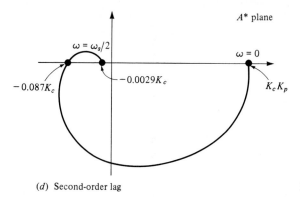

(*d*) Second-order lag

FIGURE 19.10 (*cont.*)

On the C_{out} contour, $z = re^{i\pi} = -r$ as r goes from 1 to ∞. Substituting $-r$ for z in Eq. (19.64) gives

$$A_{(z)} = \frac{K_c K_p(1 - b)}{-r - b} \qquad (19.67)$$

As r goes from 1 to ∞, $A_{(z)}$ goes from $-K_c K_p(1 - b)/(1 + b)$ to 0. See Fig. 19.10*b*.

On the C_R contour, $z = Re^{i\theta}$ where $R \to \infty$ and θ goes from π through 0 to $-\pi$. Substituting into Eq. (19.64) gives

$$A_{(z)} = \frac{K_c K_p(1 - b)}{Re^{i\theta} - b} \qquad (19.68)$$

As $R \to \infty$, $A_{(z)} \to 0$. Therefore, the infinite circle in the z plane maps into the origin in the $A_{(z)}$ plane.

The C_{in} contour is just the reverse of the C_{out} contour, going from the origin out along the negative real axis. The C_- contour is just the reflection of the C_+ contour over the real axis.

So, just like in a continuous system, we only have to plot the C_+ contour. If it goes around the $(-1, 0)$ point, this sampled-data system is closedloop unstable since $P = 0$.

The farther away the $A^*_{(i\omega)}$ curve is from the $(-1, 0)$ point, the more stable the system. We can use exactly the same frequency-domain specification that we used for continuous systems: phase margin, gain margin, and maximum closedloop log modulus. The last is obtained by plotting the function $A^*_{(i\omega)}/(1 + A^*_{(i\omega)})$. For this process (with $\tau_p = K_p = 1$) with a proportional sampled-data controller and a sampling period of 0.5 min, the controller gain that gives a phase margin of 45° is $K_c = 3.43$. The controller gain that gives a $+2$ dB maximum closedloop log modulus is $K_c = 2.28$. The ultimate gain is $K_u = 4.08$.

Example 19.10. If a deadtime of one sampling period is added to the process considered in the previous example,

$$A_{(z)} = \frac{K_c K_p(1 - b)}{z(z - b)} \qquad (19.69)$$

We substitute $e^{i\omega T_s}$ for z in Eq. (19.69) and let ω go from 0 to $\omega_s/2$. At $\omega = 0$ where $z = +1$, the Nyquist plot starts at $K_c K_p$ on the positive real axis. At $\omega = \omega_s/2$ where $z = -1$, the $A^*_{(i\omega)}$ curve ends at

$$A^*_{(i\omega)} = \frac{K_c K_p(1 - b)}{-1(-1 - b)} = \frac{K_c K_p(1 - b)}{1 + b} \qquad (19.70)$$

This is on the positive real axis. Figure 19.10c shows the complete curve in the A plane. At some frequency the curve crosses the negative real axis. This occurs when the real part of A^* is equal to -0.394 for the numerical case considered in the previous example. Thus the ultimate gain is

$$K_u = \frac{1}{0.394} = 2.54$$

Note that this is smaller than the ultimate gain for the process with no deadtime. The controller gain that gives $+2$ dB maximum closedloop log modulus is $K_c = 1.36$.

The final phase angle (at $\omega_s/2$) for the first-order lag process with no deadtime was $-180°$. For the process with a deadtime of one sampling period, it was $-360°$. If we had a deadtime that was equal to two sampling periods, the final phase angle would be $-540°$. Every multiple of the sampling period subtracts $180°$ from the final phase angle.

Remember that in a continuous system, the presence of deadtime made the phase angle go to $-\infty$ as ω went to ∞. So the effect of deadtime on the Nyquist plots of sampled-data systems is different than its effect in continuous systems.

Example 19.11. As our last example, let's consider the second-order two-heated tank process studied in Example 19.5.

$$G_{M(s)} = \frac{2.315}{(s + 1)(5s + 1)}$$

Using a zero-order hold, a proportional sampled-data controller and a sampling period $T_s = 0.5$ min gives

$$D_{(z)} H G_{M(z)} = \frac{0.0479 K_c(z + 0.8133)}{(z - 0.607)(z - 0.905)} \qquad (19.71)$$

We substitute $e^{i\omega T_s}$ for z and let ω vary from 0 to $\omega_s/2$.

$$\omega_s = \frac{2\pi}{T_s} = \frac{2\pi}{0.5} = 4\pi$$

At $\omega = 0$, the Nyquist plot starts at $A^*_{(0)} = 2.315 K_c$. This is the same starting point that the continuous system would have. At $\omega = \omega_s/2 = 2\pi$ where $z = -1$, the Nyquist plot ends on the negative real axis at

$$\frac{0.0479 K_c(-1 + 0.8133)}{(-1 - 0.607)(-1 - 0.905)} = -0.0029 K_c$$

The entire curve is given in Fig. 19.10d. It crosses the negative real axis at $A^*_{(i\omega)} = -0.087 K_c$. So the ultimate gain is $K_u = 1/0.087 = 11.6$, which is the same result we obtained from a root locus analysis.

The controller gain that gives a phase margin of 45° is $K_c = 2.88$. The controller gain that gives $+2$ dB maximum closedloop log modulus is $K_c = 2.68$.

19.4.3 Approximate Method

To generate the $A^*_{(i\omega)}$ Nyquist plots discussed above, the z transform of the appropriate transfer functions must first be obtained. Then $e^{i\omega T_s}$ is substituted for z, and ω is varied from 0 to $\omega_s/2$.

There is an alternative way to generate the $A^*_{(i\omega)}$ Nyquist plots that is often more convenient to use, particularly in high-order systems. Equation (18.13) gives a doubly infinite series representation of $A^*_{(i\omega)}$.

$$A^*_{(i\omega)} = \frac{1}{T_s} \sum_{n=-\infty}^{+\infty} A_{(i\omega + in\omega_s)} \tag{19.72}$$

where $A_{(s)}$ is the transfer function of the original continuous elements before z-transforming. Remember $A^*_{(s)}$ is a pulse transfer function. $A_{(s)}$ is a continuous transfer function. For example, the chromatograph system has an $A_{(s)}$ that is $B_{(s)} G_{M(s)} H_{(s)}$. The deadtime is included in the $G_{M(s)}$ term for convenience.

If the series in Eq. (19.72) converges in a reasonable number of terms, we can approximate $A^*_{(i\omega)}$ with a few terms in the series. Usually two or three are all that are required.

$$A^*_{(i\omega)} = [G_M BH]^*_{(i\omega)} = \frac{1}{T_s} \sum_{n=-\infty}^{+\infty} G_{M(i\omega + in\omega_s)} B_{(i\omega + in\omega_s)} H_{(i\omega + in\omega_s)} \tag{19.73}$$

$$\begin{aligned}
A^*_{(i\omega)} \simeq \frac{1}{T_s} [&G_{M(i\omega)} B_{(i\omega)} H_{(i\omega)} + G_{M(i\omega + i\omega_s)} B_{(i\omega + i\omega_s)} H_{(i\omega + i\omega_s)} \\
&+ G_{M(i\omega - i\omega_s)} B_{(i\omega - i\omega_s)} H_{(i\omega - i\omega_s)} \\
&+ G_{M(i\omega + i2\omega_s)} B_{(i\omega + i2\omega_s)} H_{(i\omega + i2\omega_s)} \\
&+ G_{M(i\omega - i2\omega_s)} B_{(i\omega - i2\omega_s)} H_{(i\omega - i2\omega_s)}] \tag{19.74}
\end{aligned}$$

This series approximation can be easily generated on a digital computer.

The big advantage of this method is that the analytical step of taking the z transformation is eliminated. You just deal with the original continuous transfer functions. For complex, high-order systems, this can eliminate a lot of messy algebra.

Example 19.12. The first-order system with a one sampling-period deadtime considered in Example 19.3 had the continuous openloop transfer function

$$G_{M(s)} = \frac{K_p e^{-T_s s}}{\tau_p s + 1}$$

If we use a continuous PI controller and a zero-order hold

$$B_{(s)} = K_c \frac{\tau_I s + 1}{\tau_I s} \qquad H_{(s)} = \frac{1 - e^{-T_s s}}{s}$$

Suppose $T_s = 0.2$ min. Then the sampling frequency ω_s is $2\pi/T_s = 31.4$ radians per minute.

$$A_{(i\omega)} = G_{M(i\omega)} B_{(i\omega)} H_{(i\omega)} = \frac{K_p e^{-i\omega T_s}}{1 + i\omega\tau_p} K_c \frac{1 + i\omega\tau_I}{i\omega\tau_I} \frac{1 - e^{-i\omega T_s}}{i\omega} \tag{19.75}$$

$$A^*_{(i\omega)} \simeq \frac{1}{T_s} [A_{(i\omega)} + A_{(i\omega + 31.4i)} + A_{(i\omega - 31.4i)} + A_{(i\omega + 62.8i)} + A_{(i\omega - 62.8i)}]$$

The approximate curve using only three terms is essentially the same as the rigorous curve.

PROBLEMS

19.1. Find the maximum value of K_c for which a proportional sampled-data controller with zero-order hold is closedloop stable for the 3-CSTR process.

$$G_{M(s)} = \frac{\frac{1}{8}}{(s + 1)^3}$$

Use sampling times of 0.1 and 1 min.

19.2. Repeat Prob. 19.1 using a sampled-data PI controller.

$$D_{(z)} = \frac{K_c}{\alpha} \frac{z - \alpha}{z - 1} \quad \text{where} \quad \alpha \equiv \frac{\tau_I}{\tau_I + T_s}$$

Use values of τ_I of 0.5 and 2 min.

19.3. Make Nyquist plots of $A^*_{(i\omega)}$ for the process of Prob. 19.1 and find the values of gain that give the following specifications:
(a) Gain margin of 2
(b) Phase margin of $45°$
(c) Maximum closedloop log modulus of $+2$ dB

19.4. Using a value of $\tau_I = \tau_p$ in the system of Example 19.6, show that the ultimate gain of the closedloop system is $K_u = \tau_p/(K_p T_s)$ by:
(a) Direct substitution in the $\mathcal{W}$ plane
(b) Root locus plot in the z plane
(c) Nyquist plot of $A^*_{(i\omega)}$

19.5. Make a root locus plot of the system in Prob. 19.1 and find the value of gain that gives a closedloop damping coefficient ζ equal to 0.3.

19.6. A distillation column has an approximate transfer function between distillate composition x_D and reflux flow rate R of

$$G_{M(s)} = \frac{0.0092}{(5s + 1)^2}$$

Distillate composition is measured by a chromatograph with a deadtime equal to the sampling period. If a proportional sampled-data controller is used with a zero-order hold, calculate the ultimate gain for $T_s = 2$ and 10.

19.7. Grandpa McCoy has decided to open up a new Liquid Lightning plant in the California goldfields. He plans to stay in Kentucky, and he must direct operation of the plant using the pony express. It takes two days for a message to be carried in either direction, and a rider arrives each day.

The new Liquid Lightning reactor is a single, isothermal, constant-holdup CSTR in which the concentration of ethanol, C, is controlled by manual changes in the feed concentration, C_0. Ethanol undergoes an irreversible first-order reaction at a specific reaction rate $k = 0.25/\text{day}$. The volume of the reactor is 100 barrels, and the throughput is 25 barrels/day.

Grandpa will receive information from the plant every day telling him what the concentration C was 2 days earlier. He will send back instructions on how to change C_0. What is the largest change Grandpa can make in C_0 as a percentage of C without causing the concentration in the reactor to begin oscillating?

19.8. A process has the following transfer function relating the controlled and manipulated variables:

$$G_{M(s)} = \frac{-s + 1}{s + 1}$$

(a) If a zero-order hold and a proportional digital controller are used with sampling period T_s, determine the openloop pulse transfer function $HG_{M(z)}$.
(b) Calculate the value of controller gain that puts the system right at the limit of closedloop stability.
(c) Calculate the controller gain that gives a closedloop damping coefficient of 0.3.

19.9. A process is controlled by a proportional digital controller with zero-order hold and sampling period $T_s = 0.25$ and has the openloop pulse transfer function

$$HG_{M(z)} = \frac{-z + 1.2212}{z - 0.7788}$$

If a unit step change is made in the setpoint, calculate the closedloop response of the process at the sampling times if a controller gain of 0.722 is used.

19.10. Make a root locus plot for the process considered in Prob. 19.9
(a) In the z plane
(b) In the log-z plane

19.11. A first-order lag process with a zero-order hold is controlled by a proportional sampled-data controller.
(a) What value of gain gives a critically damped closedloop system?
(b) What is the gain margin when this value of gain is used?
(c) What is the steadystate error for a unit step change in setpoint when this value of gain is used?

19.12. A pressurized tank has the openloop transfer function between pressure in the first tank and gas flow from the second tank

$$G_{M(s)} = \frac{0.2386}{s(0.7137s + 1)}$$

If a zero-order hold and a proportional sampled-data controller are used with $T_s = 1$.
(a) Make root locus plots in the z, log z, and $\mathcal{W}$ planes.
(b) Find the value of controller gain that gives a damping coefficient of 0.3.
(c) Find the controller gain that gives 45° of phase margin.
(d) Find the gain that gives a maximum closedloop log modulus of +2 dB.

19.13. Repeat Prob. 19.12 for a process with the openloop transfer function

$$G_{M(s)} = \frac{-s + 1}{(s + 1)^2}$$

19.14. For the process considered in Prob. 19.12, generate Nyquist and Bode plots by the rigorous method and by the approximate method using several values of n.

CHAPTER
20

DESIGN
OF
DIGITAL
COMPENSATORS

In the last two chapters we have developed the tools to analyze sampled-data systems, both openloop and closedloop. The controllers we have considered have been mostly proportional and the processes have been openloop stable.

In this chapter we look at more complex sampled-data feedback controllers and at some interesting processes: openloop unstable and deadtime processes.

In a digital computer-control system, the feedback controller $D_{(z)}$ has a pulse transfer function. What we need is an equation or algorithm that can be programmed into the digital computer. At the sampling time for a given loop, the computer looks at the current process output $x_{(t)}$, compares it to a setpoint, and calculates a current value of the error $e_{(t)}$. This error, plus some old values of the error and old values of the controller output or manipulated variable that have been stored in computer memory, are then used to calculate a new value of the controller output $m_{(t)}$.

These algorithms are basically difference equations that relate the current value of m to the current value of e and old values of m and e. These difference equations can be derived from the pulse transfer function $D_{(z)}$.

Suppose the current moment in time is the nth sampling period $t = nT_s$. The current value of the error $e_{(t)}$ is $e_{(nT_s)}$. We will call this e_n. The value of $e_{(t)}$

at the previous sampling time was e_{n-1}. Other old values of error are e_{n-2}, e_{n-3}, etc.

The value of the controller output $m_{(t)}$ that is computed at the current instant in time $t = nT_s$ is $m_{(nT_s)}$ or m_n. Old values are m_{n-1}, m_{n-2}, etc.

Suppose we have the following difference equation:

$$m_n = a_0 e_n + a_1 e_{n-1} + a_2 e_{n-2} + \cdots + a_M e_{n-M}$$

$$- b_1 m_{n-1} - b_2 m_{n-2} - b_3 m_{n-3} - \cdots - b_N m_{n-N} \quad (20.1)$$

$$m_{(nT_s)} = a_0 e_{(nT_s)} + a_1 e_{(nT_s-T_s)} + a_2 e_{(nT_s-2T_s)} + \cdots + a_M e_{(nT_s-MT_s)}$$

$$- b_1 m_{(nT_s-T_s)} - b_2 m_{(nT_s-2T_s)} - b_3 m_{(nT_s-3T_s)} - \cdots - b_N m_{(nT_s-NT_s)} \quad (20.2)$$

Limiting t to some multiple of T_s,

$$m_{(t)} = a_0 e_{(t)} + a_1 e_{(t-T_s)} + a_2 e_{(t-2T_s)} + \cdots + a_M e_{(t-MT_s)}$$

$$- b_1 m_{(t-T_s)} - b_2 m_{(t-2T_s)} - b_3 m_{(t-3T_s)} - \cdots - b_N m_{(t-NT_s)} \quad (20.3)$$

If each of these time functions is impulse-sampled and z-transformed, Eq. (20.3) becomes

$$M_{(z)} = a_0 E_{(z)} + a_1 z^{-1} E_{(z)} + a_2 z^{-2} E_{(z)} + a_3 z^{-3} E_{(z)} + \cdots$$

$$- b_1 z^{-1} M_{(z)} - b_2 z^{-2} M_{(z)} - \cdots - b_N z^{-N} M_{(z)} \quad (20.4)$$

Putting this in terms of a pulse transfer function gives

$$D_{(z)} = \frac{M_{(z)}}{E_{(z)}} = \frac{a_0 + a_1 z^{-1} + a_2 z^{-2} + \cdots + a_M z^{-M}}{1 + b_1 z^{-1} + b_2 z^{-2} + \cdots + b_N z^{-N}} \quad (20.5)$$

A sampled-data controller is a ratio of polynomials in either positive or negative powers of z. It can be directly converted into a difference equation for programming into the computer.

20.1 PHYSICAL REALIZABILITY

Continuous transfer functions are physically realizable if the order of the polynomial in s of the numerator of their transfer function is less than or equal to the order of the polynomial in s of the denominator.

The physical realizability of pulsed transfer functions uses the basic criterion that the current output of a device (digital computer) cannot depend upon future information about the input. We cannot build a gadget that can predict the future.

If $D_{(z)}$ is expressed as a polynomial in negative powers of z, as in Eq. (20.5), the requirement for physical realizability is that there must be a "1" term in the denominator. If $D_{(z)}$ is expressed as a polynomial in positive powers of z, as shown in Eq. (20.6) below, the requirement for physical realizability is that the order of the numerator polynomial in z must be less than or equal to the order of the denominator polynomial in z. These two ways of expressing physical realizability are completely equivalent, but since the second is analogous to continuous transfer functions in s, it is probably used most often.

$$D_{(z)} = \frac{M_{(z)}}{E_{(z)}} = \frac{a_0 z^M + a_1 z^{M-1} + a_2 z^{M-2} + \cdots + a_M}{z^N + b_1 z^{N-1} + b_2 z^{N-2} + \cdots + b_N} \tag{20.6}$$

Multiplying numerator and denominator by z^{-N} and converting to difference-equation form give the current value of the output m_n:

$$m_n = a_0 e_{n+M-N} + a_1 e_{n+M-N-1} + \cdots + a_M e_{n-N}$$
$$- b_1 m_{n-1} - b_2 m_{n-2} - \cdots - b_N m_{n-N} \tag{20.7}$$

If the order of the numerator M is greater than the order of the denominator N in Eq. (20.6), the calculation of m_n will require future values of error. For example, if $M - N = 1$, Eq. (20.7) tells us that we need to know e_{n+1} or $e_{(t+T_s)}$ in order to calculate m_n or $m_{(t)}$. Since we do not know at time equal t what the error $e_{(t+T_s)}$ will be one sampling period in the future, this calculation is physically impossible.

20.2 FREQUENCY-DOMAIN EFFECTS

Sampled-data controllers can be designed in the same way continuous controllers are designed. Root locus plots in the z plane or frequency-response plots are made with various types of $D_{(z)}$'s (different orders of M and N and different values of the a_i and b_i coefficients). This is the same as using different combinations of lead-lag elements in continuous systems.

These parameters are varied to achieve some desired performance criteria. In the z-plane root locus plots, the specifications of closedloop time constant and damping coefficient are usually used. The roots of the closedloop characteristic equation $1 + D_{(z)} HG_{M(z)}$ are modified by changing $D_{(z)}$.

In the frequency domain, the conventional criteria of phase margin, gain margin, or maximum closedloop log modulus are used. The shape of the $A_{(i\omega)}^*$ or $D_{(i\omega)}^* HG_{M(i\omega)}^*$ curve is modified by changing $D_{(i\omega)}^*$.

The simplest form of a $D_{(z)}$ sampled-data controller is

$$D_{(z)} = \frac{z - z_1}{z - p_1} \tag{20.8}$$

where z_1 is the zero of $D_{(z)}$ and p_1 is the pole. Selecting different values of z_1 and p_1 gives different compensation. If lead action (positive phase-angle advance) is required in the frequency domain, the pole must be chosen to lie to the *left* of the zero, as shown in Fig. 20.1a. The explanation for this necessary location of the pole and zero is given below.

Figure 20.1a shows the angles ϕ_p and ϕ_z. These are the angles between the real axis and straight lines from the pole or zero to an arbitrary location z on the unit circle where the frequency is ω_1.

$$\phi_p = \arg (z - p_1) \quad \text{and} \quad \phi_z = \arg (z - z_1) \tag{20.9}$$

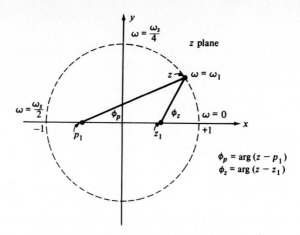

$$\phi_p = \arg(z - p_1)$$
$$\phi_z = \arg(z - z_1)$$

(a) Location of poles and zeros in the z plane

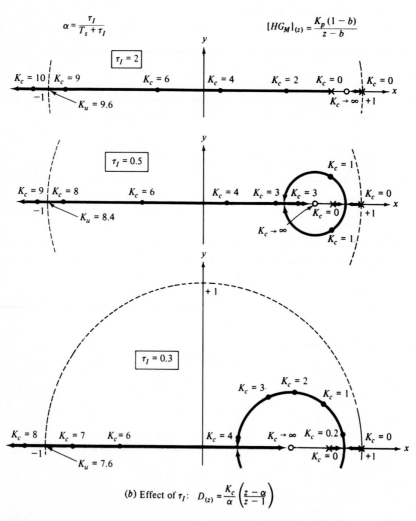

$$\alpha = \frac{\tau_I}{T_s + \tau_I} \qquad [HG_M]_{(z)} = \frac{K_p(1 - b)}{z - b}$$

(b) Effect of τ_I: $D_{(z)} = \dfrac{K_c}{\alpha}\left(\dfrac{z - \alpha}{z - 1}\right)$

At this frequency (ω_1), the argument of $D_{(z)}$ is [using Eq. (20.8)]:

$$\arg D_{(z)} = \arg (z - z_1) - \arg (z - p_1) \qquad (20.10)$$

As long as the pole p_1 is located to the left of the zero z_1, the arg $D_{(z)}$ will be positive since ϕ_z is bigger than ϕ_p over some range of frequencies. Tou (*Digital and Sampled-Data Control Systems*, p. 440, McGraw-Hill, 1959) presents curves showing how the selection of the zero and pole changes the magnitude and phase angle of $D_{(z)}$.

Example 20.1. Suppose we use a $D_{(z)}$ that approximates a continuous PI controller, as discussed in Chap. 19.

$$D_{(z)} = \frac{K_c}{\alpha} \frac{z - \alpha}{z - 1} \qquad (20.11)$$

where

$$\alpha \equiv \frac{\tau_I}{T_s + \tau_I} \qquad (20.12)$$

This pulse transfer function has a zero at $z = \alpha$ and a pole at $z = +1$. It cannot produce any phase-angle advance since the pole lies to the right of the zero (α is less than 1). The pole at $+1$ is equivalent to integration (pole at $s = 0$ in continuous systems) which drives the system to zero steadystate error for step disturbances.

Varying τ_I is equivalent to varying the location of the zero α. Figure 20.1*b* shows the effect of changing τ_I on the root locus plot of the system. Stability decreases as τ_I is decreased. Ultimate gain and damping coefficient at a given gain both decrease with decreasing τ_I.

20.3 MINIMAL-PROTOTYPE DESIGN

One of the most interesting and unique approaches to the design of sampled-data controllers is called minimal-prototype design. It is one of the earliest examples of model-based controllers.

20.3.1 Basic Concept

The basic idea is to specify the desired response of the system to a specific type of disturbance and then, knowing the model of the process, back-calculate the controller required. There is no guarantee that the minimal-prototype controller will be physically realizable for the given process and the specified response. Therefore, the specified response may have to be modified to make the controller realizable.

FIGURE 20.1
Sampled data controller $D_{(z)}$.

Let us consider the closedloop response of an arbitrary system with a sampled-data controller.

$$X_{(z)} = \frac{D_{(z)}HG_{M(z)}}{1 + D_{(z)}HG_{M(z)}} X_{(z)}^{\text{set}} + \frac{G_L L_{(z)}}{1 + D_{(z)}HG_{M(z)}} \tag{20.13}$$

If we consider for the moment only changes in setpoint,

$$\frac{X_{(z)}}{X_{(z)}^{\text{set}}} = \frac{D_{(z)}HG_{M(z)}}{1 + D_{(z)}HG_{M(z)}} \tag{20.14}$$

If we specify the form of the input $X_{(z)}^{\text{set}}$ and the desired form of the output $X_{(z)}$ and if the process and hold transfer functions are known, we can rearrange Eq. (20.14) to give the required controller. We will define this controller that is designed for setpoint changes $D_{S(z)}$.

$$D_{S(z)} = \frac{X_{(z)}}{HG_{M(z)}(X_{(z)}^{\text{set}} - X_{(z)})} \tag{20.15}$$

If all of the terms on the right side of this equation have been specified, the controller can be calculated.

Example 20.2. The first-order lag process with a zero-order hold has a pulse transfer function

$$HG_{M(z)} = \frac{K_p(1 - b)}{z - b} \tag{20.16}$$

where $b \equiv e^{-T_s/\tau_P}$. Suppose we want to derive a minimal-prototype controller for step changes in setpoint.

$$X_{(z)}^{\text{set}} = \mathcal{L}\left[\frac{1}{s}\right] = \frac{z}{z - 1} \tag{20.17}$$

We know that it is impossible to have the output of the process respond instantaneously to the change in setpoint. Therefore, the best possible response that we could expect from the process would be to drive the output $X_{(z)}$ up the setpoint in one sampling period. This is sketched in Fig. 20.2a. Remember, we are specifying only the values of the variables at the sampling times.

The output at $t = 0$ is zero. At $t = T_s$, the output should be 1 and should stay at 1 for all subsequent sampling times. Therefore the desired $X_{(z)}$ is

$$X_{(z)} \equiv x_{(0)} + x_{(T_s)}z^{-1} + x_{(2T_s)}z^{-2} + x_{(3T_s)}z^{-3} + \cdots$$
$$= 0 + z^{-1} + z^{-2} + z^{-3} + \cdots \tag{20.18}$$

$$X_{(z)} = \frac{z^{-1}}{1 - z^{-1}} = \frac{1}{z - 1} \tag{20.19}$$

Plugging these specified functions for $X_{(z)}$ and $X_{(z)}^{\text{set}}$ into Eq. (20.15) gives

$$D_{S(z)} = \frac{X_{(z)}}{HG_{M(z)}(X_{(z)}^{\text{set}} - X_{(z)})} = \frac{1/(z - 1)}{(HG_{M(z)})\{[z/(z - 1)] - 1/(z - 1)\}}$$

$$D_{S(z)} = \frac{1}{HG_{M(z)}(z - 1)} \tag{20.20}$$

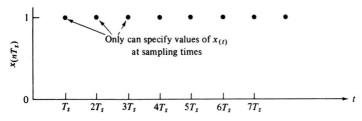

(a) Desired response to unit step change in setpoint

FIGURE 20.2
Minimal-prototype responses.

Now for this first-order process, Eq. (20.16) gives $HG_{M(z)}$. Plugging this into Eq. (20.20) gives the minimal-prototype controller.

$$D_{S(z)} = \frac{1}{HG_{M(z)}(z-1)} = \frac{1}{[K_p(1-b)/(z-b)](z-1)} = \frac{z-b}{K_p(1-b)(z-1)} \quad (20.21)$$

This sampled-data controller is physically realizable since the order of the polynomial in the numerator is equal to the order of the polynomial in the denominator. Therefore the desired setpoint response is achievable for this process.

Before we leave this example, let's look at the closedloop characteristic equation of the system.

$$1 + D_{S(z)} HG_{M(z)} = 0$$

If we substitute Eqs. (20.21) and (20.16) we get

$$1 + \frac{z-b}{K_p(1-b)(z-1)} \frac{K_p(1-b)}{z-b} = 0$$

$$1 + \frac{1}{z-1} = 0 \quad \Rightarrow \quad z = 0$$

Thus the closedloop root is located at the origin. This corresponds to a critically damped closedloop system ($\zeta = 1$). The specified response in the output was for no overshoot, so this damping coefficient is to be expected.

Example 20.3. If we have a first-order lag process with a deadtime that is equal to one sampling period, the process transfer function becomes

$$HG_{M(z)} = \frac{K_p(1-b)}{z(z-b)} \quad (20.22)$$

Suppose we specified the same kind of response for a step change in setpoint as we did in Example 20.2: the output is driven to the setpoint in one sampling period. Substituting our new process transfer function into Eq. (20.20) gives

$$D_{S(z)} = \frac{1}{HG_{M(z)}(z-1)} = \frac{1}{\{K_p(1-b)/[z(z-b)]\}(z-1)}$$

$$= \frac{z(z-b)}{K_p(1-b)(z-1)} \quad (20.23)$$

This controller is not physically realizable because the order of the numerator is higher (second) than the order of the denominator (first). Therefore we cannot achieve the response specified. This result should really be no surprise. The dead-time will not let the output even begin to change during the first sampling period, and we cannot drive the output up to its setpoint instantaneously at $t = T_s$.

Let us back off on the specified output and say we will allow two sampling periods to drive the output to the setpoint.

$$X_{(z)} \equiv x_{(0)} + x_{(T_s)} z^{-1} + x_{(2T_s)} z^{-2} + x_{(3T_s)} z^{-3} + \cdots$$
$$= 0 + (0)z^{-1} + z^{-2} + z^{-3} + \cdots \tag{20.24}$$

$$X_{(z)} = \frac{z^{-2}}{1 - z^{-1}} = \frac{1}{z(z-1)} \tag{20.25}$$

Now the minimal-prototype controller for step changes in setpoint is

$$D_{S(z)} = \frac{X_{(z)}}{HG_{M(z)}(X_{(z)}^{\text{set}} - X_{(z)})}$$

$$= \frac{\dfrac{1}{z(z-1)}}{\left(\dfrac{K_p(1-b)}{z(z-b)}\right)\left(\dfrac{z}{z-1} - \dfrac{1}{z(z-1)}\right)}$$

$$= \frac{z(z-b)}{K_p(1-b)(z^2-1)} \tag{20.26}$$

The controller is physically realizable since $N = 2$ and $M = 2$. Note that there are two poles, one at $z = +1$ and the other at $z = -1$, and there are two zeros (at $z = 0$ and $z = b$).

20.3.2   Other Input Types

Minimal-prototype controllers are designed for a specific type of input. A controller that is designed for a step change will perform quite differently if the change actually has some other shape. The following example illustrates the point.

Example 20.4. Let's go back to the first-order lag process but now we will specify that the change in the setpoint is a ramp with a slope of one.

$$X_{(s)}^{\text{set}} = \frac{1}{s^2} \quad \Rightarrow \quad X_{(z)}^{\text{set}} = \frac{T_s z}{(z-1)^2} \tag{20.27}$$

Now we must specify the desired $x_{(z)}$. It is unreasonable to expect that the controller could make the output be equal to the setpoint at the first sampling period because the controller will not see any error until this point in time. Remember the ramp setpoint started at zero at $t = 0$, so there was no error. Let us specify that the output should follow the ramp setpoint after two sampling periods.

$$X_{(z)} \equiv x_{(0)} + x_{(T_s)}z^{-1} + x_{(2T_s)}z^{-2} + x_{(3T_s)}z^{-3} + \cdots$$

$$= 0 + (0)z^{-1} + 2T_s z^{-2} + 3T_s z^{-3} + 4T_s z^{-4} + \cdots \qquad (20.28)$$

This series can be expressed in closed form.

$$X_{(z)} = \frac{T_s(2z - 1)}{z(z - 1)^2} \qquad (20.29)$$

With these specifications for $X^{set}_{(z)}$ and $X_{(z)}$, we can calculate the minimal-prototype controller for ramp setpoint changes using Eq. (20.15).

$$D_{S(z)} = \frac{X_{(z)}}{HG_{M(z)}(X^{set}_{(z)} - X_{(z)})} = \frac{\dfrac{T_s(2z - 1)}{z(z - 1)^2}}{\left(\dfrac{K_p(1 - b)}{z - b}\right)\left(\dfrac{T_s z}{(z - 1)^2} - \dfrac{T_s(2z - 1)}{z(z - 1)^2}\right)}$$

$$= \frac{(z - b)(2z - 1)}{K_p(1 - b)(z - 1)^2} \qquad (20.30)$$

Comparing this controller with that designed for a step input [Eq. (20.21)], we can see that the ramp setpoint design yields a controller that contains a double integrator.

Let's look at the closedloop characteristic equation of this system

$$1 + D_{S(z)}HG_{M(z)} = 0$$

$$1 + \frac{(z - b)(2z - 1)}{K_p(1 - b)(z - 1)^2} \frac{K_p(1 - b)}{z - b} = 0$$

$$1 + \frac{2z - 1}{(z - 1)^2} = 0$$

$$z^2 - 2z + 1 + 2z - 1 = 0 \qquad \Rightarrow \qquad z^2 = 0 \qquad (20.31)$$

So the roots of the closedloop characteristic equation are at the origin where the system is critically damped.

Now suppose this controller that was designed for ramp setpoint changes is in service, but the setpoint change is really a step: $z/(z - 1)$. What will the response of the system be? Using Eq. (20.14) gives

$$X_{(z)} = \frac{D_{(z)}HG_{M(z)}}{1 + D_{(z)}HG_{M(z)}} X^{set}_{(z)}$$

$$X_{(z)} = \frac{(2z - 1)/(z - 1)^2}{1 + (2z - 1)/(z - 1)^2}\left(\frac{z}{z - 1}\right) = \frac{2z - 1}{z^2 - z} \qquad (20.32)$$

$$X_{(z)} = 2z^{-1} + z^{-2} + z^{-3} + \cdots$$

Note that there is a 100 percent overshoot of the setpoint! And this occurs despite the fact that the damping coefficient of the system is unity.

The reason for this overshoot is the design of the controller. It is expecting a ramp setpoint input, so when it sees an error it moves the manipulated variable more than if it were expecting a step.

20.3.3 Load Inputs

Load inputs can also be designed for using minimal-prototype methods. For load disturbances only, Eq. (20.13) gives

$$X_{(z)} = \frac{G_L L_{(z)}}{1 + D_{(z)} H G_{M(z)}} \tag{20.33}$$

If the form of the load disturbance is specified and the desired output is fixed, the digital compensator (which we will call $D_{L(z)}$) can be calculated from

$$D_{L(z)} = \frac{G_L L_{(z)} - X_{(z)}}{H G_{M(z)} X_{(z)}} \tag{20.34}$$

Example 20.5. Suppose we have a process with the following openloop process transfer functions:

$$G_{M(s)} = G_{L(s)} = \frac{K_p}{\tau_p s + 1} \tag{20.35}$$

If a zero-order hold is used,

$$H G_{M(z)} = \frac{K_p(1 - b)}{z - b} \tag{20.36}$$

Suppose we want to design for a step change in load ($L_{(s)} = 1/s$).

$$G_L L_{(z)} = \mathscr{Z}\left[\frac{K_p}{(\tau_p s + 1)s}\right] = \frac{K_p(1 - b)z}{(z - b)(z - 1)} \tag{20.37}$$

Now we must specify the desired load response of the output. The ideal response would be to keep the output at zero, but this of course is not possible. The best that we could do would be to detect the error at the first sampling period and drive the process back to zero at the second sampling period. During the first sampling period, the system responds in an openloop manner to the load disturbance. So at $t = T_s$ the output will be

$$x_{(t = T_s)} = K_p(1 - e^{-T_s/\tau_p}) = K_p(1 - b)$$

Therefore the specified output for the load disturbance is

$$X_{(z)} = 0 + K_p(1 - b)z^{-1} + (0)z^{-2} + (0)z^{-3} + \cdots$$

$$= \frac{K_p(1 - b)}{z} \tag{20.38}$$

Now the digital compensator can be calculated.

$$D_{L(z)} = \frac{G_L L_{(z)} - X_{(z)}}{HG_{M(z)} X_{(z)}}$$

$$= \frac{\dfrac{K_p(1 - b)z}{(z - b)(z - 1)} - \dfrac{K_p(1 - b)}{z}}{\left[\dfrac{K_p(1 - b)}{z - b}\right]\left[\dfrac{K_p(1 - b)}{z}\right]}$$

$$D_{L(z)} = \frac{(1 + b)z - b}{(1 - b)K_p(z - 1)} \qquad (20.39)$$

This controller is physically realizable and contains integration to drive the error to zero. Now, however, the zero in the controller is located at $z = b/(1 + b)$. Remember that for step changes in setpoint, the controller had the same form [Eq. (20.21)] but the zero was located at $z = b$, which is farther to the left of the pole at $z = +1$. Thus the $D_{S(z)}$ controller should have more *lag* action than the $D_{L(z)}$ controller. This is because the step changes in setpoint are seen immediately by the controller, while the load changes are not detected for one sampling period.

The closedloop characteristic equation for this system is

$$1 + D_{L(z)} HG_{M(z)} = 0$$

$$1 + \frac{(1 + b)z - b}{K_p(1 - b)(z - 1)} \frac{K_p(1 - b)}{z - b} = 0$$

$$1 + \frac{(1 + b)z - b}{(z - 1)(z - b)} = 0$$

$$z^2 - (1 + b)z + b + (1 + b)z - b = z^2 = 0 \qquad (20.40)$$

Thus the system is critically damped as expected for minimal-prototype design.

Before we leave this example, let's see what happens if we use this $D_{L(z)}$ controller, which has been designed for step changes in load, but the real disturbance is a step change in setpoint.

$$X_{(z)} = \frac{D_{(z)} HG_{M(z)}}{1 + D_{(z)} HG_{M(z)}} X_{(z)}^{set}$$

$$= \left[\frac{\dfrac{(1 + b)z - b}{(1 - b)K_p(z - 1)} \dfrac{K_p(1 - b)}{z - b}}{1 + \dfrac{(1 + b)z - b}{(1 - b)K_p(z - 1)} \dfrac{K_p(1 - b)}{z - b}}\right] \frac{z}{z - 1}$$

$$= \frac{(1 + b)z^2 - bz}{z^2(z - 1)} = (1 + b)z^{-1} + z^{-2} + z^{-3} + \cdots \qquad (20.41)$$

Note the overshoot at $t = T_s$, despite the fact that the system is critically damped. This is due to the location of the zero in the controller.

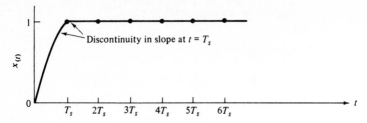

FIGURE 20.2   (*b*) Response of first-order process

20.3.4   Second-Order Processes

If the process is first-order and a setpoint change is made, we can drive the output to the setpoint in one sampling period and hold its output right on the setpoint even between sampling periods. This is possible because we can change the slope of a first-order process response curve, as shown in Fig. 20.2*b*.

If the process is second- or higher-order, we will not be able to make a discontinuous change in the slope of the response curve. Consequently we would expect a second-order process to overshoot the setpoint if we forced it to reach the setpoint in one sampling period. The output would oscillate between sampling periods and the manipulated variable would change at each sampling period. This is called *rippling* and is illustrated in Fig. 20.2*c*.

Rippling is undesirable since we do not want to keep wiggling the control valve. We may want to modify the specified output response in order to eliminate rippling. Allowing two sampling periods for the process to come up to the setpoint gives us two switches of the manipulated variable and should let us bring a second-order process up to the setpoint without rippling. This is illustrated in Example 20.6 below. In general, an Nth-order process must be given N sampling periods to come up to the setpoint if the response is to be completely ripple-free.

Since we know only the values of the output $x_{(nT_s)}$ at the sampling times, we cannot use $X_{(z)}$ to see if there are ripples. We can see what the manipulated variable $m_{(nT_s)}$ is doing at each sampling period. If it is changing, rippling is occurring. So we choose $X_{(z)}$ such that $M_{(z)}$ does not ripple.

$$M_{(z)} = D_{(z)} E_{(z)} = D_{(z)}[X^{set}_{(z)} - X_{(z)}] \tag{20.42}$$

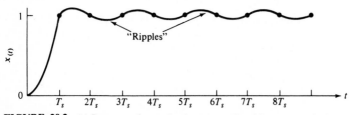

FIGURE 20.2   (*c*) Response of second-order system when driven to setpoint in one sampling period

If the controller is designed for setpoint changes [Eq. (20.15)]

$$D_{(z)} = \frac{X_{(z)}}{HG_{M(z)}(X_{(z)}^{\text{set}} - X_{(z)})}$$

$$M_{(z)} = \frac{X_{(z)}}{HG_{M(z)}} \tag{20.43}$$

Let's check the first-order system from Example 20.2.

$$HG_{M(z)} = \frac{K_p(1 - b)}{z - b} \quad \text{and} \quad X_{(z)} = \frac{1}{z - 1}$$

$$M_{(z)} = \frac{X_{(z)}}{HG_{M(z)}} = \frac{1/(z - 1)}{K_p(1 - b)/(z - b)} = \frac{z - b}{K_p(1 - b)(z - 1)} \tag{20.44}$$

Long division to see the values of $m_{(nT_s)}$ gives

$$M_{(z)} = \frac{1}{K_p(1 - b)} + \frac{1}{K_p} z^{-1} + \frac{1}{K_p} z^{-2} + \frac{1}{K_p} z^{-3} + \cdots \tag{20.45}$$

Thus the manipulated variable holds constant after the first sampling period, indicating no rippling.

Example 20.6. A second-order process considered in Example 19.5 has the following openloop transfer functions:

$$G_{M(s)} = \frac{K_p}{(\tau_1 s + 1)(\tau_2 s + 1)} \tag{20.46}$$

Using a zero-order hold gives an openloop transfer function

$$HG_{M(z)} = \frac{K_p a_0(z - z_1)}{(z - p_1)(z - p_2)} \tag{20.47}$$

where

$$p_1 = e^{-T_s/\tau_1} \quad \text{and} \quad p_2 = e^{-T_s/\tau_2} \tag{20.48}$$

$$a_0 = 1 + \frac{p_2\tau_2 - p_1\tau_1}{\tau_1 - \tau_2} \tag{20.49}$$

$$z_1 = \frac{p_1 p_2(\tau_2 - \tau_1) + p_2\tau_1 - p_1\tau_2}{\tau_1 - \tau_2 + p_2\tau_2 - p_1\tau_1} \tag{20.50}$$

We want to design a minimal-prototype controller for a unit step setpoint change. The output is supposed to come up to the new setpoint in one sampling period. Substituting Eq. (20.47) into Eq. (20.20) gives

$$D_{S(z)} = \frac{1}{HG_{M(z)}(z - 1)} = \frac{1}{\dfrac{K_p a_0(z - z_1)}{(z - p_1)(z - p_2)}(z - 1)}$$

$$= \frac{(z - p_1)(z - p_2)}{K_p a_0(z - z_1)(z - 1)} \tag{20.51}$$

This controller is physically realizable. Therefore minimal-prototype control should be attainable. But what about intersample rippling? Let us check the manipulated variable.

$$M_{(z)} = \frac{X_{(z)}}{HG_{M(z)}} = \frac{\dfrac{1}{z-1}}{\dfrac{K_p a_0(z - z_1)}{(z - p_1)(z - p_2)}} = \frac{(z - p_1)(z - p_2)}{a_0(z - 1)(z - z_1)} \tag{20.52}$$

Long division shows that the manipulated variable changes at each sampling period, so rippling occurs.

For a specific numerical case ($K_p = \tau_1 = 1$; $\tau_2 = 5$; $T_s = 0\cdot2$), the parameter values are $p_1 = 0.8187$, $p_2 = 0.9608$, $a_0 = 0.0037$, and $z_1 = -0.923$.

$$M_{(z)} = 270 - 460z^{-1} + 427z^{-2} - 392z^{-3} + 364z^{-4} - 334z^{-5} + \cdots \tag{20.53}$$

This system exhibits rippling.

To prevent rippling we will modify our desired output response to give the system two sampling periods to come up to the setpoint. The value of $x_{(t)}$ at the first sampling period, the x_1 shown in Fig. 20.2d, is unspecified at this point. The output $X_{(z)}$ is now

$$X_{(z)} = x_1 z^{-1} + z^{-2} + z^{-3} + \cdots$$

$$= x_1 z^{-1} + z^{-2}(1 + z^{-1} + z^{-2} + z^{-3} + \cdots)$$

$$= x_1 z^{-1} + z^{-2}\frac{1}{1 - z^{-1}}$$

$$X_{(z)} = \frac{x_1 z + 1 - x_1}{z(z - 1)} \tag{20.54}$$

The setpoint disturbance is still a unit step:

$$X^{set}_{(z)} = \frac{z}{z - 1}$$

The new controller is

$$D_{(z)} = \frac{X_{(z)}}{HG_{M(z)}(X^{set}_{(z)} - X_{(z)})} = \frac{\dfrac{x_1 z + 1 - x_1}{z(z - 1)}}{\left[\dfrac{K_p a_0(z - z_1)}{(z - p_1)(z - p_2)}\right]\left(\dfrac{z}{z - 1} - \dfrac{x_1 z + 1 - x_1}{z(z - 1)}\right)}$$

$$= \frac{(z - p_1)(z - p_2)(x_1 z + 1 - x_1)}{a_0(z - z_1)(z - 1)(z + 1 - x_1)} \tag{20.55}$$

The manipulated variable is

$$M_{(z)} = \frac{X_{(z)}}{HG_{M(z)}} = \frac{(z - p_1)(z - p_2)(x_1 z + 1 - x_1)}{K_p a_0(z - z_1)z(z - 1)} \tag{20.56}$$

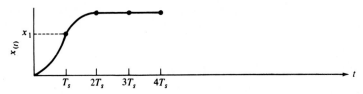

FIGURE 20.2 (*d*) Modified response of second-order system to take two sampling periods to reach setpoint without rippling

Rippling will occur whenever the denominator of $M_{(z)}$ contains any terms other than z or $z - 1$. Therefore, the $z - z_1$ term must be eliminated by picking x_1 such that the term $z - z_1$ cancels out.

$$\frac{1 - x_1}{x_1} = -z_1 \qquad \Rightarrow \qquad x_1 = \frac{1}{1 - z_1} \qquad (20.57)$$

Then $M_{(z)}$ becomes

$$M_{(z)} = \frac{x_1(z - p_1)(z - p_2)}{K_p a_0 z(z - 1)}$$

$$= \frac{x_1}{K_p a_0} + \frac{(1 - p_1 - p_2)}{K_p a_0} z^{-1} + z^{-2} + z^{-3} + z^{-4} + \cdots \qquad (20.58)$$

Thus there is no rippling.

20.3.5 Shunta Dual Algorithm

The problem of having two different controllers for setpoint or load disturbances can be very easily solved by using a "dual algorithm" proposed by Shunta (*Chemical Engineering Science*, Vol. 27, 1972, p. 1325). The structure is sketched in Fig. 20.3*a*.

The basic controller in the feedback loop is designed for load disturbances: $D_{L(z)}$. A precompensator is used on the setpoint signal. This element is designed so that the response to setpoint changes, with the $D_{L(z)}$ compensator in service, is the desired one. The precompensator is defined as $D_{SS(z)}$.

The procedure is to first design the load compensator $D_{L(z)}$ in the normal way. Then the precompensator is designed using the new closedloop servo transfer function

$$X_{(z)} = \frac{D_{SS(z)} D_{L(z)} HG_{M(z)}}{1 + D_{L(z)} HG_{M(z)}} X_{(z)}^{\text{set}} \qquad (20.59)$$

$$D_{SS(z)} = \frac{X_{(z)}}{X_{(z)}^{\text{set}}} \frac{1 + D_{L(z)} HG_{M(z)}}{D_{L(z)} HG_{M(z)}} \qquad (20.60)$$

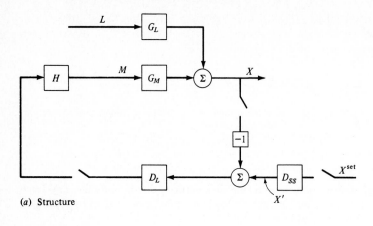

(a) Structure

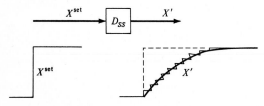

(b) Setpoint precompensator

FIGURE 20.3
Shunta dual algorithm.

Example 20.7. For the first-order process, the transfer functions for the openloop process and for the step load compensator [Eq. (20.39)] are

$$HG_{M(z)} = \frac{K_p(1 - b)}{z - b}$$

$$D_{L(z)} = \frac{(1 + b)z - b}{(1 - b)K_p(z - 1)}$$

This compensator will give the desired response in the output for step changes in load [Eq. (20.38)].

For step changes in setpoint $X^{set}_{(z)} = z/(z - 1)$, the desired response in the output is $X_{(z)} = 1/(z - 1)$. Using Eq. (20.60) gives

$$D_{SS(z)} = \frac{X_{(z)}}{X^{set}_{(z)}} \frac{1 + D_{L(z)} HG_{M(z)}}{D_{L(z)} HG_{M(z)}}$$

$$= \left[\frac{\dfrac{1}{z - 1}}{\dfrac{z}{z - 1}}\right]\left[\frac{1 + \dfrac{(1 + b)z - b}{(1 - b)K_p(z - 1)} \dfrac{K_p(1 - b)}{z - b}}{\dfrac{(1 + b)z - b}{(1 - b)K_p(z - 1)} \dfrac{K_p(1 - b)}{z - b}}\right]$$

$$D_{SS(z)} = \frac{z}{(1 + b)z - b} \tag{20.61}$$

This is the transfer function for a first-order digital filter. Thus the precompensator slows down the input to the $D_{L(z)}$ controller so that it does not see a step change in the error signal (see (Fig. 20.3b).

20.3.6   Dahlin Algorithm

As our final minimal-prototype controller, let us consider the Dahlin algorithm. The basic notion is to specify a desired step setpoint response that looks like a deadtime followed by an exponential rise up to the setpoint. See Fig. 20.4.

The deadtime is set equal to whatever the deadtime is in $G_{M(s)}$. The time constant of the specified response is a tuning factor and represents the closedloop time constant.

Let the deadtime be equal to k sampling periods and the specified time constant be τ_c. The values of $x_{(t)}$ at the sampling periods are:

$$x = 0 \qquad\qquad \text{for} \qquad t < kT_s$$

$$x = 1 - e^{-T_s/\tau_c} \qquad \text{for} \qquad t = (k + 1)T_s$$

$$x = 1 - e^{-2T_s/\tau_c} \qquad \text{for} \qquad t = (k + 2)T_s$$

$$x = 1 - e^{-3T_s/\tau_c} \qquad \text{for} \qquad t = (k + 3)T_s$$

Therefore the specified response of the output is

$$X_{(z)} = (1 - e^{-T_s/\tau_c})z^{-k-1} + (1 - e^{-2T_s/\tau_c})z^{-k-2} + (1 - e^{-3T_s/\tau_c})z^{-k-3} + \cdots$$

$$= z^{-k-1}[(1 + z^{-1} + z^{-2} + \cdots) - e^{-T_s/\tau_c}(1 + e^{-T_s/\tau_c}z^{-1} + e^{-2T_s/\tau_c}z^{-2} + \cdots)]$$

$$= z^{-k-1}\left[\frac{1}{1 - z^{-1}} - \frac{e^{-T_s/\tau_c}}{1 - e^{-T_s/\tau_c}z^{-1}}\right] \qquad (20.62)$$

$$X_{(z)} = \frac{z^{1-k}(1 - e^{-T_s/\tau_c})}{(z - 1)(z - e^{-T_s/\tau_c})} \qquad (20.63)$$

This is the specified output response for a step change in setpoint.

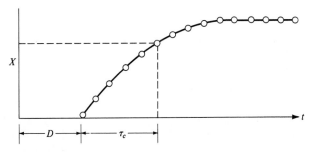

FIGURE 20.4
Dahlin algorithm.

Example 20.8. For the first-order process with a one-sampling period deadtime, the value of k in Eq. (20.63) is 1. The input is a step in setpoint:

$$X^{set}_{(z)} = \mathscr{L}\left[\frac{1}{s}\right] = \frac{z}{z-1}$$

The process openloop transfer function is

$$HG_{M(z)} = \frac{K_p(1-b)}{z(z-b)}$$

Substituting these quantities and Eq. (20.63) (with $k = 1$) into Eq. (20.15) gives the controller transfer function.

$$D_{S(z)} = \frac{X_{(z)}}{HG_{M(z)}(X^{set}_{(z)} - X_{(z)})}$$

$$= \cfrac{\cfrac{(1 - e^{-T_s/\tau_c})}{(z-1)(z - e^{-T_s/\tau_c})}}{\cfrac{K_p(1-b)}{z(z-b)}\left[\cfrac{z}{z-1} - \cfrac{(1 - e^{-T_s/\tau_c})}{(z-1)(z - e^{-T_s/\tau_c})}\right]}$$

$$D_{S(z)} = \frac{z(z-b)(1 - e^{-T_s/\tau_c})}{K_p(1-b)(z-1)(z + 1 - e^{-T_s/\tau_c})} \tag{20.64}$$

This controller is physically realizable. It contains integration, as we would expect, so that the error is driven to zero.

20.4 SAMPLED-DATA CONTROL OF PROCESSES WITH DEADTIME

Up to this point we have dealt with processes which give worse control performance when sampled-data control is used than when continuous control is used. We have demonstrated this effect quantitatively by showing that ultimate gains and gains that give a desired closedloop damping coefficient decrease as T_s is increased.

If the process contains a significant amount of deadtime, your intuition might tell you that a sampled-data controller *might* possibly give *better* control than a continuous controller because it might make sense to wait a little after the manipulated variable has been moved to see what its effect has been. In this section we will explore this interesting idea.

Suppose we have a first-order lag process with deadtime.

$$G_{M(s)} = \frac{K_p e^{-Ds}}{\tau_p s + 1} \tag{20.65}$$

We express the deadtime D as an integer of the sampling period plus a fraction of it.

$$D = kT_s + (1 - m_1)T_s \tag{20.66}$$

where k is an integer and m_1 is a fraction between 0 and 1. Using a zero-order hold gives

$$HG_{M(z)} = \mathcal{L}\left[\frac{1 - e^{-T_s s}}{s}\frac{K_p e^{-Ds}}{\tau_p s + 1}\right] = (1 - z^{-1})\mathcal{L}\left[\frac{K_p e^{-[kT_s + (1 - m_1)T_s]s}}{s(\tau_p s + 1)}\right]$$

$$= \left(\frac{z - 1}{z}\right)z^{-k}K_p\,\mathcal{L}\left[\frac{e^{-(1 - m_1)T_s s}}{s(\tau_p s + 1)}\right]$$

We now use modified z transforms [see Eq. (18.128)] to handle the fractional deadtime.

$$HG_{M(z)} = \left(\frac{z - 1}{z}\right)z^{-k}K_p\,\mathcal{L}_m\left[\frac{1}{s(\tau_p s + 1)}\right]_{m = m_1}$$

$$= \left(\frac{z - 1}{z}\right)z^{-k}K_p\left[\frac{1}{z - 1} - \frac{e^{-mT_s/\tau_p}}{z - e^{-T_s/\tau_p}}\right]_{m = m_1}$$

$$HG_{M(z)} = \left(\frac{z - 1}{z}\right)z^{-k}K_p\left[\frac{1}{z - 1} - \frac{b_m}{z - b}\right] \tag{20.67}$$

where $b = e^{-T_s/\tau_p}$

$$b_m = e^{-m_1 T_s/\tau_p} \tag{20.68}$$

If a proportional sampled-data controller is used, the ultimate gain can be found by substituting $e^{i\omega T_s}$ for z in Eq. (20.68), varying frequency ω and finding where the $HG_{M(i\omega)}$ curve crosses the negative real axis. The reciprocal of the magnitude of $HG_{M(i\omega)}$ at this point is the ultimate gain K_u.

Figure 20.5 shows how the ultimate gain varies with the sampling period for the specific numerical case: $K_p = \tau_p = 1$ and $D = 0.5$. For a sampling period T_s of zero (the continuous case), the ultimate gain is 3.8.

As the sampling period is increased, the ultimate gain initially decreases. But the ultimate gain begins to increase for T_s greater than 0.5 [where the k in Eq. (20.67) is 0], going through a peak of 4.8 at about $T_s = 1.15$. This is *higher* than the ultimate gain for continuous control. Thus it might appear that the use of a sampled-data controller could give better control in this deadtime process.

Unfortunately the sampled-data controller is really *not* better than the continuous. The controller gain that gives +2 dB maximum closedloop log modulus for the continuous controller is 1.95. The corresponding gain for a sampled-data controller with a sampling period of 1.1 is only 1.73. In addition, the sampled-data controller will not detect load disturbances as quickly as the continuous. So the performance of the sampled-data controller is not as good as the continuous.

There is a special type of controller, called a *Smith predictor* or *deadtime compensator*, that can be applied in either continuous or discrete form. It is basically a special type of model-based controller, in the same family as IMC. Figure 20.6a gives a sketch of a conventional feedback control system. Let's break up the total openloop process $G_{M(s)}$ into the portion without any deadtime $G_{ND(s)}$ and deadtime e^{-Ds}.

$$G_{M(s)} = G_{ND(s)}e^{-Ds} \tag{20.69}$$

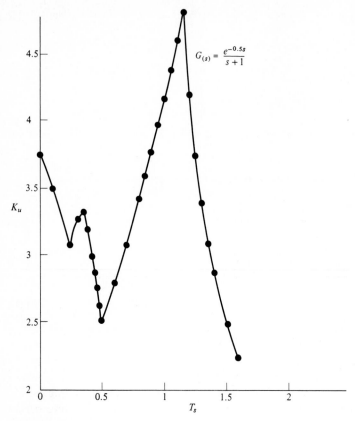

$$G_{(s)} = \frac{e^{-0.5s}}{s+1}$$

FIGURE 20.5

If a conventional feedback controller is used, the closedloop characteristic equation is

$$1 + G_{M(s)} B_{(s)} = 1 + G_{ND(s)} e^{-Ds} B_{(s)} \tag{20.70}$$

We know that the existence of the deadtime will mean that the controller will have to be detuned to maintain closedloop stability.

But if the Smith predictor (sketched in Fig. 20.6b) is used, the closedloop characteristic equation is changed. First, let's consider the inside feedback loop. The closedloop relationship between M and E is

$$\frac{M}{E} = \frac{B}{1 + BG_{ND}(1 - e^{-Ds})} \tag{20.71}$$

Now, looking at the outside loop, we see that

$$X = G_{ND(s)} e^{-Ds} M = G_{ND(s)} e^{-Ds} \frac{B}{1 + BG_{ND}(1 - e^{-Ds})} (X^{set} - X)$$

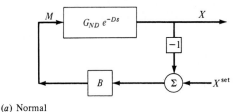

(a) Normal

(b) Predictor

FIGURE 20.6
Smith predictor.

Solving for the closedloop servo transfer function gives

$$\frac{X}{X^{\text{set}}} = \frac{BG_{ND}e^{-Ds}}{1 + BG_{ND}}$$ (20.72)

The closedloop characteristic equation of the system has been changed so that the deadtime has been removed.

$$1 + BG_{ND} = 0$$ (20.73)

Thus tighter controller settings can be used than in the conventional system.

In theory, the Smith predictor gives significant improvement in control. In practice, only modest improvement can be achieved in many processes. This is due to the sensitivity of the stability of the system to small changes in system parameters. If the controller is tightly tuned and there is a small shift in the actual deadtime of the process, the system can go unstable. Therefore, most of the successful applications have been in processes which have gains, time constants, and deadtimes that are well known and constant. Examples include paper machines, steel rolling mills, and textile manufacturing.

20.5 SAMPLED-DATA CONTROL OF OPENLOOP UNSTABLE PROCESSES

Processes with RHP poles in their openloop transfer functions are openloop unstable. The irreversible, exothermic chemical reactor is the classical chemical engineering example. These systems can sometimes be stabilized by using feedback control. Continuous controllers were applied to these systems in Chap. 11. The use of sampled-data controllers is studied in this section.

As we will show quantitatively, sampling periods must be kept quite small in order to stabilize this type of system because of the inherent degradation of performance in discrete control systems.

Example 20.9. The simplest openloop unstable process has the openloop transfer function

$$G_{M(s)} = \frac{K_p}{\tau_p s - 1}$$

Using a zero-order hold gives a pulse transfer function

$$HG_{M(z)} = \mathscr{Z}\left[\frac{1 - e^{-T_s s}}{s}\frac{K_p}{\tau_p s - 1}\right] = (1 - z^{-1})\mathscr{Z}\left[\frac{K_p}{s(\tau_p s - 1)}\right]$$

$$= \frac{z - 1}{z}K_p\mathscr{Z}\left[-\frac{1}{s} + \frac{1}{s - 1/\tau_p}\right]$$

$$= \frac{z - 1}{z}K_p\left[-\frac{z}{z - 1} + \frac{z}{z - e^{+T_s/\tau_p}}\right]$$

$$HG_{M(z)} = \frac{K_p(b^{-1} - 1)}{z - b^{-1}} \tag{20.74}$$

where $b \equiv e^{-T_s/\tau_p}$. Using a proportional sampled-data controller gives the closed-loop characteristic equation

$$1 + K_c HG_{M(z)} = 1 + K_c \frac{K_p(b^{-1} - 1)}{z - b^{-1}} = 0$$

$$z = b^{-1} - K_c K_p(b^{-1} - 1) \tag{20.75}$$

There is a single root as shown in the root locus plot given in Fig. 20.7. The path begins outside the unit circle at $z = b^{-1} = (e^{-T_s/\tau_p})^{-1} = e^{T_s/\tau_p}$. This shows the openloop instability.

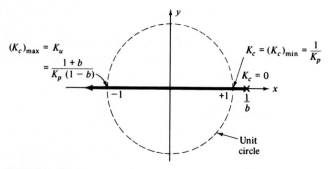

FIGURE 20.7
Openloop unstable process.

As K_c is increased, the path moves to the left on the real axis, entering the unit circle at $z = +1$. The value of controller gain at this point is found by substituting $z = +1$ into Eq. (20.75).

$$K_{min} = \frac{1}{K_p} \tag{20.76}$$

The system is closedloop unstable for gains less than K_{min}. The root locus leaves the unit circle at $z = -1$ when the gain has been increased to K_{max}.

$$K_{max} = \frac{1 + b}{K_p(1 - b)} \tag{20.77}$$

For values of gain greater than K_{max}, the system is closedloop unstable. Thus the system is conditionally stable: the controller gain must lie between K_{min} and K_{max}.

Example 20.10. Suppose we add a one-sampling period deadtime to the first-order system. The openloop system transfer function becomes

$$G_{M(s)} = \frac{K_p e^{-T_s s}}{\tau_p s - 1} \tag{20.78}$$

$$HG_{M(z)} = \frac{K_p(b^{-1} - 1)}{z(z - b^{-1})} \tag{20.79}$$

The root locus plot for a proportional controller is given in Fig. 20.8a. There are two loci, one starting outside the unit circle at $z = b^{-1}$ and the other starting at the origin. The minimum gain occurs when the one path enters the unit circle at $z = +1$. The maximum gain occurs when the two paths leave the unit circle.

Of course, if b is too small (so that b^{-1} is greater than 2), one of the paths will never enter the unit circle. Thus a proportional controller is unable to stabilize this openloop unstable system if

$$b^{-1} = e^{T_s/\tau_p} > 2$$

$$\left(\frac{T_s}{\tau_p}\right)_{max} = \ln 2 = 0.693 \qquad \Rightarrow \qquad (T_s)_{max} = 0.693\tau_p \tag{20.80}$$

This establishes an upper limit on the sampling period.

Example 20.11. A second-order openloop unstable process has the pulse transfer function given below:

$$HG_{M(z)} = \mathscr{Z}\left[\frac{1 - e^{-T_s s}}{s} \frac{1}{(\tau s - 1)(a\tau s + 1)}\right]$$

$$= \frac{a_0(z - z_1)}{(z - p_1)(z - p_2)} \tag{20.81}$$

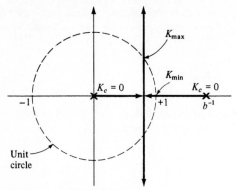

(a) First-order with deadtime

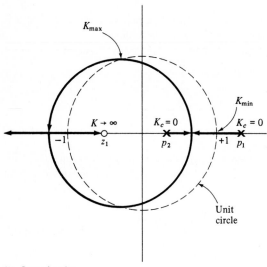

(b) Second-order

FIGURE 20.8
Openloop unstable.

where a = ratio of the negative pole to the positive pole

$$a_0 = \frac{p_1 + p_2 a - 1 - a}{1 + a} \tag{20.82}$$

$$z_1 = -\frac{p_2 + ap_1 - p_1 p_2 - ap_1 p_2}{p_1 + p_2 a - 1 - a} \tag{20.83}$$

$$p_1 = e^{+T_s/\tau} \tag{20.84}$$

$$p_2 = e^{-T_s/a\tau} \tag{20.85}$$

The pole p_1 is outside the unit circle. The zero z_1 is inside the unit circle as long as a is greater than unity. Figure 20.8b gives a root locus plot for $a < 1$.

PROBLEMS

20.1. A process has the following openloop transfer function relating the controlled and manipulated variables:

$$G_{M(s)} = \frac{-s + 1}{s + 1}$$

If a zero-order hold is used with sampling period T_s, the openloop pulse transfer function is

$$HG_{M(z)} = \frac{-z + 2 - b}{z - b}$$

Design a sampled-data minimal-prototype digital compensator for step changes in setpoint that does not ripple.

20.2. A process has openloop load and manipulated variable transfer functions

$$G_{M(s)} = \frac{e^{-T_s s}}{s + 1} \qquad G_{L(s)} = \frac{e^{-2T_s s}}{s + 1}$$

The digital compensator that gives minimal prototype setpoint response for step changes is

$$D_{s(z)} = \frac{z(z - b)}{(z^2 - 1)(1 - b)}$$

Determine the closedloop response of the system when this controller is used and a unit step change in the load input occurs.

20.3. An openloop-unstable, first-order process has the transfer function

$$G_{M(s)} = \frac{K_p}{\tau s - 1}$$

A discrete approximation of a PI controller is used.

$$D_{(z)} = \frac{K_c}{\alpha} \frac{z - \alpha}{z - 1}$$

where

$$\alpha \equiv \frac{\tau_I}{\tau_I + T_s}$$

(a) Sketch a root locus plot for this system. Show the effect of changing the reset time τ_I from very large to very small values.

(b) Find the maximum value of controller gain (K_{max}) for which the system is closedloop stable as a general function of τ, τ_I, and T_s.

(c) Calculate the numerical value of K_{max} for the case $\tau = K_p = \tau_I = 1$ and $T_s = 0.25$.

20.4. Design a minimal prototype sampled-data controller for the pressurized tank process considered in Prob. 13.27 that will bring the pressure up to the setpoint in one sampling period for a unit step change in setpoint.

(a) Calculate how the manipulated variable changes with time to test for rippling.

(b) Repeat for the case when the controller brings the pressure up to the setpoint in two sampling periods without ripple.

20.5. Design a minimal-prototype sampled-data controller for a first-order system with a deadtime that is three sampling periods. The input is a unit step change in setpoint.

20.6. Design minimal prototype controllers for step changes in setpoint and load for a process that is a pure integrator.

$$G_{M(s)} = G_{L(s)} = \frac{1}{s}$$

(a) For setpoint changes, the process should be brought up to the setpoint in one sampling period.

(b) For load changes, the process should be driven back to its initial steadystate value in two sampling periods.

(c) Calculate the controller outputs for both controllers to check for rippling.

20.7. Design a minimal prototype sampled-data controller for a process with an openloop process transfer function that is a pure deadtime.

$$G_{M(s)} = e^{-kT_s s}$$

where k is an integer. Design the controller for a unit step change in setpoint and the best possible response in the controlled variable.

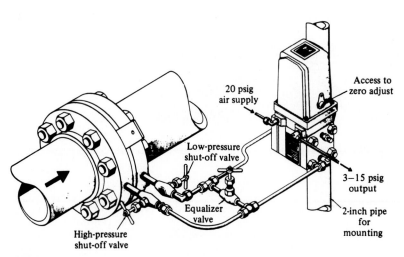

FIGURE A.1
Pneumatic differential pressure transmitter. Typical installation with orifice plate to sense flow rate.
(*Courtesy of Fischer and Porter Company.*)

FIGURE A.2
Electronic differential-pressure transmitter. (*Courtesy of Honeywell.*)

FIGURE A.3
Filled-bulb temperature transmitter. (*Courtesy of Moore Products.*)

FIGURE A.4
Control valve. (*Courtesy of Honeywell*)

FIGURE A.5
Butterfly control valve with positioner. (*Courtesy of Foxboro.*)

FIGURE A.6
Pneumatic control station. (*Courtesy of Moore Products.*)

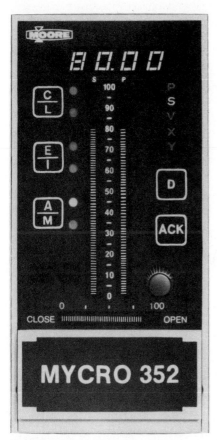

FIGURE A.7
Single-station microprocessor controller. (*Courtesy of Moore Products.*)

FIGURE A.8
Microprocessor control system (TDC 3000). (*Courtesy of Honeywell.*)

FIGURE A.9
Typical control room with computer control. (*Courtesy of Honeywell.*)

FIGURE A.10
Computer control console (CRT display). (*Courtesy of Honeywell.*)

INDEX